# CAMBRIDGE LIBRARY COLLECTION

*Books of enduring scholarly value*

## Technology

The focus of this series is engineering, broadly construed. It covers technological innovation from a range of periods and cultures, but centres on the technological achievements of the industrial era in the West, particularly in the nineteenth century, as understood by their contemporaries. Infrastructure is one major focus, covering the building of railways and canals, bridges and tunnels, land drainage, the laying of submarine cables, and the construction of docks and lighthouses. Other key topics include developments in industrial and manufacturing fields such as mining technology, the production of iron and steel, the use of steam power, and chemical processes such as photography and textile dyes.

## The International Exhibition of 1862

Replete with detailed engravings, this four-volume catalogue was published to accompany the International Exhibition of 1862. Held in South Kensington from May to November, the exhibition showcased the progress made in a diverse range of crafts, trades and industries since the Great Exhibition of 1851. Over 6 million visitors came to view the wares of more than 28,000 exhibitors from Britain, her empire and beyond. Featuring explanatory notes and covering such fields as mining, engineering, textiles, printing and photography, this remains an instructive resource for social and economic historians. The exhibition's *Illustrated Record*, its *Popular Guide* and the industrial department's one-volume *Official Catalogue* have all been reissued in this series. Volume 3, given over to the Colonial Division and the Foreign Division, includes copious examples of manufactured and agricultural goods from India, occupying nearly 300 pages. Also featured are 91 exhibitors of Jamaican rum.

Cambridge University Press has long been a pioneer in the reissuing of out-of-print titles from its own backlist, producing digital reprints of books that are still sought after by scholars and students but could not be reprinted economically using traditional technology. The Cambridge Library Collection extends this activity to a wider range of books which are still of importance to researchers and professionals, either for the source material they contain, or as landmarks in the history of their academic discipline.

Drawing from the world-renowned collections in the Cambridge University Library and other partner libraries, and guided by the advice of experts in each subject area, Cambridge University Press is using state-of-the-art scanning machines in its own Printing House to capture the content of each book selected for inclusion. The files are processed to give a consistently clear, crisp image, and the books finished to the high quality standard for which the Press is recognised around the world. The latest print-on-demand technology ensures that the books will remain available indefinitely, and that orders for single or multiple copies can quickly be supplied.

The Cambridge Library Collection brings back to life books of enduring scholarly value (including out-of-copyright works originally issued by other publishers) across a wide range of disciplines in the humanities and social sciences and in science and technology.

# The International Exhibition of 1862

### The Illustrated Catalogue
### of the Industrial Department

## VOLUME 3: COLONIAL AND FOREIGN DIVISIONS

ANONYMOUS

CAMBRIDGE
UNIVERSITY PRESS

# CAMBRIDGE
## UNIVERSITY PRESS

University Printing House, Cambridge, CB2 8BS, United Kingdom

Published in the United States of America by Cambridge University Press, New York

Cambridge University Press is part of the University of Cambridge.
It furthers the University's mission by disseminating knowledge in the pursuit of
education, learning and research at the highest international levels of excellence.

www.cambridge.org
Information on this title: www.cambridge.org/9781108067300

© in this compilation Cambridge University Press 2014

This edition first published 1862
This digitally printed version 2014

ISBN 978-1-108-06730-0 Paperback

This book reproduces the text of the original edition. The content and language reflect
the beliefs, practices and terminology of their time, and have not been updated.

Cambridge University Press wishes to make clear that the book, unless originally published
by Cambridge, is not being republished by, in association or collaboration with, or
with the endorsement or approval of, the original publisher or its successors in title.

*THE*

ILLUSTRATED CATALOGUE

OF THE

INTERNATIONAL EXHIBITION.

*COLONIAL AND FOREIGN DIVISIONS.*

THE INTERNATIONAL EXHIBITION *of* 1862.

# THE ILLUSTRATED CATALOGUE

OF THE

## INDUSTRIAL DEPARTMENT.

VOL. III.

*COLONIAL AND FOREIGN DIVISIONS.*

PRINTED FOR HER MAJESTY'S COMMISSIONERS.

*Printed for Her Majesty's Commissioners by*

CLAV, SON, & TAYLOR, Bread Street Hill.     PETTER & GALPIN, La Belle Sauvage Yard

CLOWES & SON, Stamford Street.     SPOTTISWOODE & CO., New Street Square.

# CONTENTS.

## VOLUME III.

## COLONIAL DIVISION.

| | PAGE |
|---|---|
| AUSTRALIA, SOUTH | 1 |
| AUSTRALIA, WESTERN | 3 |
| BAHAMAS | 5 |
| BARBADOS | 6 |
| BERMUDA | 7 |
| BORNEO | 8 |
| BRITISH COLUMBIA | 9 |
| CANADA | 10 |
| CAPE OF GOOD HOPE | 16 |
| CEYLON | 17 |
| CHANNEL ISLANDS (JERSEY AND GUERNSEY) | 19 |
| DOMINICA | 20 |
| HONDURAS, BRITISH | 21 |
| JAMAICA | 22 |
| MALTA | 26 |
| MAURITIUS | 27 |
| NATAL | 28 |
| NEW BRUNSWICK | 29 |
| NEWFOUNDLAND | 31 |
| NEW SOUTH WALES | 32 |
| NEW ZEALAND | 44 |
| NOVA SCOTIA | 47 |
| PRINCE EDWARD'S ISLAND | 49 |

# CONTENTS.

|  |  | PAGE |
|---|---|---|
| QUEENSLAND | . . . . . . . . . . . . . . . . . . . . . . | 50 |
| ST. HELENA | . . . . . . . . . . . . . . . . . . . . . . | 54 |
| ST. VINCENT | . . . . . . . . . . . . . . . . . . . . . . | 55 |
| TASMANIA | . . . . . . . . . . . . . . . . . . . . . . | 56 |
| TRINIDAD | . . . . . . . . . . . . . . . . . . . . . . | 60 |
| VANCOUVER | . . . . . . . . . . . . . . . . . . . . . . | 61 |
| VICTORIA | . . . . . . . . . . . . . . . . . . . . . . | 62 |

INDIAN POSSESSIONS—      *See Index at the end of "India."*

# FOREIGN DIVISION.

| AFRICA, CENTRAL | . . . . . . . . . . . . . . . . . . . . . . | 5 |
|---|---|---|
| AFRICA, WESTERN | . . . . . . . . . . . . . . . . . . . . . . | 5 |
| BELGIUM | . . . . . . . . . . . . . . . . . . . . . . | 7 |
| BRAZIL | . . . . . . . . . . . . . . . . . . . . . . | 35 |
| CHINA | . . . . . . . . . . . . . . . . . . . . . . | 43 |
| COSTA RICA | . . . . . . . . . . . . . . . . . . . . . . | 45 |
| DENMARK | . . . . . . . . . . . . . . . . . . . . . . | 47 |
| ECUADOR | . . . . . . . . . . . . . . . . . . . . . . | 59 |
| FRANCE | . . . . . . . . . . . . . . . . . . . . . . | 61 |
| FRANCE, COLONIES OF | . . . . . . . . . . . . . . . . . . . . . . | 156 |

COLONIAL DIVISION.

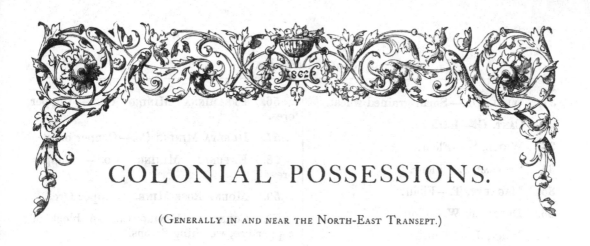

# COLONIAL POSSESSIONS.

(Generally in and near the North-East Transept.)

— ⋅◇⋅ —

## AUSTRALIA, SOUTH.

### NORTH-EAST TRANSEPT, WEST SIDE.

1. MacDonnell, His Excellency Sir R., C.B.—A case of insects; a specimen of malachite.

2. General Committee.—Native woods; a collection of stuffed birds; photographs of public buildings in Adelaide.

3. Dutton, F. J.—Malachite, and other minerals; volumes of debates, votes, and proceedings of the Parliament, and Acts of the Legislature of the Colony.

#### WINES.

4. Evans, H.—Shiraz, 1857-8; Reisling, 1857; Espanoir, 1857 and 1860; Muscatel, 1860.

5. Gilbert, J.—Shiraz, 1861; Verdeilho and Reisling, 1852.

6. Green, W.—Pineau, 1858.

7. Hector, E.—Montura, 1858.

8. Auld, P.—Palomino Blanco, 1859; Verdeilho and Donzelinho, 1860.

9. Wark, Dr.—Black Portugal, 1859.

10. Kingston, G. S.—Molle negro, 1861, &c.

11. Beaseley, F.—Black Portugal and Verdeilho, 1860.

12. Wyman, J.—Shiraz, 1860.

13. Hills, C.—Verdeilho, 1858.

14. Barnard, G. H.—Mixed, 1860.

15. Randall, D.—Shiraz and carbonet, 1860.

16. Davis, A. H.—Moore Farm, 1859.

17. Overbury, T.—Muscat, of Alexandria, 1860.

18. Ind, G. F.—Wines.

19. Warne, J. B.—Wines.

20. Davis, A. H.—Wines.

———

21. Cant, Griffin.—Wheat.

22. McDougall, J.—Wheat.

23. Dunn, J.—Wheat.

24. Hay, J.—Wheat.

25. Bell, A.—Wheat.

26. Buttfield, J. B.—Wheat.

27. Wedd, W.—Wheat.

28. Stevens, J.—Wheat.

29. Wehl, E.—Wheat.

30. Waddell, J.—Wheat.

( 1 )

31. WEHL, DR.—Small-grained wheat.

32. CANT, G.—Barley.

33. WHITE, S.—Flour.

34. STEVENS, J.—Flour.

35. MAGAREY, T.—Flour.

36. DUFFIELD, W.—Flour.

37. DUNN, J.—Flour.

38. HARRISON, BROS.—Flour.

39. HART, J. & Co.—Flour.

40. BEEBY & DUNSTAN.—Flour.

41. MURRAY, A.—Biscuits.

42. CAMPBELL, J.—Muscatel raisins, and soft shell almonds.

43. GRAVES, T.—Dried apricots.

44. DAVIS, F. C.—Jams.

45. SKIPPER, —.—A nautilus shell.

46. NEALES, J. B.—A box of gold specimens.

47. PRIEST, T.—Slabs of slate.

48. ENGLISH & AUSTRALIAN COPPER CO.—Refined copper in cake and tile.

49. S. AUSTRALIAN MINING ASSOCIATION.—Copper ore and other minerals from the Burra Burra and Karkulta mines.

50. WHEAL ELLEN MINING CO.—Copper and lead ores.

51. CUMBERLAND MINING CO. — Copper and lead ores.

52. GT. NORTHERN MINING CO.—Copper ores.

53. WALLAROO MINING CO.—Copper ores.

54. WIRRA WILKA MINING CO.—Copper ores.

55. WORTHING MINING CO.—Copper ores.

56. PREAMINNA MINING CO. — Copper ores.

57. DUREYA MINING CO.—Copper ores.

58. KAPUNDA MINING CO. — Copper ores.

59. MOUNT ROSE MINE.—Copper Ores.

60. CORNWALL MINING CO.—A block of copper ore, weighing 6 tons.

61. ROLLINSON, W.—Collections of minerals.

62. ENGLEHART, DR.—Collections of minerals.

63. MAURAU, DR.—Collections of minerals.

64. RODDA, R. V.—Copper ore smelted by his patent process.

65. KELLETT, J.—Polished marble and slate.

66. CRABB, R. S.—Malachite.

67. MELLOR, J. — A reaping machine; model of do.; cart-wheels and felloes; samples of wood; a plank of blue gum timber.

68. BENDA, A.—Native woods.

69. HAIGH, J. F. — Alpaca and angora wool.

70. PEACOCK, W. & SONS. — Wool in fleeces.

71. ANDERSON, A.—Wool.

72. HAWKER, HON. G.—Wool.

73. KELLY, P.—Wool.

74. BOWMAN, BROS.—Wool.

75. MURRAY, J.—Wool.

76. BURFORD, W. H.—Soap.

77. CHAMBERS, J.—Curiosities brought by J. M. STUART from the centre of Australia.

# AUSTRALIA, WESTERN.

## NORTH-EAST COURT, NEAR THE NAVE.

1. SAMSON, L. *Fremantle.*—Copper and lead ore from Wheal Fortune Mine.

2. SCOTT, D. *Fremantle.* — Copper and lead ore from Wheal Fortune Mine.

3. W. A. MINING ASSOCIATION, *Perth.*—Copper ore from Wanernooka Mine.

4. DRUMMOND, J. N. *Champion Bay.*—Copper ore from Gillireah Mine.

5. HORROCKS, J. L. *Champion Bay.*—Copper from Gwalla Mine.

6. SHENTON, G. *Perth.*—Surface copper ore from Wheal Arrino Mine.

7. SHENTON, A. AND CENTRAL COMMITTEE, *Perth.*—Fossils from the Greenough, York, and Kojenup districts.

8. MEADE, THE REV. W. *Albany.*—Black metallic sand.

9. ELLIOT, G. *Bunbury.*—Surface iron ore and black sand from the Bunbury district.

10. SHENTON, A. —Specimens from the Northern district.

11. BARKER, S. A. *Guildford.* — Magnetic iron ore from Cotes.

12. CARSON, J. *Perth.*—Porcelain clay, from the Darling Range, and crucible made therefrom.

13. HABGOOD, R. M. *London.*—Lead ore from the Geraldine Mine.

14. WALSH, J. *Perth.*—Modelling clay.

15. CENTRAL COMMITTEE, *Perth.*—"Wilgi" clay, with which the Aborigines rub themselves.

16. MILLARD, J. *Toodyay.*—Stone used instead of Turkey-stone.

17. R. ENGINEER DEPARTMENT, *Fremantle.*—A bar reduced from magnetic iron ore.

18. DRUMMOND, J. *Toodyay.*—Mineral, containing asbestos..

19. NEWMAN, E. *Fremantle.*—A pile of Jarrah wood.

20. CARR, J. G. C. *Perth.* — A log of sandal wood, weighing 454 lbs. ; fibrous rush.

21. KENWORTHY, J. *York.* — A log of sandal wood.

22. KING, C. *Perth.*—A wheel of native gum wood ; planks of Jarrah wood.

23. SMITH, T. *Perth.*—Jarrah wood.

24. JOYCE, W. *Perth.*—Shea oak shingles for roofing.

25. CENTRAL COMMITTEE, *Perth.* — Sections of excrescences from mahogany trees; cabinet-work and turnery ; bark of the tea-tree, reducible to pulp, for paper, &c. ; wool from York district.

26. R. ENGINEER DEPARTMENT, *Fremantle.*—Mahogany posts, 17 years immersed in water and mud ; a cabinet and slab.

27. CARSON, J. *Perth.*—A cask made of casuarina.

28. CLIFTON, W. P. *Leschenault.*—Logs, a slab, and parts of wheels.

29. JOHNSON, H. *Perth.*—An inlaid dumb waiter.

30. DURLACHER, A. *Newcastle.*—A sandal wood pedestal.

31. SLOANE, W. *Perth.*—White and red gum, for shipwrights' work.

32. LOCAL COMMITTEE, *York.* — Native woods, and parts of wheels.

33. WHITFIELD, G. *Toodyay.*—Varieties of eucalyptus wood.

34. RANFORD, B. B. *Perth.* — Colonial leathers: sole leather, kip, calf, kangaroo, &c.; barks used for tanning.

35. HOMFRAY, R. *Perth.*—Native spears, shields, kylies, head dresses, nose bones, hair girdles, opossum wool, &c.

36. YELVERTON & Co. *Vasse.*—A collection of timbers.

37. SAMSON, L.—A beam of Jarrah wood.

38. PADBURY, W. *Perth.*- Silver wattle bark.

39. McKNOE, MRS.—Feather flowers.

40. O'NEIL, MISS.—Feather screens.

41. LUKIN & KNIGHT, MISSES, *Perth.*—A muff of parrot feathers.

42. BURGES, S. *Tipperary.* — Skins of native animals.

43. FARMANER, F. *Perth.*—Emu skin.

43A. GREGORY, F. T.—Pearls and pearl oysters from Nichol Bay.

44. HILLMAN, A. *Perth.*—Native silk.

45. FLEAY, J. *Beverley.*—Wool.

46. CLIFTON, G. *Fremantle.*—A collection of shells, pearls; a fishing net, rope, &c. from Nichol Bay.

47. JACKSON & Co. *Perth.*—An inlaid work-table.

48. LEAK, MRS. G. M. *London.*—An inlaid chess table.

49. DU CANE, MRS. F. F. *London.*—Pressed flowers from Swan River.

50. DU CANE, THE REV. A.—Hortus siccus of flowers, from Swan River.

51. LITTLE, T. *Dardanup.*—Frontignan wine, of 1860 and 1861.

52. MACGUIRE, J. *Dardanup.*—Frontignan wine of 1861, from the Wellington district.

53. CENTRAL COMMITTEE.—Wine of 1860, from Pyrton.

54. CLIFTON, W. P. *Leschenault.*—Pedro Ximenes, white Frontignan, and black St. Peter's wines; olive oil; wheat reported to yield 27 to 28 bushels per acre, each weighing nearly 70 lb.

55. CLIFTON, MRS.—Australian Madeira, of 1860.

56. BARLEO, F. P. *Crawley.* — Black cluster grape wine.

57. JECKS, T. *Guildford.*—Olive oil.

58. PARKER, S. S. *York.*—3 bushels of wheat.

59. BAIN, J. *Fremantle.*—Preserved meat and fish.

60. HARDEY, J. *Perth.*—Preserved fruits.

61. CLIFTON, G.—Jelly seaweed (*euchemia speciosa*), for culinary purposes.

62. CARR, J. G. C.—Gum from the manna eucalyptus, and blackboy, or xanthorhœa.

63. BARKER, S. A. *Guildford.* — Gum from the eucalyptus resinifera.

64. DRUMMOND, J. *Toodyay.*—Gum from the "boro blackboy" tree; and wheat.

65. SHENTON, A. *Perth.*—Native hops.

65A. SANDFORD, H. A. *London.*—Furs and dressed skins; native weapons, from the Murchison, taken in action; baskets of blackboy, or xanthorhœa; emu eggs; minerals from Victoria district.

66. BOSTOCK, C. *Fremantle.*—Wheat in ear.

67. SHENTON, G. *Perth.*—Wheat in ear.

68. MUIR & SONS, *Forest Hill.* – Wheat, 66 lb. the bushel.

# BAHAMAS.

NORTH-EAST COURT, WEST SIDE OF NORTH-EAST TRANSEPT.

1. LOCAL COMMISSIONERS.—Native woods, viz.: Yellow wood, green ebony, iron wood, naked wood, Braziletto, cedar, Madura, and horseflesh mahogany.

Fibre of the pita plant, pine-apple leaf and forest pine, and specimens of cordage; indigenous cotton and seed.

Coarse grain and blown salt, produced by solar evaporation.

Arrow-root and starch.

Shell-work, palmetto hats, fans, and walking sticks; baskets and ornaments from the Jumbu Bean.

2. GEORGE, J.S., M.C.P. — Cascarilla bark, turtle shells, conch pearls, ambergris, &c.

3. HARRIS, THE HON. G. D., M.C.P.— Several kinds of sponges, and a collection of shells.

4. GENERAL APOTHECARIES' Co.—Cascarilla bark, canella alba, surgical sponges.

5. HARRIS, SAMUEL. — Conch pearls, sponges, shells, and shell-work.

# BARBADOS.

## EAST SIDE OF NORTH-EAST TRANSEPT.

S. Cave, M.P. (*Commissioner for the Colony*):—

1.  Sugar made by common process on Drax Hall, the first estate on which sugar was grown in Barbados.

2.  Sugar from Hannay's Estate, made by the oscillating process—a new method of stirring the sugar after it has been poured from the copper into the cooler. The sugar crystallizes in larger grains, and sells for nearly 2s. per cwt. more than sugar made by the old process.

3.  Sugar from Maxwell's Estate, made by same process.

4—15.  Various specimens of sugar made by the ordinary process.

16.  Two samples of rum from Sunbury and Hampton Estates.

17.  Vine cotton (*Gossypium vitifolium*).

18.  Arrowroot.

19.  Starch made from the sweet potato (*Batatas edulis*).

20.  Indigo.

21.  Barbados tar (petroleum), with two samples of lubricating oil made from it; these last sent by L. R. Valpy, Esq.

22.  Fibre of the ochro plant.

23.  Silk cotton (*Eriodendron anfractuosum*).

24.  The flower of the sugar-cane arrow.

25.  Basket of fruit, &c. of the island, in wax.

26.  Vase of flowers of the island, made of feathers.

27.  Box of flowers, made of shells.

28.  Two gourds of aloes, one split.

29.  Guinea corn (*Sorghum vulgare*).

Simmons, Mrs. J. A.

30.  Flowers of the island, in wax.

31.  Another case of the same.

32.  Two cases of fruit of the island, in wax.

33.  The flower of the night blowing cereus (*Cereus grandiflorus*) in wax.

34.  St. John, Mrs.—Case of shells, moss, and seaweed, arranged in vase.

35.  Chambers, G. H.—Specimen of white coral, which forms the basis of the island of Barbados.

# BERMUDA.

## EAST SIDE OF NORTH-EAST TRANSEPT.

TUCKER, W. C. FAHIE (*Commissioner for Bermuda*) :—

Cedar furniture, &c.

Specimens of woods.

Samples of work in palmetto, straw, grasses, and flowers.

Cotton, fibre.

Preserves.

Lime-juice.

Seeds.

Pepper.

Honey.

Beeswax.

Tannic acid.

Starch.

Models.

Pumice and brain stones.

Petrifactions.

Sponges.

Marine specimens.

Coral.

Turtle shell.

Lime-stone.

Lime, &c.

# BORNEO.

NORTH EAST COURT, NEAR ENTRANCE TO THE HORTICULTURAL GARDENS.

---

1. ST. JOHN, SPENCER :—

Native arms.

Mats.

Block of antimony ore.

Antimony paint.

Coal.

Native cloths.

Vegetable tallow.

Sago.

Dammar.

Rubber.

2. GRANT, C. T. C.—Gold and silver ornaments of native manufacture.

# BRITISH COLUMBIA.

CENTRE OF NORTH-EAST TRANSEPT.

EXECUTIVE COMMITTEE AT NEW WESTMINSTER: —

1. Model of a log hut.
2. Shingles from cedar wood.
3. Pales of cedar wood.
4. Fence post of cedar wood.
5. Stakes of cedar wood.
6. Horizontal section of Douglas fir.
7. A rent slab of Douglas fir.
8. Log of yew.
9. Log of dogwood.
10. Log of cherry tree.
11. Specimens of birds stuffed.
11A. Indian curiosities.
12. Sample of wheat.
13. Sample of peas.
14. Sample of hops.
15. Pickled fish.
16. Drawing of a tree of Douglas fir, 309 feet high.
17. Ten horizontal sections of Douglas fir.

18. DR. J. W. GILBERT, Esq. F.R.S.—A walking-stick of curled maple, with engraved head of Cariboo gold.

19. BANK OF BRITISH NORTH AMERICA.—A case of specimens of native gold.

20. HYDROGRAPHER'S OFFICE, ADMIRALTY.—Map of the south coast, and of Vancouver.

21. COMMANDER MAYNE, R.N.—Photographs of New Westminster.

# CANADA.

## CENTRE AND EAST SIDE OF NORTH-EAST TRANSEPT.

1. LARUE, A. & Co. *Three Rivers.*—Bog iron ore, with samples of wrought and cast iron obtained from the same at Radnor Forges.

2. SEYMOUR, G. *Madoc.*—Magnetic iron ore.

3. COWAN, A. *Kingston.*—Magnetic iron ore; phosphate of lime; mica; plumbago; friable sandstone; feldspar.

4. ORTON, J. *Hastings Road.*—Magnetic iron ore.

5. CLOSTER, C. C. *Gaspé Basin.*—Lead ore from Indian Cove, Gaspé.

6. WRIGHT, J. & Co. *Montreal.*—Lead ore from Upton.

7. FOLEY & Co. *Montreal.*—Lead ore and pig lead from Ramsay Mine, with a plan of the mine.

8. MONTREAL MINING Co., *Montreal.*—Copper ore, undressed and dressed, from Bruce Mine, Lake Huron, with plans of the mine; copper from Mamainse, Lake Superior.

9. WEST CANADA MINING Co. *Wellington Mine, Lake Huron.*—Copper ore, undressed and dressed, from Wellington Mine, Lake Huron, with a plan of the mine.

10. DAVIES, W. H. A., & C. DUNKIN, *Montreal.*—Copper ore, undressed and dressed, from Acton Mine, Actonvale, with plans of the mine.

11. MOORE, G. B. & Co. *Montreal.*—Copper ore from Upton Mine.

12. POMROY, ADAMS, & Co. *Sherbrooke.*—Copper ore from Wickham & Yale's Durham Mines, with plans of the mines.

13. SHAW, BIGNOL, & HUNT, *Quebec.*—Copper ore from Black River Mine, St. Flavien.

14. ENGLISH AND CANADIAN MINING Co. *Quebec.*—Copper ore, undressed and dressed, from Harvey's Hill Mine, with plans of the mine.

15. FLOWERS, MACKIE, & Co. *Montreal.*—Copper ore from St. Francis Mine, Cleveland and Coldspring Mine, Melbourne, with plans of the mines.

16. GRIFFITH & BROTHERS, *Cleveland.*—Copper ore from Jackson's Mine.

17. SWEET, S. & Co. *North Sutton.*—Copper ore from Sweet's Mine, N. Sutton, with a plan of the mine.

18. ROBERTSON, G. D. & Co. *St. Hyacinthe.*—Copper ore from Craig's Range Mine, Chester.

19. McCAW, T. *Montreal.*—Copper ore from Ascot or Haskell Hill Mine, Ascot.

20. FLETCHER, R. H. *Bruce Mines.*—Smelted copper.

21. WALTON, B. *Montreal.* — Roofing slates, serpentine, chromic iron.

22. BROWN, A. S. *Brockville.*—Cobaltiferous pyrites from Elizabethtown, and phosphate of lime from North Elmsley.

23. CANADIAN OIL Co. *Hamilton.*—Crude and refined rock oil from Enniskillen.

24. WATKINS & INGLIS, *Hamilton.* — Crude and refined rock oil.

25. RUSSELL & Co. *Kingston.* — Plumbago from Pointe du Chêne Graphite Mine, county of Argenteuil.

26. FINLAY, M. *Quebec.*—Fire clay.

27. CHEESEMAN, C. R. *Phillipsburg.* — Building stones and marble.

28. O'DONNELL, H., C.E. *Quebec.* — Building stone (gneiss) and sewerage pipe tile.

29. BROWN, T. *Thorold.* — Hydraulic cement, crude and prepared.

30. PELL & COMPTE, *Montreal.*—Bricks.

31. BULMER & SHEPPARD, *Montreal.*— Bricks.

32. TREADWELL, P. C. *L'Original.* — Drain tiles.

33. MISSISQUOI DRAIN-TILE Co. *Missisquoi.*--Drain tiles.

34. MARTINDALE, T. *Oneida.* — Crude and prepared plaster (gypsum).

35. DONALDSON, J. *Oneida.*—Crude and prepared plaster (gypsum).

36. TAYLOR, A. *York, Grand River.*— Crude and prepared plaster (gypsum), with a plan of the mine.

37. CARON, E. *St. Anne de Montmorenci.* —Iron ochre.

38. GIBB, T. *Toronto.*—White bricks and drain tiles.

39. BELL, R., M.P. *Ottawa.*—Building stones used in the construction of the Parliament House, Ottawa.

40. KNOWLES, W. *Arnprior.* — Marble from Arnprior.

41. GEOLOGICAL SURVEY OF CANADA, *Montreal.* Collection of ores of iron, lead, copper, nickel, silver, gold, platinum, indormine, chromic iron, molybdenite, dolomite, magnesite, bituminous shale, soapstone, potstone, mica rock, mica, plumbago, asbestus, fire clay, building stones, marbles, serpentines, slates, bricks, flagstones, hydraulic limestones, common limestones, whetstones, grindstones, millstone, buhrstone, freshwater shell marl, iron ochres, sulphate of barytes, lithographic stones, agates, albite, orthoclase, jasper conglomerate, epidosite sandstone for glass making, moulding sand, peat, a collection of the crystalline rocks, with a geological map of Canada.

42. PROVANCHER, ABBÉ, *St. Joachim.*— Specimens of woods, branches, leaves and flowers.

43. PRIEUR, F. X.—Specimens wood, obtained in County St. John, southern extremity Lower Canada.

44. LEPAGE, J. B.— Specimens woods, obtained at Rimouski.

45. DUBORD, DR.—Specimens of woods, collected in county St. Maurice.

46. COULTÉE, L. M.—Specimens of woods, obtained in the county of Ottawa.

47. PRICE, D.—Specimens of woods, collected in the county of Chicoutimi.

48. PATTON, DUNCAN, & Co.—Collection of commercial woods.

49. GINGRAS, G. *Quebec.*—Nine pieces of sawn wood.

50. GIROUX, O. *Quebec.*—Vegetable extracts.

51. TURGEON & OUELLET, *Quebec.*—Preserved fish.

52. TÊTU, C. H. *River Ouelle.*—Skins of white porpoise and seal; oils of white porpoise, shark, and cod liver.

53. COTÉ, O. *Quebec.*—Collection of furs.

54. VAN ALLEN, D. R. *Chatham.*—Sections of trees and planks.

55. SHARP, S., of the G. W. R. *Hamilton.* —Sections of Trees, planks, and specimens polished woods.

56. SKEAD, J. *Ottawa.*—Sections of trees and foliage; and planks.

57. LAURIE, J. *Markham.* Sections of trees and planks.

58. DICKSON, A. *Pakenham.*—Specimens polished woods.

59. McKee, H. *Norwich.* — Specimens of shrubs, twigs, and leaves.

60. Choate, J. *Ingersoll.*—Planks.

61. McCracken, A. *London.*—Planks.

62. McKellar, A. *Chatham.*— Sections, trees.

63. Burrows *Simcoe.*—Sections, trees.

63½. Trembicke, A. L. Engineer, *G. T. Railway.*—Sections of trees.

64. Bronson, A. *Bayham.*—Plank.

65. Fruit Growers' Association, U.C. —Coloured plates of fruit grown in Upper Canada (open air).

66. Passmore, S. W. *Toronto.*--Ducks, birds, fishes.

67. Thompson, J. *Montreal.* — Case of 103 birds.

68. Crooks, Miss K. *Hamilton.*—Native plants, flowers, leaves of trees, &c. from vicinity of Hamilton.

69. Fleming, J. *Toronto, U.C.*—Wheat, oats, barley; seeds of tare, millet, peas, turnips, flax, red onion, &c.

70. The Agricultural Society of Beauharnois, Lower Canada.—Barley, oats, peas, rye, wheat, flax, and grass seed.

71. The Agricultural Society of Huntingdon, L. C. -- Samples of barley, Indian corn or maize, oats, peas, wheat.

72. Boa, W. *of St. Laurent, Island of Montreal, L. C.*—Barley, beans, Indian corn, meal, oats, peas, wheat, buckwheat, and oat straw.

73. Beaudry, P. *St. Damase, L.C.*— Barley and wheat.

74. Logan, J. *Petite Côte, Island of Montreal, L. C.*—Barley, beans, butter, maize or Indian corn, oats, and wheat.

75. Malo, P. *St. Damase, L.C.*—Barley, wheat.

76. McKinnon, D. *Somerset, Megantic, L.C.*—Samples of barley and wheat.

77. Rochelau, A. *St. Bruno, L. C.*— Barley, flax.

78. Wilkins, C. *Rougemont, L. C.* — Barley, flax, Indian corn, maple sugar.

79. Evans, W. *Montreal, L.C.* — Beans, Indian corn, peas, Timothy seed.

80. Brown, D. *Nelsonville, L. C.* — Cheese, maple sugar.

81. Lymans, Clare, & Co. *Montreal, L.C.* —Clover seed, Timothy seed, flax seed.

82. Peel (County) Agricultural Society, U. C.—Barley, wheat, peas.

83. Denison, R. L. *Toronto, U.C.* — Indian corn stalks.

84. Shaw, A. *Toronto, U.C.* — Indian corn, rye, peas.

85. Martin, P. dit Ladouceur, *St. Lourent, L. C.*—Indian corn.

86. Dawes & Sons, *Lachine, Island of Montreal, L. C.*—Hops.

87. McKee, H. *Norwich. U. C.* — Honey.

88. The Agricultural Society of Wentworth and Hamilton, U. C. — Oats, wheat.

89. Badham, T. *Drummondville, L. C.* —Oats.

90. Matthieu, H. *St. Hyacinthe, L. C.* —Oats.

91. Cumming, H. *Megantic, L. C.* — Peas.

92. L'Heureux, Rev. F. *Contrecœur, L. C.*—Maple sugar.

93. Alix, J. B. *St. Cesaire, L. C.* — Maple sugar.

94. Sharon, H. *Southwick, U. C.* — Maple sugar.

95. The Agricultural Board of Upper Canada.—Samples of wheat.

96. Robertson J. *Nepean, U.C.*—Wheat.

97. Beardman, J. *Nepean, U.C.*—Wheat.

98. Brunelle, L. *St. Hyacinthe, L.C.* — Buck wheat.

99. Drummond, J. *Petite Côte, Island of Montreal, L. C.*—Wheat.

100. Lamonde, J. *St. Damase, L.C.*— Wheat.

101. Stewart, D. *Inverness, L.C.*— Wheat.

102. PILOTE, REV. F. *St. Anne's College, L. C.*—Wheat.

103. EAST BRANT AGRICULTURAL SOCIETY. —Wheat.

104. BLAIKIE & ALEXANDER, *Toronto, U.C.*—One sample of flax, and four samples of straw.

105. McNAUGHTON, E. A. *Newcastle, U. C.*—Prepared arrowroot, flax, seed, &c.

106. REINHARDT, G. *Montreal, L.C.*—Smoked meats and sausages.

107. PIGEON, N. *Montreal, L.C.*—Wine from native Canadian grapes.

108. PAULET, MDME, *Montreal, L.C.*—Native wine.

109. CRAWFORD, D. *Toronto, U.C.*—Canadian mustard.

110. LAIDLEY & TORREY, *Toronto, U.C.*—Box of wool.

111. LYMAN & Co. *Montreal, L.C.*—Arctusine; Canadian yellow wax.

112. WHEELER, J. *Montreal, L.C.*—Toilet soap.

113. LARUE & Co. *Three Rivers, L.C.*—Railway wheels from Radnor forges, St. Maurice.

114. LOWE, J. *Grand Trunk Railway. L.C.*—Model of direct action, self-balanced, oscillating cylinder for steam-engine, and a steam gauge.

115. MARTIN, J. *Toronto, U.C.*—Model of steam superheater for locomotive.

116. SHARP, S. *G.W. Railway of Canada, Hamilton, U.C.*—Models of sleeping and freight cars.

117. LEDUC, C. *Montreal, L.C.*—A four-wheeled open carriage.

118. BAWDEN, W. *Hochelaga, Montreal, L.C.*—A brick and tile making machine, with model of pug mill.

119. RICHARD, E. O. *Quebec, L.C.*—Model of an improved water wheel.

120. MOORE, T. *Etobicoke.*—Wooden handles for tools, &c.

121. TONGUE & Co. *Ottawa, U.C.*—A large collection of tools.

122. WASHBURN, S. *Ottawa, U.C.*—Axes.

123. JEFFERY, J. *Côte des Neiges, Montreal, L.C.*—An iron plough.

124. PATERSON, J. *Montreal, L.C.*—An iron swing plough.

125. COLLARD, H. *Gananoque, U.C.*—A cultivator, with wheels.

126. COMER, L. *Hinchinbrooke, Frontenac, U.C.*—Model of an improved beehive.

127. GASKIN, CAPT. R. *Kingston, U.C.*—Agricultural implements, &c.

128. McSHERRY, J. *St. David's, U.C.*—An iron plough.

129. MORLEY, J. *Thorold, U.C.*—An iron swing plough.

130. MYERS & SON, *Toronto, U.C.*—A patent churn.

131. WHITING & Co. *Oshawa, U.C.*—A collection of agricultural implements.

132. MAYNARD, REV. MR. *Toronto, U.C.*—Model of fish-tail submarine propeller.

133. OATS, R. H. *Toronto, U.C.*—Model of patent instantaneous reefers.

134. KING, T. D. *Montreal, L.C.*—Diagram of the mean diurnal changes of temperature of air and water of the St. Lawrence, Montreal.

135. THOMPSON, J. E. *Toronto, U.C.*—Heating and other apparatus.

136. NOTMAN, W. *Montreal, L.C.*—Two portfolios, and a collection in frames of photographs.

137. BONALD, G. S. D. *McGill University, Montreal, L.C.*—An apparatus for detecting consumption, &c.

138. PALMER, DR. H. *London, U.C.*—A medical magnetic instrument.

139. DUNPHY, MRS. P. *St. Malachi, L.C.*—Woollen yarn.

140. STEPHEN & Co. *Montreal, L.C.*—Woollen cloths.

141. FAIRBANK, E. *Clifton, U.C.*—Screens, mats, plumes, &c.

142. THOMPSON, T. *Toronto, U.C.*—A Shaftoe saddle.

143. ANGUS & LOGAN, *Montreal, L.C.*—Paper.

144. COUNCIL OF PUBLIC EDUCATION IN LOWER CANADA: Superintendent—Hon. P. J. O. CHAUVEAU.—A large collection of educational works.

145. ALLEN, W. *Montreal.*—School furniture.

146. EDWARDS, J. *Toronto.*—Specimens of penmanship.

147. NELSON & WOOD, *Montreal, L.C.*—Brooms and brushes.

148. EDDY, E. B. *Ottawa, U.C.*—Tubs, pails, washboards.

149. MCILROY, T. *Brampton, Peel County, U. C.*—A walnut invalid bedstead.

150. SNELL, W. H. *Victoria Iron Works, Montreal, L. C.*—Nail plates, &c.

151. BULLOCK, W. *Toronto, U. C.*—Stained glass.

152. MILLS, J. *Hamilton, U.C.*—Ornamental tiles.

153. GIBB, T. *Toronto, U.C.*—Drain tiles.

154. MISSISQUOI DRAIN TILE COMPANY, L. C.—Drain tiles.

155. PALSGRAVE, C. T. *Montreal.*—Type and impressions.

156. HENRY, P. *Montreal, L.C.*—Cigars, &c.

157. BRIDGE, A. *Westbrook, Kingston, U. C.*—Small fancy tub of Canadian woods.

158. HAYCOCK & Co. *Ottawa, U.C.*—Canadian walnut box, containing specimens of building stones used for New Parliament Houses, Canada.

159. LEWIS, C. *Ingersoll, U.C.*—A fancy keg of Canadian woods.

160. ROBERTSON, G. *Kingston, U.C.*—A case of blacking.

161. BENSON & ASPDEN, *Edwardsburgh, U. C.*—Samples of Indian corn starch.

162. MCNAUGHTON, E. A. *Newcastle, Durham County, U.C.*—Flour and potato starch.

163. HOPKINS, J. W. Architect, *Montreal, L. C.*—Architectural drawings.

164. LAWFORD & NELSON, *Montreal, L.C.*—Interior view of a building for skating during the winter.

165. HOPKINS, LAWFORD, & NELSON, *Montreal, L.C.* — Photograph of building erected by them.

166. JACOBI, O. R. *Montreal, L.C.* — Views in oil of local scenery.

167. WESTMACOTT, S. *Toronto, U.C.* — Two landscapes—Canadian scenery.

168. WHALE, R. *Burford, U.C.*—Landscapes, &c.

169. RODDEN, W. *Montreal, L.C.*—Plantagenet water.

170. SOVEREIGN, L. L. *Simcoe, C.W.*—A combined plough, drill, and harrow. A garden drill.

171. THOMPSON, MISS.—Wreath of Canadian autumn leaves.

172. HODGES, MESSRS.—Pictures of Canadian scenery and life, by Kreighoff.

172. ARMSTRONG, D. *Owen Sound.*—Spring wheat.

173. BENNING, D. *Beauharnois.*—English oats.

174. AGRICULTURAL SOCIETY, *Beauharnois.*—Flax seed.

175. BRODIE, J. *Beauharnois.*—Late peas.

175. CAREY, J. *Flamboro West, U.C.*—Spring wheat.

176. CARROLL, AMBRIDGE, & Co. *Hamilton, U. C.*—Crude and refined kerosene.

177. DRUMMOND, J. *Petite Côte, L.C.*—Spring wheat.

178. FILIATREAU, J. B. *Beauharnois, L. C.*—Late rye.

179. GALBRAITH, J. *Beauharnois.*—Canadian barley.

180. GENDRON, J. *Beauharnois.*—Early peas.

181. GERIE, A. *Ancaster, U.C.*—Soule's wheat.

182. MALO, G. *St. Damase, L.C.*—Black Sea wheat.

183. CANADA OIL WORKS, *Hamilton.*—Rock oil.

184. M‘DONALD, *Beauharnois, L. C.*— Wheat.

185. M‘GAW, T. *East Whitby, U. C.* — Spring wheat.

186. MUIR, J. *Huntingdon, L. C.*—Oats, 80 bush. to acre.

187. PERCIL, J. *Huntingdon, L. C.* — Peas.

188. ROSE, E. H.*Chatham, U.C.*—Walnut veneers.

189. SAUNDERS, W. *London, U.C.*— Medicinal herbs and fluids ; perfumery.

190. SCHUYLER, S. *Huntingdon, L.C.*— Indian corn.

191. TAIT, C. *Beauharnois, L.C.*—Black Sea wheat.

192. THOMPSON, D. *Beauharnois.*—Two-row barley.

193. TREMBICKI, A. L. *Montreal.*—Sections of wood.

194. WILSON, J. *Wellington, U.C.*—Barrel oatmeal.

———

195. BILLINGS, E. *Montreal.* — Figures and descriptions of Canadian organic remains.

196. HUNT, T. STERRY, *Montreal.* -- A collection of the crystalline rocks of Canada.

# CAPE OF GOOD HOPE.

NORTH-EAST COURT, NEAR WEST SIDE OF NORTH-EAST TRANSEPT.

---

GHISLIN, T. G. :—

Novel applications of South African algæ, as a substitute for horn, as handles in cutlery, whips, &c.

A collection of African vegetable fibres, adapted to textile fabrics, brushes, rope, paper, &c.

Various kinds of wood.

Curiosities, natural and artificial.

Specimens of aboriginal industry.

---

BOWLER, J. W. *Cape Town.*—View of Cape Town from Blue Berg (water-colour).

BOWLER, J. W. *Cape Town.*—View of Keeskamma river, British Kaffraria.

# CEYLON.

UNDER GALLERY, EAST SIDE OF NORTH-EAST TRANSEPT.

1. CEYLON COMMITTEE:—

Sponge (*N. Prov.*); oyster shells (*E. Prov.*); chenaub (*W. Prov.*).

Oils of gingelley, margosa, elleppy, and dugong (*N. Prov.*); cocoa-nut oil (*N. S. W. & N.W. Provs.*).

Medical substances (*N. Prov.*).

Green, black, and gingelly, gram, rice, honey, vinegar, fruits, and spices (*N. Prov.*); rice, &c. (*W. & Cent. Provs.*); cocoa-nuts (*E. W. Provs.*); palmyra flour (*E. & N. Provs.*).

Sunn fibres, and a variety of gums (*N. Prov.*); fibres (*Cent. Prov.*); Tagoolacoody sap and root (*E. Prov.*); hemp (*E. Prov.*); cotton (*N. E. & Cent. Provs.*).

Stuffing moss lichen (*Maritime Provs.*).

Tobacco (*N. & N. W. Provs.*).

Beeswax, vegetable wax, and birds' nests (*N. Prov.*).

Dye stuffs (*N. S. E. W. N.W. Central & Maritime Provs.*).

Tanning substances (*N. E. W. N.W. & Maritime Provs.*).

Monkey skins (*N. & E. Provs.*); sheep, goat, and bullock skins (*W. N. & E. Provs.*); cheetah and tiger skins (*N. Prov.*).

A musical instrument called "Nagoola" (*E. Prov.*).

Lace and gold embroidery (*W. Prov.*).

Carved cocoa-nut shells; manufactured articles in wood (*W. Prov.*).

Articles made of brass (*E. Prov.*).

Mats, baskets, &c. of date and screw pine leaves, pooswel and coir mats (*W. Prov.*); coir mats, native figures, cigars, and cheroots (*N. Prov.*); talipot leaves, a mat and tent of talipot (*Cent. Prov.*); towels, napkins, and a table cloth (*E. Prov.*).

Rope and cordage made of a variety of substances (*N. & Cent. Provs.*); deerskin rope (*E. Prov.*).

A bullock-pack, a casting-net, a peacock-feather fan, a pingo, a harpoon, bows and arrows, yokes, a scoop, a pitchfork, and ploughs (*E. Prov.*).

A dalada shrine (*W Prov.*).

2. RATTEMAHATMEYA OF LOWER DOOM-BERA. — Iron, rice, kittool fibres, Singhalese axes, knives.

3. RATTEMAHATMEYA OF UPPER DOOMBERA. — Iron, rice, and cotton.

4. RATTEMAHATMETA OF UDUNUWERA.— Rice, kittool fibres, cotton, Singhalese axes, bill-hooks, walking-sticks.

5. RATTEMAHATMEYA OF YATINUWERA.—Rice, Singhalese axes, cocoa-nut fibre ropes, a mammoty, bill-hooks, walking-sticks, Kandyan whips.

6. RATTEMAHATMEYA OF LOWER HEWAHETTE.—Rice, bill-hooks, ploughs.

7. POWER, T. C.—Plumbago and iron, cotton, areca-nut cutters, Singhalese axes, knives, a sickle, a mammoty, and a tevelum pack, a cotton-cleaning machine (*W. Prov.*).

8. STEELE, T.—Plumbago, red and white coral (*S. Prov.*); kittool fibre (*N. S. & Cent. Provs.*).

9. ILLANGAKOON, J. D. S. *Modliar.*—Plumbago, gums (*S. Prov.*).

10. BRODIE, A. O.—Iron (*Cent. Prov.*); neganda fibre (*N. & Cent. Provs.*): neganda mats (*Cent. Prov.*).

11. WIJESINHE, J. D. S. *Modliar.*—Iron, painted boards, Singhalese axes, knives (*S. Prov.*).

12. ISMAIL, C. L. M. *Lebbe, Merikan.*—Oils of citron and lemon grass (*S. Prov.*); citronella (*N. Prov.*).

13. BREARD.—Cinnamon-oil (*W. Prov.*).

14. COREA, D. C. *Modliar.*—Oils of cinnamon and camphor (*S. Prov.*); kittool fibre (*N. S. & Cent. Provs.*).

15. OBEYESEKERE, T. F. S. *Modliar.*—A large collection of oils and medicinal substances (*S. Prov.*).

16. PIERIS, T. A.—Paddy and rice, fibres, neganda mats, ropes of various substances, two inlaid daggers (*N. W. Prov.*); Cadjie gum (*N.W. & W. Provs.*).

17. KARUMARATINE, A. R. *Modliar.*—Paddy and rice, spices (*S. Prov.*).

18. SMITH, D.—Cinnamon (*W. Prov.*).

19. DANIEL, J. B.—Coffee (*W. Prov.*).

20. WRIGHT, W. H.—Fruits, &c. (*Cent. Prov.*).

21. SHAND & Co.—Coir fibre (*W. Prov.*).

22. THWAITES, G. H. U.—Vanilla (*Cent. Prov.*).

23. TISSA, SIRISUMANA.—Jackwood, for dyeing (*S. Prov.*).

24. FORBES, W. G.—Lace (*S. Prov.*).

25. DISSANAIKE, A. T. *Modliar.*—Carved cocoa-nut shell lamp and shells (*S. Prov.*).

26. WICKREMERATINE, A. B. *Modliar.*—Carved cocoa-nut shells; porcupine-quill desk (*S. Prov.*).

27. WEIRALASIRINAYANA ABEYARATINE, DON C. & I.—Carved ebony Davenport and jewel-case (*S. Prov.*).

28. DE COSTA, DON A.—Ebony table and chairs (*S. Prov.*).

29. KARUNANAIKE, DON F.—Ebony footstool (*S. Prov.*).

30. WIMALASIRIRIAYANA, A. DE S.—Pair of lyre tables (*S. Prov.*).

31. LAYARD, C. P.—Pair of ebony flower-stands (*W. Prov.*).

32. SIM, CAPT.—Pair of tamarind card tables (*W. Prov.*).

33. DE SILVA, DON S.—Singhalese axes.

34. RATNAIKE, W. B. *Modliar.*—Rush mats (*S. Prov.*).

35. DE SOYZA, I. *Modliar.*—Manufactured articles from the cocoa-nut tree (*W. Prov.*).

36. DE SOYZA, L. *Modliar.*—Table napkins (*S. Prov.*).

37. MENDIS, A. *Modliar.*—Walking-sticks (*W. Prov.*).

38. TEMPLER, MRS.—Caltura baskets (*W. Prov.*).

39. GIBOORE, JOS.—Models of Ceylon boats.

40. CROFT, J. McGRIGOR, M.D.—Dugong oil.

41. D'OYLEY, REV. C. T.—Dagger worn by the last of the kings of Kandy.

# CHANNEL ISLANDS.

NORTH CENTRAL COURTS, NEAR THE STAIRCASE.

---

## JERSEY.

1. BÉNEZIT, MME.—Artificial flowers.

2. FOTHERGILL, MRS.—Algæ.

3. LABALASTIERE, P.—Eau de Cologne.

4. MULLINS, H.—Photographs.

5. PEACOCK, R. A.—Model of patent dock gates.

## GUERNSEY.

1. ARNOLD, A.—Iodine, and chemical products obtained from sea-weed.

2. BISHOP, A.—Model of an improved paddle-wheel steam-boat.

# DOMINICA.

### WEST SIDE OF NORTH-EAST TRANSEPT.

---

DOMINICA COMMISSION: HON. J. IMRAY, M.D., CHAIRMAN:—

Cotton.

Coffee.

Cocoa.

Indian corn meal.

Arrowroot and other starches.

Sugar.

Spices.

Preserves.

Gums.

Honey.

Beeswax.

Specimens of sugar-cane.

Guavas.

Bananas, &c.

Cork and other woods.

About 170 specimens of indigenous woods polished, and their foliage.

Seeds.

Grasses.

Tortoiseshell and ornaments.

Specimens of natural history.

Building stone.

# HONDURAS, BRITISH.

### WALL NEAR HORTICULTURAL GARDENS, NORTHERN ENTRANCE.

COLLECTION of woods.

Aloe fibre.

Hammock made of ditto.

Samples of sugar.

Rum, coloured and uncoloured.

Tobacco.

Rice.

Pimento.

Black beans.

Turmeric.

Arnatto.

Pepita or cucurbit seed.

# JAMAICA.

EAST SIDE OF NORTH-EAST TRANSEPT, UNDER GALLERY.

---

1. McCRINDLE, J.—Ovum of the nurse shark; nereis.

2. HARRIS, R.—Sea-horse; caterpillar of an ocellated phalæna; carapaces of hawk's-bill turtle and green turtle, and parrot fish; liqueurs; chemical preparations; meals; manufactures.

3. BOWERBANK, DR. L. Q.—Saw of the Sawfish; pottery; sticks; chemical preparations; meals; oils; manufactures; ladies' ornamental work.

4. POTTS, DR.—Animated nature; woods; basts; sticks; botanical specimens; seeds; coffee; chemical preparations; starches; oils; manufactures.

5. BARRETT, L.—Minerals.

6. FYFE, HON. A. G.—Minerals.

7. SAWKINS, J. G.—Minerals.

8. BELL, J. C.—Minerals: including the ores of iron, copper, lead, zinc, cobalt, manganese.

9. R. SOC. OF ARTS, JAMAICA.—Minerals; woods; botanical specimens; rums; liqueurs; sugars; chemical preparations; starches; oils; fine arts.

10. COOPER, CAPT. R.N.—Minerals; chemical preparations; starches.

11. SAILMAN, J.—Manures.

12. SOC. INDUSTRY, HANOVER.—Woods; sticks; rums; liqueurs; sugars; chemical preparations; manufactures.

13. GORDON, R. M.—Woods; basts; coffee; chemical preparations; manufactures, &c.

14. HAUGHTON, —.—Woods.

15. CAMPBELL, REV. J.—Sea shells; woods; basts; botanical specimens; seeds; coffee; chemical preparations; starches; manufactures.

16. WILSON, N.—Vegetable fibres and basts; fruit in spirits.

17. NASH, MRS. J. — Basts; botanical specimens; seeds; ladies' ornamental work.

18. CAMPBELL, DR. C.—Basts; seeds; oils.

19. MAXWELL, DR. J.—Basts; botanical specimens; seeds; chemical preparations; starches; oils; manufactures.

20. BROWNE, W.—Basts; coffee; manufactures.

21. WILSON, E. F. — Basts; seeds; coffee; oils.

22. HYAMS, REV. —.—Sticks.

23. McINTYRE, —. — Botanical specimens.

24. HEPBURN, R.—Botanical specimens.

25. PAINE, W. S.—Botanical specimens; coffee.

26. CLARK, —GENERAL PENITENTIARY.— Botanical specimens; manufactures.

27. VIAN, THE MISSES.—Wax models.

28. VICKARS, HON. B.—Seeds.

29. PHILLIPS, H.—Seeds.

30. JAMES, MISS E.—Seeds.

31. TRENCH, J. L.—Seeds chemical preparations.

32. DARLING, HIS EXCEL. THE GOVERNOR. —Fruits in spirits; manufactures.

33. GORDON, R.—Coffee.

34. 'SHEPHERDS' HALL.'—Coffee.

35. MILLER, W.—Coffee.

## EXHIBITORS OF RUM.

36. ROBERTS & GRIFFITHS.

37. DINGWELL, J.

38. SHORTRIDGE, S.

39. CALLAGHAN, D. & HARRISON, J.

40. MITCHELL, J. W.

41. ESPEUT, P.

42. SHOUBORG, A. A.

43. EAST, HON. H.

44. MCKAY, W.

45. RUSSELL, R.

46. SOLOMON, HON. G.

47. LAWRENCE, J. F.

48. HAMMETT, J.

49. 'HALL-HEAD ESTATE.'

50. WESTMORLAND, HON. H.

51. BERRY, W. SEN. &

52. WALLACE, J.

53. 'SEVEN PLANTATION ESTATE.'

54. MITCHELL, J. H.

55. CHILD, W. D.

56. MARTIN, G. L.

57. WESTMORELAND, W.

58. 'GIBRALTAR ESTATE.'

59. CAMPBELL, T.

60. SINCLAIR, D.

61. VICKARS, W.

62. FOGARTY, D.

63. VICKARS, HON. B.

64. 'FONTABELLE ESTATE.'

65. COLVILLE, E.

66. CAMPBELL, MISS E.

67. THARPE, J.

68. FISHER, J. W.

69. WETZLER, D. B. & CO.

70. WETZLER, D. N. & CO.

71. SEWELL, W.

72. GORDON, J. W.

73. HOSTOP, L.

74. BODDINGTON & CO.

75. TYSON, H.

76. BARRETT, C. G.

77. CLARKE, E. H.

78. GRANT, MISS.

79. CLARKE, T. O.

80. FLETCHER, J. & PERYER, J.

81. HIND, R. & STIRLING, W.

82. MILNER, T. H.

83. ELMSLIE, J. & SHORTRIDGE, S.

84. HOSSACK, HON. W.

85. 'HAYWOOD HALL.'

86. DAWKINS, LIEUT.-COL. W. G.

87. 'RAYMOND ESTATE.'

88. 'NEW RAMBLE ESTATE.'

89. MELVILLE, J. C.

90. PENNANT, COL. THE HON. E. G. D.

91. WRAY, J. & NEPHEW.

92. HARVEY, J.

93. GEORGES, W. P.

94. VAZ, I. N.

95. BARCLAY, A.
96. PAINE, W. S.
97. 'BLUE MOUNTAIN VALLEY.'
98. 'DOVOR ESTATE.'
99. GARRIGUES, H. L.
100. JARRETT, J.
101. DEWAR, R.
102. LAWSON, G. M.
103. ATKINS, G. W.
104. SHAWE, R. F.
105. McGRATH, G.
106. JARRETT, H. N.
107. WILLIAMS, J. L.
108. FAYERMORE, G.
109. DOD, F.
110. HOLT & ALLAN.
111. BODDINGTON, DAVIS, & CO.
112. HINE, R.
113. STEPHEN, A.
114. HOSKINS, J. A.
115. HUTTON, W.
116. PARKINS, DAVIS, & SELBY.
117. WHITELOCK, HON. H. A.
118. HEAVEN, W. A.
119. TOD, F.
120. VAZ, I. N.
121. SHARP, E.
122. HAWTHORN & WATSON.
123. WRAY, J. & CO.
124. DENOES, P.
125. GADPAILE, C.
126. ARNABOLDI, G.

## LIQUEURS, &c.

127. DERBYSHIRE, J.
128. MELVILLE, J. C.
129. COOKE, A.
130. WRAY & CO.

131. DENOES, P.
132. DAVISON, MRS. J.
133. HENRY, H. G.
134. TRENCH, J.
135. DICKSON, R.
136. GRANT, C.

## SUGAR.

137. GARRIGUES, H. L.
138. RUSSELL, R.
139. McKENNON, HON. L. M.
140. PAINE, W. S.
141. HOLT & ALLAN.
142. WILLIAMS, J. L.
143. HIGHINGTON, J.
144. WHITELOCK, HON. H. A.
145. PARKINS, DAVIS, & SELBY.
146. HARVEY, J.
147. SHARP, E.
148. HENRY, W.
149. DOD, F.
150. WHITE, R.
151. GRANT, J.
152. BODDINGTON & CO.
153. VAZ, I. N.
154. STEPHEN, A.
155. HOSKINS, J. A.
156. HIND, R.

---

157. VICKARS, HON. B.—Coffee, &c.
158. SAWKINS, MRS.—Chemical preparations.
159. AARON.—Chemical preparations.
160. A'COURT, WM.—Chemical preparations.
161. SEGUEIRA, E. G.—Chemical preparations.
162. GREY, C.—Chemical preparations.
163. BALL, T.—Chemical preparations.

164. EDWARDS, E. B.—Chemical preparations. Meals; manufactures.

165. BRASS, J.—Chemical preparations. Meals; manufactures.

166. GALL, J.—Chemical preparations. Starches; oils; manufactures.

167. BROWN, MRS. W.—Chemical preparations; starches.

168. NUGENT.—Chemical preparations.

169. KENTISH, MRS. S.—Starches.

170. JOHNSON, MISS M.—Starches.

171. BATLEY, D. W.—Meals.

172. KEMBLE, HON. H. J.—Oils.

173. ARNABOLDI, G.—Oils, seeds.

174. 'BROWNSVILLE.' — Oils: manufactures.

175. O'HALLORAN, J. — Manufactures; ladies' ornamental work.

176. CHITTY, E.—Manufactures; ladies' ornamental work; fine arts.

177. BELL, J.—Manufactures.

178. LAWES & Co.—Manufactures.

180. CAREY, W. O.—Manufactures.

181. FINGZIES, J. K.—Manufactures.

182. 'AN AFRICAN.'—Manufactures.

183. THOMPSON, R.—Manufactures.

184. RANKINE, MRS.—Ladies' ornamental work.

185. HARRISON, MRS. & MISS.—Ladies' ornamental work.

186. JAMES, MISS.—Ladies' ornamental work.

187. ARNABOLDI, MISS. — Ladies' ornamental work.

188. POOLE, MRS. P. — Ladies' ornamental work.

189. TEAP, MISS C.—Ladies' ornamental work.

190. POTTS, MRS. DR.—Ladies' ornamental work.

191. SAVAGE, J. A.—Fine arts.

192. MAULL & POLYBLANK.—Fine arts.

193. JAMAICA COTTON Co.—Basts, &c.

194. CHITTY, E. & RIDGEWAY, A. F.—Botanical specimens; manufactures.

195. RIDGWAY, MRS. G. & SISTERS.—The banner,—"Arms of Jamaica."

196. BOWERBANK, DR.—Woods.

197. COOPER, CAPT., R.N.—Woods.

198. CAMPBELL, DR. J.—Woods.

199. PAINE, W. S.—Woods.

200. HANKEY, THOMSON.—Rum.

201. MOWATT, W.—Cotton.

202. HAMILTON, HON. DR.—Coffee.

203. CASSON, J.—Coffee.

204. ORGILL, HERBERT.—Cotton.

205. HARMAN, REV. J.—Rum.

206. SCOTT, A.—Manufactures.

207. GALL, J.—Cotton.

208. LEVIEN.—Chemical preparations.

209. CASELEY, R.—Chemical preparations.

210. ADONIO, MARCUS.—Chemical preparations.

211. HARVEY, MRS. WM. — Chemical preparations.

212. JAMAICA COTTON Co. — Manufactures.

213. DUBUISSON.—Liqueurs, &c.

214. ABRAHAMS, J. — Engraving on copper.

215. SIMONS.—Animated nature.

216. WILSON, N.—Woods; cotton.

# MALTA.

EAST SIDE OF NORTH-EAST TRANSEPT, UNDER GALLERY.

1. BORG, P. P. & Co.—Black silk lace articles.

2. MUNNERO, V.—Black silk lace articles.

3. GRECH, G.—Embroidered handkerchiefs.

4. AZZOPARDI, F.—Black silk lace articles.

5. MEILACH, A.—Black silk lace articles.

6. MICALLEF, S.—Black and white lace articles.

7. LEONARDIS & BELLIA.—Black lace articles.

8. BELLIA, M.—Lace collar, and a specimen of broad white lace.

9. BELLIA, P.—Black lace articles, silver filigree work.

10. MUIR, G.—Silver and silver gilt filigree work; stone vases; flower pots, &c.

11. VELLA, BROS.—Cotton counterpanes, quilts and table covers, and straw hats.

12. AGUIS, P. P.—Cottonina cloth.

13. MUSCAT, A.—Cottonina cloth.

14. ZAMMIT, G. B.—Cottonina cloth.

15. VELLA, A.—Woollen quilts and table cloths.

16. SCHEMBRI, DR. S.—Specimens of paste; seeds; honey; wax; and orange-flower water.

17. FRANCALANZA, L.—Stone vases.

18. TESTA, F.—Stone vase.

19. SEGOND & BROTHER.—Carved frame.

20. MELI, G.—Samples of leather, hides, &c.

21. DARMANIN & SONS.—Specimens of marble mosaic work; and stone work.

22. DIMECH, C.—Stone work.

23. TONNA, G.—A counterbass.

24. CATANIA, A.—Carved wood-work, and paper pattern.

25. POLITO, Rev. CANON.—Statuettes of saints and knights of Malta.

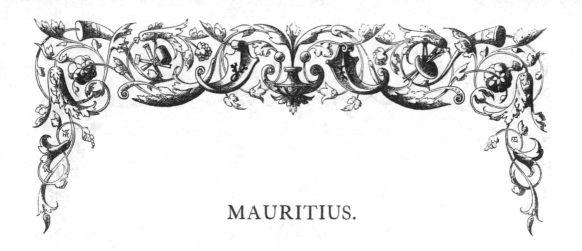

# MAURITIUS.

NEAR CENTRAL ENTRANCE FROM HORTICULTURAL GARDENS.

1. WIEHE & Co. *Labourdonnas Estate.* —Raw and manufactured sugar, candy, and syrup.

2. ICERY, E. & Co. *La Gaité Estate.*— Manufactured sugar, &c.

3. BARLOW, H. *Lucia Estate.*—Pure and impure sugar.

4. BULLEN, R. *Gros Cailloux.*—Ordinary sugar.

5. BELZIM & HAREL, *Trianon.*—Sugar for British market, and for Australian market.

6. GUTHRIE, MESSRS. *Beauchamp.* — Sugar.

7. CURRIE, J., *Beau Séjour.*—Sugar to meet English duty, and turbined with water.

8. STEIN & Co. *Gros Bois.*—Sugar.

9. BREARD, F. *Savannah Estate.*—Sugar.

10. DUNCAN, J. *Director of the Botanic Gardens, Pamplemousses.*—A collection of fibres prepared at the gardens from indigenous plants. Arrowroot, spices, and litchi fruit.

11. D'ESMERY, P.—Manure as active as guano.

12. WICHÉ, P. A.—Coral lime, reef and inland coral, and a case of colonial spirits.

13. BERGICOURT & Co.—Cigars and snuff.

14. MURPHY, W. *Long Mountain.*— Twenty-five specimens of native woods; cotton.

15. FRESQUET, A.—Vacoa tree, the roots made into various articles; hats of date tree leaves, &c.

16. MARTINDALE, E.—Finest honey; a treefern walking-stick.

17. BARCLAY, LADY.—Arrowroot.

18. BONIEUX, *Arsenal.* — Specimens of lalo fibre (*Abelmoschus esculentus*), biscuits, crystallised manioc, and arrowroot flour.

19. CLOSETS, M. DE.—Iron ore.

20. MAUGENDRE, C. A.—A barrel of disinfecting powder.

21. DUMONT, MDLLE.—Embroidered handkerchief.

22. BROUSSE, M.—Vanilla.

23. LANGLOIS, C.—Vanilla.

23. BEDINGFIELD, HON. F.—A walking-stick.

24. MORRIS, MRS. J. — Bag made of acacia seeds; articles from 'Seychelles Isles,' of cocoa-nut leaves and grasses.

25. BEDINGFIELD, HON. F.—A walking-stick.

# NATAL.

NATAL COMMISSIONERS (HON. SEC. R. J. MANN, M.D.).—Extensive collections of—

1. Food Substances: Sugar, arrowroot, coffee, cereals, pepper, roots, fruits, and preserves; tea, cheese, spirits, honey, cured meat, &c.

2. Horns, skins, carosses, tusks, &c., of native animals, the produce of the chase; feathers of the ostrich, crane, &c.; samples of wools, fleeces, &c., and sponge from the Umgeni.

3. Textile and other materials: Cotton, Kafir cotton, flax seed, hemp, &c.; specimens of a variety of woods, bark, fibres, &c.

4. Colonial manufactures: Tanned skins and articles of leather, horns, soap, candles, tallow, cigars, tobacco, bricks, tiles, models, &c.

5. Kafir manufactures, illustrating native industry and domestic economy: Shields, assegais, clubs, musical instruments, ornaments, implements, models, &c.

6. Mineral substances: Freestone, granite, plumbago, quartz, limestone, fossils, sulphuret of lead, mineral water, &c.

7. Specimens of natural history: Snakes, birds, insects, shells, &c.; Kafir medicines.

8. A counter and frame made of native woods; a map of Natal; charts illustrating the climate of Natal, from observations taken by Dr. MANN; water-colour drawings of colonial scenery, with photographs of scenery, portraits of natives, &c.

# NEW BRUNSWICK.

## CENTRE OF NORTH-EAST TRANSEPT.

1. NEW BRUNSWICK COMMISSIONERS. — Specimens of wheat, oats, buckwheat, rye, barley, beans, Indian corn, wheat flour, barley meal, buckwheat meal, rye flour, oat-meal, and hulled barley.

Albertite coal, freestone, and granite.

Native woods, with twigs and leaves.

Native woods, unmanufactured, and manufactured into doors, ballusters, and articles of furniture.

A drill harrow, a mould-board plough, a horse-rake, a saw frame, a double sleigh, a travelling wagon, ships' blocks, &c.

A sample of leather engine-hose and discharge pipe.

Models of steering apparatus; of capstan and windlass; of suspension-bridge over falls of St. John's River; of road and railway bridge over Hammond River; of railway bridge over Salmon River; of a N. B. railway train; snow plough and flange cleaner; of railway engine-house, St. John; of passenger locomotive and tender; and of a saw-mill.

Chilled locomotive car-wheels.

A collection of edge-tools, hammers, &c.

Homespun cloth, rug, socks, and mitts; Indian bead-work and dress, basket-work;

transparent shop window blinds; dried grasses.

Preserved salmon and lobster.

Photograph views in the colony.

2. SCRYMGEOUR, J. — Horse-shoes.

3. HEGAN, J. & J. — Sattinet; union tweed and flannel.

4. SCOVIL, N. H.—Nails, ship-spikes, &c.

5. PHILPS BROS. — Bunting, wrapping and sheathing paper.

7. ADAMS, W. H.—Carriage and railroad springs.

8. MAGEE, A. — Beaver cap, gauntlets, and coat; bear and lynx robe; silk hats, &c.

9. RANKINE, A. — Biscuits.

10. MCMILLAN, J. & A.—Bookbinding.

11. PEARCE, C. — Stand of brass castings.

13. FOSTER, T. A. D.—Case of dentistry.

14. PRICE & SHAW.—Single-horse sleigh.

15. MATTHEWS, G. F. — Case of minerals.

16. SPURR, DR. WOLFE. — Oils from coal.

17. CHAPMAN, J.— Hearth-rug.

18. WESTMORELAND AND ALBERT MINING AND MANUFACTURING Co. — Fossil-fish from their mine.

19. POTTER, E.— Specimen frames; box-wood moulds.

20. BLACKTIN, C.—Circular saws, variety of saws, knives, &c.

21. BOWREN & COX. — Photographic views.

22. FLOOD & WOODBURN.—Ditto.

23. JARDINE, MISSES. — Model summer-house, made from cones.

24. THOMSON, MRS. R.— Watch-pocket, made from cones.

25. STEVENS, MRS. D. B.—What-not, made from cones.

27. FLEMING & HUMBERT, *Phœnix Foun-dry.*—Oscillating steam-engine.

30. McFARLANE, P.—Forks and hoes.

31. REID, J. H.—Pair of moose horns.

32. GILBERT, S. H.—Model stone picker.

33. LAMONT, M.—Indian dress.

35. HARRIS, J. — Cast-iron enamelled mantle-piece.

36. PROVINCIAL PENITENTIARY.—Wooden ware, brushes, &c.

# NEWFOUNDLAND.

UNDER EASTERN GALLERY, NORTH-EAST TRANSEPT.

1. NEWFOUNDLAND GOVERNMENT. — Copper, galena, silver lead, silver ores, marble, and iron; skins of the silver grey, patch, and red fox, the martin, otter, beaver, weasel, and bear; stuffed birds, &c.

A model screw propeller.

2. GISBORNE, F. N.—A map of the colony; a Carriboo stag's head; Esquimaux carvings, and feather bags; patent fire-damp indicator; patent ship-steering signals, and a patent railway ticket-date cutter.

3. O'BRIEN, HON. L.—-Wheat, barley, and oats.

4. DOYLES, E.—Fish oils.

5. FOX, C.—Cod-liver-oil, seal and cod manure, deodorised.

6. O'DWYER, HON. R.—Cold-drawn seal oil.

7. ANGEL.—Cold-drawn seal oil.

8. PUNTON & MUNN.—-Seal stearine, seal and cod oils.

9. McBRIDE & KERR.—Cod-liver oil.

10. STABB, E.—Cod-liver oil.

11. DEARIN, J. J.—Cod-liver oil.

12. MOORE.—Cod-liver oil.

13. MORRIS, L.—Cod-liver oil.

14. McPHERSON.—Cod-liver oil.

15. KNIGHT, S.—Preserved salmon and lobsters.

16. NORMAN, N.—Preserved curlew.

17. TILLY, MESSRS.—Preserved salmon.

18. McMURDO, T.—Cochineal colouring.

19. DEARIN & FOX.—Advertising plates.

20. PAGE & Co.—Silver ores.

21.—LEAMON, J.—Flax.

22. A LADY.—Snake-root, and poplar blossoms.

# NEW SOUTH WALES.

UNDER STAIRS OF NORTH GALLERY, NEAR THE EAST DOME AND NAVE.

MACARTHUR, SIR W.—Woods in variety, of Southern districts. 193 specimens.

MOORE, C. *Sydney.*—Woods in variety, of Northern districts. 115 specimens.

1. THOMPSON, HENRY, *Camden.*—21 specimens of woods adapted for posts, spokes, felloes, &c.

2. HOLDSWORTH, J. B. *George Street.*—1-in. board of Wellington pine.

3. LENEHAN, A. *Castlereagh Street.*—Specimen of rosewood.

4. TRICKETT, J. *Royal Mint.*—Specimen of Wellington pine, with bark on.

5. CUTHBERT, —.—Twenty-one woods for ship-building; iron-bark knee weighing 19 cwt.

6. RILEY, J. *Glenmore.*—Log of Myall timber.

7. JOLLY W. & Co.—Bundle of forest oak shingles.

8. FOUNTAIN,—*Newtown.*—Three specimens of tree-fern stems from Lane Cove.

9. MURRAY, HON. T. A.—Specimens of Myall and Boree woods.

10. HALL, L. MISS, *Sharwood.*—Small specimen of pine from the Billybong.

11. WARD, MR. *Maitland.*—Specimen of iron-bark wood that has been under ground for 25 years.

12. WILLIAMS, J. *Pitt Street, Sydney.*—Staves of mountain ash.

13. SAMUEL, S. *Pitt Street, Sydney.*—Specimen of Myall.

14. DAWSON, A. *Sydney.*—Twenty-three woods adapted for building purposes.

15. CHAPMAN, C. *Sydney.*—Undressed staves.

16. CASEY, J. B. *for McLeay River Committee.*—Seventeen specimens of woods.

RUDDER, E. W. *Kempsey, McLeay River:*—
17. Iron-bark timber cut in 1836, constantly exposed to the atmosphere for 25 years.

18. Two specimens of ash exposed to all weathers for 15 years.

19. Blood-wood that has been under ground 23 years.

20. Turpentine wood that has been exposed for 29 years.

21. Blackbutt that has been exposed to weather for 15 years.

22. Mahogany that has been exposed for 17 years.

23. Pine that has been used in a mill for 15 years.

24. Three specimens of forest oak.

25. Three specimens of fustic.

26. Palm tree.

27. Spotted gum wood.

28. White blossoming acacia, has been exposed to weather 5 years.

29. Monthly rose.

30. Cherry-tree wood.

31. Yellow root.

32. Bastard Myall.

33. Gigantic American Aloe.

34 & 35. Yellow cedar, (two specimens).

36. Tulip wood.

—————

37. MACARTHUR, SIR W. — Two posts of stringy bark 46 years old.

38. MACARTHUR, SIR W.—Narrow leaved iron-bark posts 46 years old; section of post 40 years old; butt of post 38 years old.

39. MANNING, EDYE.—Spotted gum timber from hull of vessel built in N.S. Wales in 1830; vessel still plying.

40. MORT, T. S.—Log of Brigalow timber.

41. CLARKE, S.—Six specimens of fancy cedar.

42. DEAN, A. *Liverpool Street.*—Two pieces of blackbutt timber, one of blue gum.

43. MACARTHUR, SIR W.—Specimen of blue gum of Camden.

44. HILL, E. S.—Specimen of iron bark of Paramatta, and tea tree of Brisbane Water.

44A. DAWSON, A.—20 specimens of building wood.

44B. HOWARD & Co. *Berner Street.*— Tables and cabinet work from woods of New South Wales.

45. GOSPER, J. *Colo. Hawkesbury River.* —Maize in cob.

46. GOSPER, J.— Maize shelled.

MACARTHUR, J. & W. *Camden Park: —*

47. Six kinds of maize in cob.

48. Four kinds of shelled maize.

49. Four bundles of maize.

—————

50. PECK, M. *Hunter River.* — Yellow maize, arrived at perfection in 100 days, 70 bushels per acre.

51. BOWDEN, C. *Hunter River.*— Yellow maize in cob.

52. BATTEN, C. *Frederickton.*—Maize in cob.

53. ANDERSON, CAMPBELL, & Co. *Sydney.* — Two samples of maize in cob.

54. AYTOM & BOURNE, *near Stroud.*— Maize in cob.

55. DOYLE, A. J. *Midlorn, Hunter River.* — Maize in cob.

56. PATERSON, T. *Lorn, Hunter River.*— Maize in cob.

57. OAKS, MONTAGUE C. *Seven Oaks, McLeay River.*—Maize in cob, yielding 100 bushels an acre.

58. CLEMENTS, INGHAM S. *Bathurst.*-- Wheat.

59. MACARTHUR, J. & W. *Camden Park.* — Wheat.

61. CHAPPELL, T. *Mudgee.*—Wheat.

62. FUTTER, Mr. *Lumley, Bungonia.*— Wheat.

63. THOMPSON, H. *Camden.*— Wheat.

64. MACARTHUR, J. & W. *Camden Park.* — White wheat, three samples, 68 lbs., 66 lbs., and 64 lbs. to the bushel.

65. THOMPSON, H. *Camden.* — Wheat flour.

66. SOLOMON, VINDEN, & Co. *West Maitland.*—Wheat flour.

67. SPRING, J. & SONS, *Surrey Hills Mills.*—Wheat flour.

68. ANDERSON, CAMPBELL, & Co. *Sydney.* —Wheat flour.

69. PEMELL, J. and SONS, *Paramatta Street.*—Wheat flour.

D

70. Hays, J. *Burrowa.*—Wheat flour.

Thompson, H. *Camden :* —

71. Wheat flour.

72. Maize flour.

73. Solomon, Vinden, & Co. *West Maitland.*—Maize flour.

74. Anderson, Campbell, & Co. *Sydney.*—Two samples of maize flour.

75. Stroud Mills.—Maize flour.

76. Havenden, J. *Grafton.* — Cotton : the seeds were sown in August, and picked in May.

77. Peck, M. *Hunter River.*— Cotton.

78. Fogwell, R. *Hunter River.*—Cotton.

79. Hickey, E. *Osterley, Hunter River.* —Cotton.

80. Moss, H. *Shoolhaven.*—Cotton.

81. Nowlan, J. B. *Hunter River.*— Cotton, various samples.

82. Vinden, G. *West Maitland.*—Cotton grown by J. M. Ireland, Williams River.

83. Goodes, C. W. *Sydney.* — Cotton grown at Clarence River.

84. Caldwell, J. *Pitt Street.*—Feejee Island cotton.

85. Peck, M. *Hunter River.*—Arrowroot from Dalwood roots.

86. Neale, Mrs. *Elizabeth Farm, Paramatta.*—Arrowroot.

87. Hassall, Miss, *Camden.*—Arrowroot.

88. Robertson, W. *Hermitage, Grafton, Clarence River.* Arrowroot.

89. Filmer, W. *West Maitland.*—Arrowroot.

90. Filmer, W. *West Maitland.*—Arrowroot.

91. Beaumont, W. *Botany.*—Arrowroot.

92. Thornton, W. *McLeay River.*—Arrowroot pulp for paper stuff.

92A. Gunst, Dr. J. W. *Clarence River.*—Arrowroot : 42 square feet of land produced 436 lbs. of flour.

93. Lardner, H. *Clarence District.*—Nettletree fibre (*urtica gigas*). This tree sometimes reaches 60 feet in height : but the tree seems to consist chiefly of water, and is useless, even as firewood : the bark, however, contains abundance of fibre.

94. Gosper, J. *Colo, Hawksbury River.*—Kurrajong bark.

95. Bowler, C. E. *Newtown.*—Fibre of Miranda reed.

96. Blacket, E. T.—Kurrajong bark.

97. Bawden, T. *Lawrence, Clarence River.*—Nettletree and sycamore barks.

98. Moss, H. *Shoalhaven.*—Burrawang fibre.

99. Calvert & Castle, *Cavan, near Yass.*—Kangaroo grass, and another grass.

100. Greaves, W. A. B. *Dovedale, Clarence River.*—Sycamore fibre.

101. Lardner, A. *Grafton, Clarence River.*—Nettletree fibre and Kurrajong bark.

102. Lardner, A. *Grafton, Clarence River.*—Dilly bag of brown Kurrajong, made by aborigines of Clarence River.

103. Moore, C. *Botanic Gardens.*—Kurrajong bark.

104. Garrard, *Richmond River.*—Six bags made by aborigines, of native fibres.

105. De Mestre, A. *Terrara, Shoalhaven.*—Two kinds of bark used by the blacks, as fish poisons : one of them is an excellent tanning substance.

106. Hill, E. S. *Wollahra.*—Tea-tree bark (*phormium tenax*), and fish wattle bark.

107. Macarthur, J. & W. *Camden Park.*—Native flax.

108. Scott, T. W. *Brisbane Water.*—Tea-tree bark.

109. Snape, P. *Stroud.*—Wattle bark.

110. Thompson, *Camden.*—Wattle bark.

111. Krefft, G. *Australian Museum.*—Grass-tree gum.

112. Simmons, C. *Waverly.*—Grass-tree gum.

113. Rudder, E. W. *McLeay River.*—Collection of gums, dyes, varnish, &c., and 200 specimens of dyeing.

114, 115.   Bosch, J.—Varnishes and gums.

116, 117.—Smith, Capt.—White wine and red wine.

Macarthur, J. & W. *Camden*:—

118.   White wine of 1858.

119.   White wine of 1858.

120.   White wine of 1856.

121.   White wine of 1849.

122.   White wine of 1848.

123.   Muscat of 1845.

124.   Muscat of 1853.

125.   G. Muscat of 1851.

126.   Red wine of 1853.

127.   Red wine of 1851, bottled in 1855.

128.   White wine of 1851, bottled in 1855.

129.   White wine of 1853, bottled in 1856.

130.   Muscat, 1853, bottled in 1856.

131.   White, 1858, bottled in 1862.

132.   White, 1858, bottled in 1862.

133.   White, 1858, bottled in 1862.

134.   Red, 1854, bottled in 1862.

135.   Red, 1854, bottled in 1862.

136.   Muscat, 1854, bottled in 1862.

136A.  Red, 1849, bottled in 1861.

————

137.   Carmichael, H.—White wine.

138.   Ireland, J. M.—White wine.

139.   Pile, G.—Amontillado.

140, 141.   Bettington, Mrs. — White, vintages 1858 and 1859.

142.   Bettington, Mrs.—Claret, 1860.

143.   Blake, J. E. —Red wine.

144, 145.   McDougall, A. L. —White wine 1860 ; red wine, Malbec, 1860.

146, 147.   Windeyer, A.—Red Hermitage and white Madeira, 1858.

148, 149.   Lindemann, H. L.—Red and white Cawarra.

150.   McDougall, A. L.—White wine, 1851.

151.   Cowper, Hon. C.—Wivenhoe Madeira, 1853.

152.   Mort, T. S. — Irrawang white, bottled 1850.

153, 158.   Cooper, Sir D.—Dalwood red and white Lindemann's Cawarra (two kinds), Docker's Burgundy, and Lawson's Burgundy.

159.   Aspinall, T.—Australian wine.

160, 161.   Sangar, J. M. — Rosenberg Shiraz and brown Muscat of 1860.

162—164.   Sangar, J. M.—Riesling, Tokay, and Aucarot wines of 1860.

165.   Smith, J.—Red Kyamba, 1859.

166.   Frauensfilder, J. P.—White wine, 1858.

167, 168.   Moyse, V.—Australian wines.

169, 170.   Rodd, B. C. — Australian wines.

171.   Aspinall, T.—White wines.

172.   Farquhar, H. M.—Muscat of Camden.

173.   Smith, J.—White Kyamber, 1859.

174.   Blake, J. E. — White Kaludah, 1856.

175.   Blake, J. E.—Red Kaludah, 1856.

176.   Blake, J. E.—White Kaludah, 1858.

177.   Wright, J. *Balgowrie, Wollongong.* —Cayenne pepper, from chillies of Nepaul kind.

178.   Macarthur, J. & W. *Camden Park.*—Cayenne from chillies of Nepaul kind.

179.   Bolland, Mr. *West Maitland.*—Cayenne pepper from chillies of the Nepaul kind.

180.   Thornton, Capt. W. — Cayenne pepper from chillies grown on the McLeay River.

181.   Church, J. *West Maitland.*—Tobacco, leaf and manufactured.

182.   McCormack, J. *West Maitland.*—Leaf-tobacco.

183. HAMILTON & Co. G. & J. *Hunter Street.*—Biscuits of various kinds from flour, and arrowroot.

184. WILKIE, G. & Co., *George Street.*—Biscuits of various kinds.

185. JACQUES, MISS L. *Balmain.*—Four samples of native currant jelly.

186. SKILLMAN, MR. *near Stroud.*—Box of dried peaches.

187. CAPORN, W. G., *Port Street, Sydney.*—Orange Wine.

188. MILLER, P. *Paramatta.*—Raspberry and mulberry wines.

189. LAVERS, J. V. *Sydney.*—Ginger wine.

190—195. MACARTHUR, J. & W. *Camden Park.*—Capers, sorghum, broom-millet, imphee, beans and carob pods.

196. MOSS, H. *Shoalhaven.*—Sponge from Jerrimgong Beach.

197. PRESCOTT, H. *Sydney.*—Italian Rye-grass seed.

198. THORNTON, CAPTAIN.—Sarsaparilla grown on McLeay River.

199. MACARTHUR, J. *Camden Park.*—Walnuts.

200. SCOTT, J. W. *Point Clare, Brisbane Water.*—Sugar-cane.

201. LEDGER, C. *Sydney.*—Alpacas.

202. JONES, D. & Co. *George Street, Sydney.*—Native cat-skin rug.

203, 204. HARBOTTLE, W. *George Street.*—Whales' teeth, tortoiseshell, and beeswax.

205. NORRIE, J. S.—*Pitt Street.*—Beeswax.

206. ROSE, MRS. *Campbell Town.*—Honey.

206A. NORRIE, J. S.—Honey.

207. ROBB, MRS. JAS. *Kiama.*—Feathers.

208. CRAWLEY, T. W. *Market, Sydney.*—Skins of platypus.

209. CHILD, W. *Mount Vincent, West Maitland.*—Cochineal from the acacia.

210. BELL, H. *Pitt Street, Sydney.*—Bone Manure.

211—213. HARBOTTLE, W. *George Street, Sydney.*—Sperm, southern whale, and Dugong oils.

214. KIRCHNER, W. *Grafton, Clarence River.*—Oleine, or tallow oil.

215. YOUDALE, J. *West Maitland.*—Neat's-foot oil.

216. BELL, H. *Pitt Street, Sydney.*—Neat's-foot oil.

217. RUDDER, E. W. *Kempsey, McLeay's River.*—Purified neat's-foot oil.

218. BATTLEY, J. *Castlereagh Street.*—Shank bones.

219. SKINNER, T. *Darling Point.*—Silk.

220. TURNER, C. *West Maitland.*—Silk.

221. SANDROCK, G. F.—Silk.

222. LORD, MRS. *Double Bay.*—Silk.

223. KELLICK, J. JUN. *Philip Street, Sydney.*—Silk.

224. LEE, MRS. SEN. *Parramatta.*—Silk.

225. WHITING, J. *Hanley Street, Woolloomooloo.*—Silk.

226. LORD, MRS. *South Head Road.*—Silk.

227. BELL, H., *Pitt Street.*—Beef.

228. MANNING, J. *Kamareuka.*—Beef.

229. BATTLEY, J. *Castlereagh Street.*—Ox tongues.

230. MYERS, P. *Pitt Street, Sydney.*—Fish.

231. NOTT, J. *West Maitland.*—Tallow.

232. NOTT, J. *West Maitland.*—Alpaca tallow.

233. COMMISSIONERS OF SOUTH WALES EXHIBITION OF 1862.—Alpaca pomade.

234, 235. BATTLEY, J. *Sydney.*—Beef and mutton tallow.

236. CHILD W. *Mount Vincent, West Maitland.*—Cheese.

237. HOLDEN, A. *Gresford, Paterson River.*—Cheese.

238, 239. RILEY & BLOOMFIELD, MESSRS.—Fleece wool, washed on sheep's back.

240. Cox, E.—Fleece wool, washed on sheep's back.

241. Marlay, E.—Fleece wool, washed on sheep's back.

242. Lord & Ramsay.—Fleece wool, washed on sheep's back; from sheep the progeny of merino stock of Messrs. Macarthur.

243. Macansh, J. D.—Wool, washed on sheep's back.

244. Ebsworth & Co. *Sydney.*—Wool, washed on sheep's back.

244A. Daugar & Co.—Four fleeces washed on sheep's back.

245. Hayes, T.—Scoured wool.

246. Riley & Bloomfield, Messrs.—Scoured wool.

247. Cox, E. K.—Scoured wool.

248. Clive, Hamilton, & Rowland J. Trail.—Scoured wool, 1st, 2nd, 3rd, and 4th quality, and lambswool.

249. Bettington, Mrs.—Scoured wool.

250. Lord & Ramsay.—Scoured wool.

251. Cox, G. H. & A. B.—Scoured wool.

252. Ebsworth & Co. *Sydney.*—Scoured wool.

253. Bell, H. *Sydney.*—Scoured wool.

254. Riley & Bloomfield, Messrs.—Wool in grease.

255. Cox, E. K.—Wool in grease.

256. Clive, Hamilton, & Rowland J. Trail.—Wool in grease.

257. Macansh, J. D.—Wool in grease.

258—260. Cox, G. H. & A. B.—Wool in grease.

261. Ledger, C.—Fleece of Cotswold merino ram, 3 years old, 10 months wool.

262. Ledger, C.—Fleece Leicester merino ram, 2 years old, 10 months wool.

263. Donaldson, Sir S. A.—Six fleeces in grease.

264. Ledger, C. *Sydney.*—Alpaca wool.

265. Cooper, Lady.—Case of specimens of gold.

266. Wilson, D. *Adelong.*—Seven specimens of quartz and mundic.

267. Thomas, J., C. E. *Railway Department.*—Iron ore, coal, and quartz.

268. Bawden, T. *Clarence River.*—Carbonate of iron.

269. Lynch, R. Esq.—Auriferous quartz.

270. Gordon, Mrs.—Lead, silver, and copper ores.

271. Larkin, E. Esq.—Stalagmitic deposit from Wianamatta rocks.

272. Samuel S. Esq. & Christoe, J. P.—Copper ores and copper.

273. Loder, A. Esq.—Combustible schist.

274. Williams, Capt. D.—Silicate of magnesia.

275. Levien, A. Esq.—Specimens illustrative of strata encountered in working a rich claim.

276. Blacket, T. Esq.—Two kinds of building stone.

277. Macarthur, Sir W.—Nine specimens of building stones.

278. Croaker, C. W.—Limestone, sulphate of barytes, and green carbonate of copper from Bathurst.

279. Australian Agricultural Co.—Iron ore from Stroud.

280. Moss, H. Esq.— Native alum from Shoalhaven.

281. Laidlaw, T. Esq.—Lead ore from Jobbin's mine.

282. Hume, H.—Iron, copper, and lead ores.

283. Brown, Mrs. W.—Silicified rock, with leaves, &c.

284. Pearson, R. W.—Copper ore from Good Hope Mines.

285. Patten, W. Esq.—Polished Devonian marble.

286. Brown, Thos.—Bituminous schist, granite, ironstone, sandstone, alum, sulphate of magnesia, slate and limestone.

287. Christoe, J. P.—Copper and its ores.

288. DAWSON, A.—Stones used in building.

289. WILSON, A.—Silicate of magnesia.

290. BLAXLAND, A. — Gypsum, emery powder, coal, and geological specimens.

291. THOMPSON, H. *Camden.*—Ironstone and iron ore of Camden district.

292. MOREHEAD & YOUNG, MESSRS. — Copper ores.

293. HOLDEN, A.—Sulphide of antimony.

294. DANGAR, T.—Auriferous quartz.

295. CLARKE, REV. W. B.—White porcelain clay.

296. KEENE, W. inspector of coal fields. — Coal from eleven seams in the colony.

297. JOPLIN, C.—Tooth and bones of fossil kangaroo.

298. SOLOMON, S. & H. *Eden, Twofold Bay.*—Copper and lead ores, and auriferous sand.

299. VYNER, CAPT. A.—Polished marble.

300. McMURRICK, MR. — Auriferous quartz from new diggings.

301. SNAPE, P.—Iron ore, coal, clay, and limestone from Stroud.

302. HOWELL, MISS.—Two specimens of stone.

303. RUDDER, E. W. *McLeay River.*—Earthy cobalt.

304. SAMUEL, S. *Pitt Street.*—Marble, &c.

305. SNAPE, P. *Stroud.* — Ironstone, &c.

306. McCULLUM, A. *Woolgarlow.* — Copper ore.

307, 308. SAMUEL, S.—Rock killas from Ophir mine; and iron ore.

309. AUSTRALIAN AGRICULTURAL Co.— Coal from bore-hole seam, Newcastle.

310. AUSTRALIAN AGRICULTURAL Co.— Coal from Bellambi.

311. DALTON, F.—Specimens of minerals from Rocky River District: including petrified wood, topaz, opal, &c.

312. ROYAL MINT, *Sydney.*—Samples of gold, characteristic of the gold fields of the Colony.

313. ROYAL MINT, *Sydney.*—Auriferous quartz from some of the veins in N.S. Wales, now being worked, or capable of being worked, with profit.

314. ROYAL MINT, *Sydney.*—Two cases, illustrating the various deposits encountered in sinking for gold in this Colony, and the character of gold thus obtained.

315. CLARKE, REV. W. B.—Thirty genera of mesozoic fossils, from Wollumbilla Creek and the Fitzroy Downs.

316. KEANE, W. *Government Examiner of Coal Fields.*—Series of palæozoic fossils, in 40 compartments.

317. IRONSIDE, ADELAIDE E.—The marriage in Cana of Galilee, painted at Rome in 1861 by the Exhibitor.

318. INGELOW, G. K. — Water-colour drawings of Entrance to Sydney Harbour, and Manly Beach.

319. MARTENS, C.—Sydney Head (water colour).

320. COOPER, LADY.—Two drawings by C. Martens, and one by THOMAS, water colour.

321. COMMISSIONERS OF N.S. WALES EXHIBITION.—Six drawings, &c., by E. B. Boulton.

322. HAMILTON, E. *First Provost of University.*—The University of Sydney.

322A. DENISON, A.—Senate House of the University.

323. NICHOLL, W. G.—Plaster bust, allegorical of Australia.

324. TENERANI.—Photograph of marble statue of W. C. Wentworth, by Tenerani, of Rome.

324A. BLACKET, E. T.—Photographs in variety.

325. JOLLY & Co. *Sydney.*—Five photographic views.

326. PATERSON, J.—A. S. W. Co.'s patent ship Pyrmont—photograph.

327. FREEMAN, BROS. *George Street.*—Collection of photographs.

328. PLOMLEY, JENNER, *Paramatta River.*
—Fifty stereographic views.

329. BLACKWOOD & GOODES, *George Street.*
—Eight photographic views.

330. GALE, F. B. *Queanbyan.*—Photographic portraits of aborigines and half castes.

331. DALTON, E. *George Street.*—Fifteen frames of photographs.

332. WINGATE, MAJOR.—Panoramic view from Pott's Point—photograph.

333. MORT, T. S.—Five photographs.

334. CUTHBERT, J.—Ship-building yard—photograph.

335. WILLIAMS, J. *Pitt Street.*—Masonic officers.

336. HOBBS, J. T.—School of Arts, by an amateur—photograph.

337. JOLLY & CO. MESSRS.—Five saw mills—photograph.

338. YOUNG, RIGHT HON. SIR J. and LADY.—Kangaroo and emu of Australian gold, by Hogarth.

[338.] KANGAROO, OF AUSTRALIAN GOLD.

339. FLAVELLE, BROS. *George Street.*—Mounted inkstand.

340. FINCK & BACKEMANN, *Market Street.*—Bracelet and brooch of Australian gold.

341. HOGARTH, J. *New South Wales.*—Natives in precious metals.

342. BRUSH & MACDONNELL, *George Street.*—Two emu eggs mounted.

342A. M'LEAY, MRS. G.—Brooch of white topaz, from the Murrumbidgee

342B. M'LEAY, MRS. G.—Table ornaments, made from the seed vessels of plants near Paramatta, by the nuns of Subiaca, New South Wales.

343. RIDLEY, REV. W. *Rushcutter's Bay.* —Two primers of aboriginal language.

344. COWPER, MRS.—Key to aboriginal language.

345. BERNICKE, C. L. *Kent Street.*— Bookbinding.

346, 347. SHERRIFF & DOWNING. — Account books and bookbinding.

348. RICHARDS, T. *Government Printing Office, Sydney.*—Bookbinding and publishing.

349. FURBER, A. G.—Bookbinding.

350. SANDS & KENNY, *George Street.*— Account books and printed books.

351. REES, G. H. *Castlereagh Street.*— Books illustrative of Colonial binding.

352. WAUGH, J. W. *George Street.* — Bookbinding and publishing.

353. DEGOTARDI, G.—Specimens illustrative of the advancement of printing in Australia.

354. MOSS, L. *Hunter Street.* — Five specimens of music publishing.

355. CLARKE, J. R. *George Street.*—Book of printed music, &c.

[338.] EMU, OF AUSTRALIAN GOLD.

356. ANDERSON, J. R. *George Street, Sydney.*— Two specimens of printed music.

357. CLARKE, H. T. *Castlereagh Street.*— Two gig whips, four stock whips, and piece of hide.

358. HINTON, BROS. *George Street.* — Gentleman's saddle.

359. M'CALL, D. *Hunter Street.*—Gentleman's saddle, &c.

360. BOVIS, C. *King Street, Sydney.*— Three pairs of boots and pair of slippers.

361. VICKERY, J. 375 *George Street, Sydney.*—Four pairs of rivet boots; two pairs of slippers.

362. HALL & ANDERSON, *Pitt Street.*— Harness leather, sole and kip leather, saddles, &c. &c.

363. BRUSH, J. *George Street.*—Lady's side saddle.

364. LOBB, J. *Pitt Street.*—Boots.

365. BEGG, J. E. *Glenmore Tannery.*—Sole leather.

366. ROW, J. *Camden.*—Kip leather.

367. GOODLUCK, J. G. *Camden.*—Leather.

368. SMITH, J. *Botany Tannery.*—Skins and leather; improved graining board.

369. VINDEN, G. *West Maitland.*—Case of Colonial rosewood.

370. REYNOLDS, A. *Balmain.*—Cedar boat.

371. RENNEY, W. *Pitt Street.*—Work table.

TUCKEY, W.:—

372. Chest of drawers.

373. Library door.

374. Carved cedar font.

375. Carved Elizabethan frame.

---

376. ORAM, E., *Liverpool Street.*—Carved cedar truss.

377. MORT, T. S.—Carving of Moreton Bay staghorn fern.

378. FULLER.—Window blind.

379. MILGROVE, H. *Park Street.*—Two spiral-turned candlesticks.

380. JONES, W. & SON, 21 *Surrey Street.*—Cabinet of cypress and other woods.

381. COOPER, LADY.—Set of drawers, table, book-stand, &c.

382. GILLMAN, MRS.—Table top of cypress pine from Northern districts.

383. COMMISSIONERS OF N. S. W. EXHIBITION OF 1862.—Fourteen pieces of cabinet work from Colonial woods.

384. EBSWORTH, F. E. & Co. *Bridge Street.*—Reel cotton, manufactured from cotton of N. S. Wales.

385. PATERSON, MRS. *Manton Creek.*—Socks from opossum wool.

386. ZIONS, H. *Castlereagh Street.*—Coats and vests made from a Colonial invention.

387. RUSSELL, CAPT. *Regentville.*—Colonial tweed.

388. FARMER & PAINTER.—Two suits of Colonial tweed.

389. CAMPBELL, M. M. *Sussex Street.*—Colonial tweed.

390. LONDON COMMISSIONERS OF N. S. WALES.—(A.) Merino wool manufactures: series under ten consecutive numbers of light and heavy fabrics, manufactured by Joseph Craven, Benjamin Gott & Co. Barker & Co. Verity of Bramley, Paton & Co. Pease & Co. Carr & Co. of Twerton; also sundry goods from the general committee of Bradford. (B.) Alpaca wool manufactures: series of cloths manufactured at Bradford from Australian alpaca wool.

390A. BARKER & Co. *Leeds.*—Case showing various stages of manufacture between raw wool and cloth.

390B. BRADFORD COMMITTEE.—Case showing various stages of manufacture in merino and alpaca wool.

391. DIRECTORS OF RANDWICK ASYLUM.—Cabbage tree plait.

392. DUFFIN, J. *Sussex Street.*—Cabbage tree manufactures.

393. ENGLISH, MISS KATE, *Jamberoo.*—Cabbage tree and plait.

394. PRESCOTT, H. *Sydney.*—Cabbage tree 100 hands.

395. NEW SOUTH WALES COMMISSIONERS.—Plait work, cap, belt, and mat, of cabbage tree.

396. FARMER & PAINTER, *Pitt Street, Sydney.*—Two hats of cabbage tree.

397. HASSALL, REV. T. *Berrima.*—Cabbage tree hat.

398. GREGORY & CUBITT, *Aldermanbury.*—Cabbage tree hats.

399. BOUSFIELD, F. *Crystal Palace.*—Cabbage tree hat.

400. BIDDELL, BROS. *George Street.*—Confectionery.

401. CATES, W. G. *George Street.*—Confectionery.

402. AUSTRALIAN SUGAR Co. *Sydney.*—Sugar and spirits, manufactured in the Colony.

403. SACHER & JOSSELIN, *Sydney.*—Confectionery.

404. WELLAM, N. *Burwood.*—Pottery.

405. ENEVIS, W. *Bathurst.*—Soap and candles.

406. KIRCHNER, W. *Grafton.*—Candles and soap.

407. KENSETT BROS. *Campbell Street.*—Blacking.

408. MONK, D. J. *Pitt Street, Redfern.*—Blacking.

409. BLAND, DR. *Sydney.*—Model of ship, showing mode of extinguishing fire in hold, &c.

410. BLAND, DR.—Model of atmotic ship.

411. DAWSON, R. *Lower George Street.*—Model of coffer dam.

412. COWPER, MISS.—Model of St.Philip's Church.

413. BLACKET, E. T. *Sydney.*—Model of St. Mary's Church, Maitland.

414. THACKERAY, B.—Model of horse railroad, invented by P. Brawen.

415. WOORE, T. *Sydney.*—Model of supporting rails.

416. KIRKWOOD, D. S. *Bega.*—Model of bridge.

417. FRANCIS, H. *Balmain.*—Model gas retort.

418. COOKE, A. *Randwich.*—Model of St. John's church, Darlinghurst.

419. WOORE, T.—Model of bridge.

420. LOW, J. C. *Pitt Street.*—Model of Sofala gold diggings.

420A. CRYSTAL PALACE CO.—Model of Government dry dock in Sydney Harbour: length, 300 feet; depth, 25 feet.

420B. CRYSTAL PALACE CO.—Model of Mort dry dock in Sydney Harbour: length, 350 feet; depth, 26 feet. The 'Simla,' one of the largest steam ships in the fleet of the P. & O. Co. has been repaired in this dock.

421. COWPER, S. S., H. MOSS, E. HERBORN, AND MISS MACARTHUR.—Aboriginal implements and weapons, and work-box.

422. HALL, J. B. *Richmond Terrace.*—Case of birds.

423. CAPORN, W. G. *Fort Street.*—Two mats, rock sea-weed.

424. HENSLEY, MRS. *Hunter River.*—Sea-weed.

425. BATE, J. E. *Merrimbula.*—Collection of sea-weeds.

426. KREFFT, G. *Australian Museum.*—Seven cases of reptiles.

427. BECKER, A. *Australian Museum.*—Fishes, &c.

428. RUDDER, E. W. *Kempsey.*—Sixty-eight specimens of birds.

429. CRAWLEY, T. W. *Market, Sydney.*—Stuffed birds (120), sea-weeds, &c.

430. SAWYER, H. *Derwent Street, Glebe.*—Insects.

431. BARNES, H. *Australian Museum.*—Casts of reptiles.

432. COOPER, SIR D.—Various geological specimens.

433. KREFFT, G. *Australian Museum.*—Reptiles in variety.

434. GIPPS, LADY.—Case of birds.

435. LAVERS, G. V. *George Street.* Noyeau and orange wine.

436. MONK, D. J. *Pitt Street, Redfern.*—Vinegar.

437. LEYCESTER, A. A. *Singleton.*—Fishing rod.

438. FULLER, *Pitt Street.*—Dwarf Venetian blinds.

439. CHAPMAN, C. *Sydney.*—Colonial staves.

440. FLETCHER, D. *Wynyard Square.*—Dental work.

441. COOPER, J. *Woolloomooloo.*—Church glazing.

442. JENNINGS, W. *George Street, Sydney.*—Cutlery.

443. WACEY, G. *William Street, Woolloomooloo.*—Intersection ornaments, &c.

444. THE SURVEYOR GENERAL.—Two maps of Victoria and N. S. Wales.

445. BRYCE, J. *Laurence, Clarence River.*—Casks of different woods.

446. HILL, E. S. *Woolhara.*—Peg tops of various woods.

447. MACARTHUR, SIR W. — Cherry-brandy.

447A. MACARTHUR, SIR W.—Axe-handles.

448. MANN, G. K.—Hone stones.

449. NORRIE, J. S. *Pitt Street.*—Fossil cedar.

450. DALGLEISH, D. *Sydney.*—Fossil.

451. MANNING, SIR W. M.—Vase made of grass tree.

452. PATERSON, F. *Market Street.* — Twenty-six herbals.

453. SELFE, H. *Pitt Street.*—Broom of cabbage tree.

454. HERTZHAUMER, C.—Surgical instruments.

455. LACKERSTEIN, A.—Cayenne pepper.

456. HAYES, T.—Sheep-skin rugs.

457. NORRIE, J. S. *Pitt Street.*—Cajeput oil, and a variety of ottos and essential oils.

458. THOMSON, H. *Camden.* — Lime, bricks, bullock yoke, &c.

459. SCHULTE, R. *Woolloomooloo.*—Dyes and dyeing.

460. SHAW, G. B.—Two engravings.

461. WAINWRIGHT, J.—Flute of Myall wood.

462. JOLLY, W. & Co.—Bullock yoke, &c.

463. MACARTHUR, SIR W. — Walking-sticks.

464. ENEVER, W. *Sydney.*—Coach wheel illustrating woods used in the trade.

465. GOODSELL, F. J. *Newtown.*—Bricks.

466. HALE, T. *Bellambi.*—Quartz gold.

467. HERTZHAUMER, C. *King Street.*—Surgical instruments, invented by Dr. Bland.

468. BOUSFIELD, F. *Crystal Palace.*—A portion of Captain Cook's MS. journal, describing the discovery of Botany Bay; letters, medals, &c.

469. MACARTHUR, GENERAL.—Gold medal struck in 1856 to commemorate the establishment of Constitutional Government in Victoria.

# NEW ZEALAND.

UNDER GALLERY FRONTING NAVE, NEAR EASTERN DOME.

---

## Province of Auckland.

1. NEW ZEALAND, BANK OF.—Otago gold.

2. HUNTER.—Gold.

3. READING, J. B. — Sample of gold from Terawiti.

4. HEAPY, C. & EWEN.—Gold and auriferous quartz.

5. HEAPY, C.—Minerals, ores, auriferous deposits, fossils, views, maps, frames, shells, magnetic sand.

6. JONES, A.—Copper ores.

7. GREAT BARRIER COPPER MINING CO. —Copper ore.

8. HOLMAN, J. — Stalactite, building stone.

9. SMALES, REV. G.—Stalactite, lignite, lava, and quartz in ditto, war weapons, paddles, obsidian.

10. HANCOCK, J.—Specimens of limestone, trachytic stone.

11. CADMAN, J.—Colouring pigment, slab of mottled kauri; volcanic and building stones.

12. GILBERT, H. Pumice stone.

13. POLLOCK, T.—Ironstone; fire clays.

14. ARROWSMITH, W.—Iron sand.

15. BURNETT, J.—Marble.

16. BUCHANAN, F.—Moss agate; spar.

17. WHITE, J.—Agates, cornelians; native adzes; sinkers for fishing.

18. BRIGHTON, W.—Silicate.

19. ELLIOTT, G. E.—Sulphur, silicious incrustations of.

20. ANDREWS, H. F.—Sulphur; shells.

21. WAIHOIHOI COAL CO.—Coal.

22. COLE, G.—Petrified wood; soap stone.

23. PREECE, REV. G.—Petrified rimu (wood).

24. SCOTT, A.—Kauri gum.

25. THIERRY, E. D.—Petrified Kauri gum.

26. WELLS, S.—Blue obsidian; petrified wood.

27. OTAMA-YEA.—Specimens of petrifaction.

28. ROE & SHALDERS.—Slabs of Kauri.

29. Ring, C.—Slab mottled Kauri, rimu-matai, auriferous earth.

30. Brown, T. W.—Aleake wood.

31. Gibbons, Messrs.—Woods (Reware, wa and Hinau).

32. Commissioners International Exhibition for Auckland.—Woods from Museum at Auckland.

33. Ellis.—Vase, cotton stand, with specimens of New Zealand woods.

34. Ninnes, J.—Taraire wood.

35. Mason, J.—Loo table, made from New Zealand wood; Pohutuhawa wood.

36. Mannakan Saw Mills.—Specimens of woods.

37. Reid, Rev. A.—Mixed breed wool, dried apples.

38. Morgan, Rev. J.—Mixed breed wools, bark; war weapons, native garment.

39. Runciman, J.—Ewe Hoggett wool, cross (Leicester and Merino) wool; perennial rye grass seed.

40. Sellers, Mrs.—Wool-cross, Merino and Leicester.

41. West, J.—Long wool, mixed breed wool (Merino, Cotswold, and Leicester).

42. Church Mission School, *Otaki.*—Various samples of wool.

43. Barton.—Various samples of wool.

44. Taylor, Watt, & Co. — Various samples of wool.

45. Hunter.—Various samples of wool.

46. Ludlam.—Various samples of wool.

47. Moore.—Samples of wether fleece.

48. Shepherd, T. Jun.—Wool (Hogg, Leicester).

49. Lloyd, Neil.—Flax, ropes, lines.

50. Purchas, Rev. A. & Ninnes, J.—Patent flax.

51. Matthews, W.—Door mats, fibres.

52. Probert, J.—Flax, Pikiareo plant.

53. Smith, J. S.—Flax basket.

54. Turnbull, T.—Dressed flax and tow.

55. Therry, Baron de.—Flax and New Zealand fibre.

56. Webster, G.—Flax, Kiwi egg, vegetable caterpillars, land shells, fishing hooks.

57. Taylor, Rev. R.—Textile materials, warlike and domestic implements.

58. Innes, J. & Purchas, G. A.—Coil rope, dressed by patent machinery.

59. Holt, C.—Netting, wove by loom and shuttle.

60. James, Captn.—Rigging, made from New Zealand flax.

61. McEwen, A. — Californian Prairie grass.

62. Horne, Dr.—Ferns.

63. New Zealand Society.—Native robe.

64. Owen, G. B.—Native garment.

65. Owen, G. B.—One cabinet made entirely from New Zealand woods, height 10 ft., 5 ft. broad, 3 ft. deep.

66. White, W.—Carved tiger and alligator.

67. King, E.—Stuffed birds, dried apples, gold, limestone, cotton.

68. Goodfellow, J.—Soap, candles.

69. Blearzard, R.—Buckets.

70. Volckner, Rev. C.—War canoe, mats, garments, baskets.

71. Chamberlin, H.—Huni-Huni.

72. Chamberlin, H.—Four native New Zealand fish-hooks.

73. Combes, Daldy, & Burtt.—Guano from South Sea Islands.

74. Burtit.—Guano.

75. Fox, Mrs.—Drawing of New Zealand lora.

76. Combes & Daldy.—Coffee; cotton from S. Sea Islands; Kauri gum; iron sand.

77. Crombie, J. N.—Photographic views of local scenery, groups of members of House of Representatives.

70. Johns, F. L. Views of local scenery.

79. Barraud, C. F.—Views of local scenery.

80. Martyn, A.—Views of local scenery.

## Province of Nelson.

1. NELSON COMMISSIONERS. — A library table, cloth, photographs and stereoscopes.

2. NELSON PROVINCIAL GOVERNMENT.—A collection of gold specimens, each weighing 50 ounces, coal, and maps.

3. NELSON INSTITUTE COMMITTEE.—Native copper, rock with scales of gold, red hæmatite and chrome ore.

4. NELSON CHAMBER OF COMMERCE. — Dressed flax.

5. DUN MOUNTAIN Co.—Copper ore and crome ore.

6. MORSE, N. G.—Fleece of wool, long Leicester.

7. BLICK BROS.—Scoured wool, Hinau bark for dyeing cloth.

8. NATTRASS, L.—New Zealand flax for paper, and blue colour from Nelson chrome.

9. CURTIS, BROS.—Plumbago, from Paka-wau.

10. HACKET, T. R.— Coal from Buller River and Waimangaroha, W. coast.

11. LEWTHWAITE, J.—Coal from Paka-wau.

12. McGEE, C.—A stick of Rata.

13. EVERETT, E.—Chrome ore from Marsden's set.

14. WIESENHAVEN, C.—Iron sand, from Blind Bay.

15. REDWOOD, H. JUN.—Wheat.

16. MONRO, D.—Oats.

17. BAIGENT, SEN. — Timber, native woods, foliage of trees.

18. HARLEY, C.—Nelson hops.

19. ELLIOT, C.—Tea-poy of native wood.

20. ANDREWS, T.—Chrome ore from Ben Nevis.

## Province of Otago.

1. HOLMES, M.:—

500 ozs. of gold specimens from mines in Otago.

Views of local scenery.

Provincial newspaper printed on satin.

Samples of grasses, corn, and wool.

## Province of Wellington.

1. HUNTER.—Various samples of wool.

2. TAYLOR, WATT, & Co.—Various samples of wool.

3. LUDLAM.—Various samples of wool.

4. SELLERS.—Various samples of wool.

5. MOORE.—Various samples of wool.

6. BARTON.—Various samples of wool.

7. CHURCH MISSION SCHOOL. — Various samples of wool.

8. BARRAUD, C. F.—Sketches showing the growth of the rata tree, &c.

9. BARRAUD, C. F.—Photographs.

10. TAYLOR, REV. E.—Maori implements and numerous textile materials.

11. HUNTER.—Terawiti gold.

12. RENDING, J. B.—Wairiki gold.

13. NEW ZEALAND SOCIETY. — Basket made of 'flax.'

14. ELLIS.—Cotton-stand of New Zealand woods.

# NOVA SCOTIA.

## WEST SIDE OF NORTH-EAST TRANSEPT.

1. PROVINCIAL GOVERNMENT.—A collection of mineral specimens:—Gold from the quartz workings at Tangier, Sherbrooke, Wine Harbour, Allan's Mill, the Ovens, &c.; washings from the auriferous sands at the Ovens; gold bars, &c.

Iron and iron ore from the Londonderry Mines, and other localities.

Coal from the Sydney Mines, the Glass Bay Mines, and the Joggins; and oil-coal from Fraser's Mine.

2. SCOTT, JAMES, ESQ.—A column of coal 34 feet in height, from the Albion Mines.

3. HOWE, PROFESSOR.—224 specimens of minerals, including barytes, copper, manganese, &c., freestone, granite, ironstone, &c., marbles, clay, slate, anhydrite, clays and mineral paints, infusorial earth and cements, iron and garnet sand, amethysts, jaspars, agates, stilbite, calc spar, ankerite, selenite, and topaz.

4. HONEYMAN, THE REV.—A large collection of specimens illustrating the geology of the Colony.

5. DOWNS, A.—A stuffed bull-moose; a case of game birds and wild ducks.

6. FALES, A. JUN. — 84 varieties of polished woods, leaves, cones, &c., and a collection of native plants.

6A. HOWE, DR. — Medicinal and other plants.

7. BESSONET, MISS.—Water-colour paintings of native flowers.

8. HODGES, MISS. — Baskets made of cones.

9. LAWSON & PILLSBURY, MISSES.—Forest leaves, varnished.

10. BLACK, MRS. W. — Wax fruits and flowers.

11. CHASE, W.— Photograph of Nova Scotia vegetables.

12. COLEMAN, W.—Nova Scotian furs.

13. HALIBURTON, R. G.—Bayberry, or myrtle wax.

14. JONES, J. M.—Native fish, prepared in large glass jars, under direction of PROFESSOR AGASSIZ.

15. WILLIS, J. W.—A collection of edible shell-fish.

16. NOVA SCOTIA COMMISSIONERS.—Dried, pickled and preserved fish, as prepared for export.

17. FRAZER, R. G.—Fish oils of the province.

18. FRAZER, R. G.—A large collection of fruits and vegetables preserved in alcohol; grain, garden, and field seeds.

19. HARRIS & McKAY.—Flower seeds.

20. CURRY & Co. *Windsor.* — Patent axles.

21. DONALD & WATSON, *Halifax.*—Castings in brass, gaseliers, sleigh bells, &c.

22. CONNELLY, G. *Picton.*—Axes.

23. GRANT, P. *St. Croix.*—Horse-shoes.

24. SULLIVAN, J. *Halifax.*—Horse-shoes and a curd chopper.

25. LONDONDERRY MINING Co.—Bar iron.

26. BILL & KERRY, S. *Liverpool.*—Edge-tools, hay and manure forks, and skates.

27. CORNELIUS, J.—Jewellery, manufactured of native gold, pearls, amethysts, &c.

28. SCARFE, F.—Common and pressed bricks, and drain tiles.

29. MALCOLM, R. — Fire-bricks, drain pipes, and pottery.

30. WALLACE.—Grindstone.

31. PICTON.—Grindstone.

32. JOHNSTONE, W.—Carving in Wallace freestone.

33. HOLLOWAY, T.—Purchase blocks.

34. MOSHER, J.—Purchase blocks.

35. WILSON, W.—Purchase blocks, log reel, dead eyes, and belaying pins.

36. McEWEN & REED, *Halifax.*—Sofas, chairs, and a cabinet, &c. of native woods.

37. GORDON & KEITH, *Halifax.*—Furniture, and a ship's wheel.

38. MOORE, J. *Truro.*—Ox yokes.

39. DICKIE, J.—Patent harrow.

40. FRAZER, W. & SON, *Halifax.*—A piano of native wood.

41. BROCKLEY, MEISNER, & BROCKLEY.—A piano of native wood.

42. WYMAN & FREEMAN, *Milton.*—Laths.

43. O'BRIEN, G. L. (late).—A pony phaeton.

44. CURRY, E. & Co. *Windsor.* — A sleigh.

45. CAMERON, J. *New Glasgow.*—Model of a steamer.

46. MOSELEY, E.—Two working models on a new system.

47. McCURDY, MISS E. *Onslow.*- Woollen cloth, frilled and sewing thread.

48. DUNLOP, J. *Stewincke.*—Home-spun cotton and wool.

49. LAQUILLE MILLS.—Black and grey satinet.

50. CREED, G.—Grey homespun.

51. BEALS, MRS. *Bedford.* — Women's hose.

52. COWIE & SONS, *Liverpool.*—Skiding leather, hogskins, sole and harness leather.

53. SCOTT, MISS. — A leather picture frame.

54. PHILLIPS, N.—Bookbinding.

55. BLAIR, MRS. J. F. *Onslow.*—Sewing thread.

56. BEGG, MISS E.—A bonnet and hat.

57. CAMPBELL & McLEAN. — Tobacco, and maple sugar.

58. LYTTLETON, CAPT. — Three water-colour drawings.

59. WOODS, J.—A pencil drawing.

60. HARDING, C.—A pen and ink drawing.

61. COGSWELL, DR.—A set of artificia teeth; two bottles of silex.

62. O'CONNELL, J.—Salmon and trout flies.

63. SARRE, N.—Hair tonic.

64. CROSSKILL, J.—Bear's grease, eau de Cologne, and native cordials.

65. DUPÉ, G.—Cider and bitters.

# PRINCE EDWARD'S ISLAND.

## NORTH-EAST TRANSEPT.

---

LOCAL COMMITTEE (H. HASZARD, *Commissioner in London*):—

Corn, pulse, agricultural seeds.

Flour, meal and pearl barley.

Pork and dairy produce.

Linen and woollen manufactures.

Furniture and screens of native wood.

Agricultural machine and implements.

Harness and leather work.

Native canoes and baskets.

Patent ship's tackle.

Horse-shoes.

Preserved fish.

Samples of textile materials.

Osiers for basket-work.

Oil painting.

Hats, &c.

Bay-tree wax.

Honey, &c.

# QUEENSLAND.

NORTH-EAST COURTS, NEAR NORTH-EAST TRANSEPT.

---

*Commissioners in London :—*

M. H. MARSH, M.P., ALFRED DENISON, AND ARTHUR HODGSON.

THE Colony of Queensland, the most northern of the settlements of Australia, was only separated from New South Wales and formed into an independent colony at the close of 1859; the present is, therefore, its first appearance as a competitor for honours in International Exhibitions.

The wide extent of territory, and great variety of climate and elevation, renders the colony well suited either for pastoral occupation, for agriculture, or for raising tropical products. The population of Queensland, by the census of 1861, was 311,500, and from the immigration that has taken place, must have largely increased during the last year.

Along the coast-line, from the Clarence to the northern boundary of occupation, comprising some eight degrees of latitude, most, if not all, the productions of the Indies, South America, and not a few of those of Africa, may be successfully and profitably cultivated. The hill-slopes, from the base to the summit, are found to be admirably adapted for the cultivation of the vine, olive, indigo, cinchona, cinnamon, cocoa, allspice, tamarind, nutmeg, clove, tea, coffee, orange, cotton, &c.; and upon the rich extensive lands in the glens or valleys of the rivers near the coast, the sugar-cane, arrowroot, ginger, tobacco, banana, &c., can be produced in the highest perfection, as the samples exhibited prove.

The arrangement and grouping of the products of the colony in the Queensland Court has attracted, during the Exhibition, a large amount of public attention; and the importance and interest of most of the articles shown is demonstrated by the awards of the various juries, which comprise 26 medals and 20 honourable mentions. The four beautiful cases of stuffed birds, the opossum skins and rugs, and other objects of natural history, illustrate the animal life of the colony. Wool has hitherto been the chief staple, the number of sheep exceeding 3,500,000, and the export being upwards of 5,000,000 lbs. per annum. The beautiful fleeces shown by Messrs. Marsh, Hodgson, Watts, and Bigge, all rewarded with medals, also those of Messrs. Deuchar and Davidson; the wool in different stages of manufacture, and the broad-cloths manufactured at Leeds from wool of Mr. Marsh's, and at Westbury from Messrs. Bigge's flocks, prove the importance and value of this industry to the colony. Mess beef and tallow, honey and beeswax, leather, silk, and the oil of the dugong, a substitute for cod-liver oil, are other useful animal products.

The vegetable products are far more numerous and of equal importance in a commercial and industrial point of view. First in interest is the cotton, the quality and character of which indisputably demonstrates the fact that a vast cotton field is open to enterprise, capable, with capital and labour, of furnishing a supply of this staple equal to any demand. Sea-Island cotton, it is shown, can be successfully grown in Queensland, not only on the coast, but on the tablelands far in the interior. Four medals and five honourable mentions testify to the quality of the produce; whilst the cashmere, the warp of which is made from 250 lace thread, spun from very fine cotton, and the woof from beautifully fine wool, by Mr. T. Bazley, M.P., further demonstrate the value of these two staples.

The wealth of Queensland in forest productions is scarcely to be estimated, most of her timbers being of a kind exceedingly valuable for building and manufacturing purposes. There is a fine scientifically-named collection of 130 specimens of indigenous wood shown by Mr. W. Hill, and 26 specimens from the northern regions by M. Thozet. For both of these collections medals were given. The woods are well described, and the value of most of them has been scientifically tested. The magnificent sugar-canes, honoured with a medal, are unequalled by any others shown in the Exhibition. The arrowroot made in Queensland will also compare favourably with that of West India origin, and is much cheaper. It has carried off three prize medals, and will become of great local importance. Maize is a crop that succeeds well in Queensland; and the fine heads of corn shown by Mr. Fitzallen, for which he has received a prize medal, testify to the quality of this grain. Rice and wheat are also honourably mentioned by the jurors. Whilst flour, made in the colony, also receives honourable mention. There are two medals given for tobacco, and the samples of wine made in the colony are most favourably spoken of, one from pine-apple receiving a medal. Coal and other mineral products are exhibited, but to enumerate these would be impossible in a brief compass. It is, however, most satisfactory to find Queensland taking so prominent a place in the list of Australian colonies for raw materials and manufactures, as is evidenced by the appended list of exhibits; collected too under many disadvantages, arising from restricted time and great distance of transport.

1. ATTORNEY - GENERAL, THE HON. — Arrowroot and walking-canes.

2. ALDRIDGE, MRS.—Arrowroot; Rosella and pine-apple jams, Granadilla jelly, citron marmalade, and seeds of the *Cycas media.*

3. AUSTIN.—Wood fossils.

4. ARCHER, W.— Sandal-wood, Leichhardt tree; and geological specimens.

5. BALFOUR, J.—Twenty-two fleeces of wool.

6, 7. BAZLEY, T., M.P.—Cashmere manufactured from 250 lace-thread warp, spun from very fine cotton and wool, native produce.

8, 9. BIGGE, F. & F.—Wool and cloth manufactured by A. Laverton, from wool grown by F. & F. Bigge (Medal).

10. BIGGE, F. & F.—Log of the Bunya Bunya.

11. BARTLETT, N.—Mineral specimens; pictures; and engravings.

12. COXEN, C. — Specimens of Myall wood.

13. COXEN, MRS.—Arrowroot (Medal); beeswax, honey, and candied lemon-peel.

14. CAIRNCROSS, W.—Sea-Island cotton (Medal).

15. CAMPBELL, J.—Mess beef; beef tallow; coal.

16. CADDEN, W.—Rosella jam.

17. CANNAN.—Aboriginal implements.

18. CURPHY, MR.—Cypress pine-root.

19. CHILDS, T.—Sea-Island cotton, Rosella jam and vinegar.

20. COMMISSIONERS FOR EXHIBITION. — Wheat (Hon. Men.), dugong oil, colonial rum, 30 over-proof, leather.

21. CHAPMAN, T. T.—Arrowroot.

22. COSTIN, W. J.—Dugong oil, beeswax, and arrowroot.

23. COSTIN, T. — Colonial saddle, and stock-whip.

24. CARMODY, W.—Maize.

25. CHALLINOR, G.—Photographic views (Hon. Men.).

26. COCKBURN.—Specimens of silk.

27. COOPER, LADY.—Moreton Bay pearl, set in Australian gold.

28. CLARKE, J. & T.—Broadcloth manufactured at Troubridge from Queensland wool.

29. DAVIDSON, G. & W.—Fleeces of Queensland wool.

29A. DAY, S.—Arrowroot.

30. DUDGEON, S. V.—Sample of silk.

31. DOUGLAS, R.—Colonial soap.

32. DENISON, A.—Two cones of the Bunya Bunya, or *Araucaria Bidwillii.*

33. FITZALLEN.—Maize (Medal).

34. FLEMING, J.—Flour (Medal).

35. FAIRFAX, W.—Specimens of colonial printing.

36. GREGORY, C.—Aboriginal weapon.

37. GREGORY, A. C.—Fibre from pine-apple leaves.

38. GRAY, T.—Colonial leather and boots (Hon. Men.).

39. GAMMIE, G.—Plaid manufactured from Queensland wool.

40. HARTENSTEIN, A. T.—Arrowroot.

41. HOLDSWORTH, W. A. EXECUTORS OF.—Arrowroot.

42. HAYNES, M.—Geological specimens, cotton, fibre from indigenous plants, and aboriginal decorations.

43. HOPE, HON. L.—Large sugar-canes (Medal), rice (Hon. Men.), varieties of Banana fibre, Sea-Island cotton, and flax.

44. HOCKINGS, A. J.—Maize and preserves.

45, 46. HILL, W. *Botanical Gardens.*—120 specimens of woods, water-lily seeds, arrowroot, preserved tamarinds and ginger, medicinal and tanning barks, sarsaparilla, dye-woods, tobacco (Medal), rice (Hon. Men.), cotton, sugar-cane and rattans, walking-sticks, fibres, gums, stock-whip handles, and aboriginal weapons, implements, and ornaments; framed collection of the foliage of the indigenous woods.

47. HOLMES.—Wool in grease.

48. HODGSON & WATTS —Sixteen fleeces of washed wool (Medal).

49. HODGSON, MRS.—Bracelet of quandong seeds (*Fusanus acuminatus*) set in Australian gold, gold wine labels (Medal).

50. HODGSON, A.—Case of stuffed birds; stock-whip; gold Australian nugget; and map of Queensland.

51. IVORY, J.—Cayenne pepper; specimens of natural history.

52. ILLIDGE, R.— Scented soap from dugong oil (Hon. Men.).

53. JOHNSON, R. & J.—Arrowroot.

54. JOHNSON, J.—Honey, beeswax, and orange marmalade.

55. LOVE, G.—Arrowroot.

56. LOVE, E.—Arrowroot.

57. LADE, T.—Grape wine (Hon. Men.) and pine-apple wine (Medal).

58. LAIDLEY, J.—Opossum rugs; and photographs.

59. LUTWYCHE, MR. JUSTICE. — Table and chessmen made of Moreton Bay woods (Medal).

60. MARSHALL, W. H.—Arrowroot, beeswax (Medal), and honey.

61. MARSHALL & DEUCHAR. — Twelve fleeces of wool.

62. MARVONEY, M.—Arrowroot.

63. MARSH, M. H.—Wool in fleece (Medal); cloth manufactured at Leeds from his flocks; stock-whip; indigenous flax seed; Moreton Bay chestnut (*Castanospermum*) (Hon. Men.); Queensland flag.

64. MARSH, MRS. — Sachet made of Queensland woods; sapphire ring; brooch of white topaz found at Moreton Bay, and bracelets of Myall wood.

65. MARSH, MISS. — Eucalyptus manna (Hon. Men.).

66. MACDONALD, C.—Wool in fleece, three cases of stuffed birds, opossum rug, aboriginal weapons, and stuffed native animals.

67. NORTH BRITISH AUSTRALIAN Co.—Wool in fleece.

68. MEEKS, N.—Black wattle bark.

69. O'CONNEL, CAPT.—One ton of copper ore, and specimens of copper.

70. PRATTEN, J.—Sea-Island cotton (Hon. Men.).

71. PATHEN, MRS.—Arrowroot.

72. PATTERSON, S.—Cypress pine board.

73. PUGH, T.—Queensland almanack.

73A. PAULEY, W.—Specimens of wood, and two turned bed-posts.

74. RODE.—Sea-Island and upland cotton (Medal).

75. PETRIE, J.—Building stones (Hon. Men.).

76. PETTIGREW.—Model of a ship.

77. STEWART, J. — Arrowroot (Medal) and maize.

78. STEWART, H.—Two varieties of arrowroot (Medal), and Sea-Island cotton (Medal).

79. SUTHERLAND, MRS.—Sea-Island cotton (Hon. Men.); silk.

80. SHOLL, CAPT.—Natural curiosities.

81. SLAUGHTER, A. SEN. — Arrowroot, beeswax (Medal); honey.

82. SHEEHAN, N.—Specimen of silk.

83. THOZET.—Bitter, spice and cascarilla barks; tobacco leaf (Medal); cigars; Sea-Island and N.O. cotton (Medal); ebony, fibre, and plum-wine.

84. THOMPSON, P. W.—Sea-Island cotton (Hon. Men.), maize and banana fibre.

85. THORNTON, W.—Stone tomahawk and calabash.

86. VOWLES, G.—Specimens of silk.

87. WILDER, J. W. — Photographs and views (Hon. Men.).

88. WHITE, J. C.—Gum from Myall tree, &c.

89. WAY, E. — Walking-canes, rosella jams, beeswax, and ginger root.

90. WARNER, J.—Banana fibre, and map of Brisbane.

91. QUEENSLAND GOVERNMENT.—400 lbs. cleaned Sea-Island cotton, grown in Queensland.

92. WILDASH, J. — *Ornithorhynchus* (water mole); *Echidna* (Australian hedgehog).

---

93. NORTH BRITISH AUSTRALASIAN Co., 49, *Moorgate Street, E.C.*—Wool.

# ST. HELENA.

NORTH CENTRAL COURT, NEAR ENTRANCE TO HORTICULTURAL GARDENS.

---

1. St. Helena Committee :—
Cotton.
Coffee.

Specimens of wood with foliage, bark, &c.
Native birds.
Specimens of building stones.

# ST. VINCENT.

## EAST SIDE OF NORTH-EAST TRANSEPT.

1. CROPPER, R.—Specimens of the *bulimus rosaceus*, from the egg to the adult shell.

Sloughs of the common rock crab (*grapsus*).

Inner and outer bark of the mountain mahoe, *hibiscus elatus*, and rope made of the inner bark.

2. HAWTAYNE, G. H.—Arrowroot, guava jelly, coffee, cacao, plantain meal.

Gum from the G (?) tree.

Ginger.

Oil of the cocoa nut, benna, canole nut, castor and ground nut.

Spices.

3. ANDERSON, F.—A cask of pozzolano.

4. STEWART, C. D. & CLOKE, E. J.—Arrowroot from the Fancy Estate.

# TASMANIA.

## NORTH-EAST TRANSEPT.

1—14. ABBOTT, J. — Coal, ores, fancy woods, palings, staves, tanning bark, vegetable fibre, music.

15—17. ALLISON, N. P. — Wool, skins, shell necklace.

18. ALLISON, W. R.—Wedge-tailed eagle.

19, 20. ALLPORT, MR.—Walnuts, filberts.

21—30. ALLPORT, MRS.—Preserves, vinegar, Tasmanette, water-colour painting, topaz brooch.

31—34. ALLPORT, MR. MORTON.—Shells, Huon pine, stereographs.

35. ALLPORT, MRS. M.—Fancy plait of rush pith.

36. BACKHOUSE, R.—Flax.

37. BAKER, I.—Coal, New Town.

38. BALF, J. D.—Platypus skin.

39. BARCLAY.—Freestone, Glenorchy.

40, 41. BARTLEZ, T. B.—Manna, native bread.

42. BARNET, G.—Bituminous coal, Mersey.

43. BLYTHE, W. L.—Wool.

44—46. BOUTCHER, W. R.—Wheat, wine, and vinegar.

47—113. BOYD, J.—Clay, bricks, pottery, fruit, vinegar, shells, fancy woods, ship-building and railway timbers, palings, skins, fern trees, 230 feet spar.

114. BRYANT, MISS S.—Flying opossum skin.

115, 116. BURDON, MR.—Carriage wheels, blue gum plank.

117—120. BURGESS, MRS.— Embroidery.

121. BUTCHER, MRS.—Potter's clay.

122. BUTTON, W. S.—Glue.

123—184. CALDER, J. E.—Cubes of sandstones, marble, limestone, sea-weed, fancy woods, deer horns, photographs, opossum mouse in spirit.

185, 186. CAMERON, A. L.—Skins.

187. CARTER, W.—Bituminous coal, 12-ft. seam, Fingal.

188. CHATFIELD, W.—Chestnut-faced owl.

189. CHILTON, R. — Strong bituminous coal, $4\frac{1}{2}$-ft. seam.

190. CLARK, G. C.—Wools.

191. CLARK, MISS C. A. C.—Railway rug, fur; work-box of fancy woods.

192. CLIFFORD, S.—Stereoscopic views in Tasmania.

193. COLLINS, MISS.—Fancy basket.

194—330. COMMISSIONERS FOR TASMANIA. —Coal (bituminous and anthracite), ores, marble, freestone, grindstones, hones, gold-dust, whalebone, timber-trophy, whale-boats, casks, implements, vegetable fibre for manufacture of paper, barks for tanning and medicinal uses, models of fruits, furs, skins, furniture, guano.

331. COOK, MRS. H.—Myrtle-wood vase.

332. COX, E.—Native bread.

333, 334. COX, F. — Peppermint wood, 25 years cut, cantharides-beetle.

335—337. CRESSWELL, C. F. —Wheat, Talavera and Tuscan ; preserves.

338—341. CROUCH, MRS. S.—Preserved meat, hams, penguin skin.

342—347. CROWTHER, MRS. B.—Sinews, opossum fur, ornamental feather work, bird-skins.

348—379. CROWTHER, W. L.—Spermaceti, oils, whale's jaws, split timber, sawn timber, railway sleepers, ship planking, blue gum 45 years in use.

380. CRUTTENDEN, T. — Opossum wool, gloves.

381. DALGETTY, F. G.—Oil painting of Hobart Town.

382. DENNY, H.—Bituminous coal.

383. DOBSON, A.—Opalized wood, Syndal.

384. DOOLEY, M.—Native bread.

385, 386. DOUGLAS, Opalized wood, 'La Perouse's' tree.

387—391. DOYLE.—Leather of sorts.

392. DU CROZ, MRS.—Couch rug, black native cat.

393. DU CROZ, F. A.—Rug of grey and black opossum fur.

394. DYSODILE Co.—Resiniferous shale : Mersey river.

395. EMMETT, S.—Gold dust ; Hellyer river, Tasmania, N.W.

396. FAWNS, J. A.—Table of fancy woods of Tasmania.

397. FENCHIER, H.—Iron ore.

398. FINLAYSON, A. H.—Cabin bread in Huon pine case.

399. FLETCHER, D. S.—Wheat.

400. GERRARD, REV. T.—Wool.

401—404. GELL, P. H.—Clay, wheat, wools.

405. GIBSON, J.—Oats.

406. GILLON.—Granite from Clark's Island.

407—409. GLEDHILL. — Boots, sorts, hones.

410—436. GOULD, C.—Ores, dysodile, alum, porphyry, granites, limestones, topazes, bituminous coals, marbles, skins, geological maps, table of Huon pine and muskwood.

437, 438. GOURLAY, F. R.—Porpoise oil, heart of fern tree.

439. GOWLAND & STANARD.—Freestone.

440. GRAY.—Hickory wood knee.

441. GREENBALGH, M —Rug.

442. GROOM, F.—Strong bituminous coal, Mount Nicholas, 12 feet seam.

443—445. GRUBB & TYSON. — Bench screws, office rulers, shingles.

446. GUNN, RONALD.—Aromatic wood.

447, 448. HAES, FREDERICK.—Ornamental plat of rush-pith, and fern-tree vase.

449—452. HALL, R.—Granite, kangaroo, and platypus skins.

453. HAWKINS.—Palings.

454—456. HILL, R. — Blackwood log, palings, and shingles.

457—460. HORNE, A. J.—Skins ; wool.

461. HULL, J. F.—Leather of elephant seal skin.

462. HULL, H. M.—Manna, insects.

463. HULL, H.—Skins of tiger-cat of Tasmania.

464. IRWIN, D.—Muskwood.

465. JUDD, H.—Native bread.

466—468. JOHNSON, T.— Model of apparatus for conveying salmon over to Tasmania.

469. KERMODE, R. Q.—Fine wools.

470—476. LETTE, R. L.—Minerals, shells, woods, Xerotes fibre.

477—479. LEWES, J. L.—Lavender, she-oak, bronze pigeon.

480. LLOYD, MAJOR.— Wattle gum.

481. MACCRACKEN, R.—Beef in canisters.

482. MACCRACKEN, MISS.—Bronze-wing beetles.

483. MACDONALD, W.—Asbestos in serpentine.

484. MACFARQUHAR.—Rug of native furs.

485—488. MCGREGOR, J.—Ship timbers of blue gum.

489. MACLANACHAN, J.—Fine wool.

490—492. MARSHALL, J.—Wheat, oats, barley.

493—495. MEREDITH, C. — Bituminous coal, fibre, native bread.

496—507. MEREDITH, MRS.—Water-colour paintings of flowers of Tasmanian trees, shrubs and plants, framed in muskwood.

508. MEREDITH, J.—Wool.

509—546. MILLIGAN, J.—Topazes, jacinths, beryl, cairngorm, rock crystal, opal, cornelian, garnet, schorl, hornstone, auriferous quartz, iron ores, galena, obsidian, pumice, alum, Epsom salts, aboriginal baskets and necklaces, shells, loo-table of muskwood inlaid.

547—549. MOORE, DR.—Palings, staff, gate.

550. MORRISON, A.—Busts (2) of aborigines of Tasmania.

551. MORRISON, J. A.—Grindstone, kangaroo point.

552. NIXON, RIGHT REV. DR., BISHOP OF TASMANIA.—Photograph of groups of Tasmanian aborigines.

553. NOAKE, E.—Flour.

554. OFFICER, R.—Cajeput oil, distilled from the leaves of blue gum.

555—558. OLDHAM, T.—Draywheels, piles, and ship planking of blue gum.

559. PINK.—Beef.

560—566. POWELL, W.—Table of myrtle, and black-wood, vases; walking-sticks, office rulers, turned of jaw and teeth of sperm whale.

567. PROCTOR, W.—Peppermint wood, 29 years in use.

568—570. PYBUS, R.—Arrowroot, grass-tree gum; peppermint wood, 35 years cut and exposed.

571. RANSOM, J.—Bituminous coal, Mount Nicholas 12-ft. seam.

572, 573. RITCHIE, R.—Oatmeal, groats.

574—576. ROGERS. — Vegetable fibre-barks.

577, 578. ROSS, J.—Ship-timber, of blue-gum.

579, 580. SANDERSON, M. — Teeth of sperm-whale, and walking-stick of jaw of the same.

581, 582. SCOTT, J.—Flours; she-oak timber.

583. SEARLE.—Flour.

584, 585. SHARLAND, F. W.—Wattle-tree gum, grindstone.

586. SHARLAND, W. S.—Golding hops.

587. SHAW, M.—Native bread.

588. SHOOLRIDGE, R.—Fossiliferous limestone.

589. SMITH, J. L.—Oats.

590. SMITH, P. T.—Wools.

591. STEVENSON.—Lard.

592—604. STUART, J. W.—Ink, basketwork, knee-caps, carriage-mats, ladies' boots.

605—615. STUTTARD, J.—Views in water colours of scenery on the north coast of Tasmania.

616, 617. SWIFT, A. H.—Bituminous coal, east coast of Tasmania, seam 5-ft. to 6-ft. 10-in.

618. TUPFIELD, MISS.—Feathers of Tasmanian emu.

619. THOMAS, MR.—Skin of penguin, Bass's Strait.

620, 624. THOMPSON, MR. H.—Minerals, gloves of opossum fur, skins, emu's egg.

625. THORNE, J.—Red ochre.

626—628. TULLY, W. A.—Paraffin oil, native bread, backgammon board.

629. WADE, MR.—Products of botanic garden.

630. WALCH & SONS.—Bookbinding.

631. WALKER, J. C.—Almonds grown by exhibitor.

632, 633. WEAVER, W. G.—Alcohol, skin of musk duck.

634. WEDGE, J.—Paraffin oil.

635. WEDGE, J. H.—Vegetable fibre : a grass.

636, 637. WHITING, G.—Opossum skin rugs.

638. WHITING, J.—Stalactites.

639. WHYTE, J.—Fresh-water limestone.

640. WILKINSON.—' More pork,' skins, &c.

641, 642. WILSON, G.—Wheat, oats.

643. WILSON, J. J.—Preserved fruits of 1861.

644—648. WRIGHT, I.—Wheat, leather of sorts, ' Cape Barren ' goose.

649. YOUL, J. A.—Busts of two aborigines of Tasmania.

650—654. YOUNG, LADY.—Book-stand, writing ditto, paper knife, paper-weight, casket.

655. SMITH, L.—Wheat.

656. COOPER, A. H.—Sketches of South Sea Whale fisheries.

657. MARSHALL, G.—Wheat.

658. SMITH, J.—Wheat.

659, 660. LINDLEY, G. H.—Wheat, barley.

661. WILLIAMS, W.—Flour.

662, 663. NUTT, R. W.—Fine wools from Malahide.

664—668. ARCHER, W.—Wools.

# TRINIDAD.

## UNDER EASTERN GALLERY, NORTH-EAST TRANSEPT.

TRINIDAD EXHIBITION COMMITTEE :—

A collection of minerals : Asphalte from the Pitch Lake, glance pitch, iron ore from Maracas, gypsum from St. Joseph, tertiary coal from the eastern coast, and lignite from the Irois coast.

Chemical and pharmaceutical products.

A collection of food substances : Rice, in the husk and cleaned ; ground nuts ; gingili or sesamum ; varieties of cacao or cocoa ; coffee, nutmegs, akee seeds, Brazil nuts, tea ; flour of the bread fruit, plaintain, yams, tania, sweet potatoe, cushcush, (a kind of yam), bitter cassada, sweet cassada ; ochro ; starch from the cassada, arrowroot, and toloman (canna) ; cloves, nutmegs, black pepper, and vanilla.

Among substances used in manufactures : Oil from the cocoa-nut, pressed and boiled : whale, castor, Avocado-pear, and carap oils ; balsam of copaiva ; a collection of ornamental seeds, timite fruit and seed, vegetable ivory, grugru-nuts, rough leaves of curatella, skins of sharks, cauto bark, sponges, mamure ; timite, raw and prepared ; arnatto, from fresh and fermented seeds.

Textile materials : Wild and cultivated cotton ; ochroma or corkwood cotton ; leaf and fibres of œnocarpus batawa, of carata, macerated and unmacerated, of langue de bœuf or agave vivipara, sanseviera, of wild cane or heliconia, of musas, plantains, and various other plants.

141 specimens of native woods.

Rope from sterculia caribea and malachra radiata.

Indian jugs, pots, and garglets made of clay mixed with the ashes of cauto bark.

Plantains stewed in syrup, and fruit preserves.

Indian wicker-work made of tirite, a species of calathea ; Indian impermeable baskets of the same. Plain and ornamented calabashes ; razor strops made of various vegetable substances ; fancy baskets of luffa fruit ; Indian fans ; cocayes, and ornamented articles made of seeds.

Patent fuel, manufactured from pitch, by HAMILTON WARNER, *San Fernando.*

Walking-sticks of native woods.

Photographs made by MR. WILLIAM TUCKER, *Port of Spain.*

# VANCOUVER.

### NORTH-EAST TRANSEPT.

---

EXECUTIVE COMMITTEE, *Victoria* :—

1. Gold ; copper and iron ore ; coal, limestone, cement stone, slate, sandstone, granite.

A spar of Douglas fir for the International Exhibition flag-staff, 220 feet long.

White and Douglas pine, silver fir, spruce, yellow cypress, cedar, oak, yew, hemlock, maple, dogwood, alder, white pine and cypress cones and twigs.

Wheat, barley, oats, peas, timothy, and potatoes of field growth. A bunch of barley, of timothy, and of hemp nettle. Garden vegetables.

A bundle of kelp. A specimen of the rock crab.

Oils of whale, seal, dog-fish, and oulachan.

A sample of wool.

A pair of antlers ; a buck.

Indian manufactures : Hemp and net, from the hemp nettle ; rope and mantle, from the bark of yellow cypress ; hats, a basket, whaling tackle, a harpoon, float and line, halibut fish-hooks.

Models of a stern-wheel steamer, of a side-wheel steamer, of a centre-board schooner.

Specimens of red bricks, manufactured near Victoria.

Three small kegs, a claret-jug, and a drinking-cup, made of native oak.

2. HENLEY, *Clover Point*. — Fifty-two varities of kitchen-garden seeds.

3. DRIARD, S.—Prepared meat, concentrated soup ; apples, and other fruit ; sardines and anchovies.

4. FOUCAULT.—Halibut and salmon.

5. FARDON.—Photograph views and portraits.

6. A COLONIAL AMATEUR.—Sketches of scenery near and in the town of Victoria, and wild flowers from Fern Wood.

# VICTORIA.

## CENTRE AND WEST SIDE OF NORTH-EAST TRANSEPT.

1. ABEL, PROF. J. *Ballarat.*—Meteorite found at Cranbourne; collection of minerals; wine.

2. ALBION QUARTZ MINING CO.—Five cwt. of auriferous quartz.

3. BANK OF AUSTRALASIA.—Samples of gold, 50 in number, from the various gold fields in Victoria.

4. BANK OF VICTORIA DIRECTORS.—Specimens of gold, for the most part alluvial.

5. BANK OF NEW SOUTH WALES, DIRECTORS OF VICTORIA BRANCH.—Specimens of the occurrence of gold in the matrix.

6. BANK ORIENTAL.—Two specimens of quartz rich in gold.

7. BRISTOL REEF MINING CO.—Section showing the strata cut through in reaching the quartz reef; bag of quartz.

8. BACK CREEK LOCAL COMMITTEE.—Sections, showing strata cut through in reaching Cornubian Reef, All Nations Reef, and alluvial sinking.

9. BAILLIE & BUTTERS.—Large specimens of quartz, studded with gold; samples of wash-dirt; copper ore, tin ore, &c.

10. BLUCHER'S REEF CO. *Maryborough.*—Sample bag of quartz.

11. BLAND, W. H. CLUNES. — Twenty ounces of silver taken from Victorian gold; gold in various forms; sample of arsenic.

12. BENDIGO GOLD MINING CO.—Section, showing strata and workings of their reef, Bendigo.

13. BUCHANNAN'S REEF, *Inglewood.* — Piece of sandstone, showing quartz veins therein; piece of slate, &c.

14. BURKITT, A. H. *Beechworth.*—Sample of analysis of black sand, gold, &c. from the Middle Woolshed, Ovens District.

15. BENYON, J. *Tarnagulla.* — Alluvial specimen from Doctor's-gully, quartz studded with gold, various specimens of quartz, rich in gold.

16. BLIGH & HARBOTTLE, *Melbourne.*—Two samples of antimony.

17. BLACK HILL QUARTZ MINING CO.—Twenty-five tons of ordinary quartz for crushing.

18. BEECHWORTH LOCAL COMMITTEE. — Samples of granite and other building stones, wheat, Indian corn, and flour.

19. BARKLY, SIR H.—Specimen of meteoric iron from Western Port, and horseshoe made therefrom.

20. BREADING, P. G. *Castlemaine.*—Graptolite, found in forming Barker Street.

21. BANNERMAN, A. *Sandhurst.*—Twenty-seven specimens of auriferous quartz; sundry small specimens from the Eagle Hawk Reef.

22. BRIGHT, BROS. *Melbourne.*—Sample of iron ore from the Ovens District.

23. CLARK & SONS, *Melbourne.*—Tin-ore; six casks of auriferous quartz, from Ajax Mine, Castlemaine.

24. CASTLEMAINE LOCAL COMMITTEE.—Sandstone, slate, fossils, &c.

25. CAMPBELL, *Back Creek.* — Fourteen fossils; two precious stones; quartz, with mundic and small crystal.

26. COTOWORTH & WOOD, *Morse's Creek.*—Specimens of gold in quartz, from Oriental Reef.

27. CHAMBERS & GITCHELL, *Beechworth.*—Gold in quartz, from reefs in the Ovens District.

28. COLLES, J. *Back Creek.*—Volcanic specimens from Mount Greenock, an extinct volcano.

29. CLUNES ALLIANCE MINING Co.—Section, showing strata in sinking shaft.

30. CLUNES MINING Co. — Twenty-five tons of quartz, to be crushed by machine in Exhibition.

31. CATHERINE REEF MINING Co.—Three samples of auriferous quartz; one sample of quartz tailings.

32. CAIRNS, WILSON, & AMOS, *Melbourne.*—Antimony, reduced from the ore; bar iron, rolled from scraps.

33. CAKEBREAD, G. *Geelong.*— Block of limestone, polished.

34. GREAT REPUBLIC GOLD MINING Co.—Cask of auriferous wash-dirt from a depth of 290 feet through two layers of basaltic rock.

35. ROYAL SAXON GOLD MINING Co. *Inkerman Lead.*—Auriferous wash-dirt from a depth of 300 feet through two layers of basaltic rock.

36. PRINCE OF WALES GOLD MINING Co. —*Cobbler's Lead, Ballarat.* — One cask of wash-dirt.

37. NELSON GOLD MINING Co. *Sebastopol Hill.*— Wash-dirt, auriferous cement, and lignite found at a depth of 378 feet through four layers of basaltic rock.

38. TEMPERANCE GOLD MINING Co. *Band of Hope Reef, Lit. Bendigo, Ballarat.*—One cask of quartz (Schist Reef).

39. RED JACKET GOLD MINING Co.—Auriferous wash-dirt.

40. CLIFFORD, G. P. *Melbourne.*—Twenty-five small surface stones containing gold, from Fryer's Town; quartz road metal containing gold.

41. COGDON, J. *Ballarat.*—Small specimen from Hiscocks prospecting claim, Buningong, where gold was first discovered.

42. MAJESTIC MINING Co. *Black Hill, Ballarat.*—Sample of quartz (Schist Reef).

43 INDEPENDENT MINING Co. *Little Bendigo, Ballarat.*—Sample of quartz (Schist Reef).

44. CAMP Co. *Cobbler's Lead, Ballarat.*—Wash-dirt 400 feet from surface through three layers of basaltic rock.

45. COMMISSIONERS OF VICTORIA EXHIBITION.—Three specimens of auriferous quartz, five specimens of quartz with gold.

46. CASTLEMAINE LOCAL COMMITTEE. —Two quartz crystals.

47. DARTMOUTH REEF, *Inglewood.* — Quartz stone with crystals, taken ten feet from surface.

48. DALY'S REEF, *Inglewood.* — Specimens of quartz.

49. DARCY, *Heathcote.* — Two quartz crystals found at 120 feet from surface.

50. DOWDING & Co. *Sandhurst.*—Specimens of quartz gold, and other metals, from Johnson's Reef, Bendigo.

51. DYER & Co. *Melbourne.*—Limestone and lime from Geelong and Point Nepean.

52. DALGETTY & Co.—Sample of tin ore.

53. EASTWOOD CAPT. *Sandhurst.*—Seven specimens of auriferous quartz, sample of conglomerate from the White Hills.

54. FOORD, G. *Melbourne.*—Collection of minerals associated with gold; meteoric iron etched to exhibit the structure; titanic iron sand; sample of coal and coke from New Caledonia.

55. GETHING, G. *Ballarat.*—Specimen of basaltic rock containing zeolithes.

56. GUILFORD, MR. *Loddon.*—One bag of upper or below wash-dirt, one bag alluvial wash-dirt.

57. HART, G. H. *Sandhurst.* — Five samples of auriferous drift from the neighbourhood of Huntly.

58. HALL, J. *Emerald Hill.*—Sample of iron ore from Sandhurst, crude and reduced to pig iron.

59. HEFFERNAN, J. *Sandhurst.*—Iron ore from Sandhurst.

60. INDEPENDENT GOLD MINING CO. *Amherst, Back Creek.*—Section of the Company's claim at Rocky Flat.

61. JOSKE, PAUL, *Melbourne.*—A collection of specimens of quartz rich in gold, from the exhibitor's claim at Sandhurst.

62. JOSEPH, H. *Sandhurst.*—Specimen of quartz from Wellington Claim, Golden Gully; two pieces of quartz, road metal containing gold; specimen of gold in cement.

63. KNIGHT, J. G. *Melbourne* (Architect Secretary to Victoria Department of Exhibition).—Specimen of building stones at present known in Victoria, and treatise thereon; drain pipes, bricks, tiles, &c.

64. KIDD, P. R. *Fryer's Creek.*—Iron ore producing 70 per cent. of metal.

65. KER, R. *Melbourne*—Sample of red granite from Western Port.

66. LEWIS, J. *Whroo.*—Specimens of gold-bearing quartz and other minerals, from Balaclava Hill, Whroo District.

67. LEVIATHAN REEF MINING CO. *Maryborough.* — Section, showing strata cut through in reaching quartz reef.

68. LEVY & SONS, *Melbourne.*—Sample coal from Cape Patterson.

69. LEICESTER, C. *Melbourne.*—Case of minerals; specimens and illustrations of various methods of extracting gold.

70. MARINER'S REEF, *Maryborough.*—Section, showing strata cut through in getting to the reef; samples of quartz.

71. MAXWELL'S REEF (Laidlaw and Party, *Inglewood*).—Eighteen specimens of quartz containing gold, five containing sulphurets.

72. MEADS, R. G.—Specimens of galena, found in the Tullarook Ranges, near Goulburn River.

73. McNAIR, J. — Clunes nugget, from Clunes, weighing 16 oz.

74. MITCHELL, A. *Avoca.*—Specimens of gold in calcined quartz.

75. MITCHELL, M. *Melbourne.* — Quartz, from the first opening of McIvor Caledonia Reef, McIvor.

76. MARYBOROUGH LOCAL COMMITTEE.—Samples of quartz, various stones, and quartz crystals.

77. MALAKOFF REEF CO. *Steglitz.*—Sample of auriferous sulphides.

78. NUGGETY MINING CO. *Campbell's Creek, Castlemaine.*—Specimens of quartz with gold; yield of reef, 25 oz. to the ton.

79. MIXON, WM. *Geelong.* — Sample of coal found at the surface, about 10 miles from Geelong.

80. POLKINGHORNE, J. *McIvor.*—Samples of tin ore, bar tin, antimony, oxide of calcium.

81. PARKINS, H. *Sandhurst.*—Two sections, showing strata in deep sinking at Huntly.

82. PRESHAW, W. J. *Castlemaine.*—Boulder taken from a freestone quarry; three quartz crystals.

83. POOLE, A. *Castlemaine.* — Fossils found at Talbot Quarry, Tarradale.

84. ROBERTSON, J. S. *Inglewood.*—Three casks of auriferous wash-dirt, and quartz from Inglewood District.

85. ROBERTS & JONES, *Castlemaine.*—Slate flag; sample of granite.

86. RODDA, R. N.—Case of minerals and metals, operated upon by a patent process.

87. RICHARDS, A. *Scotchman's Gully, Bendigo.*—Section, showing distribution of alluvial deposits in connexion with gold.

88. RIGBY, E. *McIvor.*—Washing of gold and black sand from McIvor Creek.

89. SPECIMEN HILL MINING CO. *Eagle Hawk, Bendigo.*—Bottle containing quicksilver and alluvial gold, rough gold, fine gold, &c.

90. SANDHURST (BENDIGO) LOCAL COMMITTEE.—Quartz with gold, flagging building stones, slate.

91. ST. MUNGO GOLD MINING CO. *Bendigo.*—Large quartz stone, containing gold; two small ditto.

92. SMYTH, BROUGH, *Melbourne.*—Collection of rocks and fossils relating to the geology of Victoria.

93. SELWYN, A. C. R. Government Geologist.—Six cases of minerals, rocks, and fossils, relating to the geology of Victoria; gypsum, coal, &c.

94. STIELING, G. F. *Richmond.*—Sample of modelling clay, fine clay, &c.

95. TURNER, W. J. *Beechworth.* — Gold, alluvial gold in slate, bar tin gold in crystallised quartz; precious stones, jewellery.

96. TRIUMPHANT GOLD MINING CO. *Rocky Flat, Back Creek.*—Petrified wood found at 120 feet from surface.

97. VICTORIA GOVERNMENT, per HON. C. HAINES, Minister of Finance—8,000 ounces of alluvial gold.

98. VICTORIA KAOLIN CO. *Bulla Bulla.*—Block of kaolin, and specimens of its manufactures in various forms.

99. WATSON, J. F. *Back Creek.*—Iron ore from the ranges between the Bet Bet and Adelaide Lead.

100. WALL, DR.—Specimen of magnesian limestone.

101. WALTERS & WRIGHT, *St. Arnauds.*——Two quartz stones from a cross spar 90 feet deep; specimens containing gold and silver.

102. WRIGHT, G. E. *Inglewood.*—Large quartz boulder, and numerous specimens of gold in quartz.

103. WELLINGTON CLAIM, *Maryborough.*—Section showing strata in reaching the great reef.

104. WILKINSON, R. *Back Creek.*—Fossils, and three precious stones.

105. WILSON, E. *Back Creek.*—Sample of blue stone.

106. ASKUNAS & CO. *Melbourne.*—Guano from Flat Island.

107. BARNARD, J. *Kew.* — Hyoscyamus leaves, extract and tincture.

108. BOSISTO, J. — Oils, tinctures, varnishes, drugs, &c.

109. CONNOR, D. *Bunyip Creek.*—Resin, palm nuts, extracts, &c.

110. COLE, MR. *Murray River.*—Resins of the *Eucalypti.*

111. CAULFIELD, E. *Toorak.*—Olive oil.

112. DAINTREE, H. *Melbourne.*—Resins.

113. DENNY, E. & J. *Geelong.* — Meat manure.

114. FLETCHER, G. —Resins of various kinds.

115. GRAY, H. *Ballarat.*—Essential oils; pyroligneous acid.

116. HARRIS, *South Yarra.*—Resin of *Eucalyptus Viminalis.*

117. HOLDSWORTH, *Sandhurst.*—Sample of pyroxylic spirit.

118. JOHNSON, W. *St. Kilda.* — Oils, resins, &c.

119. KRUSE, J. *Melbourne.*—Fluid magnesia, mineral waters, beeswax, leeches from Murray River.

120. MECKMERKAN & CO. *Flemington.*—Superphosphate of lime.

121. MACDONALD, MR. *Wickliffe.*—Samples of salts and crystals from Lake Bolac.

122. MORTON, W. L. *Melbourne.* — Sandarac from *Callitris Verrucosa.*

123. MULLER, DR.—Resins and oils from various indigenous trees and plants.

124. PRAGST, G. *Williamstown.*—Charcoal, tar, and the residue from wood leaves in the manufacture of vegetable gas.

125. ROBERTSON, DR. *Queen's Cliff.* — Resins and essential oils.

126. WOODWARD, G. *Kew.*—Two samples of Victorian guano.

127. AITKEN, T. *Melbourne.*—One barrel of ale.

128. BEECHWORTH LOCAL COMMITTEE.—Small samples of wheat, Indian corn, and flour.

129. BENCRAFT, G. *Melbourne.* — Two barrels of oatmeal.

130. BAYLES & Co. *Melbourne.* — Two bags of wheat.

131. BOWLES, J. B. *Back Creek.* — Two small cases of biscuits.

132. CASTLEMAINE LOCAL COMMITTEE.— Small sample bag of barley.

133. CLARK, R. *Benalla.* — Sample of wheat, barrel of flour from Ovens District.

134. DOEPPER, H. *Richmond.* — Samples of maccaroni and vermicelli.

135. DANELLI, S. *Richmond.* — Samples of maccaroni and vermicelli.

136. DOCKER, REV. J. *Wangaruta.* — One bag of wheat.

137. DENNYS, C. & J. *Geelong.* — Charqui; preserved meats.

138. DEWAR, J. *Gisborne.* — One bag of wheat.

139. ELLIOT & FAWNS, *Sandhurst.* — One barrel of ale.

140. FINLAY, J. *Emerald Hill.* — Two small sample bags of oats.

141. FALLON, J. F. — One bag of wheat.

142. FRY, J. *Ascot Mills.* — Flour.

143. FORDHAM. — Bottles of fruit; assorted jams.

144. GREEN, RAWDON. — Mess beef in tierces.

145. GRANT, T. *Melton.* — Victorian prize barley.

146. GREEN, *Warnambool.* — Bag of wheat; barrel of flour.

147. GIRAUD, L. *Collingwood.* — Liqueur, confectionery.

148. HODGKINSON, W. *Prahran.* — Two bottles of honey; one bottle of mead; beeswax.

149. HADLEY, T. H. & Co. *Melbourne.* — Flour from wheat weighing 69 lbs. per bushel.

150. JOHNSON, J. *Newburn Park, Port Albert.* — Four tierces of mess beef and pork.

151. KRUSE, J. *Melbourne.* — Sample of Sorghum sugar.

152. KINNERSLEY, D. *Burrambeet.* — Victorian prize oats, 49 lbs. per bushel.

153. LAWRENCE, W. *Merri Creek.* — Three Stilton cheeses.

154. MCKENZIE & Co. *Melbourne.* — Oatmeal.

155. MUELLER, DR. — Tea; ginger; bark.

156. RAMSDEN, S. *Carlton Mills.* — Flour, bran, and wheat.

157. REYNOLDS & Co. *Melbourne.* — Seeds of agricultural produce.

158. RICHARDS, MR. *Albert River.* — Arrowroot grown at South Gipps Land.

159. SWALLOW & Co. — Two cases of biscuits.

160. SANDHURST LOCAL COMMITTEE.— Samples of wheat, Indian corn, and tobacco leaf.

161. STEWART, R. *Geelong.* — Three tins of biscuits; jam, and marmalade.

162. SMITH, T. *Collingwood.* — Two bags of wheat.

163. VICTORIA EXPLORATION COMMITTEE. — Dried beef and meat; Nardoo flour, on which Burke, Wills, and King, the explorers, for a long time subsisted.

164. WILKIE, J. & Co. *Melbourne.* — Prize wheat, weighing 69 lbs. 4ozs., grown by William Thompson, Gisborne.

165. ABEL, A. T. *Ballarat.* — Wine: colonial.

166. ALBURY & MURRAY RIVER AGRICULTURAL SOCIETY. — Wine: colonial.

167. BRYDEN & HENDRICK, *Geelong.* — Wine: colonial.

168. BLAKE, J. A. *Melbourne.* — Wine: five cases.

169. BREQUET, F. *Geelong.* — Wine: Australian, Sauterne, Burgundy, Claret, white Sauterne.

170. COOPER, R. *Melbourne.* — Wine: red Victoria, white Victoria.

171. DUNOYER, J. *Geelong.* — Wine: white Pineau, Gris.

172. DIXON, P. G. *Melbourne.* — Orange bitters, ginger wine, ginger brandy, sodawater.

173. EVERIST, T. J.—Wine: white Carignan, white Gouais.

174. FALLON, J. F.—Wine: Aucarot, Carbeitrel, Sauvignon, Muscat, Riesling, red Scyras.

175. GROSMANN, *Melbourne.*—Wine: Burgundy.

176. HIRSCHI, F. *Castlemaine.*—Mount Alexander (red); ditto (white).

177. LEMME & Co. *Castlemaine.*—Red Castlemaine wine.

178. MCMULLEN, W. *Geelong.*—Wine: Hermitage; brandy.

179. MATE & Co.—Wine: Aucarot, white Muscat of Alexandria, white tokay, Reisling (white).

180. NIFFENECKER BROS. *Barabool Hills.* —Wine: Auvernat black cluster, Burgundy, sparkling Chasselas; brandy.

181. PASSELAIGUE.—Wine: Hermitage.

182. SIDEL, B. *Barrabool Hills.*—Burgundy.

183. WEBER, BROS. *Batesford, Geelong.* —Wines: Chasselas, Burgundy, sweetwater.

184. WALSH, H. S. *Hawthorne.*--Wine: white Longfield.

185. ZORNE, E. *Oakleigh.*—Five bottles of tomato sauce.

186. VICTORIA EXHIBITION COMMISSIONERS. —Coloured plaster casts of fruits grown in Victoria, comprising 57 varieties of apples, 45 pears, 10 cherries, plums, strawberries, figs, oranges, melons, and a large assortment of vegetables.

187. ACCLIMITIZATION SOCIETY OF VICTORIA.—Hair from llamas, alpacas, camels, Angora goats, &c.

188. BARKER, J. & R. *Melbourne.*—Raw silk from worms fed on the black mulberry.

189. CASTLEMAINE LOCAL COMMITTEE.— Native cochineal.

190. CROPPER, W. H. *Melbourne.*—Raw silk.

191. CHUCK, T. *Melbourne.*—Small sample of native cotton and fibres.

192. CROFTS, MR. *Melbourne.*—Samples of raw silk.

193. DARDANELLI, SIG. *Melbourne.*—Raw silk.

194. DOWNIE & MURPHY, *Melbourne.*— Mixed, purified, and common tallow.

195. GASKELL, J. *Melbourne.*—Pure emu oil; raw silk.

196. GOUGH & Co. *Richmond.*—Bag of malt from Victorian barley; one bag of malt from Californian barley.

197. HAYTER, H. H. *Melbourne.*—Specimens of *Cryptostemma Calendulaceum.*

198. LAMBERT, T. *Richmond.* — Four samples of bark wood for tanning.

199. LOUGHMAN & Co. *Melbourne.*—Tobacco leaf.

200. MEARS, J. & A. *Collingwood.*—Medical herbs and roots (14 varieties).

201. MACKMEIKAN & Co. *Flemington.*— Glue pieces and bone dust.

202. MURPHY, F. M. *Castlemaine.*—Native flax.

203. MUELLER, DR.—Fibres; various plants; lichens.

204. QUIRK, H. B. *Maryborough.*—Native silk.

205. REED, J. *Collingwood.*—Rope, &c. made of New Zealand flax, grown at the Botanical Gardens, Melbourne.

206. RIDGE, MRS. *Melbourne.*—Hair of the first cross with the Angora and common goat.

207. STABER, F. *Collingwood.*—Fibre of the *Yucca gloriosa*, from leaves grown at the Botanical Gardens, Melbourne.

208. SADDLER, T. *St. Kilda.*—Silk, from worms reared at Caulfield.

209. WILSON, E. *Melbourne.*—Hair of the Poiteau ass.

210. BAYLDON & GRAHAM, *Geelong.* — Scoured wool.

211. CLOUGH & Co. *Melbourne.*—Sixteen bales of choice wool of various brands; ninety-six fleeces.

212. CORRIGAN, S. B. *Geelong.*—Combing, clothing, and lambs'-wool.

213. COMMISSIONERS OF VICTORIAN EXHIBITION.—Two samples of wool.

214. DOUGLASS, A. & Co. *Geelong.*— Scoured combing, clothing, and lambs'-wool.

215. DEGRAVES, WM. *Coliban Park.*— Spanish merino wool.

216. ELDER & SON, *Kuruc Kuruc.*— Merino fleece wool.

217. GOLDSBOROUGH & Co. *Melbourne.*—Thirteen bales of choice wool of various brands, forming one half of the trophy in conjunction with Clough & Co.

218. LEARMONTH, MESSRS. *Ercildoun.*— Washed fleece.

219. MARSHALL, T. *Geelong.*—Scoured wool.

220. RUSSELL, P. *Carngham.* — Cross-bred wool.

221. RUSSELL, T. *Wanook.*—Fleece wool.

222. ROWE, E. *Melbourne.*—First cross between merino and Cotswold wool.

223. SIMSON, R. *Langi Kal Kal.*—Beaufort fleece wool.

224. SPIRO, F. *Melbourne.*—Three bales of scoured wool.

225. TONDEUR, O. & Co. *Melbourne.*— Trophy, containing 70 samples of wool, various brands.

226. VICTORIAN EXHIBITION COMMISSIONERS.—Specimens of Victorian timber, in all 447 pieces, the greater portion being in slabs, 8 feet in length and 4 inches thick; collected under the direction of Dr. Mueller, Government Botanist. The collection also comprises specimens contributed by Messrs. Beveridge, Allitt, Kidd, McHaffie, Levy Bros. Williams & Little, Rodgers, Weatherhead, and Dr. Backhaus.

227. PURCHAS, A. *Melbourne.*—Working models of a railway carriage, with self-acting brake, and of gas tender, for lighting railway trains.

228. RANDALL, WM. *Melbourne.*—Working model of a locomotive engine and tender.

229. HACKETT & Co. *Collingwood.*—An 'Albert' street car.

230. WILLIAMS, WM. *Melbourne.*—Pair-horse carriage.

231. NICOL, D.—An improved saw-set.

232. KAY, J. A. *Melbourne.*—Sewing machine.

233. MACINTOSH, *Melbourne.* — Mining picks, hammers, drills, and gadges.

234. ROBARDT, O.—Model of puddling machine, prismatic cross tramel, parallel ruler, and beam compass.

235. BROWN, W. *Fitzroy.*—Model of a road-scraper by horse-power; model of a battery of stampers.

236. COMMISSIONERS OF VICTORIAN EXHIBITION.—Working battery of 12 stampers for crushing quartz and amalgamating gold with ripples, and amalgamation complete; manufactured by the Port Philip Gold Mining Company, at Clunes.

237. GROLEY, W. *East Collingwood.*— Model of a quartz-grinding and amalgamating mill.

238. MERIDETH, J. *Castlemaine.*—Model of an improved gold amalgamator.

239. HARPER, R.—Working model of an automatic coffee roaster.

240. McNAUGHT, *Chewton.*—Models of horse-puddling machines, &c.

241. STRACHAN, W. *Melbourne.*—Model of engine for extinguishing bush-fires.

242. THOMSON, R. & W.—Mercurial filter for separating gold amalgam from the liquid mercury.

243. HENDERSON & BETT, *South Yarra.* Swing plough.

244. ROBINSON & Co. *Melbourne.*—Victorian prize reaping machine with side delivery.

245. CHAMP, WM. *Pentridge.*—Model of a pump made by a Chinese prisoner at the penal establishment.

246. LAMBERT & CURTIS, *Collingwood.*— Perforated gratings, for stamper boxes connected with quartz-crushing machinery.

247. LOVE, R. A. *Sandhurst.*—Model of a new compound truss suspension-bridge.

248. White, J. H. *Melbourne.*—An improved fire-hose director, with revolving nozzles of different sizes; an improved lever hose.

249. HENSON, A. W. *Melbourne.*—Fowling-piece, bullet-moulds, ramrods.

250. FERGUSON, CAPT. *Williamstown.*—Model of a life-boat.

251. HEATH & JACKSON, *Geelong.*—Model of yacht, ' Southern Cross.'

252. HIDDLE, J. *Melbourne.*—Model of an improved shackle for heavy chains.

253. SKINNER, MR. *Melbourne.*—Model of the iron steamer, ' Phantom.'

254. VAIL, MR. *Melbourne.*—Model of wreck escape.

255. WHITE, MESSRS. *Williamstown.*—Working models of vessels built by exhibitors.

256. WILKIE, HON. D. *Melbourne.*—Model of a new form of propeller for steam navigation.

257. BOLTON, J. *Williamstown.*—A gravitating dial.

258. ROBARDT, OTTO, *Melbourne.*—Prismatic cross-trammel parallel ruler and beam-compass.

259. GRIMOLDI, J. *Melbourne.*—Barometer and thermometers.

260. AMHERST MUNICIPAL COUNCIL.—Photographs of views and buildings in the municipality and suburbs.

261. BALLARAT MUNICIPAL COUNCIL.—Views and buildings in the town and district of Ballarat.

262. BELFAST MUNICIPALITY.—Views of Belfast.

263. BEECHWORTH MUNICIPALITY.—Views.

264. CASTLEMAINE MUNICIPAL COUNCIL.—Photographs of views and buildings.

265. COX & LUCKEN, *Melbourne.*—Photographs of stores and buildings in Melbourne, &c.

266. CARLTON MUNICIPAL COUNCIL.—Panoramic view of Carlton.

267. DUNOLLY MUNICIPAL COUNCIL.—Photographs of views and buildings.

268. DAINTREE, R.—Photographs of panoramic views of Ballarat, Castlemaine, &c.; geological sections and views.

269. DAVIS.—Photographs of buildings in Melbourne and Fitzroy.

270. GEELONG CORPORATION. — Photographs of public buildings in Geelong.

271. GEELONG. — Photographs of banks and private buildings, presented by the owners of the property.

272. HAIGH, E.—Photographs of views and buildings in and around Melbourne.

273. JOHNSON, MESSRS.—A collection of photographic views.

274. KILMORE MUNICIPAL COUNCIL. — Views and buildings in the district.

275. KYNETON MUNICIPAL COUNCIL. — Views of Kyneton.

276. MOONAMBEL MUNICIPAL COUNCIL.—Photographs of views of the district.

277. MELBOURNE CITY COUNCIL.—General views of the city, photographed by Littleton.

278. NETTLETON, *Melbourne.* — Photographs of buildings.

279. PUBLIC WORKS DEPARTMENT. — Photographs of public buildings in the neighbourhood of Melbourne.

300. RICHMOND MUNICIPAL COUNCIL.—Views, &c., in the municipality.

301. SMYTHESDALE MUNICIPAL COUNCIL. — Photographs of general views in the districts.

302. SANDRIDGE MUNICIPAL COUNCIL.—Photographs of views and buildings in Sandridge.

303. St. KILDA MUNICIPAL COUNCIL. — Photographs of views and buildings in the municipality.

304. SPIERS & POND, MESSRS.—Photograph of racket ground, showing the ' All England ' match.

305. SANDHURST MUNICIPAL COUNCIL.—Views and buildings at Sandhurst.

306. VICTORIA VOLUNTEERS. — Photograph by Batchelder and O'Neil, Melbourne.

307. WILLIAMSTOWN MUNICIPAL COUNCIL.—Photographs of views and buildings in Williamstown.

308. OSBORNE, S.W.—Specimens of photolithography, the process invented and patented by exhibitor.

309. MATTHIAS, J. R. *Melbourne.* — A bass drum, constructed on a new principle.

310. THORNE, J. *Melbourne.*—New kind of silver strings for violins, tenors, violoncellos, &c.

311. WITTON, H. *Collingwood.*—Case of clarionet reeds.

312. BEANEY, J. G., F.R.C.S.—An improved fracture apparatus.

313. CHRINSIDE, T. *Werribee.* — Basket and nets made by natives on the Grampians; baskets and nets made of reeds.

314. HOPWOOD, H. *Echuca.*—Nets made of Victoria flax.

315. MACKENZIE, J. *Swan Hill.*—Fishing net made by natives of the Murray, from fibre cyperus vaginatus.

316. CHAMP, W. *Pentridge.* — Woollen door mats.

317. HOLLINGS & CHAMBERS, *Melbourne.* —Wool flock for upholsterers.

318. McLENNAN & Co. *Castlemaine.*—Six pair of socks; one pair of gloves.

319. POTTS, MRS. R. *Melbourne.*—Point lace.

320. BEECHWORTH LOCAL COMMITTEE.—An opossum skin rug.

321. CLARK, J. *Melbourne.*—Kangaroo skins; opossum skins; flying squirrel; native cat skins.

322. FITZGERALD, RYAN, *Portland.* — Kangaroo skins; opossum skins.

323. GRAY, MRS. *Portland.* — Emu feathers: wool and skin of native cat, dyed with sea-weed.

324. HART, J. *Melbourne.*—Large rug of native cat skins.

325. ROBERTSON, J. *Melbourne.*—Dressed and dyed feathers of Australian birds.

326. WILLIAMSON, J. *Collingwood.* — Sample of curled horse hair.

327. BREARLEY, BROS. *Geelong.* — Two crop butts, four crop sides, four sides dressed curried shoe leather.

328. CLARK, J. *Melbourne.*—Dressed and curried hides, shoe leather, harness leather, waxed and brown kangaroo dressed saddle leather curried.

329. CHIRNSIDE, T.—Stock whips, saddle girths, hide rope.

330. COMMISSIONERS OF VICTORIAN EXHIBITION.—Pack saddle, as made for the Burke and Wills exploring expedition.

331. CHRISTIAN, H. *Kew.*—Halters.

332. DOCKER, REV. J. *Wangaratta.*—Stock whip, with handle of Myall wood.

333. FORD, BROS. *Melbourne.*—Camel shoes made for the Burke and Wills exploring expedition.

334. LADE & SANDERS, *Melbourne.*—Ladies' and gentlemen's riding saddles.

335. McFARLANE, *Melbourne.* — Stock whip, 17 ft. long.

336. MUELLER, DR.—Pair of saddle bags, with wire and leather covers, as used by Exhibitor for drying plants when travelling in the bush.

337. CHAMP, W. *Pentridge.*—Cabbage tree hats; boots; uniform; prisoners' clothes, made at penal establishment.

338. FORD, BROS. *Melbourne.*—Patent washing hats.

339. GALVIN, J. *Melbourne.*—Case of hats.

340. KING, MRS. *Collingwood.*—Bonnets made of colonial straw.

341. OXLEY, G. W. *Talbot.*—Black Creek volunteer uniform, in Sydney tweed.

342. THOMAS & MURPHY, *Melbourne.*—Pair of jockey boots of kangaroo leather.

343. WALLWORTH, *Melbourne.*—Military hats (Busbies).

344. ANGUS & ELLERAY, *Castlemaine.*—Specimen of printing.

345. BURNIE, J. D. *Warnambool.*—Copy of Warnambool Sentinel.

346. CLARSON, SHALLARD, & Co.—Printing in colours.

347. COOK & FOX, *Melbourne.*—Super royal folio ledger, bound in vellum, double Russia bands.

348. CLOUGH & Co. *Melbourne.*—Vol. Clough's Circular, and loose copies of ditto.

349. DETMOLD, WM. *Melbourne.*—Specimens of bookbinding—3 vols. Shakespeare.

350. EVANS & SOMMERTON, MESSRS.—Vol. Maryborough and Dunolly Advertiser.

351. FRANKLYN, F. B. *Melbourne.*—Vols. of Melbourne Herald, Bell's Life, and Illustrated Post.

352. FERRES, JOHN, Government Printer. —Copies of official books and acts of parliament; vellum binding; stereotyping.

353. HEATH & CORDELL, *Geelong.* — Almanack for 1862.

354. LEVEY, WM. *Melbourne.*—Copies of Victorian Ruff, Stud Book, Fistiana.

355. MELBOURNE PUBLIC LIBRARY, TRUSTEES.—Progress catalogue; Melbourne Benevolent Asylum; Reports.

356. MELBOURNE UNIVERSITY.—Copy of Library Catalogue Calendar, 1861-1862.

357. MASON & FIRTH, MESSRS. *Melbourne.*—Copies of various pamphlets and papers published by them.

358. PROPRIETORS OF ARARAT ADVERTISER.—Copy of journal.

359. PROPRIETORS OF BALLARAT STAR.— Volume of journal.

360. PROPRIETORS OF BUNINGONG TELEGRAPH.—Copy of journal.

361. PROPRIETORS OF CASTLEMAINE ADVERTISER.—Copy of journal.

362. PROPRIETORS OF CRESWICK & CLUNES ADVERTISER.—Copies of journal.

363. PROPRIETORS OF ECONOMIST.— Volume of journal.

364. PROPRIETORS OF CHRISTIAN TIMES. --Volume of journal.

365. PROPRIETORS OF FEDERAL STANDARD. —Copies of journal.

366. PROPRIETORS OF ILLUSTRATED AUSTRALIAN MAIL.—Copies of journal.

367. PROPRIETORS OF MOUNT ALEXANDER MAIL.—Volume of journal.

368. PROPRIETORS OF OVENS AND MURRAY ADVERTISER.—Copies of journal.

369. PROPRIETORS OF REVIVAL RECORD. —Copies of journal.

370. PROPRIETORS OF SOUTH BOURKE STANDARD.—Volume of journal.

371. PROPRIETORS OF TALBOT LEADER.— Copies of journal.

372. PROPRIETORS OF WARNAMBOOL EXAMINER.—Copies of journal.

373. ROYAL SOCIETY OF VICTORIA.—Copies of transactions.

374. SANDS & KENNY.—Copies of Melbourne Illustrated, Kelly on the Vine, Cricketer's Guide, map of Australia.

375. SMYTH, R. B.—Mining Surveyors' Reports, 2 vols.

376. SUPREME COURT OF VICTORIA.—Copy of Catalogue of Law Library.

377. SYME & Co.— *Melbourne.* —Vol. Daily Age, Weekly Age, Leader, and Farmers' Journal.

378. STATISTICS IN ILLUMINATED WRITING, showing the rise and progress of the corporations and municipalities in Victoria.

379. TURNER, J. *Melbourne.*—Letter dispatch box, pocket-books, card cases.

380. VICTORIAN COMMISSIONERS.—Copies of Melbourne Exhibition Certificates, and season tickets.

381. WILSON, E. — Vols. of Examiner and Yeoman newspapers.

382. WILSON & MACKINNON. — 2 vols of Daily Argus, Weekly Argus.

383. BERNARD, W. H. *Beechworth.*—An improved scale diagram.

384. MARSH S. D. *Melbourne.*—Portion of MS. Opera, ' The Gentleman in Black.'

385. O'BRIEN, J. *Hotham.*—Ornamental writing, in frame of the oak.

386. TOLHURST, G. P. *Melbourne.*—Music compositions.

387. WILKIE & Co. *Melbourne.* — Music published by Exhibitors.

388. ALCOCK & Co. *Melbourne.* — Billiard table made of myrtle wood, specimens of turned work, in wood and ivory.

389. BARRY, SIR REDMOND. — Portable wardrobe, of Colonial woods, by Thwaites & Son.

390. BOWIE, DR. *Yarra Bend Lunatic Asylum.* — A loo table, made of colonial woods.

391. CHAMP, W. *Pendridge.* — 2 fire-screens, 1 work-table, 1 flower-stand, chess-board and men, solitaire board.

392. CARR & SON. — Model of Venetian window-blind.

393. MCKENZIE, D. *South Yarra.*—Rustic seat.

394. PASLEY, CAPTAIN, R. E. — Inlaid desk of Australian woods, with gold mountings.

395. PIETOICHE, F. — Potichiomanie chimney-piece and ornament.

396. VICTORIAN EXHIBITION COMMISSIONERS. — A Gothic case, 9 feet by 9 feet, and 15 feet high, for exhibiting gold and precious stones, writing-desks, work-boxes of inlaid wood.

397. WHITEHEAD, J. *Melbourne.*—Picture frames.

398. CAIRNS, WILSON, & AMOS, *Melbourne.* —Bar-iron rolled out of scraps and refuse.

399. HUGHES & HARVEY, *Melbourne.*— Fletcher's anti-agitator or milk-preserving cans.

400. JENKINSON, W. *Fitzroy.*—Portable oven for use in the bush.

401. LAMBLE, S. *Geelong.*—A set of horse-shoes mounted on hoofs.

402. WARD, *Melbourne.*—A small sample of plain cutlery.

403. BRUCE, J. V.—Gold inkstand, with granite pedestal, presented to Exhibitor by the workmen on Melbourne and Murray River Railway.

404. COMMISSIONERS OF VICTORIAN EXHIBITION.—A pyramid, 44 feet, $9\frac{1}{4}$ inches high, and 10 feet square at the base, representing the quantity of gold exported from Victoria from 1st of October 1851, to 1st of October 1861, viz., 26,162,432 ounces troy, equal to 1,793,995 lbs. avoirdupois, or 800 tons 17 cwt. 3 qrs. 7 lbs., equal in solid measurement to $1,492\frac{1}{2}$ cubic feet of gold, of the value of 104,649,728*l.* sterling. Designed by J. G. Knight, Fel. R.I.B.A., Secretary to the Commissioners of Victorian Department.

405. RHODES, MRS. *Melbourne.* — Topaz found at Kangaroo Flat, near River Loddon.

406. TURNER, W. J. *Beechworth.* — 7 brooches, 2 bracelets, 1 diamond ring, gold and precious stones.

407. WIPER, J. *Melbourne.*—Specimens of cut and bent glass, frame of coloured cut glass.

409. HIRSCHI, F. *Castlemaine.*—Pottery.

410. KELLY, T. *Brunswick.*—Stone ware drain pipes.

411. STIELING, G. F. *Richmond.*—Pottery.

412. ARNOLD, C. *Melbourne.*—Five cases of myall wood pipes, case of myall wood.

413. ANDERSON, SHARP, & WRIGHT.—Machine-made window sash and frame.

414. ADAMSON, H. A. *Queen's Cliff.* — Two cases of sea-weeds.

415. BEALE, MRS. *Emerald Hill.*—Marine bouquet made of sea-weeds and shells.

416. BEALE, W. J. *Prahran.* — Three cases of fancy soap.

417. BOEHMN, J. *East Collingwood.*—Soap.

418. BALLARAT COMMITTEE. — Three pieces of steam-bent timber.

419. BROWN, W. *Melbourne.* — Sample gold and produce bags.

420. CHATFIELD, C. M. *Melbourne.*—Australian earth soap.

421. CHAMP, W. *Pentridge.*—Coir matting, door-mats, paper-knives, walking-sticks, imitation books, made of Victoria woods, granite fountain.

422. CROMPTON, *Beechworth.* — Cigars made from tobacco grown at Beechworth.

423. COMMISSIONERS OF VICTORIAN EXHIBITION.—Fifty-one paper-knives of different Victorian woods.

424. DOWNIE & MURPHY, *Hobson's Bay, Candle Works.*—Sample boxes of candles, soap, refined tallow, &c.

425. DOTHERS, E. *Castlemaine.*—Sample of prepared and common bricks.

426. EWING, J. A. *Fitzroy.*—Yeast powder.

427. FRAZER, MR. *Back Creek.*—Six bottles of hair dye and atmospheric oil.

428. FITTS, C. *Melbourne.*—Sample of Victorian glue.

429. FOORD, G. *Melbourne.*—Two samples of hydraulic cement made from septara.

430. GRAY, W. *Philiptown.*—Bricks and tiles.

431. GRAY & WARING, *Melbourne.*—Butter churn.

432. GOERNAMANN, F. *Melbourne.*—Steam-bent timber.

433. HEATH, R. *Geelong.*—Case of artificial teeth in gold and vulcanite.

434. HODGKINS, W. *Emerald Hills.*—Brushes.

435. HODGKINSON, G. *Prahran.*—Bricks, tiles, and terra cotta.

436. HEWETH, *Beechworth.*—Soap.

437. KELSALL, J. *Buningong.*—Case of soap.

438. KNIGHT, J. G. *Melbourne.*—Bricks and tiles from various parts of the colony.

439. LEVY BROS. *Melbourne.*—Myall wood pipes mounted in gold.

440. LEE, P. *St. Kilda.*—Cigars.

441. MERCER, MRS. *Geelong.*—Frame of fancy leather work.

442. MILLER, F. McD. *Melbourne.*—Cartridge and compressed bullets of various kinds.

443. MONTGOMERY, R. *Collingwood.*—Samples of cork cuttings.

444. MEMMOTT, W. *Collingwood.*—Specimen of comb making, made on the goldfields.

445. MELBOURNE ASPHALT CO.—Eleven small specimens of asphalt.

446. MOURANT, G. W. *Collingwood.*—Wooden taps and spiles.

447. McILWRAITH, *Melbourne.*—Sheet lead and piping.

448. McCAPE, A. *Chiltern.*—Patent powder-proof lock.

449. NIGHTINGALE, E. *Melbourne.*—Milliners' boxes.

450. PERRY, J. *Melbourne.*—Bent timber for carriages.

451. QUELCH, BROS. *Prahran.*—Case containing various kinds of candles.

452. SKEATS & SWINBOURNE, *Melbourne.*—Machine-wrought mouldings for joiners' work.

453. SHIP & CO. *Melbourne.*—Samples of wire work.

454. SHIEBLACH, C. *Back Creek.*—Candles and soap.

455. SANDHURST LOCAL COMMITTEE.—Bricks and tiles.

456. STIELING, G. F. *Richmond.*—Fire bricks.

457. SANSOM, H. *St. Kilda.*—A curious chain cut in wood.

458. SHECKEL, T. *Geelong.*—Fancy basket and willow work.

459. TALLERMANN, W. *Collingwood.*—Samples of india-rubber manufacture.

460. WILKIE, MRS.—Two cases of seaweeds.

461. WILLIAMS, W. *Melbourne.*—Machine-wrought timber for carriage building.

462. WHITELAW, G. *Ballarat.*—Four specimens of graining in imitation of wood.

463. WARD & CO. *Melbourne.*—Soap powder.

464. WOOD, W. J. *Toorak.*—Blacking.

465. WESTALL, MRS. *Melbourne.*—Case of wax flowers.

466. CROUCH & WILSON, *Melbourne.*—Design for Town-hall, Prahran; design for Wesleyan chapel, Collingwood.

467. BILLING, N. *Melbourne.*—Design for Presbyterian church.

468. KNIGHT, J. G. *Melbourne.*—Design for Government House.

469. PURCHAS & SAWYER.—Designs for Melbourne Savings' Bank, Bank of Australasia, Temple Court, and premises, Bourke Street.

470. TERRY, L. *Melbourne.*—Design for Flower, Macdonald & Co.'s warehouse.

471. BATEMAN, *Melbourne.*—Designs for woollen fabrics, introducing indigenous flowers and foliage.

472. CALDER, J. *Ballarat.*—Oil painting: View in the Pyrenees, Victoria.

473. DE GRUCHY & LEIGH, *Melbourne.*—Specimen of chromo-lithography, ' The Three Marys.'

474. EATON, MISS, *Melbourne.*—Three water-colour paintings of indigenous flowers.

475. GUERARD, E. VON, *Melbourne.*—Oil paintings of Victorian scenery. No. 1. Fern Tree Gully. No. 2. Stony Rises. No. 3. View of Geelong. No. 4. Mount William. No. 5. Forest scene. No. 6. Sydney Heads.

476. ROWE, G.—Six water-colour paintings of scenery in Victoria.

477. STRUTT, W. *Melbourne.*—Portrait of the late Colonel Neil, Portrait of Major-General McArthur, Maoris driving off Settler's Cattle, and six other subjects.

478. TAYLOR, T. G.—Painting, ' View at Fryer's Creek, in 1852 ;' ' Road Making in Black Forest'—a sketch taken in 1852.

479. ARNOLDI, X. *Melbourne.*—Seal of the Commissioners of the Victorian Exhibition.

480. POOLE, MRS. G. H. *Melbourne.*—Small medallion in wax of Burke and Wills, the explorers.

481. SCURRY & MACKENNAL, *Melbourne.*—Two designs for fountains ; design for chimney-piece.

482. SUMMERS, C. *Melbourne.*—Collection of medallion portraits ; design for the seal for Victoria Exhibition ; cast of the head of Wonga Wonga, a native chief.

483. THOMAS, MISS, *Richmond.*—Bust of Dr. Barnett ; figure, ' Napea.'

484. COSMOPOLITAN GOLD MINING CO. *Ballarat.*—Drawing showing the workings of the Company's claim.

485. DAVIDSON, MR.—Large geological map of Ballarat, showing the principal leads of gold in that district.

486. McCOY, PROFESSOR, *Melbourne University.*—Three plates of Decades of the Memoirs of the Museum of Victoria.

487. SELWYN, A. R. C., GOVERNMENT GEOLOGIST.—Progress map of the geology of Victoria.

488. TOCKNELL, W. *Melbourne.*—Steel-plate engraving.

489. TOCKNELL, W. *Melbourne.*—View of Melbourne in 1839.

491. BLACKHAUS, DR. *Sandhurst.*—Platypus.

492. BEECHWORTH COMMITTEE. — Two small Murray River turtles, porcupine, small platypus.

493. CHAMP, W. INSPECTOR-GENERAL OF PENAL ESTABLISHMENTS, *Pentridge.*—Native weapons.

494. DUNN, E. H. *Beechworth.*—Seventeen bottles containing snakes, lizards, &c.

495. DENNYS, M. L. *Geelong.*—A collection of Victorian quails.

496. HOWITT, W. LEADER OF CONTINGENT EXPLORING PARTY.—Native bag containing pitcherry.

497, 498. HOWARD, THE REV. W. C. AND BROKEY, P. LE P.—A collection of birds shot in the Ovens District.

499. HENSON, W. A. *Melbourne.*—Collection of native birds.

500. JAMIESON.—Native weapons used by the Murray tribe.

501. JOHNSON, T. *Back Creek.*—Native tomahawk.

502. MURRAY RIVER FISHING CO. — Preserved specimens of Murray cod and other fish.

503. McCOY, *Melbourne University.*—A collection of insects (six cases).

504. MUELLER, DR.—Native weapons.

505. MARTINDALE, MRS. *Back Creek.*—Two bottles containing snakes.

506. PRESHAW, W. J. *Castlemaine.*—Native tomahawk.

507. PAULL, J. M. *Back Creek.*—Two bottles containing reptiles.

508. SANBRIDGE, H. A.—Native peaches.

509. THOMAS, W. *Melbourne.*—A kur-bur-er, or native bear ; native te work, used by natives to produce fire ; native basket.

510. COOP.—Lead piping.

511. McILWRAITH. — Milled lead and piping.

512. CRISP. — Diamond and precious stones.

513. SMYTH, W.—A collection of quartz specimens, rich in gold, from Inglewood.

514. COMMISSIONERS OF VICTORIAN EXHIBITION.—Sixteen cases of cereals.

515. ATTWOOD, S.—Sample of wheat.

516. BUTCHER, BENJ. N. — Sample of wheat.

517. BUCHANAN, R. & J.—Two samples of wheat.

518. M'CASKILL.—Two samples of wheat.

519. COCKRANE.—Sample of wheat.

520. CONNOR, J. H.—Sample of wheat.

521. DYER, R.—Oats.

522. DARCY, M.—Sample of wheat.

523. FORREST, J.—Sample of wheat.

524. GAUL & LUMSDEN.—Sample of wheat.

525. HANCOCK, A. & B.—Sample of wheat.

526. HADLEY & Co.—Bag of flour.

527. JOHNSON, D.—Sample of wheat.

528. KINNERSLEY, D.—Sample of wheat.

529. KITSON, S.—Sample of wheat.

530. MORGAN, J.—Sample of wheat.

531. M'ALPINE, W.—Sample of wheat.

532. MORRISON, W. J.—Sample of wheat.

533. HALL, MR.—Box of beans.

534. PATTERSON, A.—Sample of wheat.

535. PATTEN, A.—Sample of wheat.

536. PORTER, B. C.—Sample of wheat.

537. SIMPSON, G. H.—Sample of wheat.

538. THOMSON, W.—Three samples of wheat.

539. VEARING, T.—Sample of wheat.

540. WESTLAKE, A.—Sample of wheat.

541. DALGETTY, F. G.—A collection of Victorian birds in three cases.

542. BENJAMIN, D. — Presentation gold cup, gold brooch.

PRINTED FOR HER MAJESTY'S COMMISSIONERS

BY

SPOTTISWOODE AND CO., NEW-STREET SQUARE, LONDON

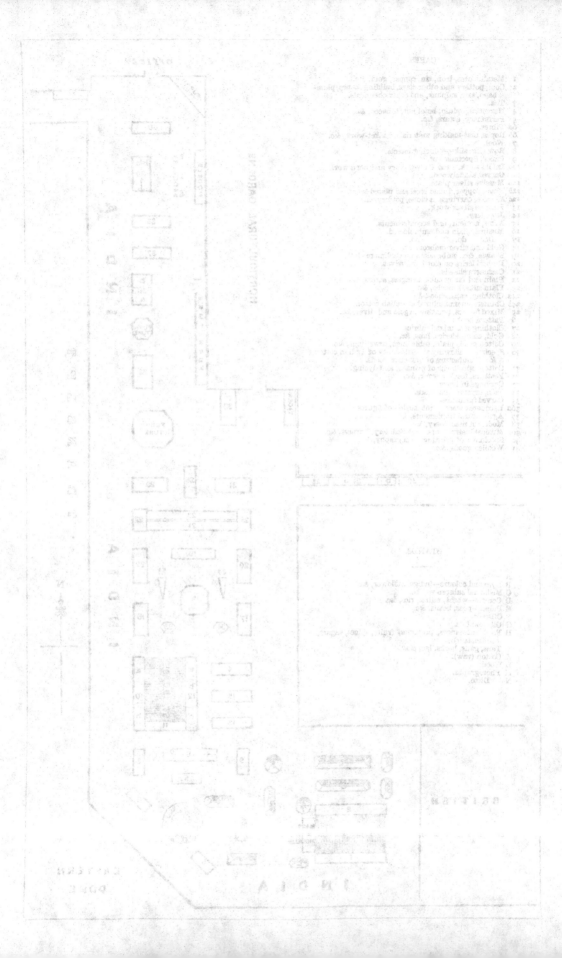

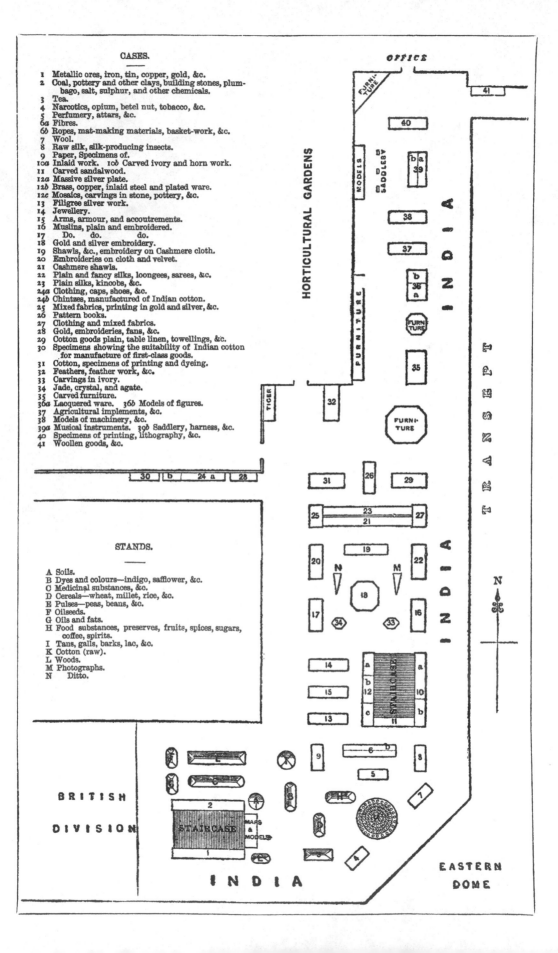

THE INTERNATIONAL EXHIBITION *of* 1862.

# A CLASSIFIED

### AND

# DESCRIPTIVE CATALOGUE

### OF THE

# INDIAN DEPARTMENT.

*BY J. FORBES WATSON, A.M., M.D.*

*F.L.S., F.C.S., F.R.A.S.*

*Reporter on the Products of India ; Director &c. Indian Department, International Exhibition.*

PRINTED FOR HER MAJESTY'S COMMISSIONERS.

PRINTED FOR HER MAJESTY'S COMMISSIONERS

BY

SPOTTISWOODE AND CO., NEW-STREET SQUARE, LONDON

# PREFACE.

———◆———

IN preparing for Her Majesty's Commissioners the Catalogue of the Indian Department, my object has been to supply not only a correct guide to the articles actually exhibited, but at the same time to afford as much information as possible within the space at my disposal.

To the Committees established in India for the purpose of collecting specimens illustrative of the districts represented by them, as well as to individual exhibitors, the public are indebted for much information which finds a permanent record here, as well as in MR. DOWLEANS' valuable catalogue of the articles forwarded through the CENTRAL COMMITTEE for *Bengal*.

For the numerous statistical tables, which add so materially to the value of this compilation, I am indebted to the Statistical Department of the India Office.

I have offered but few remarks on the quality of the various articles exhibited in the department over which I have the honour to preside : these have necessarily been reserved for my final report to the Indian Government. It may, however, be fairly assumed that, in a commercial point of view, the value to India and to this country of the present Exhibition will be considerable, and that the substantial advantages accruing therefrom will far exceed those which followed the display of Indian articles on the two previous occasions of 1851 and 1855.

JOHN FORBES WATSON.

# INDIAN COMMITTEES.

# CENTRAL COMMITTEE FOR BENGAL.

## CALCUTTA.

### PRESIDENT.

The Honorable Sir Bartle Frere, K.C.B., *Member of the Supreme Council of India.*

### MEMBERS.

Honorable Sir Mordaunt Wells, Kt., *Puisne Judge of the Supreme Court of Judicature*

Honorable H. B. Harington, *Member of the Legislative Council of India*

Honorable W. FitzWilliam, Esq., *Member of the Legislative Council of India*

Honorable D. Cowie, Esq., *Member of the Legislative Council of India, and Sheriff of Calcutta*

A. Grote, Esq., *President of the Income Tax Commission, and Member of the Board of Revenue*

R. Temple, Esq., *Currency Commissioner for India*

W. S. Seton-Karr, Esq.

Col. C. B. Young, *Secretary to the Government of Bengal, Department Public Works*

Professor T. Oldham, *Superintendent of the Geological Survey of India*

Dr. T. Anderson, *Superintendent of the Botanical Gardens*

Dr. F. N. Macnamara, *Chemical Examiner, and Professor of Chemistry at the Calcutta Medical College*

W. S. Atkinson, Esq., *Director of Public Instruction*

Col. Baird Smith, C.B., *late Mint Master of Calcutta*

G. Brown, Esq., *of the firm of Messrs. Jardine, Skinner, & Co.*

Seth Apcar, Esq., *of the firm of Messrs. Apcar & Co.*

F. Jennings, Esq., *Master of the Trades' Association*

Rajah Kali Krishna Bahadoor

Manockjee Rustomjee, Esq.

Syud Keramut Ali.

### SECRETARY.

A. M. Dowleans, Esq.

( v )

# CENTRAL COMMITTEE FOR THE PUNJAB.

## LAHORE.

### PRESIDENT.

D. F. McLeod, Esq., C.B., *Financial Commissioner of the Punjab*

### MEMBERS.

T. D. Forsyth, Esq., C.B., *Commissioner of the Lahore Division*

Dr. Burton Brown, *Chemical Examiner, and Professor of Chemistry in the Lahore Medical College*

### SECRETARY

T. H. Thornton, Esq., C.S.

# DISTRICT COMMITTEES.

## UMRITSUR.

Major Farrington
H. Cope, Esq.
T. C. Vaughan, Esq.
D. FitzPatrick, Esq., *Secretary*

## LOODIANA.

Captain McNeile

## PESHAWUR.

Captain Coxe
Lieut. Shortt, *Secretary*

## DELHI.

Fred. Cooper, Esq., C.B.
Dr. B. Smith, *Secretary*

## MOOLTAN.

Lieut.-Colonel Hamilton
Colonel Voyle
Lieut. Lane, *Secretary*

## KANGRA.

P. Egerton, Esq.
Sir A. Lawrence, *Secretary*

# CENTRAL COMMITTEE FOR THE N.W. PROVINCES.

## ALLAHABAD.

### PRESIDENT.

Rowland Money, Esq.

### MEMBERS.

C. B. Thornhill, Esq.

Dr. Beatson

Captain Hodgson

Captain Peile

G. Sibley, Esq.

Cecil Stephenson, Esq.

D. W. Lorne Campbell, Esq.

J. Middleton, Esq.

Nazir Ali Khan, Esq.

Buldeo Narain *alias* Chotee Loll

### SECRETARY.

G. H. M. Batten, Esq.

# SUB-COMMITTEES.

### ALLAHABAD.

W. Johnstone, Esq.

W. Young, Esq.

### BANDA.

H. W. Dashwood, Esq.

C. W. Carpenter, Esq.

Ahmed Hossein

Ooodey Ram Séth

### CAWNPORE.

C. A. Elliott, Esq.

E. S. Robertson, Esq.

— Maxwell, Esq.

— Palmer, Esq.

Lalla Sheopershaud

### FUTTEHPORE.

J. Power, Esq.

W. Tyrrell, Esq.

Ahmed Buksh

Lall Bahadoor

SUB-COMMITTEES—*continued.*

## ROHILCUND DIVISION.

*President.*

W. Roberts, Esq.

*Members.*

J. Inglis, Esq.
Dr. Corbyn
Dr. Cunningham
Captain Unwin
Rajah Byjnath
Moulvie Khyrooddeen Ahmed Bahadoor
Baboo Gunga Pershad
Rajah Khyratee Lall
Lalla Luchmee Narain
Moonshee Buldeo Singh
Nawab Neamut Khan

*Secretary.*

H. R. Wilson, Esq.

## AGRA DIVISION.

*President.*

G. F. Harvey, Esq.

*Members.*

A. L. M. Phillipps, Esq.
John Murray, Esq., M.D.
G. R. Playfair, Esq., M.D.
Captain L. Macbel
Lieut. J. L. Watts
J. T. Pritchard, Esq.
W. Birks, Esq.
Lalla Motee Lall
Moulvie Syud Ameer Alli Shah
Hukeem Syud Nusseerooddeen
Lalla Jyaram Dass
Chowbey Muttrapershad

## JUBBULPORE DIVISION.

*President.*

Major W. C. Erskine.

*Members.*

Captain W. Nembhard
Captain A. Impey
Captain Pearson
J. W. Williams, Esq.
F. Macbell, Esq.
Kooshall Chund Seth
Innarjun Pundit

*Secretary.*

A. M. Russell, Esq.

## BENARES DIVISION.

*President.*

F. B. Gubbins, Esq., C.B.

*Members.*

W. Edwards, Esq.
J. H. Bax, Esq.
Rajah Deonarain Singh
Baboo Gooroodas Mitter
Baboo Narain Dass
Baboo Shevapershad
Baboo Dabeepershad

## MEERUT DIVISION.

*President.*

Colonel Huyshe

*Members.*

Colonel Hogge, C.B.
Major Ross, H.M. 36th Regiment
Captain Brown, H.M. 35th Regiment
G. D. Turnbull, Esq., C.S.
H. C. Cutcliffe, Esq., F.R.C.S.
E. E. Watson, Esq., C.S.
Koour Ouzeer Alli Khan
Mahomed Alli Khan
Oomer Alli Khan
Lalla Bankey Raie

### SUB-COMMITTEES—*continued.*

## JHANSIE DIVISION.

*President.*

Major Ternan

*Members.*

C. Daniell, Esq.
Captain Davidson
Captain Baillie
Doctor Annesley
Rajah of Kutera

*Secretary.*

W. R. N. James, Esq.

## MIRZAPORE DISTRICT.

H. P. Fane, Esq.
C. B. Denison, Esq.

## MIRZAPORE DISTRICT—*continued.*

J. H. Loch, Esq., M.D.
C. McLean, Esq.
Mulanut Jyramgeer
Ukbur Alee Khan

## GORUCKPORE DISTRICT.

*President.*

A. Swinton, Esq.

*Members.*

F. M. Bird, Esq.
G. Osborne, Esq.
M. Cooke, Esq.
Chedeeloll Muhajun
Ramsuroop Muhajun
Meer Zuhoor Alee

# CENTRAL COMMITTEE FOR OUDE.

## LUCKNOW.

### PRESIDENT.

G. Campbell, Esq.

### MEMBERS.

Colonel Abbott
Major Evans

Major Crommelin, C.B.
Captain Chapman

### SECRETARY.

Dr. Bonavia

# LOCAL COMMITTEES IN BENGAL AND ITS DEPENDENCIES.

## RANGOON.

*President.*

Major F. P. Sparks

*Members.*

Captain Browne
R. S. Edwards, Esq.
R. S. Steel, Esq.
Thomas Christien, Esq.
G. Bullock, Esq.
E. Fowle, Esq.

---

## CUTTACK.

*President.*

A. Shore, Esq.

*Members.*

Colonel Maclean
T. De B. Armstrong, Esq.
W. L. Heeley, Esq.
R. Pringle, Esq., M.D.
Baboo Juggomohun Rai
Baboo Luckheenarain Rai Chowdry

*Secretary.*

W. C. Lacey, Esq.

---

## TIPPERAH.

M. Lecolier, Esq.
H. G. Leicester, Esq.
Baboo Hurkally Ghose
Baboo Hurrish Chunder Bose

## AKYAB.

*President.*

Major G. Verner

*Members.*

J. Bullock, Esq.
A. L. McMillan, Esq.
Mahomed Bux
T. Shepherd, Esq.
Captain Ripley
H. W. Beddy, Esq.
Captain Twynam
J. H. O'Donel, Esq.
Phat-May

*Secretary.*

Dr. Graham

---

## CHITTAGONG.

*President.*

V. H. Schalch, Esq.

*Members.*

J. D. Ward, Esq.
J. E. Bruce, Esq.
— Hollingsworth, Esq.
Dr. W. B. Beatson
Captain Magrath
W. J. Brown, Esq.
Baboo Obhoy Chunder Doss
Baboo Ramkinoo Dutt
Baboo Brijo Mohun Roy
Baboo Hur Chunder Roy
Baboo Thacoorbux Tewaree

LOCAL COMMITTEES IN BENGAL AND ITS DEPENDENCIES — *continued.*

## BULLOOAH.

Dr. H. M. Davis
Baboo Ramgopal Roy
Baboo Gungapersaud Sein
Baboo Gungadass Sein

———

## BACKERGUNGE.

— Harvey, Esq.

———

## SYLHET.

— Davies, Esq.

———

## MYMENSING.

R. Abercrombie, Esq.

———

## FURREEDPORE.

— Tottenham, Esq.

———

## CACHAR.

Lieut. Stewart

———

## ARRAH, IN SHAHABAD.

C. P. Leycester, Esq.
R. F. Hutchinson, Esq.
S. Bayley, Esq.

———

## GYA, IN BEHAR.

T. C. Trotter, Esq.
E. F. Lautour, Esq.
James Farrell, Esq.
J. B. Allan, Esq.

## CHUPRAH, IN SARUN.

H. Nelson, Esq.
J. Jackson, Esq.

———

## DACCA.

*President.*

C. T. Davidson, Esq.

*Members.*

— Dodgson, Esq.
— Wise, Esq.
Dr. Simson.
J. G. Pogose, Esq.
Khajeh Abdool Gunny
— Reynolds, Esq.
— Davy, Esq.
— Brennand, Esq.
Baboo Gobind Chunder Dutt

———

## PATNA.

*President.*

H. D. H. Fergusson, Esq

*Members.*

H. Wake, Esq.
J. Sutherland, Esq.
J. K. Walter, Esq.
C. J. Muller, Esq.
R. F. Chisholm, Esq.
Syed Lootf Ally Khan
Baboo Hurry Kishen

———

## MOOTEEHAREE, IN CHUMPARUN.

W. Macpherson, Esq.
J. M. Coates, Esq.

## MOZUFFERPORE, IN TIRHOOT.

A. E. Russell, Esq.
T. B. Lane, Esq.
N. C. Macnamara, Esq.
J. Forlong, Esq.
Rai Nundeeput Mabta Bahadoor

## CHOTA NAGPORE.

*President.*

Captain E. T. Dalton

*Members.*

Major W. H. Oakes
Captain J. S. Davies
Captain G. C. Depree
Captain A. P. S. Moncrieff
Bura Lall Oopendurnath Sahee Deo
Thukoorai Kishendeal Singh Rae Bahadoor
Thukeorai Rughooburdeal Singh Rae Bahadoor
Lieut.-Col. T. Simpson
M. de W. George, Esq.
W. F. Clark, Esq.
C. G. Wray, Esq.
Baboo Kallie Dass Paulit
Koomar Ramnath Sing
M. Liebert, Esq.
Captain G. N. Oakes
Dr. W. J. Ellis
R. C. P. Perry, Esq.
Baboo Nundookoomar Aikat
Baboo Umbikachurn Mookerjee
Captain R. C. Birch
Dr. A. J. Meyer
Rajah Chunderhur Sing Deo Bahadoor
Rajah Chillresser Dhul
Rajah Bindhessuree Pershad Sing Deo Bahadoor
Rajah Protap Narain Sing Bahadoor.

*Secretary.*
Lieut. R. C. Money

## BHAUGULPORE.

*Vice-President.*

C. B. Skinner, Esq., *Officiating Commissioner*

*Members.*

W. R. Davies, Esq.
Baboo Dwarkanauth Chatterjee
Captain Layard

## MONGHYR.

*President.*

W. H. Henderson, Esq.
T. Duka, Esq., M.D.
J. Bean, Esq.

## PURNEAH.
*President.*

W. L. T. Robinson, Esq.

*Members.*
F. J. Earle, Esq.
W. M. Beaufort, Esq.
Baboo Kalyprosonno Roy Chowdry
T. J. Shillingford, Esq.

## BURDWAN.
*President.*

G. Plowden, Esq., *Commissioner*

*Members.*

The Maharajah of Burdwan
Pierce Taylor, Esq.
Captain Short
S. S. Hogg, Esq.
Baboo Sharoda Pershad Roy

*Secretary.*
J. Tweedie, Esq.

LOCAL COMMITTEES IN BENGAL AND ITS DEPENDENCIES—*continued.*

## RAJSHAHYE.

*President.*

H. M. Reid, Esq.

*Members.*

L. S. Jackson, Esq.
S. Taylor, Esq.
Dr. Shircore
James Cockburn, Esq.
C. R. Jennings, Esq.
D. T. Gordon, Esq.
Rajah Prosunno Nath Roy
Koour Anund Nath Roy
Baboo Mothooranath Bannerjee

## GOWHATTY, IN ASSAM.

*Members.*

H. Bainbridge, Esq.
Colonel D. Reid
J. N. Martin, Esq.
W. Becher, Esq.
Colonel H. Vetch
Captain R. Campbell
Dr. De Fabeck

*Secretary.*

Captain Lloyd

## DARJEELING.

Dr. A. Campbell

## BOGRAH.

— Larkins, Esq.

## GOWALPARRAH.

Lieutenant Morton

## NOWGONG.

Lieutenant Sconce

## DURRUNG.

Lieutenant W. Phaire

## SEEBSAGUR.

W. O. A. Beckett, Esq.

## LUCKIMPORE.

Captain H. S. Bivar

## BERHAMPORE.

*Members.*

A. Pigou, Esq.
A. W. Russell, Esq., *Judge*
H. A. Cockerell, Esq., *Collector*
J. Guise, Esq., *Civil Surgeon*
Sir W. Wemys, Bart.
Rajah Prosononarain Deb Bahadoor
Baboo Rajkisto Sein
Baboo Poolin Beharee Sein
Baboo Punchanun Banerjee

*Secretary.*

Major H. C. James

# CENTRAL COMMITTEE FOR MADRAS.

## PRESIDENT.

The Honorable E. Maltby, Esq., *Member of Council*

## MEMBERS.

W. R. Arbuthnot, Esq. (*Messrs. Arbuthnot & Co.*)

Surgeon Major Edward Green Balfour, *Examiner of Accounts, Medical Department*

W. H. Crake, Esq. (*Messrs. Parry & Co.*)

Surgeon Alexander Hunter, M.D., *Superintendent of the School of Industrial Arts*

Captain A. H. Hope, *Superintendent of Army Clothing*

## SECRETARIES.

C. Collett, Esq., *Acting Collector of Sea Customs* | C. J. Shubrick, Esq., *Collector of Sea Customs*

# CENTRAL COMMITTEE FOR MYSORE.

APPOINTED BY

## C. B. SAUNDERS, ESQ.

*Officiating Commissioner for the Government of the Territories of His Highness the Maharajah of Mysore.*

## PRESIDENT.

Major R. S. Dobbs, *Officiating Judicial Commissioner.*

## MEMBERS.

Colonel Lawford, *Chief Engineer of Mysore*

Major Sankey, *Assistant to Chief Engineer of Mysore*

Captain J. Pearse, *Secretary to the Mysore Commission*

J. Kirkpatrick, Esq., *Surgeon to the Mysore Commission*

Captain Puckle, *Assistant to Superintendent of Bangalore Division*

J. Garrett, Esq., *Director of Public Instruction, Mysore*

B. Kristniengar, *Sheristadar of Judicial Commissioner's Office*

S. B. Krishnasawmy Iyengar, *Naib Sheristadar of Chief Commissioner's Office*

## SECRETARY.

Lieutenant R. A. Cole, *Assistant to Judicial Commissioner*

# CENTRAL COMMITTEE FOR BOMBAY.

## PRESIDENT.

The Honorable Sir Joseph Arnould, Knight.

## MEMBERS.

Lieut.-Col. H. Barr

Bhawoo Dajee, Esq.

G. Birdwood, Esq. M.D.

W. R. Cassels, Esq.

Cowasjee Jehangeerjee, Esq.

J. N. Fleming, Esq.

A. K. Forbes, Esq., *Civil Service*

Sir Alexander Grant, Bart.

Sir Jamsetjee Jejeebhoy, Bart.

Juggonath Sunkersett, Esq.

Munguldass Nathoobhoy, Esq.

H. Newton, Esq., *Civil Service*

John Peet, Esq., M.D.

Lieut.-Col. John Pottinger, C.B.

Lieut.-Col. H. Rivers

Colonel Commandant H. B. Turner

Commodore G. G. Wellesley, C.B., R.N.

M. R. Westropp, Esq.

Rev. John Wilson, M.D.

## SECRETARY.

Dr. G. Birdwood.

# COMMITTEE FOR PENANG & PROVINCE WELLESLEY.

## PRESIDENT.

The Honorable Major Man, *Resident Councillor.*

## MEMBERS.

G. W. Earl, Esq.

J. J. E. Brown, Esq.

L. Nairne, Esq.

G. Scott, Esq.

## HONORARY SECRETARY TO SINGAPORE COMMITTEE.

Colonel Collyer, *Chief Engineer, Straits Settlements.*

NOTE.—Lists of the Committees for Singapore, Sinde, and some districts in Southern India, have not yet been received. It should, however, be added that many important contributions, and valuable information connected therewith, have been collected and forwarded by resident Government Collectors, acting under instructions from the Central Committees.

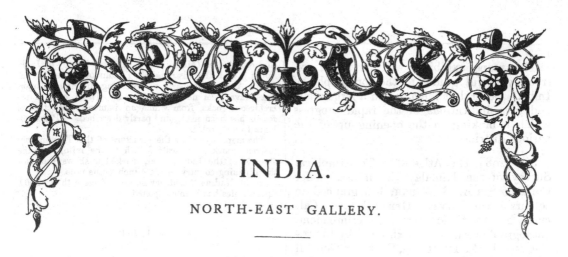

# INDIA.

## CLASS I.

# MINING, QUARRYING, METALLURGICAL OPERATIONS, AND MINERAL PRODUCTS.

## I.—MINING AND QUARRYING OPERATIONS.

*No representatives of this division.*

## II.—GEOLOGICAL MAPS, PLANS, AND SECTIONS.

1. [5967] Topographical Model of India constructed by R. MONTGOMERY MARTIN, Esq. *From the India Museum, Fife House, Whitehall Yard.*

Gold double lines represent railways finished.
Gold single lines represent railways in progress or sanctioned.
Distinctive colouring indicates the varied fluvial drainage; each shade of colour shows the extent of country drained by the main river, which flows through it and discharges its waters into the sea.
Horizontal scale 1 inch to 15 miles.
Vertical scale about 1 inch to 3,000 feet. For the lesser heights of the Himalaya range the horizontal scale slightly differs. The white ridge on the culminating parts of the Himalaya represents the line of perpetual snow.
The sandy tract near the Indus shows the great desert, the limits of which are very imperfectly known.

2. [5968] Topographical Model of India, recast by F. PULMAN from the original by R. MONTGOMERY MARTIN, Esq.; drawn and coloured by EDWARD STANFORD.

The MOUNTAINS and HIGHLANDS are displayed in relief. The Indian mountains are divided into two distinct groups by the valley of the Ganges. On the north of the river extends the snow-capped chain of the Himalaya mountains, separating India from the Chinese dominions, and drained on all sides by the rivers Indus, Ganges,

Brahmaputra and their affluents. South of the Ganges is spread out an elevated table land including the Deccan. This Highland is bordered by the Western Ghauts, rising abruptly from the shores of the Arabian Sea, — by the Eastern Ghauts towards the Bay of Bengal, — and by several chains of hills skirting the Ganges and its feeders: among the last the Aravulli Mountains are the most westerly group and the Rajmahal Hills the most easterly. The Vindhya Mountains separate the affluents of the Ganges from the basin of the Nerbudda, and thus divide the watersheds of the Bay of Bengal and the Arabian Sea.

Two other groups of mountains border India on the east and west. On the east are the snowy ranges between Assam, China, and Burmah; also the Yumadung coast range between Arracan and Burmah. On the west are the Suliman and Hala ranges, rising abruptly from the right bank of the river Indus and forming the frontiers of Kabul, Affghanistan and Beloochistan.

Northward the Hindoo Koosh separates Kabul from the Highlands of Central Asia.

RIVER SYSTEM.—All the principal rivers of India flow into the Indian Ocean, most of them being tributary to the Bay of Bengal, and the remainder to the Arabian Sea. The water parting between these two divisions or watersheds is distinguished by a bi-coloured line (blue and red).

The watershed of the Bay of Bengal is divided into minor basins (distinguished by a red line). It includes the Ganges, Brahmaputra, Mahanuddy, Godavery, Kishna, Pennar, Cauvery, &c.

The watershed of the Arabian Sea is similarly divided (distinguished by a blue line). It includes the rivers Indus, Loonee, Nerbudda, Tapty, and the numerous torrents descending from the Western Ghauts.

The maps exhibited will be found extremely useful guides to the topography of India. Several of them are on a large scale, and give indications of the rapid progress which is making in the opening up of the various districts.

3. [3298] The Atlas of a Topographical Survey of the Himalaya Mountains, under COLONEL SIR A. S. WAUGH, lithographed in colours at the Surveyor General's office, Calcutta, by MR. H. M. SMITH, Superintendent, Lithographic Branch. Exhibited by LIEUT.-COLONEL H. L. THUILLIER, Surveyor General of India.

This atlas is not only intended as a specimen of the Topographical Survey of a portion of the Himalaya Mountains ranging from 22,832 feet above mean sea level, but also to exhibit a new style of representing such difficult hilly ground and intricate details by a combination of chalk and ordinary transfer, drawing, and printing in colours from separate stones, as first introduced by the exhibitor in the publication of the results of the Indian Survey, at the Surveyor General's Office, Calcutta, whereby the details on the ¼ inch geographical scale may be more easily read than by line engraving.

Maps of the Geological Survey of India. Exhibited by THOMAS OLDHAM, Esq., Superintendent of the Geological Survey of India.

The Geological Survey of India was first established on a systematic basis in 1856. Previously to that year several detached districts had been examined and reported on, but these being isolated and unconnected, no general results of any value could be looked for. In 1856 a systematic and continuous examination of the entire country was commenced on a plan suggested by the present Superintendent of the Survey, and this plan has been steadily maintained since.

Very large portions of the Indian territories have not as yet been topographically surveyed, and of these no trustworthy maps exist. The Geological Survey is, therefore, compelled to confine its examination, for the present, to such districts as have been mapped. In some cases where maps did not exist, topographical sketch surveys have been made by the Geological Survey as the examination of the country progressed.

Wherever maps on a large scale could be procured, they have invariably been used as the records of the Geological Survey, but for many districts the only maps available are on the small scale of four miles to the inch or $\frac{1}{253440}$. This is the scale of the Indian Atlas Sheets, the only permanent or engraved maps issued by the Government of India, all others being only lithographed, and a limited number of copies printed.

The following statement gives a brief outline of the progress already made in the examination of the country: —

### MADRAS.

4. [6619] Geological Map of the Trichinopoly District.

5. [6620] Do. of the adjoining country.

The examination of the Madras Presidency was commenced in 1857. The Nilghiri Hills have been mapped

and reported on (Memoirs of Geological Survey of India, vol. i.), and the districts of Trichinopoly, Salem, and South Arcot, with parts of adjoining districts (Tanjore &c.) have been nearly completed. These contain the cretaceous rocks, from which an immense collection of fossils has been made, and partly described. — Palæontologia Indica, vol. i.

The maps sent show the structure of the Trichinopoly district, marked A (scale ½ inch to the mile); and the sheet 79 of the Indian Atlas, marked B, shows that and the adjoining country; scale ¼ inch to the mile.

In the Madras Presidency an area of more than 18,500 square miles has been completed.

### CENTRAL INDIA.

6. [6621] Geological Map of a part of the Nerbudda Valley.

7. [6622] Do. of Bundelkund.

The geological map of part of the Nerbudda Valley, marked C, represents about 8,200 square miles; that of Bundelkund about 5,000 square miles. In addition to the geological mapping of the Nerbudda Valley, the greater portion of the district was also topographically surveyed. The examination of the country adjoining this on the east, and along the valley of the Soane River, had completed (up to 1861) about 5,000 square miles in addition to that now published.

### BENGAL AND THE NORTH-WESTERN PROVINCES.

In Bengal the following districts have been completed, and in every case (where the maps were available) on the scale of 1 inch to the mile: of these only a few are sent as specimens, as the size of the maps, if combined, would be far greater than space could be obtained for.

| AREA IN SQUARE MILES. | | | | DISTRICTS. |
|---|---|---|---|---|
| 2,698 | . | . | . | Pooree, ⎫ |
| 3,062 | . | . | . | Cuttack, ⎬ Orissa |
| 1,876 | . | . | . | Balasore, ⎭ |
| 600 | . | . | . | Talcheer Coal Field, Orissa |
| 5,032 | . | . | . | ⎰ Midnapore ⎱ Hidgellee |
| 1,349 | . | . | . | Bancoorah |
| 2,692 | . | . | . | Burdwan |
| 2,007 | . | . | . | Hooghly |
| 3,114 | . | . | . | Beerbhoom |
| 2,634 | . | . | . | Moorshedabad |
| 7,804 | . | . | . | Bhaugulpore |
| 3,035 | . | . | . | Rajshahye |
| 1,288 | . | . | . | Maldah |
| 5,878 | . | . | . | Purneah |
| 3,599 | . | . | . | Monghyr |
| 1,829 | . | . | . | Patna |
| 48,495 | | | | |

In addition to the above, large portions of Dinagepore, Tirhoot, Behar, Shahabad, Mirzapore, Benares, Goruckpore, &c. (in all about 15,000 square miles) have been examined, although the entire districts are not yet completed.

In *Eastern Bengal* the Khasia Hills and parts of adjoining districts have also been examined.

In the *North-West of India* the country extending from Hurdwar on the Ganges to the Beeas along the Sewalik and Sub-Himalaya ranges has been mapped, embracing an area of about 6,000 square miles.

In *Pegu*, the examination of which was commenced in 1860-61, 3,000 square miles have been completed, while portions of the Burmese Empire and of the Tenasserim Provinces had also been reported on.

In addition to the foregoing, preliminary examinations of large areas have been completed (Guzerat &c.)

A total area, therefore, of more than 94,000 square miles has been completed and geologically mapped ; *an area much larger than the whole of Great Britain.*

8. [6623] Maps of the Raneegunge Coal Field.

9. [6624] Atlas Sheets.

10. [6617] Geological Map of Bancoorah.

11. [6616] Geological Map of Burdwan.

12. [6615] Do. Beerbhoom.

13. [6618] Do. Monghyr.

The maps of the Raneegunge Coal Field on the larger scale of 1 inch to the mile, marked E, and the Atlas Sheets Nos. 112, 113, 114F, which embrace a large portion of the country which has been examined in Bengal, will show the amount of detail and care which has been applied ; while the full maps of a few of the districts are sent to show the character of the original field work. (Bancoorah G, Burdwan H, Beerbhoom I, Monghyr J.)

Owing to the peculiar mode of publication of these topographical maps, no general or combined map of the whole area can be exhibited on the larger scale.*

## III.—ORES AND METALLURGICAL OPERATIONS.

Ores of the more common metals, as of iron, copper, zinc, tin, lead :—

### IRON ORES.

The iron ores of India are of great interest and value. Many are noted for their singular purity, especially those belonging to the magnetic iron-ores. Some are distinctly titaniferous, being in this respect similar to the Taranaki iron-sand from New Zealand, which has lately attracted so much attention in this country.

*Series illustrating the Iron-manufacture, exhibited by the East India Iron Company, through E. J. BURGESS, Esq. Secretary.*

16. [9845] Laterite stone, of which the shell of the blast furnaces are built, and all other permanent buildings at Beypore.

17. [9847] Shells used as a flux in the Beypore blast furnace instead of lime.

18. [9848] Lime used at Trincomallee blast furnace.

WOODS USED FOR MAKING CHARCOAL.

19. [9849] Vella Murdah.

20. [9850] Kurrah Murdah.

21. [9851] Errool.

22. [9852] Indian Gooseberry.

23. [9853] Poohum.

24. [9854] Nux vomica.

25. [9855] Cassan.

SAMPLES OF CHARCOAL EMPLOYED.

------

ORES.

26. [9857] Magnetic iron ore from Nynamullay, Salem.

27. [9858] Do. Tullamullay, Salem.

28. [9859] Do. Shadamunglam, Salem.

29. [9860] Do. Moorakcully, Salem.

30. [9861] Do. Rajakcullay, Salem.

31. [9862] Hydrous peroxide of iron, Arreanattum, South Arcot.

32. [9863] Magnetic iron ore, Palamcotta, South Arcot.

33. [9864] Do. Penatoor, South Arcot.

34. [9865] Do. Tondamum, Poodoocola District.

35. [9866] Do. Poonpara, Beypore.

36. [9867] Do. Honore, Beypore.

37. [9868] Do. Edewannah, Beypore.

The bulk of these ores are rich magnetic oxides, and when freed from earthy matter, and ready for the blast furnace, contain about 72 per cent. of iron. They are found

* Note by PROFESSOR OLDHAM.

in mountain masses, and are obtained by quarrying with a crowbar. The quantity is so large, that it is not necessary to have recourse to underground operations. They are quite free from sulphur, arsenic, and phosphorus, and upon a large average have been found to yield 68 per cent. of metal in the blast furnace.

The Poonpara and Honore ores are those chiefly used at present for the production of steel by the Bessemer process.

38. [9856] Two pieces of bamboo used for supports and rafters in sheds, floating rafts for timber, &c.

39. [9846] Clay found at Beypore 20 to 30 feet below the surface, used for fire-bricks and for lining furnaces.

40. [9874] Specimen of native steel.

41. [9875] Two samples of steel made direct from the ore in Malabar native furnace.

42. [9873] Three samples of hammered bar iron, manufactured by native process.

43. [9870] Pair of goat-skin bellows for blowing smiths' fires.

44. [9871] Cast iron mamooty.

45. [9869] Nine specimens of East Indian charcoal pig iron.

46. [9876] Samples of Indian Bessemer steel, made at Beypore, direct from blast furnace.

47. [9877] Three jungle knives, manufactured by native smiths at Beypore Works, from Indian Bessemer steel, which had been made direct from blast furnace.

48. [9878] Three sickles, do.

49. [9879] Four felling axes, do.

50. [9880] Razor, do.

51. [9881] Two penknives, do.

52. [9882] Six pocket knives, manufactured from Indian Bessemer steel, which had been made direct from blast furnace, by ANTHONY ROTHERHAM, *Sheffield.*

53. [9883] Four desk knives, do.

54. [9884] One pruning knife, do.

The plan adopted for the production of Indian cast-steel at the Beypore Works, by the Bessemer process, is similar to that pursued in Sweden, but differs essentially from the Sheffield method.

At Sheffield and elsewhere in this country, where the process is in operation, pig-iron is melted in a reverberatory furnace, and run thence into the converter or Bessemer vessel, which is mounted on axes; but in Sweden, and at the Beypore Works in Madras, the crude metal is run direct from the blast-furnace into an

ordinary founder's ladle, which is raised to a sufficient height by means of a travelling crane, and then poured into the converter, which is a *fixed* vessel, lined with a mixture of native fireclay and sand, and pulverized English firebrick. Steam is raised to about 50 lbs. in the boilers, giving a pressure of blast of about 6½ or 7 lbs. per square inch, and the air is driven into the converter through 11 tuyeres of ¾-inch diameter, placed horizontally at the bottom of the vessel. No manganese or other metal is added to temper the steel, the quality of the metal required being regulated by the pressure of blast and the time of blowing.

As soon as the metal is sufficiently decarbonised, the vessel is tapped, and the fluid steel run into a ladle provided with an outlet in the bottom. This ladle is swung round over the cast-iron ingot moulds, the fireclay-plug withdrawn, and the steel allowed to flow in a clear stream into the moulds beneath.

These ingots are then cogged down, under a Nasmyth hammer, and drawn into finished steel bars of various sizes.

The native tools exhibited were made from steel produced at some of the early experiments, and drawn down under considerable disadvantages; but with the appliances that will soon be at command, improved steam tilt-hammers by Messrs. Kitson & Hewitson, and Messrs. Hudswell & Clarke of Leeds, having been despatched to Madras, it is expected that nicely-finished bars of Indian cast-steel, of excellent quality, will soon be supplied from the Beypore Works.

55. [9885] Samples of Bessemer steel, made from East Indian charcoal pig iron, various sizes, tilted and rolled.

56. [9885] Two pieces of ingots, broken to show fracture, made by Bessemer process from Indian charcoal pig iron, at the works of Messrs. JOHN BROWN & Co., Atlas Steel and Iron Works, *Sheffield.*

57. [9886] Three steel sheets, do.

58. [9887] Small steel circular saw, do.

59. [9888] Piece of steel bridge rail, do.

60. [9889] Piece of bright shafting, do.

61. [9890] A pocket knife, made from Indian steel by ANTHONY ROTHERHAM, *Sheffield.*

62. [9891] Four table knives, do.

63. [9892] Two bread knives, do.

64. [9893] Two carvers, do.

65. [9872] Small and large native lamp, used in pagodas &c.

66. [9894] Three razors, and case of razors, made from Indian iron and steel.

67. [9896] Two gun barrels and gun lock, do.

68. [9894] Sword blade, do.

69. [9898] Two pieces of bright shafting, do.

70. [9897] Small spring, made from Indian iron and steel.

71. [9899] Walking stick, do.

72. [9895] Carving knife and fork, and bread and table knives, do.

73. [9900] Nails (various) and screws, do.

74. [9898] Small spindle, do.

75. [9901] Five bundles of wire, do.

76. [9902] Two small malleable castings, do.

77. [5975] Magnetite in lustrous flakes, Salem.

78. [5978] Do. massive and singularly pure, Salem.

79. [5979] Do. rusting on the surface, Salem.

80. [5976] Do. in octohedral crystals, Salem.

81. [5977] Do. rusting on the surface, first sort from Salem.

82. [5974] Do. rusted on the surface, second sort from Salem.

The iron ore of the Salem districts of the Madras Presidency is a rich magnetic oxide of iron, very heavy and massive. It is commonly known as loadstone. The yield averages 60 per cent. of metallic iron. Much of the ore being a pure black magnetic oxide would doubtless yield 73 per cent. The ore is, however, often mixed with quartz, which is a very refractory material in the blast furnace. Limestone, and, in some places, shell lime, is employed as a flux; and the charcoal of some kind of acacia is the fuel.

83. [7222] Iron ore. Very fine and massive hæmatite. DR. HUNTER.

IRON SANDS EXHIBITED BY CAPT. J. MITCHELL.

84. [5041] Magnetic iron sand, village of Pankam, Vellore Talook.

85. [5042] Do. village of Pulleputt, Vellore Talook.

86. [5043] Do. village of Anchenamput, Vellore Talook.

87. [5044] Do. village of Vannanthangal, Vellore Talook.

88. [5045] Do. village of Vennembutt, Vellore Talook.

89. [5046] Do. village of Catharercooppum, Vellore Talook.

90. [5047] Magnetic iron sand, village of Vamembady, Vellore Talook.

91. [5048] Do. village of Nimmanapully, Santooroo, Congoondy Talook.

92. [5049] Iron ore, from Santghum Talook.

93. [5050] Iron sand, from Talook of Streevellypootoor, in district of Tinnevelly.

94. [5304] Iron ore, Dhenkanal, Cuttack.

95. [5302] Iron ore of the class of hæmatites, both micaceous and siliceous, Dhenkanal.

96. [5303] Do. do. Talchere, Cuttack. GOVERNMENT.

An abundance of this ironstone is found in the district of Sumbulpore, and it is plentiful in the Cuttack Tributary States of Talchere, Dhenkanal, Pal-Lahara, and Ungool, and indeed throughout the hilly country bordering the settled districts of this province on the north-west. The whole of the iron used for various purposes in this division is supplied from these local sources. In Sumbulpore, according to Dr. Shortt, of the Madras Army, who passed through that district in 1855, the crude iron is sold at one anna per seer, which is equivalent to about three-fourths of a penny per English pound. From a report by the same observer, the following information relative to the method of smelting is gathered. No flux is used; the broken ironstone is mixed with charcoal, which can be prepared in any required quantity on the spot, and the mixture is then, probably in alternate layers, put into the furnace,—a kiln in miniature, standing about 4 feet high, and made of clay. The top is open, and the bottom and sides thoroughly closed. The fire is maintained by an artificial blast, introduced through a fire-clay pipe, which is sealed up with clay after the insertion of the nozzle of the bellows. The slag escapes, or more properly is raked out, through an aperture made in the ground, and which runs up into the centre of the furnace base. Three men — one to serve the fire, and two to work the bellows — are required to tend each furnace. Nearer home, this ore abounds, as has been observed, in Ungool, Talchere, Pal-Lahara, and Dhenkanal. The specimens sent are from Talchere and Dhenkanal. These are a red-ochry ore, said to produce very excellent metal, without the aid of a flux. The method of smelting here is very similar to that already described, the main difference being that the slag is passed out through an arched opening in the base of the furnace. The charcoal used is made from the Sal or *Shorea robusta.* Limestone in calcareous nodules is abundant on the spot, in Ungool at least, but is nowhere used in smelting. The price of the crude iron in Ungool is a trifle less than one anna per seer. It is, as might be expected, mixed with impurities. A specimen of the Ungool ore, taken from the ground where it had lain exposed to sun and rain, gave 66 per cent. of the peroxide of iron, equal to 46 per cent. of metallic iron. A sample from Pal-Lahara gave 60½ per cent. of the protoxide of iron, equivalent to 47 per cent. of metal. These results are given on the authority of Mr. Piddington, late Curator of Economic Geology, Calcutta. The native method of smelting is, however, rude and wasteful.*

Although there is abundance of mineral coal in South Mirzapore, in Palamow, Singrowlie, and Rewah, native smelters use only wood charcoal prepared by themselves.

* Local Committee of *Cuttack*. W. C. LACEY, Esq., Secretary.

and as their furnaces and tools are small, they can all be constructed and arranged by one man in half a day; this fuel and ore are close at hand to the furnace, the latter being remade farther in the jungle to suit their main requirements, while the wretched hut in which they live may well be prepared in the half day remaining. The process employed by the smelters is a very simple one. To each furnace there are two men, and it is kept in full play all day. Each day, if the smelters have wives and children to break up the ore into ½ or ¼-inch cubes, and bring charcoal, they will charge the furnace four times, and the day's work will be 4 or 5 small malleable pigs of 2 or 2½ seers each, or in all 12 annas to a rupee's worth of iron. They employ no flux, and the slag runs off first in pipe-like lumps. The furnace is emptied at each charge. The metal never runs liquid from the furnace, but falls to the bottom, below the blast tube, from whence it is taken in a flaming mass by a pair of iron tongs, and while incandescent it is hammered on a hard stone, or if the smelter be rather rich, on a rough iron anvil, into a double-wedge-shaped pig, and so on, the labour being divided between the smelter and his family, who think themselves fortunate if they can earn 1½ anna per head.

### 97. [6504] Iron ore, Monghyr. E. B. HARRIS, Esq. Found in excavating the Monghyr tunnel.

### 98. [7968] Iron ore, Shahabad District. R. W. BINGHAM, Esq., Hon.-Assistant Magistrate, Chynepore.

The part of the Vhyudhya Hills forming the southern portions of Shahabad, and of Mirzapore, north and north-west of the Soane River, together with Mirzapore, south of the Soane, Rewah, Palamow, and, in fact, the whole chain and spurs of the Vhyudhya range in this neighbourhood, is full of mineral wealth of various kinds, and will doubtless, in the course of a few years, when railways run down the valley of the Soane, connecting the Gangetic valley with that of the Nerbuddah, be found to yield products of immense value. Abundant quarries of the peroxide and proto-peroxide of iron, as also of iron-pyrites, abound in the most accessible portions of the Kymore range. The Kymore range is the north-easterly spur of the Vhyudhya range, and fills all Southern Mirzapore and Shahabad. Most of the ores are peculiarly rich in metal, some of them even yielding 70 to 75 per cent. of pig iron, but without accessible coal they are comparatively useless. Considerable quantities of iron, and that some of the best in India, are annually produced in Palamow, Rewah, Bidjugghur, and Singrowlie. The iron from the latter place in particular bears a high character in the market, being tough, flexible, and easily worked, while English iron, having originally been smelted from an inferior ore (the clay ironstone) and with mineral coal, is almost unworkable by native blacksmiths.

The ores are extremely rich, and the cost merely nominal, probably not more than 2 per cent. upon the cost of quarrying; and the ores being all above ground, would reduce the cost of quarrying to a minimum. One rupee per ton for royalty and cost of quarrying would give an ample margin for all contingencies, allowing rates of labour to remain as at present. Charcoal, as used by native smelters, may be obtained at 10 or 11 maunds per rupee, say 2½ to 3 rupees per ton, in the forest, to which, of course, must be added cost of carriage to site. Native charcoal is, however, made in open kilns in a most wasteful manner. Burnt in close kilns more than double the quantity, and that of a much better quality, would be obtained, while the tar and wood vinegar obtained at the same time would materially diminish the cost.*

### 99. [1476] Iron ore, Chundeyrie, Jhansee.

### 100. [983] Do. hydrous peroxide in flakes and powder (Dhaoo), Gwalior. H.H. the MAHARAJAH.

Iron ore (Dhaoo) is produced in the land lying between Mouzah Sathoo Nurwaree of the Gwalior District and Punehar, *i. e.* about 4 coss (8 miles) from east to west, and 1 coss (2 miles) north to south; also in the hills adjoining. In that neighbourhood people dig for the ore: after digging 20 cubits deep and 50 yards square, a description of earth called dhaoo (the ore), which is like small stones, but very soft, is found. This earth is loaded on bullocks, and taken to Dhoa and Bugrowlee and other places, where it is smelted and iron made from it.

The cost of digging and refining the ore is 12 annas per maund, as per following detail:—

|  | R. | A. | P. |
|---|---|---|---|
| Digging of ore . . . . . . | 0 | 0 | 5½ |
| Duty do. . . . . . | 0 | 0 | 3 |
| Carriage do. . . . . | 0 | 1 | 0 |
| Duty paid to Zemindars . . . | 0 | 0 | 0½ |
| Charcoal . . . . . . | 0 | 8 | 6 |
| Wages for one blacksmith for three hours | 0 | 0 | 4½ |
| Do. of men for working the bellows . | 0 | 1 | 1½ |
| Do. of Bhistee (water carrier) . . | 0 | 0 | 3 |
| Total . . . | 0 | 12 | 0 |

At this rate, a piece of iron about 20 seers in weight is made in three hours, at the cost of 12 annas. The price in the bazaar of 20 seers of iron is 14 annas, thus a profit of 2 annas is derived by the manufacturer.*

### 101. [10251] Specular iron ore, of a peculiar character. Very fine specimen. Kumaon. KUMAON IRON COMPANY.

### 102. [10252] A peroxide, mixed with much gangue, Kumaon. KUMAON IRON COMPANY.

### 103. [10277] Massive hæmatite, Kumaon. KUMAON IRON COMPANY.

### 104. [10215] Do. Kumaon. KUMAON IRON COMPANY.

### 105. [1465] Iron ore. Siliceous peroxide. First quality. Tendookhera, Nursingpore.

### 106. [1466] Do. Second quality. Tendookhera, Nursingpore.

The ore actually worked is a large vein or lode in the limestone of the great schist formation of the Indian Geological Survey, and the only rock in its immediate vicinity is hard grey and blue crystalline limestone. It occurs to the north of Nerbudda in the open flat country between the river and the Vhyudhya Hills. Only one mine is worked at present, but ore of a similar quality has been found at one or two other places in the neighbourhood. The only fuel used is charcoal, which is of very fair quality. Some coal mines also are situated at Mohpanee, not far from Tendookhera. The distance of fuel from mines is from

---

* Note by MR. BINGHAM.

* MAJOR R. J. MEADE, Political Agent, *Gwalior*.

5 to 10 miles. The iron is smelted in small clay furnaces, blown by goat-skin bellows, worked by the hand. It is obtained in small lumps or blooms called 'cutcha,' or raw iron, and is afterwards reheated and hammered, and then sold as 'pucka,' or finished iron. Intermixed with the raw iron as it comes from the furnace, is a sort of crude steel, which is carefully selected and used for the manufacture of tools and agricultural implements. The ore is largely smelted by the natives at the town of Tendookhera, about 2 miles from the mines, where, during the eight dry months of the year, about sixty furnaces are worked, but the mines have now been leased to the Nerbudda Coal and Iron Company formed in London. About 5 tons 3 cwt. of iron ore, and 5 tons 12 cwt. of charcoal, are used for the manufacture of 2 tons of 'pucka,' or finished iron. The ore contains, upon an average, about 40 per cent. of iron; it is of a calcareous nature, very fusible, and somewhat resembles the ores of the Forest of Dean. The ore is obtained by means of pits sunk from 30 to 40 feet, through the alluvium of the valley, to the ore. They are washed in during the rains, and require to be resunk yearly. The iron is obtained at a very small cost, as the government do not demand at present any royalty from the smelters. The fuel or charcoal is sold at from 3 to 3½ buffaloe-loads per rupee, which is equivalent to about 8 shillings per ton. The pucka iron sells at from 5 to 6 rupees, 12 shillings, per goan or bullock-load of 3 maunds, equal to 24 bundles, or from 4l. 10s. to 5l. 8s. per ton. The iron is sent by bullock carts and buffaloes to all fairs of any consequence within 100 miles, and in some instances is sent even 200 and 250 miles. From the iron of these mines, several years ago, a very good suspension bridge was built near Saugor.*

## 107. [1467] Iron ore, Azureea, Jubbulpore.

The geological description of these mines will best be obtained, as well as those at Tendookhera, from the Memoirs of the Geological Survey of India, vol. ii., part 2. It may here be said, however, that the Azureea mines are situated on a hill consisting of iron ore found at 1½ feet from the surface, and extending over an area of about 60,000 yards square and 30 feet deep. The ore exists in thin *flakes* of a grey iron colour and metallic lustre. The nature of fuel used is common wood 'charcoal, and for refining the metal, bamboo charcoal: the fuel is brought from a distance of about 5 miles from the mines. The ore and charcoal are thrown in small quantities every half hour into an earthen furnace 5 feet high and 2 feet square; a part of the bottom of the furnace is filled with fuel only; this being kindled, a pair of bellows is applied to raise the heat, and a passage made at the side of the furnace for the melted metal to run out. Four maunds (320 lbs.) of ore and 2½ maunds of charcoal are daily used in a furnace; the fuel is used in the proportion of 5-8ths or 62 per cent. of the ore for smelting, and 1-5th more for refining the metal. A furnace furnishes daily 2 maunds (160 lbs.) or 50 per cent. of the crude iron from 4 maunds of the ore; this, when forged, yields 30 seers, or nearly 19 per cent. of wrought iron. The ore is simply dug out with pickaxes; it costs 6 pie per maund for excavating and carrying to the furnace. The fuel or charcoal costs Rs. 1-1-6 per every maund of wrought iron. The entire cost of the pure metal obtained amounts to Rs. 1-13 per maund, including labour and materials. The ore is generally sold at the works and conveyed on bullocks to different markets. When brought to Jubbulpore, the nearest market, it costs 2 annas 9 pie per maund, exclusive of duty. The specimen exhibited from Azureea, the village where it is found, is not far from the proposed branch line of the East Indian Railway to Allahabad.†

## 108. [10212] Iron dust, Purulia, Chota Nagpore.

## 109. [10213] Do. Purulia, Chota Nagpore.

## 110. [10210] Magnetite, near Ranchee, Chota Nagpore.

Iron ore, in this form, or granulated, is found lying on the surface of almost all the high grounds in Chota Nagpore; where it lies, the soil is only surface, the rocks primary formation.*

## 111. [7979] Hæmatite, Assam. LIEUT. W. PHAIRE.

## 112. [6578] Iron ochre, Gholagat, Assam. — BECKETT, Esq.

Iron is obtained from this earth, which is found in Shoroo Cacharree Mehaul, in Gholagat Sub-Division, about 15 or 18 feet under the surface of the ground.

## 113. [6579] Iron sand containing iron prepared therefrom. — BECKETT, Esq.

## 114. [6580] Iron in the rough prepared therefrom. — BECKETT, Esq.

## 115. [6581] Refuse after extracting the iron. — BECKETT, Esq.

## 116. [10211] Rusty magnetic iron ore, Ranchee, Chota Nagpore.

## 117. [10212] Do. Maubhoom.

Found in beds of streams.

## 118. [6582] Iron clay, from Gellaka Mouzah, in Seebsaugor, Assam.

## 119. [5305] Iron ore, Pegu.

These specimens were brought from the base and western slope of a mountain called 'Popah,' about 3,000 feet in height, 15 miles inland from the river Irrawaddy, on its left bank. The ore is found in large quantity in nodules, and is collected and smelted by the inhabitants of the neighbouring villages. They use no flux of any description in smelting. Price 4 rupees per ton, delivered at the neighbouring furnaces.†

## 120. [7226] Hæmatite. Makoom, Upper Assam. MAJOR H. S. BIVAR.

## 121. [6747] Ironstone (Laterite?), Malacca.

## 122. [10195] Iron sand, Moulmein.

## 123. [5305] Bog-iron ore, Rangoon.

### CHROME.

## 124. [5980] Chrome-iron ore, Salem.

---

* Local Committee, *Jubbulpore.* A. M. RUSSEL, Esq., Secretary.
† Local Committee, *Jubbulpore.*

* Local Committee. *Chota Nagpore.*
† Local Committee, *Rangoon.* MAJOR T. P. SPARKS, President.

The chrome-iron ore of Salem is exceedingly pure, and of great value as a source of chromic acid. It is now largely imported into England.

### COPPER.

Copper ores are found in several parts of India.

125. [5320] Copper ore, specimen of carbonate, Ulwar.

126. [5322] Copper ore. A carbonate. Ulwar.* H.H. the MAHARAJAH.

Copper is found but in one or two places, and its working cost has not been ascertained, as it is under the direct management of the State.

127. [5889] Copper ore. An impure carbonate, very rich. One specimen weighs nearly ¼ cwt. From Singhboom. D. C. MACKAY, Esq.

The fuel used for smelting, is charcoal made from the extensive forests in the immediate vicinity of the mines and works in Landoo, in Dalbhoom, and Singhboom, in the south-west frontier of Bengal. The distance of the works from Calcutta is about 140 miles, and may be reached *via* Midnapore or *via* Raneegunge and Purulia.
By assay this sample is said to give copper 31½ per cent., and silver 2 oz. 5 dwts. 17 grs. per ton of ore.

128. [5890] Kunkur flux, Landoo in Dalbhoom. D. C. MACKAY, Esq.

129. [5891] Lime flux, Landoo in Dalbhoom. D. C. MACKAY, Esq.

130. [5892] Copper slag, Landoo in Dalbhoom. D. C. MACKAY, Esq.

131. [1051] Copper-pyrites, Hills of Beloochistan, Upper Sindh.

132. [1844] Carbonate of copper. LAHORE COMMITTEE, *Punjab.*

133. [5022] Copper ore. A silicate of copper, Nellore. CAPTAIN J. MITCHELL.

### ANTIMONY.

134. [1474] Antimony ore from Candahar *via* Umritsur.

Tersulphide of antimony is said to be found in the salt range near the Keura salt mine. Vast quantities of antimony have been found by Major Hay in the Himalayan ranges of Spite. The ore is also imported from Cabul.†

135. [1844] Antimony ore, Punjab.

136. [7230] Antimony, Sarawak.

### LEAD.

The only ore of lead is the common sul-

phide or galena. It frequently contains silver. The specimens exhibited are few, but good.

137. [4410] Galena, very pure, Kandahar.

138. [5040] Do. in quartz. Several large specimens. Contains no silver. Kurnool. CAPTAIN J. MITCHELL.

139. [5051] Do. massive. Contains silver. Catloor, near Cuddapah.

140. [1050] Do. massive. Picked from the surface of the soil in the hills. Beloochistan.

141. [1052] Do. massive. From the mines. Beloochistan.

### TIN.

The tin-ore of India and of the Archipelago is the same as that of Cornwall. It is the ordinary tin-stone, or binoxide of tin. It occurs in veins, and also in rounded masses or grains. It is often beautifully crystallized, interspersed with decomposing granite, and is generally free from sulphur and arsenic.

142. [4164] Tin ore from Larut, Malay Peninsula. FORBES BROWNE, Esq.

143. [4165] Do. from Junk Ceylon. FORBES BROWNE, Esq.

144. [4075] Do. diffused through quartz &c., Martcham Booboo, province Wellesley.

145. [4076] Do.

146. [4077] Tin in bar, obtained from the tin-stone of Macham, province Wellesley.

147. [9362] Twelve samples of tin ore, with prepared tin. Contains capital specimens of stream-tin. British possessions, Malacca.

148. [4833] Kensac tin, Burmah.

149. [5973] Tin-stone, picked specimen in more or less perfect crystals, Malacca. F. F. GEACH, Esq.

*Illustrations of Tin produce from Malacca.* HON. CAPTAIN BURN.

150. [6352] 1. Alluvial soil containing tin ore, Malacca.

151. [6353] 2. Quartz from which the tin ore is extracted, Malacca.

152. [6354] 3. Tin ore, Malacca.

153. [6355] 4. Tin when smelted, Malacca.

154. [6358] 5. Tin goblet and cup, Kassang.

---

155. [6712] Tin-ore from Kassang. J. MONIOT, Esq.

156. [6718] Do. from Tringanu, Malay Peninsula. JOSE D'ALMEIDA, Esq.

157. [6719] Tin from Pattani, Malay Peninsula. Do.

158. [6720] Do. from Johore, Malay Peninsula. Do.

159. [6721] Do. from Lanwan, Malay Peninsula. Do.

160. [5972] Do. from Lingie, Malay Peninsula. Do.

161. [6722] Do. from Pahang, Malay Peninsula. Do.

162. [6723] Do. from Kassang. F. F. GEACH, Esq.

*Nature of Fuel.*—Charcoal made from the Gomposs tree, is the only description of fuel employed. Its distance from the mines varies according to circumstances. If a mine is opened in the primeval forest the distance is not great, abundance of charcoal can be procured within the radius of a mile; but in some old mines in the valley of Kassang, where all the jungle has been cleared, the distance varies from 4 to 6 miles.

*Method in Smelting.* — A funnel-shaped blast furnace, 6 feet high and 4 feet diameter at the mouth. The sides of the trunk and funnel-hole are shaped and backed with clay. The fused matters escape from the cavity and flow continually into an exterior reservoir, hollowed out for that purpose, from which the liquid metal is ladled out into moulds, shaped in moist sand. The trunk is filled with charcoal, and combustion is accelerated by a cylindrical blowing machine, worked by eight men, of which the nozzle is introduced by an aperture. When the whole mass is brought to a red heat, the crude ore is sprinkled on top of the burning embers and kept constantly fed, by successive charges of charcoal and mineral.

*Quantity of Metal obtained from the Ore.* — Each charge consists of 30 piculs of washed ore containing from 45 to 60 per cent. of tin.

*Nature of Mining Operations.* — The ground being marked out and cleared of vegetation, a square or oblong pit is sunk, varying in depth from 40 to 80 feet, through an alluvial deposit and the ore extracted by a series of stream works. The stanniferous deposits occur in the form of regular beds, in which the binoxide of tin is associated with coarse sand and decomposed quartz, which are removed in baskets by the Chinese coolies, arranged in heaps on the surface, and exposed to sun and rain for a month or two. The washing is conducted in wooden jutters, through which a stream of water is made to flow, the dirty ore or 'work' thrown into coarse wicker baskets immersed in water in the wooden trough and shaken about : the metallic ore and finer particles of sand and decomposed quartz are washed through the crevices of the basket into the wooden trough, through which the stream of water flows, and is there kept in constant motion by several coolies with spades, by which means all the dirt and lighter particles of sand are carried off by the stream, and the heavy ore collected in the heap when the flow of water is stopped, and the metallic ore conveyed to the smelting shed.

*Cost of Ores.* — Fifty per cent. of the reduced metal.

The annual importation of tin from these mines to Singapore is, on the average, as follows : —

| | | | | |
|---|---|---|---|---|
| Tringanu | 150 piculs | or | 178 | cwt. |
| Pattani | 100 | „ | 118 | „ |
| Johore | 250 | „ | 296 | „ |
| Lanwan | 2,000 | „ | 2,375 | „ |
| Lingie | 1,000 | „ | 1,187½ | „ |
| Pahang | 3,000 | „ | 3,562½ | „ |
| Kassang | 3,000 | „ | 3,562½ | „ |

### b. *Native Metals.*

The only representative of this subdivision is gold, which is found in a number of districts throughout India, and attempts are, it is said, about to be made in the Belgaum district to turn to account the gold resources of the South Maharatta country.

163. [6447] Gold sand. From Purulia, Chota Nagpore.

164. [6448] Do. From Purulia, Chota Nagpore.

Gold dust is extracted from sand in the beds of rivers in Maunbhoom and Palamow, but not in large quantities.[*]

165. [3042] Gold-washers' sand. Sumbulpore, Cuttack.

Gold-washers sand from Sumbulpore. It is a matter of regret that a more ample specimen has not been forwarded. Such as it is, the committee submit it. The tools and vessels used in washing are so primitive and simple, that it has not been thought worth while to forward specimens.[†]

Gold deposits from Rangoon, viz. :

166. [10056] Deposit from which gold is washed on the Meh-Tyne stream, a tributary of the Shoay Gyeen River.

167. [10055] Deposit taken from the bottom of Shoay Gyeen River, containing gold.

168-70. [6466-7-8] Sand containing gold, deposited by the Nars Rivers at the mouth of the Martaban stream, a tributary of the Shoay Gyeen River.

171. [3043] From Meh-Tyne stream, a tributary of the Shoay Gyeen River.

172. [3048] From the sewers of the Meh-wine stream, a tributary of the Beeling River.

173. [2102] Gold dust from Jubbulpore.

---

[*] Local Committee, *Chota Nagpore.*
[†] Local Committee, *Cuttack.*

The gold dust is found in the Paiqdhur Nullah, in the Seonee district. The little stream rises in the Konye range of hills, and falls into the river Wyne-Gungah. The gold is obtained by washing the sand, and the natives say they never get more than four annas worth by a day's work, and would consider it unlucky if they did, as the goddess who is supposed to make it would then leave their locality.*

**174. [4491] Gold and dust. Luckimpore, Assam. LIEUT. W. PHAIRE.**

**175. [2101] Gold washings from Peshawur.**

Gold is found in minute scales in the sandstone of the Salt range, a lower range of hills running parallel to the Himalayan chain, between the rivers Indus and Jhelum; it is also found in small quantities in the sands of the Indus, Jhelum, Beas, and Sutlej; but the occupation of gold-washing is not very remunerative, amounting on an average to not more than from 3*d.* to 6*d.* a day, and the proceeds of the annual lease of gold-washing amounted last year to but 84*l.* Gold dust is also imported from Elaché, in Khoktan.†

**176. [3045] Gold dust found in beds of rivers at Purnalia, Chota Nagpore.**

**177. [3046] Gold manufactured from the same, Chota Nagpore.**

**178. [9363] Auriferous rock, from Candahar.**

**179. [4451] Burmese gold dust. Messrs. HALLIDAY, FOX, & Co.**

----

### Subdivision IV.—ADAPTATION OF METALS TO SPECIAL PURPOSES.

#### *a.* and *b.*

**180. [5307] Refined iron from Rangoon.**

From iron of this description, blacksmiths manufacture all the implements for common use throughout Burmah. In Pegu, English iron and tools of British manufacture are rapidly supplanting the native articles. Price 12*l.* per ton on the spot.‡

**181. [5314] Crude iron from Assam.**

**182. [10201] Do. KUMAON IRON COMPANY.**

**183. [10253] Welded iron from Wuzeeree Hills.**

**184. [10254] Iron from Peshawur.**

**185. [10255] Iron slag from Kangra Hills. GOVERNMENT.**

Magnetic iron ore is found in considerable quantity, and of a very fine description, closely resembling the Swedish, in the Himalayas, about 30 miles north-east of Dhurmsala, in the Kangra district; close to the sanitarium of Dalhousie; in the native states of Maudi and Kotkai; in the Sulymani range near Kolachi, on the western frontier of the Punjab. In all these places mines are worked, but iron is also to be found in the Salt range, in the Afidi hills to the west of Peshawur, and the Mewatti hills of the Goorgaon district. But there is one great obstacle to the successful working of iron mines in all these localities, viz., the absence of coal. In the year 1858 sixty bars of Kangra iron were sent to England, in order to ascertain the quality of the metal and its value in the European market. On being tested at the Atlas works of Messrs. Sharp, Stewart, & Co., of Manchester, while the best English iron yielded at a pressure of about 56,000 lbs. to the square inch, the Kangra iron, in the state in which it was received, required a force of 61,300 lbs. per square inch to break it; and, after being hammered in Manchester sustained a pressure of 71,800 lbs. The quality was considered 'equal to that of Yorkshire iron.' At present, however, its cost in Kangra, about 30 miles from the mines, is no less that 14*l.* a ton.*

**186. [7222] Iron for railway purposes. DR. HUNTER, *Madras.***

**187. [5298] Crude iron, Cuttack.**

**188. [5299] Do.**

**189. [5299] Do. Pal Lahara.**

**190. [5304] Do. Dhenkanal.**

**191. [5306] Do. Talchere.**

**192-3. [1477-78] Bloom, or pig iron, Gwalior. H.H. the MAHARAJAH.**

**194. [10214] Iron, Chota Nagpore.**

**195. [5823] Iron smelting, Shahabad.**

**196. [7949] Pig iron from Sirkee, Rohtass Spur, Kymore range. R. W. BINGHAM, Esq.**

**197-9. [7950-53, 7959] Do. from Biggeryghur, Rohtass Spur, Kymore range. R. W. BINGHAM, Esq.**

**200-2. [7958, 7961-3] Do. from Singrowlee, Rohtass Spur, Kymore range. R. W. BINGHAM, Esq.**

**203-6. [7954-5-7, 7960] Do. from Sirkee, Rohtass Spur, Kymore range. R. W. BINGHAM, Esq.**

**207. [9361] Iron from Gholagat. — BECKETT, Esq.**

**208. [10199] Cast-iron railway bars. KUMAON IRON COMPANY.**

**209. [10200] Do. bed. KUMAON IRON Co.**

**210. [2425] Do. plate. KUMAON IRON Co.**

**211. [6726] Ingot of iron. Cochin China. G. ANGUS, Esq.**

**212-3. [5970-71] Sixteen bars of iron. Chittledroog, Mysore.**

**214. [5969] Ingots of steel. Chittledroog, Mysore.**

----

* Local Committee, *Jubbulpore.*
† Central Committee, *Lahore.*
‡ Local Committee, *Rangoon.*

----

* Central Committee, *Lahore.*

215. [5948] Wire rope, 1¾ inch, galvanized. CALCUTTA GOVERNMENT.

216. [5263] Iron rope, galvanized, from Shalimar Patent Rope Works, Calcutta. Messrs. AVISHER & Co.

217. [6508] Steel wire (used as strings for the 'Citar'), Cuttack.

218. [2583] Do. Bangalore, Mysore.

219. [5893] Slab copper, made from the ores of Landoo, in Dalbhoom. D. C. MACKAY, Esq.

220. [5894] Sheet copper, rolled in Calcutta mint from the above. D. C. MACKAY, Esq.

221. [4834] Copper from Rangoon. Messrs. HALLIDAY, FOX, & Co.

222. [5300] Lead from Rangoon. GOVERNMENT.

Exported by sea from Rangoon, in the year 1860-61, to the value of 12,000*l*.*

223. [4836] Lead from Rangoon. Messrs. HALLIDAY, FOX, & Co.

## IV.—NON-METALLIC MINERAL PRODUCTS.

Subdivision I.—MINERALS USED AS FUEL.

### *a. Coal and derived products.*

Series of Indian coals, collected and exhibited by PROFESSOR OLDHAM, Superintendent Geological Survey of India:—

#### LOCALITIES ETC.

224. [1464] *Kurhurbalee* is in the district of Hazareebaugh. It contains several valuable seams of coal varying from 7 to 16 ft. in thickness, and is worked by the East Indian Railway Company. In 1860–61, 275,256 maunds of coal were raised. This coal is superior to any of the coals raised elsewhere in Bengal. A comparative trial in the locomotives of the East Indian Railway, continued for three months, showed a superiority, amounting to 13 per cent., over the good steam coals of the Raniganj field.

225. [1463] *Kasta* is situated to the north of the Adjai River, in the extreme north of the great Raniganj field. Here an immense seam of upwards of 30 ft. in thickness crops to-day, and is worked in open quarries. The lower 11½ ft. of this are of superior quality, and from these the specimen has been selected, which is of a quality much above the *average* of this coal. At Kasta 11,892 maunds were raised in 1860–61. It is less accessible than other collieries.

226. [1480] At *Chokidanga*, the most northerly of these, a fine seam of 15½ ft. is worked. The average production of three years has been 360,000 maunds.

227. [1431] At *Toposi*, a seam (higher in the series of rocks) of 22 ft. is worked. In 1860–61, 300,000 maunds were raised.

228. [1487] *Bansra* is another seam still higher in the series, of about 7 ft. in thickness. In 1860–61, 70,000 maunds were raised.

229. [1462] *Mangalpur*, a long-established colliery: a seam of 15½ ft. (including 9 in. of *shale*) is worked, yielding 1,000,000 maunds in 1860–61.

The Chokidanga, Toposi, Bansra, Mangalpur, Babùsol, and Harispur collieries are all on the Singàrun, a feeder of the Damùda River, and in the eastern portion of the great Raniganj coal field.

230-1. [4697-4695] *Babùsol* and *Madhubpur* (or Harispur) are situated in the lower portion of the Singàrun stream, and are the most eastern collieries in the field. From *Babùsol* 84,000 maunds, and from *Harispur* 440,000 maunds, were raised in 1860–61.

232. [1462] *Rogonathchuk* is on the banks of the Damùda River, and is one of the oldest collieries in the field. The bed is 12½ ft. thick, and yielded, in 1860–61, 300,000 maunds of coal.

233-4. [4702, 4696] *Raniganj.* — The most extensive workings in the field are near the Damùda River. The entire seam is 13 ft. in thickness, divided by a band of shale into two seams of 9 ft. and 3 ft. From each of these, specimens are sent. The Raniganj

* Local Committee, *Rangoon.*

workings yielded 1,600,000 maunds in 1860–61.

235. [4700] *Bhangaband* is in the same neighbourhood, and yielded, in 1860-61, 250,000 maunds.

236. [1208] *Banali* is a recently opened colliery, where a fine seam of 12 ft. is worked at a depth of 43 ft. below the surface.

237. [1484] *Futtehpur* is on the Grand Trunk Road. The bed is of 10 ft. in thickness, and of excellent quality. In 1860–61, 150,000 maunds were raised.

The Rogonathchuk, Raniganj, Bhangaband, Banali, and Futtehpur collieries are in the middle of the Raniganj field.

238. [1209] *Hattinal.*—This colliery is in the west of the Raniganj field, near the junction of the Baràkar and Damùda. The seam is 8½ ft. thick, the pits only 42 ft., the out-turn in 1860–61, 200,000 maunds.

239. [4694] *Chinakùri* (Cheenacooree) is close to Hattinal, and coal has long been worked here. In 1860-61, its out-turn was 3,290,000 maunds.

240. [4701 A] *Dùmarkùnda* lies to the west of the Baràkar, and is the most westerly colliery now worked in the field.

These three collieries, Chinakuri, Hattinal,

and Dùmarkùnda, are all in the western portion of the Raniganj coal field.

241. [3195] At *Panchbyni* a 7 ft. seam has been worked to some extent in open quarries.

### ALUBERA COLLIERIES.

242. [1212] At *Chilgo* a 5 ft. seam yielded 20,000 maunds in 1860-61.

243. [1211] At *Oormoo,* two seams of 7 ft. and 3 ft. produced 30,000 maunds in 1860-61, and at *Bankijora,* a thick bed of 19 ft., worked in open quarries, produced 30,000 maunds. The Chilgo, Oormoo, and Bankijora collieries are often spoken of as the Alubera collieries.

244. [1485] At *Bhorah,* a thick seam of 17 ft. produced, in 1860-61, 700,000 maunds. This colliery is only 20 miles from the Ganges. It is worked in open quarries.

These collieries are all in the Rajmahal hills. That of Panchbyni is on the Brahmini stream, at the extreme south of the hills. The Alubera collieries are near the Bansloi stream in the centre of the hills; and the Bhorah colliery to the north end.

The foregoing 21 specimens give a fair average representation of the coals of Bengal. They are from three distinct districts: 1st,—the detached coal field of Kurhurbalee: 2nd,—the great coal field of Raniganj, or, as it is not uncommonly called, the Burdwan field: and 3rd,—the Rajmahal hills. The following is the general classification of all these coals, with the names of the proprietors, and arranged in the order of the relative amounts of fixed carbon which they contain, which may be taken as a fair index of their relative value as fuel.

| Names of Collieries | Thickness of seam in feet | COMPOSITION OF COAL | | | PROPRIETORS |
| | | Carbon | Volatile matter | Ash | |
|---|---|---|---|---|---|
| Kurhurbalee . . | 7 to 16 | 66·70 | 24·80 | 8·45 | East Indian Railway Company |
| Futtehpùr . . | 10 | 63·80 | 25 00 | 11·20 | Messrs. Apcar & Co. |
| Dùmarkùnda . | 10 | 62·40 | 22·60 | 15·00 | Bengal Coal Company |
| Kasta . . . | 30 | 61·40 | 28·00 | 10·60 | East Indian Coal Company, and Messrs. Nicol & Sage |
| Chokidanga . . | 15½ | 56·80 | 34·00 | 9·20 | Messrs. Nicol & Sage |
| Chinakùri . . | 10½ | 53·20 | 35·50 | 11·30 | Bengal Coal Company |
| Hattinal. . . | 11 | 52·60 | 33·00 | 14·40 | Beerbhoom Coal Company |
| Madubpur . (Harispur) | 17 | 51·10 | 35·40 | 13·50 | Bengal Coal Company |
| Raniganj . . | 9 | 50·80 | 36·00 | 13·20 | Do. |
| Do. . . . | 3 | 50·30 | 36·30 | 13·40 | Do. |
| Toposi . . . | 22 | 49·20 | 35·40 | 15·40 | East Indian Coal Company |
| Bansra . . . | 13 | 47·00 | 40·00 | 13·00 | Do. |
| Rogonathchuk . | 10½ | 46·90 | 35·00 | 18·10 | Beerbhoom Coal Company |
| Babùsol . . . | 17 | 46·00 | 35·40 | 18·60 | Bengal Coal Company |
| Chilgo . . . | 5 | 45·50 | 43·50 | 11·00 | Messrs. Eaton & Browning |
| Oormoo . . . | 7 & 3 | 45·00 | 44·60 | 10·40 | Do. |
| Panchbyni . . | 7 | 44·20 | 34·10 | 21·70 | Messrs. Mackey & Co. |
| Mangalpur . . | 15½ | 43·90 | 38·40 | 17·70 | Beerbhoom Coal Company |
| Bankijora . . | 19 | 43·50 | 42 00 | 14·50 | Messrs. Eaton & Browning |
| Banali . . . | 12 | 42·60 | 44·20 | 13·20 | Beerbhoom Coal Company |
| Bhangaband . . | 7 | 40·30 | 28·40 | 31·30 | Bengal Coal Company |
| Bhorah . . . | 17 | 25·20 | 37·20 | 37·60 | East Indian Railway Company |

If, on the other hand, these coals were arranged according to the relative amounts of ash in each, which for many purposes is a more useful classification, they would stand as follows:—

AMOUNT OF ASH.

| | |
|---|---|
| Kurhurbalee | 8·45 |
| Chokidanga | 9·20 |
| Oormoo | 10·45 |
| Kasta | 10·60 |
| Chilgo | 11·00 |
| Futtehpur | 11·20 |
| Chinakùri | 11·30 |
| Bansra | 13·00 |
| Banali | 13·20 |
| Raniganj, (average of 2 seams) | 13·30 |
| Madubpur (Harispur) | 13·50 |
| Hatinal | 14·40 |
| Bankijora | 14·50 |
| Dùmarkùnda | 15·00 |
| Toposi | 15·40 |
| Mangalpur | 17·70 |
| Rogonathchuk | 18·10 |
| Babùsol | 18·60 |
| Panchbyni | 21·70 |
| Bhangaband | 31·30 |
| Bhorah | 37·60 |

Some curious *Ball-Coal* from the Dùmarkùnda mines is also sent.

Full statistics of the amount of coal raised in the years 1858-9-60, are given in the Memoirs of the Geological Survey of India.*

The total returns give an average of coal yield for the past three years of 87,37,454 maunds, or about 320,631 tons. But it is scarcely just to consider this as giving a fair mean of the present production, for during the first of these years there were, as is well known, disturbing causes at work tending to injure the regular trade of the country—and a fairer average, though determined by too small a number of years, will be obtained by taking the mean of the last two years' produce. This will give 100,25,020 maunds, or about 367,890 tons in the twelve months.

The returns also show one important and interesting fact, namely, that however the local out-turn may have increased or diminished, as affected by local causes, the general out-turn has steadily and markedly increased, apparently indicating a healthy and sound extension of trade and commerce.

The total out-turn for 1860 (that is, for the twelve months ending October 1860) was 100,88,113 maunds, or 370,206 tons, an amount about only the 200th part of the coals raised annually in Great Britain, viz. 72 millions of tons, but still evidencing a large and increasing commerce, and the spread of many of the arts of civilisation.

**245.** [3196] Coal, indurated and rendered columnar by the intrusion of trap rock, Lower Damùda.

The coals in the lower portion of the Damùda Coal Field are very frequently found intersected with basaltic trap, and in most cases the structure of the coal is entirely changed. The coal has become beautifully prismatic or columnar, and this may be seen over large areas. The columns are often not more than half an inch diameter, and generally are so completely separated, that it is ex-

ceedingly difficult to procure a specimen which will show more than one single prism.*

**246.** [4701B] Ball Coal, Dumarkhunda.

**247.** [9322] Coal from Tirop, Assam.

**248.** [5578] Do. Bancoorah. Two samples. A steam coal. BABOO GOVIND, Pundit of *Bancoorah*.

**249.** [1468] Do. Mohpanee, Nursingpore, Jubbulpore.

The coal mines in Nursingpore are entirely confined to the south side of the Nerbudda Valley, where they form a strip or band of irregular width, along the foot of the Puchmurree hills. Thin seams of inferior coal, from 18 inches to 3 feet thick, have been found also on the Shere River, but the only workable seams are at Mohpanee on the Seeta-Rewah River. At this point three seams, respectively 10 feet, 6 feet, and 3 feet 6 inches thick, are found. The coal is of very fair quality, resembling that of Bengal, and small quantities that have been used experimentally by the Great Indian Peninsular Railway Company and the Indian Navy, have been very favourably reported of. These mines have been leased to the Nerbudda Coal and Iron Company.†

**250.** [10281] Coal, Hills, Manbhoom, Chota Nagpore. A surface coal.

**251.** [10285] Do. Hills near Hazareebaugh, Chota Nagpore. A surface coal.

**252.** [10279] Do. near Purulia, Chota Nagpore. A surface coal.

**253.** [6713] Do. from Bintaloo, Borneo. H. C. READ, Esq.

**254.** [6356] Do. Malacca. HON. CAPTAIN BURN.

*b. Lignite and Peat.*

**255.** [5576] Lignite, Sumbulpore.

**256.** [5576] Do. Talchere.

**257.** [17] Do. Rajmahal.

**258.** [6503] Do. Chittagong.

This specimen was found up the Kurnafulloo river, amongst the hills; exact locality unknown.‡

**259.** [5577] Do. Talchere.

**260.** [10202] Do. Assam. — BECKETT, Esq.

**201.** [1024] Do. 'Dalajit,' Kangra, Punjab.

* Also exhibited by PROFESSOR OLDHAM.
† Local Committee, *Jubbulpore*.
‡ Local Committee, *Chittagong*.

*c. Bituminous Bodies and native Naphtha.*

262. [5183] Petroleum, Yuynanyoung, Burmah.

Supply unlimited, but price high, being a close monopoly of the King of Burmah.*

263. [5705] Petroleum, Assam. J. N. MARTIN, Esq.

264. [5706] Do. Assam. J. N. MARTIN, Esq.

265. [5708] Cheduba petroleum, Akyab.

266. [5707] Do. Akyab.

It is used by natives for burning; by Europeans for medical purposes; by both for varnish and to preserve wood. Also put on the bottoms of boats, it being an excellent preservative of wood from insects and worms. In the Island of Ramree there are 13 wells, in Cheduba 22 wells. Each well produces about 2 maunds per season, the aggregate produce of all the wells being 70 maunds per annum. The produce might be increased some 10 to 20 maunds by digging more wells. No petroleum is exported from the province. The petroleum is thick and dark coloured.†

267. [5729] Ramree naphtha, Akyab.

Used as varnish for oiling boats, posts, &c. There are 2 wells; they each produce about 14 maunds per annum. No great increase is expected. The naphtha is clear and bright, and none is exported. Price from 6 to 7 rupees per maund of 80 lbs.‡

268. [5716] Petroleum.

269. [4810] Petroleum, Rangoon. HALLIDAY, FOX, & Co.

270. [10702] Do. Mangalore. V. P. COELHO.

271. [6337] Do. Minia Kayo (?) Malay. G. ANGUS, Esq.

272. [10702] Do. Mangalore. V. P. COELHO.

---

Subdivision II.—*a.* FOR PURPOSES OF CONSTRUCTION GENERALLY — SILICEOUS OR CALCAREOUS FREE-STONES AND FLAGS, GRANITES, PORPHYRITIC AND BASALTIC ROCKS, SLATES.

Although India affords abundant illustrations of every stone belonging to this subdivision, the present Exhibition, from the absence of their international commercial value, does not attempt to display them in anything like adequate proportions. Never-

theless, many of the specimens supplied are of great interest. Most of them need only to be named: whenever a description is required, it will be found in its proper place.

273. [10232] Moss sandstone, Banda.

274. [10233] Do. Banda.

275. [7917] Sandstone, Chynepore Spur, Kymore Range, of Vhyndhya Hill. R. W. BINGHAM, Esq.

| | | |
|---|---|---|
| 276. [7926] Do. | Do. | Do. |
| 277. [7918] Do. | Do. | Do. |
| 278. [7930] Do. | Do. | Do. |
| 279. [7921] Do. | Do. | Do. |
| 280. [7919] Do. | Do. | Do. |
| 281. [7932] Do. | Do. | Do. |
| 282. [7922] Do. | Do. | Do. |
| 283. [7920] Do. | Do. | Do. |
| 284. [7925] Do. | Do. | Do. |
| 285. [7927] Do. | Do. | Do. |

MR. BINGHAM adds the following remarks:—The sandstones of this range have a high commercial value at Chunar and Mirzapore, being used as flagstones, and for ornamental purposes. The stones at those places owe their advantage to the proximity of the Ganges, which affords an easy river carriage; otherwise they are the worst and most destructible description of stone in the range. The millstones of Chynepore, Sasseram, and Tilowlhoo (perhaps also Ackbarpore), are famous, but must always be dear in a distant market for want of river carriage. The Soane causeway and the Koylwan railway bridge are built of the dense sandstone of Sasseram, while even little quantities are found in the higher portions of the range towards Rohtass. The best stone, while easily workable, is almost as hard as granite, and may be had of any colour, viz., white, crystalline, blue, grey, and all shades to a dark red.

286. [5875] Flexible sandstone, Ulwar.

287. [7844] Do. Jhend.

288. [4275] Do. Darjeeling. DR. CAMPBELL.

289. [4276] Moss sandstone, Darjeeling. Do.

290. [6727] Sandstone, Singapore. COL. COLLYER.

291. [7946] Old red sandstone from the Sasseram spur of the Vhyndhya Hills.

292. [10280] Slate found near Purulia, Chota Nagpore.

293. [6524] Do. Monghyr.

294. [5400] Slate found near Cossia Hills.

295. [7150] Grey slate, 'Teluck matee,' Cuttack.

Grey slate from Nilgiri, in Orissa, used for the purpose of making the marks on the forehead, nose, arms, and breast, more particularly affected by Hindoo devotees, and also by high-class natives in the Madras Presidency, and by Stirling called 'Meerschaum.'*

296. [5301] Kharee or slate stone, Cuttack.

Specimen of what is locally called 'Kharee,' which is used, among other purposes, for the manufacture of pencils and balls for writing on the ground or floor, being so used in all rural schools, and by native accountants.†

297. [5399] Slate pencils, Cossia Hills.

298. [5295] Moongnee stone or chlorite slate, Orissa.

Specimen of what is locally known as *Moongnee Stone*, apparently a kind of chlorite slate. According to locally received accounts, this stone, when freshly quarried, is comparatively soft and easily workable, but by long weathering becomes highly indurated, black, and bright. It comes from the hill state of '*Nilgiri*,' in Orissa, where extensive quarries are said to exist. This stone is used principally for the manufacture of various utensils. Idols are also made of it, and if the popular assertion that it is the true '*Moongnee*' be accepted, this stone is that on which the finest specimens of native sculpture extant in the province are executed, to wit, the '*Aroon Khumba*,' a polygonal column of considerable grace and beauty now standing before the principal entrance of the Pooree Temple; also the elaborately carved and figured slabs that adorned the top and sides of the doorways of the old Temple of the Sun at *Kanarac*, in the same district, and the gigantic figures of certain native deities of *Jajpore*, in the Cuttack district. It is probable, however, that '*Moongnee*' is a general term confined, not to one species of stone, but applying to several, and that the specimen is what is called the '*Kharee*' or slate '*Moongnee*.'‡

299. [7912] Limestone deposit overlying the old red sandstone, Rohtass Spur, Kymore range. R. W. BINGHAM, Esq.

300. [7931] Limestone from the Mussaye quarries, Chynepore Spur. R. W. BINGHAM, Esq.

301. [7914] Do. Do. Do.

302. [10283] Do. from hills near Hazareebagh. R. W. BINGHAM, Esq.

303. [10284] Do. Do. Do.

304. [7945] Mountain limestone underlying the old red sandstone of the Rohtass range. R. W. BINGHAM, Esq.

305. [7933] Mountain limestone underlying the old red sandstone of the Rohtass Pass. R. W. BINGHAM, Esq.

306. [7913] Limestone from the Mussaye quarries, Kymore. R. W. BINGHAM, Esq.

307. [7915] Do. from the Mussaye quarries, Kymore. R. W. BINGHAM, Esq.

308. [7916] Do. Do. Do.

The so-called mountain limestone underlies the whole of the Kymore range in Shahabad, and it also shows itself along the valley of the Soane as far at least as Mungeysur peak in Mirzapore. In some parts, as in Rohtass, it crops up boldly to 200 or 300 feet, forming a sloping base to the precipitous sandstone rock. In these places there appear to be three well-defined strata, viz. an upper one of a yellowish blue mixed with disintegrated sandstone, iron-pyrites, and chalk, — all in thin plates. Below that a more bluish-grey limestone with occasional calcspar crystals again is found, but generally of the same nature as a German lithographic stone.

Under the aforesaid strata lies a very dense bluish-grey limestone mixed with veins of calcspar. It is not used by native lime-burners, as being intractable. This is the lowest stratum, and would be an almost indestructible building or flooring stone from its great hardness, much harder than granite, and approaching to porphyry. It may be had in large blocks, and, if sawn into slabs, would be a very handsome building stone, bluish-grey with white streaks, and moreover it would probably make a superior kind of lime. Immense quantities of lime are made from the quarries of the western bank of the Soane, and exported down the Soane and the Ganges as far as Monghyr. Perhaps 300,000 to 400,000 tons are made annually, and the material is inexhaustible. The same limestone rock crops out on the northern face of the range at intervals, between the Soane river and Mirzapore; and again, especially in the singular and interesting limestone caverns of Goopteswar in the valley of the Doorgowtee River, at Beetree Band, in Khawah Koh at Mussaye, on the Sooreh River, and near Mirzapore. With canals and tramways, these quarries could supply all Northern India with the finest lime in the world. The cost of the lime at these quarries varies from 6 to 16 rupees per 100 maunds, or, say 5 to 14 shillings per ton. The present system of lime-burning is a very imperfect one, and indeed only suited to native wants, but with European supervision, although the material could not perhaps be produced cheaper, it could be produced with much more certainty and evenness in quality.*

309. [1471] Limestone from the vicinity of Jubbulpore.

310. [1472] Do. Do.

311. [1473] Do. Do.

312. [1469] Do. Do.

313. [1470] Do. Do.

There is close to Jubbulpore a range of low hills within a circumference of about ten miles, interspersed with masses of limestone both above and below the surface. The fuel generally used and most available for burning

---

* Local Committee, *Cuttack.*
† Local Committee, *Cuttack.*
‡ Local Committee, *Cuttack.*

* Note by R. W. BINGHAM, Esq.

the lime is brushwood. It is cut and brought from a distance of 7 or 8 miles. The stone is broken into fragments of 6 to 12 inches in size, then piled like a dome over a hole of about 9 feet diameter dug in the ground, and a passage left for introducing the fuel. This kiln is kept burning continually for the whole of the day, and the lime removed on the following morning. The fuel is used in the proportion of 40 maunds to every 75 maunds of limestone. Seventy-five maunds of the stone yield about 50 maunds of well-burnt lime. The stone is simply collected and broken up by manual labour, and the cost of collecting and putting it in the kiln amounts to 3 rupees for every 100 maunds of lime. The fuel costs from 5 to 8 rupees for every 100 maunds of lime. The entire cost of preparing the lime varies from 8 to 10 rupees per 100 maunds. The lime is at present only used in the city and station of Jubbulpore, and the locality is leased by Government to a farmer from year to year for a trifling sum; but the railway works shortly to commence will enhance its value. The hills are conveniently situated both as regards the line of railway to Bombay as well as to Mirzapore.*

314. [9800] Magnesite, Salem. DR. HUNTER.

315. [9801] Dolomite, Cuddapah. DR. HUNTER.

316. [9802] Mountain limestone, much used in Madras. DR. HUNTER.

317. [9803] Grey hydraulic limestone, banks of Godavery. DR. HUNTER.

318. [9804] Slaty grey limestone, containing magnesia, Kurnool. DR. HUNTER.

319. [9805] Nodular limestone or kunkur, Bellary. DR. HUNTER.

320. [9806] Nodular stalactitic limestone, Bangalore. DR. HUNTER.

321. [9807] Black slaty limestone, Cuddapah. DR. HUNTER.

322. [9808] Brown slaty limestone, Guntoor. DR. HUNTER.

323. [9809] Blue modular limestone, Ennore, near Madras. DR. HUNTER.

324. [6502] Limestone, Chittagong.

325. [7965] Limestone deposit, Kymore range.

326. [9786] Twelve varieties of building stones used for railway purposes, Madras. DR. HUNTER.

327. [6724] Granite, Singapore. COL. COLLYER.

328. [6725] Do. Do. Do.

329. [7934] Porphyritic granite from the

upheaved range of isolated rocks at Burnmonee, eight miles north of Sasseram, and ten miles from the nearest hills of the Rohtass range.

330. [7937] Porphyritic granite from do.

331. [7932] Do.

332. [7935] Do.

333. [7929] Do.

334. [7924] Do.

335. [7938] Quartzose conglomerate from the upheaved Plutonic rocks at Sonar, near Rohtass range.·

336. [9225] Rough slabs of slate from Karakambady, in the Chendraghiri Talug, North Arcot.

This description of slate is found in abundance in quarries in the Karakambady jungles, which belong to the Poligar of that place. When first taken from the quarry it is very soft, and can easily be cut into slates, or otherwise, as required.

*b.   Massive Minerals used for Ornament, Decoration, and the Fine Arts.*

The marbles are few in number, and give an imperfect notion of the varieties existing in India. The collection of marbles made by Messrs. SCHLAGINTWEIT, under the direction of the late East India Company, and deposited at the India Museum, Fife House, together with others there exhibited, furnish proof of the varieties to be met with.

337. [9791] White saccharine marble, from the banks of the Nerbudda. DR. HUNTER.

Bhera Ghât on the Nerbudda near Jubbulpore (ten miles), on the line of the railway to Bombay. The marble is plentiful and easily accessible. It has been used in a limited degree at Jubbulpore, sometimes to make lime, and other times for metalling roads. It is made up into images by natives, but does not take a good polish. A block was sent to the late Paris Exhibition, and pronounced to be equal to Italian marble for statuary purposes.*

338. [9821] Three slabs of polished marble.

339. [9822] Six do.

340. [9789] Crystalline marble for statuary, Tinnevelly. DR. HUNTER.

341. [9790] Do. in large grains, Rangoon. DR. HUNTER.

---

* Local Committee, *Jubbulpore.*

* Local Committee, *Jubbulpore.*

342. [9796] Green marble passing into serpentine, Cuddapah. DR. HUNTER.

343. [9792] White calcareous spar, Masulipatam.

344. [9793] Pink calcareous spar, Travancore.

345. [9794] Rhomb spar, Nellore.

346. [9795] Satin spar, Hyderabad.

347. [1129] Alabaster, Boogtee Hills, near Jacobabad.

348. [9797] Serpentine, Nellore.

349. [9798] Steatite, Madras. DOCTOR HUNTER.

350. [9799] Potstone for carving images, Chittoor. DR. HUNTER.

351. [4782] Steatite, Burmah. Messrs. HALLIDAY, FOX, & Co.

352. [5297] Do. Do.

353. [5294] Boulmala stone, Cuttack.

Procured from the hill state of *Dhenkanal*, in Orissa; this is used to make the little tripods on which sandal-wood is ground, and the small mortars used by natives.

354. [5296] Dalimba stone, Cuttack.

A hard, granulated coarse stone, very common, and worked into utensils of various kinds.*

355. [4838] Burmese stone.

356. [10144] Carbonate of lime, Calcutta. KANNY LOLL DEY.

357. [1816] Iceland spar (Surma Safed), Kabul.

This mineral is found in rocks in Kabul, and is extracted and broken into crystalline fragments, more or less opaque, which belong to the rhombohedral system. It is employed by the natives as an astringent in ophthalmia, gonorrhœa, and other fluxes, in doses, internally, of $7\frac{3}{16}$ grains, and also externally as a local application. It is called Surma Safed, or white antimony, from being thought to be similar to black antimony, the common tersulphide of that metal. Price 3*d.* per lb.

358. [7911] Indurated potstone, from the Mussaye quarries, Spur of Kymore range, Chynepore. R. W. BINGHAM, Esq.

359. [7940] Do. Quoindee quarries. R. W. BINGHAM, Esq.

360. [7936] Do. from old red sandstone of the Rohtass Pass, Kymore range. R. W. BINGHAM, Esq.

361. [7941] Indurated potstone, from the Quoindee quarries. R. W. BINGHAM, Esq.

362. [7944] Do. Do. Do.

363. [7943] Do. Do. Do.

364. [7942] Do. Do. Do.

365. [7910] Do. Do. Do.

It is found in several parts of the range, and from some quarries can be had in large slabs, and in great varieties of colour. It takes a high polish, and might be used for a great deal of ornamental work. Want of cheap carriage is against its being much worked. It is at present simply taken to Benares for the purpose of making images of gods for temples of private worship, or used locally. Although it has many of the qualities of, and is called 'indurated potstone' by Colonel Sherwill in his Geological Survey, it is not potstone, but rather a fine sandstone, stained of a dark hue, varying from greenish to dark brown, (and, when polished, black) by some mineral oxide, and hardened by igneous action. It is a useful stone, and deserves to be better known.*

366. [9811] Black mica.

367. [9812] Mica.

368. [9813] Do. from Behar.

369. [10282] Do. Chota Nagpore.

Mica is exported from Bombay in considerable quantities. In the year 1861, 5 tons of the value of 146*l.* were exported.

370. [7152] Talc slate. Cuttack.

*c. Cements and Artificial Stones.*

With the exception of gypsum for plaster, the only farther illustrations of this subdivision are supplied by—

371. [2927] Two large slabs of cement and concrete, Bhagulpore. J. SANDYS, Esq.

372. [4787] White lime, Burmah. Messrs. HALLIDAY, FOX, & Co.

373. [4762] Red lime. Do. Do.

374. [6745] White coral, for making lime. Malacca. T. NEUBRONNER, Esq.

Subdivision III. — MINERALS USED IN THE MANUFACTURE OF POTTERY AND GLASS.

Of the clays used for bricks, tiles, and the various kinds of pottery and porcelain, a large and interesting collection, embracing

---

* Local Committee, *Cuttack.*

* R. W. BINGHAM. Esq.

Nos. 375 to 410, is exhibited by DR. HUNTER, of *Madras.* The varieties of colours are unusual, and all are worthy of careful study.

375. [9751] White kaolin from decaying Clevelandite, Bimlipatam. DR. HUNTER.

376. [9752] Do. from decaying Albite, Vizianagarum.

377. [9753] Do. from decaying Pegmatite, Bangalore.

378. [9754] Do. from decaying granite, Bimlipatam.

379. [9755] Do. do., Bangalore.

380. [9756] Do. from decaying Felspar, Kercumbada.

381. [9757] Cream-coloured do. do. Bangalore.

382. [9758] Kaolin from Beder, near Secunderabad.

383. [9759] Lavender-coloured do., approaching Lithomarge, from near Tanjore.

384. [9760] White do. from Cuddapah.

385. [9761] Do. from railway cuttings near Cheyar Bridge, Cuddapah District.

386. [9762] Do. from Madura.

387. [9763] Cream-coloured do. from Angoomly, Nugger Division, Mysore.

388. [9764] Greenish-yellow do., Bangalore.

389. [9765] Buff-coloured do. Do.

390. [9766] Do. Yercand, Salem.

391. [9776] Fawn-coloured do. Do.

392. [9767] White pipe-clay (occurs below kaolin), Cuddapah.

393. [9768] White ball-clay, Conjevaram.

394. [9769] Do. Bangalore.

395. [9770] Ball-clay, Coopoor, near Madras.

396. [9771] Do. Awady, near Madras.

397. [9772] Do. Mount Capper, Cuddalore.

398. [9773] Silty clay (bed of river), near Cuddapah.

399. [9774] Blue clay, Bangalore.

400. [9775] Fire clay or aluminous shale, Streepermatoor.

401. [9777] Yellowish clay, Bangalore.

402. [9778] Greenish clay, Bangalore.

403. [9779] Yellowish clay, Salem.

404. [9780] Pink-coloured clay, Bangalore.

405. [9781] Soft red clay, accompanying kaolin, from Bangalore.

406. [9782] Yellow ochrey clay, Bangalore.

407. [9783] Brown potter's clay (three qualities), Madras.

408. [9784] Black clay (containing manganese), Bangalore.

409. [9785] Black clay, Madras.

410. [5053] Series of pottery clays, from Madras.

411. [6576] White earth, banks of Dunseerie Nuddee, Assam. R. W. BINGHAM, Esq.

412. [6574] Washed earth, banks of Nambur Nuddee, Assam. R. W. BINGHAM, Esq.

413. [6573] White earth, Seel Chitta, Assam. R. W. BINGHAM, Esq.

414. [6575] Do. unwashed, from banks of Nambur Nuddee, Assam. R. W. BINGHAM, Esq.

415. [6577] Black earth, Assam. R. W. BINGHAM, Esq.

416. [6714] Blue clay, Singapore. COL. COLLYER.

417. [6715] Do. Do. Do.

418. [6716] Yellow clay Do. Do.

419. [6717] Pipe clay Do. Do.

420. [6730] Clay No. 6 Do. Do. Hydraulic cement made therefrom.

421. [4166] Pipe clay, Penang.

422. [4167] Do. Do.

423. [4168] Do. Do.

424. [2863] Chooee mutty or pipe clay, Raepore.

425. [2703] Samples of clays for brick,

pottery, and china, from Bangalore, Mysore. CAPT. PUCKLE.

> *a* Clays.
> *b* Do.
> *c* Do.
> *d* Do.
> *e* Quartz powder.
> *f* Powdered grit.

426. [2723] Vase, ordinary native pottery, from potter's clay, intended to illustrate the clays (No. 2703), Bangalore. CAPT. PUCKLE.

427. [2724] Four vases, made of the finer clays, at the Bangalore jail and industrial school, in illustration of clays (No. 2703). These clays had not been previously used. Bangalore. CAPT. PUCKLE.

428. [6748] Tannah Mala (white clay), Malacca.

429. [6729] Fire clay bricks, Singapore.

These bricks were manufactured at the government convict brick establishment, by native convicts from various parts of India, mostly transported to Singapore for life. The bricks are made in the slop method on a slab table, with water trough, &c. Each moulder turns out in a working day of eight hours 2,500 bricks. Eight tables can be worked daily, and the out-turn has been 500,000 per month. The cost of the establishment last year was 26,727 rupees, and the value of manufacture produced 41,526 rupees, showing a balance of profit to the government of 1,500*l*.*

430. [5935] Bricks, Government works, Calcutta.

431. [5936] Do. made at Akra.

432. [4111] Fire brick, made at Avarendang, Prov. Wellesley, by J. C. THOMPSON, Manager of the Malakoff estate. The bed of clay extends for several miles along the north bank of the Pry river, and its depth has not been ascertained.

Subdivision IV.—MINERALS USED FOR PERSONAL ORNAMENTS, OR FOR MECHANICAL AND SCIENTIFIC PURPOSES.

In this subdivision but few specimens are exhibited. Agates are fairly represented. Rubies are almost entirely absent, and sapphires altogether.

TABLE SHOWING THE VALUE OF PRECIOUS STONES EXPORTED FROM INDIA AND EACH PRESIDENCY TO ALL PARTS OF THE WORLD, FROM 1856-57 TO 1860-61.

| YEARS | WHENCE EXPORTED | COUNTRIES WHITHER EXPORTED | | | | | | | TOTAL EXPORTED TO ALL PARTS |
|---|---|---|---|---|---|---|---|---|---|
| | | UNITED KINGDOM | FRANCE | OTHER PARTS OF EUROPE | SUEZ AND ADEN | CHINA | ARABIAN AND PERSIAN GULFS | OTHER PARTS | |
| | | Value | Value | Value | Value | Value | Value | Value | Value |
| | | £ | £ | £ | £ | £ | £ | £ | £ |
| 1856-57 | Bengal | 16,776 | 272 | 1,122 | 1,800 | 250 | .. | 971 | 21,191 |
| | Madras | 8,205 | 70 | 35 | .. | .. | .. | 290 | 8,600 |
| | Bombay | .. | 13,000 | .. | 84,908 | 3,546 | 5,195 | 1,784 | 108,433 |
| | ALL INDIA | 24,981 | 13,342 | 1,157 | 86,708 | 3,796 | 5,195 | 3,045 | 138,224 |
| 1857-58 | Bengal | 9,618 | 261 | 2,125 | 5,060 | 50 | .. | 416 | 17,530 |
| | Madras | 14,950 | .. | .. | .. | .. | .. | 134 | 15,084 |
| | Bombay | 22 | 3,000 | .. | 61,304 | 4,459 | 6,839 | 2,591 | 78,215 |
| | ALL INDIA | 24,590 | 3,261 | 2,125 | 66,364 | 4,509 | 6,839 | 3,141 | 110,829 |
| 1858-59 | Bengal | 16,296 | .. | .. | 8,530 | 1,958 | .. | 746 | 27,530 |
| | Madras | 9,927 | 1,000 | .. | .. | 60 | .. | 536 | 11,523 |
| | Bombay | 12,129 | .. | .. | 66,318 | 8,865 | 1,486 | 397 | 89,195 |
| | ALL INDIA | 38,352 | 1,000 | .. | 74,848 | 10,883 | 1,486 | 1,679 | 128,248 |
| 1859-60 | Bengal | 7,645 | 710 | .. | 10,990 | 6,542 | .. | 698 | 26,585 |
| | Madras | 2,243 | 470 | .. | .. | 280 | .. | 1,947 | 4,940 |
| | Bombay | 18 | 1,800 | .. | 90,206 | 15,018 | 3,623 | 644 | 111,309 |
| | ALL INDIA | 9,906 | 2,980 | .. | 101,196 | 21,840 | 3,623 | 3,289 | 142,834 |
| 1860-61 | Bengal | 7,955 | 2,840 | .. | 7,490 | 1,317 | .. | 2,363 | 21,965 |
| | Madras | 1,578 | 115 | .. | .. | 150 | .. | 913 | 2,756 |
| | Bombay | 1 | 600 | .. | 108,309 | 9,367 | 9,963 | 787 | 129,027 |
| | ALL INDIA | 9,534 | 3,555 | .. | 115,799 | 10,834 | 9,963 | 4,063 | 153,742 |

433. [10272] Jasper, Banda.

434. [10271] Bloodstone, Do.

435. [988] Piece of fossil wood and bamboo, Jubbulpore.

436. [10266] Petrified bamboo, Banda.

437. [10270] Bloodstone, Do.

---

* H. A. M'NAIR, Esq., Superintendent.

c 2

438. [987] Piece of fossil bamboo, Jubbulpore.

439. [986] Piece of fossil wood, Do.

440. [985] Do. Do.

441. [10263] Jasper, Banda.

442. [10264] Do. reddish, Do.

443. [10265] Petrified palm tree, Do.

444. [7964] Box of pebbles, Soane River. R. W. BINGHAM, Esq.

445. [4461] Unpolished rubies, Burmah. Messrs. HALLIDAY, FOX, & Co.

446. [7992] Pieces of malachite, C. India. INDIA MUSEUM.

447. [7991] Turquoise (large specimens), Thibet. Do.

448. [7990] Garnets, cut and polished. Do.

449. [2143] Turquoise, Jubbulpore.

450. [2370] Garnets, Mysore.

451. [2139] Specimen of Tombra stone, Peshawur.

452. [7989] Garnets, rough, Viziangram. INDIA MUSEUM.

453. [10278] Goree Soleymance stone, Kane River, Banda.

454 [1479] Rock crystal, Jubbulpore.

455. [5884] Do. Ulwar. Presented by H. H. the MAHARAJAH.

456. [616] Do. Vizagapatam. GOVERNMENT.

457. [4837] Large piece rock crystal, Burmah. Messrs. HALLIDAY, FOX, & Co.

458. Rough agate stones, Cambay.

459–61. [5311–12–13] Specimens of jade, Rangoon.

462. [10268] Polished Goodurreea stone, Banda.

463. [10269] Polished shells, Do.

464. [10262] Goodurreea stone, Do.

Subdivision V.—*a.* SIMPLE BODIES, OR COMPOUNDS CONTAINING THE ALKALIES OR ALKALINE EARTHS.

Of those used principally for culinary purposes or for medicine, salt and mineral waters require special consideration.

Salt is very well represented, as the following varieties prove.

465. [6176] Scinde Kurkutch salt, 1860-61, Cuttack. Salt produced by solar evaporation.*

466. [6184] Chilka Kurkutch salt. Cuttack. Salt produced by solar evaporation.

467. [6188] Chilka Pungah salt. Do. Do. Do.

468. [6189] Khoredah Pungah salt. Do. Do. Do.

469. [6185] Balasore Pungah salt. Do. Do. Do.

470. [6181] Cuttack Pungah salt. Do. Do. Do.

471. [6195] Pungah salt. (Ghaut Narainpore). Tumlook. Do.

472. [6200] Ghaut Pooreeghattah salt, 1267 S. S. Hidgelee agency. Boiled salt.

473. [6174] Ghaut Russoolpore salt. Do. Do. Do.

474. [6173] Ghaut Ramnuggur salt. Do. Do. Do.

475. [6197] Ghaut Kalinuggur salt. Do. Do. Do.

476. [6192] Ghaut Kissennuggur salt. Do. Do. Do.

477. [6186] Salt, Narainpore. Manufactured under excise during season 1860-61, at the factory of Messrs. Collins & Hills, at Narainpore, within the limits of the Barripore Salt Chokey Superintendency, Zillah 24-Pergunnahs, Bengal.

478. [6178] Do. Barripore. Manufactured under excise during season 1860-61, by Baboo Mohendronarain Dutt, in Lot No. 122, within the limits of the Barripore Superintendency, Zillah 24-Pergunnahs.

479. [6198] Do. Saugor Island. Manufactured under excise during season 1860-61,

* Contributed by GOVERNMENT.

at Saugor Island, by H. Frazer, Esq., within the limits of the Barripore Superintendency, Zillah 24-Pergunnahs.

**480.** [6175] Pungah salt, Southern Arungs, manufacture of 1267. Chittagong.

**481.** [6196] Pungah salt, Nezampore Arungs, manufacture of 1267. Do.

**482.** [6179] Salt, Midnapore.

**483.** [6183] Common salt, Pungah salt. Cuttack.

**484.** [6177] Gravel do., Kurkutch. Do.

Two specimens are sent, furnished by W. J. MONEY, Esq. Salt Agent of Pooree. The first, called *Pungah* salt, is obtained by boiling to a residuum highly concentrated brine. The second sample is the *Kurkutch*, or gravel salt. As the word signifies, it is produced by the aid of solar evaporation only, from sea-water. The water is introduced into small beds prepared with a smooth bottom of clay, rightly depressed in the ground, and surrounded by a slight ridge of earth. A few hours' exposure in the burning sun of March and the two following months, is sufficient to evaporate the water in these beds, which deposits the salt it held in solution. A fresh supply is then let in, and the process of total or only partial evaporation is continued, till the bottom of the beds is covered with a layer of this salt, more or less thick, which is then scraped up and is the salt of the sample.

Both these kinds of salt are produced all along the seaboard of the province, from February to June, and under what is practically a Government monopoly. The last season's manufacture amounted to 50,000 tons of the one, and 44,000 tons of the other. The production is considered handsomely to remunerate the petty contractors, who engage with Government for its supply at 10 annas and 4 annas per maund for each respectively, which is equivalent to 35s. 5d. per ton for the one, and 14s. 5d. per ton for the other, in English money and measure. To the Pungah must be added about 50 per cent., to the Kurkutch about 25 per cent., for expenses of superintendence &c. To the more extended manufacture of the white salt, the present insalubrity of the manufacturing localities and the consequent difficulty of procuring labour, as well as the insufficiency of the fuel supply, are obstacles. The coarser kind may be manufactured *ad infinitum*, but is nowhere appreciated so much as locally, that is, in the district, as, not being 'cooked' like the Pungah, it is more acceptable to the scrupulous caste prejudices of the Oryahs. The bulk of both kinds is exported to Calcutta. The local retail price at Cuttack in the shops of the bazaar for Kurkutch is 6s. 3d. per maund of 100 English lbs. Pungah is sold at the Government depôts at 8s. 3d. per maund.*

---

* Local Committee, *Cuttack*, which also supplies the following note:—

"This committee, in specifying the local prices of articles, do so with this explanation, that it is impossible to give any rates which are average ones, or which are in any degree equally applicable all over the district. Where any particular article is produced, there it is cheapest: and the cost and difficulty of transport, and the want of competition, are such, that a distance of 50 miles between the place of production and the central market, makes a difference in price of 50 or 100 per cent. in the rates at which some articles are bought and then sold. This is especially the case with the oil seeds of the province, cotton, &c. Under these circumstances it would have been impossible for this committee to have stated any average rates. The prices which have been generally mentioned in the catalogue are the local, that is, the Cuttack bazaar

**485.** [6568] Common salt, Cha. Ramree.

Used with food by the inhabitants; price, 1 rupee per maund; annual production, 100,000 maunds, but could be extended to 500,000, if necessary. The surplus salt not required for consumption in the province, is exported by Government to Chittagong. The salt is manufactured by boiling the salt sea-water after it has been allowed to stand for some days on land prepared to receive it.*

**486.** [6187] Salt in lumps on wood, Isadaing. Ramree.

Used with food by the inhabitants; price, 1 rupee per maund. The quantity annually manufactured is 400 maunds.†

**487.** [6182] Salt, Racee Nimuck. Lucknow.

This is an impure chloride of sodium. Formerly the greater part of the salt consumed by natives of Oude was made in this province; now it comes from other provinces, and this is the only kind made at present in Oude.‡

**488.** [10259] Do. From salt mines of the Shahpore district.

There are five salt mines worked by Government in the salt range; one at Kalabagh, across the Indus, and several in the Kohat district, and the supply from these sources may be said to be inexhaustible. An excise duty of 3 rupees per maund of 80 lbs. is now charged upon all salt sold, the rate having been lately increased (two years ago it was but 2 rupees); and the revenue derived from this source amounted to upwards of 280,000l. The salt mines are the means of supplying the traders of the Punjab with a kind of paper currency. By payment of the regulated price at any of the Punjab treasuries, a warrant for the delivery of so much salt at the mines may be obtained; these documents are transferable, and pass from hand to hand like bank-notes.§

**489.** [9824] Crystals of common salt from old salt works, Patree.

**490.** [9825] Do. Do. Do.

**491.** [10259] Rock salt from the salt range, Punjab.

**492.** [4786] Salt, Burmah. Messrs. HALLIDAY, FOX, & Co.

**493.** [1849] Salt, No. 4416.

**494.** [9826] Crystals of common salt found in salt-pits, Futtehpore.

**495.** [2175] Black salt, Padah noon, Calcutta.

**496.** [2176] Do. Do. Do.

---

retail prices. It may be as well to state, that all articles coming from Sumbulpore or elsewhere, by river carriage, are cheapest in the months of July and August, when the rivers first admit of navigation, and that between July and January the rates for the same articles may vary as much as 50 per cent."

* Local Committee. *Akyab.*
† Local Committee. *Akyab.*
‡ Central Committee, *Oude.*
§ Central Committee, *Lahore.*

## MINERAL WATERS.

497. [2103] Mineral water, Deoree, Jubbulpore.

498. [2104] Do.     Do.     Do.

499. [2105] Do.     Kosunghat.     Do.

500. [2106] Do.     Do.     Do.

501. [2119] Do.     Surar.     Do.

502. [2120] Do.     Do.     Do.

503. [2107] Mineral water, Koodra, Jubbulpore.

504. [2108] Do.     Do.     Do.

These four springs produce a good deal of water; they bubble up, and are supposed to contain no sulphur; the natives do not attribute to them any beneficial effect in disease, but after recovery from sickness drink the waters, as they are supposed to create an appetite, and thereby give strength.

505. [2121] Mineral water, Sonachur.

506. [2122] Do.     Do.

Does not bubble up, but produces a good deal of water; the villagers drink it daily. Comes out of black earth.

507. [2109] Mineral water, Kooslee.

508. [2110] Do.     Do.

509. [2111] Do.     Churgaon.

510. [2112] Do.     Do.

511. [2113] Do.     Bilba.

512. [2114] Do.     Do.

513. [2115] Do.     Bumhee Boomba.

514. [2116] Do.     Do.

There is very little water in the last-named spring; it bubbles up mixed with sand. It contains no sulphur, and is not supposed to be beneficial in disease.

515. [2117] Mineral water, Nugur Moha.

516. [2118] Do.     Do.

Very little water; does not bubble up; contains no sulphur. Dries up in hot weather.

517. [2123] Mineral water. Artesian well near Jubbulpore.

518. [2124] Do.     Do.     Do.

A chalybeate used as a tonic by convalescents in fever cases.

519–20. [5718–19] Water from mineral springs from the Singphoo country, Khouang, Assam. H. L. JENKINS, Esq.

As sulphur and borax belong to the same subdivision as the mineral waters, they are inserted. The specimens of sulphur are very few.

521. [1049] Sulphur from the mountains of Beloochistan.

522. [1128] Do. from mines near Shoruns, Beloochistan.

523. [1848] Do. from salt range, Punjab.

524. [2300] Do. from Rangoon. Messrs. HALLIDAY, FOX, & Co.

### BORAX OR BIBORATE OF SODA.

*Sohaga,* or *Borax,* is also called Tincal. It is obtained in large quantities in the valley of Puga, in Ladakh, and in Thibet; it is collected on the borders of the lakes as the water dries up, then smeared with fat to prevent loss by evaporation, and transported across the Himalayas on the backs of sheep and goats; refined at Umritsur and Lahore by washing with lime water. It is employed by the natives as a tonic for loss of appetite; also as a deobstruent and diuretic in ascites; and also to promote labour. It is used in the arts to clean metals before soldering, to form a glaze on earthenware, and in the preparation of varnishes. It is employed as a chemical flux in experiments with the blowpipe. It is in composition a biborate of soda. Price of raw borax, 4*d.* per lb.; of refined borax, 6*d.* per lb.*

525. [1793] Sohaga or tincal, Umballah.

526. [9834] Crystals of borax, Thibet.

527. [1794] Sohaga, Umballah.

528. [1802] Do. Thibet.

529. [1803] Do. Do.

530. [9835] Crystals of native borax, Thibet.

531. [10140] Borax, Calcutta. KANNY LOLL DEY.

### b. *Earthy and Semi-crystalline Minerals.*

Minerals useful for grinding and polishing abound in many parts of India. Although its grindstones and honestones are as yet but little known, its corundum is attracting attention. Several of the specimens are beautifully crystallized.

532. [9814] Corundum stone, Chittledroog, Mysore.

533. [9815] Corundum, Salem.

* Note by DR. BROWN, Lahore.

534. [6505] Emery, Moonghyr.

535. [9816] Corundum.

536. [7978] Do. Mysore.

537. [9817] Do.

538. [9818] Red corundum.

539. [9819] Corundum stone.

540. [9820] Corundum cake, made of stick lac and corundum powder, as used by jewellers.

541. [9226] Honestones, North Arcot.

542. [9364] Twelve rough honestones.

Of LITHOGRAPHIC STONES there are only the specimens exhibited under

543. [9823] Lithographic stones. R. W. BINGHAM, Esq.

MR. BINGHAM says: "I had stones for lithographic purposes made from the grey limestone, and it was used in the office of the Surveyor-General. The first stone was used in the press of Shah Kubeerooddeen Ahmed of Sasseram; it answered admirably for the purpose, but the stone must be freshly quarried or it chips, as after exposure to the atmosphere it grows intensely hard, and could then only be sawn into shape. Outside stone of limited sizes can only be obtained, owing to the ages of débris and decay which cover the main strata, but after quarrying some feet into the living rock, I feel satisfied from observation that lithographic stones of any size can be obtained."

PLUMBAGO, graphite, or black lead, is found in various parts of India. A deposit of some extent exists at Trevandrum, near Travancore. The following samples are exhibited:

544. [6357] Plumbago, Malacca. HON. CAPT. BURN.

545. [617] Do. Vizagapatam. H. H. the RAJAH.

546. [10260] Do. Goorgaon. DR. THORNTON.

This specimen is taken from a plumbago mine, discovered by Dr. W. J. Thornton, Civil Assistant Surgeon, Goorgaon, in October 1861. It is found in masses of variable sizes, and in general quite detached; though, in some cases, the rock all round is full of plumbago mixed with finely divided micaceous particles. Provision has been made in the budget of 1862-63 to admit of farther enquiries and examination of the deposits being carried on.*

547. [7232] Plumbago, Travancore. INDIA MUSEUM.

548. [1583] Plumbago (cake), Umritsur.

549. [4274] Plumbago brick, Darjeeling. DR. CAMPBELL.

Used to make ink with rice-water.

Of earthy and other minerals used as pigments or for staining, dyeing, and colouring, there are some few illustrations.

550. [10157] Geroo, from Rohtass Spur. R. W. BINGHAM, Esq.

551. [2849] Geroo, from Raepore.

552. [10160] Kirimchee or Geroo, Rohtass Spur. R. W. BINGHAM, Esq.

553. [1822] Multani mitt, Mooltan.

554. [6392] Yellow ochre, Malacca.

Belonging to the same class and subdivision are various minerals used in manufacture, such as alum schist, fuller's earth, French chalk, &c.

554a. [693] Alum schist. H. H. the RAO of KUTCH.

555. [7977] Fuller's earth, Chittledroog, Mysore.

556. [9836] Do. Scinde.

---

## Subdivision VI. — SOILS AND MINERAL MANURES.

Of the latter there are none exhibited. But of the former, a portion has been selected from the large collection made by the Messrs. SCHLAGINTWEIT, and since then submitted to analysis. The accompanying Table (pp. 24-6) will convey useful information† to persons interested in agriculture. The soils which form a portion of the present Exhibition have been transferred from the India Museum.

---

* Central Committee, *Lahore.*
† A complete report on these and other Indian soils has been prepared for separate publication.

# INDIAN SOILS.

### ANALYSED UNDER THE DIRECTION OF THE REPORTER ON THE PRODUCTS OF INDIA.

| Number | TERRITORY &c. | LOCALITY | REMARKS | COMPOSITION PER CENT. | | | | | | | | | | | | |
|---|---|---|---|---|---|---|---|---|---|---|---|---|---|---|---|---|
| | | | | Water at 212° F. | Water above 212°, and organic matter | Silica, free and combined | Alumina | Peroxide of iron | Lime as carbonate | Lime as sulphate | Lime in other forms | Magnesia | Potash | Sodium as chloride | Soda in other forms | Phosphoric acid |
| | **ALLUVIAL SOILS** | | | | | | | | | | | | | | | |
| | GANGES RIVER SYSTEM | | | | | | | | | | | | | | | |
| 1 | Bundelkund (14) | Bomori, 93 miles north-west from Saugur | Light brown with much silex. Extends over a large surface. Good, but stony | 1·082 | 1·247 | 84·000 | 9·180 | 1·350 | trace | 0·036 | 1·405 | 0·100 | 0·614 | 0·023 | 0·590 | 0·041 |
| 2 | Oude (13) | Cawnpore | Drab-coloured, micaceous. Cultivated, not manured | 1·870 | 0·900 | 79·978 | 9·600 | 3·060 | 0·909 | 0·097 | 0·441 | 0·120 | 1·790 | 0·037 | 1·181 | 0·107 |
| 3 | Oude (19) | Lucknow | 'Very fertile soil' | — | 2·400 | 28·290 | 2·870 | 3·270 | 59·550 | trace | 3·057 | 0·270 | 0·193 | trace | 0·050 | trace |
| 4 | North-Western Provinces (31) | Agra, near the military cantonments | Subsoil taken 16 feet below surface. Light grey with tinge of yellow. Much finely-divided silex | — | 3·725 | 68·400 | 3·827 | 3·657 | 18·727 | 0·087 | 0·493 | 1·181 | 0·568 | 0·048 | 0·087 | 0·110 |
| 5 | North-Western Provinces (33) | Agra, near the military cantonments | Very light grey, with but few pebbles. Subsoil taken 32½ feet below surface | — | 3·720 | 81·177 | 5·109 | 3·778 | 4·964 | 0·051 | none | 0·574* | 0·422 | 0·079 | 0·155 | 0·181 |
| 6 | North-Western Provinces (32) | Agra, near the military cantonments | Yellowish, lumpy, containing much fine sand. Subsoil 44 feet below surface | — | 4·325 | 67·742 | 6·462 | 4·334 | 14·466 | trace | 1·582 | 0·660 | 0·558 | trace | 0·079 | 0·289 |
| 7 | Bahar (44) | Nandialpur, 3 miles east of Monghyr | A red soil, good, extensively cultivated, and not manured | 6·363 | 0·618 | 69·082 | 12·810 | 8·670 | trace | 0·036 | 0·261 | 0·390 | 1·500 | 0·041 | 0·188 | 0·043 |
| 8 | Bahar (49) | Gaya, 55 miles south of Patna | Light brown soil, very fine, not impalpable. Good soil, very extensively cultivated | 3·470 | 0·960 | 69·860 | 14·210 | 9·310 | 0·079 | 0·013 | 0·259 | 0·290 | 0·550 | 0·053 | none | 0·153 |
| 9 | Bengal (39) | Buksar, on right bank of Ganges | Drab-coloured, micaceous, very finely divided. Cultivated, but not manured | 8·170 | 0·370 | 72·842 | 13·560 | 8·770 | 2·727 | 0·005 | 2·401 | 0·240 | 0·445 | 0·051 | 0·582 | 0·121 |
| 10 | Bengal (42) | Rajpur, 3 miles east of Bhagalpur | Light brown, almost impalpable. Good soil, extending over a large surface. Cultivated, but not manured | 1·470 | 1·130 | 78·325 | 9·100 | 5·410 | 0·670 | 0·081 | 0·195 | 0·400 | 1·570 | 0·041 | 1·162 | 0·088 |
| 11 | Bengal (45) | Bhagalpur | Snuff-coloured soil, with much lumpy carbonate of lime and pebble. Good soil, extensively cultivated | 3·210 | 1·520 | 61·000 | 10·360 | 2·710 | 17·045 | 0·010 | 1·781 | 0·772 | 0·933 | 0·020 | 0·392 | 0·107 |
| 12 | Bengal (52) | Khanjarpur factory, near Bhagalpur | Good indigo soil, not manured | 1·650 | 5·216 | 55·194 | 10·683 | 3·700 | 17·925 | 0·018 | 3·700 | 1·146 | 0·800 | 0·014 | 1·100 | 0·095 |
| 13 | Bengal (54) | Bhagalpur | Do. do. | 1·470 | 4·700 | 57·100 | 10·800 | 4·000 | 16·943 | 0·012 | 3·446 | 1·250 | 0·832 | none | 1·012 | 0·122 |
| 14 | Bengal (26) | Calcutta, from the bottom of river Hughly | Drab-coloured micaceous soil, almost impalpable | 2·750 | 1·960 | 66·000 | 15·690 | 8·640 | 0·680 | 0·029 | 0·414 | 0·871 | 2·020 | 0·106 | 0·944 | 0·102 |
| | **BRAHMAPUTRA RIVER SYSTEM** | | | | | | | | | | | | | | | |
| 15 | Assam (11) | Debrugurh, neighbourhood of | A peculiar micaceous soil of a red colour. Used for tea cultivation | 5·700 | 0·020 | 65·046 | 16·550 | 7·850 | trace | 0·002 | 2·530 | 0·740 | 0·820 | 0·022 | 0·758 | 0·115 |
| 16 | Assam (12) | Debrugurh | Drab-coloured, approaching white. Only occasionally cultivated, not manured | 3·330 | 0·180 | 68·700 | 17·470 | 4·530 | 0·014 | 0·005 | 0·681 | 1·450 | 1·810 | 0·029 | 1·795 | 0·042 |
| 17 | Assam (1) | Debru, east of Debrugurh | A fine micaceous soil, of a light reddish-brown colour. A regularly inundated but indifferent soil | 3·930 | 1·440 | 69·660 | 12·830 | 9·660 | 0·091 | 0·010 | 0·173 | 0·500 | 0·700 | 0·038 | 0·043 | 0·063 |
| 18 | Assam (35) | Machuli island in the Brahmaputra. Near Salamara village | Stiff drab-coloured soil, approaching whiteness. Cultivated | 5·830 | 3·500 | 55·300 | 29·630 | 3·570 | trace | 0·007 | — | 1·210 | 0·620 | 0·025 | 0·167 | 0·058 |

* Carbonate of magnesia, 0·107.

## ALLUVIAL SOILS — *continued*

### BRAHMAPUTRA RIVER SYSTEM

| Locality | Description | | | | | | | | | | | | | | Ref. |
|---|---|---|---|---|---|---|---|---|---|---|---|---|---|---|---|
| Eastern Bengal (10) — Jairampur, left side of Brahmaputra | A ferruginous clay. Cultivated and inundated | 6·750 | 1·950 | 62·505 | 22·240 | 5·230 | trace | 0·084 | none | 0·240 | 0·664 | 0·053 | 0·513 | 0·015 | 19 |
| Eastern Bengal (25) — Jumalpur, in most northern part of Delta of Brahmaputra | Yellowish brown, with much fine silex. Cultivated, but not manured. Inundated at the time the sample was obtained | 4·700 | 1·780 | 71·350 | 14·010 | 7·040 | 0·144 | 0·009 | — | 0·232 | 0·450 | 0·041 | 0·253 | 0·125 | 20 |
| Eastern Bengal (3) — Bajaganj, on the Surma | Very light brown, somewhat lumpy. Cultivated and inundated | 2·930 | 1·730 | 72·000 | 16·270 | 3·460 | trace | 0·007 | 0·883 | 1·240 | 0·786 | 0·016 | 0·686 | 0·082 | 21 |
| Eastern Bengal (9) — Bairab bazaar, one mile north of, near the junction of the Surma and Brahmaputra | Cultivated, not manured | 1·155 | 3·892 | 60·443 | 15·835 | 12·662 | trace | 0·414 | 2·160 | 0·295 | 1·830 | 0·791 | 0·022 | 0·250 | 22 |
| Eastern Bengal (7) — Koralea, 5 miles east of, near the Megna | Drab-coloured with reddish tint. Regularly inundated and cultivated eight months of the year | 4·160 | 1·190 | 67·000 | 17·440 | 4·890 | trace | 0·088 | 1·404 | 1·430 | 0·598 | 0·090 | 1·656 | 0·132 | 23 |

### INDUS RIVER SYSTEM

| Locality | Description | | | | | | | | | | | | | | Ref. |
|---|---|---|---|---|---|---|---|---|---|---|---|---|---|---|---|
| Punjab (2) — Near Rorki | Whitish soil called 'pakka zamin' (hard soil,) obtained just above the limit of the regular perennial overflow of the Indus. | 2·900 | 6·593 | 52·950 | 15·347 | 8·153 | 8·557 | 0·022 | 1·046 | 1·388 | 1·220 | 0·038 | 0·976 | 0·025 | 24 |
| Punjab (38) — Near Raulpindi Sindh Sager Duab | Drab-coloured, dense soil. Ordinary cultivated soil | 4·430 | 2·840 | 59·500 | 17·890 | 5·110 | 1·164 | 0·069 | 6·778 | 0·720 | 1·038 | 0·074 | 0·404 | 0·107 | 25 |
| Punjab (22) — Numbal, Sindh Sager Duab | Reddish drab, distinctly calcareous. Cultivated, but not manured | 3·940 | 1·010 | 50·490 | 11·890 | 7·420 | 16·660 | 0·839 | 3·915 | 1·218 | 1·102 | 0·315 | 0·943 | 0·069 | 26 |
| Punjab (34) — Kot Isa Shah, in the Jech Duab between the Jhilam and Chinab | Cultivated, but not manured | 1·750 | 2·410 | 66·640 | 12·260 | 6·310 | 9·810 | trace | — | 1·330 | 0·520 | trace | 0·213 | 0·043 | 27 |
| Punjab (37) — Lahor, between the Ravi and the Sutlej | A light-drab-coloured, impalpable soil. Carefully irrigated and cultivated with rice. The best soil near Lahor | 3·930 | 0·930 | 63·800 | 12·660 | 10·000 | 4·000 | 0·041 | 1·033 | 1·910 | 1·029 | 0·086 | 0·722 | 0·182 | 28 |
| Punjab (4) — Lahor | Light drab-coloured, almost impalpable. Subsoil to (37); 5 feet below the surface of a rice field | 2·160 | 0·500 | 66·337 | 14·630 | 4·100 | 5·459 | 0·051 | 1·590 | 2·010 | 1·820 | 0·331 | 1·318 | 0·097 | 29 |
| Punjab (23) — Lahor. Do. | Light drab-coloured subsoil from 10 feet below the surface | 1·260 | 3·700 | 65·800 | 12·814 | 6·280 | 4·690 | 0·155 | — | 3·070 | 1·244 | 0·080 | 0·900 | 0·092 | 30 |
| Punjab (24) — Between Rorki and Adoisha in the Bari Duab | Whitish, retentive of moisture. Very good siliceous soil. Cultivated, but not manured. Occasionally inundated by the Indus | 1·800 | 7·860 | 49·783 | 15·450 | 8·600 | 9·160 | 0·209 | 1·466 | 3·235 | 1·287 | 3·026 | 1·180 | 0·064 | 31 |
| Punjab (5) — Serdarpur in the Bari Duab, between the Ravi and the Sutlej | Cultivated soil | 3·980 | 0·930 | 63·800 | 12·660 | 10·000 | 4·000 | 0·041 | 1·033 | 1·910 | 1·029 | 0·036 | 0·722 | 0·182 | 32 |
| Punjab (6) — Multan, Bari Duab | A light drab-colour. Almost impalpable. Covered with grass | 1·860 | 1·520 | 70·000 | 12·980 | 8·010 | 1·637 | 0·070 | 1·204 | 0·443 | 2·240 | 0·055 | 0·801 | 0·117 | 33 |
| Punjab (3) — Multan, in the Bari Duab | Light drab subsoil, 3 feet below surface | 1·090 | 1·870 | 68·670 | 14·150 | 6·340 | 3·182 | 0·090 | 6·610 | 0·590 | 2·130 | 0·078 | 1·232 | 0·109 | 34 |
| Punjab (20) — Naushera, in Bhawulpore | Stone-coloured calcareous subsoil, 9ft. 9 in. below the surface | 2·120 | 4·080 | 57·900 | 15·000 | 3·500 | 9·620 | 0·452 | 3·130 | 2·300 | 0·830 | 0·262 | 1·000 | 0·140 | 35 |
| Punjab (21) — Ahmedpur, in Bhawulpore | Whitish, somewhat calcareous. Cultivated | 1·300 | 2·300 | 63·800 | 12·300 | 6·250 | 5·145 | 5·198 | 1·290 | 1·022 | 0·086 | 0·862 | 0·063 | 36 |
| Sindh (18) — Larkhana | Good cultivated soil | — | | 54·940 | 15·850 | 14·720 | 9·563 | trace | none | 0·560 | 1·103 | trace | 0·620 | 0·660 | 17 |

### RIVER OF CENTRAL INDIA

| Locality | Description | | | | | | | | | | | | | | Ref. |
|---|---|---|---|---|---|---|---|---|---|---|---|---|---|---|---|
| Orissa (43) — Puoke Par, near the Godaveri, to the north of Rajamundri | Grey siliceous soil, with very slight retentiveness for moisture. It is the predominant soil | 0·235 | 1·188 | 97·298 | 0·604 | 0·147 | trace | 0·142 | 0·091 | 0·060 | 0·028 | 0·017 | 0·109 | 38 |

### BLACK COTTON SOILS

| Locality | Description | | | | | | | | | | | | | | Ref. |
|---|---|---|---|---|---|---|---|---|---|---|---|---|---|---|---|
| Gujrat (27) — Purderi, north-west of Rajkot | Coarse pebbly soil of pigeon-grey colour. A good soil. Cultivated, but not manured | 4·323 | 0·117 | 74·202 | 8·000 | 3·200 | 5·545 | 0·922 | 1·396 | 1·780 | 0·594 | 0·065 | 0·749 | 0·127 | 39 |
| Gujrat (28) — Between Dassa and Daruka. | Dark grey, very tenacious. Good ordinary soil, generally cultivated | 7·960 | 1·870 | 36·980 | 16·410 | 9·340 | 22·600 | 0·018 | 1·210 | 1·020 | 0·710 | 0·065 | 0·236 | 0·025 | 40 |
| Gujrat (30) — Near Kotra | Fine specimen of cotton soil | 5·767 | 6·194 | 49·588 | 12·089 | 19·860 | 0·454 | 0·374 | 4·750 | 0·804 | 0·287 | 0·033 | 0·383 | 0·151 | 41 |

| TERRITORY &c. | LOCALITY | REMARKS | Water at 212° F. | Water above 212°, and organic matter | Silica, free and combined | Alumina | Peroxide of iron | Lime as carbonate | Lime as sulphate | Lime in other forms | Magnesia | Potash | Sodium as chloride | Soda in other forms | Phosphoric acid | Number |
|---|---|---|---|---|---|---|---|---|---|---|---|---|---|---|---|---|
| **BLACK COTTON SOILS—*continued*** | | | | | | | | | | | | | | | | |
| Malwa (15) | Between Panagurh and Jubbulpur | A brown lumpy soil, with much silex. 'Good black cotton soil, of great extent. Cultivated, but not manured' | 10·700 | none | 61·500 | 16·860 | 6·000 | 0·454 | 0·121 | 1·377 | 0·720 | 0·573 | 0·063 | 1·719 | 0·166 | 42 |
| Malwa (36) | Surki, south of Sauger | Dark grey, coarse and powdery, with few fibres. Good black cotton soil. Cultivated, but not manured | 7·500 | 2·560 | 63·820 | 8·867 | 12·629 | trace | 0·061 | 1·824 | 0·447 | 0·184 | 0·066 | 0·087 | 0·063 | 43 |
| Berar (16) | Near Shironcha | Good black cotton soil, extensively cultivated | 3·459 | 3·099 | 76·510 | 6·194 | 6·885 | 0·795 | 0·382 | 0·929 | 0·492 | 0·744 | 0·058 | 0·273 | 0·080 | 44 |
| Dekhan (50) | Navi Ghat, between Rematpur and Pusesauli, near the Krishna | Deep brown, with few visible particles of chalk, and but little silex. Good black cotton soil. Cultivated, but not manured | 20·000 | 0·330 | 48·100 | 21·500 | 10·000 | 2·000 | 0·004 | 0·240 | 0·570 | 0·695 | 0·048 | 1·700 | 0·025 | 45 |
| Dekhan (53) | Khamlapur, on the right side of the Krishna | Deep brown, with some coarse silex. Good black cotton soil. Cultivated, but not manured | 17·000 | 0·460 | 41·730 | 19·930 | 11·400 | 2·228 | 0·024 | 4·353 | 0·143 | 0·819 | 0·053 | 0·779 | 0·088 | 46 |
| Mysur (29) | Bellari | Deep slate colour, with much silex and visible chalk particles. Good black cotton soil. Cultivated, not manured | 12·000 | 0·260 | 57·330 | 13·330 | 5·230 | 9·231 | 0·057 | 1·154 | 0·290 | 0·382 | 0·055 | 0·774 | 0·101 | 47 |
| **SOILS DERIVED FROM ROCKS IN SITU** | | | | | | | | | | | | | | | | |
| Nilgiris (47) | Paikara, west of Utakamand | Bright ochreous soil, with great affinity for water. 'A very good specimen of laterite' | 5·070 | 10·080 | 31·450 | 14·910 | 36·823 | 0·090 | 0·017 | 0·448 | 0·195 | 0·760 | 0·016 | 0·360 | 0·005 | 48 |
| Nilgiris (48) | Kunnur, SSE. from Utakamand | Bright ochreous soil, with great affinity for moisture. 'Laterite very frequently met with' | 7·610 | 10·190 | 35·817 | 27·986 | 17·265 | — | 0·014 | 0·024 | 0·086 | 0·660 | 0·020 | 0·120 | 0·007 | 49 |
| Orissa (51) | Hills near Rajamandri | Light reddish brown, very lumpy, as if burnt. Highly siliceous. 'Finely grained laterite, extensively cultivated, but not manured' | 2·240 | 0·826 | 86·183 | 5·807 | 2·659 | trace | 0·037 | 0·932 | 0·034 | 0·450 | 0·018 | 0·110 | 0·486 | 50 |
| Berar (17) | Deolapur, north of Nagpur | Sandy soil of decomposed granite. Cultivated, but not manured' | 1·042 | 2·507 | 76·083 | 10·989 | 5·481 | trace | 0·898 | 1·104 | 0·461 | 1·444 | 0·024 | 0·505 | 0·102 | 51 |
| **SOILS OF MIXED ORIGIN** | | | | | | | | | | | | | | | | |
| Bundelkund (46) | Teri, 72 miles north-west from Sauger | 'Good soil for wheat cultivation : covers a large area.' | 2·860 | 0·140 | 81·430 | 11·930 | 1·400 | 0·454 | 0·002 | 0·018 | 0·308 | 0·578 | 0·033 | 0·883 | 0·023 | 52 |
| Malwa (40) | Nursinghpur | A deep pigeon-grey soil, with much coarse silex. Rich in peroxide of iron : it contains no alumina. 'Ordinary soil capable of cultivation' | 12·000 | 0·530 | 54·730 | trace | 25·000 | 2·454 | 0·004 | 3·286 | 0·960 | 0·412 | 0·081 | 0·302 | 0·066 | 53 |
| Malwa (41) | Gurchamer, north of Nursinghpur | An olive-brown soil : gritty from silex. 'A good cotton soil, not manured' | 16·000 | 2·000 | 41·930 | 28·292 | 13·438 | 0·454 | 0·005 | 1·194 | 1·440 | 0·720 | 0·055 | 0·529 | 0·008 | 54 |

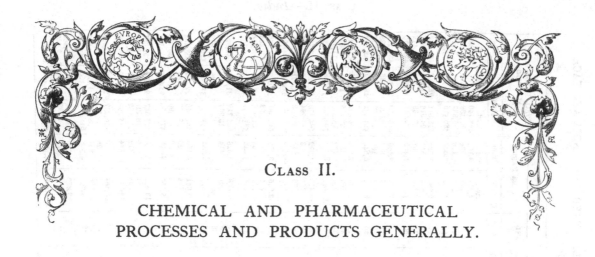

# Class II.

## CHEMICAL AND PHARMACEUTICAL PROCESSES AND PRODUCTS GENERALLY.

---

### Section A.—CHEMICAL SUBSTANCES AND PRODUCTS.

Division I. — Chemical Substances used in Manufacture :—Alkalies, Earths, and their Compounds.

#### IMPURE CARBONATE OF SODA.

This occurs as an efflorescence in some part or other in almost every district in India. It is largely employed in the manufacture of native soap. Specimens from various localities, as indicated below, have been forwarded.

557. [1895] Impure carbonate of soda, Calcutta. Kanny Loll Dey.

558. [9831] Crude carbonate of soda, Madras.

559. [1935] Crystals of carbonate of soda. C. B. Wood, Esq.

560. [10256] Sajji, from the common Sajji mutti (soda earth), Mooltan.

561. [10257] Do. Jung district.

562. [10258] Do. Do.

#### SALTPETRE.

Bengal supplies the largest portion of this important salt sent into the European market. The Punjab is said to possess large resources in this respect, which have still to be developed. The subjoined Table indicates the extent and value of the Indian export trade in saltpetre from the various Presidencies, whilst the samples exhibited embrace representatives from a number of districts. The notes entered below the various samples have been furnished by the different exhibitors or committees, and, although chiefly of local application, are possessed of considerable general interest.

563. [6180] Nitre or saltpetre (black), from Cuttack.

564. [6193] Do. (white) Do.

Nitre is known locally as 'Khai jabkhai.' The black specimen is obtained by a process of solution and filtration of the salt, which is found effloresced on old mud walls. The white is the same salt more carefully prepared for the most part in the hill tracts, from a similar efflorescence found in the cold months on the base of cow-house walls, and there generated, it is to be supposed, by the oxydation of the ammonia thrown off from the urine of the cattle. Neither kind is manufactured extensively enough for commercial purposes; still the local manufacture furnishes a good deal of the saltpetre, if not the bulk of it, used in native gunpowder for shooting and for fireworks.*

565. [10164] Saltpetre, first quality, from Ulwar. Presented by H. H. the Maharajah of Ulwar.

566. [10165] Do. second quality, Do.

567. [1947] Saltpetre from Jhansi.

568. [6194] Saltpetre (Shorah), Lucknow.

For gunpowder and frigorific mixtures; 2 seers per rupee; made in various parts of Oude. This is purified nitre, and is the kind only used for making gunpowder.†

569. [9827] Saltpetre, Nellore.

---

\* Local Committee, *Cuttack.*
† Central Committee, *Lucknow.*

( 27 )

# TABLE SHOWING THE QUANTITIES (AS FAR AS CAN BE ASCERTAINED) AND THE VALUE OF SALTPETRE EXPORTED FROM INDIA AND EACH PRESIDENCY TO ALL PARTS OF THE WORLD FROM 1850-51 TO 1860-61

COUNTRIES WHITHER EXPORTED

| Years | Whence Exported | United Kingdom Qty (cwt.) | United Kingdom Value (£) | France Qty (cwt.) | France Value (£) | Other Parts of Europe Qty (cwt.) | Other Parts of Europe Value (£) | America Qty (cwt.) | America Value (£) | China Qty (cwt.) | China Value (£) | Arabian and Persian Gulfs Qty (cwt.) | Arabian and Persian Gulfs Value (£) | Other Parts Qty (cwt.) | Other Parts Value (£) | Total Exported Qty (cwt.) | Total Exported Qty (tons) | Total Exported Value (£) |
|---|---|---|---|---|---|---|---|---|---|---|---|---|---|---|---|---|---|---|
| 1850-51 | Bengal | 260,893 | 202,489 | 49,031 | 88,392 | 11,529 | 9,027 | 105,961 | 82,954 | 32,282 | 25,278 | .. | .. | 4,044 | 8,166 | 463,740 | 23,187 | 361,306 |
| | Madras | 16,759 | 11,345 | 249 | 167 | .. | .. | .. | .. | .. | .. | 8 | 6 | 428 | 294 | 17,439 | 872 | 11,812 |
| | Bombay | 4,886 | 2,228 | 114 | 91 | .. | .. | .. | .. | .. | .. | 84 | 47 | 258 | 148 | 6,887 | 267 | 2,814 |
| | ALL INDIA | 282,538 | 216,062 | 49,394 | 38,650 | 11,529 | 9,027 | 105,961 | 82,954 | 32,282 | 25,278 | 92 | 53 | 4,680 | 3,608 | 486,516 | 24,326 | 375,612 |
| 1851-52 | Bengal | 224,548 | 175,908 | 21,046 | 16,476 | 9,707 | 7,600 | 241,787 | 192,288 | 19,578 | 15,322 | .. | .. | 2,280 | 1,785 | 518,941 | 25,947 | 409,369 |
| | Madras | 23,267 | 15,641 | 541 | 364 | .. | .. | .. | .. | .. | .. | 3 | 2 | 805 | 654 | 24,616 | 1,231 | 16,561 |
| | Bombay | 6,855 | 5,304 | .. | .. | .. | .. | .. | .. | .. | .. | 98 | 78 | 116 | 67 | 7,068 | 353 | 5,449 |
| | ALL INDIA | 254,670 | 196,848 | 21,587 | 16,840 | 9,707 | 7,600 | 241,787 | 192,288 | 19,573 | 15,322 | 101 | 80 | 3,200 | 2,406 | 550,625 | 27,531 | 431,379 |
| 1852-53 | Bengal | 318,064 | 245,169 | 42,282 | 33,089 | 8,518 | 6,786 | 148,788 | 116,043 | 31,013 | 24,284 | .. | .. | 7,855 | 6,140 | 556,510 | 27,325 | 431,461 |
| | Madras | 11,313 | 7,601 | 4,665 | 8,202 | .. | .. | .. | .. | .. | .. | 18 | 11 | 524 | 314 | 16,590 | 826 | 11,128 |
| | Bombay | 10,067 | 6,017 | .. | .. | .. | .. | .. | .. | .. | .. | 11 | 9 | 384 | 190 | 10,462 | 523 | 6,216 |
| | ALL INDIA | 339,444 | 258,787 | 46,947 | 36,291 | 8,518 | 6,736 | 148,788 | 116,043 | 31,013 | 24,284 | 29 | 20 | 8,763 | 6,644 | 583,492 | 28,674 | 448,805 |
| 1853-54 | Bengal | 357,057 | 279,449 | 43,647 | 18,316 | 336 | 263 | 220,679 | 172,705 | 28,742 | 22,527 | .. | .. | 10,146 | 7,941 | 660,607 | 33,080 | 501,201 |
| | Madras | 19,042 | 12,807 | 3,049 | 2,080 | .. | .. | .. | .. | .. | .. | 19 | 33 | 705 | 476 | 22,815 | 1,141 | 15,396 |
| | Bombay | 17,844 | 11,568 | 13 | 11 | .. | .. | .. | .. | .. | .. | 192 | 234 | 294 | 162 | 18,343 | 917 | 11,875 |
| | ALL INDIA | 393,943 | 303,824 | 46,709 | 20,407 | 336 | 263 | 220,679 | 172,705 | 28,742 | 22,527 | 211 | 267 | 11,145 | 8,579 | 701,765 | 35,088 | 528,572 |
| 1854-55 | Bengal | 240,151 | 187,095 | 57,555 | 45,045 | 10,693 | 8,370 | 216,308 | 170,229 | 44,405 | 34,983 | 17 | 12 | 15,817 | 11,992 | 584,429 | 29,221 | 457,714 |
| | Madras | 12,850 | 8,649 | 723 | 486 | .. | .. | .. | .. | .. | .. | 68 | 28 | 704 | 473 | 14,294 | 715 | 9,620 |
| | Bombay | 26,370 | 15,257 | 66 | 40 | .. | .. | .. | .. | 3,342 | 2,018 | .. | .. | 176 | 115 | 30,032 | 1,501 | 17,458 |
| | ALL INDIA | 279,371 | 211,001 | 58,344 | 45,571 | 10,693 | 8,370 | 216,308 | 170,229 | 47,747 | 37,001 | 85 | 40 | 16,197 | 12,580 | 628,745 | 31,437 | 484,792 |
| 1855-56 | Bengal | 355,204 | 278,018 | 30,765 | 24,073 | .. | .. | 145,674 | 114,095 | 5,503 | 4,308 | .. | .. | 3,835 | 3,000 | 540,981 | 27,049 | 423,489 |
| | Madras | 10,378 | 7,211 | .. | .. | .. | .. | .. | .. | .. | .. | 71 | 50 | 214 | 144 | 10,592 | 529 | 7,355 |
| | Bombay | 31,688 | 18,392 | .. | .. | .. | .. | .. | .. | 90 | .. | .. | .. | 179 | 104 | 31,918 | 1,596 | 18,546 |
| | ALL INDIA | 397,250 | 303,616 | 30,765 | 24,073 | .. | .. | 145,674 | 114,095 | 5,593 | 4,308 | 71 | 50 | 4,228 | 3,248 | 583,491 | 29,174 | 449,390 |
| 1856-57 | Bengal | 266,250 | 226,833 | 82,337 | 70,102 | 10,449 | 8,864 | 249,732 | 212,525 | 18,276 | 15,472 | 12 | 8 | 7,668 | 6,528 | 634,712 | 31,736 | 540,324 |
| | Madras | 13,149 | 8,849 | 950 | 570 | 196 | 182 | .. | .. | .. | .. | 46 | 33 | 607 | 414 | 13,964 | 698 | 9,403 |
| | Bombay | 44,100 | 25,958 | .. | .. | .. | .. | .. | .. | .. | .. | .. | .. | .. | .. | 4,519 | 226 | 26,619 |
| | ALL INDIA | 323,499 | 261,640 | 83,287 | 70,672 | 10,645 | 8,996 | 249,732 | 212,525 | 18,276 | 15,472 | 58 | 41 | 8,275 | 6,942 | 653,195 | 32,660 | 576,346 |
| 1857-58 | Bengal | 246,393 | 192,837 | 28,480 | 22,259 | 6,612 | 5,175 | 139,629 | 109,326 | 24,935 | 19,515 | .. | .. | 5,641 | 4,186 | 451,640 | 22,582 | 358,298 |
| | Madras | 8,844 | 5,943 | 138 | 187 | 203 | 137 | .. | .. | .. | .. | .. | .. | 174 | 120 | 9,359 | 468 | 6,288 |
| | Bombay | 52,006 | 27,993 | 4,587 | 2,459 | 508 | 240 | 287 | 172 | .. | .. | .. | .. | 2 | 2 | 57,390 | 2,869 | 30,796 |
| | ALL INDIA | 307,243 | 226,703 | 33,155 | 24,811 | 7,323 | 5,552 | 139,916 | 109,498 | 24,935 | 19,515 | .. | .. | 5,817 | 4,308 | 518,389 | 25,919 | 390,387 |
| 1858-59 | Bengal | 353,402 | 276,597 | 32,171 | 24,485 | 3,732 | 2,920 | 154,741 | 121,109 | 28,190 | 22,090 | .. | .. | 13,311 | 9,786 | 585,547 | 29,277 | 457,687 |
| | Madras | 13,193 | 8,860 | 4,880 | 3,946 | .. | .. | .. | .. | .. | .. | .. | .. | 84 | 56 | 18,107 | 905 | 12,162 |
| | Bombay | 107,486 | 55,927 | 2,443 | 1,225 | .. | .. | .. | .. | 1,177 | 683 | .. | .. | 636 | 412 | 111,742 | 5,587 | 58,247 |
| | ALL INDIA | 474,081 | 341,384 | 39,494 | 29,656 | 3,732 | 2,920 | 154,741 | 121,109 | 29,367 | 22,773 | .. | .. | 14,031 | 10,254 | 715,396 | 35,769 | 528,096 |
| 1859-60 | Bengal | 325,878 | 281,456 | 4,096 | 3,206 | 849 | 278 | 147,704 | 115,599 | 32,771 | 25,807 | .. | .. | 5,464 | 4,446 | 516,257 | 25,818 | 480,587 |
| | Madras | 2,111 | 1,419 | 3,287 | 2,209 | .. | .. | .. | .. | .. | .. | .. | .. | 34 | 28 | 5,432 | 274 | 3,656 |
| | Bombay | 88,349 | 45,962 | .. | .. | .. | .. | 605 | 575 | 292 | 184 | 39 | 28 | 296 | 234 | 89,581 | 4,479 | 46,983 |
| | ALL INDIA | 416,333 | 328,817 | 7,383 | 5,415 | 849 | 273 | 148,309 | 116,174 | 33,063 | 25,991 | 39 | 28 | 5,794 | 4,708 | 611,270 | 30,566 | 531,226 |
| 1860-61 | Bengal | 216,956 | 288,581 | .. | .. | 4,991 | 6,795 | 168,326 | 227,739 | 44,829 | 60,972 | .. | .. | 3,380 | 4,380 | 438,482 | 21,974 | 598,467 |
| | Madras | 4,376 | 8,027 | .. | .. | .. | .. | .. | .. | .. | .. | .. | .. | 16 | 18 | 4,399 | 220 | 8,040 |
| | Bombay | 83,786 | 64,152 | .. | .. | .. | .. | 1,336 | 1,111 | .. | .. | .. | .. | 2 | 2 | 85,124 | 4,256 | 65,265 |
| | ALL INDIA | 305,118 | 360,760 | .. | .. | 4,991 | 6,795 | 169,662 | 228,850 | 44,829 | 60,972 | .. | .. | 3,398 | 4,395 | 527,998 | 26,450 | 661,777 |

570. [9828] Pure nitre, gunpowder factory, Madras.

571. [4785] Saltpetre, Burmah. Messrs. HALLIDAY, FOX, & Co.

572. [9829] Do. purified in the laboratory, Calcutta.

573. [10586] Do. Salem. Messrs. FISHER & Co.

574. [9837] Do., a very beautiful well-crystallized specimen, Bengal.

575. [9838] Do. Nellore.

576. [9839] Do. Bengal.

577. [10105] Do. Do. BABOO NEWGEE.

578. [1814] Do. Mooltan.

Saltpetre is found in considerable quantities in many parts of the Punjab, especially on the sides of old villages and towns, but it is produced in greatest abundance in the districts of Shahpore, Gujeerat, Multan, and Jhung, from which districts a brisk export trade with Kurachi is carried on.*

579. [6199] Nitrate of potash (Shorah), Lucknow.

This is nitre of the first crystallization. Nitre is found in the earth of old buildings and ruins which has been long exposed to the air; water filtered through this earth is then boiled and concentrated. This kind is only used for frigorific mixtures.†

580. [6191] Do. Do. Do.

For frigorific mixtures and gunpowder; 2 seers per rupee; made in various parts of Oude. This is nitre of the second crystallization, and is only used for frigorific mixtures.‡

581. [10162] Saltpetre, Shahabad. R. W. BINGHAM, Esq.

582. [10163] Do.

Saltpetre and salt are produced abundantly in some parts of Shahabad, but crude saltpetre is prepared at from 6 to 7 rupees per local maund by the Nooneahs; this in its crude state would be 15l. to 18l. per ton, while the salt produced with the saltpetre is of a coarse kind, and only sold to the poorest of the community, under the name of *Kharree Nimuck*. It can, however, easily be purified by boiling, and then is a good and pure salt. §

583. [6190] Sal-ammoniac, or chloride of ammonium, Rausadar, Lucknow.

For tinning and frigorific mixture; 12 annas per seer; made in various parts of Oude. It is manufactured from the contents of cesspools, and is used for tinning copper pots and pans; mixed with common salt, nitre, &c., forms a frigorific mixture.‖

* Central Committee, *Lahore.*
† Central Committee, *Lucknow.*
‡ Central Committee, *Lucknow.*
§ R. W. BINGHAM, Esq.
‖ Central Committee, *Lahore.*

584. [10142] Chloride of ammonium, Calcutta. KANNY LOLL DEY, Sub-Assistant Surgeon, Calcutta.

585. [9833] Sulphate of magnesia, or Epsom salts, prepared from magnesite, or carbonate of magnesia, at Calcutta.

ALUM.

This salt has not hitherto been produced to any considerable extent in India. Large quantities of it are produced in Cutch. The chief source of supply, however, is China, from which it is largely imported into Bombay.

586. [10146] Alum (potash-alum), Calcutta. KANNY LOLL DEY.

587. [1847] Alum from Dera Ismael Khan.

Alum is manufactured from a black shale, principally at Kalabag on the Indus, where some 430 tons are annually sold at the rate of 7l. 16s. per ton. The process of manufacture is almost identical with that employed in European alum works.*

c. *Compounds of Metals Proper, as Salts of Iron, Copper, Lead, &c.*

588. [10159] Green vitriol, Shahabad. R. W. BINGHAM, Esq.

589. [1857] Do., or sulphate of iron, procured at Pind Dadun Khan.

It is known as *Heera kasis*, and is said to be dug out of the ground in large masses.

590. [1898] Do. Calcutta. KANNY LOLL DEY.

591. [6455] Do. Do. Messrs. BATHGATE & Co.

592. [9840] Blue vitriol, or sulphate of copper, Umritsur.

Made at Umritsur by boiling sheet copper in oil of vitriol. Sells at 8d. per lb.

593. [1825] White lead, or basic carbonate of lead, Punjab.

594. [10139] Do. Do., Calcutta. KANNY LOLL DEY.

Subdivision III. — MANUFACTURED PIGMENTS, DYES, AND MISCELLANEOUS CHEMICAL MANUFACTURES.

595. [9841] Litharge, or oxide of lead, Calcutta Bazaar.

* Central Committee, *Lahore.*

596. [9842] Vermilion, or protosulphide of mercury, Calcutta.

597. [5309] Realgar, or bisulphide of arsenic, Burmah.

598. [5310] Orpiment, or sulph-arsenious acid, Pegu.

599. [4783] Do. Burmah. Messrs. HALLIDAY, FOX, & Co.

600. [4772] Do. Do. Do.

### Subdivision V. — MANUFACTURES FOR SANITARY PURPOSES.

This subdivision contains a curiosity, in the shape of a camphor cup, for use after the manner of the quassia-wood, or ' bitter cup,' in vogue in this country, Calcutta Bazaar.

601. [5130] Camphor Cup, Baboot Gopaul Chunder Goopta.

## SECTION B. — CHEMICAL SUBSTANCES USED IN MEDICINE AND PHARMACY.

### Subdivision I. — FROM THE MINERAL KINGDOM.

Non-metallic substances and their compounds; alkalies, earths, and their compounds. These have been already enumerated.

#### METALLIC PREPARATIONS.

Several of these have been described. The following belong especially to this group, and have been carefully collected by BABOO KANNY LOLL DEY, of Calcutta, as representatives of the metallic preparations employed in medicine : —

602. [1897] Oxide of lead, Calcutta.

603. [10141] Do. Do.
Much used in ointments. Administered also internally as a tonic.

604. [1919] Protosulphide of mercury, or vermilion, Calcutta.

Used externally as well as in fumigations.

605. [10116] Subchloride of mercury, or calomel, Calcutta.

606. [10145] Tersulphide of antimony, Do.

It is used by native practitioners as an astringent in hæmorrhagia. Dose 5 to 10 grs.

607. [10107] Arsenious acid, or white arsenic, Calcutta.

608. [10109] Do. known by the name Semulkhur, Do.

609. [10110] Orpiment, or sulph-arsenious acid, Do.

610. [10127] Do. known as Jorode Sanko, Do.

611. [10108] Realgar, bisulphide of arsenic, or Darmooj, Do.

612. [10111] Do., or Mansul, Do.

Arsenical preparations enter largely into the composition of the native drugs; they are used to cure leprosy, snake-bites, intermittent fevers and other diseases. They are also employed all over India for suicidal and criminal purposes.

#### MEDICINAL SUBSTANCES FROM THE VEGETABLE AND ANIMAL KINGDOMS.

It has been found impossible to arrange these according to the directions in the Jury Catalogue. So many medicinal substances employed in India are unknown in any European pharmacopœia, and indeed so little generally known, that no more satisfactory arrangement could be arrived at than by placing them in the following order.

It is to be regretted that very many of the specimens were so damaged as to be beyond the power of recognition, and that so many of the bottles were broken, and the labels completely destroyed. These accidents have diminished the value of the collection, and have much reduced the numbers sent by various enterprising exhibitors. Many specimens are also too small in quantity to admit of any careful examination.

A large collection has been sent from Moulmein; of these only about twenty are missing owing to the above-named causes. It comprises the following : —

613. [7305] Khekya-poo Khadat Kouk (*Cinnamomum* sp.).

This is used in flatulency and in impurity of blood.

614. [7187] Khadat Kouk, Anan Kouk, Three-leaved Caper (*Capparis* sp).

This medicine is given in debility, and as an external application in swellings and dropsical diseases.

615. [7348] Nuag Yan (*Menispermaceæ*).

This medicine is given in fever, and also dropsical diseases and dysentery.

616. [7192] Tse Yoa.

This medicine is given in fever, and also dropsical diseases and dysentery.

617. [7301] Oopathetgah.

This is a cooling remedy, dried, ground into powder, and applied to the skin during pyrexia.

618. [7189] Boo-yet (*Clerodendron* sp).

For dysentery and dropsical diseases.

619. [7329] Gay-donk (*Connarees*).

This is used as an external application on swollen œdematous parts.

620. [7287] Hoan-myet.

This root, pounded, is used for poisoning fishes.

621. [7123] Than-mat (*Sesamum indicum*).

A decoction used in tooth-ache.

622. [7281] Than-ma-ka (*Poirea*).

This root, pounded and mixed with honey, is given in hysterical complaints, and the expressed juice of the leaf is given as an antidote to poison. It acts as an emetic.

623. [7356] Kurway.

This seed is used in eruptive swellings and in impurity of blood, ground into a paste with water.

624. [7322] Kaboung-gyee (*Strychnos Nux Vomica*).

The root of this tree is given in dyspepsia and in fever; it is ground on a flat stone with water and a little salt, and then rubbed on the tongue.

625. [7255] Thaman Kya.

For dyspepsia and hysterical complaints.

626. [7153] Phwa-bet.

This medicine is given as a purgative.

627. [7690] Thettyenggee (*Acacia concinna*).

This is given in impurity of blood, dyspepsia, and also as a purgative.

628. [7294] Danzeekoo.

Used in all diseases of impurity of the blood, the pounded root being mixed with water and drank off.

629. [7311] Ngabyey Jin.

This is given in dyspepsia and to promote secretion of milk. The decoction of the root is taken internally.

630. [7341] Nga Poung Tsag.

This is a valuable medicine in flatulency.

631. [7323] Bon-ma-yaya.

This is given internally in dyspepsia. The powder is also applied to the fauces in sore throat, also (mixed with aloes) given as a purge to horses.

632. [6583] Thenwen.

This is given in cough; the powdered root, mixed with lime juice, is given as a draught.

633. [6590] Kyoungban Myit (*Chaste Root*).

This is given in flatulency.

634. [7300] Gin-dine-tsaynee.

This medicine is used in impurity of blood.

635. [7307] Mahagah.

This is given in febrile diseases.

636. [7303] Yo-doan.

This is used in rheumatism, externally and internally.

637. [7256] Toa-tha-ngay Myouk.

This root is ground on a flat stone, with water and a little salt, and then rubbed on the tongue in fever.

638. [7347] Kinpoon Myit.

This medicine is used in hysterical complaints.

639. [7317] Makee Tuka.

It is given to those that are attacked with leprosy; used internally.

640. [6588] Rangoon Croton.

Used as a cathartic and in rheumatism of the joints.

641. [6587] Gon-ga.

This root is ground on a flat stone, with water and a little salt, and then rubbed on the tongue in fever.

642. [7310] Kapmhat.

This medicine is used in impurity of blood.

643. [7332] Tsey-ma-khan.

This is used on sores rubbed as an unguent.

644. [7336] Thamay.

It is given in hysterical complaints.

645. [7331] Yin Bya.

This medicine is used after child-birth as a draught, the patient being kept close to the fire, when puerperal fever threatens.

646. [7324] Than Thet Ngui.

In flatulency and impurity of the blood.

647. [7325] Thee-ha-yaza.

This medicine is used in fever by being rubbed on the tongue.

648. [7327] Get Myit (*Acacia* sp.)

Root used in leprosy.

649. [7314] Kouk-kho.

Used in weakness and general debility.

650. [7319] Tseet-doan.

This medicine is used as an external application in œdematous swellings.

651. [6594] Tsau Bah.

Used in impurity of the blood.

652. [7305] Tha Bya Kouk (*Eugenia jambolana*).

Used in flatulency and impurity of blood.

653. [7312] Karamet.

Used in flatulency, impurity of blood, and as pearl powder by Burmese females.

654. [7316] Kyee-pya.

For dysentery.

655. [7337] Than-zeyet Kyee.

Used in flatulency and in impurity of blood.

656. [7289] Nga-ra-nin.

This is given in weakness and general debility, and acts as a tonic.

657. [7288] Thag-yey Zin.

This is used in enlargement of liver.

658. [6596] Tonk Tsa (*Vitex arborea*).

This is used in enlargement of liver.

659. [6586] Tha Ta Tsa.

Used in flatulency.

660. [6591] Kyouk Tsha Gouk (Trumpet Flower Bark).

This is used by the Burmese, by grinding and rubbing on swollen parts, to keep down the swelling in recent wounds.

661. [6598] Na Bai (*Odina Wodier*).

Used with Kyouk Tsha Gouk.

662. [6593] Oak Shet Myit (*Ægle marmelos*).

This is used in hysterical complaints and dysentery.

663. [6599] Mee Young Noay.

Used in weakness and debility.

664. [7313] Thaphan Myit (*Ficus glomerata*).

Used in dropsical diseases.

665. [7284] Ak Nyah.

This medicine is given in hysterical fits.

666. [7296] Kadoan Nga-nhat.

This medicine is given in fever and bowel complaints.

667. [7290] Kankan Myit (*Mesua*).

Used in impurity of blood.

668. [6595] Kya Yin.

Used as a cathartic.

669. [7318] Ngat Kysap.

Used as a cathartic.

670. [7334] Bagyee Myit (*Clerodendron* sp).

This is given in hysterical complaints.

671. [7308] Tha Min Ya-pya.

Used in dropsical diseases.

672. [7346] Theling Katha.

This wood is used in fever by being rubbed on the tongue.

673. [7295] Tsin Bagoo.

Used in fever.

674. [7349] Shin Ka See.

Used in fever.

675. [7339] Mayo Myit.

This medicine is given in impurity of blood, and when put in milk causes it to curdle.

676. [7321] Padai Kungya.

This is used in cough.

677. [7304] Myat Meway.

This is used in impurity of blood.

678. [7355] Thasa-gya-yoo.

This medicine is given in bowel complaints.

679. [7328] Yoosa-na Bah (*Linum catharticum*).

This is given in fever as sudorific.

680. [6589] Mingagah.

It is given as a cathartic.

681. [7689] Yee-byoo-thee (*Emblica officinalis*).

This is given in cough.

682. [7254] Thit Shain-thee (*Terminalia belerica*).

Used in fever.

683. [7190] Kan Gyouk Nee.

Used in flatulency and impurity of blood.

684. [7324] Kya Nee, Red Lily.

Used in impurity of blood.

685. [7353] Kya Phu, White Lily.

Used in impurity of blood.

686. [7158] Kha Padoou Mah.

Used in impurity of blood and fever.

687. [6585] Kya-thee (*Nelumbium speciosum*).

Used in fever.

688. [6592] Kayapun (*Acanthus* sp.)

This is given in impurity of blood.

689. [7193] Tin-bwen (*Nauclea* sp.)

This is used in impurity of blood, causing vertigo &c.

690. [7188] Thyet Maouk-thee (*Ardisia* sp.)

Used in flatulency and impurity of blood.

691. [7280] Pait Khai-thee, Long Pepper (*Chavica officinalis*).

This is given in fever and flatulency.

692. [7259] Yoay-gyee-thee (*Adenanthera pavonina*).

Used in fever.

693. [7293] Nga-rok-goung (*Piper nigrum*).

Used in bowel complaints.

694. [7289] Tsayootha-tong.

Used in weakness of any kind before or after illness.

695. [7258] Kyee-thee (*Semecarpus anacardium*).

Used in leprosy.

696. [7257] Aik Moug thee (*Embelia ribes*).

Used in impurity of blood.

697. [7291] Yoay-pyoo Tsan.

This medicine is given to children for colds and dyspepsia. The kernel of the fruit is used.

698. [7298] Caraway-thee.

Used in impurity of blood.

699. [6584] Kakanwoot Tsan.

Used in impurity of blood.

700. [7354] Noày Gyo (*Glycyrrhiza officinalis*).

This medicine is given in palpitation of the heart; it depresses circulation &c.

701. [7333] Nanwen-det (*Curcuma* sp.)

This medicine is given when the body is heated and feverish.

702. [7330] Tsaythangaigyet (*Helicteres isora*).

Used in impurity of blood.

703. [7343] Lam Nai.

Used in bowel complaints.

704. [7326] Ga Moungmhoag.

Used in impurity of blood.

705. [7350] Kanaka Tsan (*Croton tiglium*).

Used as a cathartic.

706. [7282] Dundoongoup.

Used in hysteria.

707. [7335] Thetdoupkouk.

Used in dropsy.

708. [7345] Kalani Myit.

Used in leprosy.

709. [7340] Aigareet Myit.

This root is said to deprive spirituous liquor of all its strength. A decoction given to an intoxicated person is said to render him immediately sober.

710. [7342] Tset-ga-doung-youk.

Used in fever and on external swelling.

711. [7352] Kap Boo.

Used in impurity of blood.

712. [7155] Poung Myit.

Used in debility and weakness.

713. [7344] Pya Noung.

Used in fever.

714. [7292] Alan Bai.

This root is used in fever, rubbed on the tongue.

715. [7302] Oung Maiphu.

This is used in sore-eyes.

716. [7315] Kyet-thaheng.

This is given as a purgative.

717. [7351] Kalaia Tsan (*Guilandina bonduc*).

Used in leprosy.

718. [7320] Pahla (*Amomum* sp.).

Cardamom seeds, used in fever; the powder mixed with Nga'rok Goung acts as stomachic.

719. [7297] Marayan.

Used in impurity of blood.

720. [7299] Ban ya Khin.

This is given as a cathartic.

721. [7154] Ngoo thee (*Cassia fistula*).

This is given in debility, and as a cathartic.

Among individuals, the largest collection of drugs has been made by BABOO KANNY LOLL DEY, Sub-Assistant Surgeon of Calcutta. Unfortunately the greater part of the specimens sent by BABOO KHETTUR MOHAN GOOPTU, also a graduate of the Medical College of Calcutta, was so injured as to be incapable of exhibition. As many of the drugs sent by individuals and committees are of the same kind, most of them are enumerated together, but the names of the senders are, as far as possible, always given.

722. [7017] Atees (*Aconitum heterophyllum*), Bombay.

723. [9932] Do. Calcutta. KANNY LOLL DEY.

The root has long been celebrated as a tonic and valuable febrifuge; it is intensely bitter and slightly astringent, with an abundance of farina. There are two kinds of atees, the black and the white, both equally valuable.

724. [10117] Bish (*Aconitum ferox*), Calcutta.

725. [7023] Butchnab (*Aconitum ferox*), Bombay.

The root is highly poisonous, and in the northern parts of India is used for poisoning arrows. Native practitioners administer it in cases of chronic rheumatism. It grows in the Himalayas, and is imported in considerable quantities into the plains, and sold at the rate of one rupee the seer.

726. [10119] Aconite (*Aconitum* sp.), Calcutta.

727. [7156] Mishmee Teeta (*Coptis teeta*), Luckimpore. W. G. WAGENTRIEBER, Esq.

728. [7025] Do. Bombay.

729. [7048] Kulie Koothie (*Helleborus niger*), Bombay.

730. [9923] Do. Calcutta. KANNY LOLL DEY.

Roots brought from Nepaul, used as a powerful cathartic in mercurial and dropsical cases. Price 8 annas per lb.

731. [6165] Black cumin (*Nigella sativa*), Bombay.

*Nigella sativa* is not found wild, but is extensively cultivated and employed both as a condiment and a medicine.

732. [6143] Seetaphul (*Anona squamosa*), Ahmedabad.

The seeds contain a highly acrid principle, fatal to insects, on which account the natives of India use them powdered and mixed with the flour of *Cicer arietinum* for occasionally washing the hair.

733. [6112] Badian Khutai (*Illicium anisatum*), Bombay.

The natives employ the seeds most extensively as a stomachic and carminative.

734. [9947] Rusot (extr. *Berberis lycium*), Calcutta. KANNY LOLL DEY.

Is prepared by digesting in water sliced pieces of the root, stem and branches in an iron vessel, boiling for some time, straining and then evaporating to a proper consistence. This extract is much employed in Hindoo medicine, especially as an external application in ophthalmia. It is likewise considered an extremely valuable febrifuge. Dose 3 ʒss diffused in water. Price 4 annas per lb.

735. [7039] Goorwail or Gooluncha (*Tinospora cordifolia*), Bombay.

Is extensively used as a tonic and febrifuge. The stems, roots and leaves are bitter, and afford a decoction; much used as a bitter tonic in convalescence from fevers and acute diseases generally. Dose ʒij to ʒss in decoction. Price 2 annas per lb.

**736.** [1914] Colomba (*Cocculus palmatus*), Calcutta. KANNY LOLL DEY.

**737.** [7021] Do. Bombay.

This is the well-known Columba root. It is cultivated in some localities and employed as a tonic.

**738.** [7020] Cocculus Indicus (*Anamirta cocculus*), Bombay.

The fruits known as 'Cocculus Indicus' are exported from Malabar and Travancore, and shipped for the London market. From 1852 to 1856, 5,817 cwt. were exported from the Madras Presidency. In India they are powdered and employed for destroying pediculi, or made into an ointment and applied externally in itch.

**739.** [6166] Nilofer (*Nymphæa lotus*), Bombay.

**740.** [6043] Kumul ka phool (*Nymphæa lotus*), Do.

**741.** [6607] Rumalgatta (*Nelumbium speciosum*), Calcutta. KANNY LOLL DEY.

This is a very common plant in tanks all over India. The roots and seeds are eaten, the young shoots as an ingredient in curries, the seeds either raw, roasted or boiled. In medicine the root is said to be demulcent and diuretic, and the stalks, leaves, and flowers as cooling and tonic.

**742.** [6040] Post or Poppy capsules (*Papaver somniferum*), Bombay.

In Upper India an intoxicating liquor is prepared by heating the capsules of the poppy with jaggery and water. In medicine they are employed for their anodyne properties as fomentations for inflamed surfaces, and a syrup prepared from them to allay cough &c.

**743.** [4950] *Cratæva Roxburghii*, Chingleput. DR. SHORTT.

The bark macerated in water, and mixed with ginger, long pepper, milk and gingelly oil is applied as a liniment for drying up humours. An infusion of the bark is also given in flatulency.

**744.** [6148] Bendee seed (*Hibiscus esculentus*), Ahmedabad.

The mucilaginous seeds of the Ochro are added to soups, and in medicine are employed chiefly as a demulcent.

**745.** [7167] Patwah seeds (*Hibiscus cannabinus*), Lucknow.

These seeds are also mucilaginous.

**746.** [6562] Mooskh dana (*Abelmoschus moschatus*), Calcutta.

The seeds are cordial and stomachic, and when bruised have been given to counteract the effects of bites from venomous reptiles.

**747.** [4946] Indian mallow (*Abutilon indicum*), Chingleput. DR. SHORTT.

The seeds are considered laxative, and the fruits entire are employed in decoction for the sake of their mucilage.

**748.** [7376] Mallow flowers (*Althæa* sp.), Bombay.

Employment similar to the marsh mallow flowers of Europe.

**749.** [7052] Murad sing (*Helicteres Isora*), Bombay.

The natives of India, like those of Europe in former times, believing that external signs point out the properties possessed by plants, consider that the twisted fruit of *Helictires Isora* indicates that it is useful, and they therefore prescribe it in pains of the bowels.

**750.** [7032] Kakadee (*Citrus medica*), Bombay.

The citron is cultivated in the Peninsula, and is used as in Europe.

**751.** [7038] Narungee-ke-chal (*Citrus aurantium*), Bombay.

**752.** [7049] Bael (*Ægle marmelos*), Bombay.

**753.** [10747] Belpatri (*Ægle marmelos*), Mangalore. V. P. COELHO, Esq.

**754.** [1835] Bael fruit (*Ægle marmelos*), Punjab.

Belgiri or the Indian Bel fruit (dried fruit of the *Ægle marmilos*). Grown everywhere in the hills to the south of the Himalayan mountains. Used as a sherbet; the pulp also is eaten. Greatly used by the natives as an astringent in diarrhœa and dysentery, in doses of gr. xlv. Has been employed in the latter disease at the Medical College Hospital, Calcutta, with success. The clear jelly-like substance from the seeds forms a substitute for gum. Dose, one fruit. Price ½d. each. *

**755.** [6132] Wood apple (*Feronia elephantum*), Ahmedabad.

**756.** [9911] Neem bark (*Azadirachta indica*), Calcutta. KANNY LOLL DEY.

The tree is very common in India. The decoction of the leaves is used for clearing foul ulcers; leaves are used for making poultices. The bark is used as a most valuable tonic and febrifuge, also as a vermifuge; it can be used as a substitute for cinchona. Dose, ʒj. to ʒij. in infusion. Price 4 annas per lb.

**757.** [7033] Dek ke phul (*Melia azedarach*).

**758.** [6242] Mangosteen rind (*Garcinia mangostana*), Bombay.

Rind of the fruit imported from Singapore; used with much effect in chronic hæmorrhagic dysentery. Dose, ʒj. to ʒij. in infusion. Price 1 rupee per lb.

**759.** [6245] Gookhroo (*Tribulus terrestris*), Poona.

* DR. BROWN, Lahore Medical College.

760. [6207] Gookhroo (*Tribulus lanuginosus*), Calcutta. KANNY LOLL DEY.

The powdered seeds are given in decoction in dropsy and gonorrhœa.

761. [7014] Tejbul bark (*Xanthoxylon* sp.), Bombay.

This aromatic bark is employed as a condiment; in medicine for its supposed efficacy in strengthening the stomach.

762. [1833] Jujubes or Unab (*Zizyphus jujuba*), Punjab.

The dried fruit of the Lotus tree. Common all over the Punjab, ripens in February, and is then plucked and dried. The best comes from Kabul. Used by the natives as a refrigerant in fevers, and as a local application for pain in the ear; also as a laxative. Employed as an infusion. Price 2*s.* per lb.*

763. [9369] Kakra-singee (*Rhus kakrasinghee*), Kangra.

KAKRA-SINGEE.

These horn-like excrescences, formed probably in consequence of the deposition of the ova of some insect, have long constituted a famed article of Hindoo medicine: they are found in the Deyra Dhoon and everywhere in the hills, at moderate elevations.

DR. ROYLE refers the specimens contained in his collection, which are identical with the above, to *Rhus kakrasinghee*, but it has been doubted whether these galls are produced by a species of *Rhus*, as they are nearly identical with those found on *Pistacia terebinthus*.

764. [9366] Bhela (*Semecarpus anacardium*), Calcutta. KANNY LOLL DEY.

The nuts are in general use for marking cotton cloths; the colour is improved and prevented from running by the mixture of a little quick lime and water. The acrid juice of the shells is given in small doses in leprous and scrofulous affections. Price 2 annas per lb.

765. [6038] Balessan (fruits of *Balsamodendron carpobalsamum*), Bombay.

These fruits have probably no medicinal use, but are sent on account of their botanical interest.

766. [6241] Hudgah (*Balanitis ægyptiaca*), Bombay.

The unripe drupes are bitter and violently purgative.

767. [1831] Root of *Hedysarum gangeticum*, Punjab.

*Saloporni*, believed to be the *Hedysarum gangeticum*, one of the *Leguminosæ*, a bitter tonic used in fever.

768. [4973] Legumes of *Agati grandiflora*, Chingleput. DR. SHORTT.

These legumes are eaten by the natives in their curries.

769. [6025] Pulas ke beej (*Butea frondosa*), Bombay.

The large flat seeds deprived of their outer covering used as an anthelmintic in cases of tapeworm and also as deobstruent. Dose gr. iij. to Эj.

770. [6554] Areca nuts (*Areca catechu*), Midnapore.

771. [9935] Do. Calcutta. KANNY LOLL DEY.

772. [6555] Sooparee (*Areca catechu*), Cuttack.

The Goobak of the Sungscreet classics—the nut of the *Areca catechu*. This graceful tree is cultivated to any extent only in the Pooree district. The nut is used in this province almost exclusively in conjunction with other ingredients, with the leaf of the 'Piper Betel,' forming together the well-known masticatory of the East, '*Pan.*' Locally at Cuttack the best nuts are retailed at about 2 annas per English lb. weight. The tree is grown in company with the cocoa nut in plantations in moist situations, and, like the cocoa nut, is productive and valuable. An astringent extract is also, the Committee believe, obtained from old areca nuts, which is used in dyeing processes.*

The betel nuts are much relished by the natives, being chewed with the leaf of betel pepper (*Chavica Betel*) spiced with chunam (lime), but they sadly discolour the teeth. They are considered to be astringent and tonic. Roasted and powdered they make an excellent charcoal powder for the teeth, and are much used as an antiseptic tooth powder.

773. [7993] Babool bark (*Acacia arabica*.)

Babool bark is used as a tonic in infusion, and a strong decoction is employed as a wash for ulcers, and finely powdered and mixed with gingelly oil it is recommended externally in cancerous affections.

774. [7016] Fruits of a species of *Astragalus.*

775. [1840] Mulathi (*Glycyrrhiza glabra*), Punjab.

776. [1911] Liquorice (*Glycyrrhiza glabra*), Calcutta. KANNY LOLL DEY.

777. [4809] Liquorice, Burmah. Messrs. HALLIDAY, FOX, & Co.

The root of the liquorice plant, grown in large quantities about Peshawur; dug up, dried, and cut into pieces; used by the natives as a tonic in fever, in doses of gr. lx. Also as a demulcent in coughs &c.; also in all diseases consequent upon an undue accumulation of phlegm or bile. Price 1*d.* per lb.†

778. [7007] Cowage (*Mucuna prurita*), Bombay.

The stinging hairs of the pod are employed when mixed with honey as an anthelmintic in European practice, but it is doubtful whether the natives employ them at all.

779. [4041] Fenugrec (*Trigonella fœnum-græcum*), Madras.

780. [6469] Do. Calcutta.

781. [9936] Do. Do. KANNY LOLL DEY.

The seeds are mucilaginous and are employed in dysentery.

782. [7003] Gardul (*Entada pursœtha*), Bombay.

The kernel of the seeds is employed by the Hill people as a febrifuge.

783. [10126] Croton seed (*Croton tiglium*), Calcutta. KANNY LOLL DEY.

784. [9252] Do. Moulmein.

785. [4027] Do. Madras.

786. [10739] Croton fruits. S. Canara. V. P. COELHO, Esq.

787. [4769] Cankhosa (*Croton tiglium*), Burmah. Messrs. HALLIDAY, FOX, & Co.

788. [132*] Croton oil (*Croton tiglium*), Madras.

The seeds yield a powerfully cathartic oil. It is prepared by grinding the seeds, placing the powder in bags, and pressing them between plates of iron. The oil is then allowed to stand fifteen days, and afterwards filtered. The residue of the expression is saturated with twice its weight of alcohol, heated on the sand-bath from 120° to 140° Fahr., and the mixture pressed again. The alcohol is distilled off, the oil allowed to settle, and filtered after a fortnight. One seer (2 lb.) of seed furnishes 11 fluid ounces of oil, 6 by the first process, 5 by the second. The oil is well known in this country for its medicinal properties.

789. [6161] Bowchee (*Psoralia corylifolia*), Bombay.

790. [9944] Babchee (*Psoralia corylifolia*), Bazaar, Calcutta.

791. [6250] Babchee, Poona.

The seeds are aromatic and slightly bitter; they are used by the natives as a stomachic and deobstruent, and also in cases of leprosy.

792. [6237] *Notonia grandiflora*, Bombay.

Dr. Gibson's specific in hydrophobia.

793. [4967] Seeds of *Bauhinia pulcherrima*, Chingleput. DR. SHORTT.

794. [4960] Seeds of *Bauhinia tomentosa*, Chingleput. DR. SHORTT.

795. [4968] Seeds of *Acacia* sp. Do. Do.

796. [9367] Nickar (*Guilandina Bonduc*), Calcutta. KANNY LOLL DEY.

Seeds, a powerful tonic, and very valuable febrifuge; the kernels are very bitter; reduced to powder and mixed with black pepper, they are used in ague with the best results; powdered small with castor oil, they are applied externally in hydrocele. Dose gr. v. to gr. x. Price 12 annas per lb.

The CHITTAGONG COMMITTEE adds the following information. It grows well in waste land and in the jungles of this district. It is used as a febrifuge administered in pills, 4 grains of the pounded kernel with 4 grains of the black pepper. It is a very effectual remedy, but to make it so, the pills must be given fresh. If the seeds are not broken they will keep good for years, but if broken, and the kernel kept for four or five days, they become useless as a medicine for fever.

797. [6167] Senna (*Cassia lanceolata*), Bombay.

798. [1790] Ditto (*Cassia* sp.) Punjab.

799. [1791] Do. Do. Do.

800. [9942] Senna mookee (*Cassia obovata*), Calcutta. KANNY LOLL DEY.

Largely grown about Peshawur. The leaves are collected and dried in the sun. Used by both Europeans and natives as a purgative and carminative. Dose ℈ij. Price 6d. per lb.

801. [7012] Chemil seeds (*Cassia absus*), Bombay.

The seeds are bitter and mucilaginous; they are sometimes employed in ophthalmia.

802. [1859] Poar seeds (*Cassia tora*), Punjab.

The seeds are employed in preparing a blue dye. They are also used medicinally.

803. [9915] Beekeedanna (*Cydonia vulgaris*), Calcutta. KANNY LOLL DEY.

804. [7010] Do. Bombay.

Seeds used as a valuable demulcent, tonic, and a restorative remedy. Price 12 annas per lb.

805. [6201] Chaulmoogra (*Hydnocarpus odorata*), Chittagong.

Seeds imported from Sylhet: yield by expression about 10 per cent. of a thick fixed oil, of unpleasant flavour and rather offensive smell; used extensively in the treatment of cutaneous diseases; also given internally in *Lepra tuberculosa*. Dose m iij. to m vj. of oil. Price for oil, 2 rupees per lb.; seed, 2 annas per lb.

806. [9931] Potole root (*Trichosanthes dioica*), Calcutta. KANNY LOLL DEY.

The root is extensively used as a hydragogue cathartic in dropsy. Fruits and leaves are used extensively in curry. Dose gr. ij. to gr. vj. in powder. Price 5 annas per lb.

807. [4962] Root of *Coccinia indica*, Chingleput. DR. SHORTT.

This is a favourite medicine with the native practitioners in syphilitic disorders.

808. [7009] Katairee (*Cucurbita* sp.), Bombay.

809. [9937] Ghosal paul (*Momordica dioica*), Calcutta. KANNY LOLL DEY.

The root is employed in the form of an electuary in hæmorrhoids.

810. [1890] Colocynth (*Citrullus colocynthis*), Calcutta. KANNY LOLL DEY.

811. [6956] Bitter apple (*Cucumis colocynthis*), Chingleput. DR. SHORTT.

The pulp and seeds of the colocynth (*Indrewan*) produce exceedingly powerful cathartic effects; extract made from the pulp is equal to English extract of colocynth. Dose gr. v. to gr. x. Price 4 annas per lb.

812. [6075] Singara (*Trapa bispinosa*), Poona.

813. [6072] Do. Bombay.

These farinaceous fruits are chiefly employed as an article of food—they are also mixed with hermodactyl as an adulteration. In China the similar fruits of *Trapa bicornis* are much sought after as food.

TRAPA BICORNIS.

814. [4977] *Hydrocotyle asiatica*, Chingleput. DR. SHORTT.

815. [1882] Thalkin (*Hydrocotyle asiatica*), Calcutta. KANNY LOLL DEY.

The plants grow wild in Bengal; the leaves are bitter; are toasted and given in infusion to children in bowel complaints and fevers. They are also applied as anti-inflammatory to bruises; it is said to be an excellent remedy in leprosy on the Malabar Coast. Dose ʒj. to ʒij. of fresh juice. Price 2 annas per lb.

816. [6240] Fitura (*Prangos pabularia*), Bombay.

817. [6141] Gajur (*Daucus carota*), Ahmedabad.

818. [6991] Sowa (*Anethum Sowa*).

The seeds of the Indian dill are to be met with in every bazaar. They not only form an ingredient in curries, but are also employed as a carminative, and are bruised and applied externally in rheumatic and other swellings of the joints.

819. [7036] Bhaphullee (*Ptychotis montana*), Bombay.

820. [9916] Ajwan (*Ptychotis ajowan*), Bazaar, Calcutta.

821. [6470] Do. Do. Do. KANNY LOLL DEY.

The dried seeds are used as a substitute for aniseed, both as an aromatic and in colic. Also as a deobstruent in ischaria and dysmenorrhœa, and as a stimulant in catarrh and hemicrania. Employed as an infusion. Dose two drachms. Price 2s. per lb.*

822. [1805] Black Carraway (*Carum nigrum*), Punjab.

823. [9929] Carraway (*Carum nigrum*), Bazaar, Calcutta.

824. [9903] Sha jeera (*Carum album*), Do. KANNY LOLL DEY.

The dried seeds are used as a substitute for carraway seeds, similarly to those of the *Ptychotis*. Dose 30 grains. Price 1s. per lb.

825. [1829] Anisbu (*Ptychotis involucrata*), Kabul.

826. [9919] Bonjewan (*Ligusticum diffusum*), Calcutta. KANNY LOLL DEY.

Seeds used as a carminative. Price 3 annas per lb.

827. [9906] Khetpapra (*Oldenlandia biflora*), Calcutta. KANNY LOLL DEY.

Plants extensively cultivated in Bengal. The whole plant used in infusion as excellent tonic and febrifuge in chronic fever. Dose ʒj. to ʒij. Price 8 annas per lb.

828. [7011] Pieraloo (*Randia dumetorum*), Bombay.

The fruits are used as a fish poison, and also in medicine as an emetic.

829. [2840] Deekamul (*Gardenia lucida*), Raepore.

The young branches of the tree which yields the Dikamali resin, and with which they are impregnated.

830. [1841] Baman Lal (*Centaurea Behen*), Punjab.

* DR. BURTON BROWN.

**831.** [7028] Lal Behmen, Bombay.

The dried roots of two varieties of a composite plant, chiefly obtained from Kabul. Used by the natives as a tonic in debility, in doses of 4 drachms; also in impotence as a deobstruent. Price 1s. per lb. Not at present used in European medicine, but was formerly employed as an aromatic stimulant.

**832.** [1837] Baman sufed, Kabul.

This is the white Behen root of the old pharmacopœias, now rarely met with and not employed medicinally in Europe.

**833.** [7034] Kasnee seeds of *Cichorium intybus,* Bombay.

**834.** [7046] Akurkura (*Pyrethrum indicum*), Do.

**835.** [9912] Do. (*Anacyclus pyrethrum*), Calcutta. Kanny Loll Dey.

Indian pellitory; the roots imported; used as an external as well as an internal stimulant and sialogogue. Dose 3j. to 3j. in infusion. Price 12 annas per lb.

**836.** [7045] Nak chiknee (*Artemisia sternutoria*), Bombay.

The powdered plant, which is called the 'sneezer,' is used as a sternutatory.

**837.** [1867] Assautin (*Artemisia indica*), Kabul.

It has a very bitter taste. Grows chiefly in Kabul. Used by the natives as a febrifuge, in asthma, in diseases of the brain, and also in dyspepsia. Leaves are much used in scents for its strong odour; it yields a volatile essential oil when distilled. Price 5 annas per lb.

**838.** [6435] Baahiring (*Embelia ribes*), Chota Nagpore.

**839.** [9920] Borunga (*Embelia ribes*), Calcutta. Kanny Loll Dey.

**840.** [7008] Baibarung (*Embelia ribes*), Poona.

The dry berries come from Sylhet; are pungent; given in infusion, they are anthelmintic; also heating and stimulant. Dose gr. v. to 3j. in powder. Price 5 annas per lb.

**841.** [4972] Mudar bark (*Calotropis gigantea*), Chingleput. Dr. Shortt.

**842.** [10125] Do. Calcutta. Kanny Loll Dey.

The root, bark, and inspissated juice are used as powerful alteratives; the natives use the powder of the roots in almost all cutaneous affections, especially in syphilitic complaints, and in leprosy; it is used in place of ipecacuanha. Dose gr. iv. to gr. viij. Price 4 annas per lb.

**843.** [5121] Anta Moole (*Hemidesmus indicus*), Cuttack.

**844.** [9917] Anta Moole, Calcutta. Kanny Loll Dey.

Plant very common in Bengal; the roots largely used as a substitute for sarsaparilla. Dose 3j. to 3iij. in decoction. Price 3 annas per lb.

**845.** [9913] Shamalata (*Ichnocarpus frutescens*), Calcutta.

The plant is occasionally used as a substitute for sarsaparilla. Price 12 annas per lb.

**846.** [6238] Inderjan (*Wrightia antidysenterica*), Bombay.

The seeds very bitter, are used medicinally, being boiled in milk, and given in hæmorrhoids and dysentery, and in decoction in fever and gout; also used as an anthelmintic. Dose gr. v. to 3j. Price 5 annas per lb.
The bark of the root is astringent and febrifuge, and is used as a specific in dysentery and bowel complaints. Dose 3j. to 3iij. in decoction. Price 8 annas per lb.

**847.** [2835] Kechlab (*Strychnos nux vomica*), Raepore.

**848.** [10742] Do. Mangalore. V. P. Coelho, Esq.

**849.** [10113] Do. Calcutta. Kanny Loll Dey.

**850.** [10112] Do. Do.

**851.** [10114] Do. Do.

**852.** [4915] Do. Chingleput. Dr. Shortt.

Employed chiefly as an infusion; the trees are very common in the Bengal jungle; the seeds are sold in great abundance in the bazaar; it is used as a stimulant and tonic in debility and rheumatism; as an aphrodisiac; also as a poison. Price 4 annas per lb.

**853.** [1855] Clearing-nuts (*Strychnos potatorum*), Punjab.

**854.** [9905] Nermali (*Strychnos potatorum*), Calcutta. Kanny Loll Dey.

The clearing nut. By rubbing the nuts round the inside of water-pots containing muddy water, the impurities very soon fall to the bottom, leaving the water clear and perfectly wholesome. Price 2 annas per lb.

**855.** [1845] Chiretta (*Agathotes chirayta*), Punjab.

**856.** [6570] Do. Calcutta.

**857.** [7047] Do. Bombay.

**858.** [9943] Do. Calcutta.

This is an exceedingly valuable tonic and febrifuge. The whole plant is pulled up at the time that the flowers begin to decay, and is dried for use. It is much employed in India as a substitute for gentian, by which name it is known amongst the European residents. The root is considered the bitterest part, and it is best administered in the form of a tincture ; the nuts of *Guilandina bonduc* are sometimes pounded and given with it.

859. [6247] Moota (*Pedalium murex*), Bombay.

The seeds are diuretic, and are employed in dropsy.

860. [6205] Tooti (*Martynia diandra*), Calcutta. KANNY LOLL DEY.

861. [5032] Do. Chingleput. DR. SHORTT.

862. [7022] Turbeth or Teoree (*Ipomea turpethum*), Bombay.

The bark of the root is used as a hydragogue cathartic ; the root being free from a nauseous taste and smell, possesses a decided superiority over jalap, for which it might be substituted. Dose gr. x. to ʒss. Price 12 annas per lb.

863. [7005] Kala dana (*Pharbites nil*), Bombay.

864. [9930] Do. Calcutta. KANNY LOLL DEY.

Seeds used as an effectual and safe cathartic. Dose ʒss. to ʒj. Price 4 annas per lb.
The plant is cultivated in Bengal ; the powder of the seeds is used as an excellent tonic purgative, and is very useful in *Lepra tuberculosa* &c. The infusion of the seed is demulcent. Dose ɘij. to ɘiss. in powder.

865. [1832] Dodder (*Cuscuta reflexa*), Punjab.

Grows about Lahore, but the best specimens are obtained from the Sub-Himalayan country. The whole leafless herb is gathered and dried ; it is employed either in mixture or infusion as a laxative in fever, as an anthelmintic, and also as an alterative in cancer. Dose ʒij. of the powder. Price 1s. per lb.*

866. [1247] Gaozuban (*Onosma bracteata*), Bombay.

867. [6145] Wairitak (*Solanum melongena*), Ahmedabad.

The fish boiled in gingelly oil, is used as a remedy for toothache.

868. [1904] Bakoor (*Solanum indicum*), Calcutta.

The root is used in infusion as stimulant in cases of fevers and coughs ; the juice of the leaves boiled with the juice of fresh ginger is administered to stop vomiting. Dose ʒij. to ʒj. in infusion. Price 4 annas per lb.

869. [9910] Kantee karee (*Solanum Jacquini*), Calcutta.

The whole plant is used in decoction as expectorant in coughs and consumptive complaints, also in humoral

asthma. Vinum made from root is used to check vomiting. Dose ʒij. to ʒvj. in infusion. Price 2 annas 6 pie per lb.

870. [6243] Neesah (*Physalis somnifera*), Bombay.

The root is deobstruent and diuretic, the fruit diuretic. All parts of the plant are employed by native practitioners.

871. [1789] Henbane (*Hyoscyamus niger*), Punjab.

872. [9909] Do. Calcutta. KANNY LOLL DEY.

873. [7030] Do. Bombay.

The seeds are used as a slight stimulant. Price 3 annas per lb. They are obtained chiefly from Kabul and Khorassan.

874. [4952] Datura (*Datura alba*), Chingleput. DR. SHORTT.

The plant is known for the intoxicating and narcotic properties of its fruit. Price 5 annas per lb.
The root is used in cases of violent headache, and in epilepsy. In India it is a very common practice to employ Datura seeds for stupefying and even poisoning those whom they are at enmity with. Price 5 annas per lb.

875. [6049] Do. (*Datura metel*), Bombay.

876. [4974] Datoora root (*Datura fastuosa*), Chingleput. DR. SHORTT.

877. [6818] Wild mint seed, Burmah, Messrs. HALLIDAY, FOX, & CO.

878. [4955] Basil seed (*Ocymum basilicum*), Chingleput. DR. SHORTT.

879. [9921] Babooi Toolsi (*Ocymum basilicum*, var. *pilosum*), Calcutta. KANNY LOLL DEY.

The whole plant has a grateful smell. The juice of the leaves are used in catarrhal affections in children. This shrub is considered sacred in India. Dose ʒj. to ʒij. warm. Price 12 annas per lb.

880. [7004] Balungoo (*Dracocephalum Royleanum*), Bombay.

881. [1785] Ustakhudas (*Lavendula stœchas*), Punjab.

Chiefly from Kashmir. Employed either in powder or infusion as a laxative and tonic in diseases of the heart and brain ; especially as a nervine tonic in epilepsy and headache. Dose ʒiv. Price 1d. per lb.

882. [1888] Kala megh (*Andrographis paniculata*), Calcutta.

883. [6214] Kala megh (*Andrographis paniculata*), Calcutta. KANNY LOLL DEY.

Under the name of Creyat this plant has a wide reputation for its tonic and stomachic properties. It is employed in cholera and dysentery. It is found wild in the North, and is cultivated in the South.

884. [1879] Boree Gopum (*Ruellia litibrosa*), Calcutta.

885. [4975] *Asteracantha longifolia*, Chingleput. DR. SHORTT.

The roots are considered tonic and diuretic, and the leaves boiled in vinegar are used as a diuretic.

886. [7035] Lal chitra (*Plumbago rosea*), Bombay.

887. [6988] Lall chitta, Chittagong.

Is a species of plumbago, with bright red flowers; the Mugs use the roots of this as an irritant; they bruise a piece of the root and lay it on any part of the human body, and in a very short time a blister rises, but the action is so violent and painful that it could not be generally employed in practice.*

888. [9904] Isabgool (*Plantago ispaghula*), Calcutta.

889. [7002] Ispagool (*Plantago ispaghula*), Bombay.

The seeds highly mucilaginous and very much used as a demulcent in dysentery. Dose ℨij. to ℨss. in infusion. Price 3 annas per lb.

890. [9907] Azareh (*Achyranthes aspersa*), Calcutta. KANNY LOLL DEY.

891. [7015] Do. Bombay.

The flowering spikes, rubbed with a little sugar, are made into pills and given in hydrophobia, and in cases of snake-bites. Fresh leaves made into a pulp applied externally to the bites of scorpions with great effect. Dose gr. v. to gr. xx. Price 4 annas per lb.

892. [1795] Rhubarb (*Rheum emodi*), Punjab.

893. [1796] Do. (*Rheum sp.*), Kangra.

894. [9939] Do. (*Rumex tuberosus?*), Calcutta. KANNY LOLL DEY.

The inferior quality of rhubarb comes from China and is used as a gentle purgative. Dose ℈ss. to ℨj. in powder. Price 12 annas per lb.

895. [7006] Maida luckri bark (*Tetranthera Roxburghii*), Bombay.

Mildly astringent, used by the Hill people in diarrhœa.

* Local Committee, *Chittagong.*

896. [1854] Bhera (*Terminalia belerica*).

The fruit is collected and called the Beleric Myrobalan. It is administered by the natives either in powder or infusion, and is used as an astringent tonic in loss of appetite and debility. The kernels are said to be intoxicating. It is distinguished from other Myrobalans by being slightly pentangular, with rounded angles and having a smooth surface. Dose ℨj. of the powder. Price 3d. per lb.

897. [1817] Harhur (*Terminalia chebula*).

FRUIT.

The fruit is used both in powder and infusion as a laxative and tonic in fever and in brain disease. Dose ℨj. Price 3d. per lb.

898-9. [1800-1] Pakhan bed, Kangra.

A root obtained from Thibet, believed to be an antidote for opium, and used as such in cases of poisoning by that drug, either in powder, in doses of 15 grains, or in infusion.

900. [7042] Keeramar (*Aristolochia bracteata*), Bombay.

Every part of the plant is nauseously bitter. An infusion of the dried leaves is given as an anthelmintic. The leaves beaten up with water are given internally in cases of snake-bites. The plant grows abundantly in the Deccan.

901. [1839] Juramina mudharo (*Aristolochia longa*), Kabul.

Is used both in powder and mixture; employed as a tonic in diseases of the chest and brain, and especially in headache. Dose ℨij. Price 2s. per lb.

902. [10123] (*Aristolochia indica*), Calcutta. KANNY LOLL DEY.

The root is nauseously bitter; it possesses a powerful emmenagogue property; it is used for procuring abortion; and is also considered to be a valuable antidote to snake-bites, being applied both externally and internally. Price 4 annas 6 pie per lb.

903. [1868] Asarabacca (*Asarum europœum*), Kangra.

904. [1869] Do. Do. Do.

905. [1830] Asarum (*Asarum europœum*), Do.

The dried leaves are used as an expectorant, and employed also by natives as a deobstruent and tonic in diseases of the brain, stomach, and liver. Dose ℨiss. Price 3d. per lb.

**905a.** [1786] Ukkulbeer (*Datisca cannabina*), Punjab.

*Ekolbir*, used as an expectorant in catarrh in doses of gr. xlv.; also as a local application for carious teeth.

**906.** [9934] Kamala (*Rottlera tinctoria*), Calcutta. KANNY LOLL DEY.

A red powder formed on the outside of the capsules of the *Rottlera tinctoria*, called the monkey's nut, as these animals are believed to use the powder instead of rouge, is obtained from the hills about Chumba. It is used in powder mixed with the curd of milk as a purgative and anthelmintic in doses of ʒij. Price 1s. per lb. Has lately been introduced into European medicine, and used with great success as an anthelmintic.

**907.** [9941] Amleka (*Phyllanthus emblica*), Calcutta. KANNY LOLL DEY.

Seeds given in infusion as a cooling remedy in bilious affections and nausea, also in diarrhœa; used also by native females for cleaning their hairs. Dose ʒj. to ʒss. Price 2 annas per lb.

**908.** [1828] Hirbi (*Euphorbia verrucosa*), Kangra.

**909.** [1850] Bhang (*Cannabis sativa*), Punjab.

**910.** [1846] Churrus (*Cannabis sativa*), Punjab. The resinous exudation of hemp.

Charus is a resinous exudation from the hemp plant of the tracts north of the Punjab. It is brought into the Punjab from Ladakh *viâ* Kulu, Kangra, and Cashmere; also from Yarkand and Persia *viâ* Peshawur and Dera Ishmael Khan on the western frontier of the Punjab. A small quantity placed in the hookah and smoked, produces almost immediately an intoxicating effect. Price, 8s. 4d. per lb.*

The dried hemp plant which has flowered, and from which the resin has not been removed, is called Gunjah; it is used for smoking alone. The alcoholic extract of it is used medicinally as antispasmodic and anodyne, very useful in spasmodic coughs, in tetanus, in hydrophobia, &c. Dose — Extract ¼ gr. to gr. j. Price, 8 annas per lb.

The larger leaves and capsules without the stalks are called *Bhang*, *Subjee*, or *Siddhee*. They are used for making an intoxicating drink, for smoking, and in the conserve or confection termed *Majoom*. *Bhang* is cheaper than *Gunjah*, and though less powerful, is sold at such a low price, that for one pice enough can be purchased to intoxicate a person habituated to it. Dose, ʒj. in infusion for intoxication. Price, 8 annas per lb.

The resinous juice that exudes and concretes on the leaves, slender stems, and flowers, when separated, constitutes the churrus; it is collected in great abundance in Nepal, also in Goruckpore; it is used for smoking, mixed with a little tobacco. Price, 2 rupees per lb.

The majoom or hemp confection is a compound of sugar, butter, flour, milk, and siddhee or barley. Dose ʒj to ʒiij. for intoxication. Price, 2 rupees per lb.

**911.** [7051] Cubab cheenee (*Piper cubeba*), Bombay.

**912.** [9918] Kabob chenee (*Piper cubeba*), Calcutta. KANNY LOLL DEY.

**913.** [1906] Do.     Do.     Do.

**914.** [6341] Do.     Do. Java. JOSE D'ALMEIDA, Esq.

Employed among other purposes in gonorrhœa. Dose, ʒj. to ʒij. Price 1 rupee 4 annas per lb.

**915.** [6164] Pipilee mool (*Chavica Roxburghii*), Bombay.

**916.** [1798] Baami (*Taxus baccata*), Kangra.

The leaves are used by the natives as an expectorant in catarrh; administered in powder in doses of gr. xlv. or in decoction.

**917.** [1886] Aloes (*Aloë indica*), Calcutta. KANNY LOLL DEY.

**918.** [9914] Zafran (*Crocus cashmerianus*), Calcutta. KANNY LOLL DEY.

**919.** [7031] Keyswo (*Crocus cashmerianus*), Bombay.

**920.** [6068] Punjabee Peaz or Podshap salep, Bombay.

PUNJABEE PEAZ OR PODSHAP SALEP.

* Central Committee, *Lahore.*

921. [7018] Soorungan tulli (*Bitter hermodactyls*), Bombay.

922. [7044] Suffaid Mooslee (*Murdannia scapiflora*).

923. [1836] Orris root (*Iris* sp.), Kashmir.

924. [7037] Cairoofsha (*Iris florentina*), Bombay.

925. [1808] Kuth metha (*Costus arabicus*), Punjab.

926. [1807] Do. Kabul.

The root is brought down from Kashmir in large quantity, and sold to the Chinese, who burn it as incense. Used by the natives in infusion in rheumatism as an anodyne. Dose, gr. xxx. Price, 3s. per lb.

927. [9908] Huldee (*Curcuma longa*), Calcutta.

Plant is common in Bengal, where the root is extensively used as a condiment. Price, 1 anna 6 pice per lb.

928. [7026] Amba Hulda (*Curcuma zedoaria*), Ahmedabad.

929. [7027] Kutchoora (*Curcuma zerumbet*), Bombay.

930. [7019] Kupoor Kuchree (*Hedychium spicatum*), Bombay.

931. [1834] *Calamus aromaticus*, Kangra.

932. [9922] Do. Calcutta.

The dried rhizome is employed either in powder or infusion as an aromatic tonic and carminative, in doses of 3iss. Price, 1s. per lb.

933. [9926] Mootha (*Cyperus longus*), Calcutta.

934. [6569] Moothoo (*Cyperus rotundus*), Do.

935. [9938] Moothe (*Cyperus rotundus*), Do. KANNY LOLL DEY.

In medicine it is used as tonic and stimulant. In the fresh state given in infusion as a demulcent in fevers, and also used in cases of dysentery and diarrhœa. Dose, gr. v. to gr. xv. Price, 8 annas per lb.

936. [7041] Nagur mootha (*Cyperus pertennis*), Bombay.

937. [1810] Bothesi (*Nardostachys Jatamansi*), Punjab.

An aromatic root, common in the Hills; employed chiefly as an expectorant in cough and colds. Dose, gr. xlv. of the powder. Price, 9d. per lb.

A kind of spikenard, imported from Nepal, and is used in scents for its strong odour; it is also used as refrigerant. Dose, gr. v. to 3j. in infusion. Price, 6 annas per lb.

938. [5343] Bamboo fungus, Rangoon.

This fungus is formed at the roots of bamboos. It is a favourite vermifuge with the Burmese, and has been used with much success by our own medical officers in their practice. Dose, a piece about the size of a large cherry pounded and administered in any convenient vehicle.*

939. [6428] Bunslochun, Chota Nagpore.

940. [2127] Do. white, Mundla.

941. [2125] Do. pink. Do.

942. [2126] Do. blue, Jubbulpore.

943. [10118] Do., Calcutta. KANNY LOLL DEY.

Found in the hollow bamboo, obtained from the pith of some bamboos. It is said to be found in old wood only, and about one bamboo in three produces it. Used by natives as a stimulant and tonic, in doses of about five grains; though what purpose it can serve would be difficult to understand, as it is a very pure kind of silica or silex. Some of it approaches opal in appearance and composition. Price, 1 rupee 8 annas per lb.†

944. [35*] Massoy bark (cosmetic), New Guinea.

945. [2343] Cosmetic root, Burmah. Messrs. HALLIDAY, FOX, & CO.

946. [2342] Pitch wood. Do. Do.

947. [6351] Tuba root (poison), Singapore. COL. COLLYER.

948. [1827] Kanochar or Konachan, Kangra.

Supposed to be a species of Barrera; used as an aromatic expectorant in doses of gr. xxx.

949. [1858] Bistang or Bartang, Kangra.

An astringent used in diarrhœa and dysentery. Dose of the powder, gr. ij.

950. [6249] Sicilian manna (*Ornus europœa*), Bombay.

Imported into Bombay from the South of Europe.

* Local Committee, *Rangoon.*
† Local Committee, *Jubbulpore.*

**951. [1864] Shirkhist, Kabul.**

The collector states that this is obtained from the *Ononis rotundifolia.* It is found on the plant in small, rounded, reddish grains, and is used by the natives similarly to Sicilian manna. Price, 2*s.* per pound. Dose ℥j.

DR. ROYLE says :—' *Sheerkhisht,* a kind of manna, is said to be procured from a tree of Khorasan, perhaps a species of *Fraxinus* or *Ornus.*' Perhaps the collector has mistaken *Ornus rotundifolia* for *Ononis rotundifolia,* as the former is a much more probable source than the latter.

**952. [1864] Alhagi manna or Taranjibin, Punjab.**

A sort of manna found on the *Hedysarum Alhagi,* one of the *Leguminosæ,* produced principally in Kabul. It occurs in unequal, dark brown pieces, of a bitter, sweetish taste, and is mixed with the stalks of the plant on which it is produced. It is used by the natives as a laxative and refrigerant in fever, in doses of ℥j. Price, 1*s.* 6*d.* per pound.

In the latter it is much used as a substitute for sugar. It is imported into India from Kabul and Khorasan.

### MISCELLANEOUS.

**953. [5189] Essence of Chiretta, Calcutta. Messrs. BATHGATE & Co.**

**954. [6416] Powder for sore throat, Singapore. TAN KIM SING.**

**955. [6417] Powder for extracting teeth. Do. Do.**

**956. [2253] Bremma Dundoo Veray Ennye, an ointment prepared from Argemone seed, Salem. CHEDUMBARA PILLAY.**

**957. [3746] Kottanguchee Thylum, oil of cocoa-nut shells. Do.**

**958. [3748] Meza Rajanga Thylum, a mixture of Gingellie, Iloopa, Vappum, Poonga, and Poonnay oils. Do.**

**959. [3747] Myel Ennye, an ointment of peacock's fat. Salem.**

**960. [7*] Blistering beetles (*Mylabris,* various sp.), Bombay. DR. BIRDWOOD.**

About 180 lbs. forwarded last year by DR. BIRDWOOD, to test market value in England, were sold at 5*s.* 8*d.* per lb.

**961. [10690] Pills for diabetes, Mangalore. V. P. COELHO, Esq.**

# Class III.

## SUBSTANCES USED AS FOOD.

### Section A. — AGRICULTURAL PRODUCE.

A large and interesting series is exhibited — one which embraces almost every kind of grain used as food in India.

The collection of grains in the ear comprises the following: —

962. [9948] Maize, Mangalore. V. P. Coelho, Esq.

963. [9949] Mulloo Navane. Do. Do.

964. [9950] Save. Do. Do.

965. [9951] White and black Navane. Do. Do.

966. [9952] Raghy. Do. Do.

967. [9953] Samugga paddy. Do. Do.

968. [9954] Maskatta do. Do. Do.

969. [9955] Guddoo Maskatta do. Do. Do.

970. [9956] Kalane white do. Do. Do.

971. [9957] Sangula paddy. Do. Do.

972. [9958] Somasale do. Do. Do.

973. [9959] Kalune red do. Do. Do.

974. [9960] Amate paddy Do. Do.

975. [9961] White Nerrir paddy Do. Do.

976. [9962] Kini bitter paddy Do. Do.

977. [9963] Chokambally do. Do. Do.

978. [9964] Kukuma do. Do. Do.

979. [9965] Jarsali do. Do. Do.

980. [9966] White Maskatta paddy, Mangalore. V. P. Coelho, Esq.

981. [9967] Black paddy Do. Do.

982. [9968] Kojari do. Do. Do.

983. [9969] Paddy (*Oryza sativa*), Elooppapoo Simba, Madras. Dr. Hunter.

984. [9970] Do. Cada. Do. Do.

985. [9971] Do. Oodicha. Do. Do.

986. [9972] Do. Muvadasu. Do. Do.

987. [9973] Do. Peseme. Do. Do.

988. [9974] Do. Nuroolly Sumba. Do. Do.

989. [9975] Do. Manakata. Do. Do.

990. [9976] Do. Vellay Car. Do. Do.

991. [9977] Do. Vellay Manacartha Car. Do. Do.

992. [9978] Common do. Perroo Car. Do. Do.

993. [9979] Paddy, Koondo Sumba. Do. Do.

994. [9980] Do. Vaday Sumba. Do. Do.

995. [9981] Do. Sumba. Do. Do.

996. [9982] Do. Mathoo Sumba, Do. Do.

997. [9983] Paddy, Segapoo Windoo Car. Madras. DR. HUNTER.

998. [9984] Do. Moja Manacartha Car. Do. Do.

999. [9985] Do. Seeroomanee. Do. Do.

1000. [9986] Do. Poompada. Do. Do.

1001. [9987] Do. Pall Sunba. Do. Do.

1002. [9988] Little millet (*Panicum miliaceum*), Shama. Do. Do.

1003. [9989] Italian millet (*P. italicum*), Tenney. Do. Do.

1004. [9990] Raggy (*Eleusine stricta*), Rolla Kevara. Do. Do.

1005. [9991] Little millet (*P. italicum*), Waragoonura var. Do. Do.

1006. [9992] Do. (*P. miliaceum*), Waragoo. Do. Do.

.1007-8. [9993-94] Raggy (*Eleusine stricta*), Kararoo or Kehuwaragoo. Do. Do.

1009. [9995] Paddy (*Oryza sativa*), Carapoo Manacartha Car. Do. Do.

1010. [9996] Common paddy, (*do.*), Coodum Sumba. Do. Do.

1011. [9997] Paddy (*do.*) Parroon Sumba. Do. Do.

1012. [9998] Do. Muvadasse, blue var. Do. Do.

1013. [9999] Do. Essara Cova. Do. Do.

1014. [10000] Red paddy, Legapoo. Do. Do.

1015. [192] Paddy (*O. sativa*), Sacree Sumba. Do. Do.

1016. [193] Do. Aroonjothee. Do. Do.

1017. [194] Do. Vulloor Sumba. Do. Do.

1018. [195] White Cholum, or great millet (*Sorghum vulgare*). Do. Do.

1019. [196] Cholum, or great millet (*Sorghum vulgare* var.) Do. Do.

1020. [197] Do. var. Do. Do.

1021. [198] Do. var. Do. Do.

1022. [199] Yellow Cholum, or great millet (*Sorghum vulgare* var.), Murja Cholum, Do. Do.

## CEREALS.

### WHEAT

Is cultivated and employed as an article of food throughout almost the whole of India. The analyses of the samples first entered, as well as those of the other grains which follow, are by the compiler of the present Catalogue.

1023. [55] Wheat, Broach, Bombay.

COMPOSITION.

|  | Per cent. |
|---|---|
| Moisture | 12·40 |
| Nitrogenous matter | 14·68 |
| Starchy matter | 69·78 |
| Fatty or oily matter | 1·16 |
| Mineral constituents (ash) | 1·98 |
| TOTAL | 100·00 |

1024. [56] Wheat (*Triticum vulgare*), Guzerat.

COMPOSITION.

|  | Per cent. |
|---|---|
| Moisture | 10·88 |
| Nitrogenous matter | 13·30 |
| Starchy matter | 73·23 |
| Fatty or oily matter | 1·29 |
| Mineral constituents (ash) | 1·30 |
| TOTAL | 100·00 |

1025. [57] Wheat, Guzerat.

COMPOSITION.

|  | Per cent. |
|---|---|
| Moisture | 13·28 |
| Nitrogenous matter | 13·19 |
| Starchy matter | 70·87 |
| Fatty or oily matter | 1·20 |
| Mineral constituents (ash) | 1·45 |
| TOTAL | 100·00 |

1026. [58] Wheat, Bombay Bazaar.

COMPOSITION.

|  | Per cent. |
|---|---|
| Moisture | 13·41 |
| Nitrogenous matter | 12·84 |
| Starchy matter | 70·99 |
| Fatty or oily matter | 1·17 |
| Mineral constituents | 1·59 |
| TOTAL | 100·00 |

1027. [59] Wheat, Bombay Bazaar.

COMPOSITION.

|  | Per cent. |
|---|---|
| Moisture | 13·32 |
| Nitrogenous matter | 14·90 |
| Starchy matter | 68·54 |
| Fatty and oily matter | 1·14 |
| Mineral constituents (ash) | 2·10 |
| TOTAL | 100·00 |

1028. [60] Wheat, Madras.

COMPOSITION.

| | Per cent. |
|---|---|
| Moisture . . . . . | 10·80 |
| Nitrogenous matter . . . | 12·98 |
| Starchy matter . . . . | 73·51 |
| Fatty or oily matter . . . | 1·03 |
| Mineral constituents (ash) . . . | 1·68 |
| TOTAL | 100·00 |

1029. [61] Wheat, Calcutta.

COMPOSITION.

| | Per cent. |
|---|---|
| Moisture . . . . . | 11·78 |
| Nitrogenous matter . . . | 12·73 |
| Starchy matter . . . . | 72·58 |
| Fatty or oily matter . . . | 1·01 |
| Mineral constituents (ash) . . | 1·90 |
| TOTAL | 100·00 |

1030. [62] Wheat, Bombay Bazaar.

COMPOSITION.

| | Per cent. |
|---|---|
| Moisture . . . . . | 12·56 |
| Nitrogenous matter . . . | 14·26 |
| Starchy matter . . . . | 70·26 |
| Fatty or oily matter . . . | 1·06 |
| Mineral constituents (ash) . . . | 1·86 |
| TOTAL | 100·00 |

Several of the samples of wheat forwarded on the present occasion were destroyed by weevil; and a number of the other products had suffered from insects to a similar extent.*

The specimens of wheat enumerated below will be found on the stand along with those before named.

1031. [4980] Wheat, grown at Bangalore, Madras.

1032. [6082] Do. Poona, Bombay.

1033. [2368] Do. Nuggur Division, Mysore.

1034. [6086] Do. Ahmedabad.

1035. [6087] Do. Do.

1036. [9368] Do. Gilghit, N.E. of Peshawur, now being grown experimentally in Punjab.

1037. [1942] Do. Jubbulpore, Bengal.

This beautiful wheat is produced in the Jubbulpore,

Nursingpore, and Hoshungabad districts, all along the line of the railway to Bombay. The average selling price at this place is about 30 to 35 seers per rupee (about 2*d.* per bushel), and other descriptions of wheat can be had cheaper.*

1038. [1943] Wheat, Jubbulpore, Bengal.

1039. [6307] Do. Rangoon. Messrs. HALLIDAY, FOX, & Co.

1040. [4790] Do.† Burmah. Messrs. HALLIDAY, FOX, & Co.

## BARLEY

Is extensively grown in the north of India, and occasionally in the hill regions in the south.

1041. [63] Barley (unhusked), Bombay Bazaar.

COMPOSITION.

| | Per cent. |
|---|---|
| Moisture . . . . . | 8·00 |
| Nitrogenous matter . . . | 10·94 |
| Starchy matter . . . . | 77·14 |
| Fatty or oily matter . . . | 1·65 |
| Mineral constituents (ash) . . . | 2·27 |
| TOTAL | 100·00 |

1042. [64] Barley (husked like pearl-barley), Nepal.

COMPOSITION.

| | Per cent. |
|---|---|
| Moisture . . . . . | 12·90 |
| Nitrogenous matter . . . | 11·46 |
| Starchy matter . . . . | 72·30 |
| Fatty or oily matter . . . | 1·25 |
| Mineral constituents (ash) . . | 2·09 |
| TOTAL | 100·00 |

1043. [6084] Barley (*Hordeum hexastichon*), Ahmedabad.

1044. [2808] Do.

1045. [6910] Do. Calcutta.

1046. [6911] Do. Lucknow.

Two kinds of barley are grown in Oude on light soils, and not irrigated. The one kind is called 'Jan,' and is grown everywhere; the other is called 'Dasawree,' and is grown on the banks of rivers. Sown in October. Sells for 2 or 2½ maunds for 1 rupee.‡

## OATS

Are not indigenous. Some years ago they were introduced into Patna and Moonghyr,

---

* With but few exceptions the whole collection of grains from Mysore was destroyed by insects.

* Local Committee, *Jubbulpore.*
† Local Committee, *Rangoon.*
‡ Central Committee, *Lucknow.*

but the cultivation is not carried on to any extent. The sample the analysis of which is given below was not a good specimen.

1047. [65] Oats (*Avena sativa*). Mixed sample from Patna and Moonghyr, 1850.

COMPOSITION.

| | Per cent. |
|---|---|
| Moisture | 13·52 |
| Nitrogenous matter | 10·13 |
| Starchy matter | 68·79 |
| Fatty or oily matter | 3·63 |
| Mineral constituents (ash) | 3·93 |
| TOTAL | 100·00 |

1048. [7371] Oats, Moonghyr, 1861.

### MAIZE (*Zea Mays*).

Extensively cultivated in many parts of India. Only two representatives are shown.

1049. [69] Maize (*Zea Mays*), Bombay Bazaar.

COMPOSITION.

| | Per cent. |
|---|---|
| Moisture | 12·90 |
| Nitrogenous matter | 9·23 |
| Starchy matter | 74·63 |
| Fatty or oily matter | 1·59 |
| Mineral constituents (ash) | 1·66 |
| TOTAL | 100·00 |

1050. [4799] Maize, Burmah. Messrs. HALLIDAY, FOX, & CO.

It is common in Burmah. Maize is also extensively cultivated in Oude. The stem and leaves, when dry, are chopped up and given to cattle. The seed is ground for bread, and eaten under the name of *Chabena.* It is sold at 30 seers for the rupee.*

### MILLETS.

JOWAR, JOWAREE (*Sorghum vulgare*), (*Holcus Sorghum*), INDIAN MILLET. This and the next mentioned species, the *Penicillaria spicata*, or Bajra, constitute the chief grain food of a considerable proportion of the people of India. Taking the country as a whole, it is only the rich or well-to-do classes who can afford to employ rice and wheat. The native names are those which have been attached to the different samples by the exhibitors.

1051. [66] Jowaree (white) (*Sorghum vulgare*), Patna, Bengal.

COMPOSITION.

| | Per cent. |
|---|---|
| Moisture | 12·70 |
| Nitrogenous matter | 9·18 |
| Starchy matter | 74·53 |
| Fatty or oily matter | 1·90 |
| Mineral constituents (ash) | 1·69 |
| TOTAL | 100·00 |

1052. [67] Jowaree (red) Bombay Bazaar.

COMPOSITION.

| | Per cent. |
|---|---|
| Moisture | 12·00 |
| Nitrogenous matter | 9·51 |
| Starchy matter | 74·71 |
| Fatty or oily matter | 2·15 |
| Mineral constituents (ash) | 1·63 |
| TOTAL | 100·00 |

1053. [6080] Jowaree Do. Ahmedabad.

1054. [4940] Mootoo cholum (*Sorghum vulgare*), Madras. DR. HUNTER.

1055. [5004] Cumboo. Do. Do. Do.

1056. [2814] Jowaree (yellow).

Various samples of this grain from Oude and other parts had become unfit for exhibition.

BAJRA (*Penicillaria spicata; Holcus spicatus*), SPIKED MILLET.

1057. [70] Bajra (*Penicillaria spicata*), Bellary, Madras.

COMPOSITION.

| | Per cent. |
|---|---|
| Moisture | 12·40 |
| Nitrogenous matter | 10·14 |
| Starchy matter | 73·37 |
| Fatty or oily matter | 2·20 |
| Mineral constituents (ash) | 1·89 |
| TOTAL | 100·00 |

1058. [71] Bajra, Bombay Bazaar.

COMPOSITION.

| | Per cent. |
|---|---|
| Moisture | 9·82 |
| Nitrogenous matter | 10·90 |
| Starchy matter | 74·27 |
| Fatty or oily matter | 3·05 |
| Mineral constituents (ash) | 1·96 |
| TOTAL | 100·00 |

1059. [72] Bajra, Narespore.

COMPOSITION.

| | Per cent. |
|---|---|
| Moisture | 11·80 |
| Nitrogenous matter | 10·00 |
| Starchy matter | 71·45 |
| Fatty or oily matter | 4·62 |
| Mineral constituents (ash) | 2·13 |
| TOTAL | 100·00 |

* Central Committee, *Lucknow.*

1060. [106] Bajra, Bengal.

1061. [4060] Do.

1062. [4807] Do. Burmah. Messrs. HALLIDAY, FOX, & CO.

*Panicum miliaceum,* common throughout India.

1063. [73] *Panicum miliaceum,* Bengal.

COMPOSITION.

|  |  | Per cent. |
|---|---|---|
| Moisture | . . . . . | 12·00 |
| Nitrogenous matter | . . . | 12·60 |
| Starchy matter | . . . | 70·43 |
| Fatty or oily matter | . . . | 3·62 |
| Mineral constituents (ash) | . . | 1·35 |
| TOTAL | | 100·00 |

1064. [108] *Panicum miliaceum,* Travancore.

1065. [5003] Chamay (*Panicum miliaceum*), Madras. DR. HUNTER.

1066. [111] Do. Travancore.

1067. [6083] Chenoo (*Panicum miliaceum*), Ahmedabad.

1068. [107] Koda (*Paspalum scrobiculatum*), Travancore.

1069. [113] Do. Benares.

1070. [1967] Do. Bengal.

1071. [2805] Do.

1072. [6310] Saurwah (*Panicum frumentaceum*), Bengal.

1073. [5838] Shegapoo Thenee (*Panicum frumentaceum*), Madras.

1074. [6308] Kakum (*Panicum italicum*), Lucknow.

1075. [6085] Do. Ahmedabad.

1076. [6553] Do. Cuttack.

1077. [5002] Hicana (*Panicum italicum*), Madras. DR. HUNTER.

1078. [74] Millet, called wild rice (*Panicum colorium*), Gaujam.

COMPOSITION.

|  |  | Per cent. |
|---|---|---|
| Moisture | . . . . | 11·96 |
| Nitrogenous matter | . . . | 9·64 |
| Starchy matter | . . . | 75·76 |
| Fatty or oily matter | . . . | 0·60 |
| Mineral constituents (ash) | . . | 2·04 |
| TOTAL | | 100·00 |

1079. [4051] Little millet (*Panicum miliare*), Madras.

1080. [75] Raggee (*Eleusine coracana*), Bombay Bazaar.

COMPOSITION.

|  |  | Per cent. |
|---|---|---|
| Moisture | . . . . | 11·16 |
| Nitrogenous matter | . . . | 5·76 |
| Starchy matter | . . . | 79·94 |
| Fatty or oily matter | . . . | 0·50 |
| Mineral constituents (ash) | . . | 2·64 |
| TOTAL | | 100·00 |

1081. [76] Raggee, Bombay Bazaar.

COMPOSITION.

|  |  | Per cent. |
|---|---|---|
| Moisture | . . . . | 12·00 |
| Nitrogenous matter | . . . | 6·00 |
| Starchy matter | . . . | 78·69 |
| Fatty or oily matter | . . . | 1·20 |
| Mineral constituents (ash) | . . | 2·11 |
| TOTAL | | 100·00 |

1082. [109] Raggee (black), Bombay Bazaar.

1083. [5007] Do. Do. Madras. DR. HUNTER.

1084. [6552] Do. Do. Cuttack.

1085. [4966] Mutlinga pilloo (*Eleusine ægyptiaca*), Madras. DR. SHORTT.

1086. [110] Kiery (*Amaranthus frumentaceus*).

1087. [6309] Kiery (*Amaranthus anardhana*), Lucknow.

1088. [7485] Kiery Rajura (*A. anardhana*), Ahmedabad.

RICE (*Oryza sativa*).

This is the favourite food-grain of the people; but, except in Arracan, and a few other districts, in which it constitutes the chief and almost only article cultivated, its use is confined to the richer classes throughout the country.

The extent to which it is exported from India to this and other countries will be gathered from an inspection of the Table pp. 50 and 51. A very large collection of paddy (rice in the husk) and cleaned rice has been forwarded on the present occasion. The space at disposal has, however, only admitted of the exhibition of the following samples:—

## QUANTITIES (AS FAR AS CAN BE ASCERTAINED) AND VALUE OF RICE EXPORTED

COUNTRIES WHITHER

| Years | Whence Exported | United Kingdom Quantity | United Kingdom Value | France Quantity | France Value | Other Parts of Europe Quantity | Other Parts of Europe Value | America Quantity | America Value | China Quantity | China Value | Arabian and Persian Gulfs Quantity | Arabian and Persian Gulfs Value | Aden and Africa Quantity | Aden and Africa Value | Cape of Good Hope Quantity | Cape of Good Hope Value |
|---|---|---|---|---|---|---|---|---|---|---|---|---|---|---|---|---|---|
| | | qrs. | £ | qrs. | £ | qrs. | £ | qrs. | £ | qrs. | £ | qrs. | £ | qrs. | £ | qrs. | £ |
| 1850-51 | Bengal | 84,762 | 95,277 | 6,597 | 7,578 | 6,632 | 5,665 | 5,076 | 4,328 | 7,223 | 6,223 | 59,533 | 48,408 | .. | .. | 3,550 | 5,446 |
| | Madras | 47,894 | 36,190 | 6,941 | 5,817 | .. | .. | .. | .. | .. | .. | 58,576 | 54,589 | .. | .. | 328 | 275 |
| | Bombay | 77 | 165 | .. | .. | .. | .. | .. | .. | 5 | 10 | 14,751 | 25,632 | 2,788 | 4,373 | .. | .. |
| | ALL INDIA | 132,733 | 131,632 | 13,538 | 13,395 | 6,632 | 5,665 | 5,076 | 4,328 | 7,233 | 6,233 | 132,860 | 128,629 | 2,788 | 4,373 | 5,878 | 5,721 |
| 1851-52 | Bengal | 82,674 | 82,965 | 4,418 | 5,268 | 12,426 | 10,038 | 6,485 | 5,725 | 1,936 | 1,652 | 47,407 | 42,768 | .. | .. | 14,419 | 15,513 |
| | Madras | 29,761 | 22,363 | 2,224 | 1,934 | .. | .. | .. | .. | .. | .. | 93,502 | 85,583 | .. | .. | 6,057 | 5,196 |
| | Bombay | .. | .. | .. | .. | .. | .. | .. | .. | 7 | 12 | 9,584 | 16,553 | 2,145 | 4,003 | .. | .. |
| | ALL INDIA | 112,435 | 105,328 | 6,642 | 7,202 | 12,426 | 10,038 | 6,485 | 5,725 | 1,943 | 1,664 | 150,493 | 144,904 | 2,145 | 4,003 | 20,476 | 20,709 |
| 1852-53 | Bengal | 107,922 | 113,574 | 16,124 | 17,553 | 2,650 | 2,749 | 9,914 | 7,931 | 3,827 | 3,570 | 49,616 | 41,188 | .. | .. | 5,018 | 4,843 |
| | Madras | 89,015 | 63,533 | 13,357 | 10,068 | .. | .. | .. | .. | .. | .. | 71,077 | 64,956 | .. | .. | .. | .. |
| | Bombay | .. | .. | .. | .. | .. | .. | .. | .. | 7 | 13 | 9,858 | 17,517 | 4,888 | 9,576 | .. | .. |
| | ALL INDIA | 196,937 | 177,107 | 29,481 | 27,621 | 2,650 | 2,749 | 9,914 | 7,931 | 3,834 | 3,583 | 130,551 | 123,661 | 4,888 | 9,576 | 5,018 | 4,843 |
| 1853-54 | Bengal | 847,116 | 328,665 | 57,210 | 50,657 | 70,610 | 55,608 | 3,620 | 2,127 | 38,783 | 15,559 | 49,753 | 33,334 | 4,966 | 3,005 | 8,116 | 5,542 |
| | Madras | 79,875 | 80,678 | 13,883 | 11,720 | .. | .. | .. | .. | 3,345 | 2,040 | 83,861 | 91,272 | .. | .. | 7 | 12 |
| | Bombay | 1,052 | 1,596 | .. | .. | .. | .. | .. | .. | 1 | 1 | 9,034 | 13,615 | 10,576 | 16,084 | .. | .. |
| | ALL INDIA | 928,043 | 410,939 | 71,093 | 62,377 | 70,610 | 55,608 | 3,620 | 2,127 | 42,129 | 17,601 | 142,648 | 138,221 | 15,542 | 19,089 | 8,123 | 5,554 |
| 1854-55 | Bengal | 11,507,584 | 538,959 | 1,982,248 | 73,819 | 2,204,936 | 114,782 | 144,889 | 5,024 | 342,561 | 13,276 | 1,448,292 | 53,730 | 5,228 | 2,520 | 192,096 | 7,778 |
| | Madras | 21,508 | 21,270 | 7,882 | 6,041 | .. | .. | .. | .. | .. | .. | 80,914 | 99,275 | .. | .. | 8 | 14 |
| | Bombay | 1,477 | 2,240 | .. | .. | .. | .. | .. | .. | 82 | 117 | 31,458 | 21,700 | 5,101 | 7,862 | .. | .. |
| | ALL INDIA | 11,530,569 | 562,469 | 1,990,130 | 79,860 | 2,204,936 | 114,783 | 144,889 | 5,024 | 342,643 | 13,393 | 1,560,664 | 174,705 | 10,329 | 10,382 | 192,104 | 7,792 |
| 1855-56 | Bengal | 13,936,612 | 1,181,079 | 3,736,701 | 192,736 | 1,169,815 | 169,947 | 257,908 | 10,211 | 2,063,521 | 87,209 | 1,177,860 | 48,755 | 49,022 | 1,672 | 227,361 | 8,954 |
| | Madras | 111,308 | 82,975 | 47,204 | 39,427 | .. | .. | 2,283 | 1,696 | 2,296 | 3,850 | 71,125 | 90,454 | 5 | 6 | .. | .. |
| | Bombay | 2,784 | 4,107 | 885 | 1,343 | 34 | 52 | .. | .. | 332 | 504 | 19,449 | 26,597 | 3,286 | 4,978 | .. | .. |
| | ALL INDIA | 14,050,704 | 1,268,161 | 3,784,790 | 233,506 | 1,169,849 | 169,999 | 260,191 | 11,907 | 2,066,149 | 91,563 | 1,268,434 | 165,806 | 52,313 | 6,656 | 227,361 | 8,954 |
| 1856-57 | Bengal | 7,436,159 | 662,096 | 1,800,688 | 162,281 | 236,196 | 41,279 | 251,700 | 14,954 | 4,611,622 | 224,453 | 1,689,480 | 88,969 | 14,060 | 3,828 | 312,556 | 17,654 |
| | Madras | 102,420 | 86,260 | 18,966 | 17,129 | 100 | 129 | .. | .. | .. | .. | 40,077 | 52,546 | 562 | 818 | 16 | 21 |
| | Bombay | .. | .. | .. | .. | .. | .. | .. | .. | 3,381 | 3,315 | 17,906 | 27,267 | 4,744 | 7,320 | .. | .. |
| | ALL INDIA | 7,538,579 | 748,356 | 1,819,654 | 179,410 | 236,296 | 41,408 | 251,700 | 14,954 | 4,615,003 | 227,768 | 1,747,463 | 168,782 | 19,366 | 11,966 | 312,572 | 17,675 |
| 1857-58 | Bengal | 5,392,635 | 1,181,419 | 906,807 | 135,397 | 93,552 | 41,556 | 391,368 | 31,541 | 3,515,860 | 593,595 | 1,676,842 | 111,844 | 113,750 | 6,548 | 344,249 | 25,568 |
| | Madras | 126,663 | 102,717 | 13,672 | 11,098 | 1,698 | 1,553 | .. | .. | .. | .. | 50,146 | 76,678 | 928 | 1,762 | .. | .. |
| | Bombay | .. | .. | .. | .. | 1,892 | 2,867 | .. | .. | 3,757 | 5,619 | 11,947 | 11,906 | 5,796 | 7,872 | .. | .. |
| | ALL INDIA | 5,519,298 | 1,284,136 | 920,479 | 146,495 | 97,142 | 45,976 | 391,368 | 31,541 | 3,519,617 | 599,214 | 1,738,935 | 200,425 | 119,546 | 14,420 | 344,245 | 25,568 |
| 1858-59 | Bengal | 496,893 | 461,662 | 13,748 | 14,891 | 161 | 219 | 28,468 | 44,792 | 168,966 | 159,944 | 113,357 | 156,944 | 6,224 | 9,586 | 1,457 | 23,356 |
| | Madras | 34,891 | 22,412 | 21,261 | 13,164 | .. | .. | .. | .. | 1,713 | 2,230 | 36,271 | 64,717 | 279 | 414 | .. | .. |
| | Bombay | .. | .. | .. | .. | .. | .. | .. | .. | 245 | 371 | 9,150 | 13,539 | 4,342 | 6,569 | .. | .. |
| | ALL INDIA | 531,784 | 484,074 | 35,009 | 28,055 | 161 | 215 | 28,468 | 44,792 | 170,924 | 162,545 | 158,778 | 235,200 | 10,845 | 16,569 | 1,457 | 23,356 |
| 1859-60 | Bengal | 499,920 | 518,799 | 18,571 | 24,691 | 17,663 | 7,171 | 20,027 | 31,792 | 44,261 | 44,654 | 101,854 | 146,975 | 35,375 | 7,401 | 16,740 | 33,197 |
| | Madras | 57,379 | 46,589 | 11,243 | 9,477 | .. | .. | .. | .. | 43 | 89 | 34,678 | 66,537 | 928 | 1,762 | .. | .. |
| | Bombay | .. | .. | .. | .. | .. | .. | .. | .. | 70 | 117 | 9,656 | 14,792 | 2,314 | 3,533 | .. | .. |
| | ALL INDIA | 557,299 | 565,388 | 29,814 | 34,168 | 17,663 | 7,171 | 20,027 | 31,792 | 44,374 | 44,862 | 146,188 | 228,304 | 38,617 | 12,696 | 16,740 | 33,197 |
| 1860-61 | Bengal | 838,196 | 769,060 | 49,206 | 83,441 | 31,300 | 27,824 | 20,468 | 6,886 | 128,962 | 134,696 | 135,153 | 131,527 | 77,964 | 15,958 | 13,341 | 22,039 |
| | Madras | 54,602 | 38,339 | 31,805 | 25,918 | .. | .. | .. | .. | 99 | 155 | 44,604 | 84,892 | 585 | 1,080 | 33 | 50 |
| | Bombay | 1,796 | 3,476 | .. | .. | .. | .. | .. | .. | 70 | 132 | 31,945 | 49,396 | 4,154 | 6,231 | .. | .. |
| | ALL INDIA | 894,594 | 810,875 | 81,011 | 109,359 | 31,300 | 27,824 | 20,468 | 6,886 | 129,131 | 134,983 | 211,702 | 265,815 | 82,701 | 13,269 | 13,374 | 22,089 |

## FROM INDIA AND EACH PRESIDENCY TO ALL PARTS OF THE WORLD FROM 1850-51 TO 1860-61.

EXPORTED

| Ceylon | | Mauritius and Bourbon | | New South Wales | | Straits Settlements | | Turkey | | West Indies | | Other Parts | | Total Exported to All Parts | |
|---|---|---|---|---|---|---|---|---|---|---|---|---|---|---|---|
| Quantity | Value | Quantity | Value | Quantity | Value | Quantity | Value | Quantity | Value | Quantity | Value | Quantity | Value | Quantity | Value |
| qrs. | £ | qrs. | £ | qrs. | £ | qrs. | £ | qrs. | £ | qrs. | £ | qrs. | £ | qrs. | £ |
| 5,607 | 3,417 | 141,311 | 119,467 | 1,228 | 1,253 | 9,463 | 7,669 | .. | .. | 3,006 | 2,405 | 5,361 | 4,367 | 341,352 | 311,503 |
| 245,545 | 190,648 | 28,210 | 23,017 | 164 | 138 | 1,137 | 947 | .. | .. | 20,037 | 15,890 | 3,631 | 3,023 | 412,463 | 330,534 |
| 18 | 29 | .. | .. | .. | .. | .. | .. | .. | .. | .. | .. | 118 | 192 | 17,757 | 30,401 |
| 251,170 | 194,094 | 169,521 | 142,484 | 1,392 | 1,391 | 10,600 | 8,616 | .. | .. | 23,043 | 18,295 | 9,110 | 7,582 | 771,572 | 672,438 |
| 1,201 | 5,175 | 163,432 | 140,873 | 4,434 | 5,520 | 6,622 | 6,518 | .. | .. | 4,114 | 3,322 | 4,081 | 4,960 | 353,649 | 330,297 |
| 336,008 | 236,449 | 75,914 | 62,926 | 1,636 | 1,496 | 1,353 | 1,218 | .. | .. | 7,223 | 6,028 | 1,370 | 1,178 | 555,048 | 424,371 |
| 13 | 27 | 325 | 638 | .. | .. | 3 | 5 | .. | .. | .. | .. | 781 | 1,125 | 12,858 | 22,363 |
| 337,222 | 241,651 | 239,671 | 204,437 | 6,070 | 7,016 | 7,978 | 7,741 | .. | .. | 11,337 | 9,350 | 6,232 | 7,263 | 921,555 | 777,031 |
| 5,965 | 8,780 | 220,586 | 135,456 | 10,600 | 10,184 | 8,031 | 6,947 | .. | .. | 3,213 | 2,870 | 5,430 | 4,363 | 448,296 | 355,008 |
| 301,251 | 221,923 | 64,101 | 51,146 | 196 | 179 | 2,038 | 1,742 | .. | .. | 7,247 | 6,464 | 1,138 | 733 | 549,420 | 420,744 |
| 8 | 16 | .. | .. | 56 | 112 | .. | .. | .. | .. | .. | .. | 1,020 | 1,484 | 15,837 | 28,718 |
| 307,224 | 225,719 | 284,687 | 166,602 | 10,852 | 10,475 | 10,069 | 8,689 | .. | .. | 10,460 | 9,334 | 7,588 | 6,580 | 1,013,553 | 804,470 |
| 143,939 | 5,588 | 265,267 | 155,649 | 6,142 | 4,543 | 343,376 | 69,882 | .. | .. | 31,820 | 20,347 | 60,707 | 6,852 | 1,931,425 | 757,358 |
| 255,104 | 206,300 | 67,290 | 68,313 | 34 | 54 | 2,523 | 2,354 | .. | .. | 5,370 | 9,004 | 546 | 505 | 511,838 | 472,252 |
| 10 | 16 | 21 | 32 | .. | .. | .. | .. | .. | .. | .. | .. | 442 | 552 | 21,136 | 31,896 |
| 399,053 | 211,904 | 332,578 | 223,994 | 6,176 | 4,597 | 345,899 | 72,236 | .. | .. | 37,190 | 29,351 | 61,695 | 7,909 | 2,464,399 | 1,261,505 |
| 582,497 | 18,911 | 4,304,316 | 159,883 | 236,562 | 8,557 | 2,245,595 | 105,518 | .. | .. | 834,332 | 30,630 | 673 | 482 | 26,431,809 | 1,143,870 |
| 248,361 | 226,582 | 29,260 | 30,241 | 354 | 593 | 261 | 259 | .. | .. | 405 | 680 | 507 | 803 | 389,460 | 385,758 |
| 48 | 74 | .. | .. | .. | .. | 1 | 1 | .. | .. | .. | .. | 622 | 698 | 39,789 | 32,692 |
| 830,906 | 245,567 | 4,333,576 | 190,124 | 236,916 | 9,150 | 2,245,857 | 105,778 | .. | .. | 834,737 | 31,310 | 1,802 | 1,983 | 26,861,058 | 1,562,320 |
| 819,230 | 29,585 | 5,170,340 | 213,895 | 642,756 | 23,166 | 1,067,117 | 74,566 | .. | .. | 786,500 | 28,398 | 338,049 | 11,175 | 31,442,792 | 2,081,348 |
| 289,640 | 240,060 | 17,355 | 17,759 | .. | .. | 914 | 942 | .. | .. | 836 | 700 | 804 | 835 | 543,767 | 478,704 |
| .. | .. | 38 | 57 | .. | .. | .. | .. | .. | .. | .. | .. | 853 | 62 | 27,661 | 38,694 |
| 1,108,870 | 269,645 | 5,187,733 | 231,711 | 642,756 | 23,166 | 1,068,031 | 75,508 | .. | .. | 787,336 | 29,098 | 339,706 | 12,072 | 32,014,220 | 2,598,746 |
| 770,936 | 30,729 | 6,667,617 | 303,838 | 785,872 | 69,432 | 1,207,292 | 91,019 | .. | .. | 1,633,948 | 71,104 | 554,957 | 15,964 | 27,773,083 | 1,797,600 |
| 295,585 | 253,053 | 15,299 | 14,718 | .. | .. | 316 | 288 | 30,641 | 38,544 | 1,542 | 940 | 182 | 66 | 505,706 | 464,512 |
| .. | .. | 52 | 80 | .. | .. | .. | .. | .. | .. | .. | .. | 840 | 1,089 | 26,923 | 39,071 |
| 1,066,521 | 283,782 | 5,682,968 | 318,636 | 785,872 | 69,432 | 1,207,608 | 91,307 | 30,641 | 38,544 | 1,635,490 | 72,044 | 555,979 | 17,119 | 28,305,712 | 2,301,183 |
| 446,776 | 27,459 | 5,412,321 | 410,699 | 608,123 | 51,203 | 1,055,395 | 232,969 | .. | .. | 552,394 | 38,853 | 114,617 | 14,492 | 20,624,689 | 2,903,143 |
| 288,468 | 243,995 | 38,381 | 37,376 | 180 | 187 | 4,926 | 5,018 | 22,175 | 33,708 | 3,277 | 2,997 | 2,110 | 114 | 549,696 | 515,438 |
| 289 | 438 | 215 | 281 | .. | .. | 85 | 129 | .. | .. | .. | .. | 1,274 | 1,480 | 25,255 | 30,592 |
| 735,533 | 271,892 | 5,450,917 | 448,356 | 608,303 | 51,390 | 1,060,406 | 138,116 | 22,175 | 33,708 | 555,671 | 41,850 | 118,001 | 15,086 | 21,199,640 | 3,449,173 |
| 22,428 | 164,956 | 320,298 | 507,645 | 96,935 | 158,480 | 290,792 | 260,906 | .. | .. | 32,940 | 46,300 | 21,008 | 17,303 | 1,613,675 | 2,026,984 |
| 236,423 | 229,119 | 34,750 | 34,863 | 23 | 42 | 1,116 | 1,170 | 6,884 | 12,395 | 1,230 | 1,875 | 107 | 177 | 374,948 | 382,578 |
| 64 | 97 | 1,164 | 1,764 | .. | .. | .. | .. | .. | .. | .. | .. | 1,079 | 1,244 | 16,044 | 23,584 |
| 258,915 | 394,172 | 356,212 | 544,272 | 96,958 | 158,522 | 291,908 | 262,076 | 6,884 | 12,395 | 34,170 | 48,175 | 22,194 | 18,724 | 2,004,667 | 2,433,146 |
| 95,959 | 66,035 | 376,044 | 511,526 | 84,758 | 139,912 | 248,667 | 112,650 | .. | .. | 38,341 | 135,022 | 184,845 | 35,309 | 1,783,025 | 1,815,134 |
| 287,440 | 258,186 | 33,287 | 38,355 | 3 | 5 | 489 | 497 | 7,628 | 14,912 | 3,974 | 4,405 | 7 | 11 | 437,099 | 440,825 |
| .. | .. | 77 | 117 | .. | .. | .. | .. | .. | .. | .. | .. | 623 | 1,777 | 13,740 | 20,336 |
| 383,399 | 324,221 | 409,408 | 549,998 | 84,761 | 139,917 | 249,156 | 113,147 | 7,628 | 14,912 | 42,315 | 139,427 | 185,475 | 37,097 | 2,233,864 | 2,276,295 |
| 290,906 | 123,250 | 420,080 | 483,175 | 60,941 | 103,432 | 460,673 | 63,903 | .. | .. | 29,197 | 37,741 | 281,695 | 27,191 | 2,838,082 | 2,030,123 |
| 340,598 | 314,031 | 83,457 | 89,274 | .. | .. | 306 | 312 | 11,966 | 22,018 | 4,280 | 5,812 | 271 | 446 | 572,604 | 582,327 |
| 15 | 23 | 5 | 10 | .. | .. | 2 | 4 | .. | .. | .. | .. | 1,041 | 1,618 | 39,028 | 60,890 |
| 631,519 | 437,304 | 503,542 | 572,459 | 60,941 | 103,432 | 460,981 | 64,219 | 11,966 | 22,018 | 33,477 | 43,553 | 283,007 | 29,255 | 3,449,714 | 2,673,340 |

E 2

1089. [77] Rice (*Oryza sativa*), Pegu.
INDIA MUSEUM.

COMPOSITION.

|  | Per cent. |
|---|---|
| Moisture . . . . . | 13·50 |
| Nitrogenous matter . . . | 7·41 |
| Starchy matter . . . | 78·10 |
| Fatty or oily matter . . . | 0·40 |
| Mineral constituents (ash) . . | 0·59 |
| TOTAL | 100·00 |

1090. [78] Rice, Bombay Bazaar. Do.

COMPOSITION.

|  | Per cent. |
|---|---|
| Moisture . . . . . | 13·00 |
| Nitrogenous matter . . . | 7·44 |
| Starch matter . . . . | 77·63 |
| Fatty or oily matter . . . | 0·70 |
| Mineral constituents (ash) . . . | 1·23 |
| TOTAL | 100·00 |

1091. [79] Rice, Broach, Bombay. Do.

COMPOSITION.

|  | Per cent. |
|---|---|
| Moisture . . . . . | 13·10 |
| Nitrogenous matter . . . | 7·12 |
| Starchy matter . . . . | 78·70 |
| Fatty or oily matter . . . | 0·49 |
| Mineral constituents (ash) . . . | 0·66 |
| TOTAL | 100·00 |

1092. [81] Rice, Bareilly. Do.

COMPOSITION.

|  | Per cent. |
|---|---|
| Moisture . . . . . | 12·80 |
| Nitrogenous matter . . . . | 8·24 |
| Starchy matter . . . . | 77·80 |
| Fatty or oily matter . . . . | 0·64 |
| Mineral constituents (ash) . . | 0·52 |
| TOTAL | 100·00 |

1093. [80] Pulut rice, Malacca. Do.

COMPOSITION.

|  | Per cent. |
|---|---|
| Moisture . . . . . | 12·90 |
| Nitrogenous matter . . . | 7·24 |
| Starchy matter . . . . | 78·56 |
| Fatty or oily matter . . . | 0·60 |
| Mineral constituents (ash) . . | 0·70 |
| TOTAL | 100·00 |

1094. [82] Rice, Arracan. Do.

1095. [83] Do., Bengal. Do.

From Arracan, which exports rice to a larger extent than any other district, the undermentioned samples and information are forwarded by the Local Committee at *Akyab.*

1096. [9253] Lak-taw-ree-tha-jung-thau, Akyab.

1097. [9254] Gua-kreen-thee, Akyab.

1098. [9255] Loong-phroo, Akyab.

1099. [9256] Lak-roong, Akyab.

1100. [9257] Toung-phroo, Akyab.

1101. [9258] Byah, Akyab.

Nos. 9253—9258 are exported: No. 9254 in very large quantities; Nos. 9255, 9257, and 9258 are varieties of No. 9254, 'Gua-kreen-thee.' It is not possible to state the quantity exported of each. Previous to 1845–46, Nos. 9253 and 9256, Lak-taw-ree. and Lak-roong, were more extensively grown than Gua-kreen-thee, as they were preferred in China, in the Straits, and Coast markets, but Gua-kreen-thee being preferred in the European markets, and their yield per acre being greater, it has almost entirely supplanted the other kinds. The quantity produced may be estimated at, 'Gua-kreen-thee,' and varieties, 200,000 tons, and Lak-roong and Lak-taw-ree at 20,000 tons. Last year 125,000 tons rice were exported from Arracan, of which not more than about 5,000 tons were Lak-roong and Lak-taw-ree. The average annual export of rice from Arracan, during the last eight years, has been, to Europe 112,000 tons, to the East and Indian ports about 4,000 tons. The wholesale price of rice varies considerably, according with the demand : formerly it might be purchased at from 2*l.* 10*s.* to 3*l.* 10*s.* the ton. The average rate, for the last ten years, may be set down at 5*l.* per ton. In consequence of the rise in the price, China and other Eastern countries have been nearly driven out of the market.

With regard to the desirability of endeavouring to cause an extension of the cultivation of any particular kind of rice, the natives will readily see which gives the best return, and there being a very large extent of waste land in this province, were it desirable, with increase of population, the cultivation of rice might be increased ten-fold; but taking the population into consideration, the export of rice from Arracan is very great, owing, no doubt, to the excellent water-communication throughout the province: could, however, the inhabitants be induced to cultivate other articles to a greater extent than is now done, such as cotton, jute, tea, &c., it would be very desirable.

Most of the information contained in the above has been furnished by J. BULLOCK, Esq., of the firm of Messrs. Halliday, Bullock, & Co., of Akyab, Member of the Akyab Committee.

The specimens of rice submitted are of last year's produce, and are consequently not of as good colour as if they were of this year's produce.*

1102. [5732] Rice, Kurak-thor, Akyab.

1103. [9259] Paddy (rice in the husk), Akyab.

1104. [5733] Gua-mounk-way, Akyab.

1105. [5736] Lak-taw-ree-tha-jung-thau, Akyab.

1106. [5735] Gua-kreen-thee, Akyab.

1107. [5737] Loong-phroo, Akyab.

1108. [9260] Lak-roon, Akyab.

1109. [5734] Toung-phroo, Akyab.

* Local Committee, *Akyab.*

1110. [9261] Byab, Akyab.

**Paddy** (*Rice in Husk*) *from* **Moulmein.**

1111. [6292] Paddy, Ahphet or Koukgyee, Moulmein.

This grain is cultivated in June and reaped in December. This is the daily food of the people.

1112. [6286] Paddy, Shangalay, Moulmein.

1113. [6285] Do., Konk Yen, Do.

The sowing of rice takes place in June, and it is reaped in October; it is very quick in its growth, and requires little culture.

1114. [6306] Paddy, Konk Mhoag, Moulmein.

1115. [6297] Do., Nhat Tsa Bah, Do.

1116. [6287] Do., Yahine, Do.

1117. [6289] Do., Wetssee, Do.

1118. [6288] Do., Thootpaya, Do.

1119. [6296] Do., Than Bah, Do.

1120. [6284] Do., Konk Mhoag, Do.

1121. [6280] Do., Khakabong, Do.

1122. [9262] Do., Tsankouk Ngen, Do.

1123. [6278] Do., Konk Ya, Do.

1124. [6305] Do., Na Kheit, Do.

1125. [6281] Do., Toung Byan, Do.

1126. [6279] Do., Tsa Tha, Do.

1127. [9263] Do., Ain Thee, Do.

1128. [6291] Do., Yui Gnay, Do.

1129. [6293] Do., Toung Aw, Do.

1130. [6303] Do., Tsin Soay, Do.

1131. [9264] Do., Nga Tin Thoay, Do.

1132. [6302] Do., Kyet Thai, Do.

1133. [9265] Do., Thakapheit, Do.

1134. [6300] Do., Tsin Thee, Do.

1135. [6298] Do., Gkakha Yine, Do.

1136. [6282] Do., Nnagkheit Nhet, Do.

1137. [6299] Do., E—thine, Do.

1138. [6283] Do., Thetkouppan, Do.

1139. [6290] Do., Gkalaon, Do.

1140. [9266] Do., Myai-thuay, Do.

All the varieties of rice in these provinces are sown in May or June at the setting in of the south-west monsoon; some are cultivated in high land, and others in low land.

A few samples showing the extremes of quality of the rices produced in the Cuttack district have been forwarded, accompanied with the remarks by the Local Committee here appended: —

Rice is the staple of this division. It is used for food for man, beast, and bird; for the manufacture of starch, the distillation of spirits, &c. Its varieties are as numerous as its uses. There are in this province three distinct crops; the first, grown on somewhat high ground, is the early crop, sown for the most part in June, and reaped in August and September. The second is the main crop, sown in June and July, and cut from November to January. It requires a great deal of moisture, some varieties growing in several feet of water. The third is a dwarf crop, cultivated in the months of March, April, and May, on low-lying land, generally on the sides of marshes and pools, where irrigation is easy. The ratio of productiveness is said to be, in a good season, as 1 to 35. The market value of this grain varies so much according to locality and season, as is indeed the case with all other raw products of this part of the country, that it would only tend to mislead to give any rates purporting to be of general application.

1141. [5738] Rice in the husk, Meeshay, Rangoon.

1142. [5746] Do. Natsieng, Do.

1143. [5749] Cargo rice husked by the Oriental Rice Company's steam mill, Do.

1144. [5745] Cleaned rice, Natsieng, dressed by the Burmese, Do.

1145. [5750] Do. dressed by the Oriental Rice Company's mill, Do.

1146. [5747] Rice in the husk, Meedo, Do.

1147. [5748] Cargo rice husked by the Oriental Rice Company's mill, Do.

1148. [6275] Cleaned rice dressed by the Oriental Rice Company's mill, Do.

1149. [5741] Rice in the husk, Beeat, Do.

1150. [6276] Do. black rice, Do.

1151. [5744] Do. 'winged' rice, Do.

1152. [5742] Cleaned rice, red Koungnyeen (*Oryza glutinosa*), Do.

1153. [5743] Rice in the husk, white Koungnyeen (*O. glutinosa*), Do.

1154. [5740] Cleaned rice, white Koungnyeen, Do.

The Burmese recognise nearly a hundred varieties of rice, but the principal distinctions between the different kinds are as follows:—Hard grain, soft grain, glutinous rice.

The 'Natsieng' is the hardest grain, and is the rice which is accordingly principally exported to Europe.

The 'Meedo' is the chief of the soft grain varieties.

It is much preferred by the Burmese to the hard-grained sorts, and it is certainly superior in taste when cooked; but the hard-grained rice is chiefly purchased by the merchants for export, as it keeps better, and the soft-grained rice is too much broken by European machinery in cleaning. Latterly, on the Continent, this last objection appears to have been overcome, and a greater demand is consequently springing up for the 'Meedo' rice for the markets of foreign Europe.

The 'Koungnyeen,' or hill rice, is called 'glutinous' rice by Europeans, from the property it possesses, when cooked, of the grains all adhering in a thick glutinous mass. It is the chief article of food with the Karens and other hill tribes, but is not much eaten by the inhabitants of the low swampy plains, where the common rice is grown.

Nos. 5749, 5750, 5748, and 6275, have been husked and cleaned respectively by the steam mills erected at Rangoon by the Oriental Rice Company.

Price of rice in the husk 50 rupees per 100 baskets of 52 lbs.

Cargo rice 95 rupees per 100 baskets of 63 lbs.

Cleaned rice 150 rupees per 100 baskets of 70 lbs.

Of the special collection furnished by Messrs. HALLIDAY, FOX, & Co., the under-mentioned samples are exhibited:—

1155. [4210] Baw-gyn.

1156. [4211] Shwai thway, Burmah.

1157. [4213] Tha-za, Do.

1158. [4214] Cargo thang-bong, Do.

1159. [4217] Na-pangyu-cow nyeing, Do.

1160. [4219] Tha-za, Do.

1161. [4231] Mian ma kla dow, Do.

1162. [4233] Kya-nee-young, Do.

1163. [4234] Cow nyien ka nghet (glutinous rice), Do.

1164. [4247] Tann dwing, Do.

1165. [4819] Necranzie, Do.

1166. [4824] Byack, Arracan.

1167. [4825] Do. Do.

LIEUT. W. PHAIRE, Deputy-Commissioner, Assam, sends a large collection, of which the subjoined are a few representatives:—

1168. [6277] Rice, unshelled, Durrung, Assam.

1169. [6758] Do. shelled, Hukcoahbac, Do.

1170. [6796] Do. unshelled (Paddy), Kuchareehalle, Do.

1171. [6797] Do. shelled, do. Do.

1172. [6775] Do. unshelled, Bugaguha, Do.

1173. [6776] Do. shelled, do. Do.

1174. [6781] Rice, unshelled, Homzul, Assam.

1175. [6782] Do. shelled, do. Do.

1176. [6794] Do. unshelled, Hurra Pooa, Do.

1177. [6795] Do. shelled, do. Do.

1178. [6770] Ditto, unshelled, Burhooagmony, Do.

1179. [6771] Ditto, unshelled, Dulkoosee, Do.

1180. [6772] Do. shelled, do. Do.

1181. [6752] Do. unshelled, Maneekeemadoovy, Do.

1182. [6751] Do. shelled, do. Do.

Of the specimens of rice forwarded by the Central Committee, Allahabad, the following are exhibited:—

1183. [6856] Rice, Bansmuttee, Philibheet District, Rohilcund.

1184. [6850] Do. Dhanee, Do.

1185. [6853] Do. Roymoonia, Do.

1186. [6857] Do. Gantanee, Do.

1187. [6868] Do. Hirrunj, Do.

1188. [6861] Do. Rutnee, Do.

1189. [6848] Do. Surhee, Do.

1190. [6862] Do. Hunsa, Do.

1191. [6867] Do. Hunsraj, Do.

1192. [6859] Do. Bunkee, Do.

1193. [6866] Do. Jhilma, Do.

1194. [6872] Do. Sammaloo, Do.

The collection forwarded from Lucknow comprises the following:—

1195. [6817] Dhan (*Oryza sativa*), Lucknow, Oude.

Eighteen seers per rupee. Grows all over Oude extensively. This is considered as one of the best sorts of rice, and is produced from the *bateesa* paddy.

1196. [6818] Dhan (*O. sativa*), Lucknow, Oude.

Twenty seers per rupee. Grows all over Oude. This is considered a superior kind, and produces a white rice called *bateesa.*

1197. [9267] Dhan (lamba) (*O. sativa*), Lucknow, Oude.

Sixteen seers per rupee. Grows all over Oude extensively. This is also a superior kind, and is called *lamba chawl.*

1198. [6819] Kala Dhan (*Oryza sativa*), Lucknow, Oude.

Twenty-five seers per rupee. Grows all over Oude. This is a specimen of brown rice produced from the black paddy; it is an inferior kind, and used by the poorest people; it is called *bagree chawl.**

1199. [6816] Dhan (lamba) (*O. sativa*), Lucknow, Oude.

Twenty seers per rupee. Grows all over Oude extensively. This is considered one of the good kinds of paddy, and produces the long white rice.

1200. [6820] Dhan (*O. sativa*), Lucknow, Oude.

Twenty-eight seers per rupee. Grows all over Oude. This is an inferior kind of paddy, and produces, when husked, a reddish rice; the natives call this kind *bagree*.
Very many varieties of rice are grown in Oude. A heavy soil and plenty of water suits them best. There are five kinds which are considered among the best; 'Mihee' and 'Bansee' are foremost. The peculiarity in the cultivation of these two kinds is, that they are transplanted and placed about 5 inches apart. And, by this method, if the soil is good, they grow to the height of an ordinary sized man, and produce a much larger quantity than if otherwise treated. The odour and flavour of these two kinds, when cooked, are superior to those of any other kind. They are only used by those who can afford to buy them.
As the labour in cultivating them makes them dearer than the other sorts, the three other varieties which are considered good are the 'Bateesa,' the '——' and the 'Phool-Biring.' They are sown broad-cast in June, and left so, and they are the kinds mostly used by natives. The first two mentioned, when new, sell for 10 or 12 seers per rupee, and become dearer according as they become older. The other three kinds sell for about 19 seers per rupee, and are dearer if older. Some consider 'Phool-Biring' the best, as it swells in boiling, and has an agreeable odour.**

The collection of rices from Chota Nagpore, as forwarded by the Local Committee there, is partly represented by the following:—

1201. [6883] Rice, Pershad Chogdan, first sort, Chota Nagpore.

1202. [6884] Do. Shamzeerer Dhan, Do.

1203. [6885] Do. Rajnath Dhan, Do.

1204. [6886] Do. Seetul Cheenie Dhan, Do.

1205. [6887] Do. Sikhee Dhan, Do.

1206. [6888] Do. Ramghurria Dhan, Do.

1207. [6889] Do. Chundunphul Dhan, Do.

1208. [6890] Do. Kussoor Sal Dhan, Do.

1209. [6891] Do. Gehu Hurree, Do.

1210. [6892] Do. Raichoonee Doshawdar Dhan, Do.

1211. [6893] Rice, Siree Kumul, Chota Nagpore.

1212. [6894] Do. Chundunphul Dhan, Do.

1213. [6895] Do. Bansmuttee Dhan, Do.

From Madras the following examples are exhibited:—

1214. [4045] Rice (*Oryza sativa*), Seeroomanen aresee, Madras.

1215. [4046] Do. Hoonda sumba aresee, Do.

1216. [4048] Do. Coodum sumba aresee, Do.

1217. [4048] Do. Paswell, Do.

1218. [4054] Do. Seeraga sumba, Do.

1219. [4056] Do. Peroo car aresee, Do.

1220. [4057] Do. Segapoo aresee, Do.

1221. [4061] Do. Nelloo var. Monakata, Do.

1222. [4248] Do. Medong Cargo, Do.

1223. [4979] Do. Red aresee, Do.

1224. [4982] Do. Majay aresee, Do.

1225. [4984] Do. Carah aresee, Do.

1226. [4985] Do. Piegoh aresee, Do.

1227 [4986] Do. Baulah aresee, Do. DR. HUNTER.

1228. [4988] Do. Pootoo aresee, Do. Do.

1229. [4989] Do. Pallow aresee, Do. Do.

1230. [4990] Do. Curpoo pootoo aresee, Do. Do.

1231. [4991] Do. Car aresee, Do. Do.

1232. [4992] Do. Chodum Chumbah, Do. Do.

1233. [4994] Do. Kaddy Khythan, Do. Do.

1234. [4999] Do. Palin Chumbah, Do. Do.

Rice is not extensively cultivated in the Bombay Presidency. The following are examples from Poonah and Ahmedabad.

1235. [7410] Rice (*Oryza sativa*), Chowl Dakosal, Poonah.

1236. [7411] Do. Chowl Bhasud, Do.

1237. [7412] Do. Chowl Dhowlasal, Do.

1238. [7413] Rice, Chowl Umbemoher, Poonah.

1239. [7415] Do. Bhat Danger allahurry, Ahmedabad.

1240. [7416] Do. Bhat Danger satee, Do.

1241. [7417] Do. Bhat Camood, Do.

1242. [7418] Do. Bhat Danger pakalee, Do.

1243. [7419] Do. Bhat Danger satee sat, Do.

## PULSES.

Pulses occupy an important position in the food vocabulary of the people of India. They are eaten with, and supply to rice, and some other cereals, the nitrogenous or 'flesh-forming' material in which these are defective.

Of the PEA TRIBE, Gram (*Cicer arietinum*), or chick pea, occupies an important position. It is largely used by the people, and constitutes, besides, the great horse-food of Northern and Western India. It can be used for this purpose for a length of time without causing 'heating,' or the other deleterious effects, ordinarily produced by the too exclusive employment of peas and beans in this country.

A comparison of the various analyses which follow will show that the proportion of nitrogenous or 'flesh-forming' matter in *gram* is, with one exception, *less* than in any of the other pulses enumerated : —

1244. [84] Gram (*Cicer arietinum*), Bengal. INDIA MUSEUM.

COMPOSITION.

| | Per cent. |
|---|---|
| Moisture . . . . . | 10·80 |
| Nitrogenous matter . . . | 19·32 |
| Starchy matter . . . . | 62·20 |
| Fatty or oily matter . . . | 4·56 |
| Mineral constituents (ash) . . | 3·12 |
| TOTAL | 100·00 |

1245. [85] Gram, Bombay Bazaar. Do.

COMPOSITION.

| | Per cent. |
|---|---|
| Moisture . . . . | 12·24 |
| Nitrogenous matter . . . | 18·05 |
| Starchy matter . . . . | 61·70 |
| Fatty or oily matter . . . | 4·95 |
| Mineral constituents (ash) . . | 3·05 |
| TOTAL | 100·00 |

1246. [86] Gram, Bombay. INDIA MUSEUM.

COMPOSITION.

| | Per cent. |
|---|---|
| Moisture . . . . . | 10·86 |
| Nitrogenous matter . . . | 21·17 |
| Starchy matter . . . . | 60·11 |
| Fatty or oily matter . . . | 4·47 |
| Mineral constituents (ash) . . | 3·39 |
| TOTAL | 100·00 |

1247. [87] Gram, Bombay Bazaar. Do.

COMPOSITION.

| | Per cent. |
|---|---|
| Moisture . . . . ·. | 9·25 |
| Nitrogenous matter . . . | 20·64 |
| Starchy matter . . . . | 63·62 |
| Fatty or oily matter . . . | 4·11 |
| Mineral constituents (ash) . . | 2·38 |
| TOTAL | 100·00 |

1248. [88] Gram, Bombay. Do.

COMPOSITION.

| | Per cent. |
|---|---|
| Moisture . . . . . | 10·80 |
| Nitrogenous matter . . . | 21·23 |
| Starchy matter . . . . | 60·30 |
| Fatty or oily matter . . . | 4·77 |
| Mineral constituents (ash) . . | 2·90 |
| TOTAL | 100·00 |

1249. [90] Gram (husked), Madras. Do.

COMPOSITION.

| | Per cent. |
|---|---|
| Moisture . . . . | 11·30 |
| Nitrogenous matter . . . | 21·04 |
| Starchy matter . . . . | 60·45 |
| Fatty or oily matter . . . | 4·31 |
| Mineral constituents (ash) . . | 2·90 |
| TOTAL | 100·00 |

1250. [91] Gram (white variety), Saharunpore. Do.

COMPOSITION.

| | Per cent. |
|---|---|
| Moisture . . . . , | 12·20 |
| Nitrogenous matter . . . | 20·13 |
| Starchy matter . . . . | 60·24 |
| Fatty or oily matter . . . | 4·63 |
| Mineral constituents (ash) . . | 2·80 |
| TOTAL | 100·00 |

1251. [6081] Chenna peela (*Cicer arietinum*), Ahmedabad.

1252. [4032] Chick pea (*C. arietinum*), Madras.

1253. [5039] Do.

1254. [4032] Do. (husked), Madras.

1255. [89] Gram, husked, (*C. arietinum*).

1256. [4805] Gram, Burmah. Messrs. HALLIDAY, FOX, & Co.

### PIGEON PEA (*Cajanus indicus*).

This pea is a particular favourite. When husked and split, it constitutes the kind of 'dhol' which, when procurable, most commonly enters with rice into the formation of the vegetable curry of the Hindoo.

The three samples which follow show its composition. The others are a portion of those forwarded for exhibition on the present occasion.

1257. [92] Pigeon pea, Jaffrabad, Bombay. INDIA MUSEUM.

COMPOSITION.

|  | Per cent. |
|---|---|
| Moisture . . . . | 10·77 |
| Nitrogenous matter . . . | 20·19 |
| Starchy matter . . . . | 64·32 |
| Fatty or oily matter . . . | 1·32 |
| Mineral constituents (ash) . . | 3·40 |
| TOTAL | 100·00 |

1258. [93] Pigeon pea, Calcutta Bazaar. Do.

COMPOSITION.

|  | Per cent. |
|---|---|
| Moisture . . . . | 12·80 |
| Nitrogenous matter . . . | 20·38 |
| Starchy matter . . . . | 61·90 |
| Fatty or oily matter . . . | 1·52 |
| Mineral constituents (ash) . . | 3·40 |
| TOTAL | 100·00 |

1259. [96] Pigeon pea, (husked), Broach. Do.

COMPOSITION.

|  | Per cent. |
|---|---|
| Moisture . . . . | 12·30 |
| Nitrogenous matter . . . | 19·83 |
| Starchy matter . . . . | 63·12 |
| Fatty or oily matter . . . | 1·86 |
| Mineral constituents (ash) . . | 2·89 |
| TOTAL | 100·00 |

1260. [4934] Thoraray (*Cajanus indicus*), Madras. DR. HUNTER.

1261. [4927] Do., Madras. DR. HUNTER.

1262. [4929] Mullay thoraray (*C. indicus*), Do.

1263. [4034] Pigeon pea (*C. indicus*), Do.

1264. [7165] Arhar kala, dark variety (*C. indicus*), Lucknow.

1265. [7166] Arhar safaid, white variety (*do.*), Do.

1266. [7182] Parbuttee (*Cajanus indicus*), Beerbhoom.

1267. [7179] Do. Hooghly.

1268. [4034] Pigeon pea, husked, (*C. indicus*), Do.

1269. [97] Pigeon pea, Do.

1270. [117] Do.

### COMMON PEA (*Pisum sativum*).

1271. [94] Pea (*P. sativum*), Benares. INDIA MUSEUM.

COMPOSITION.

|  | Per cent. |
|---|---|
| Moisture . . . . | 12·70 |
| Nitrogenous matter . . . | 25·20 |
| Starchy matter . . . . | 58·38 |
| Fatty or oily matter . . . | 1·10 |
| Mineral constituents (ash) . . | 2·53 |
| TOTAL | 100·00 |

1272. [98] Pea, Benares. Do.

COMPOSITION.

|  | Per cent. |
|---|---|
| Moisture . . . . | 12·60 |
| Nitrogenous matter . . . | 21·80 |
| Starchy matter . . . . | 62·19 |
| Fatty or oily matter . . . | 1·12 |
| Mineral constituents (ash) . . | 2·29 |
| TOTAL | 10·000 |

1273. [95] Pea.

1274. [4933] Buttanee (*P. sativum*), Madras. DR. HUNTER.

1275. [4040] Do. Madras.

1276. [4764] Do. Burmah. Messrs. HALLIDAY, FOX, & Co.

1277. [4817] Paisailong (*P. sativum*), Do. Do.

1278. [2819]

### LENTILS (*Ervum lens*).

Cultivated in many parts, but not generally held in high repute.

1279. [101] Lentils (*E. lens*), Calcutta Bazaar. INDIA MUSEUM.

COMPOSITION.

|  | Per cent. |
|---|---|
| Moisture . . . . | 12·70 |
| Nitrogenous matter . . . | 24·57 |
| Starchy matter . . . . | 59·43 |
| Fatty or oily matter . . . | 1·01 |
| Mineral constituents (ash) . . | 2·29 |
| TOTAL | 100·00 |

1280. [102] Lentils (husked), Bombay Bazaar. INDIA MUSEUM.

COMPOSITION.

|  | Per cent. |
|---|---|
| Moisture | 12·50 |
| Nitrogenous matter | 24·65 |
| Starchy matter | 59·34 |
| Fatty or oily matter | 1·14 |
| Mineral constituents (ash) | 2·37 |
| TOTAL | 100·00 |

1281. [103] Lentils (whole), Calcutta Bazaar. Do.

COMPOSITION.

|  | Per cent. |
|---|---|
| Moisture | 11·40 |
| Nitrogenous matter | 26·18 |
| Starchy matter | 59·43 |
| Fatty or oily matter | 1·00 |
| Mineral constituents (ash) | 1·99 |
| TOTAL | 100·00 |

1282. [104] Lentils, Bombay Bazaar. Do.

COMPOSITION.

|  | Per cent. |
|---|---|
| Moisture | 10·72 |
| Nitrogenous matter | 25·20 |
| Starchy matter | 59·96 |
| Fatty or oily matter | 1·92 |
| Mineral constituents (ash) | 2·20 |
| TOTAL | 100·00 |

1283. [4753] Lentils, Burmah. Messrs. HALLIDAY, FOX, & Co.

1284. [4031] Do. (split), Madras.

1285. [4939] Lentils, Madras. DOCTOR HUNTER.

## VETCH (*Lathyrus sativus*).

This pulse is cultivated in many parts of the country; it is not, however, considered a particularly wholesome article of food, for either man or beast. It is too rich in nitrogenous matter, and therefore requires to be largely *diluted*.

1286. [105] Vetch (*L. sativus*), Calcutta Bazaar. INDIA MUSEUM.

COMPOSITION.

|  | Per cent. |
|---|---|
| Moisture | 10·10 |
| Nitrogenous matter | 31·50 |
| Starchy matter | 54·26 |
| Fatty or oily matter | 0·95 |
| Mineral constituents (ash) | 3·19 |
| TOTAL | 100·00 |

1287. [6597] Khesaree (*L. sativus*), Cuttack.

## BEAN TRIBE.

Beans are largely cultivated and employed similarly to the foregoing. Of the *Dolichos* species, the two first enumerated are those chiefly used as articles of human food. Of the first mentioned, the *Dolichos Lablab*, there are a number of varieties, all of them favourites.

1288. [133] Whal (*Lablab vulgaris*), Bombay Bazaar. INDIA MUSEUM.

COMPOSITION.

|  | Per cent. |
|---|---|
| Moisture | 10·81 |
| Nitrogenous matter | 24·55 |
| Starchy matter | 60·81 |
| Fatty or oily matter | 0·81 |
| Mineral constituents (ash) | 3·02 |
| TOTAL | 100·00 |

1289. [136] Ghot wall (*L. vulgaris* var.), Bombay Bazaar. Do.

COMPOSITION.

|  | Per cent. |
|---|---|
| Moisture | 12·02 |
| Nitrogenous matter | 22·45 |
| Starchy matter | 60·52 |
| Fatty or oily matter | 2·15 |
| Mineral constituents (ash) | 2·86 |
| TOTAL | 100·00 |

1290. [4926] Velley Mochay (*L. vulgaris*), Cottay, Madras. DR. HUNTER.

1291. [4928] Segapoo (*L. vulgaris*), Madras. DR. HUNTER.

1292. [4800] Do. Burmah. Messrs. HALLIDAY, FOX, & Co.

## CHOWLEE (*Dolichos sinensis*).

Extensively cultivated. There are three varieties, white, brown, and black.

1293. [132] Chowlee (*D. sinensis*), Bombay Bazaar. INDIA MUSEUM.

COMPOSITION.

|  | Per cent. |
|---|---|
| Moisture | 12·44 |
| Nitrogenous matter | 24·00 |
| Starchy matter | 59·02 |
| Fatty or oily matter | 1·41 |
| Mineral constituents (ash) | 3·13 |
| TOTAL | 100·00 |

1294. [6079] Chowlee, Ahmedabad.

1295. [6158] Do. Poona.

1296. [135] Katjang pootayam (*D. sinensis*), Malacca.

**COOLTEE,** 'Madras Horse Gram,' (*Dolichos uniflorus*).

Chiefly cultivated in the Madras territory. Steeped and used for horses; occasionally eaten by man.

1297. [122] Cooltee (*D. uniflorus*), Bombay Bazaar. INDIA MUSEUM.

COMPOSITION.

| | Per cent. |
|---|---|
| Moisture | 11·30 |
| Nitrogenous matter | 23·47 |
| Starchy matter | 61·02 |
| Fatty or oily matter | 0·87 |
| Mineral constituents (ash) | 3·34 |
| TOTAL | 100·00 |

1298. [129] Cooltee, Bellary, Madras. Do.

COMPOSITION.

| | Per cent. |
|---|---|
| Moisture | 11·50 |
| Nitrogenous matter | 23·03 |
| Starchy matter | 61·85 |
| Fatty or oily matter | 0·76 |
| Mineral constituents (ash) | 2·86 |
| TOTAL | 100·00 |

1299. [138] Cooltee (*D. uniflorus*).

1300. [4938] Caroopoo colloo (*D. uniflorus*), Madras. DR. HUNTER.

1301. [4035] 'Horse gram' (*D. uniflorus*). Madras.

1302. [128] Do.

1303. [4696] Colloo (*D. uniflorus*), Madras. DR. HUNTER.

**GUWAR** (*Dolichos fabæformis*).

A hard refractory bean, cultivated in the Deccan, steeped, and then almost entirely used for animals.

1304. [139] Guwar (*D. fabæformis*), Poona. INDIA MUSEUM.

COMPOSITION.

| | Per cent. |
|---|---|
| Moisture | 11·75 |
| Nitrogenous matter | 29·80 |
| Starchy matter | 53·89 |
| Fatty or oily matter | 1·40 |
| Mineral constituents (ash) | 3·16 |
| TOTAL | 100·00 |

1305. [4044] Muthee puttay (*Cyamopsis psoralioides*), Madras.

1306. [6078] Guour, Ahmedabad.

**BHOOT,** *Soja hispida*, (*Dolichos Soja*).

Cultivated in many parts to the north of India. This is the same as the well-known Chinese bean, which constitutes such a large article of trade between the northern and southern ports of China. Of all vegetable substances, it is richer in nitrogenous or 'flesh-forming' matter than any yet discovered.

1307. [99] Salmca bean, white (*Soja hispada*). INDIA MUSEUM.

COMPOSITION.

| | Per cent. |
|---|---|
| Moisture | 8·12 |
| Nitrogenous matter | 40·63 |
| Starchy matter | 29·54 |
| Fatty or oily matter | 17·71 |
| Mineral constituents (ash) | 4·00 |
| TOTAL | 100·00 |

1308. [100] Salmca bean, white. Do.

COMPOSITION.

| | Per cent. |
|---|---|
| Moisture | 7·96 |
| Nitrogenous matter | 37·74 |
| Starchy matter | 31·08 |
| Fatty or oily matter | 18·90 |
| Mineral constituents (ash) | 4·32 |
| TOTAL | 100·00 |

1309. [130] Salmca bean (black), Sumatra. Do.

COMPOSITION.

| | Per cent. |
|---|---|
| Moisture | 10·40 |
| Nitrogenous matter | 41·54 |
| Starchy matter | 30·82 |
| Fatty or oily matter | 12·31 |
| Mineral constituents (ash) | 4·93 |
| TOTAL | 100·00 |

Of the **PHASEOLI,** the *P. radiatus* is one of the principal favourites, although it is doubtful if any of these are in such general repute as some of the preceding.

1310. [118] *Phaseolus radiatus*, Bombay Bazaar. INDIA MUSEUM.

COMPOSITION.

| | Per cent. |
|---|---|
| Moisture | 11·00 |
| Nitrogenous matter | 22·48 |
| Starchy matter | 62·15 |
| Fatty or oily matter | 1·46 |
| Mineral constituents (ash) | 2·91 |
| TOTAL | 100·00 |

1311. [4937] Patchay pyaroo (*P. radiatus*), Madras. DR. HUNTER.

1312. [4936] Do. Do.

1313. [4964] Thoolokah pyaroo (*Phaseolus radiatus*), Chingleput. DR. SHORTT.

1314. [4816] Paidee (*P. radiatus*), Burmah. Messrs. HALLIDAY, FOX, & Co.

1315. [116] Do.

1316. [120] Do.

1317. [123] Do.

1318. [7184] Oord (*P. radiatus*), Lucknow.

1319. [4043] Do. (husked), Madras.

1320. [125] Green gram (*Phaseolus mungo*), Bombay Bazaar. INDIA MUSEUM.

COMPOSITION.

| | Per cent. |
|---|---|
| Moisture . . . . . | 9·20 |
| Nitrogenous matter . . . | 24·70 |
| Starchy matter . . . . | 60·36 |
| Fatty or oily matter . . . | 1·48 |
| Mineral constituents (ash) . . | 3·26 |
| TOTAL | 100·00 |

1321. [126] Green gram (husked as 'dholl'), Bellary, Madras. Do.

COMPOSITION.

| | Per cent. |
|---|---|
| Moisture . . . . . | 12·90 |
| Nitrogenous matter . . . | 23·54 |
| Starchy matter . . . . | 59·38 |
| Fatty or oily matter . . . | 1·11 |
| Mineral constituents (ash) . . | 3·07 |
| TOTAL | 100·00 |

1322. [4037] Black gram (*P. mungo*), Madras.

1323. [4935] Oolendoo (*P. mungo*), Madras. DR. HUNTER.

1324. [4033] Do. Madras.

1325. [4802] Painaut (*P. mungo*), Burmah. Messrs. HALLIDAY, FOX, & Co.

1326. [4806] Paikinant (*P. mungo*), Burmah. Do.

1327. [4033] Moong, husked (*P. mungo*), Madras.

1328. [124] *Phaseolus aconitifolius,* Calcutta Bazaar.

COMPOSITION.

| | Per cent. |
|---|---|
| Moisture . . . . | 11·22 |
| Nitrogenous matter . . . | 23·80 |
| Starchy matter . . . . | 60·78 |
| Fatty or oily matter . . . | 0·64 |
| Mineral constituents (ash) . . | 3·56 |
| TOTAL | 100·00 |

It is extensively cultivated in Oude. When split, it forms one of the '*Dâls*,' and ground into flour is used for bread by natives. It is also sometimes used mixed up with wheaten flour. Bullocks, sheep, goats, and many of the native cavalry horses are fed on it. Two varieties are cultivated, white and black. 22 seers per rupee.

1329. [4930] Cawramanee (*P. vulgaris*), Madras. DR. HUNTER.

1330. [4931] Do. Do.

1331. [6239] Kiwanch-ka-beej (*Mucuna prurita*), Bombay.

1332. [105*] Dhoolaconda (*M. prurita*), Do.

1333. [4941] Poona kalie (*M. prurita*), Madras. DR. HUNTER.

1334. [4039] Agathee (*Agate grandiflora*), Madras.

The undermentioned is a peculiar grain-like substance from Malacca, the botanical source of which has still to be determined. The appearance of the seed does not afford any very distinct clue to this; its composition leads, however, to the conclusion that it is of leguminous origin.

1335. [38*] Ejin or Ee-gin, Malacca.

COMPOSITION.

| | Per cent. |
|---|---|
| Moisture . . . . | 12·60 |
| Nitrogenous matter . . . | 23·06 |
| Starchy matter . . . . | 59·40 |
| Fatty or oily matter . . . | 0·89 |
| Mineral constituents (ash) . . | 4·05 |
| TOTAL | 100·00 |

# Section B. — GROCERY OR PREPARATIONS OF FOOD AS SOLD FOR CONSUMPTION.

## FRUITS.

The specimens of dried fruits are good, and their condition, generally speaking, excellent. They comprise many which are but little known in England.

1336. [4190] Mangosteen (*Garcinia Mangostana*), Rambutan; Rambai; Limon Kandangsa; Langsat; Kumbule; Rumania. Penang.

1337. [4191] Custard apples; Mangosteen; Guava; Buahbell, Pulasan, Amrah, Limonpurut, Blimbing-buloh, Blimbing-saga, Assam-glugor, Katapang and Blinjow. Penang.

1338. [4192] Durian, Rambutan, Rambai, Sillooh, Assam Kulubi, Pulasan, and Blimbing (*Averrhoa Bilimbi*). Penang.

1339. [4193] Mangosteen, Jamboo (*Eugenia Jambolana*), Papaya (*Carica Papaya*), Blinjow, Buah-dulee-mah, Buah-bungah Siam, Buahblingai, Assam Glugor. Penang.

1340. [4194] Durian, Mangosteen; Amrah, Nutmegs and Rambai. Penang.

1341. [4359] Durian, Mangosteen, Lemon kaya, Namnam, Rambutan Langoat, Konajang. Singapore.

A number of specimens of imitation fruits have also been sent for exhibition. These may now be fitly introduced. They are all from Lucknow, with one exception.

1342. [5131] Imitation squash.

1343. [5132] Bael fruit.

1344. [5134] Mango.

1345. [5136] Guava.

1346. [5138] Custard apple.

1347. [5140] Orange.

1348. [5142] Kumralk.

1349. [5144] Plantain.

1350. [5147] Cucumbers.

1351. [5150] China peach.

1352. [5151] Lime.

1353. [10052] Certain imitation fruits from Bareilly.

## PICKLES, PRESERVES, DRIED FRUITS, ETC.

1354. [5730] Preserve of *Asparagus racemosus*, Lucknow.

1355. [5672] Limes, Do.

1356. [5673] Imitation fish, cut out of *Benincasa cerifera*, Do.

1357. [5670] White mango, Do.

1358. [5669] Preserve of *Momordica charantia*, Do.

1359. [5667] Carrot, Do.

1360. [5663] Petha (*Benincasa cerifera*), Do.

1361. [5665] Chillies, Do.

1362. [5668] Green mango, Do.

1363. [5671] Oranges, Do.

1364. [5677] Nauratan pickle, of nine ingredients, among which are mangoes, figs, &c., Do.

1365. [5674] Pickles of *Averrhoa Carambola*, Do.

1366. [5676] Tamarind chutney (*Tamarindus indica*), Do.

1367. [5675] Mango chutney (*Mangifera indica*), Do.

1368. [10169] Bael (*Ægle marmelos*), Calcutta.

1369. [4777] Gee Thee, Burmah. Messrs. HALLIDAY, FOX, & Co.

1370. [4066] Niangis burong, Penang.

1371. [5664] Preserve of *Artocarpus lacoosha*, Lucknow.

1372. [10193] Pickles of *Artocarpus lacoosha*, Do.

1373. [10171] Mowha flowers (*Bassia latifolia*), Moonghyr.

1374. [7177] Do., Lucknow.

1375. [6456] Mowha flowers, Allahabad.

1376. [6110] Oomla (*Phyllanthus emblica*), Ahmedabad.

1377. [5182] Dried plantains (*Musa paradisiaca*), Ramree.

1378. [6107] Figs (*Ficus Carica*), Bombay.

1379. [6106] Kokum (*Garcinia purpurea*), Poona.

1380. [6103] Raisins (*Vitis vinifera*), Bombay.

1381. [6113] Sultanas, Do.

1382. [7100] Cabul grapes, Calcutta.

1383. [6138] Raisins, Bombay.

1384. [1861] Do. Punjab.

1385. [1910] Tamarinds (*Tamarindus indica*), Bengal.

1386. [4780] Do. Burmah. Messrs. HALLIDAY, FOX, & Co.

1387. [6102] Do. Ahmedabad.

1388. [6101] Do. Poona.

1389. [6108] Dates (*Phœnix dactylifera*), Bombay.

1390. [6099] Jujubes (*Zizyphus jujuba*), Do.

1391. [6134] Do. Do.

1392. [6248] Sebestens (*Cordia latifolia*), Do.

1393. [6100] Amboolie (*Mangofera indica*), Almedabad.

1394. [6034] Shah bulloot (edible acorns), Bombay.

1395. [6071] Pounded do. Do.

1396. [6074] Chilgoza (*Pinus Gerardiana*), Do.

1397. [6109] Badam (*Amygdalus communis*), Do.

1398. [6133] Peroo (*Psidium pomiferum*), Do.

1399. [4795] Kya Tsai (*Nelumbium speciosum*), Burmah.

### SPICES, ETC.

1400. [6690] Black pepper (*Piper nigrum*), Singapore. JOSE D'ALMEIDA, Esq.

1401. [6689] White pepper, Singapore. JOSE D'ALMEIDA, Esq.

1402. [6608] Black pepper (*Piper nigrum*), Rhio. Do.

1403. [6685] White do. Do. Do.

1404. [1*] Black pepper (*P. nigrum*), Travancore.

1405. [6105] Do. Bombay.

1406. [4*] Wild pepper (*P. trioicum*), Travancore.

1407. [4145] Soosoo pepper (*P. nigrum*), Penang.

1408. [4146] Trang pepper (*do.*), Do.

1409. [8*] White pepper (*do.*), Bengal.

1410. [10730] Do. Mangalore. V. P. COELHO, Esq.

1411. [4808] Black pepper (*P. nigrum*), Burmah. Messrs. HALLIDAY, FOX, & Co.

1412. [6104] Pepile (*Chavica Roxburghii*), Bombay.

1413. [6691] Long pepper (*C. officinalis*), Java. JOSE D'ALMEIDA, Esq.

1414. [6210] *C. Roxburghii*, Calcutta.

1415. [10170] Do. Assam. LIEUT. PHAIRE.

1416. [6486] Sha morich (*P. nigrum*), Calcutta. KANNY LOLL DEY.

1417. [4779] Chillies (*Capsicum* sp.), Burmah. Messrs. HALLIDAY, FOX, & Co.

1418. [6114] Muchai (*Capsicum frutescens*), Poona.

1419. [6116] Do. Ahmedabad.

1420. [6213] Capsicums (*C. fastigiata*), Calcutta.

1421. [6212] Do. (*C. frutescens*), Do.

1422. [6124] Assalia (*Lepidium sativum*), Ahmedabad.

1423. [10737] Kavate (*Xanthoxylum* sp.), Mangalore. V. P. COELHO, Esq. Used by the natives in their curries.

1424. [10678] Cardamoms (*Elettaria cardamomum*), Do.

1425. [4788] Do. seeds (*Amomum* sp.), Burmah. Messrs. HALLIDAY, FOX, & Co.

1426. [70*] Cardamoms (*Elettaria cardamomum*), Calcutta.

1427. [65*] Do.

1428. [66*] Do. Travancore.

1429. [6218] Ginger (*Zinziber officinalis*), Midnapore.

1430. [1908] Do. Calcutta.

1431. [6216] Do.  KANNY LOLL DEY.

1432. [6092] Green ginger (*do.*), Poona.

1433. [6093] Soont ginger (*do.*), Ahmedabad.

1434. [6094] Udruck ginger (*do.*), Bombay.

1435. [6111] Turmeric(*Curcuma longa*), Ahmedabad.

1436. [4144] Cloves (*Caryophyllum aromaticum*), Penang.

CLOVES.

1437. [6089] Do. (mature fruits), Bombay.

1438. [4195] Do. (preserved), Penang.

1439. [6088] Do. Bombay.

1440. [44*] Cassia buds (*Cinnamomum* sp.), Canara.

1441. [7181] Cinnamon(*Laurus nitida*), Calcutta.  KANNY LOLL DEY.

1442. [6090] Do. Bombay.

1443. [6091] Do. (*C. cassia*), Bombay.

1444. [36*] Wild cinnamon, Coorg.

1445. [33*] Cinnamon, Travancore.

1446. [6344] Do. Singapore.  JOSE D'ALMEIDA, Esq.

CINNAMON.

1447. [34*] Sassafras bark, Martaban.

1448. [98*] Malabathrum leaves (*Cinnamomum cassia*).

1449. [4139] Nutmegs (*Myristica moschata*), Penang.

1450. [4194] Do. (preserved).  Do.

1451. [4195] Do.      Do.

1452. [4140] Do.      Do.

1453. [4138] Do (in shell), Do.

1454. [6073] Jaiputree (*Myristica* sp.), Bombay.

1455. [6097] Jawntree mace (*Myristica tomentosa*), Bombay.

1456. [4142] Mace (*M. moschata*), Penang.

1457. [6096] Jaiputree mace (*Myristica* sp.), Bombay.

1458. [6119] Sowa (*Anethum Sowa*), Ahmedabad.

1459. [6118] Warcalee (*Fœniculum panmorium*), Poona.

1460. [6117] Coriander (*Coriandrum sativum*), Poona.

1461. [6122] Sowa (*Anethum Sowa*), Poona.

1462. [6123] Jeera (*Cuminum cyminum*), Ahmedabad.

1463. [6120] Carraway (*Carum nigrum*), Bombay.

1464. [6121] Ajowan (*Ptychotis ajowan*), Poona.

1465. [6115] Poppy seed (*Papaver somniferum*), Ahmedabad. Used as a condiment.

SUBSTANCES USED IN THE PREPARATION OF DRINKS :—

**TEA.**

The collection is one of the greatest interest, both as to magnitude and the quality of the article displayed. No fewer than 153 different samples of tea are exhibited, and many of these are of excellent quality. All of them are unadulterated with colour, and show at least what we ought to receive in 'green' and 'black' teas.

An exhibition of teas from localities other than the well-known ones in China possesses more than ordinary interest. There can be no doubt that tea will be extensively cultivated in India — land spread over a district exceeding 1,000 miles in length being more or less adapted to its growth. The production of tea in Assam has taken very firm root, and is spreading with almost unexampled rapidity. But the cultivation of tea is not confined to Assam; the Government of India having succeeded, through the able agency of Dr. Jameson, in introducing its cultivation into Dhera Dhoon, Kumaon, Gurhwal, and Kangra. As the result of this, private enterprise — as represented by a number of individuals and several companies — is now engaged in extending its cultivation in the districts in question — the Government still, however, continuing to foster its growth, not with the view to immediate pecuniary profit, but simply to encourage so important an article of commerce. Of what is being done, a very good notion is conveyed by the samples displayed.

**Tea from Assam.**

1466. [7243] Congou, No. 1, Assam. ASSAM TEA COMPANY.

1467. [7244] Congou, No. 2. Assam. ASSAM TEA COMPANY.

1468. [7245] Flowery Pekoe, Do. Do.

1469. [7246] Orange Pekoe, Do. Do.

1470. [7247] Pekoe Souchong, Do. Do.

The Assam Company has been established since 1839, and now supplies the markets of London and Calcutta with no fewer than 1,000,000 lbs. of tea annually. The plant, which is reared in nurseries until it is matured, was first obtained from the shrubs indigenous to the country. It begins to yield in its third year, and attains its maximum production in the seventh. One and the same plant affords the following varieties:—Pekoe, Flowery Pekoe, Orange Pekoe, Souchong, Congou, and Bohea. The kind of tea is determined simply by the number of the sieve through which the dried leaves will pass.

1471. [4703] Fine flowery Pekoe, Do. Pekoe, Do. Souchong, Green tea, Chundypore Factory (season 1861), Assam. Messrs. JOHN BORRADAILE & Co., *Calcutta.*

1472. [4704] Fine Congou, Do. Do.

1473. [4705] Fine flowery Pekoe, Do. Pekoe, Do. Souchong, Do. Congou, Soorispore Ballicandy Factory (season 1861), Assam. Do.

1474. [2023] Fine Souchong, Coocheela Factory (season 1861), Assam. Do.

1475. [2024] Fine Pekoe, Do. Do.

1476. [4404] Fine flowery Pekoe, Do. Do.

1477. [2085] Pekoe tea, Do. T. MORGAN, Esq., *Debrooghur, Assam.* Maijan Plantation.

1478. [2086] Souchong, Do. Do.

1479. [2087] Congou, Do. Do.

1480. [2095] Flowery Pekoe, Do. W. STRAFORD, Esq., *Jorehaut, Assam.* Noakharee Plantation.

1481. [2096] First class Pekoe, Do. Do.

1482. [2097] Unsorted Black, Do. Do.

1483. [2098] Greenish tea (finest), Do. Do.

1484. [2034] Pekoe, Do. G. WILLIAMS, Esq., *Seebsagur,* Assam. Benganakooah Tea Plantation.

1485. [2035] Do. Do. Singlo, Assam.

1486. [2036] Do. Do. Nowgong. Diffloo, Assam.

1487. [2037] Do. Do. Seebsagur. Gilliedary, Assam.

1488. [3970] Orange Pekoe (Chah or Phalap), Dibrooghur. H. E. S. HANNAY, Esq.

1489. [3967] Pekoe, Do., Mankottah, Upper Assam. Messrs. HIGGS, SEVENOAKS, & MELANY.

1490. [3968] Hyson, Do. Do. Do.

1491. [3969] Scented Pekoe, Do. Do. Do.

1492. [3971] Pekoe, Do., Nagaghooli, Upper Assam. Messrs. BARRY & WAGENTRIEBER.

These teas have been manufactured from China plants, in July 1861.

1493. [3972] Souchong (Chah or Phalap), Nagaghooli, Upper Assam. Messrs. BARRY & WAGENTRIEBER.

1494. [3973] Congou, Do. Do. Do.

1495. [3974] Pekoe, Do. Do. Do.

Manufactured from indigenous plants, in July 1861.

1496. [3975] Souchong, Do. Do. Do.

1497. [3976] Congou, Do. Do. Do.

1498. [3977] Green, Mankottah Sesa Barie, Upper Assam. DHATOORAM JEMADAR.

The manufacture of this tea differs from China tea. The Assam green teas are unfaced, and not coloured in any degree.

1499. [2092] Pekoe (Chah or Phalap), Maijam, Upper Assam. C. H. MORGAN, Esq.

1500. [2093] Souchong, Do. Do. Do.

1501. [2094] Congou, Do. Do. Do.

1502. [2088] Flowery Pekoe, Khowang, Assam. CAPTAIN LLOYD, Commissioner of *Assam.*

Manufactured by H. L. JENKINS, Esq.

1503. [2089] Orange Pekoe, Gowhatty. H. G. BAINBRIDGE, Esq., *Gowhatty.*

Manufactured from Assam leaf.

1504. [2090] Flowery Pekoe, Do. H. G. BAINBRIDGE, Esq., *Gowhatty.*

Manufactured from China leaf.

1505. [2091] Do. Do.

## Tea from Cachar.

A number of new plantations have been started in this district. Already upwards of fifty are said to exist, and some thousands of acres are under cultivation.

1506. [3963] Flowery Pekoe, Cachar. The produce of the Goongoor Pahar Plantation, of the season of 1861, belonging to the Cachar Company. Superintendent, R. STEWART, Esq.

1507. [3964] Orange Pekoe, Do. Do.

1508. [3965] First-class Souchong manufactured from leaf taken from China plant, Do. Do.

1509. [3966] First-class Souchong manufactured from the plant indigenous to Cachar, Do. Do.

1510. [4400] Flowery Pekoe. Messrs. TYDD, FORBES, & Co., *Cachar.* Victoria Tea Garden Plantation.

1511. [4401] Pekoe, Do. Do.

1512. [4402] First quality Souchong, Do. Do.

1513. [4403] Second do. Do. Do.

1514. [3717] Souchong. MR. PATERSON, *Tullee Cheera*, at *Sylhet.*

1515. [3718] Congou, Do.

1516. [3730] Flowery Pekoe, Do.

1517. [3731] Pekoe, Do.

1518. [6566] Tea, Chittagong.

Tea thrives vigorously in this place. It is good, though gathered from trees little cultivated, and not properly dried and prepared.*

1519. [2149] Flowery Pekoe, Hazareebaugh. From Seetagurah Plantation.

1520. [2148] Orange Pekoe, Do. Do.

1521-24. [2144-7] Pekoe, Do. Do.

1525. [3734] Indigenous tea plant, Do.

Found in the jungles, being of wild indigenous growth.†

## Tea from Darjeeling.

The following remarks on the cultivation of tea and coffee in the Darjeeling territory are by DR. CAMPBELL, to whom we are chiefly indebted for the excellent display of Indian teas on the present occasion.

The first trial of the tea plant at Darjeeling was made in 1841, with a few seeds grown in Kamaon from China stock. It was quite successful as to its growth, and the quality was approved of by the Assam tea planter who visited Darjeeling in 1846, and made the first tea here.

* Local Committee. *Chittagong.*
† Local Committee, *Chota Nagpore.*

The original plants are now to be seen. All are of gigantic size: one is a bush 50 feet in circumference and 20 feet high.

Nevertheless 7000 feet, the elevation of Darjeeling, is too great for profitable planting; the frost kills the seedlings, and there is not a sufficiently rapid succession of leaf in the warm season to make the manufacture pay. At 7,500 feet the plant does not thrive at all. Elevations of 4,500 feet and under that to 2000, are the best for tea, and from 3000 to 1200 feet will probably be found the best for coffee. Tea and coffee plantations at higher elevations than these may eventually come into occasional use to secure high-flavoured produce without reference to profit.

Although experiments continued to be made on the growth of the tea plant, and seed from Assam and Kumaon was distributed gratuitously by Government, it was not till 1856 that the first plantation was started at Kursing, and another near Darjeeling, by Captain Samler, who was also the first to try coffee. The success in both cases has been complete, and others have followed in the same path. Indeed all that is now required, is careful and liberal encouragement by the Government, to render these hitherto valueless mountains a rich and productive field for European enterprise, a profitable source of pleasant labour to the Hill tribes, and through these channels a source of strength and stability to our power.

The following table exhibits the quantity of land taken up for tea and coffee planting in Darjeeling up to 1861, with the extent of cultivation, the numbers of plants out of nursery, the probable quantity of tea for the season of 1861, and the labour employed.

RETURN OF TEA AND COFFEE PLANTATIONS IN THE DARJEELING TERRITORY, 1861.

| Name of Planter or Company | Extent of Grant in Acres | Acres in Tea Cultivation | No. of Plants not in Nursery | No. of lbs. of Tea to be made in 1861 | Labourers |
|---|---|---|---|---|---|
| Kursing and Darjeeling Tea and Coffee Company, Limited | 2,000 | 421 acres in tea 112 in coffee | At Darjeeling 318,000. At Kursing, 184,000. At Kursing, coffee 91,800 | 20,000 lbs. of tea at Darjeeling and Kursing. At Kursing, 2,000 lbs. of coffee, 1,200 tea seed | 300 |
| Tuckvor Tea Company, Limited | 1,500 | 500 tea, 20 coffee | 950,000 | 3,200 | 360 |
| Hopetown Tea Association, Limited | 1,500 | 500 | 900,000 | 4,800 | 360 |
| Balasan Tea Company, Limited | 1,000 | 130 | 200,000 | None | 150 |
| J. Perry, Kenwick . . . | | 25 | 30,000 | | 30 |
| J. Cleene, Hopetown . . | | 25 | 20,000 | | 25 |
| W. Taylor, Hopetown . . | 300 | 8 Coffee. Nursery 300 acres. | | | 25 |
| Mr. Vaughan . . . . . | | Seed in Nursery | | | 10 |
| Darjeeling Tea Company. Limited | 4,150 | 200 | 240,000 | 4,800 | 700 |
| Ging Park and Ambotia . | 3,000 | 50 | 60,000 | | |
| Mr. Truetter . . . . . | 26 | 9 | 13,000 | 1,600 | 24 |
| Minchu . . . . . | 81 | 8 | 20,000 | | 10 |
| Singell . . . . . | 1,500 | 240 | | None | 200 |
| Wardroper and Ames :— | | | | | |
| Kursing . . . . . | 100 | 80 | 132,000 | 2,000 | 50 |
| Tuckvor . . . . | | 15 | 36,000 | | |
| Torikoolla . . . . . | 360 | 200 | 8,000 | 1,000 | 30 |
| Mr. Grant . . . . . | 150 | 12 | 20,000 | 400 | 20 |
| Mr. Barnes . . . . | 300 | 13 Tea. 3 Coffee | | | 40 |
| Castleton . . . . . | 300 | 200 | 300,000 | | 100 |
| Dooturia . . . . . | 4,000 | 500 | 800,000 | 48 oz. | |
| Darjeeling Tea Company, Limited | 1,598 | | | | |
| Total | 21,865 | 3,251 | Tea plants, 4,303,000 Coffee, 91,800 | 42,600 Tea 20,000 Coffee | 2,534 |

The manufacture of tea in Darjeeling begins in April and ends in October. During the period twenty pickings of leaves are reckoned on. The tea of April, May, and October is the finest. The coffee is cured from October to January.

The number of tea plants per acre varies from 1860 to 2700, according as they are placed at 5 or 4 feet apart. The produce of tea per acre looked for from the *first year of manufacture* to the fourth or fifth, when a plantation is at maturity, cannot be correctly estimated. The produce per plant in the fourth year of age is variously estimated at ½ to 3 ounces. Captain Massan in a memorandum of his operations at Tuckvor, states ' he got last season from a few indigenous Assam plants grown at an elevation of 5000 feet above the sea *one pound* of manufactured tea from *each tree*. The trees were seven years old.' This is an immense return, and not to be reckoned on, on a large scale.

Labour is still abundant, and is likely to continue so from the absence of demand in Eastern Nepaul, the great source of supply. The plantations give steady employment to about 3000 persons, with extra hands occasionally. Wages of Coolies 4s. 8d. to 5s. per mensem.

Employment on tea and coffee is preferred to that on roads and buildings. Wages of manufacturing Coolies amount to 8 rupees per mensem; that of European assistants to from 100 to 150 rupees; of managers from 200 to 400, with house, &c.

The capital involved in tea and coffee plantations in Darjeeling up to 1861, is 1,600,000 rupees or 160,000*l.*

The difficulty of communication with Rajmahal is still the greatest drawback to the progress of tea and coffee planting, as it is to the general advancement of Darjeeling. For the full developement of its resources, a continuation of the rail to the foot of the Hills is required.

1526. [3732] Pekoe tea, *Darjeeling.* Forwarded by Messrs. WOOD, OLIFFE, & Co., agents of the Kursiong and Darjeeling Tea Company. CAPTAIN SAMLER.

1527. [3733] Do. Kursiong. Do.

1528. [3718] Pekoe. Kursiong Plantation. *Darjeeling.* P. H. SCANLAN, Esq.

1529. [3719] Souchong. Do.

1530. [3720] Congou. Do.

1531. [3721] Souchong, Kursiong Plantation. P. H. SCANLAN, Esq.

1532. [3722] Pekoe. Do.

1533. [3723] Congou. Do.

1534. [3724] Pekoe. Do.

1535-8. [3725-28] Four samples of mixed black tea. Tuckvor Tea Plantation. *Darjeeling.* CAPTAIN MASSON.

1539. [2047] Souchong. *Darjeeling.* MAHOMED TONIKOOLLA.

1540. [2040] Do. *Darjeeling.* HOPE TOWN TEA ASSOCIATION.

1541. [2039] Pouchong. Do.

1542. [2038] Souchong. Do.

1543. [2041] Hyson. Do.

1544. [4415] Pekoe. Do.

1545. [4416] Pouchong. Do.

1546. [4417] Souchong. Do.

1547. [4418] Hyson. Do.

1548. [4711] Fine Souchong. Koalagire Tea Plantation. *Darjeeling.* GOVERNMENT.

1549. [4412] Souchong. Do.

1550. [4411] Pouchong. Do.

1551. [4410] Bohea. Do.

1552. [4708] Young Hyson. Do.

1553. [4414] Hyson. Do.

1554. [4709] Imperial gunpowder. Do.

1555. [4707] Gunpowder. Do.

1556. [4706] Hyson skin. Do.

1557. [4413] Pekoe. Koalagire Tea Plantation. *Darjeeling.* GOVERNMENT.

1558. [2045] Hyson. Do.

1559. [2046] Souchong. Do.

1560. [3729] Young Hyson. Do.

1561. [3994] Brick tea, Sikkim. DR. CAMPBELL.

### Tea from Deyrah Dhoon.

The following samples have been forwarded from this important district:—

1562. [4407] Pouchong. Hurbunswala, Arcadia, and Hope Town Plantation, *Deyrah Dhoon.* NORTH-WEST TEA COMPANY.

1563. [2099] Young Hyson. Do.

1564. [4710] Hyson. Do.

1565. [4405] Pukh Hyson. Do.

1566. [4406] Gunpowder. Do.

1567. [4409] Company's Mixture. Do.

1568. [2100] Fine Souchong. Do.

1569. [4408] Souchong. Do.

### Tea from Kumaon and Gurwhal.

Both in Kumaon and Gurwhal, and the Deyrah Dhoon, Government plantations,* as well as those of individuals and companies, are to be found. In the year 1861 alone, the Government distributed to private planters throughout Kumaon, Gurwhal, Deyrah Dhoon, and the Punjab upwards of 130 tons of seed, and 2,400,000 seedlings.

1570. [3986] Souchong. Megree Tea Plantation. *Kumaon.* C. R. TROUP, Esq.

1571. [3987] Pouchong. Do.

1572. [3988] Bohea. Do.

1573. [3989] Souchong. Konsamire Tea Plantation. *Kumaon.* KONSAMIRE TEA COMPANY.

1574. [3990] No. 1 Young Hyson. Do.

1575. [3991] No. 2 Do. Do.

1576. [3992] Hyson. Do.

1577. [3993] Imperial Gunpowder. Do.

* The Government tea factories and plantations in Kumaon and Gurwhal, as well as those in Kangra, are about to be notified for sale by public auction.

1578. [4385] No. 1 Imperial Gunpowder. Kumaon Hawalbagh Tea Plantation. GOVERNMENT.

1579. [4386] No. 2 Do. Do.

1580. [4387] Gunpowder. Do.

1581. [4388] No. 1 Young Hyson. Do.

1582. [4389] No. 2 Do. Do.

1583. [4390] No. 1 Hyson. Do.

1584. [4391] No. 2 Do. Do.

1585. [4392] Hyson Skin. Do.

1586. [4393] Souchong. Do.

1587. [4394] Bohea. Do.

1588. [4395] Souchong. Kumaon Agartola Tea Plantation. GOVERNMENT.

1589. [4396] Pouchong. Do.

1590. [4397] Bohea. Do.

1591. [4398] Souchong. Kumaon Bhurtpore Tea Plantation. GOVERNMENT.

1592. [4399] Pouchong. Do.

1593. [7242] Bohea. Kumaon Blimthal Tea Plantation. GOVERNMENT.

1594. [3980] Fine Souchong. *Gurwhal.* GOVERNMENT.

1595. [3981] Souchong. Do.

1596. [3982] Pouchong. Do.

1597. [3983] Bohea. Do.

1598. [3984] Souchong. Warrand Field Tea Plantation, *Gurwhal.* T. WARRAND, Esq.

1599. [3985] Souchong. Willow Bank Tea Plantation, *Gurwhal.* G. RICHARDS, Esq.

### Tea from Kangra, Punjab.

Upwards of 1416 maunds, or more than 52 tons, of the seed were obtained from the Government plantations at Kangra for distribution in 1861. The present exhibition includes the following samples:—

1600. [2026] Hyson. Kangra Tea Plantation. *Kangra.* GOVERNMENT.

1601. [2027] Young Hyson. Do.

1602. [2028] Gunpowder. Do.

1603. [2029] Imperial Gunpowder. Do.

1604. [2030] Hyson Skin. Kangra Tea Plantation. GOVERNMENT.

1605. [2031] Souchong. Do.

1606. [2032] Pouchong. Do.

1607. [2033] Bohea. Do.

*Note.*—The above samples were those first forwarded. The eight which follow are superior in point of quality, but these did not come to hand until the jury had all but concluded its labours.

1608. [7860] Souchong, Kangra. GOVERNMENT.

1609. [7861] Hyson, Do. Do.

1610. [7862] Imperial Gunpowder, Do. Do.

1611. [7863] Young Hyson, Do. Do.

1612. [7864] Souchong, Do. Do.

1613. [7865] Do. Do. Do.

1614. [7866] Hyson, Do. Do.

1615. [7867] Souchong, Do. Do.

### Sample of Tea from Burmah.

1616. [4758] Brick tea, Burmah. Messrs. HALLIDAY, FOX, & Co.

### COFFEE.

This important berry is being extensively cultivated in the high-lands of Southern India, and large tracts of country are available for the extension of its growth.

The amount and value of the coffee exported from India will be gathered from the Table subjoined. (*See next page.*)

1617. [5065] Coffee (*Coffea arabica*), Mysore. COLONEL ONSLOW.

Known in the market as 'Cannon's Mysore.'

The subjoined remarks on coffee-planting in Mysore are by COLONEL ONSLOW:—

According to the traditions of the country, the coffee plant was introduced in Mysore by a Mussulman pilgrim, named Baba Booden, who came from Arabia about 200 years ago, and took up his abode as a hermit in the uninhabited hills in the Nuggur Division named after him, and where he established a muth or college, which still exists endowed by Government. It is said that he brought seven coffee berries from Mocha, which he planted near to his hermitage, about which there are now to be seen some very old coffee trees. However this may be, there is no doubt that the coffee plant has been known in that neighbourhood from time immemorial; but the berry has never come into general use among the people for a beverage. It is only of late years that the coffee trade of these districts has become of any magnitude, or that planting has been carried to any important extent. There is no record of either farther back than the year 1822, when the revenue was under contract. In the year 1837

SHOWING THE QUANTITIES (AS FAR AS CAN BE ASCERTAINED) AND THE VALUE OF COFFEE EXPORTED FROM INDIA AND EACH PRESIDENCY TO ALL PARTS OF THE WORLD FROM 1850-51 TO 1860-61.

| Years | Whence Exported | United Kingdom Quantity (lbs.) | United Kingdom Value (£) | France Quantity (lbs.) | France Value (£) | Other Parts of Europe Quantity (lbs.) | Other Parts of Europe Value (£) | America Quantity (lbs.) | America Value (£) | China Quantity (lbs.) | China Value (£) | Arabian and Persian Gulfs Quantity (lbs.) | Arabian and Persian Gulfs Value (£) | Other Parts Quantity (lbs.) | Other Parts Value (£) | Total Exported to All Parts Quantity (lbs.) | Total (tons) | Total Value (£) |
|---|---|---|---|---|---|---|---|---|---|---|---|---|---|---|---|---|---|---|
| 1850-51 | Bengal | 28,787 | 353 | 691,476 | 13,137 | 128 | 1 | | | | | 1,254,692 | 11,377 | 36,848 | 453 | 757,234 | 338 | 13,944 |
| | Madras | 1,728,076 | 18,105 | 205,001 | 2,050 | | | | | | | 537,451 | 6,778 | 97,177 | 1,179 | 3,284,946 | 1,466 | 32,711 |
| | Bombay | 2,452,854 | 43,026 | 216,627 | 3,949 | | | | | 381 | 5 | | | 7,928 | 96 | 3,215,341 | 1,435 | 63,854 |
| | ALL INDIA | 4,209,717 | 61,484 | 1,113,104 | 19,136 | 123 | 1 | | | 381 | 5 | 1,792,143 | 18,155 | 141,953 | 1,728 | 7,257,421 | 3,239 | 100,509 |
| 1851-52 | Bengal | 3,043 | 59 | 336,814 | 6,458 | | | | | | | 1,272,269 | 11,463 | 1,974 | 40 | 341,831 | 153 | 6,557 |
| | Madras | 956,537 | 9,568 | 398,435 | 3,985 | | | | | | | 227,285 | 2,435 | 146,377 | 1,464 | 2,773,638 | 1,238 | 26,480 |
| | Bombay | 5,364,855 | 48,802 | | | | | | | 280 | 5 | | | 2,639 | 27 | 5,595,059 | 2,498 | 51,269 |
| | ALL INDIA | 6,324,435 | 58,429 | 735,269 | 10,443 | | | | | 280 | 5 | 1,499,554 | 13,898 | 150,990 | 1,531 | 8,710,528 | 3,889 | 84,306 |
| 1852-53 | Bengal | 19,164 | 372 | 143,773 | 2,798 | | | 136,080 | 1,215 | | | 1,668,413 | 14,896 | 73,449 | 1,043 | 236,386 | 106 | 4,213 |
| | Madras | 1,097,760 | 10,298 | 885,966 | 8,636 | | | | | | | 300,823 | 3,194 | 197,987 | 1,911 | 3,986,206 | 1,777 | 36,956 |
| | Bombay | 3,127,921 | 50,359 | 201,246 | 2,594 | | | | | | | | | 12,367 | 162 | 3,642,960 | 1,626 | 56,391 |
| | ALL INDIA | 4,244,845 | 61,029 | 1,230,985 | 14,028 | | | 136,080 | 1,215 | | | 1,969,236 | 18,090 | 283,803 | 3,116 | 7,865,552 | 3,509 | 97,490 |
| 1853-54 | Bengal | 804,033 | 8,041 | 497,753 | 5,347 | 6,720 | 130 | 448 | 6 | | | 1,835,533 | 16,472 | 827,904 | 8,283 | 3,965,223 | 1,770 | 37,774 |
| | Madras | 2,214,751 | 39,548 | 1,001,905 | 17,891 | 37,380 | 467 | | | | | 278,502 | 3,044 | 277,411 | 4,908 | 3,772,569 | 1,683 | 65,391 |
| | Bombay | .. | 141 | .. | 4,978 | | | | | | | | | .. | 1,102 | .. | .. | 6,596 |
| | ALL INDIA | 3,018,784 | 47,730 | 1,499,658 | 28,216 | 44,100 | 597 | 448 | 6 | | | 2,114,035 | 19,516 | 1,105,315 | 14,293 | .. | .. | 109,761 |
| 1854-55 | Bengal | 171,696 | 3,350 | 104,384 | 2,030 | | | | | | | 444,800 | 870 | 55,888 | 697 | 383,488 | 171 | 7,077 |
| | Madras | 771,716 | 7,717 | 339,875 | 3,399 | | | | | | | 2,089,656 | 18,716 | 477,922 | 4,779 | 3,679,169 | 1,642 | 34,611 |
| | Bombay | 2,235,815 | 27,198 | 726,594 | 9,082 | | | | | 168 | 2 | 336,066 | 4,325 | 2,471 | 32 | 3,338,494 | 1,490 | 41,106 |
| | ALL INDIA | 3,179,227 | 38,265 | 1,170,853 | 14,511 | | | | | 168 | 2 | 2,470,522 | 23,911 | 536,281 | 5,508 | 7,401,151 | 3,303 | 82,794 |
| 1855-56 | Bengal | 6,608 | 187 | 413,252 | 8,546 | | | | | 392 | 8 | | | 9,727 | 190 | 429,979 | 192 | 8,881 |
| | Madras | 1,494,542 | 14,956 | 1,107,716 | 11,077 | | | 2,464 | 25 | | | 1,757,148 | 16,679 | 368,168 | 3,541 | 4,730,038 | 2,112 | 46,278 |
| | Bombay | 3,259,688 | 53,317 | 566,293 | 8,761 | 52,628 | 861 | | | 2,960 | 43 | 148,660 | 1,854 | 15,857 | 206 | 4,046,086 | 1,806 | 65,042 |
| | ALL INDIA | 4,760,818 | 68,410 | 2,087,261 | 28,384 | 52,628 | 861 | 2,464 | 25 | 3,352 | 51 | 1,905,808 | 18,533 | 393,752 | 3,937 | 9,206,103 | 4,110 | 120,201 |
| 1856-57 | Bengal | .. | 499 | .. | 3,744 | .. | 2,277 | 40 | 1 | 603 | 1,443 | .. | 4 | .. | 1,147 | .. | .. | 9,054 |
| | Madras | 1,810,972 | 18,132 | 1,554,319 | 15,543 | 241,259 | 2,413 | | | | | 230,410 | 2,110 | 1,342,864 | 12,190 | 5,179,864 | 2,312 | 50,389 |
| | Bombay | 3,298,822 | 51,352 | 601,742 | 8,893 | 734,306 | 11,169 | | | | | 142,577 | 1,765 | 14,507 | 207 | 4,791,954 | 2,184 | 73,376 |
| | ALL INDIA | 5,109,794 | 69,913 | 2,156,061 | 28,180 | 975,565 | 15,859 | 40 | 1 | 603 | 1,443 | 372,987 | 3,879 | 1,357,371 | 13,544 | .. | .. | 132,819 |
| 1857-58 | Bengal | 8,250 | 205 | 70,188 | 1,451 | 40,068 | 829 | 84 | 2 | 392 | 9 | | | 11,887 | 246 | 130,477 | 58 | 2,738 |
| | Madras | 1,068,723 | 10,678 | 136,614 | 1,366 | 61,119 | 611 | | | | | 1,024,741 | 9,579 | 1,958,652 | 18,995 | 4,249,949 | 1,897 | 41,229 |
| | Bombay | 14,303 | 29,270 | 609,296 | 8,039 | 420,815 | 5,965 | | | | | 663,416 | 12,195 | 3,534 | 76 | 1,743,481 | 778 | 55,765 |
| | ALL INDIA | 1,091,276 | 40,153 | 816,098 | 10,856 | 522,002 | 7,401 | 84 | 2 | 392 | 9 | 1,688,157 | 21,774 | 1,974,073 | 19,317 | 6,123,807 | 2,733 | 99,727 |
| 1858-59 | Bengal | 157,164 | 3,202 | 415,884 | 8,511 | 1,643 | 16 | | | | | | | 14,896 | 7 | 573,776 | 266 | 11,729 |
| | Madras | 3,402,854 | 33,742 | 2,055,386 | 20,032 | | | 5,264 | 109 | 392 | 9 | 527,453 | 4,756 | 1,637,222 | 12,401 | 7,356,883 | 3,284 | 70,947 |
| | Bombay | 1,506,938 | 20,182 | 1,701,291 | 24,722 | 164,808 | 2,207 | 19,236 | 258 | | | 324,892 | 4,337 | 66,657 | 913 | 3,764,586 | 1,681 | 52,361 |
| | ALL INDIA | 5,066,956 | 57,126 | 4,172,561 | 53,265 | 166,451 | 2,223 | 24,500 | 367 | 392 | 9 | 852,345 | 9,093 | 1,718,775 | 13,321 | 11,695,195 | 5,221 | 135,037 |
| 1859-60 | Bengal | 896 | 19 | 456,288 | 8,595 | 12,544 | 153 | | | | | | | 14,896 | 207 | 489,888 | 219 | 9,188 |
| | Madras | 3,831,800 | 38,793 | 3,831,984 | 54,176 | | | | | | | 688,680 | 9,245 | 1,637,222 | 26,446 | 9,988,686 | 4,459 | 128,660 |
| | Bombay | 812,408 | 9,792 | 2,837,760 | 38,005 | | | | | | | 181,718 | 2,433 | 15,113 | 203 | 3,866,235 | 1,726 | 50,691 |
| | ALL INDIA | 4,645,104 | 48,604 | 7,126,032 | 100,776 | 12,544 | 153 | | | | | 870,398 | 11,678 | 1,667,231 | 26,956 | 14,345,809 | 6,404 | 188,534 |
| 1860-61 | Bengal | 3,108 | 74 | 501,760 | 11,199 | 336 | 8 | | | | | | | 1,008 | 24 | 506,212 | 226 | 11,805 |
| | Madras | 5,124,599 | 91,148 | 7,089,441 | 125,644 | 20,412 | 204 | | | 385 | 8 | 574,976 | 9,920 | 1,426,504 | 23,564 | 14,236,282 | 6,355 | 250,480 |
| | Bombay | 2,122,358 | 28,427 | 1,914,213 | 41,019 | | | 135,975 | 2,914 | | | 141,064 | 1,879 | 63,770 | 1,410 | 4,877,765 | 1,954 | 75,651 |
| | ALL INDIA | 7,250,365 | 119,649 | 9,505,414 | 177,862 | 20,748 | 212 | 135,975 | 2,914 | 385 | 8 | 716,040 | 11,799 | 1,490,282 | 24,998 | 19,119,209 | 8,535 | 337,436 |

when the country had been some years under British rule, the Raja's authority having been suspended in 1832–3, the contract system was discontinued, and a duty of one rupee per maund of 28 lbs. was fixed. From that time the production of coffee and duty is duly recorded. In 1843 the duty was reduced to half a rupee per maund on exportation, and in 1849 to a quarter of a rupee. Together with the reduction of duties, regulations for taking up and holding coffee lands were adopted. At the same time prices continued to rise. 'Cannon's Mysore' (the coffee exhibited) has risen from 48s. per cwt. in 1846-7 to an average of 96s. per cwt., and has fetched so high as 115s. Native coffee sold in the country has risen from 1 rupee per maund of 28 lbs. to 6 and 8 rupees.

The encouragement thus given to coffee planters has resulted in the great extension of planting, the prosperity of the planters, and an increase of revenue to the state.

Under the contract system the revenue averaged from 1822 to 1832, 4270 rupees annually, and from 1832 to 1837, 7472 rupees annually. The yearly average during the next six years under the duty system, the duty being 1 rupee per maund, was 15,238 rupees on that number of maunds. During the next six years, the duty being half a rupee per maund, the average yearly produce rose to 52,236 maunds, giving a revenue of 26,118 rupees yearly. During the next 12 years, that is, up to 1861 inclusive, to which time the accounts are made up, the yearly average of produce rose to 346,083 maunds, and the revenue to 86,524 rupees, the duty having been reduced to a quarter of a rupee per maund. This short statement serves to show the good effect of liberal measures.

More than 30 years ago a few Europeans were engaged in coffee planting near Chickmoogloor, a few miles from the Bababooden Hills. About 20 years ago the plantations producing the well-known coffee called 'Cannon's Mysore' and others on the Memzera, 'bad mountain,' were commenced by two enterprising gentlemen. The success of these has induced many more Europeans to plant coffee in Mysore. The consequence is, that the coffee trade of Mysore bids fair to emulate that of Ceylon. It has given also an example to other parts of India, and the plant originally taken from the Bababooden Muth is now extending over tens of thousands of acres in Coorg, the Wynaad district, the Neilgherry Hills, and along the Western Ghauts, North and South.

In Mysore the number of European coffee planters has within the last 10 years increased to 20 or 30. The number of native planters is estimated between 3000 and 4000. The quantity of land planted or taken up cannot be ascertained with any degree of accuracy. The revenue depending upon the quality of the coffee produced, not upon a tax on land, there is no regular correct system of land measurement. This way of taxing is bad; it leads to bad cultivation and smuggling. It is to be·hoped that a land tax will be adopted instead, which would have a good moral and fiscal effect. It would put an end to smuggling, and would be a great inducement to the natives to improve their cultivation, which is now very slovenly. If the tax were on the land, they would make more effort to increase the produce of it. The average produce per acre in Mysore is probably not half that of Ceylon.

The coffee districts are confined to the region of the Western Ghauts and the Bababooden Hills. Some attempts have been made to cultivate coffee in the open country, but without success; it seems to require forest land and considerable elevation and moisture. 'Cannon's Mysore' is grown on a range of hills from 3,500 to 4.000 feet above the sea, having the benefit of the south-west monsoon, which very seldom fails at all, never entirely, and of the tail-end of the north-east monsoon. It is probably to these advantages that the peculiar qualities of 'Cannon's Mysore' are attributed, viz. closeness of texture and richness of flavour. This elevation gives a pleasant climate well suited to the Europeans. During the south-west monsoon, the planter may be in his gardens all day long without oppression in the hottest weather; the thermometer in house on these plantations

rises no higher than 81° or 82° Fahrenheit. The whole of the coffee district, with here and there an exception of feverish spots, possesses a climate in which the European can live and work with comfort, and, with moderate care and prudence, with health.

Planting has of late years been carried to such an extent by Europeans and natives in Mysore, that but little available land remains. These mountain and forest wastes have been turned into rich productive gardens. From being the most wild and desolate parts of Mysore, these districts have become very prosperous, and the people have been raised from poverty to comfort, and in many instances to wealth. The natives are benefiting largely by the capital and example of European planters, and are learning the science of planting.

Mysore generally, especially the coffee districts, affords a most promising field for European capital and enterprise.

1618. [10583] Coffee, Salem. Messrs. FISHER & CO.

1619. [10799] Do. in husk, Mangalore. V. P. COELHO, Esq.

1620. [6423] Coffee, Chota Nagpore. M. LIEBERT.

1621. [6422] Do. Do.

1622. [6567] Do. Chittagong.

1623. [6098] Do. Bombay.

1624. [6387] Do. Malacca. CAPTAIN PLAYFAIR.

1625. [4771] Do. in husk (*C. arabica*), Burmah. Messrs. HALLIDAY, FOX, & CO.

1626. [4174] Do. Penang.

1627. [6699] Do. Singapore. W. J. VALBORG, Esq.

1628. [4171] Do. Penang.

1629. [4148] Do. Do.

———

1630. [6694] Cocoa (*Theobroma Cacao*), Singapore. CAPTAIN W. SCOTT.

This sample was grown experimentally in a garden at Singapore.

———

Among the substances used in the preparation of drinks the following are included:—

1631. [6095] Hura Char (*Andropogon citratum*), Ahmedabad. Used as a beverage.

1632. [5188] Dietetic Bael (*Ægle marmelos*), Calcutta. Messrs. BATHGATE & CO.

This dietetic preparation is obtained from the fruit of the Bael (*Ægle marmelos*), and is strongly recommended by the manufacturers for invalids and dyspeptic persons, and an agent has been appointed in England for its sale.

## STARCHES &c.

### Arrowroot, Tapioca, and Sago.

1633. [7148] Arrowroot (*Maranta arundinacea*), Cuttack.

ARROWROOT PLANT (*Maranta arundinacea*).*

1634. [7149] Wild arrowroot, Do.

It is not easy to decide whether this is identical with our garden arrowroot. A cup of arrowroot made of the one is not distinguishable from a cup made of the other, except, perhaps, by a slightly earthy taste and smell observable in the wild arrowroot, which is easily accounted for by its imperfect manufacture. The cultivation and more perfect manufacture of the garden arrowroot have been comparatively recently introduced into the province, so that it is neither generally grown nor its produce used by the natives. The specimen sent was made from plants of his own growing by a native Christian of 'Khundittur,' who sells his produce among the European residents of Cuttack, his price being a little under 6d. per English lb. This arrowroot is of excellent quality, and the process of manufacture as simple as can be. The tubers are taken up in the cold season, washed, put into a large wooden mortar, and mashed. The mash is then taken out, and well washed in cold water, the water drained off, and set to stand in large flat vessels, in which it deposits a large proportion of the arrowroot flour, which is rewashed in cold water, and set to dry in the sun. The wild arrowroot, known in the bazaar as 'Palooa,' grows abundantly in the jungles of the district. It is collected in the cold season by the Sahars, the tubers pounded and mashed, and the sediment dried in the sun. By these people it is eaten and sold for the manufacture of what is called 'Abheer.' In the Sumbulpore, and to a less degree also in the Cuttack District, the wild arrowroot is made

into cakes, or boiled with milk, and thus used as an article of food. This Committee had intended to send specimens of sago and tapioca meal, the trees being indigenous, but the time and the season of the year have prevented it.*

1635. [6425] Arrowroot from plant growing wild in the jungles, Chota Nagpore. M. Liebert.

1636. [6425] Arrowroot, Chota Nagpore.

1637. [6565] Starch from wild ginger, Chittagong.

The plant which furnishes this sample grows everywhere in this district; it is very difficult to eradicate it from land, as the smallest root or piece of a root that has an eye will spring up again. The plant dies off in December. A rough experiment was made with this root by the Civil Assistant-Surgeon of this place, Dr. W. B. Beatson, and the yield was estimated at 1 ounce of starch from 1 pound of the root. The experiment, however, was not precise enough to be satisfactory, and he is inclined to think that the yield would be much larger, as the microscope shows the root to be loaded with starch granules. The supply of the root being inexhaustible, any quantity of starch might be extracted from it yearly, and it might be found a valuable article of commerce. There would be no expense for cultivation, and allowing for the cost of digging the root, and manufacturing the starch by bruising and macerating the root in water and drying the deposit, the product would be cheaper than Arracan rice, which is believed to be largely exported to Europe to be used, not as food, but in manufacture for glazing linen, &c.†

1638. [22*] Arrowroot (*Curcuma*, species), Rohilcund.

1639. [18*] Do (*C. angustifolia*), Malabar.

1640. [5731] Arrowroot, Burdwan.

1641. [7147] Do. Akyab.

Large quantities can be produced if required. This description of arrowroot is prepared from the Pemban Oo root, obtainable in large quantities. Price 4 rupees per maund.‡

1642. [10694] Arrowroot flour, Mangalore. V. P. Coelho, Esq.

1643. [6684] Arrowroot, Singapore. M. T. Davidson, Esq.

1644. [6390] Kledey Poteh, sweet potato flour (*Batatas edulis*), Malacca. Captain Burn.

1645. [10697] Byne Palm flour (*Caryota urens*), Mangalore. V. P. Coelho, Esq.

---

* From Mr. R. Hardwicke, 192 Piccadilly.

* Local Committee, *Cuttack.*
† Local Committee, *Chittagong.*
‡ Local Committee, *Akyab.*

1646. [10698] Amroota Bally, Bangalore. V. P. COELHO, Esq.

1647. [1881] Singara flour (*Trapa bispinosa*), Calcutta. KANNY LOLL DEY.

1648. [34*] Sweet farina of *Parkia biglobosa*, Madras.

1649. [4073] Tapioca (*Jatropha manihot*). Province Wellesley.

TAPIOCA PLANT (*Jatropha manihot*).*

1650. [4074] Tapioca. LAWRENCE NAIRNE.

1651. [4182] Do. Alma Estate, Province Wellesley. D. C. THOMPSON, Esq.

1652. [1331] Speed's steam-made arrowroot (*Maranta arundinacea*), Calcutta. A. GEORGE, Esq.

1653. [1332] Tapioca, steam-made (*Jatropha manihot*), Calcutta. A. GEORGE, Esq.

1654. [4180] Tapioca flour, Alma Estate, Province Wellesley.

1655. [4181] Refuse of tapioca root after extracting the starch, Alma Estate, Province Wellesley.

1656. [6388] Tapioca, Malacca. T. NEUBRONNER, Esq.

1657. [190] Raggy flour (*Eleusine coracana*), Mangalore. V. P. COELHO, Esq.

1658. [6389] Sago Rombiya, Malacca. CAPTAIN BURN.

1659. [4070] Sago, Penang.

1660. [6697] Sago flour, Singapore. JOSE D'ALMEIDA, Esq.

SAGO PALM.

1661. [6687] Do. Lingee. Do.

1662. [6692] Pearl sago, Sarawak. G. ANGUS, Esq.

1663. [4071] Sago, Penang.

1664. [4069] Do.

1665. [4072] Do.

## SALEPS.

1666. [3835] Bechundie, Jubbulpore.

1667. [2838] Do. Raepore.

This substance, if pulverised, resembles arrowroot, and is made use of by natives on fast days, prepared in various ways. It is obtained from the glutinous matter which issues from the stems of a jungle plant, after being soaked in running water for some days. The Gonds prepare the Behchandee. It can be had in any quantity in the Jubbulpore bazaar, but most of it comes from Mundla and Seonee.*

The specimens seem to consist of the dried sections of a farinaceous root containing bassorin, and allied in composition to salep.

1668. [6067] Pungie salep, Bombay.

1669. [1811] Salep misree, Punjab.

---

* From MR. R. HARDWICKE.

* Local Committee, *Jubbulpore.*

1670. [6069] Salep misree (*Eulophia campestris?*), Bombay.

1671. [7040] Mooslee seah (*Murdannia scapiflora*), Bombay.

1672. [5057] Alstrœmeria root. DR. RIDDELL.

Both of these are probably the dried corms of the plant first named, and *not* of an *Alstrœmeria.*

### EDIBLE ALGÆ &c.

1673. [6677] Sea weed (*Plocaria candida*), Eastern Archipelago. G. ANGUS, Esq.

1674. [1*] Do.

1675. [2*] Agar agar (*Eucheuma spinosa*), Malacca.

1676. [3*] Do. Macassar.

1677. [6669] Do. Eastern Archipelago. G. ANGUS, Esq.

1678. [7115] Kek kieo, Ramree.

1679. [7116] Do. Do.

This is a lichen, doubtless *Alectoria jubata*, but the absence of fructification renders it difficult to decide with certainty.
Gelatinous : eaten by the natives with rice. Cost 2 annas. Not exported. Good samples not procurable during the rains.*

1680. [4162] Mushrooms collected from the stumps of trees (*Agaricus (Pleurotus) subocreatus*, n.s.), China. COL. COLLYER.

This is a new and apparently undescribed species of *Agaricus* belonging to the sub-genus *Pleurotus*. It is nearly allied to the British *Agaricus ulmarius*, from which it is separated by the volva, remains of which may be traced at the base of the stem. It is a dendrophytal species, drying readily, and is employed in the Straits Settlements as an article of food. If it proves to be really, as it appears to be, a new and undescribed species, the most suitable name would be the one here adopted.†

1681. [6688] Dried Fungi (*Hirneola auricula-Judæ*), Singapore. COL. COLLYER.

This fungus does not appear to differ from our indigenous *Hirneola auricula-Judæ*, which has a wide range, and an almost obsolete reputation in medicine. It is sent as a food product, but, we should imagine, of very little merit.‡

### SUGARS.

Although the samples of sugar on show are not numerous, a few are of excellent quality. The Tables subjoined will give the requisite information as to the extent of the Indian export trade in this important article of daily life. (*See pages* 74, 75.)

1682. [5348] Double refined loaf sugar, Shahjehanpore. Messrs. CAREW & Co.

1683. [5350] Crystallized sugar, Do. Do.

1684. [6226] Goor from sugar-cane, Hooghly.

1685. [6228] Sugar-candy, Midnapore.

1686. [7146] Do. No. 1, Calcutta.

1687. [7145] Do. No. 2, Do.

1688. [2177] Daloo, Calcutta Bazaar.

1689. [2178] Ach Borah (first quality), Do.

1690. [2179] Do. (second quality), Do.

1691. [2180] Dobarrah, Do.

1692. [2181] Cassee Chena, Do.

1693. [6227] Ook or Junnah (*Saccharum officinarum*), Lucknow.

This is the sugar extracted from the above, called by the natives *Cheenee*, partly refined. 4 seers for the rupee. Used for sugar and spirits.*

1694. [6229] Goor (*S. officinarum*), Lucknow.

This is the appearance of the sugar after the first boiling of the cane juice : the natives call it *Goor*. 11 seers per rupee. Used for sugar and spirits.†

1695. [6230] Kund (*S. officinarum*), Lucknow.

This is refined sugar, and called *Kund* by the natives ; this is what many of the Europeans use for their tea, coffee, &c. 2 seers per rupee.‡

1696. [3833] Calpee candy, Jubbulpore.

1697. [2738] Sugar (first quality). ASTAGRAM SUGAR COMPANY.

1698. [2739] Do. (second quality), Do.

---

* Local Committee, *Akyab.*
† M. C. COOKE, Esq.
‡ M. C. COOKE, Esq.

* Central Committee, *Lucknow.*
† Central Committee, *Lucknow.*
‡ Central Committee, *Lucknow.*

( 73 )

# CLASS III.—*India.*

TABLE SHOWING THE QUANTITIES (AS FAR AS CAN BE ASCERTAINED) AND THE VALUE OF SUGAR EXPORTED FROM INDIA AND EACH PRESIDENCY TO ALL PARTS OF THE WORLD FROM 1850-51 TO 1860-61.

COUNTRIES WHITHER EXPORTED

| Years | Whence Exported | United Kingdom | | France | | Other Parts of Europe | | America | | China | | Arabian and Persian Gulfs | | Aden | | Coast of Africa | | New South Wales | | Ceylon | | Other Parts | | Total Exported to All Parts | | |
|---|---|---|---|---|---|---|---|---|---|---|---|---|---|---|---|---|---|---|---|---|---|---|---|---|---|---|
| | | Quantity (cwt.) | Value (£) | Quan. (cwt.) | Value (£) | Quan. (cwt.) | Value (£) | Quan. (cwt.) | Value (£) | Qua. (cwt.) | Val. (£) | Quan. (cwt.) | Value (£) | Quan. (cwt.) | Val. (£) | Quan. (cwt.) | Val. (£) | Quan. (cwt.) | Value (£) | Quan. (cwt.) | Value (£) | Quan. (cwt.) | Value (£) | Quantity (cwt.) | (tons) | Value (£) |
| 1850-51 | Bengal | 1,190,454 | 1,505,487 | 1,508 | 1,792 | 178 | 206 | | | | | 4,942 | 5,739 | | | | | 4,225 | 4,131 | 1,785 | 2,137 | 1,539 | 2,035 | 1,204,631 | 60,232 | 1,521,527 |
| | Madras | 290,147 | 199,681 | | | | | | | | | 3 | | | | | | 159 | 81 | 1,829 | 1,943 | 1,298 | 742 | 293,436 | 14,672 | 202,448 |
| | Bombay | 8,278 | 4,575 | | | | | | | | | 80,646 | 90,505 | 2,812 | 2,682 | 1,545 | 1,777 | | | | | 283 | 275 | 93,564 | 4,678 | 99,814 |
| | ALL INDIA | 1,488,879 | 1,709,743 | 1,508 | 1,792 | 178 | 206 | | | | | 85,591 | 96,245 | 2,812 | 2,682 | 1,545 | 1,777 | 4,384 | 4,212 | 3,614 | 4,080 | 3,120 | 3,052 | 1,591,651 | 79,582 | 1,823,789 |
| 1851-52 | Bengal | 1,106,298 | 1,425,059 | 2,113 | 2,875 | 1,360 | 1,551 | 1,735 | 2,362 | | | 2,202 | 2,496 | | | | | 7,645 | 9,922 | 4,746 | 5,413 | 1,985 | 2,243 | 1,128,084 | 56,404 | 1,451,921 |
| | Madras | 399,753 | 264,093 | 2,775 | 1,489 | | | | | | | | | | | | | 7,113 | 4,222 | 2,435 | 2,107 | 61 | 52 | 412,137 | 20,607 | 271,963 |
| | Bombay | | | | | | | | | | | 63,456 | 73,088 | 1,610 | 1,825 | 1,866 | 2,282 | | | | | 355 | 581 | 67,287 | 3,364 | 77,776 |
| | ALL INDIA | 1,506,051 | 1,689,152 | 4,888 | 4,364 | 1,360 | 1,551 | 1,735 | 2,362 | | | 65,658 | 75,584 | 1,610 | 1,825 | 1,866 | 2,282 | 14,758 | 14,144 | 7,181 | 7,520 | 2,401 | 2,876 | 1,607,508 | 80,375 | 1,801,660 |
| 1852-53 | Bengal | 1,048,236 | 1,376,128 | | | 153 | 208 | 2,876 | 3,917 | | | 2,471 | 2,591 | | | | | 20,752 | 27,691 | 2,576 | 2,880 | 6,600 | 6,896 | 1,083,264 | 54,163 | 1,420,311 |
| | Madras | 307,624 | 228,346 | | | | | 461 | 206 | | | | | | | | | 20,471 | 9,738 | 3,271 | 2,781 | 115 | 134 | 331,942 | 16,597 | 241,205 |
| | Bombay | 770 | 847 | | | | | | | | | 57,023 | 62,286 | 2,164 | 2,380 | 1,999 | 2,200 | 89 | 98 | | | 396 | 435 | 62,441 | 3,122 | 68,246 |
| | ALL INDIA | 1,356,630 | 1,605,321 | | | 153 | 208 | 3,317 | 4,123 | | | 59,494 | 64,877 | 2,164 | 2,380 | 1,999 | 2,200 | 41,312 | 37,527 | 5,847 | 5,661 | 7,111 | 7,465 | 1,477,647 | 73,884 | 1,729,762 |
| 1853-54 | Bengal | 458,429 | 478,652 | 1,667 | 2,269 | | | 1,004 | 1,367 | 48 | 65 | 4,560 | 5,998 | 662 | 907 | | | 30,034 | 40,865 | 2,960 | 4,016 | 954 | 1,306 | 500,318 | 25,016 | 535,445 |
| | Madras | 493,712 | 305,982 | | | | | | | | | 1,197 | 658 | | | | | 7,774 | 4,112 | 1,035 | 709 | 46 | 28 | 503,764 | 25,188 | 311,489 |
| | Bombay | 22 | 24 | | | | | | | | | 78,633 | 85,578 | 3,067 | 3,374 | 2,078 | 2,286 | | | | | 172 | 190 | 83,972 | 4,199 | 91,452 |
| | ALL INDIA | 952,163 | 784,658 | 1,667 | 2,269 | | | 1,004 | 1,367 | 48 | 65 | 84,390 | 92,234 | 3,729 | 4,281 | 2,078 | 2,286 | 37,808 | 44,977 | 3,995 | 4,725 | 1,172 | 1,524 | 1,088,054 | 54,403 | 938,386 |
| 1854-55 | Bengal | 520,431 | 696,040 | 12,616 | 17,173 | 1,411 | 1,922 | 14,903 | 3,148 | 45 | 62 | 18,939 | 21,099 | | | | | 54,131 | 73,383 | 4,419 | 5,038 | 2,342 | 2,933 | 629,237 | 31,462 | 820,798 |
| | Madras | 187,954 | 145,825 | | | | | | | | | 41 | 32 | | | | | 17,273 | 8,744 | 1,573 | 1,180 | 13 | 2 | 206,841 | 10,342 | 155,783 |
| | Bombay | 362 | 452 | | | | | | | | | 102,121 | 127,551 | 1,661 | 2,076 | 2,262 | 2,828 | | | 27 | 34 | 429 | 535 | 106,862 | 5,343 | 133,476 |
| | ALL INDIA | 708,747 | 842,317 | 12,616 | 17,173 | 1,411 | 1,922 | 14,903 | 3,148 | 45 | 62 | 121,101 | 148,682 | 1,661 | 2,076 | 2,262 | 2,828 | 71,404 | 82,127 | 6,019 | 6,257 | 2,771 | 3,470 | 942,940 | 47,147 | 1,110,057 |
| 1855-56 | Bengal | 607,117 | 782,512 | 7,771 | 9,322 | 12,028 | 15,834 | 21,340 | 28,239 | 1,100 | 1,490 | 3,259 | 3,888 | | | | | 67,772 | 86,534 | 7,158 | 8,900 | 2,416 | 3,157 | 730,058 | 36,503 | 939,975 |
| | Madras | 430,280 | 279,897 | | | | | | | | | 352 | 377 | | | | | | | 2,499 | 1,770 | 13 | 20 | 433,144 | 21,657 | 282,064 |
| | Bombay | 28,728 | 30,004 | | | | | | | | | 79,683 | 87,832 | 3,435 | 3,721 | 1,534 | 1,688 | | | 180 | 200 | 298 | 338 | 113,858 | 5,693 | 123,783 |
| | ALL INDIA | 1,066,125 | 1,092,411 | 7,771 | 9,322 | 12,028 | 15,834 | 21,340 | 28,239 | 1,100 | 1,490 | 83,294 | 92,097 | 3,435 | 3,721 | 1,534 | 1,688 | 67,772 | 86,534 | 9,837 | 10,870 | 2,727 | 3,515 | 1,277,060 | 63,853 | 1,345,822 |
| 1856-57 | Bengal | 755,982 | 1,028,982 | 43,190 | 38,787 | 61,493 | 53,700 | 50,265 | 38,417 | 4,190 | 5,704 | 7,115 | 9,686 | 370 | 504 | | | 53,434 | 72,731 | 3,739 | 5,088 | 8,446 | 11,445 | 988,224 | 49,411 | 1,345,044 |
| | Madras | 502,281 | 355,118 | 585 | 491 | 132 | 242 | | | | | 491 | 551 | | | | | | | 1,696 | 1,123 | 33 | 38 | 505,218 | 25,261 | 357,563 |
| | Bombay | 8,929 | 5,364 | | | | | | | | | 61,922 | 73,051 | 1,323 | 1,328 | 2,586 | 3,232 | | | 69 | 86 | 400 | 408 | 75,229 | 3,761 | 83,469 |
| | ALL INDIA | 1,267,192 | 1,389,464 | 43,775 | 39,278 | 61,625 | 53,942 | 50,265 | 38,417 | 4,190 | 5,704 | 69,528 | 83,288 | 1,693 | 1,812 | 2,586 | 3,232 | 53,434 | 72,731 | 5,504 | 6,297 | 8,879 | 11,891 | 1,568,671 | 78,433 | 1,786,076 |
| 1857-58 | Bengal | 419,808 | 562,326 | 9,795 | 13,231 | 13,248 | 17,874 | 33,405 | 43,689 | | | 16,011 | 21,893 | 783 | 1,090 | | | 19,320 | 25,717 | 4,002 | 6,359 | 1,571 | 2,004 | 517,943 | 25,897 | 694,183 |
| | Madras | 329,654 | 331,116 | 23,449 | 22,458 | | | 5,035 | 5,318 | | | 48 | 80 | | | | | 6,993 | 6,339 | 934 | 895 | | | 366,164 | 18,308 | 366,196 |
| | Bombay | 2,283 | 2,740 | | | | | | | | | 68,845 | 81,749 | 3,653 | 5,919 | 2,894 | 3,473 | | | 104 | 125 | 242 | 293 | 78,021 | 3,901 | 94,299 |
| | ALL INDIA | 751,745 | 896,182 | 33,244 | 35,689 | 13,248 | 17,874 | 38,440 | 49,007 | | | 84,904 | 103,690 | 4,436 | 7,009 | 2,894 | 3,473 | 26,313 | 32,056 | 5,040 | 7,379 | 1,813 | 2,319 | 961,128 | 48,106 | 1,154,678 |
| 1858-59 | Bengal | 662,528 | 933,980 | 1,109 | 1,510 | 3,375 | 4,594 | 15,519 | 26,567 | 211 | 287 | 4,829 | 6,903 | 800 | 1,089 | | | 64,878 | 88,299 | 7,797 | 11,630 | 8,991 | 12,343 | 770,037 | 38,502 | 1,087,202 |
| | Madras | 233,175 | 211,187 | 1,572 | 1,408 | | | | | | | 373 | 450 | | | | | 1,780 | 1,569 | 928 | 1,006 | 43 | 32 | 237,871 | 11,894 | 215,652 |
| | Bombay | 445 | 476 | | | | | | | | | 98,751 | 117,946 | 2,459 | 2,811 | 3,877 | 4,653 | | | 88 | 106 | 266 | 319 | 105,886 | 5,294 | 126,311 |
| | ALL INDIA | 896,148 | 1,145,643 | 2,681 | 2,918 | 3,375 | 4,594 | 15,519 | 26,567 | 211 | 287 | 103,953 | 125,299 | 3,259 | 3,900 | 3,877 | 4,653 | 66,658 | 89,868 | 8,813 | 12,744 | 9,300 | 12,694 | 1,113,794 | 55,690 | 1,429,165 |
| 1859-60 | Bengal | 381,611 | 534,325 | | | 16,508 | 23,110 | 11,756 | 16,458 | 127 | 178 | 11,186 | 15,660 | 813 | 1,138 | | | 45,762 | 64,066 | 4,749 | 6,649 | 13,477 | 18,869 | 485,989 | 24,299 | 680,453 |
| | Madras | 304,022 | 243,917 | | | | | | | 3 | 5 | 100 | 119 | | | | | 2,935 | 1,982 | 1,933 | 2,141 | 12 | 6 | 309,002 | 15,450 | 248,170 |
| | Bombay | 1,857 | 2,228 | | | | | | | | | 58,469 | 70,040 | 3,180 | 3,817 | 1,225 | 1,470 | | | | | 279 | 336 | 65,010 | 3,251 | 77,891 |
| | ALL INDIA | 687,490 | 780,470 | | | 16,508 | 23,110 | 11,756 | 16,458 | 130 | 183 | 69,755 | 85,819 | 3,993 | 4,955 | 1,225 | 1,470 | 48,697 | 66,048 | 6,682 | 8,790 | 13,758 | 19,211 | 860,001 | 43,000 | 1,006,514 |
| 1860-61 | Bengal | 380,904 | 571,336 | 2 | 3 | 16,326 | 24,491 | 19,711 | 29,566 | 365 | 548 | 16,991 | 25,187 | 1,884 | 2,825 | | | 34,221 | 51,331 | 6,518 | 9,776 | 3,092 | 4,638 | 480,171 | 24,009 | 719,936 |
| | Madras | 313,902 | 263,410 | 223 | 112 | | | | | | | 409 | 364 | | | | | | | 2,898 | 2,637 | | | 317,432 | 15,872 | 256,594 |
| | Bombay | 1,206 | 964 | | | | | | | | | 42,343 | 49,253 | 2,990 | 3,557 | 1,327 | 1,592 | | | 150 | 180 | 342 | 410 | 48,358 | 2,418 | 55,956 |
| | ALL INDIA | 696,012 | 835,710 | 225 | 115 | 16,326 | 24,491 | 19,711 | 29,566 | 365 | 548 | 59,743 | 74,804 | 4,874 | 6,382 | 1,327 | 1,592 | 34,221 | 51,331 | 9,566 | 12,593 | 3,434 | 5,049 | 845,961 | 42,299 | 1,032,416 |

TABLE SHOWING THE QUANTITIES (AS FAR AS CAN BE ASCERTAINED) AND THE VALUE OF MOLASSES OR JAGREE EXPORTED FROM INDIA AND EACH PRESIDENCY TO ALL PARTS OF THE WORLD FROM 1857-58 TO 1860-61.

| Years | Whence Exported | COUNTRIES WHITHER EXPORTED | | | | | | | | | | | | | | TOTAL EXPORTED TO ALL PARTS | | |
| | | United Kingdom | | France | | Other Parts of Europe | | America | | China | | Arabian and Persian Gulfs | | Other Parts | | | | |
| | | Quan. | Value | Quan. | Value | Quan. | Value | Quan. | Value | Quan. | Value | Quan. | Value | Quan. | Value | Quantity | | Value |
| | | cwt. | £ | cwt. | £ | cwt. | £ | cwt. | £ | cwt. | £ | cwt. | £ | cwt. | £ | cwt. | tons | £ |
| 1857-58 | Bengal | 9,884 | 1,427 | .. | .. | .. | .. | 681 | 307 | 1 | 1 | 1,296 | 326 | *2,690 | *733 | 13,256 | 663 | 2,468 |
| | Madras | 59,122 | 25,630 | 1,270 | 578 | .. | .. | .. | .. | .. | .. | .. | 2,043 | *5,877 | *2,346 | 67,565 | 3,378 | 28,880 |
| | Bombay | .. | 14 | .. | .. | .. | .. | .. | .. | .. | .. | .. | .. | .. | 619 | .. | .. | 2,676 |
| | ALL INDIA | | 27,071 | 1,270 | 578 | .. | .. | 681 | 307 | 1 | 1 | .. | 2,369 | .. | 3,698 | .. | .. | 34,024 |
| 1858-59 | Bengal | 1,911 | 650 | .. | .. | .. | .. | .. | .. | .. | .. | .. | .. | *5,824 | *2,207 | 7,735 | 387 | 2,857 |
| | Madras | 79,769 | 35,496 | .. | .. | .. | .. | .. | .. | .. | .. | 164 | 49 | †2,584 | †751 | 82,517 | 4,126 | 36,296 |
| | Bombay | 104 | 82 | .. | .. | .. | .. | .. | .. | .. | .. | 2,302 | 893 | ‡1,050 | ‡425 | 3,456 | 173 | 1,400 |
| | ALL INDIA | 8,1784 | 36,228 | .. | .. | .. | .. | .. | .. | .. | .. | 2,466 | 942 | 9,458 | 3,383 | 93,708 | 4,686 | 40,553 |
| 1859-60 | Bengal | 1,532 | 201 | .. | .. | .. | .. | .. | .. | .. | .. | .. | .. | *11,358 | *1,546 | 12,890 | 644 | 1,747 |
| | Madras | 63,154 | 28,023 | .. | .. | .. | .. | .. | .. | .. | .. | 85 | 26 | †2,991 | †994 | 66,230 | 3,312 | 29,043 |
| | Bombay | .. | .. | .. | .. | .. | .. | .. | .. | .. | .. | 3,240 | 1,288 | ‡2,281 | ‡875 | 5,471 | 274 | 2,163 |
| | ALL INDIA | 64,686 | 28,224 | .. | .. | .. | .. | .. | .. | .. | .. | 3,325 | 1,314 | 16,580 | 3,415 | 84,591 | 4,230 | 32,953 |
| 1860-61 | Bengal | 6,639 | 2,656 | .. | .. | 392 | 157 | 7 | 3 | .. | .. | .. | .. | 36 | 14 | 7,074 | 354 | 2,830 |
| | Madras | 41,144 | 15,884 | .. | .. | .. | .. | .. | .. | .. | .. | 28 | 9 | †2,826 | †972 | 43,998 | 2,200 | 16,865 |
| | Bombay | 36 | 13 | .. | .. | .. | .. | .. | .. | .. | .. | 3,184 | 1,471 | ‡300 | ‡322 | 3,520 | 176 | 1,806 |
| | ALL INDIA | 47,819 | 18,553 | .. | .. | 392 | 157 | 7 | 3 | .. | .. | 3,212 | 1,480 | 3,162 | 1,308 | 54,592 | 2,730 | 21,501 |

* Principally to New South Wales.    † Principally to Ceylon.    ‡ Principally to Aden and Coast of Africa.

1699. [2740] Sugar (third quality). ASTAGRAM SUGAR COMPANY.

1700. [10576] Do. North Arcot.

1701. [10577] Do. Do.

1702. [4778] Jaggery, Rangoon. Messrs. HALLIDAY, FOX, & CO.

1703. [4781] Sugar, Do.

1704. [4149] Sugar from Caledonia Estate, Penang, the property of RIGHT HON. E. HORSMAN, M.P.

1705. [4150] Do.

1706. [4183] Do. from Golden Grove Estate, Province Wellesley.

1707. [4184] Cane sugar from Province Wellesley. Grown and manufactured on the Chinese system by JOW KAM MEAH, at Songhy Bakow.

1708. [4185] Do. Do. by TEO AH TOO, at Songhy Bakow.

1709. [4186] Cane sugar. Grown and manufactured on the Chinese system by JOW KAM MEAH, at Songhy Bakow.

1710. [4187] Do. Do. on Golden Grove Estate.

1711. [4188] Cane sugar. Grown and manufactured on the Chinese system by TEO AH TOO, at Songhy Bakow, Province Wellesley.

1712. [6667] Molasses (first quality), Singapore. JOSE D'ALMEIDA, Esq.

1713. [6666] Do. (second quality), Singapore. JOSE D'ALMEIDA, Esq.

The Date Palm (*Elate sylvestris*) furnishes almost the whole of the sugar exported in such considerable quantities from Calcutta.

1714. [7139] Kasee chinnee (*E. sylvestris*), Calcutta.

1715. [7140] Dhoba batta (*E. sylvestris*), Do.

1716. [7141] Suckhur (*E. sylvestris*), Do.

1717. [7144] Ball sugar, 'Oullah' (*E. sylvestris*), Do.

1718. [7142] Crushed sugar (*do.*), Do.

1719. [7143] Refined sugar (*do.*), Do.

1720. [6225] Goor from date juice (*E. sylvestris*), Do.

1721. [6231] Do. Beerbhoom.

MR. S. H. ROBINSON, of Calcutta, makes the following remarks on the date palm, and the manufacture of date sugar in Bengal:—

The date tree is met with in almost every part of Bengal Proper, but it flourishes most congenially, and is found plentifully only in the alluvial soils which cover its south-eastern portion, excepting only such tracts as suffer entire submersion annually from the overflow of their rivers, as is common in portions of the Dacca, Mymensing, and Sunderbund districts. The extent of country best suited for its growth, and over which it is found most plentifully as above indicated, may therefore be taken as within an area stretching east and west about 200 miles, and north and south about 100 miles, and comprehending by a rough estimate about 9,000 square miles, within an irregular triangular space.

When not stunted in its growth by the extraction of its juice for sugar, it is a very handsome tree, rising in Bengal from 30 to 40 feet in height, with a dense crown of leaves spreading in a hemispherical form from its summit. These leaves are from 10 to 15 feet long, and composed of numerous leaflets or pinnules about 18 inches long. The trunk is rough, from the adherence of the bases of the falling leaves; this serves to distinguish it at a glance from the smooth-trunked cocoa-nut palm, which in its leaves only it resembles. The fruit consists more of seed than of pulp, and altogether is only about one-fourth the size of the Arabian kind brought annually to Calcutta for sale, and, when fresh imported, a rich and favourite fruit there. This inferiority of the Bengal fruit may no doubt be attributed to the entire neglect of its improvement there from time immemorial, and, perhaps, in some measure, to the practice of tapping the trees for their sap, so universally followed in the districts around Calcutta, its principal range of growth.

The process of tapping and extracting the juice commences about the 1st of November, and terminates about the 15th of February. Some days previously, the lower leaves of the crown are stripped off all round, and a few extra leaves from the side of the tree intended to be tapped. On the part thus denuded, a triangular incision is made with a knife, about an inch deep, so as to penetrate through the cortex, and divide the sap vessels; each side of the triangle measuring about 6 inches, with one point downwards, in which is inserted a piece of grooved bamboo, along which the sap trickles, and from thence drops into an earthen pot suspended underneath it by a string. The pots are suspended in the evening, and removed very early the following morning, ere the sun has sufficient power to warm the juice, which would cause it immediately to ferment, and destroy its quality of crystallizing into sugar.

The cutting being made in the afternoon, next morning the pot is found to contain, from a full-grown tree, 10 seers of juice, the second morning 4 seers, and the third morning 2 seers of juice; the quantity exuding afterwards is so small, that no pot is suspended for the next four days.

Daily at sunrise, throughout the goor season, the industrious ryot may be seen climbing his trees, and collecting at a convenient spot beneath them the earthen pots containing the juice yielded during the past night. Under a rude shed, covered with the leaves of the date tree itself, and erected under the shade of the plantation, is prepared the boiling apparatus to serve for the goor season. It consists of a hole of about 3 feet in diameter, sunk about 2 feet in the ground, over which are supported by mud arches four thin earthen pans of a semi-globular shape, and 18 inches in diameter; the hole itself is the furnace, and has two apertures on opposite sides for feeding in the fuel, and for escape of the smoke. The fire is lit as soon as the juice is collected, and poured into the four pans, which are kept constantly supplied with fresh juice as the water evaporates, until the whole produce of the morning is boiled down to the required density. As the contents of each pan become sufficiently boiled, they are ladled out into other earthen pots or jars, of various sizes, from 5 to 20 seers of contents, according to local custom, and in these the boiled extract cools, crystallizes into a hard compound of granulated sugar and molasses, and is brought to market for sale as *goor.*

The subsequent processes by which the goor is deprived more or less of its molasses and impurities are too long to be detailed.

1722. [6880] Dhannee (*Nipa fruticans*), Moulmein.

This molasses is made out of a plant called *Dhannee.**

1723. [52*] Nipa sugar (*N. fruticans*).

1724. [59*] Gomuti palm sugar (*Borassus flabelliformis*), Java.

1725. [5187] Palm sugar (*do.*), Burmah.

This coarse substitute for sugar is obtained from the toddy of the Palmyra tree. The sugar-cane grows and thrives admirably in Pegu, but it is not very largely cultivated, and none but the very coarsest sugar is manufactured from it.†

## ANIMAL FOOD PRODUCTS.

### Fish Maws and Shark Fins.

As will be gathered from the tables (p. 77), these form a considerable article of export.

1726. [6703] Fish maw, Eastern Archipelago. G. ANGUS, Esq.

1727. [5579] Do. Akyab.

1728. [5580] White shark fins, Do.

1729. [5581] Black shark fins, Akyab.

1730. [5582] White shark back fins, Do.

1731. [4157] Edible birds' nests (*Collocalia nidifica*) from islands adjoining Junk Ceylon.

1732. [4158] Do.      Do.      Do.

1733. [4159] Do.      Do.      Do.

1734. [4160] Do.      Do.      Do.

1735. [4155] Beche de Mer or Trepang (*Holothuria*, species of,) Penang.

1736. [6671] Do. Eastern Archipelago.

1737. [4156] Do. Penang.

1738. [4153] Dried shell fish, Do.

1739. [4154] Do.      Do.      Do.

1740. [2367] Coorg honey, from Mysore.

---

* Local Committee, *Moulmein.*
† Local Committee, *Rangoon.*

TABLE SHOWING THE QUANTITIES (AS FAR AS CAN BE ASCERTAINED) AND THE VALUE OF FISH MAWS EXPORTED FROM INDIA AND EACH PRESIDENCY TO ALL PARTS OF THE WORLD FROM 1857-58 TO 1860-61.

| YEARS | WHENCE EXPORTED | COUNTRIES WHITHER EXPORTED | | | | | | TOTAL EXPORTED TO ALL PARTS | | |
|---|---|---|---|---|---|---|---|---|---|---|
| | | UNITED KINGDOM | | CHINA | | OTHER PARTS | | | | |
| | | Quantity | Value | Quantity | Value | Quantity | Value | Quantity | | Value |
| | | cwt. | £ | cwt. | £ | cwt. | £ | cwt. | tons | £ |
| 1857-58 | Bengal | .. | .. | 42 | 265 | 7,237 | 1,208 | 7,279 | 364 | 1,473 |
| | Madras | 2 | 4 | .. | .. | 15 | 44 | 17 | 1 | 48 |
| | Bombay | 36 | 184 | 1,587 | 7,273 | 33 | 84 | 1,656 | 83 | 7,541 |
| | ALL INDIA | 38 | 188 | 1,629 | 7,538 | 7,285* | 1,336* | 8,952 | 448 | 9,062 |
| 1858-59 | Bengal | .. | .. | 113 | 770 | 128 | 525 | 241 | 12 | 1,295 |
| | Madras | .. | .. | .. | .. | 2 | 3 | 2 | .. | 3 |
| | Bombay | 180 | 899 | 1,449 | 6,271 | .. | .. | 1,629 | 81 | 7,170 |
| | ALL INDIA | 180 | 899 | 1,562 | 7,041 | 130* | 528* | 1,872 | 93 | 8,468 |
| 1859-60 | Bengal | .. | .. | .. | 431 | 138 | 507 | .. | .. | 938 |
| | Madras | 32 | 91 | 2 | 6 | 39 | 111 | 73 | 4 | 208 |
| | Bombay | 253 | 1,247 | 1,184 | 5,054 | .. | .. | 1,437 | 72 | 6,301 |
| | ALL INDIA | 285 | 1,338 | .. | 5,491 | 177* | 618* | .. | .. | 7,447 |
| 1860-61 | Bengal | .. | .. | .. | 137 | .. | .. | .. | .. | 137 |
| | Madras | 68 | 279 | 2 | 10 | 47 | 171 | 117 | 6 | 460 |
| | Bombay | 733 | 2,876 | 854 | 2,928 | .. | .. | 1,587 | 79 | 5,804 |
| | ALL INDIA | 801 | 3,155 | .. | 3,075 | 47* | 171* | .. | .. | 6,401 |

* Comprise exports to Straits Settlements.

TABLE SHOWING THE QUANTITIES (AS FAR AS CAN BE ASCERTAINED) AND THE VALUE OF SHARK FINS EXPORTED FROM INDIA AND EACH PRESIDENCY TO ALL PARTS OF THE WORLD FROM 1857-58 TO 1860-61.

| YEARS | WHENCE EXPORTED | COUNTRIES WHITHER EXPORTED | | | | | | TOTAL EXPORTED TO ALL PARTS | | |
|---|---|---|---|---|---|---|---|---|---|---|
| | | UNITED KINGDOM | | CHINA | | OTHER PARTS * | | | | |
| | | Quantity | Value | Quantity | Value | Quantity | Value | Quantity | | Value |
| | | cwt. | £ | cwt. | £ | cwt. | £ | cwt. | tons | £ |
| 1857-58 | Bengal | .. | .. | .. | .. | .. | .. | .. | .. | .. |
| | Madras | 1 | 2 | .. | .. | 218 | 651 | 219 | 11 | 653 |
| | Bombay | .. | .. | 7,433 | 18,338 | 15 | 44 | 7,448 | 352 | 18,382 |
| | ALL INDIA | 1 | 2 | 7,433 | 18,338 | 233 | 695 | 7,667 | 363 | 19,035 |
| 1858-59 | Bengal | .. | .. | .. | .. | .. | .. | .. | .. | .. |
| | Madras | 4 | 10 | .. | .. | 229 | 676 | 233 | 12 | 686 |
| | Bombay | .. | .. | 8,062 | 16,126 | 2 | 5 | 8,064 | 403 | 16,131 |
| | ALL INDIA | 4 | 10 | 8,062 | 16,126 | 231 | 681 | 8,297 | 415 | 16,817 |
| 1859-60 | Bengal | .. | .. | .. | .. | .. | .. | .. | .. | .. |
| | Madras | .. | .. | .. | .. | 148 | 469 | 148 | 7 | 469 |
| | Bombay | .. | .. | 6,216 | 12,435 | .. | .. | 6,216 | 311 | 12,435 |
| | ALL INDIA | .. | .. | 6,216 | 12,435 | 148 | 469 | 6,364 | 318 | 12,904 |
| 1860-61 | Bengal | .. | .. | .. | .. | .. | .. | .. | .. | .. |
| | Madras | .. | .. | 84 | 222 | 101 | 362 | 185 | 9 | 584 |
| | Bombay | .. | .. | 5,564 | 11,127 | .. | .. | 5,564 | 278 | 11,127 |
| | ALL INDIA | .. | .. | 5,648 | 11,349 | 101 | 362 | 5,749 | 287 | 11,711 |

* Consisting of the Straits Settlements only.

# SECTION C.—SPIRITS; INTOXICATING OR STIMULATING DRUGS, ETC.

1741. [10689] Palmyra arrack (*Borassus flabelliformis*), Mangalore. V. P. COELHO, Esq.

Other interesting arracks sent by V. P. COELHO were lost in transit by the breaking of the bottles.

1742. [5717] Sugar-cane spirit (*Saccharum officinarum*), Lucknow.

1743. [2746] Spirits of wine. ASTAGRAM SUGAR COMPANY.

1744. [2172*] Country spirit (*Backur Khatee*), Calcutta.

1745. [2174*] Country rum, Do.

1746. [10129] Rum, Calcutta. KANNY LOLL DEY.

1747. [10128] Doasta or country spirit, Do. Do.

1748. [6415] Medicinal arrack, Singapore. TAN KIM SING.

1749. [2747] Rum, Mysore. ASTAGRAM SUGAR COMPANY.

1750. [6340] Do. Singapore. JOSE D'ALMEIDA, Esq.

1751. [5351] Do. Shahjehanpore. Messrs. CAREW & Co.

1752. [5728] Rice arrack (*Oryza sativa*), Cuttack.

A spirit distilled from rice. This is the only distilled spirit used by the natives of this province, and that only by those of the lower classes. It is the same to the use of which the wild tribes of Orissa, the Khonds, Sahars, and Coles are so addicted. It is unpalatable and nauseous. It is made 25 below London proof, 1 maund of rice making 8 gallons. An intoxicating spirit is distilled also in the Sumbulpore district, chiefly from the fruit or flower of the *Bassia latifolia*, the *Mahool, as locally called.* This tree is also met with throughout the forest jungles of this province: the sweet fruit or flower is a favourite food of wild animals, especially the bear, and it is believed that the saccharine matter, which apparently abounds in the fruit or flower, whichever it may be, might be turned to the very best account.*

1753. [4196] Rice arrack, from Penang rice, Penang.

1754. [5727] Chellee, Midnapore.

1755. [5726] Do. Do.

1756. [2168] Country spirit, Cumlaha, Calcutta.

* Local Committee, *Cuttack.*

1757. [2173] Country spirit, Pattaha, Calcutta.

1758. [2167] Do. Allachee, Do.

1759. [2169] Do. Aumish, Do.

1760. [2171] Do. Atturee, Do.

1761. [2170] Do. Joobabee, do.

These ardent spirits are distilled from sugar-cane, and used by the Hindoos of the lower order. Backerkhatee is the spirit distilled, in which cardamom is put and weakened with water, and called 'Allachee;' 'Cumlaha' with orange peel; 'Joobabee' and 'Pattaha' are adulterated with tobacco leaf, and 'Atturee' is scented with uttur. 'Aunish' is the only pure spirit distilled from aniseed.

1762. [1934] Mango spirit (*Mangofera indica*), Malda. DR. THOMPSON.

Prepared from the mango, a fruit well known, cheap, and to be had in abundance in Bengal and in many parts of India. The taste of the spirit is not unlike whisky, and far superior to anything of the sort sold in our Indian bazaar for every purpose to which the latter is applied. The specific gravity of that in the phial is about 903·5, which at a temperature of 80° F. gives about 60 per cent. of alcohol to the volume of spirit.

## OPIUM.

The extent to which this drug is exported to China and other parts, is shown by the Table. (*See next page.*)

### Series Illustrative of the Manufacture of Opium at Patna.

1763. [5378] Poppy seed (*Papaver somniferum*).

1764. [5379] Poppy-seed oil.

1765. [42*] Poppy capsules as cut by the nushtur for the purpose of collecting the opium.

1766. [6535½] Nushtur for scratching the capsules.

1767. [5380] Abkaree opium cake.

1768. [5381] Medicinal opium cake.

1769. [5382] Bottle of lewah.

1770. [    ] Tawa or iron plate.

1771. [5383] Bottle of morphia.

1772. [5384] Bottle of narcotine.

## TABLE SHOWING THE QUANTITIES (AS FAR AS CAN BE ASCERTAINED) AND THE VALUE OF OPIUM EXPORTED FROM INDIA AND EACH PRESIDENCY TO ALL PARTS OF THE WORLD FROM 1850-51 TO 1860-61.

**COUNTRIES WHITHER EXPORTED**

| Years | Whence Exported | America Qty (chests) | America Value (£) | Ceylon Qty (chests) | Ceylon Value (£) | China Qty (chests) | China Value (£) | Java and Sumatra Qty (chests) | Java and Sumatra Value (£) | Pegu and the Straits Qty (chests) | Pegu and the Straits Value (£) | New South Wales Qty (chests) | New South Wales Value (£) | Other Parts Qty (chests) | Other Parts Value (£) | Total Qty (tons*) | Total Qty (chests) | Total Value (£) |
|---|---|---|---|---|---|---|---|---|---|---|---|---|---|---|---|---|---|---|
| 1850-51 | Bengal | | | | | 28,892 | 2,777,518 | 100 | 9,206 | 3,910 | 368,350 | | | | | 2,203 | 32,902 | 3,155,074 |
| | Madras | | | | | | | | | | | | | | | | | |
| | Bombay | | | | | 19,138 | 2,296,560 | | | 62 | 7,500 | | | | | 1,286 | 19,200 | 2,304,060 |
| | All India | | | | | 48,030 | 5,074,078 | 100 | 9,206 | 3,972 | 375,850 | | | | | 3,489 | 52,102 | 5,459,134 |
| 1851-52 | Bengal | | | | | 27,921 | 2,713,469 | 50 | 5,035 | 4,335 | 419,277 | | | 1 | 110 | 2,163 | 32,306 | 3,187,781 |
| | Madras | | | | | | | | | | | | | | | | | |
| | Bombay | | | | | 28,168 | 3,368,838 | | | 73 | 8,485 | | | | | 1,891 | 28,242 | 3,377,433 |
| | All India | | | | | 56,089 | 6,082,307 | 50 | 5,035 | 4,408 | 427,762 | | | 1 | 110 | 4,054 | 60,548 | 6,515,214 |
| 1852-53 | Bengal | | | | | 31,433 | 3,482,948 | | | 4,745 | 537,146 | | | | | 2,422 | 36,178 | 4,020,094 |
| | Madras | | | | | | | | | | | | | | | | | |
| | Bombay | | | | | 24,979½ | 2,987,967 | | | 239 | 25,950 | | | ¼ | 64 | 1,689 | 25,219 | 3,013,981 |
| | All India | | | | | 56,412½ | 6,470,915 | | | 4,984 | 563,096 | | | ¼ | 64 | 4,111 | 61,397 | 7,034,075 |
| 1853-54 | Bengal | | | | | 33,941 | 3,069,895 | | | 7,970 | 618,485 | 5 | 481 | 1 | 102 | 2,807 | 41,917 | 3,688,963 |
| | Madras | | | | | | | | | | | | | | | | | |
| | Bombay | | | | | 26,113 | 2,732,574 | | | 142 | 15,380 | | | 2 | 181 | 1,758 | 26,257 | 2,748,135 |
| | All India | | | | | 60,054 | 5,802,469 | | | 8,112 | 633,865 | 5 | 481 | 3 | 283 | 4,565 | 68,174 | 6,437,098 |
| 1854-55 | Bengal | | | 6 | 390 | 43,952 | 3,150,320 | 30 | 2,025 | 7,430 | 541,858 | 2 | 142 | 1 | 81 | 3,443 | 51,421 | 3,694,816 |
| | Madras | | | | | | | | | | | | | | | | | |
| | Bombay | | | | 18 | 25,958 | 2,534,657 | | | 17 | 1,700 | | | 1 | 86 | 1,739 | 25,976 | 2,636,461 |
| | All India | | | 6 | 408 | 69,910 | 5,684,977 | 30 | 2,025 | 7,447 | 543,558 | 2 | 142 | 2 | 167 | 5,182 | 77,397 | 6,331,277 |
| 1855-56 | Bengal | 15 | 982 | 10 | 667 | 37,851 | 3,031,735 | | | 7,018 | 600,457 | 41 | 4,847 | 3 | 253 | 3,009 | 44,938 | 3,638,941 |
| | Madras | | | | | | | | | | | | | | | | | |
| | Bombay | | | | | 25,576 | 2,560,797 | | | 90 | 900 | | | 2 | 233 | 1,719 | 25,668 | 2,561,930 |
| | All India | 15 | 982 | 10 | 667 | 63,427 | 5,592,532 | | | 7,108 | 601,357 | 41 | 4,847 | 5 | 486 | 4,728 | 70,606 | 6,200,871 |
| 1856-57 | Bengal | | | 6 | 544 | 36,459 | 3,276,751 | | | 5,767 | 522,619 | 208 | 18,785 | 1 | 104 | 2,842 | 42,441 | 3,818,803 |
| | Madras | | | | | | | | | | | | | | | | | |
| | Bombay | | | | | 29,846 | 3,228,836 | | | 96 | 8,600 | | | 2 | 390 | 2,005 | 29,944 | 3,237,826 |
| | All India | | | 6 | 544 | 66,305 | 6,505,587 | | | 5,863 | 531,219 | 208 | 18,785 | 3 | 494 | 4,847 | 72,385 | 7,056,629 |
| 1857-58 | Bengal | | | | | 31,878 | 3,913,086 | | | 6,666 | 826,309 | 67 | 6,534 | | 60 | 2,586 | 38,611 | 4,745,929 |
| | Madras | | | | | | | | | | | | | | | | | |
| | Bombay | | | | | 36,125 | 4,327,945 | | | 229 | 32,700 | | | 1 | 60 | 2,434 | 36,355 | 4,360,700 |
| | All India | | | | | 68,003 | 8,241,031 | | | 6,895 | 859,009 | 67 | 6,534 | 1 | | 5,020 | 74,966 | 9,106,629 |
| 1858-59 | Bengal | 10 | 1715 | 40 | 5,775 | 33,858 | 5,050,103 | 20 | 3,217 | 746 | 112,299 | 21 | 3,235 | | | 2,322 | 34,685 | 5,174,629 |
| | Madras | | | | | | | | | | | | | | | | | |
| | Bombay | | | | | 40,849 | 5,610,548 | | | 288 | 42,464 | | | | | 2,755 | 41,137 | 5,653,012 |
| | All India | 10 | 1715 | 40 | 5,775 | 74,707 | 10,660,651 | 20 | 3,217 | 1,034 | 154,763 | 21 | 3,235 | | | 5,077 | 75,822 | 10,827,641 |
| 1859-60 | Bengal | | | | | 22,329 | 3,661,830 | 5 | 820 | 3,483 | 636,110 | 112 | 18,785 | 11 | 1,813 | 1,737 | 25,950 | 4,321,078 |
| | Madras | | | | | | | | | | | | | | | | | |
| | Bombay | | | | | 32,534 | 4,704,505 | | | 192 | 28,120 | 2 | 170 | 3 | 525 | 2,192 | 32,731 | 4,733,320 |
| | All India | | | | | 54,863 | 8,366,335 | 5 | 820 | 3,675 | 664,230 | 114 | 18,955 | 14 | 2,338 | 3,929 | 58,681 | 9,054,393 |
| 1860-61 | Bengal | | | | | 15,714 | 2,862,168 | | | 3,810 | 711,066 | 10 | 1,860 | | 20 | 1,308 | 19,534 | 3,575,114 |
| | Madras | | | | | | | | | | | | | | | | | |
| | Bombay | 2 | 2 | | | 43,691 | 6,566,719 | | | 263 | 42,598 | | | 2 | 280 | 2,943 | 43,956 | 6,609,699 |
| | All India | 2 | 1 | | | 59,405 | 9,428,887 | | | 4,073 | 753,664 | 10 | 1,860 | 2 | 300 | 4,251 | 63,490 | 10,184,773 |

\* In Bengal, about 164 lbs. go to the chest of opium; in Bombay, about 140 lbs. In calculating the amount in tons, 150 has been the factor employed.

1773. [5385] Powdered opium.

1774. [5386] Model of the manufacturer, with attendant and apparatus.

1775. [5387] Provision opium cakes.

1776. [6525] Model mutchan, with wooden balls and earthen cups.

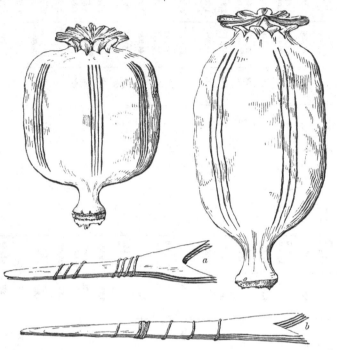

Poppy capsules and nushturs (*a*, *b*) of their natural size. The poppy capsules have been scarified in parallel lines.*

1777. [5388] Jar of opium as received from cultivator.

1778. [6527] Tin tagar.

1779. [6528] Iron lowah.

1780. [6529] Brass moulding cup.

1781. [6030] Brass lowah cup.

1782. [6531] Brass ladle.

1783. [6532] Tin puttee.

1784. [6533] Model provision chest.

1785. [6534] Model Abkara chest.

1786. [6537] Leaves for opium cake.

1787. [6526] Jar and cover for opium.

1788. [6536] Two sitoohas for collecting the opium.

1789. [6537] Press for forming the Ab-karee cake.

1790. [6539] Karhaee.

**Illustrations of Opium Manufacture from the Benares Opium Agency.**

1791. [5390] Opium cake-maker's seat.

1792. [5391] Lewah measure.

Natives engaged in scraping off the exuded juice.

* Figures illustrating opium manufacture from the *Pharmaceutical Journal.*

1793. [5391*b*] Box of poppy flowers.

1794. [5392] Box of poppy wash.

1795. [5393] Leaves for packing opium.

1796. [5394] Cake mould.

1797. [5389] Ball of provision opium.

Figure representing the section of an opium store.

1798. [5389*b*] Ball cut in section.

1799. [5414] Ball opium.

Figure showing a native engaged in the manufacture of ball opium.

1800. [7099] Opium from Lucknow.

1801. [10115] Opium, Calcutta.

———

1802. [1865] Nirmasor (antidote for opium), Punjab.

### CHURRUS OR HEMP RESIN.

1803. [10130] Bang (*Cannabis indica*), Calcutta.

1804. [10131] Do. Do.

1805. [6127] Do. Ahmedabad.

1806. [6128] Do. Bombay.

1807. [6129] Gunjah (*C. indica*) Ahmedabad.

1808. [6131] Do. Poona.

1809. [2891] Majoom.

1810. [10134] Sidee majoon, Bengal.

1811. [1853] Bhang prepared with sugar, Punjab.

1812. [10132] Churrus or hemp resin, Nepaul.

1813. [10133] Do. Gorruckpore.

1814. [6557] Do. Lucknow.

The narcotic properties of hemp become concentrated in a resinous juice, which in certain seasons and in tropical countries exudes, and concretes on the leaves, slender stems, and flowers. This constitutes the base of all the hemp preparations, to which all the powers of the drug are attributable. In Central India, the hemp resin, called *churrus*, is collected during the hot season in the following manner:—Men clad in leathern dresses run through the hemp fields, brushing through the plants with all possible violence; the soft resin adheres to the leather, and is subsequently scraped off and kneaded into balls, which sell at from 5 to 6 rupees the seer, or about 5*s*. or 6*s*. per pound. A still finer kind, the *momeca* or waxen churrus, is collected by the hand in Nepaul, and sells for nearly double the price of the ordinary kind. Dr. M'Kinnon says — 'In Nepaul, the leathern attire is dispensed with, and the resin is collected on the skin of naked coolies.' In Persia the churrus is obtained by pressing the resinous plant on coarse cloths, and then scraping it from these and melting it in a pot with a little warm water. Mirza considers the churrus of Herat the most powerful of all the varieties of the drug. The hemp resin, when pure, is of a blackish grey colour, with a fragrant narcotic odour, and a slightly warm, bitterish, acrid taste.

1815. [7000] Gunjah (a substitute, and not the true *Cannabis indica*).

1816. [10735] Betel nuts (*Areca cate-chu*), Mangalore. V. P. COELHO, Esq.

BETEL NUT PALM.*

1817. [4067] Betel nuts (*A. catechu*).

1818. [4109] Do. Penang.

1819. [10729] Do. Mangalore. V. P. COELHO, Esq.

1820. [4776] Do. Burmah. Messrs. HALLIDAY, FOX, & Co.

1821. [4792] Do. Do. Do.

1822. [2700] Do. Mysore.

1823. [5367] Betel nut cutter, Bengal.

The Areca palm, which supplies the betel nut, is known by the Malay name *Pinang*, whence also the name of the island Penang, which is now the chief emporium of the trade. There are various kinds in use, and the mode of preparation also differs. The three ingredients of the betel nut, as commonly used, are, the sliced nut, the leaf of the betel pepper in which the nut is rolled, and chunam or powdered lime, which is smeared over the leaf.

Of the absolute quantity of betel nuts which are employed no accurate estimate can be given. Prof. Johnston calculated that they are chewed by at least fifty millions of the human race.

1824. [10748] Byne seeds (*Caryota urens*) (substitute for betel nuts), Mangalore. V. P. COELHO, Esq.

1825. [3260] Imitation paun in clay, Bengal.

1826. [9388] Chunam or powdered shell lime.

1827. [6126] *Datura hammatii* var. *fastuosa*, Ahmedabad.

To facilitate theft, and for other criminal purposes, the seeds mixed with sweetmeats are very commonly administered in Bengal. The individual sinks into profound lethargy: the respiration is natural, but the pupils of the eye become greatly dilated. These symptoms have been known to continue two days.

1828. [7442] White gambir.

1829. [6662] Gambir (*Nauclea gambir*), Malay Peninsula.

1830. [5342] Catechu (*Acacia catechu*), Lucknow.

1831. [10102] Kaha kehan (spiced catechu), Calcutta.

## TOBACCO.

The Table (p. 83) shows the quantity exported to all parts.

1832. [2682] Tobacco (*Nicotiana tabacum*), Mysore.

1833. [4097] Tobacco, Province Wellesley.

1834. [4098] Leaf tobacco, Bukel Matagam, Do.

1835. [4170] Tobacco, Penang.

1836. [470] Do. Do.

1837. [1485] Do. Rangoon. Messrs. HALLIDAY, FOX, & Co.

1838. [5324] Do. Akyab, Arracan.

1839. [5325] Do. Do. Do.

1840. [5326] Do. Do. Do.

1841. [5327] Do. Rangoon.

1842. [5328] Do. Cuttack.

1843. [5335] Do. Do. Do.

1844. [5337] Do. Assam. Do. LIEUT. W. PHAIRE.

1845. [6125] Do. Ahmedabad.

1846. [6130] Do. Poona.

1847. [6556] Do. Midnapore.

1848. [6558] Do. Lucknow.

# TABLE SHOWING THE QUANTITIES (AS FAR AS CAN BE ASCERTAINED) AND THE VALUE OF TOBACCO EXPORTED FROM INDIA AND EACH PRESIDENCY TO ALL PARTS OF THE WORLD FROM 1850-51 TO 1860-61.

COUNTRIES WHITHER EXPORTED

| Years | Whence Exported | United Kingdom Qty (lbs) | United Kingdom Value (£) | France Qty (lbs) | France Value (£) | Other Parts of Europe Qty (lbs) | Other Parts of Europe Value (£) | America Qty (lbs) | America Value (£) | China Qty (lbs) | China Value (£) | Arabian and Persian Gulfs Qty (lbs) | Arabian and Persian Gulfs Value (£) | Other Parts* Qty (lbs) | Other Parts* Value (£) | Total Qty (lbs) | Total (tons) | Total Value (£) |
|---|---|---|---|---|---|---|---|---|---|---|---|---|---|---|---|---|---|---|
| 1850-51 | Bengal |  | 1,494 |  | 77 |  | 4 |  | 636 |  | 21 |  | 166 |  | 7,564 |  |  | 9,962 |
|  | Madras | 827 | 131 |  |  |  |  |  |  |  |  |  | 33 |  | 10,983 |  |  | 11,147 |
|  | Bombay |  | 260 |  |  |  |  |  |  | 394 | 16 | 103,483 | 3,869 | 16,609 | 619 | 121,313 | 54 | 4,764 |
|  | All India |  | 1,885 |  | 77 |  | 4 |  | 636 | 394 | 37 |  | 4,068 |  | 19,166 |  |  | 25,873 |
| 1851-52 | Bengal |  | 1,452 |  | 7 | 37,968 | 135 |  | 641 |  | 2 |  | 55 |  | 5,478 |  |  | 7,713 |
|  | Madras |  | 192 |  | 3 |  |  |  |  |  |  |  |  |  | 13,040 |  |  | 13,292 |
|  | Bombay | 431 | 307 |  |  |  |  | 5 | 2 | 130 | 7 | 47,579 | 1,828 | 6,549 | 308 | 54,694 | 24 | 2,452 |
|  | All India |  | 1,951 |  | 10 | 37,968 | 135 | 5 | 643 | 130 | 9 |  | 1,883 |  | 18,826 |  |  | 23,457 |
| 1852-53 | Bengal |  | 242 |  | 10 |  |  |  | 433 |  | 2 |  | 32 |  | 5,559 |  |  | 6,276 |
|  | Madras |  | 43 |  | 3 |  |  |  |  |  |  |  | 44 |  | 12,509 |  |  | 12,601 |
|  | Bombay | 432 | 152 |  |  |  |  |  |  | 261 | 3 | 2,312 | 143 | 12,885 | 652 | 15,890 | 7 | 950 |
|  | All India |  | 437 |  | 13 |  |  |  | 433 | 261 | 5 |  | 219 |  | 18,720 |  |  | 19,827 |
| 1853-54 | Bengal |  | 2,693 |  | 451 |  |  |  | 511 |  | 35 |  | 112 |  | 9,940 |  |  | 13,742 |
|  | Madras | 349 | 34 |  | 1 |  |  |  |  |  |  |  | 6 |  | 5,574 |  |  | 6,615 |
|  | Bombay |  | 175 |  |  |  |  |  |  | 227 | 9 | 1,248 | 98 | 123,355 | 4,779 | 125,179 | 56 | 5,061 |
|  | All India |  | 2,902 |  | 452 |  |  |  | 511 | 227 | 44 |  | 216 |  | 20,293 |  |  | 24,418 |
| 1854-55 | Bengal |  | 1,115 |  | 447 |  |  | 6,832 | 252 |  | 71 | 45,808 | 226 |  | 5,921 |  |  | 8,092 |
|  | Madras | 605 | 151 |  | 37 |  |  |  |  | 16 | 4 |  | 10 |  | 10,005 |  |  | 10,207 |
|  | Bombay |  | 23 |  |  |  |  |  |  |  |  | 4,698 | 364 | 57,332 | 2,212 | 62,651 | 28 | 2,600 |
|  | All India |  | 1,289 |  | 484 |  |  | 6,832 | 252 | 16 | 76 | 50,506 | 600 |  | 18,138 |  |  | 20,859 |
| 1855-56 | Bengal |  | 3,306 |  | 742 | 28,896 | 129 |  | 224 |  | 453 | 18,816 | 90 |  | 7,925 |  |  | 12,869 |
|  | Madras | 1,032 | 223 |  | 33 |  |  |  |  |  | 1 |  | 1 |  | 7,496 |  |  | 7,754 |
|  | Bombay |  | 388 | 166 | 5 |  |  |  |  |  |  | 5,016 | 464 | 13,714 | 1,008 | 19,928 | 9 | 1,865 |
|  | All India |  | 3,917 |  | 780 | 28,896 | 129 |  | 224 |  | 454 | 23,832 | 555 |  | 16,429 |  |  | 22,488 |
| 1856-57 | Bengal |  | 8,645 |  | 432 | 495,264 | 2,443 |  | 332 |  | 97 | 50,960 | 248 |  | 15,224 |  |  | 27,421 |
|  | Madras | 1,057 | 118 |  |  |  |  |  |  |  |  |  | 6 |  | 7,549 |  |  | 7,739 |
|  | Bombay |  | 361 |  |  |  | 54 |  | 4 | 11 | 2 | 298,796 | 1,998 | 13,491 | 441 | 813,355 | 140 | 2,802 |
|  | All India |  | 9,124 |  | 440 | 495,264 | 2,497 |  | 336 | 11 | 99 | 349,756 | 2,252 |  | 23,214 |  |  | 37,962 |
| 1857-58 | Bengal | 258,644 | 27,147 | 232,288 | 1,290 | 162,260 | 1,022 | 18,612 | 243 | 5,852 | 51 | 18,340 | 141 |  | 10,027 |  |  | 39,921 |
|  | Madras |  | 3,774 |  | 11 |  | 837 |  |  |  |  |  | 2 |  | 8,206 |  |  | 12,830 |
|  | Bombay |  | 3,831 | 1,665 | 12 | 19,312 | 369 |  |  | 77 | 9 | 6,614 | 363 | 8,632 | 1,001 | 294,944 | 132 | 5,585 |
|  | All India |  | 34,752 |  | 1,313 |  | 2,228 | 18,612 | 243 | 5,929 | 60 | 24,954 | 506 |  | 19,234 |  |  | 58,336 |
| 1858-59 | Bengal |  | 6,961 |  | 813 | 479,388 | 4,084 |  | 2,567 | 644 | 5 | 60,088 | 394 |  | 16,844 |  |  | 31,668 |
|  | Madras | 4,467 | 779 |  | 4 |  |  |  |  |  |  |  | 40 |  | 11,370 |  |  | 12,198 |
|  | Bombay |  | 362 |  |  | 472 | 85 |  |  | 425 | 36 | 6,349 | 476 | 5,884 | 497 | 17,597 | 8 | 1,456 |
|  | All India |  | 8,102 |  | 817 | 479,860 | 4,169 |  | 2,567 | 1,069 | 41 |  | 910 |  | 28,711 |  |  | 45,317 |
| 1859-60 | Bengal |  | 7,058 |  |  | 797,356 | 5,723 |  | 188 | 12,712 | 95 |  | 33 |  | 11,313 |  |  | 24,410 |
|  | Madras |  | 228 |  |  |  |  |  |  |  |  |  |  |  | 10,944 |  |  | 11,172 |
|  | Bombay |  | 248 |  |  |  |  |  |  | 198 | 5 | 67,919 | 1,009 | 61,641 | 812 | 130,766 | 68 | 2,074 |
|  | All India |  | 7,534 |  |  | 797,356 | 5,723 |  | 188 | 12,910 | 100 |  | 1,042 |  | 23,059 |  |  | 37,656 |
| 1860-61 | Bengal | 1,134 | 760 | 467 | 9 |  |  |  | 179 | 369 | 575 | 18,760 | 201 | 996,946 | 6,147 | 998,216 | 445 | 7,862 |
|  | Madras |  | 54 | 72 | 16 |  |  |  |  | 711 | 8 |  |  |  | 8,596 |  |  | 8,667 |
|  | Bombay | 469 | 218 |  |  | 874 | 15 |  |  |  | 21 | 1,228,247 | 9,263 | 483,183 | 3,731 |  |  | 13,264 |
|  | All India |  | 1,032 | 539 | 25 | 374 | 15 |  | 179 |  | 604 | 1,247,007 | 9,454 |  | 18,474 | 1,713,046 | 764 | 29,783 |

* Consisting chiefly of Mauritius and Bourbon, New South Wales, Aden, Africa, and Straits Settlements.

1849. [6673] Tobacco.  G. ANGUS, Esq.

1850. [10194] Snuff, Bengal.

1851. [10138] Goolee   for   smoking. Bengal.

———

1852. [6506] Carved snuff gourd, Do.

1853. [6507] Two carved gourd snuff-boxes, Peshawur.

1854. [2900] Silver hookah, with apparatus, Ulwar.   H. H. the MAHARAJAH.

1855. [2918] Tobacco pipe, Arracan.

1856. [10301] Silver pipe, Calcutta.

1857. [5492] Clay hookah, Hooghly.

1858. [5493] Cocoa nut shell hookah, Do.

# CLASS IV.

## ANIMAL AND VEGETABLE SUBSTANCES USED IN MANUFACTURES.

### SECTION A. — OILS, FATS, AND WAX.

This important section is well represented in the vegetable oil series. But in order to form a correct notion of the wealth of our Indian Empire in this respect, it will be necessary to have regard also to Section D of this class, which embraces the odorous essential oils. The subdivisions of this section have been included under the two headings animal and vegetable fats and oils, as there are so few representatives of the products manufactured from fats.

The tables inserted in this section show the quantities of linseed and other seeds exported.

### Division I. — ANIMAL OILS.

1859. [5*] Shark's liver oil, Tellicherry.

1860. [15*] Do. Do.

1861. [7*] 'Seepho' fish oil.

1862. [11*] Neat's-foot oil, Madras.

1863. [10*] Porpoise oil, Patna.

1864. [6338] Minia ekan, fish oil, Seas of Indian Archipelago. G. ANGUS, Esq.

The samples of beeswax are very good of their kind, and afford at least a general idea of the abundance of this article. (See Table, p. 87.)

1865. [2935] Beeswax, Province of Orissa.

1866. [2581] Beeswax, Chittledroog, Mysore.

1867. [2937] Yellow beeswax, Burmah.

1868. [2936] Do. Do.

1869. [6480] Beeswax, Chota Nagpore.

1870. [6481] Do. (second quality), Do.

1871. [2934] Do. Moulmein.

1872. [2938] Do. Ulwar.

1873. [2865] Do. Raepore.

### SOAP AND CANDLES.

1874. [7138] Soap, Lucknow.

1875. [1899] Country soap, Bengal. KANNY LOLL DEY.

1876. [1875] Stearine candles, manufactured by Messrs. SAINTE & Co., Cossipore.

### Division II. — VEGETABLE OIL SERIES.

1877. [47*] Almonds (*Amygdalus communis*).

1878. [49*] Almond oil (*do.*)

The almond is a native of the Himalayas, and is abundant in Cashmere. The oil is colourless or very slightly yellow, and is congealed with difficulty. It is obtained for native use in India, but does not as yet form a recognized article of export. Both varieties of almond, bitter and sweet, are imported into the northern parts of India from Ghoorbund, and into the southern parts from the Persian Gulf.

# CLASS IV.—*India.*

TABLE SHOWING THE QUANTITIES (AS FAR AS CAN BE ASCERTAINED) AND THE VALUE OF OILS EXPORTED FROM INDIA AND EACH PRESIDENCY TO ALL PARTS OF THE WORLD FROM 1852-53 TO 1860-61.

COUNTRIES WHITHER EXPORTED

| Years | Whence Exported | United Kingdom Quantity | UK Value £ | France Quan. | France Value £ | Other Parts of Europe Quan. | O.P. Europe Value £ | America Quan. | America Value £ | China Quan. | China Val. £ | Arabian & Persian Gulfs Quan. | Arab./Pers. Value £ | Aden Quan. | Aden Value £ | Coast of Africa Quan. | C. of Africa Val. £ | Straits Settlements Quan. | Straits Val. £ | Mauritius & Bourbon Quan. | Maur. Value £ | Other Parts Quan. | Other Parts Value £ | Total Exported Quantity | Total Value £ |
|---|---|---|---|---|---|---|---|---|---|---|---|---|---|---|---|---|---|---|---|---|---|---|---|---|---|
| 1852-53 | Bengal (cwt.) | 25,809 | 26,277 | 572 | 615 | 440 | 419 | 3,842 | 3,749 | | | 235 | 107 | | | | | 11,226 | 642 | 3,249 | 2,890 | 1,414 | 1,364 | 85,326 | 35,314 |
| | Madras (gallons) | 1,083,239 | 29,581 | 27,300 | 591 | | | 91,804 | 2,474 | 164 | 170 | 4,600 | 2,640 | 3,008 | 310 | | | 24 | 89 | 180,210 | 3,082 | 268,164 | 8,055 | 1,602,188 | 44,432 |
| | Bombay (") | 91,958 | 6,666 | | | | | | | | | | | | | 175 | 84 | | | 3,638 | 263 | 116 | 71 | 103,683 | 10,293 |
| | ALL INDIA | | 62,474 | | 1,206 | 440 | 419 | | 6,223 | 164 | 170 | 4,815 | 2,747 | 3,008 | 310 | 175 | 84 | 11,250 | 731 | | 6,185 | | 9,490 | | 90,039 |
| 1853-54 | Bengal (cwt.) | 9,266 | 9,364 | 218 | 401 | 296 | 283 | 1,781 | 1,713 | 10 | 15 | 2 | 2 | | | | | 100 | 152 | 5,854 | 4,468 | 1,828 | 1,616 | 19,361 | 18,024 |
| | Madras (gallons) | 1,581,412 | 44,974 | 198,473 | 4,062 | 163,390 | 3,869 | | | | | 1,447 | 102 | 10,891 | 811 | 2,975 | 606 | 16,052 | 921 | 212,179 | 5,377 | 245,797 | 6,864 | 2,418,750 | 66,159 |
| | Bombay (") | 193,745 | 14,580 | 18,635 | 1,381 | | | | | 158 | 135 | 15,910 | 1,586 | | | | | 83 | 305 | 3,552 | 341 | 1,218 | 281 | 247,097 | 19,976 |
| | ALL INDIA | | 68,918 | | 5,844 | | 4,152 | 1,781 | 1,713 | 158 | 150 | | 1,690 | 10,891 | 821 | 2,975 | 606 | | 1,378 | | 10,186 | | 8,701 | | 104,159 |
| 1854-55 | Bengal (gallons) | 1,789,814 | 17,178 | 11,551 | 233 | 57,142 | 278 | | | | 76 | 2,216 | 1 | 7,273 | 651 | 6,300 | 529 | 4,649 | 232 | 164,182 | 5,428 | 327,195 | 2,940 | 2,356,599 | 29,225 |
| | Madras (") | 528,647 | 46,095 | 22,800 | 248 | 9,733 | 1,229 | | | 16 | 45 | 17,588 | 186 | | | | | 178 | 255 | 3,359 | 3,695 | 8,582 | 8,588 | 599,476 | 60,246 |
| | Bombay (") | | 33,842 | | 1,303 | | 645 | | | | | | 3,995 | | | | | | 312 | | 218 | | 565 | | 41,487 |
| | ALL INDIA | | 96,615 | | 1,784 | | 2,152 | | 2,859 | | 121 | 19,804 | 4,132 | 7,273 | 651 | 6,300 | 529 | | 799 | | 9,341 | | 12,093 | | 130,958 |
| 1855-56 | Bengal (cwt.) | 23,251 | 24,509 | 1,847 | 943 | 571 | 692 | 10,304 | 11,335 | 631 | 205 | 6 | 8 | | | | | 236 | 128 | 3,886 | 3,427 | 4,393 | 4,225 | 45,075 | 45,402 |
| | Madras (gallons) | 1,073,790 | 33,434 | 188,812 | 3,375 | 311,609 | 6,705 | | | | | 444 | 61 | | | | | 10,984 | 700 | 189,312 | 4,246 | 287,045 | 4,397 | 2,011,996 | 52,908 |
| | Bombay (") | 540,470 | 41,691 | 41,222 | 2,944 | 20,382 | 1,332 | | | | | 29,047 | 5,136 | 7,705 | 644 | 1,197 | 358 | 262 | 665 | 49,253 | 3,280 | 2,242 | 180 | 691,780 | 56,280 |
| | ALL INDIA | | 99,634 | | 7,262 | | 8,659 | 10,304 | 11,335 | 631 | 205 | | 5,205 | 7,705 | 644 | 1,197 | 358 | | 1,493 | | 10,953 | | 8,802 | | 154,540 |
| 1856-57 | Bengal (gallons) | 1,304,977 | 80,454 | 149,288 | 1,208 | 269,810 | 1,625 | 14,086 | 14,086 | | 242 | 225 | 14 | | | | | 12,962 | 90 | 191,909 | 2,191 | 181,454 | 2,826 | 2,060,625 | 52,726 |
| | Madras (") | | 37,690 | | 3,153 | | 6,516 | | | | 199 | | 6,699 | | | | | | 707 | | 4,382 | | 4,055 | | 56,467 |
| | Bombay (") | 655,372 | 49,247 | 44,907 | 3,629 | 58,858 | 4,631 | | | | | 20,065 | 6,777 | 9,790 | 986 | 7,319 | 801 | 116 | 354 | 41,077 | 3,154 | 3,459 | 271 | 840,963 | 69,971 |
| | ALL INDIA | | 117,391 | | 7,990 | | 12,777 | 14,086 | 14,086 | | 441 | | 6,777 | 9,790 | 986 | 7,319 | 801 | | 1,151 | | 9,677 | | 7,152 | | 179,164 |
| 1857-58 | Bengal (gallons) | | 84,350 | | 1,927 | | 1,654 | | 9,178 | | 226 | | 14 | | 106 | | | | 186 | | 1,234 | | 2,268 | | 101,037 |
| | Madras (") | | 56,354 | | 3,094 | | 5,715 | | 995 | | 84 | | 136 | | 1,689 | | 2,813 | | 424 | | 5,358 | | 4,755 | | 75,942 |
| | Bombay (") | | 60,146 | | 4,703 | | 2,548 | | | | | | 8,963 | | | | | | 791 | | 5,343 | | 217 | | 88,292 |
| | ALL INDIA | | 200,850 | | 9,724 | | 9,917 | | 10,173 | | 310 | | 9,113 | | 1,795 | | 2,813 | | 1,401 | | 11,935 | | 7,240 | | 265,271 |
| 1858-59 | Bengal (gallons) | 837,171 | 26,297 | 7,707 | 560 | 16,263 | 1,170 | 184,823 | 13,427 | 9,476 | 885 | 1,608 | 90 | | | | | 6,127 | 245 | 74,289 | 4,267 | 50,794 | 3,390 | 1,188,208 | 50,331 |
| | Madras (") | 2,024,121 | 46,982 | 5,051 | 106 | 90,231 | 2,143 | | | 66 | 417 | 1,419 | 114 | | | | | 12,579 | 1,029 | 224,815 | 4,891 | 37,085 | 1,673 | 2,394,801 | 56,938 |
| | Bombay (") | 854,740 | 69,274 | 7,674 | 740 | | | | | | | 60,306 | 7,912 | 14,410 | 1,290 | 15,920 | 355 | 109 | 421 | 41,925 | 3,284 | 5,843 | 1,600 | 1,000,993 | 85,293 |
| | ALL INDIA | 3,716,032 | 142,553 | 20,432 | 1,406 | 106,494 | 3,313 | 184,823 | 13,427 | 9,542 | 1,302 | 63,333 | 8,116 | 14,410 | 1,290 | 15,920 | 355 | 18,815 | 1,695 | 340,479 | 12,442 | 93,722 | 6,663 | 4,584,002 | 192,562 |
| 1859-60 | Bengal (gallons) | 584,562 | 42,464 | 12,445 | 1,089 | 25,323 | 2,077 | 214,777 | 18,781 | 9,050 | 798 | 30 | | 329 | 29 | 126 | 6 | 7,891 | 337 | 47,690 | 3,601 | 27,694 | 2,037 | 929,817 | 71,249 |
| | Madras (") | 1,586,172 | 42,089 | 70,070 | 1,484 | 49,669 | 1,039 | | | | | 4,082 | 232 | 448 | 25 | | | 7,098 | 524 | 358,888 | 7,912 | 75,039 | 2,286 | 2,151,466 | 55,584 |
| | Bombay (") | 339,824 | 28,132 | | | | | | | 560 | 707 | 132,971 | 13,103 | 20,471 | 2,736 | 31,737 | 3,006 | 246 | 565 | 60,533 | 4,807 | 2,339 | 177 | 588,681 | 58,233 |
| | ALL INDIA | 2,510,558 | 112,678 | 82,515 | 2,573 | 74,992 | 3,116 | 214,777 | 18,781 | 9,610 | 1,505 | 137,053 | 13,365 | 21,248 | 2,790 | 31,863 | 3,012 | 15,235 | 1,426 | 467,041 | 16,320 | 105,072 | 4,500 | 3,669,954 | 180,066 |
| 1860-61 | Bengal (gallons) | 41,016 | 27,508 | 17,840 | 1,720 | 10,966 | 1,058 | 130,962 | 12,704 | 24,093 | 2,499 | 20 | 6 | 22 | 2 | | | 3,096 | 325 | 31,630 | 3,053 | 14,469 | 2,695 | 274,114 | 51,570 |
| | Madras (") | 2,219,980 | 114,309 | 176,758 | 8,254 | 72,157 | 1,747 | | | 28 | 11 | 9,900 | 493 | 2,160 | 211 | | | 4,571 | 324 | 216,022 | 17,009 | 67,616 | 3,275 | 2,768,592 | 145,638 |
| | Bombay (") | 278,361 | 25,661 | | | | | | | 233 | 397 | 36,787 | 5,615 | 6,593 | 831 | 2,985 | 318 | 175 | 390 | 45,232 | 3,653 | 2,328 | 311 | 372,644 | 37,176 |
| | ALL INDIA | 2,538,757 | 167,478 | 194,598 | 9,974 | 83,123 | 2,805 | 130,962 | 12,704 | 24,354 | 2,907 | 46,657 | 6,114 | 8,775 | 1,044 | 2,985 | 318 | 7,842 | 1,039 | 292,884 | 23,715 | 84,413 | 6,281 | 3,415,350 | 234,379 |

TABLE SHOWING THE QUANTITIES (AS FAR AS CAN BE ASCERTAINED) AND THE VALUE OF WAX AND WAX CANDLES EXPORTED FROM INDIA AND EACH PRESIDENCY TO ALL PARTS OF THE WORLD FROM 1857-58 TO 1860-61.

| Years | Whence Exported | United Kingdom | | France | | Other Parts of Europe | | America | | China | | Arabian and Persian Gulfs | | Other Parts * | | Total Exported to All Parts | | |
|---|---|---|---|---|---|---|---|---|---|---|---|---|---|---|---|---|---|---|
| | | Quan. | Value | Quan. | Value | Quan. | Value | Quan. | Value | Quan. | Value | Quan. | Value | Quan. | Value | Quantity (lbs.) | (tons) | Value |
| | | lbs. | £ | lbs. | £ | lbs. | £ | lbs. | £ | lbs. | £ | lbs. | £ | lbs. | £ | lbs. | tons | £ |
| 1857-58 | Bengal | 144,368 | 5,958 | 51,744 | 2,202 | 5,040 | 214 | .. | .. | 336 | 17 | 112 | 2 | 21,504 | 982 | 223,104 | 100 | 9,375 |
| | Madras | 251,248 | 5,992 | 97,546 | 3,525 | .. | .. | .. | .. | .. | .. | 7,796 | 279 | .. | .. | 356,590 | 159 | 9,796 |
| | Bombay | 130,085 | 5,807 | 18,974 | 847 | 751 | 33 | .. | .. | 25 | 1 | 6,126 | 313 | 9,111 | 425 | 165,072 | 73 | 7,426 |
| | ALL INDIA | 525,701 | 17,757 | 168,264 | 6,574 | 5,791 | 247 | .. | .. | 361 | 18 | 14,034 | 594 | 30,615 | 1,407 | 744,766 | 332 | 26,597 |
| 1858-59 | Bengal | 164,710 | 5,182 | 131,848 | 4,562 | .. | .. | .. | .. | 4,714 | 169 | .. | .. | 51,803 | 1,847 | 352,575 | 157 | 11,760 |
| | Madras | 133,002 | 6,074 | 18,161 | 851 | 2,148 | 75 | .. | .. | .. | .. | 683 | 43 | 1,877 | 175 | 156,871 | 70 | 7,218 |
| | Bombay | 157,552 | 6,989 | 79,178 | 3,534 | .. | .. | .. | .. | 2,270 | 113 | 1,686 | 84 | 37,154 | 1,689 | 277,840 | 124 | 12,359 |
| | ALL INDIA | 455,264 | 18,195 | 228,687 | 8,947 | 2,148 | 75 | .. | .. | 6,984 | 282 | 2,369 | 127 | 90,834 | 3,711 | 787,286 | 351 | 31,337 |
| 1859-60 | Bengal | 105,915 | 4,632 | 45,584 | 1,994 | 588 | 26 | .. | .. | 34,496 | 1,509 | .. | .. | 112,821 | 5,126 | 299,404 | 134 | 13,287 |
| | Madras | 79,184 | 3,884 | 58,908 | 2,079 | .. | .. | .. | .. | .. | .. | 1,774 | 125 | 4,310 | 153 | 144,176 | 64 | 6,241 |
| | Bombay | 149,942 | 6,694 | 17,614 | 790 | .. | .. | .. | .. | 3,755 | 187 | 4,403 | 220 | 76,844 | 3,471 | 252,558 | 113 | 11,362 |
| | ALL INDIA | 335,041 | 15,210 | 122,106 | 4,863 | 588 | 26 | .. | .. | 38,251 | 1,696 | 6,177 | 345 | 193,975 | 8,750 | 696,138 | 311 | 30,890 |
| 1860-61 | Bengal | 86,968 | 4,277 | 36,008 | 1,748 | .. | .. | 4,032 | 144 | 13,188 | 652 | .. | .. | 173,348 | 7,831 | 313,544 | 139 | 14,652 |
| | Madras | 30,285 | 1,515 | 9,916 | 560 | 2,715 | 95 | .. | .. | 3,585 | 125 | 1,234 | 78 | 15,154 | 599 | 62,899 | 29 | 2,972 |
| | Bombay | 111,101 | 5,042 | 1,127 | 50 | .. | .. | .. | .. | 703 | 35 | 4,926 | 245 | 93,093 | 4,516 | 210,950 | 94 | 9,888 |
| | ALL INDIA | 228,354 | 10,834 | 47,051 | 2,358 | 2,715 | 95 | 4,032 | 144 | 17,476 | 812 | 6,160 | 323 | 281,595 | 12,946 | 587,393 | 262 | 27,512 |

* Consisting of principally Batavia, Java, Aden, Africa, Mauritius and Bourbon, and the Straits Settlements.

1879. [177*] Dessee Akroot seeds (*Aleurites triloba*). INDIA MUSEUM.

1880. [50*] Indian walnut oil (*do.*), Hyderabad.

1881. [7084] Belgaum walnut oil (*do.*) DR. R. RIDDELL.

This oil is also called Kekui or Kekune, and Lumbang. The fruit whence it is procured is very abundantly produced in India, and the facility with which it is separable from the nut is an important recommendation. This oil is said to have been successfully employed as a drying oil.

1882. [5037] Moringa seed (*Moringa pterygosperma*), Madras. R. BROWN, Esq.

1883. [72*] Ben oil (*M. pterygosperma*).

This oil is valuable, because it does not soon turn rancid; and it might be extensively procured, as the tree is common throughout India. But though the flowers, foliage, and fruit are eaten by the natives, and the rasped root employed as a substitute for horse-radish, the oil is seldom extracted, and does not form an article of export, except in very small quantities. It is occasionally employed by the natives as an unguent in gout and rheumatism.

1884. [7358] Bonduc nut oil (*Guilandina bonduc*), Lucknow.

1885. [4959] Do. Madras. DR. SHORTT.

1886. [10118a] Bonduc nut oil, Calcutta. KANNY LOLL DEY.

This oil is medicinal only; it is considered useful in palsy and other diseases.

1887. [7172] Teorah seed (*Brassica erucastrum*), Lucknow.

1888. [5686] Teorah oil, (*do.*) Lahore.

The cost of the oil is from 3 to 10 seers per rupee. Used for burning.

It may be remarked that all oils in Oude are extracted by the native press called 'Kolhoo,' turned by means of bullocks, with the exception of the castor oil seed, the oil from which is extracted by boiling in water, and afterwards skimming.*

1889. [123*] Cabbage seed oil (*B. oleracea*), Madras.

This is a fluid oil resembling rape seed oil.

1890. [96*] Capala seed oil (*Rottlera tinctoria*), Calcutta.

The oil which is obtained plentifully from the kernels of this fruit after the removal of the celebrated *Kupli* or *Kamala* powder, promises to be of some importance medicinally as a cathartic oil, and deserves a more complete investigation.†

1891. [68*] Cardamom seed oil (*Elettaria cardamomum*), Madras.

Used medicinally.

1892. [10097] Hidglee Badam oil, (*Anacardium occidentale*), Calcutta. MESSRS. DOSS & DEY.

* Central Committee, *Lucknow.*
† Local Committee, *Cuttack.*

1893. [10733] Cashewnuts (*Anacardium occidentale*), Mangalore. V. P. COELHO, Esq.

1894. [6486] Do. Calcutta. KANNY LOLL DEY.

1895. [4022] Do. Madras.

1896. [6058] Do. Bombay.

1897. [10703] Cashew-nut oil (*do.*), Mangalore. V. P. COELHO, Esq.

1898. [5700] Do. Cuttack.

1899. [42*] Do. Madras.

This light-yellow and sweet-tasted oil is affirmed to be equal, if not superior, as an edible oil, to that of the olive or almond. The kernels have lately been met with in English commerce under the name of Cassia seeds — an evident corruption of Cashew. The tree is common in the East and West Indies. In Bengal it is found chiefly near the sea. The nuts are largely employed as a table fruit.

1900. [10683] Cassia-seed oil, Mangalore. V. P. COELHO, Esq.

A semi-solid oil with an aromatic odour of cinnamon.

1901. [10731] Castor-oil seeds (*Ricinus communis*), South Canara. V. P. COELHO, Esq.

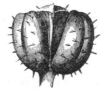

CASTOR-OIL FRUIT.*

1902. [4026] Do. Madras.

1903. [6059] Erindee (*R. communis*), Ahmedabad.

1904. [10093] Do. Calcutta. Messrs. DOSS & DEY.

1905. [5678] Do. Lucknow.

1906. [5691] Do. Do.

1907. [55*] Castor oil (*R. communis*), Calcutta.

1908. [6641] Do. Chota Nagpore.

1909. [10681] Do. Mangalore. V. P. COELHO, Esq.

The castor oil plant is extensively cultivated all over India. The following notes accompanied the samples. The plant is cultivated at Lucknow as a mixed crop. It is sown in June by almost all the villagers, not extensively, but principally for their own use. Its cultivation can be extended all over Oude. This oil is extracted by bruising the seed and then boiling it in water; the oil is afterwards skimmed off. This is the only seed out of which the oil is extracted by boiling, as in this case it is found cheaper than the method used for other seeds, which is by pressure. The cost of the seed is one rupee per maund, and the price of the oil is from 2 to 5 seers per rupee, according

to the abundance of the crop in the season. The proportion of the oil yielded is about half the weight of the seeds boiled; it is only used for burning.*

In Cuttack, the plant is grown all over the province, a good deal in patches of newly cleared land in the jungles of the Tributary States and Sumbulpore. The oil i extracted in two ways. It is used for burning and culinary purposes, and medicinally also. The local market is now 11½ peculs of the seeds per rupee. Both the native methods of extracting oil are wasteful and tedious, and therefore expensive. European oil presses, and a knowledge of some methods of clarifying the expressed oil, seem only to be required to render the oil seed crops of this extensive division of great value.†

1910. [10695] Jungle lamp oil (*Ricinus communis major*), Mangalore. V. P. COELHO, Esq.

1911. [3750] Lamp oil (*do.*), Salem. CHETUMBARA PILLAY.

This oil is pressed from the larger variety of castor oil seed, and is sometimes drawn cold, when a pale straw-coloured oil is obtained, resembling that from the small-seeded kind. It is generally extracted by heat, and forms the common 'lamp oil' of the bazaars.

1912. [7191] Chaulmoogra seed (*Hydnocarpus odorata*), Moulmein.

1913. [5680] Chaulmoogra seed oil (*do.*), Chittagong.

1914. [10094] Do. Calcutta. Messrs. DOSS & DEY.

The tree which produces the seed from which this oil is expressed, is a slender tree, with very delicate but tough branches. The leaves are lance-shaped and dark green; the fruit is produced at the ends of the branches. It is very like a wood-apple, but rough, about 3 inches in diameter, and is filled with the seeds, each about an inch long, of an irregular shape, and about 1½ inches in circumference in the thickest part of the kernel; is covered with a thin hard shell of an earthy colour. The seeds are obtainable in the north of Chittagong, but it cannot be stated how much could be procured in any one season. The oil is very successfully used by native physicians for the cure of bad sores. Dr. Beatson remarks, that 'there is no doubt that the oil expressed from the Chalmoogree seeds is a most valuable remedy in leprosy, and in obstinate ulcers of all kinds.' He has seen ulcers, both leprous and non-leprous, take on a healing action on the application of Chalmoogree, after resisting every kind of treatment. The oil obtained in the bazaar is coarse and unpleasant.

1915. [2856] Cheerongee seed (*Buchanania latifolia*), Raepore.

1916. [6061] Do. Bombay.

This oil is rarely extracted from the abundantly oleaginous seeds which are eaten by the natives to make them fat. The oil is clear, sweet, and straw-coloured; the trees grow abundantly in Mysore and Cuddapah, &c.

1917. [1238] Chemmarum seeds (*Amoora rohituka*), Calcutta.

From the seeds, where the trees are plentiful, as in Travancore and Bengal, the natives extract an oil which is useful for many economic purposes.

---

* From MR. M. C. COOKE.

* Central Committee, *Lucknow.*
† Local Committee, *Cuttack.*

1918. [9389] Circassian beans (*Adenanthera pavonina*). DR. SHORTT.

These hard red seeds are best known here by their employment for ornamental purposes; but it would appear that an oil has been obtained from them.

1919. [7180] Hoorhooryaseed (*Polanisia icosandra*), Calcutta.

The seeds of this plant, which are employed medicinally in India, and sold in the bazaars under the name of *chorie-ajooain*, yield an oil, when subjected to very powerful pressure, which is of a light olive-green colour, and very limpid. It might prove of service in cases where a very limpid oil would be an acquisition.

1920. [1936] Cotton seed (*Gossypium herbaceum*), Umritsur.

1921. [7121] Do. Lucknow.

1922. [9391] Cotton seed oil.

Manufactured in India; it is pale and limpid.

1923. [4791] Cress seed (*Lepidium sativum*), Burmah. Messrs. HALLIDAY, FOX, & Co.

From the seeds of this plant an oil has been extracted in India, as well as in this country, but samples have not been forwarded to this Exhibition.

1924. [6136] Turbooz seed (*Cucurbita citrullus*), Ahmedabad.

1925. [64*] Water-melon oil (*C. citrullus*), Madras.

The seeds of this cucurbitaceous plant are used as the source of a mild culinary oil in Western India.

1926. [7054] Cucumber-seed oil (*Cucumis sativus*).

The oil of cucumber seed is a mild edible oil, which is obtained by expression for native use only.

1927. [6147] Rheera seed (*Cucumis*, species of), Ahmedabad.

1928. [6144] Khurbooza (*C. melo*), Ahmedabad.

1929. [4025] Do. Madras.

1930. [6564] Kurbooja seed (*do.*), Calcutta.

1931. [4020] Kankur (*C. utilissimus*), Madras.

The fruit is yellow, the size of an ostrich's egg. The seeds are ground into meal, and an oil expressed from them, which is very nourishing. It is much cultivated in the Guntoor Circars.

1932. [6563] Kuddoo beeah (*Lagenaria vulgaris*).

1933. [4914] Bottle gourd seed (*L. vulgaris*), Madras. DR. SHORTT.

1934. [6135] Doodee seeds (*Lagenaria vulgaris*), Ahmedabad.

The fruit is employed for bottles. The seeds yield a bland oil. The plant is common in the East and West Indies.

1935. [6140] Toorai seed (*Luffa acutangula*), Ahmedabad.

A medicinal oil is obtained from this seed, which has a reputation among the native practitioners.

1936. [9390] Custard-apple seeds (*Anona squamosa*).

1937. [1930] Exile oil (*Cerbera Thevetia*). C. B. WOOD, Esq.

1938. [4789] Fennel-flower seed (*Nigella sativa*), Burmah. Messrs. HALLIDAY, FOX, & Co.

1939. [*] Nigella oil (*N. sativa*). INDIA MUSEUM.

A clear and colourless, but rather viscid oil is obtained from these seeds, which appears to be employed principally as a medicine. It is called *Jungle geerah oil* in Mysore.

1940. [67*] Garlic oil (*Allium sativum*).

This is only a medicinal oil. It is clear, colourless, limpid, and contains the full odour of the plant. It might be available in cookery for those who admire the flavour of garlic in their dishes, but this will evidently be the fullest extent of its application; hence it can scarcely be considered of any importance commercially.

1941. [7731] Wolley yallu seed (*Sesamum orientale*), Astagram, Mysore.

1942. [6682] Black til seed (*S. orientale*), Siam. G. ANGUS, Esq.

1943. [6670] White do., Do. Do.

1944. [6438] Gingellie (*S. orientale*), Chota Nagpore.

1945. [7117] Black til (*S. orientale*), Cuttack.

1946. [7122] White sesame, Nhan kyen (*S. orientale*), Moulmein.

1947. [7123] Black til, Nhan mai (*S. orientale*), Do.

1948. [4794] Do. Burmah. Messrs. HALLIDAY, FOX, & Co.

1949. [4804] White til, Do. Do.

1950. [10732] Gingelly, Mangalore. V. P. COELHO, Esq.

1951. [4023] Gingelly seed, Madras. DR. HUNTER.

1952. [10680] Gingelly or Sesame oil (*Sesamum orientale*), South Canara. V. P. COELHO, Esq.

1953. [2252] Do. Salem. CHEDUMBARA PILLAY.

1954. [5685] Dark til oil, Lucknow.

1955. [17*] Gingelly oil (*do.*), Patna.

1956. [5701] Sesame oil, Cuttack.

1957. [5687] Do. Bengal.

1958. [10091] Teel oil, Calcutta. Messrs. DOSS & DEY.

1959. [19*] Do. Tanjore.

Gingelly oil is largely employed by the natives of India. The seed, as well as the oil, is exported in large quantities.

The black-seeded variety (*Tillee*) affords a larger percentage of oil than the red-seeded kind. Sesamum seed has of late been exported largely to France, where it is said to be employed for mixing with olive oil.

Three varieties of sesame seed are cultivated in India—the white seeded (*Suffed-til*), the red or parti-coloured (*Kala-til*), and the black variety (*Tillee*): it is the latter which affords the greater proportion of the gingelly oil of commerce. At the commencement of 1861, white seed was worth, in the London markets, 65s.; black and brown, 58s. and 60s. per quarter.

A second sort of sesame oil, sometimes called 'Rape,' is obtained from the red-seeded variety.

*Black* sesame is sown in March, and ripens in May. *Red* sesame is not sown till June.

1960. [40*] Moneela gram oil (*Dolichos uniflorus*).

A pale yellow or almost colourless, viscid, clear oil, in Tanjore is stated to be obtained from a variety of this pulse. The variety employed for oil has nearly white seeds.

1961. [6486] Ground nuts (*Arachis hypogœa*), Calcutta. KANNY LOLL DEY.

1962. [4942] Vayr Kadalay (*A. hypogœa*), Madras. DR. HUNTER.

1963. [6339] Minia Kachang oil (*A. hypogœa*), Java. G. ANGUS, Esq.

1964. [30*] Ground-nut oil (*Arachis hypogœa*), Madras.

1965. [7085] Do. Deccan. DR. R. RID-DELL.

The ground nut is extensively cultivated in various parts of India; the kernels yield about 44 per cent. of a clear pale yellow oil, which is largely used as food and as fuel for lamps. The value of ground-nut kernels in London is about 16*l.* 10*s.* per ton, and of the oil 42*l.* to 43*l.* per ton. For ordinary purposes it is quite equal to olive oil.

1966. [6065] Bhang ka beej (*Cannabis sativa*).

1967. [6131] Hemp seed, Bombay.

The hemp plant is largely cultivated in India, but the oil from its seed is not in general use.

1968. [1957] Hemp, Madras seed (*Crotalaria juncea*), Jhansee.

1969. [7112] Do. Lucknow.

1970. [185*] Kanari fruits (*Canarium commune*).

Yield a mild oil which is mixed with food when fresh.

1971. [6066] Khaliziri seed (*Vernonia anthelmintica*), Poona.

1972. [7081] Do. Bombay.

1973. [10118] Do. Calcutta. KANNY LOLL DEY.

The oil from these seeds is semi-fluid. The plant which produces them is very common in India.

Large quantities of linseed are imported into Britain from India. The Table (p. 91), shows the amount and value of the linseed from India exported to all parts for the eleven years ending 1861, whilst the others indicate points of interest regarding the amount and price of that and other oilseeds exported from Bombay.

TABLE* SHOWING RANGE OF PRICES IN BOMBAY OF THE UNDERMENTIONED OIL SEEDS TAKEN MONTHLY FOR FOUR YEARS.

| Article | 1857 | | 1858 | | 1859 | | 1860 | |
|---|---|---|---|---|---|---|---|---|
| | Range | Average | Range | Average | Range | Average | Range | Average |
| Linseed, per cwt. Rs | 4/8 @ 6/2½ | 5/8 | 4/10 @ 5/12 | 5/3 | 4/10 @ 5/2 | 4/12 | 3/14 @ 5/1 | 4/9 |
| Teel seed, per candy . | 23/ @ 32/ | 26/13 | 23/8 @ 26/8 | 25/8 | 23/8 @ 25/ | 24/14 | 22/ @ 30/ | 26/12 |
| Rape seed, per cwt. . | 4/12 @ 6/ | 5/6 | 4/8 @ 6/ | 5/6 | 5/ @ 5/9 | 5/4 | 5/ @ 6/ | 5/8 |
| Mustard . . . . . | 5/8 @ 5/12 | 5/9 | 4/12 @ 5/8 | 5/2 | 4/8 @ 4/12 | 4/11 | 4/8 @ 5/ | 4/ |
| Ground nuts, per cdy. | 18/ @ 25/ | 21/6 | 16/ @ 21/ | 18/10 | 15/ @ 18/ | 16/7 | 16/ @ 25/ | 21/ |
| Niger seed . . . . | 15/8 @ 21/ | 18/4 | 15/8 @ 20/ | 17/12 | 15/ @ 17/8 | 15/10 | 15/8 @ 21/ | 19/10 |

* Furnished to DR. BIRDWOOD by the HON. W. B. TRISTRAM.

## TABLE SHOWING THE QUANTITIES (AS FAR AS CAN BE ASCERTAINED) AND VALUE* OF LINSEED EXPORTED FROM INDIA AND EACH PRESIDENCY TO ALL PARTS OF THE WORLD FROM 1850-51 TO 1860-61 INCLUSIVE.

### COUNTRIES WHITHER EXPORTED

| Years | Whence Exported | United Kingdom Qty (lbs.) | United Kingdom Value (£) | France Qty (lbs.) | France Value (£) | Other Parts of Europe Qty (lbs.) | Other Parts of Europe Value (£) | North America Qty (lbs.) | North America Value (£) | China Qty (lbs.) | China Value (£) | Arabian and Persian Gulf Qty (lbs.) | Arabian and Persian Gulf Value (£) | Islands and Shores of the Indian Seas Qty (lbs.) | Islands and Shores of the Indian Seas Value (£) | All Other Places Qty (lbs.) | All Other Places Value (£) | Total Exported to All Parts Qty (tons) | Total Exported to All Parts Qty (lbs.) | Total Exported to All Parts Value (£) |
|---|---|---|---|---|---|---|---|---|---|---|---|---|---|---|---|---|---|---|---|---|
| 1850-51 | Bengal | 24,521,504 | 59,601 | 3,701,712 | 8,997 | 72,688 | 177 | 34,675,760 | 84,277 | | | | | 2,016 | 5 | 15,568 | 38 | 28,120 | 62,989,248 | 153,095 |
| | Madras | 89,264 | 226 | | | | | | | | | | | | | 448 | 1 | 40 | 89,712 | 227 |
| | Bombay | 5,611,424 | 17,051 | | | | | | | | | | | | | | | 2,505 | 5,611,494 | 17,051 |
| | All India | 30,222,192 | 76,878 | 3,701,712 | 8,997 | 72,688 | 177 | 34,675,760 | 84,277 | | | | | 2,016 | 5 | 16,016 | 39 | 30,665 | 68,690,384 | 170,373 |
| 1851-52 | Bengal | 70,467,488 | 171,275 | 1,452,528 | 6,330 | 2,262,064 | 5,498 | 42,039,648 | 102,181 | | | | | | | 31,806 | 77 | 51,899 | 116,253,636 | 285,361 |
| | Madras | 114,464 | 277 | | | | | | | | | | | | | | | 51 | 114,464 | 277 |
| | Bombay | 10,822,672 | 28,989 | | | 672 | 2 | 1,066,352 | 3,332 | | | | | | | 448 | 3 | 5,308 | 11,890,144 | 32,336 |
| | All India | 81,404,624 | 200,541 | 1,452,528 | 6,330 | 2,262,736 | 5,500 | 43,106,000 | 105,513 | | | | | | | 32,256 | 80 | 57,258 | 128,258,144 | 317,954 |
| 1852-53 | Bengal | 55,871,424 | 135,798 | 573,440 | 1,394 | 50,624 | 123 | 50,433,600 | 105,916 | | | | | 3,360 | 8 | 10,640 | 26 | 47,742 | 106,943,088 | 243,265 |
| | Madras | 116,032 | 306 | | | | | | | | | | | | | 1,008 | 3 | 52 | 117,040 | 309 |
| | Bombay | 12,802,608 | 34,293 | 560 | 1 | | | | | | | | | | | 33,712 | 90 | 5,731 | 12,836,880 | 34,384 |
| | All India | 68,790,064 | 170,397 | 574,000 | 1,395 | 50,624 | 123 | 50,433,600 | 105,916 | | | | | 3,360 | 8 | 45,360 | 119 | 53,525 | 119,897,008 | 277,958 |
| 1853-54 | Bengal | 36,017,968 | 87,544 | 1,510,320 | 3,671 | 224,560 | 546 | 47,370,400 | 114,892 | 448 | 1 | | | | | 81,872 | 199 | 38,038 | 85,205,568 | 206,853 |
| | Madras | 315,952 | 933 | | | | | | | | | | | | | 1,904 | 6 | 142 | 317,856 | 939 |
| | Bombay | 5,782,672 | 69,586 | 30,128 | 362 | | | | | | | | | | | 9,408 | 113 | 2,599 | 5,822,208 | 70,061 |
| | All India | 42,116,592 | 158,063 | 1,540,448 | 4,033 | 224,560 | 546 | 47,370,400 | 114,892 | 448 | 1 | | | | | 93,184 | 318 | 40,779 | 91,345,632 | 277,851 |
| 1854-55 | Bengal | 126,898,464 | 307,460 | 3,618,832 | 8,796 | 8,176 | 20 | | | 2,576 | 15 | | | 58,576 | 107 | | | 58,298 | 130,586,694 | 316,398 |
| | Madras | 192,864 | 591 | | | | | | | | | | | | | | | 86 | 192,864 | 591 |
| | Bombay | 5,972,176 | 71,380 | 31,472 | 377 | 64,512 | 1,050 | | | | | | | | | | | 2,709 | 6,068,160 | 72,807 |
| | All India | 133,063,504 | 379,431 | 3,650,304 | 9,173 | 72,688 | 1,070 | | | 2,576 | 15 | | | 58,576 | 107 | | | 61,093 | 136,847,648 | 389,796 |
| 1855-56 | Bengal | 124,002,144 | 301,430 | 3,630,144 | 8,823 | | | 87,189,872 | 211,921 | | | | | 72,352 | 176 | 235,312 | 572 | 96,040 | 215,129,894 | 522,992 |
| | Madras | 204,064 | 650 | | | | | | | | | | | | | | | 91 | 204,064 | 650 |
| | Bombay | 7,183,792 | 116,770 | 264,992 | 4,308 | 3,360 | 54 | | | | | | | | | | | 3,327 | 7,452,144 | 121,132 |
| | All India | 131,390,000 | 418,850 | 3,895,136 | 13,131 | 3,360 | 54 | 87,189,872 | 211,921 | | | | | 72,352 | 176 | 235,312 | 572 | 99,458 | 222,786,032 | 644,704 |
| 1856-57 | Bengal | 20,463,632 | 66,071 | 2,002,000 | 4,858 | 505,344 | 1,228 | 145,506,480 | 353,667 | | | | | 3,248 | 8 | 247,296 | 601 | 75,325 | 168,728,000 | 426,433 |
| | Madras | 717,472 | 2,003 | 13,664 | 38 | 7,280 | 13 | | | | | | | | | 7,280 | 21 | 333 | 745,696 | 2,075 |
| | Bombay | 7,654,640 | 124,034 | 434,336 | 7,039 | 127,680 | 2,070 | | | | | | | | | | | 3,668 | 8,216,656 | 133,148 |
| | All India | 28,835,744 | 192,108 | 2,450,000 | 11,935 | 640,304 | 3,311 | 145,506,480 | 353,667 | | | | | 3,248 | 8 | 254,576 | 622 | 79,326 | 177,690,352 | 561,651 |
| 1857-58 | Bengal | 74,057,088 | 187,925 | 2,296,672 | 5,582 | 136,528 | 332 | 89,848,752 | 218,382 | | | | | 229,712 | 558 | 22,512 | 55 | 74,871 | 166,591,264 | 412,884 |
| | Madras | 703,696 | 2,469 | 20,384 | 62 | | | | | | | | | | | | | 323 | 724,080 | 2,681 |
| | Bombay | 8,875,776 | 177,837 | 284,256 | 5,346 | 342,048 | 7,369 | 1,737,792 | 30,792 | | | | | | | | | 5,018 | 11,239,872 | 221,844 |
| | All India | 83,636,560 | 368,231 | 2,601,312 | 10,990 | 478,576 | 7,701 | 91,586,544 | 249,174 | | | | | 229,712 | 558 | 22,512 | 55 | 79,712 | 178,555,216 | 636,709 |
| 1858-59 | Bengal | 29,942,464 | 292,956 | 1,390,310 | 13,603 | 24,436 | 240 | 27,686,152 | 290,883 | | | | | | | 1,442,134 | 14,217 | 27,002 | 60,485,496 | 611,899 |
| | Madras | 92,168 | 1,544 | 78,884 | 1,256 | | | | | | | | | | | | | 76 | 171,052 | 2,800 |
| | Bombay | 27,345,190 | 435,366 | 231,404 | 3,686 | 51,578 | 820 | 1,727,002 | 42,210 | | | | | | | | | 13,105 | 29,355,174 | 482,282 |
| | All India | 57,379,822 | 730,066 | 1,700,598 | 18,545 | 76,014 | 1,060 | 29,413,154 | 333,093 | | | | | | | 1,442,134 | 14,217 | 40,183 | 90,011,721 | 1,096,981 |
| 1859-60 | Bengal | 41,842,508 | 409,386 | 278,800 | 2,728 | 245,672 | 2,403 | 23,396,896 | 228,915 | | | | | | | | | 29,358 | 65,768,876 | 643,432 |
| | Madras | 104,878 | 1,731 | 7,134 | 117 | | | | | | | | | | | | | 50 | 112,012 | 1,848 |
| | Bombay | 26,093,024 | 415,651 | 171,216 | 2,728 | | | 3,129,694 | 49,126 | | | | | | | | | 13,128 | 29,395,984 | 467,605 |
| | All India | 68,040,410 | 826,768 | 457,150 | 5,573 | 245,672 | 2,403 | 26,526,590 | 278,041 | | | | | | | | | 44,531 | 95,271,842 | 1,112,785 |
| 1860-61 | Bengal | 15,190,840 | 446,915 | 423,976 | 1,558 | | | 11,464,712 | 24,225 | 15,260 | 56 | | | | | 12,404 | 217,718 | 12,101 | 27,107,192 | 690,472 |
| | Madras | 44,930 | 741 | | | | | | | | | | | | | | | 20 | 44,980 | 741 |
| | Bombay | 27,870,652 | 457,608 | 453,952 | 7,321 | 241,736 | 3,908 | 5,958,776 | 95,727 | | | | | | | 86 | 2 | 15,414 | 34,525,198 | 564,666 |
| | All India | 43,106,422 | 905,264 | 877,928 | 8,879 | 241,736 | 3,908 | 17,443,488 | 119,952 | 15,260 | 56 | | | | | 12,490 | 217,720 | 27,535 | 61,677,330 | 1,255,779 |

* The Indian rupee is here converted at the exchange of two shillings.

TABLE* SHOWING THE PLACE OF CHIEF PRODUCTIONS IN WESTERN INDIA—THE QUALITY, PRICE AND QUANTITY OF THE PRINCIPAL OILS AND OIL SEEDS EXPORTED FROM BOMBAY.

| Article | Place of Production | Quality | | Nominal Prices in 1861 | Export for year ending 30th April 1860 |
|---|---|---|---|---|---|
| | | In Bombay. | In London. | | |
| Linseed . | Sholapore and Khandeish . . | Best in the World | | Rs. 5/4 @ 5/6 per cwt. | |
| Do. . . | Arvee, Hingunghaut, Oomrawutty | Good | Good | 5/ @ 5/2 | 262,970 Qrs. |
| Do. . . | Scinde . . . . . | Inferior | Good | 4/14 | |
| Teel seed . | Khandeish . . . . . | Best | Best | Rs. 40/ per candy of about 5 cwt. | 40,963 do. |
| Do. . . | Scinde and Guzerat . . . | Good | Good | 38/ do. do. | |
| Rape seed . | Guzerat . . . . . | Best in the world | | 6/9 per cwt. | |
| Do. . . | Scinde, Ferozepore . . . | Good | Good | 6/ do. | 73,270 do. |
| Do. . . | Cutch . . . . . | Do. | Do. | 5/10 @ 6/ | |
| Mustard seed . | Guzerat . . . . . | Best | Good to best | 25 @ 26 per candy of 5½ cwt. | |
| Do. . . | Ghattee . . . . . | Good | Poor to good | 22 @ 23/ do. | 1,039 do. |
| Do. . . | Scinde . . . . . | Do. | Do. | do. do. | |
| Poppy seed . | Guzerat and Malwa . . . | Only description | | 6/ | 219 do. |
| Niger seed . | Nassick and Sholapore . . | Do. | | 25 @ 26 per candy of 5¼ cwt. | 458 do. |
| Ground nuts . | Vingorla, Mhar, Barsee, Nassick . | Good | Good | 26 @ 27 per do. of 5 cwt. | 4,057 do. |
| Castor oil . | Bulsar and Neighbourhood . . | Do. | Poor | 4/8 @ 5/ per maund of 28 lbs. | 11,063 Gals. |
| Teel oil . | Bombay . . . . . | Do. | Good | 5/ per maund of do. | 407,089 do. |
| Do. . . | Ghatee . . . . . | Inferior | Inferior | 4/ @ 4/4 | |
| Safflower . | Barsee . . . . . | Best | Inferior | 9/ @ 10/ per do. of lbs. 4106 | 237 cwt. |
| Do. . . | Rajapore and Compta . . | Good | Do. | 7/ @ 8/ do. do | |
| Kokum . | Goa . . . . . . | Best | Do. | 40 @ 45/ per candy of 7 cwt. | None. |
| Do. . . | Mhar, Vingorla, Compta . . | Poor | Do. | 20 @ 22/ per do. | |

* Furnished to DR. BIRDWOOD by the HON. W. B. TRISTRAM.

1974. [10172] Linseed (*Linum usitatissimum*), Lahore.

1975. [10173] Do. Jhansee.

1976. [4028] Do. Madras.

1977. [7119] Do. Cuttack.

1978. [7169] Do. Lucknow.

1979. [7168] Do. Midnapore.

1980-1. [6486, 7175] Do. Calcutta. KANNY LOLL DEY.

1982. [6062] Do. Ahmedabad.

1983. [5692] Linseed oil (*L. usitatissimum*), Lucknow.

1984. [5684] Do. Cuttack.

1985. [10095] Teese oil (boiled), Calcutta. Messrs. DOSS & DEY.

1986. [10089] Do. Do. Do.

1987. [151*] Linseed oil, Madras.

1988. [6312] Macassar oil, Macassar. G. ANGUS, Esq.

This oil is solid at ordinary temperatures.

1989. [6063] Malkungunnee seed (*Celastrus paniculatus*), Bombay.

1990. [100*] Malkungee oil (*do*).

1991. [99*] Valuluvy oil (*do.*), Madras.

The oil yielded by the seeds of this species of *Celastrus* by expression is of a deep scarlet colour; it is only employed medicinally.

1992. [6486] Marking nuts (*Semecarpus anacardium*), Calcutta. Messrs. DOSS & DEY.

The acrid and vesicating oil which is contained between the laminæ of the pericarp is employed as a preventive against the attack of the white ant, and also by native practitioners as a remedy in rheumatic and leprous affections. The oil which is obtained from the kernel of the nut is of a different character, but it has only been obtained experimentally, and it would seem to resemble the mild oil of *Cashew-nut* kernels.

1993. [7176] Bakul seeds (*Mimusops elengi*), Calcutta.

1994. [6244] Rana seeds (*M. kaki*), Ahmedabad.

Although the oil of the Bakul is obtainable in considerable quantities, it is not much used. It has some medicinal reputation.

———

1995. [9392] Pulas seed (*Butea frondosa*).

1996. [65*] Moodooga oil (*B. frondosa*).

Only obtainable in small quantities. Merely employed medicinally.

———

1997. [7725] Sasvey seed (*Sinapis dichotoma*), Astagram, Mysore.

1998. [7118] Do. Cuttack.

1999. [6452] *S. ramosa*, Chota Nagpore.

2000. [7170] Sarson (*S. juncea*), Lucknow.

2001. [7178] Raee (*S. ramosa*), Midnapore.

2002. [7183] Do. Beerbhoom.

2003. [7174] Do. Hooghly.

2004-5. [6236, 6232] Do. Ahmedabad.

2006. [6233] Do. Bombay.

2007. [6235] Do. Poona.

2008. [10696] Mustard-seed oil, Mangalore. V. P. COELHO, Esq.

2009. [5681] *S. dichotoma*, Cuttack.

2010. [5689] *S. ramosa*, Lucknow.

2011. [5712] *S. juncea*, Do.

2012. [5696] *S. ramosa*, Hooghly.

2013. [6443] *S. dichotoma*, Chota Nagpore.

Five or six species of *Sinapis* are cultivated throughout India for the sake of their oil, which is much esteemed in the country for cookery, for medicine, and for anointing the person.

———

2014. [6041] Myrabolans (*Terminalia chebula*).

The kernels of the fruit yield in small quantities an oil which is occasionally extracted in India for medicinal purposes.

2015. [1854] Beleric Myrabolans (*T. belerica*), Punjab.

2016. [97*] Myrabolan oil (*T. belerica*), Madras.

The fruit is astringent, but the kernels are eaten. They yield a small quantity of oil, which appears to be of medicinal use in India.

2017. [46*] Indian almond oil (*T. Catappa*), Calcutta.

2018. [6439] Badam oil (*T. Catappa*), Chota Nagpore.

The kernel of the fruit resembles an almond or filbert in taste and composition; hence it has been called the *wild almond* and *country almond*. It yields an excellent fixed oil, which is rather thicker and more amber-coloured than almond oil, for which it might be substituted.

———

2019. [9393] Nahor seeds (*Mesua ferrea*), Bengal.

2020. [6209] Nageshur (*M. ferrea*), Chittagong.

2021. [10738] Nago sampige (*M. ferrea*), Mangalore. V. P. COELHO, Esq.

2022. [1932] Nahor oil (*M. ferrea*). C. B. WOOD, Esq.

2023. [5709] Nagkesur oil (*M. ferrea*), Chittagong.

This promises to be a valuable oil, if it can be obtained in sufficient quantities.

The tree from which seeds are obtained grows wild in the jungles in the hills of this district, and has been planted many years ago on the sides of the road leading to some of the dwelling houses in the station. The seeds are contained in a strong brown skin, one, two, or three in each. When ripe the skin bursts and the seeds drop out. The seed is covered with a thin hard shell. The oil is an excellent cure for cutaneous diseases.*

———

2024. [88] Physic nuts (*Jatropha curcas*), Mangulpore.

2025. [6203] Arenda seeds (*J. curcas*), Chittagong.

2026. [86*] Physic nut oil (*J. curcas*), Madras.

2027. [5704] Arenda oil (*J. curcas*), Chittagong.

The oil is employed not only medicinally, but also for lamps. It is, however, very local, both in its manufacture and use.

The bush from which the seed is obtained is used for fencing ground : it is readily increased by cuttings, which rapidly take root. The seeds are three or four, contained in a thin skin, which is black; the seed is of the same colour, and grows in branches; the stems of the bushes are not strong, but they answer excellently for fences, with split bamboo tied on each side to keep them straight and together, and the great advantage is that no kind of cattle eat them. The seeds are collected and the oil expressed in the usual way.†

———

* Local Committee, *Chittagong.*
† Local Committee, *Chittagong.*

2028. [144*] Bherinda oil (*Jatropha*, a species of), Bheerbhoom.

2029. [7120] Polang seed (*Calophyllum inophyllum*).

2030. [4029] Do. Madras.

2031. [93*] Pinnaycottay oil (*C. inophyllum*), Madras.

2032. [5682] Polang oil (*C. inophyllum*), Cuttack.

Is manufactured and used at Bombay, Tinnevelly, and other parts of India, as a lamp-oil. The seeds from which it is obtained are very oleaginous, and yield about 60 per cent. of their weight of oil.

2033. [10736] Hone seeds (*C. calaba*), Mangalore. V. P. COELHO, Esq.

2034. [4021] Poonga seed (*Pongamia glabra*), Madras.

2035. [6055] Kurrunj seed (*P. glabra*), Ahmedabad.

2036. [6202] Caron seed (*P. glabra*), Chittagong.

2037. [82*] Kurrunjate oil (*P. glabra*), Sattara.

2038. [6444] Kurrunj oil (*P. glabra*), Chota Nagpore.

2039. [5694] Caron oil (*P. glabra*), Chittagong.

This oil is of a pale brownish colour, and is fluid at a temperature above 55°. It has a slight smell, which becomes more evident in the darker-coloured samples than in the pale sherry-coloured. The tree from which it is obtained is a crooked tree, grows in wet places near fresh water, very common on the sides of ditches which surround native dwellings. The seed is bean-shaped, and produced in a flat pod: the pods grow several together. The flower is pink and white, of the shape of a bean-flower or blossom. The oil is used for burning in native lamps, and in large quantities for boiling with dammer to soften it for the seams and bottoms of ships. It is also often used by native practitioners for the cure of itch. A maund of seeds costs 1R. 8A., and the extraction of the oil by heat costs 8 annas: the oil produced amounts to 6¾ seers per maund.*

2040. [10156] Poppy seed (*Papaver somniferum*).

2041. [4024] Do. Madras.

2042. [5697] Poppy seed oil (*do*), Bengal.

2043. [10101] Posto oil (*P. somniferum*), Calcutta. Messrs. DOSS & DEY.

2044. [11*] Poppy seed oil (*P. somniferum*), Vizianagram.

2045. [6955] Do. Lucknow.

The seeds would yield by expression about 50 per cent. of a bland and very valuable oil, of a pale golden colour, fluid to within 10° of the freezing-point of water. It dries easily, is inodorous, of agreeable odour, and partially soluble in alcohol. The seed is worth about 61s. in the English market. By simple exposure to the rays of the sun in shallow vessels, the oil is rendered perfectly colourless. It is expressed by means of a heavy circular stone, placed on its edge, made to revolve by a long lever, and the apparatus is worked by draught bullocks.

MR. BINGHAM adds the following note on this oil seed:—'The seed has no narcotic qualities, but has a sweet taste, and is used, parched, by the lower class of natives as a food; it is also much used by the sweetmeat-makers as an addition in their wares. This and the seed of the Teel (*Sesamum orientale*) are the only oilseeds, with the exception of the cocoanut, which are used for that purpose. It produces, under the native method, a clear limpid oil, which burns very quickly. About 30 per cent. of oil is generally extracted, and the cake is then sold as a food to the poorer classes. The oil at present sells at about 5 seers per rupee at Shahabad. The production of this seed is only limited by the production of the poppy.

'In Oude each ryot sows from 2 to 4 beegahs in the month of October. The oil is extracted by the common native press. The cost of the seed is 10 seers for the rupee, and the oil sells for 3 seers for the rupee; two-fifths of the weight of the seed employed is about the proportion of oil yielded by the native process. The poppy seed is eaten by the natives made into sweetmeats, provided the opium has been extracted from the seed vessel, otherwise it is bitter and narcotic, and under these circumstances the oil extracted is also bitter. Used for cooking and burning.'*

2046. [9394] Prickly poppy seed (*Argemone mexicana*), Chingleput. DR. SHORTT.

2047. [1933] She-al-kanta oil (*A. mexicana*), Malda. DR. R. F. THOMPSON.

2048. [5699] Do. Chittagong.

The plant from which the oil is obtained is a very common, troublesome weed, growing almost everywhere, on any abandoned heap of rubbish. The plant itself is well known, having prickly, thistle-like leaves, and bright yellow flowers. The seed yields a large quantity of oil, nearly as much as the common mustard seed. The oil is pale yellow-coloured and clear. It is mild, resembling that of the poppy, and may be taken in one-ounce doses without producing purgative effects. It is readily procurable, and so cheap that a considerable saving has been effected from its introduction by Dr. Thompson into the Malda jail for burning in place of mustard oil.

2049. [6060] Moora seed (*Raphanus sativus*), Bombay.

2050. [6150] Moola seed (*do*), Ahmedabad.

2051. [6076] Ramtil seed (*Guizotia oleifera*), Poona.

2052. [7171] Sooregoja (*G. oleifera*), Midnapore.

2053. [6442] Soorgooja oil (*G. oleifera*), Chota Nagpore.

This is a sweet-tasting edible oil, plentiful in some parts of the country, and is employed similarly to sesame or gingelly oil, but is not generally considered so good. The seeds yield about 34 per cent. of oil. In some parts of India its value is said to be about 10*d.* per gallon. It is exported under the name of *Niger* seed. Its price in the Bombay market will be seen on reference to the Table, (p. 90). It was first shipped to London experimentally in 1851. It is grown in very large quantities in Chota Nagpore, being a favourite crop, of easy cultivation, and giving good returns.

2054. [6234] Rape seed (*Brassica napa*), Ahmedabad.

Although we do not receive the oil, large quantities of seed are imported under the name of rape-seed from India, probably some of it the produce of species of *Sinapis.*

2055. [121*] Vappauley seeds (*Wrightia antidysenterica*).

The seeds of this tree are in great repute in India for their medicinal virtues, which the oil is supposed also to possess. It is employed only medicinally, and is obtained in but small quantities.

2056. [1788] Koosum or koosumba seed (*Carthamus tinctorius*), Punjab.

2057. [6070] Hurdee seed (*C. tinctorius*), Ahmedabad.

2058. [1931] Safflower oil (*C. tinctorius*), Calcutta. C. B. Wood, Esq.

2059. [5688] Do. Lucknow.

2060. [7053] Do. (refined), Do.

This is a light yellow clear oil, when properly refined or prepared; it is used in India for culinary and other purposes. This oil deserves more attention than it has hitherto received in this country; and if once fairly introduced, there is no doubt whatever of its becoming a staple import. It is used in some of the Government workshops as a 'drying oil.' It is believed to constitute the bulk of the celebrated 'Macassar oil.' The seed is exported under the name of Curdee or safflower seed. The Lucknow Committee appends the following note: — In Oude it is sown in October, either alone, or along the edge of wheat crops; both light and heavy soils are adapted to it. It is cultivated in every village, but not extensively. There would be no difficulty in farther cultivating it to any extent. The oil is extracted by pressing. The cost of the seed, which is called 'Dundi,' is 10? seers per rupee, and the cost of the oil is from 3 to 4 seers per rupee.

2061. [7113] Retha, soap berry (*Sapindus emarginatus*), Cuttack.

2062. [6064]. Reetah (*do.*), Ahmedabad.

2063. [76*] Soap-nut oil (*S. emarginatus*), Madras.

This is a pale yellow semi-solid oil, valued in medicine by the natives, but too costly to be otherwise employed.

2064. [77*] Tobacco oil (*Nicotiana tabacum*).

An oil extracted from tobacco-seed of a very dark colour, thick, and with an oppressive odour.

2065. [*] Babool seed (*Acacia arabica*).

The pods of this tree, known as the *Babool* in many parts of India, have long been employed in tanning on account of their astringency, but the extraction of oil from the seeds is only of recent date. It must be classed among experimental oils.

2066. [8*] Bassia butter in cakes (*Bassia butyracea*).

This is a beautiful, white, solid fat, the produce of the fruit of *Bassia butyracea.* It melts at a temperature above 120° Fahr., and in this respect it is superior to all other vegetable fats produced in India. The oil concretes immediately it is expressed.

2067. [4961] Mowha seeds (*Bassia latifolia*), Madras. Dr. Shortt.

2068. [6139] Do. Ahmedabad.

2069. [6457] Do. Allahabad.

2070. [6458] Do. Do.

2071. [5683] Mowha oil (*do.*), Cuttack.

2072. [6440] Do. Chota Nagpore.

2073. [7*] Do. Do.

2074. [5690] Do. Lucknow.

2075. [10100] Do. Calcutta. Messrs. Doss & Dey.

2076. [7082] Do. Madras.

This fatty substance, obtained also from the kernels of the fruit, is an article of common consumption in India, and may often be met with under the names of Mowha or Yallah oil in the London market. The tree grows wild in the Taree, and is also planted in groves in most parts of Oude, near villages, &c. Its cultivation can be extended almost indefinitely, and it thrives without any trouble. Its flowers have a thickened and enlarged tube, in which is contained a considerable amount of sugar. They are dried and eaten by the natives, and also fermented for the manufacture of *Mohwah Spirits.* The cost of the oil extracted is 3 rupees per maund. The proportion of oil yielded by native process is about half the weight of the seed; used only for burning.*

2077. [9*] Illoopa oil (*B. longifolia*), Madras.

* Central Committee, *Lucknow.*

2078. [2251] Illoopa, Chedumbara Pillay.

This solid fat is obtained from the kernels of the fruit, and is generally of a dirty white colour, and not so firm as the Bassia butter of *B. butyracea.* It melts at a temperature above 70° The tree is common everywhere in Southern India, and the fat or oil is employed largely by the natives.

2079. [5036] Cocoa nut (*Cocos nucifera*).

2080. [6326] Cold-drawn cocoa-nut oil (*do.*), Singapore. JOSE D'ALMEIDA, Esq.

2081. [4197] Cocoa-nut oil, Ayer Rajah, Penang. G. SCOTT, Esq.

2082. [9370] Do. South Canara. V. P. COELHO, Esq.

2083. [5703] Do. Cuttack.

2084. [2951] Do. Cossipore.

2085. [6001] Kokum butter (*Garcinia purpurea*), Bombay.

This solid vegetable oil melts at a temperature of 95°. It now forms an article of export. The seeds are first sun-dried, and then pounded and boiled in water; the oil collects on the surface, and on cooling concretes into a solid cake. When purified from extraneous matter, the product is of a rather brittle quality, of a pale yellowish hue, inclining to greenish, and mild to the taste. The seeds yield about one-tenth of their weight of oil. It is admirably adapted for the manufacture of healing ointments.

2086. [10740] Gamboge fruits (*Garcinia pictoria*), Mangalore. V. P. COELHO, Esq.

2087. [10693] Gamboge butter (*do.*), Do. Do.

This solid fat is obtained from the fruit of the gamboge tree of India. It is obtained by pounding the seeds in a stone mortar, and boiling the mass until the fat rises to the surface. It is used as a substitute for ghee (purified butter) by the poor.

2088. [10684] Mangosteen oil (*Garcinia mangostana*), Mangalore. V. P. COELHO, Esq.

A hard solid fat, resembling the vegetable tallow of Borneo

2089. [4954] Coodiri seed (*Sterculia fœtida*), Madras. DR. SHORTT.

2090. [141*] Kikuel oil (*Salvadora Persica*).

This oil has a somewhat aromatic odour, is of a bright green colour, and solid at a temperature below 95°. It most resembles in appearance the vegetable tallow of Borneo.

2091. [7001] Neem seed (*Azadirachta indica*), Bombay.

2092. [4951] Do. Chingleput. DR. SHORTT.

2093. [3749] Veppa ennye oil (*Azadirachta indica*),Salem. CHEDUMBARA PILLAY.

2094. [130*] Margosa oil (*A. indica*), Madras.

This is a pale yellow semi-solid oil, obtained from the fruits of a tree common in India. It is much employed by native practitioners, administered both internally and externally, and is sold in the bazaars, for illuminating purposes, under the name of *bitter oil.*

2095. [7055] Neem oil (*A. indica*).

2096. [115*] Marotty seeds (*Hydnocarpus inebrians*).

2097. [10734] Surrate fruits (*H. inebrians*), Mangalore. V. P. COELHO, Esq.

2098. [10682] Surrate oil (*do.*), Do. Do.

2099. [*] Neeradimootoo oil (*H. inebrians*).

The oil is used as a sedative, and as a remedy in scabies and ulcers of the feet.

2100. [4202] Nutmeg butter (*Myristica moschata*), Penang. G. SCOTT, Esq.

This fat is generally prepared by beating up the nutmegs, enclosing the paste in a bag and exposing it to the vapour of water, and afterwards expressing the fat by means of heated plates.

2101. [10744] Pundi kai (*M. malabarica*), Mangalore. V. P. COELHO, Esq.

2102. [11*] Piney tallow (*Vateria indica*), Canara.

2103. [191] Do. Mangalore. V. P. COELHO, Esq.

This butter is of solid consistence, and requires a higher temperature to melt it than animal tallows.

2104. [6675] Vegetable tallow, Siam. G. ANGUS, Esq.

2105. [6676] Do. Cochin China. Do.

2106. [6679] Do. Borneo. Do.

The Borneo tallow is obtained from the seeds of a Dipterocarpous tree, and is generally run whilst melted into joints of bamboo. It has a pale, greenish tint, is very hard, and approximates nearly to a vegetable wax. The other two vegetable tallows now exhibited appear to be new, and they are unaccompanied with any information.

2107. [1876] Vegetable wax. H. E. LOWER, Esq.

This is an artificial product manufactured by the exhibitor from castor oil.

# SECTION B. — ANIMAL SUBSTANCES USED IN MANUFACTURES.

## I.—FOR TEXTILE FABRICS AND CLOTHING.

### 1. WOOL.

As will be gathered from the Table (p. 98), wool has very rapidly become an important article of export from India.

The following information respecting the various wools used in the Punjab, of which specimens have been forwarded, is supplied by the LAHORE COMMITTEE :—

The following woollen substances are used in the Punjab : —

*a. Pashum,* or shawl wool, properly so called, being a downy substance, found next the skin and below the thick hair of the Thibetan goat. It is of three colours : white, drab, and dark lavender (Túsha).

The best kind is produced in the semi-Chinese Provinces of Turfan Kichar, and exported *viá* Yarkand to Kashmere. All the finest shawls are made of this wool, but as the Maharajah of Kashmere keeps a strict monopoly of the article, the Punjab shawl-weavers cannot procure it, and have to be content with an inferior kind of Pashum produced at Cháthán, and exported *viá* Leh to Umritsur, Núrpúr, Loodianah, Jelalpúr, and other shawl-weaving towns of the Punjab. The price of white Pashum in Kashmere is for uncleaned, 3*s.* to 4*s.* per lb. ; ditto cleaned, 6*s.* to 7*s.* per lb. Of Túsha ditto, uncleaned, 2*s.* to 3*s.* a lb. ; cleaned, from 5*s.* to 7*s.*

*b. The fleece of the Dumba sheep of Kabul and Peshawur.*—This is sometimes called *Kabuli Pashum.* It is used in the manufacture of the finer sorts of chogas, an outer-robe or cloak with sleeves, worn by Affghans and other Mahomedans of the Western frontier. Specimens of these are included in the collection.

*c. Wahab Sháhi,* or *Kirmani Wool.*—The wool of a sheep found in Kirman, a tract of country in the south of Persia, by the Persian Gulph. It is used for the manufacture of a spurious kind of shawl cloth, and for adulterating the texture of Kashmere shawls. Specimens of this wool will be found in the collection.

*d. The hair of a goat common in Kabul and Peshawur,* called *Pat,* from which a texture called *Pattu* is made.

*e. The woolly hair of the camel.*—From this a coarser kind of choga is made.

*f. The wool of the country sheep of the Plains.* — Regarding the production of wool in the Himalayan or Sub-Himalayan portion of the Punjab, the last year's Revenue Report states that 'there can be no doubt that the valleys of the Sutlej, Ravee, Chandrabaga (or Chenab), Namisukh, and other tributaries of the Indus, supply grazing grounds not to be surpassed in richness and suitableness in any part of the world. The population inhabiting them are chiefly pastoral, but owing to sloth and ignorance the wool they produce is but small in quantity, full of dirt and ill-cared for in every way.' The government of the Punjab have made efforts to improve the breed by the importation of Merino rams, but hitherto with little success. However, a truss of Merino wool produced at Huzara, a hill district to the north-west of the Punjab, and sent to England in 1860, was there valued at 1*s.* 6*d.* per lb.

2108. [3504] Wool, Kashmere goat's, Umritsur.

2109. [3508] Do. weft (Wahab Shahi).

2110. [3509] Wool cleaned (Wahab Shahi), Umritsur.

2111. [3510] Do. Kashmere goat's, Do.

2112. [3511] Do. Do., new, Do.

2113. [3512] Do. raw, Do.

2114. [3513] Do. first class, white Cabul goat's, cleaned, Do.

2115. [3514] Do. original colour, Do.

2116. [3515] Do. first class, brown Cabul goat's, cleaned, Do.

2117. [3516] Do. third class, black and raw, Do.

2118. [3517] Do. first class, white, Do.

2119. [3518] Do. second class, original colour, Do.

2120. [3519] Do. second class, red, Do.

2121-2. [3520-1] Do. fourth class, black, Do.

2123. [5612] Do. raw Pashum of the Thibetan goat, used in the manufacture of Kashmere shawls of the kind called Tusha, produced in Thibet, Lahore.

2124. [5613] Do. raw, white, Do.

2125. [5614] Do. cleaned, of the kind called Tusha, Do.

2126. [5656] Wool of the Dumba sheep, Do.

2127. [3505] Do. first class, for weft, Kashmere.

2128. [3506] Do. second class, do., Do.

2129. [3507] Do. third class, do., Do.

2130. [5650] Wool, Thibet.

2131. [5651] Untwisted yarn, Lahore.

2132. [5652] Pashum thread, Do.

2133. [5615] Thread used in the manufacture of Kashmere shawl, Do.

2134. [5653] Pushum thread, cleaned, Do.

2135. [5654] Do. of ordinary quality, Do.

2136. [5655] Do. of finest quality, Do.

# TABLE SHOWING THE QUANTITIES (AS FAR AS CAN BE ASCERTAINED) AND VALUE * OF WOOL EXPORTED FROM INDIA AND EACH PRESIDENCY TO ALL PARTS OF THE WORLD FROM 1850-51 TO 1860-61 INCLUSIVE.

| Years | Whence Exported | United Kingdom Qty (lbs.) | UK Value (£) | France Qty (lbs.) | France Value (£) | Other Parts of Europe Qty (lbs.) | Other Europe Value (£) | North America Qty (lbs.) | N. America Value (£) | China Qty (lbs.) | China Value (£) | Arabian and Persian Gulfs Qty (lbs.) | Value (£) | Islands and Shores of the Indian Seas Qty (lbs.) | Value (£) | All Other Places Qty (lbs.) | Value (£) | Total Exported to All Places Qty (lbs.) | Qty (tons) | Value (£) |
|---|---|---|---|---|---|---|---|---|---|---|---|---|---|---|---|---|---|---|---|---|
| 1850-51 | Bengal | | | | | | | | | | | | | 164 | 50 | | | 164 | | 50 |
| | Madras | | | | | | | | | | | | | | | | | | | |
| | Bombay | 4,492,794 | 65,525 | | | | | 189,114 | 2,758 | | | | | | | | | 4,681,910 | 2,090 | 68,285 |
| | ALL INDIA | 4,492,794 | 65,525 | | | | | 189,114 | 2,758 | | | | | 164 | 50 | | | 4,682,074 | 2,090 | 68,335 |
| 1851-52 | Bengal | 30,985 | 236 | | | | | | | | | | | | | | | 30,985 | 14 | 236 |
| | Madras | | | | | | | | | | | | | | | | | | | |
| | Bombay | 7,025,728 | 100,370 | | | | | 448 | 6 | | | | | | | | | 7,026,176 | 3,137 | 100,376 |
| | ALL INDIA | 7,056,713 | 100,606 | | | | | 448 | 6 | | | | | | | | | 7,057,161 | 3,151 | 100,612 |
| 1852-53 | Bengal | † 17,600 | 285 | | | | | † 8,000 | 123 | | | | | † 7,800 | 114 | 26,400 | 407 | † 59,800 | 27 | 929 |
| | Madras | 79,072 | 856 | | | | | | | | | | | | | | | 79,072 | 35 | 856 |
| | Bombay | 11,921,927 | 170,313 | | | | | | | | | | | | | | | 11,921,927 | 5,322 | 170,313 |
| | ALL INDIA | 12,018,599 | 171,454 | | | | | 8,000 | 123 | | | | | 7,800 | 114 | 26,400 | 407 | 12,060,799 | 5,384 | 172,098 |
| 1853-54 | Bengal | 54,096 | 875 | | | | | | | | | | | | | | | 54,096 | 24 | 875 |
| | Madras | 128,088 | 1,578 | | | | | | | | | | | | | | | 128,088 | 57 | 1,578 |
| | Bombay | 14,165,069 | 202,359 | 28,281 | 304 | | | | | | | | | | | 33,945 | 485 | 14,227,295 | 6,351 | 203,148 |
| | ALL INDIA | 14,347,253 | 204,812 | 28,281 | 304 | | | | | | | | | | | 83,945 | 485 | 14,499,479 | 6,432 | 205,601 |
| 1854-55 | Bengal | 120,762 | 1,682 | | | | | | | | | | | | | | | 120,762 | 54 | 1,682 |
| | Madras | | | | | | | | | | | | | | | | | | | |
| | Bombay | 12,964,065 | 205,196 | 13,272 | 190 | | | 13,052 | 221 | | | | | | | 2,576 | 35 | 12,990,389 | 5,799 | 205,647 |
| | ALL INDIA | 13,084,827 | 206,818 | 13,272 | 190 | | | 13,052 | 221 | | | | | | | 2,576 | 35 | 13,113,727 | 5,854 | 207,264 |
| 1855-56 | Bengal | 11,566 | 121 | | | | | | | | | | | | | | | 11,566 | 5 | 121 |
| | Madras | | | | | 114,781 | 1,947 | | | | | | | | | | | 114,781 | 51 | 1,947 |
| | Bombay | 18,084,762 | 306,795 | 273,557 | 5,352 | | | | | | | | | | | | | 18,358,319 | 8,247 | 312,147 |
| | ALL INDIA | 18,096,328 | 306,916 | 273,557 | 5,352 | 114,781 | 1,947 | | | | | | | | | | | 18,484,666 | 8,252 | 314,215 |
| 1856-57 | Bengal | 26,962 | 393 | | | 2,576 | 38 | 9,632 | 150 | | | | | | | | | 39,170 | 17 | 581 |
| | Madras | 66,663 | 962 | | | | | | | | | | | | | | | 66,663 | 30 | 962 |
| | Bombay | 16,892,920 | 351,783 | 662,607 | 13,252 | 21,840 | 370 | 952,196 | 20,165 | | | | | | | | | 18,529,563 | 8,272 | 385,560 |
| | ALL INDIA | 16,986,545 | 353,138 | 662,607 | 13,252 | 24,416 | 408 | 961,828 | 20,305 | | | | | | | | | 18,635,396 | 8,319 | 387,103 |
| 1857-58 | Bengal | 812 | 8 | | | | | | | | | | | | | | | 812 | | 8 |
| | Madras | 26,799 | 319 | | | | | | | | | | | | | | | 26,799 | 12 | 319 |
| | Bombay | 15,189,131 | 339,045 | 196,044 | 4,376 | | | 275,410 | 6,147 | | | | | | | | | 15,660,585 | 6,991 | 349,568 |
| | ALL INDIA | 15,216,742 | 339,372 | 196,044 | 4,376 | | | 275,410 | 6,147 | | | | | | | | | 15,688,196 | 7,010 | 349,895 |
| 1858-59 | Bengal | † 70,000 | 840 | | | | | | | | | | | | | | | † 70,000 | 31 | 840 |
| | Madras | | | | | | | | | | | | | | | | | | | |
| | Bombay | 15,953,942 | 270,647 | 85,786 | 1,455 | | | | | | | | | | | | | 16,039,728 | 7,161 | 272,102 |
| | ALL INDIA | 16,023,942 | 271,487 | 85,786 | 1,455 | | | | | | | | | | | | | 16,109,728 | 7,192 | 272,942 |
| 1859-60 | Bengal | | | | | | | | | | | | | | | | | | | |
| | Madras | | | | | | | | | | | | | | | | | | | |
| | Bombay | 18,688,328 | 417,151 | | | | | 874,569 | 19,521 | | | | | | | | | 19,562,897 | 8,733 | 436,672 |
| | ALL INDIA | 18,688,328 | 417,151 | | | | | 874,569 | 19,521 | | | | | | | | | 19,562,897 | 8,733 | 436,672 |
| 1860-61 | Bengal | 750 | 10 | | | | | | | | | | | | | | | 700 | | 10 |
| | Madras | | | | | | | | | | | | | | | | | | | |
| | Bombay | 21,068,289 | 470,823 | 37,968 | 1,223 | | | 270,816 | 6,045 | | | 9,632 | 43 | | | | | 21,381,705 | 9,505 | 478,134 |
| | ALL INDIA | 21,069,089 | 470,833 | 37,968 | 1,223 | | | 270,816 | 6,045 | | | 9,632 | 43 | | | | | 21,382,405 | 9,595 | 478,144 |

* The Indian rupee is here converted at the exchange of two shillings.

† The quantities of wool exported from the Bengal Presidency in the year 1852-53, and the Madras Presidency in 1855-56, owing to their not having been officially recorded, are estimated from the given values.

2137. [3522] Six samples of coloured wool thread for needle-work, Umritsur.

2138. [3523] A sample of blue wool thread, Do.

2139. [3524] Do. dark yellow, Do.

2140. [3525] Do. light do., Do.

2141. [3526] Do. scarlet, Do.

2142. [3527] Do. light blue, Do.

2143. [3528] Do. green, Do.

2144. [3529] Do. rose-coloured, Do.

2145. [3530] Do. scarlet, Do.

2146. [3531] A sample of light green thread, Do.

2147. [3532] Do. crimson, Do.

2148. [3533] Do. black, Do.

2149. [3534] Do. dark rose-coloured, Do.

2150. [3535] Do. purple, Do.

2151. [3536] Do. scarlet, Do.

2152. [1041] Raw wool, Khelat.

2153. [777] Do. Sindh.

2154. [696] Do. Kutch.

2155. [1270] Black wool, Ahmedabad.

2156. [1270] White wool, Do.

2157. [2698] Uncleaned wool, Mysore.

2158. [2699] Woollen thread, Astagram, Mysore.

2159. [4267] Thibet wool, Darjeeling.

2160. [1054] Goat's hair, Hyderabad.

2161. [1055] Camel's hair, Do.

2162. [981] Raw wool, Bengal. A. M. Dowleans, Esq.

2163. [5645] Do. Do. Do.

2164. [4484] Burmese twist, Rangoon.

## 2. SILK.

Among the textile fabrics of animal origin, silk holds a high rank. The Table (p. 100) shows the export of silk from India. The present Exhibition embraces not only a considerable number of the silks produced by the various kinds of silkworms, but also a very valuable collection of the silk-producing moths, together with drawings of the caterpillars and specimens of the cocoons. These latter have been prepared by Mr. F. Moore. They are contained in a series of trays in Case 8, and are as follows:—

*Tray I.*—Containing cocoons and moths of various species of mulberry-feeding silkworms (*Bombyx*), &c.
1. Cocoon of *Bombyx mori* (Linn.) from Cashmere stock.
2. Cocoons of do. as imported into England from the continent.
3. Cocoon of cross between Cashmere *B. mori* and Bengal *B. cræsi*, obtained by Capt. T. Hutton at Mussooree.
4. Cocoons of *B. textor* (Hutton), from Mussooree.
5. Do. of *B. sinensis* (?), from Bengal.
6. Golden-yellow cocoons, from Shanghai, as imported into England.
7. Pure white do.    do.    do.
8. Do.    do. from Canton.    do.
9. Cocoon of very large size; locality unknown.
10. Do. from the Punjab.
11. Cocoon, raw silk, and specimen of moth of *B. Huttoni* (Westw.), from Mussooree.
12. Do. do. *B. Horsfieldi* (Moore), from Java.
13. Do. with drawing of caterpillar and cocoon of *Ocinara dileetula* (Walk.), from Java.

*Tray II.*—Containing samples of raw silk, produced by various species of *Bombyx.*
1. Silk from Surdah filature, Bengal.
2. Commercolly silk.
3. Silk from Rangamutty filature, Bengal.
4, 5, 6. Punjab silk—first and third years' cultivation.
7. Madras silk.
8. Mysore silk — raised from eggs originally from Bengal.
9. Silk obtained from cross between ♀ Cashmere (*B. mori*) and ♂ Bengal (*B. cræsi?*), from Bengal.
10. Do. obtained from cross between French and Bengal moths, from Surdah filature, Bengal.
11. White silk, obtained from the Boro-poloo, from Bengal.
12. Do. do. from Mysore, raised from eggs originally from Bengal.
13, 14. Do. do. from Bangalore.

*Tray III.*—Containing raw silk, produced by various species of *Bombyx*, from several localities, viz.—
1. Pat raw silk, from Durrang, Central Assam.
2, 3. Several samples of silk obtained from the large and small 'Pat Polo' or mulberry worm of Assam, from Durrang, Central Assam.
4. Silk raised in Munguldye, Assam.
5. 'Pat Soot,' Pat silk ungummed, from Seebsagur, Upper Assam.
6. 'Na Dhoroa Pat,' silk thread unbleached, from Nowgong, Do.
7. 'Dhoroa Pat' silk thread bleached, do.
8. Silk from Sandoway, Arracan.
9. Burmese silk, from Tenasserim.
10. White silk, from Pegu.
11. Yellow silk, Do.
12. Do. silk from Cochin China.

*Tray IV.*—Containing samples of cocoons, raw, thrown, and waste silk as imported into England from China (Shanghai) and Japan. Also a compressed batch of relaxed cocoons, stated to be imported from Southern Russia.

*Tray V.*—Containing drawing of caterpillar, specimens of cocoons and moths of *Attacus atlas* (Linn.) from China, Sylhet, Mussooree, Madras, and Java.

*Tray VI.*—Containing specimens of:—
1. Male and female moths of *Attacus Edwardsii* (White), from Sikkim.

# Class IV.—*India.*

## TABLE SHOWING THE QUANTITIES (AS FAR AS CAN BE ASCERTAINED) AND VALUE* OF SILK EXPORTED FROM INDIA AND EACH PRESIDENCY TO ALL PARTS OF THE WORLD FROM 1850-51 TO 1860-61 INCLUSIVE.

| Years | Whence Exported | UK Qty (lbs) | UK Value (£) | France Qty (lbs) | France Value (£) | Other Europe Qty (lbs) | Other Europe Value (£) | N. America Qty (lbs) | N. America Value (£) | China Qty (lbs) | China Value (£) | Arabian & Persian Gulf Qty (lbs) | Arabian & Persian Gulf Value (£) | Islands & Shores Indian Seas Qty (lbs) | Islands & Shores Indian Seas Value (£) | All Other Places Qty (lbs) | All Other Places Value (£) | Total Qty (lbs) | Total Qty (tons) | Total Value (£) |
|---|---|---|---|---|---|---|---|---|---|---|---|---|---|---|---|---|---|---|---|---|
| 1850-51 | Bengal | 1,256,533 | 605,387 | 247 | 66 | .. | .. | .. | .. | .. | .. | 905 | 412 | .. | .. | .. | .. | 1,257,685 | 561 | 605,865 |
|  | Madras | .. | .. | .. | .. | .. | .. | .. | .. | .. | .. | .. | .. | .. | .. | .. | .. | .. | .. | .. |
|  | Bombay | 14,716 | 4,063 | .. | .. | .. | .. | .. | .. | .. | .. | 45,985 | 7,704 | .. | .. | 7,552 | 1,686 | 68,253 | 31 | 13,458 |
|  | ALL INDIA | 1,271,249 | 609,450 | 247 | 66 | .. | .. | .. | .. | .. | .. | 46,890 | 8,116 | .. | .. | 7,552 | 1,686 | 1,325,938 | 592 | 619,318 |
| 1851-52 | Bengal | 1,434,275 | 679,260 | .. | .. | 657 | 297 | .. | .. | .. | .. | 3,537 | 1,526 | .. | .. | .. | .. | 1,438,469 | 642 | 681,083 |
|  | Madras | 210 | 116 | .. | .. | .. | .. | .. | .. | .. | .. | .. | .. | .. | .. | .. | .. | 210 | .. | 116 |
|  | Bombay | 3,173 | 1,161 | .. | .. | .. | .. | .. | .. | .. | .. | 15,825 | 2,328 | 9,878 | 3,471 | 1,789 | 486 | 30,665 | 14 | 7,441 |
|  | ALL INDIA | 1,437,658 | 680,537 | .. | .. | 657 | 297 | .. | .. | .. | .. | 19,362 | 3,849 | 9,878 | 3,471 | 1,789 | 486 | 1,469,344 | 656 | 688,640 |
| 1852-53 | Bengal | 1,361,165 | 658,673 | 987 | 497 | 576 | 296 | .. | .. | 100 | 50 | 82 | 27 | .. | .. | .. | .. | 1,362,800 | 608 | 659,493 |
|  | Madras | 240 | 96 | .. | .. | .. | .. | .. | .. | .. | .. | .. | .. | .. | .. | 63 | 16 | 303 | .. | 112 |
|  | Bombay | 19,813 | 5,685 | .. | .. | .. | .. | .. | .. | 280 | 126 | 17,546 | 1,850 | .. | .. | 978 | 406 | 38,337 | 17 | 7,941 |
|  | ALL INDIA | 1,381,208 | 664,454 | 987 | 497 | 576 | 296 | .. | .. | 380 | 176 | 17,628 | 1,877 | .. | .. | 1,041 | 422 | 1,401,440 | 625 | 667,546 |
| 1853-54 | Bengal | 1,479,193 | 548,531 | 17,724 | 8,613 | .. | .. | 3,808 | 1,864 | .. | .. | 672 | 361 | .. | .. | 90,244 | 45,950 | 1,591,741 | 711 | 605,369 |
|  | Madras | 200 | 88 | .. | .. | .. | .. | .. | .. | .. | .. | 254 | 82 | .. | .. | 2,731 | 1,147 | 3,185 | 1 | 1,317 |
|  | Bombay | 75,555 | 29,619 | .. | .. | .. | .. | .. | .. | .. | .. | 21,690 | 3,156 | .. | .. | 7,613 | 2,888 | 105,138 | 47 | 35,789 |
|  | ALL INDIA | 1,554,948 | 578,238 | 17,724 | 8,613 | .. | .. | 3,808 | 1,864 | .. | .. | 22,616 | 3,599 | .. | .. | 100,588 | 49,985 | 1,700,064 | 759 | 642,475 |
| 1854-55 | Bengal | 1,019,648 | 446,720 | 56,468 | 4,332 | .. | .. | .. | .. | .. | .. | 2,800 | 1,385 | .. | .. | 7,280 | 7,878 | 1,086,196 | 485 | 460,310 |
|  | Madras | .. | .. | .. | .. | .. | .. | .. | .. | .. | .. | .. | .. | .. | .. | .. | .. | .. | .. | .. |
|  | Bombay | 32,826 | 13,887 | .. | .. | .. | .. | .. | .. | .. | .. | 33,809 | 9,186 | .. | .. | 40,259 | 16,728 | 106,894 | 48 | 39,796 |
|  | ALL INDIA | 1,052,474 | 460,607 | 56,468 | 4,332 | .. | .. | .. | .. | .. | .. | 36,609 | 10,571 | .. | .. | 47,539 | 24,596 | 1,193,090 | 533 | 500,106 |
| 1855-56 | Bengal | 1,322,608 | 641,118 | 47,040 | 23,407 | .. | .. | .. | .. | .. | .. | 1,344 | 699 | .. | .. | 13,216 | 6,363 | 1,384,208 | 618 | 671,587 |
|  | Madras | .. | .. | .. | .. | .. | .. | .. | .. | .. | .. | .. | .. | .. | .. | .. | .. | .. | .. | .. |
|  | Bombay | 21,814 | 8,791 | .. | .. | .. | .. | .. | .. | .. | .. | 91,546 | 9,723 | .. | .. | 79,637 | 17,604 | 192,997 | 86 | 36,118 |
|  | ALL INDIA | 1,344,422 | 649,909 | 47,040 | 23,407 | .. | .. | .. | .. | .. | .. | 92,890 | 10,422 | .. | .. | 92,853 | 23,967 | 1,577,205 | 704 | 707,705 |
| 1856-57 | Bengal | 1,155,056 | 562,093 | 171,248 | 78,255 | .. | .. | 32,144 | 1,583 | .. | .. | .. | .. | .. | .. | 94,024 | 45,734 | 1,452,472 | 648 | 687,665 |
|  | Madras | 11,250 | 3,922 | .. | .. | .. | .. | .. | .. | .. | .. | .. | .. | .. | .. | .. | .. | 11,250 | 5 | 3,922 |
|  | Bombay | 141,044 | 55,043 | 9,520 | 3,332 | .. | .. | .. | .. | .. | .. | 3,704 | 371 | .. | .. | 138,788 | 31,806 | 293,056 | 131 | 90,552 |
|  | ALL INDIA | 1,307,350 | 621,058 | 180,768 | 81,587 | .. | .. | 32,144 | 1,583 | .. | .. | 3,704 | 371 | .. | .. | 232,812 | 77,540 | 1,756,778 | 784 | 782,139 |
| 1857-58 | Bengal | 1,341,200 | 658,718 | 125,216 | 64,193 | 784 | 390 | 1,120 | 583 | .. | .. | 112 | 65 | .. | .. | 49,392 | 23,906 | 1,517,824 | 678 | 747,865 |
|  | Madras | 1,038 | 413 | 870 | 553 | 24 | 13 | .. | .. | .. | .. | .. | .. | .. | .. | .. | .. | 1,932 | 1 | 979 |
|  | Bombay | 20,737 | 8,578 | 672 | 235 | .. | .. | 800 | 324 | .. | .. | 18,378 | 2,473 | .. | .. | 20,120 | 6,229 | 60,707 | 27 | 17,839 |
|  | ALL INDIA | 1,362,975 | 667,709 | 126,758 | 64,981 | 808 | 403 | 1,920 | 907 | .. | .. | 18,490 | 2,538 | .. | .. | 69,512 | 30,135 | 1,580,463 | 706 | 766,673 |
| 1858-59 | Bengal | 911,110 | 579,978 | 110,992 | 57,241 | .. | .. | .. | .. | .. | .. | 448 | 220 | .. | .. | 33,142 | 10,761 | 1,055,692 | 471 | 648,200 |
|  | Madras | 146,136 | 70,719 | 1,440 | 720 | .. | .. | .. | .. | .. | .. | 17,386 | 3,683 | .. | .. | 6,784 | 2,333 | 171,746 | 77 | 77,465 |
|  | Bombay | .. | .. | .. | .. | .. | .. | .. | .. | .. | .. | .. | .. | .. | .. | .. | .. | .. | .. | .. |
|  | ALL INDIA | 1,057,246 | 650,697 | 112,432 | 57,961 | .. | .. | .. | .. | .. | .. | 17,834 | 3,903 | .. | .. | 39,926 | 13,094 | 1,227,438 | 548 | 727,438 |
| 1859-60 | Bengal | 1,430,324 | 702,392 | 149,240 | 73,299 | 1,008 | 496 | 168 | 80 | .. | .. | .. | .. | .. | .. | 50,176 | 22,985 | 1,630,916 | 728 | 799,252 |
|  | Madras | 30,625 | 16,001 | 960 | 518 | .. | .. | .. | .. | .. | .. | 2,845 | 389 | .. | .. | 5,352 | 1,693 | 39,782 | 18 | 18,601 |
|  | Bombay | .. | .. | .. | .. | .. | .. | .. | .. | .. | .. | .. | .. | .. | .. | .. | .. | .. | .. | .. |
|  | ALL INDIA | 1,460,949 | 718,393 | 150,200 | 73,817 | 1,008 | 496 | 168 | 80 | .. | .. | 2,845 | 389 | .. | .. | 55,528 | 24,978 | 1,670,698 | 746 | 817,853 |
| 1860-61 | Bengal | 1,572,704 | 842,503 | 212,688 | 113,933 | 1,568 | 841 | .. | .. | .. | .. | .. | .. | .. | .. | 7,195 | 4,004 | 1,794,155 | 801 | 961,281 |
|  | Madras | 143,076 | 72,195 | 2 | 1 | .. | .. | .. | .. | .. | .. | 9,937 | 1,416 | .. | .. | 8,586 | 1,835 | 161,501 | 72 | 75,447 |
|  | Bombay | .. | .. | .. | .. | .. | .. | .. | .. | .. | .. | .. | .. | .. | .. | .. | .. | .. | .. | .. |
|  | ALL INDIA | 1,715,780 | 914,698 | 212,690 | 113,934 | 1,568 | 841 | .. | .. | .. | .. | 9,937 | 1,416 | .. | .. | 15,781 | 5,839 | 1,955,656 | 873 | 1,036,728 |

* The Indian rupee is here converted at the exchange of two shillings.

2. Male *Attacus Guerini* (Moore), from Bengal.
3. *Attacus Ricini*, from Assam.
4. Eggs, caterpillars (preserved in spirits), cocoons, and male and female moths of the 'Ailanthus silkworm,' or cross between the Chinese and Bengal 'Eria;' reared in London in 1859 by Mr. F. Moore; also sample of the spun silk, and cloth manufactured in France.

*Tray VII.*—Containing specimens of eggs, drawing of caterpillar, specimens of cocoons, male and female moths of the ' Eria,' (*Attacus Cynthia* Drury), from China (Hong Kong), Nepal, Mussooree (reared in London from the cocoon), and Java.

*Tray VIII.*—Containing samples of 'Eria' or 'Arrindy' silk in several conditions (raw, dyed, flossed, &c.), with specimens of the cloth, from various districts in Assam, Bengal, &c.

*Tray IX.*—Containing —
1. Bundle of relaxed cocoons, raw and twisted silk, and cloth of the 'Mezankoorie,' (*Antheræa Mezankooria* Moore).
2. Drawing of caterpillar, specimens of cocoons and moths of *Actias Selene* M'L.), from Darjeeling and Mussooree.

*Tray X.* — Containing eggs, drawing of caterpillar, specimens of cocoons, raw silk and thread, cloth, &c., of the 'Tusseh' or 'Tussur,' (*Antheræa Paphia* Linn.), from Darjeeling, Bhagulpore, &c.

*Tray XI.*—Containing drawing of caterpillar, specimens of the cocoons, raw and floss (dyed pink and yellow), silk, cloth, and moth of the 'Moonga' or 'Moogha,' (*Antheræa Assama* Helfer), from various districts in Assam.

*Tray XII.*—Containing specimens of various species of so-called 'Tusseh ' moths.
1. *Antheræa Pernyi* (Guer.-Men.) drawing, North China.
2. *A. Perrottetti* (G. Men.) do. Pondicherry.
3. *A. Roylei* (Moore), moth and cocoon, from Mussooree.
4. *A. Helferi* (Moore), from Darjeeling.
5. *A. Jana* (Cram.), drawing, Java.
6. *A. Frithii* (Moore), Darjeeling.
7. *A. Larissa* (Westw.), drawing, Java.

*Tray XIII.*—Containing specimens of—
1. Male moth of *Saturnia pyretorum* (Westw.), from China.
2. Do. *Saturnia Grotei* (Moore), from Darjeeling.
3. Drawing of caterpillar and cocoon; male and female moths of *Loepa Katinka* (Westw.), from Java.
4. Drawing of *Neoris Huttoni* (Moore), from Mussooree.
5. Do. *Caligula Thibeta* (Westw.), and specimen of cocoon, from Mussooree.
6. Male and female moths of *Caligula Simla* (Westw.)
7. Drawing of Salassa Lola (Westw.), South-east Himalayas.
8. Drawing of caterpillar, specimens of cocoon, and moths of *Cricula trifenestrata* (Helfer), from Java.

2165. [3786] Raw silk (dyed). Reared and reeled by JAFFER ALLEE, of *Goordaspore*, under the direction of LALLA CHUMBA MULL.

2166. [3787] Do. Do.

2167. [3788] Do. Do.

2168. [3789] Do. Do.

2169. [3790] Do. Do.

2170. [3791] Do. Do.

2171. [3792] Raw silk (dyed). Reared and reeled by JAFFER ALLEE, of *Goordaspore*, under the direction of LALLA CHUMBA MULL.

2172. [3793] Do. Do.

2173. [3794] Do. Do.

2174. [3795] Do. Do.

2175. [3796] Do. Do.

2176. [3797] Do. Do.

2177. [3798] Do. Do.

2178. [3799] Do. Do.

---

2179. [7228] Bengal raw silk in its various stages. F. BASHFORD, Esq.

2180. [7229] Bengal raw silk (yellow and white), from the filatures of Messrs. LYALL, RENNIE, & Co., *Berhampore* and *Calcutta.*

2181. [2590] Silk thread (dyed), Chittledroog, Mysore.

2182. [2582] Do. Do.

2183. [5616] Floss silk (dyed), Peshawur. MAHOMED AUZUM.

2184. [815] Do. Shikarpore.

2185. [34*] Do. Cachar.

2186. [41*] Silk thread (dyed), Sattara.

2187. [38*] Do. Madras.

2188. [5100] Mezankoree silk, Assam. H. BAINBRIDGE, Esq.

2189. [8227] Do. Seebsagur, Assam. LIEUT. PHAIRE.

2190. [4775] Raw silk (dyed of various colours), Burmah. Messrs. HALLIDAY, FOX, & Co.

2191. [9245] Do. (yellow), Bengal.

2192. [9244] Do. Punjab.

2193. [5618] Do. Cashmere.

2194. [9246] Do. Mysore.

2195. [9247] Do. Pegu.

2196. [9248] Do. Arracan.

2197. [9249] Do. (white), Singapore.

2198. [5545] Do. (yellow and white), Pegu.

2199. [5543] 'Pat soota' (white floss), Luckimpore. H. L. MICHEL, Esq.

2200. [5099] Do. Balasore, Assam.

2201. [4483] Raw silk, Rangoon. Messrs. HALLIDAY, FOX, & Co.

2202. [2371] Do. (pale yellow), Mysore.

2203. [3785] Do. (white), Bokhara.

2204. [3784] Do. (yellow). Reared in Umritsur by MR. H. COPE, and reeled by JAFFIR ALLEE, of *Goordaspore.*

2205. [7060] Do., reared at Umritsur by MR. H. COPE.

2206. [816] Do. (pale yellow), Khorasan.

Almost all the raw silk used in the Punjab is produced in Bokhara, but it is not improbable that the submontane districts of the Punjab were formerly silk-producing countries, as we know that in the time of Justinian silk was produced in Sirhind. A series of experiments was made a few years ago by the government of the Punjab, with a view of ascertaining whether silkworms could be successfully reared in the plains, but the experiment was a failure, the heat and want of sufficient moist food having rendered the worms nearly unproductive. MR. COPE, of Umritsur, who originated the experiment, has, however, again tried it with considerable success.* In the month of February of the present year, the Manchester Chamber of Commerce reported thus on the samples forwarded from Umritsur by MR. COPE:— 'The silk is very well and carefully reeled in every respect; the thread is clean, round, and uniform in size, well laid on the reel, and the hank a convenient size. It winds well, and passes through the other operations of throwing in a satisfactory manner. Its elasticity is about 1 in 6, equal to good China. If the silk can be supplied in quantity in every respect equal to sample, it would be worth 25s. per lb. in the present state of the market.' The sample of yellow silk has been valued at very nearly the same amount.

2207. [9243] Raw silk, Oude. R. CARNEGIE, Esq. Forwarded by the CHIEF COMMISSIONER of OUDE.

Reared by MR. CARNEGIE at Seetapore, in Oude, from the silkworm of Cashmere. The cultivation of the silkworm in Oude is believed to be capable of great extension, as the mulberry abounds throughout the province, and the climate appears favourable to the rearing and thriving of the worms.

Of this sample Messrs. DURANT & Co., of Copthall Court, report as under:—'So far as we can judge from so small a sample, we consider it to be most creditable to the reeler, and such as affords good hope of future success.

'The colour and quality are good, in appearance strongly resembling silk of Lombardy. The thread is well made, clean, and fairly open in size; and we think that when tried on the throwing mill and in the loom, such silk will be found to yield satisfactory results. In every respect it is equal, and in some better, than the silk of Bengal, inasmuch as it has more nerve and less of the "fluffiness" and small foul which more or less seems to be an inherent condition of even the best of the European filatures of that country.

'The present value is about 25s. per lb.'

2208. [2589] Raw silk, Chittledroog, Mysore.

2209. [2591] Mulberry silkworm cocoons, Do. Do.

2210. [2708] Raw silk, No. 1, Bangalore, Mysore.

2211. [2709] Do. No. 2, Do. Do.

2212. [4005] Kora silk, Salem.

2213. [3836] Raw Tusseh silk and cocoons, from Seonee, Jubbulpore.

This is produced extensively in the Seonee district, but is not manufactured into cloth there. It is exported to Nagpore, where it is woven into native cloths, called 'Tussur Sarees,' &c. No farther information can be given at present, as the trade is carried on by the Gonds, living in scattered jungles.*

2214. [5056] Do. Chynepore, Bengal.

2215. [6517] Do., and thread in various states, Cuttack.

Locally called the '*Khosa*'—'*Khoseare.*' These are, it is presumed, the product of the *Saturnia Mylitta,* moth-caterpillar. The cocoons are found in abundance throughout the forest jungles of this division, and for the most part on the larger trees called the '*Asan*' (the *Pentaptera tomentosa*), the *Sal* or *Shorea robusta,* and less frequently on the common Indian plum or '*Barkolee*' tree. The wild cocoons are collected by the 'Sahars' and other poor and half wild castes (whose villages are often met with in the heart of the jungles), and sold so many for the pice, to the best advantage, but at no fixed rates. Each cocoon being very carefully enclosed within two leaves brought together and made to wrap around it, it is almost impossible to discover by mere sight on which trees cocoons are to be found. This is, therefore, done by observing the dung of the caterpillar under the tree. The eggs of the moth are also collected and preserved, and the caterpillar regularly reared and tended, in many parts in the Hill tracts, on trees pruned and preserved for the purpose. The cocoons vary much in size and colour, and there is also a very perceptible diversity in the texture and glossiness of the raw silk, which most probably depends on the species of tree on which the caterpillar may happen to have fed. The same variety is observable in the female moths, which are of three or four different sizes and colours. The mode of winding off the thread does not differ from that pursued in the case of the ordinary silkworm cocoon proper.†

2216. [2864] Tussur cocoons, Raepore, Bengal.

2217. [9250] Moonga silk, Assam.

2218. [7909] Do. Luckimpore, Upper Assam. BABOO CHAROO CHUNDA, *Mowzahdar.*

2219. [5094] Do. Balasore, Assam.

2220. [5095] Moonga silk, Kamroop, Assam. LIEUT.-COL. HAMILTON VEITCH.

2221. [5097] Eria silk, Assam. LIEUT.-COL. HAMILTON VEITCH.

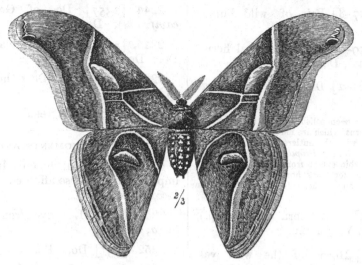

ERIA SILK MOTH.*

2222. [5098] Eria silk, Assam. H. BAINBRIDGE, Esq.

2223. [9251] Do. Assam.

2224. [7094] Yellow Eria cocoons of castor oil silkworms, No. 1, Durrung, Assam. LIEUT. W. PHAIRE.

2225. [7095] Do. No. 2, Do. Do.

2226. [7096] White do. Do. Do.

2227. [5546] Do. cocoons and silk in various states, Assam.

2228. [5542] Eria silk, Luckimpore. JAWRAM DEKA BOROOVAH PESKA.

---

II. — FOR DOMESTIC OR ORNAMENTAL PURPOSES, OR FOR THE MANUFACTURE OF IMPLEMENTS.

**BONE, HORN, ETC.**

2229. [4249] Pair of buffalo horns, Burmah. Messrs. HALLIDAY, FOX, & Co.

TABLE SHOWING THE QUANTITIES (AS FAR AS CAN BE ASCERTAINED) AND THE VALUE OF HORNS EXPORTED FROM INDIA AND EACH PRESIDENCY TO ALL PARTS OF THE WORLD FROM 1857-58 TO 1860-61.

| Years | Whence Exported | United Kingdom | | France | | Other Parts of Europe | | America | | China | | Arabian and Persian Gulfs | | Other Parts | | Total Exported to All Parts | |
|---|---|---|---|---|---|---|---|---|---|---|---|---|---|---|---|---|---|
| | | Quan. | Value | Quan. | Val. | Quan. | Val. | Quan. | Val. | Quan. | Val. | Quan. | Val. | Quan. | Val. | Quantity | Value |
| | | | £ | | £ | cwt | £ | cwt | £ | | £ | | £ | | £ | tons | £ |
| 1857-58 | Bengal .cwt. | .. | 8,263 | .. | 1,408 | .. | 314 | .. | .. | .. | .. | .. | .. | .. | 145 | .. | 10,130 |
| | Madras .cwt. | 1,363 | 5,788 | 185 | 930 | 300 | 383 | .. | .. | .. | .. | .. | .. | 921 | 1,569 | 2,769 / 138 | 8,670 |
| | Bombay .„ | .. | 7,934 | 1,643 | 919 | 561 | 316 | 330 | 187 | .. | 943 | .. | 110 | .. | 45 | .. | 10,454 |
| | ALL INDIA . | .. | 21,985 | .. | 3,257 | .. | 1,013 | 330 | 187 | .. | 943 | .. | 110 | .. | 1,759 | .. | 29,254 |
| 1858-59 | Bengal | .. | 5,873 | .. | 4,098 | .. | .. | .. | .. | .. | 17 | .. | .. | .. | 200 | .. | 10,188 |
| | Madras .No. | 413,292 | 4,929 | 114,814 | 1,082 | .. | .. | .. | .. | .. | .. | 237 | 2 | 55,558 | 1,077 | 583,901 | 7,090 |
| | Bombay .cwt. | .. | 6,986 | 1,034 | 621 | .. | .. | .. | .. | .. | 576 | .. | 79 | .. | .. | .. | 8,262 |
| | ALL INDIA . | .. | 17,788 | .. | 5,801 | .. | .. | .. | .. | .. | 593 | .. | 81 | .. | 1,277 | .. | 25,540 |
| 1859-60 | Bengal | .. | 5,900 | .. | 2,667 | .. | 522 | .. | .. | .. | 88 | .. | .. | .. | 382 | .. | 9,559 |
| | Madras . | .. | 5,704 | .. | 102 | .. | .. | .. | .. | .. | .. | .. | .. | .. | 109 | .. | 6,515 |
| | Bombay .cwt. | .. | 6,975 | 100 | 60 | .. | .. | .. | .. | .. | 1,317 | 29 | 17 | .. | .. | .. | .. |
| | ALL INDIA „ | .. | 18,579 | .. | 3,429 | .. | 522 | .. | .. | .. | 1,405 | 29 | 17 | .. | 491 | .. | 24,443 |
| 1860-61 | Bengal | .. | 10,909 | .. | 4,190 | .. | 462 | .. | .. | .. | 50 | .. | .. | .. | 203 | .. | 15,814 |
| | Madras {No. / cwt. | 406,328 / 1,971 | 6,211 | 127,675 | 1,675 | .. | .. | .. | .. | .. | .. | .. | .. | 8,549 / 323 | 426 | 542,552 / 2,294 | 8,312 |
| | Bombay .cwt. | .. | 12,138 | 758 | 455 | .. | .. | 100 | 60 | .. | 1,215 | 4 | 3 | .. | .. | .. | 13,871 |
| | ALL INDIA „ | .. | 29,258 | .. | 6,320 | .. | 461 | 100 | 60 | .. | 1,265 | 4 | 3 | .. | 619 | .. | 37,997 |

\* Received from J. H. JOHNSON, Esq.

2230. [4756] Pair of elephants' tusks, Burmah. Messrs. HALLIDAY, FOX, & Co.

2231-2. [5157–8] Pair of wild buffalo horns, Cuttack.

2233-4. [5159–60] Do. wild Gyal horns, (*Bibos cavifrons*), Do.

2235-6. [5161–2] Do. (*Antilope cervicapra*), Cuttack.

These horns have been polished by native workmen. The tame buffalo horns, which are much smaller than the wild, form, together with the antlers of the spotted axis and the *Sambur* or *Cervus Hippelaphus*, an article of export to a considerable extent from this district. The local retail rates are, for black horn, about 5 annas per seer, and for deer horn (wholesale) 6 lb. weight per shilling.*

2237. [5914] Deer horn (*Cervus* sp.), Assam. LIEUT. W. PHAIRE.

2238. [2346] Horns of the wild yak (*Poephagus grunniens*). DR. A. CAMPBELL, Darjeeling.

2239. [2348] Do. of tame yak (*do.*) Do. Do.

2240. [2353] Do. of Chiru antelope (*Kemas Hodgsoni*). Do. Do.

2241. [2360] Do. of Shou (*Cervus affinis*). Do. Do.

2242. [2354] Skull and horn of rhinoceros. Do. Do.

* Local Committee, *Cuttack.*

2243. [2356] Horns of Jharai (*Rusa* sp.). DR. A. CAMPBELL, *Darjeeling.*

2244. [2357] Do. of Gowree (*Bibos cavifrons*). Do. Do.

2245. [4271] Do. of black antelope. Do. Do.

2246-9. [2362–5] Deer horns (*Cervus* sp.). Do. Do.

2250. [2961] Shagreen.

---

## V. — FOR PIGMENTS AND DYES.

Of animal origin, the only Indian dyes of importance are those afforded by the *Coccus lacca* insect.

2251. [41*] Lac dye from the *Coccus lacca*, Calcutta.

2252. [43*] Do. Do.

2253. [10104] Do. Do.

Is the well-known colouring matter of 'lac,' and is obtained from those incrustations by the insect *Coccus lacca*. It is employed chiefly as a red dye on wool.

---

2254. [1916*] Cochineal, Calcutta.

---

2255. [10166] Purree or Indian yellow. Monghyr. W. H. HENDERSON, Esq.

Produced from the urine of horned cattle fed on mango leaves. It is used in the locality of production, and also sent to Calcutta for exportation.*

* W. H. HENDERSON, Esq.

# SECTION C.—VEGETABLE SUBSTANCES USED IN MANUFACTURES.

## I.—GUM AND RESIN SERIES.

This subdivision comprises the gums, the resins, and the gum resins. The Table (p. 105) shows the extent to which these are exported.

### GUMS.

2256. [136*] Gum ghati (*Acacia arabica*), Bombay.

2257. [6037] Babool (*A. arabica*), Do.

2258. [6208] Do. Lucknow.

2259. [6221] Babool gum (*Acacia arabica*), Calcutta.

2260. [1948] Do. Jhansee.

2261. [6050] Do. Ahmedabad.

2262. [141*] Do. Calcutta.

2263. [148*] Do.

2264. [2585] Acacia gum, Bangalore, Mysore.

2265. [2587] Do. Chittledroog, Mysore.

TABLE SHOWING THE QUANTITIES (AS FAR AS CAN BE ASCERTAINED) AND THE VALUE OF GUMS EXPORTED FROM INDIA AND EACH PRESIDENCY TO ALL PARTS OF THE WORLD FROM 1857-58 TO 1860-61.

| YEARS | WHENCE EXPORTED | UNITED KINGDOM | | FRANCE | | OTHER PARTS OF EUROPE | | AMERICA | | CHINA | | ARABIAN AND PERSIAN GULFS | | OTHER PARTS | | TOTAL EXPORTED TO ALL PARTS | | |
|---|---|---|---|---|---|---|---|---|---|---|---|---|---|---|---|---|---|---|
| | | Quan. | Value | Quan. | Value | Quan. | Value | Quan. | Value | Quan. | Value | Quan. | Value | Quan. | Value | Quantity | | Value |
| | | cwt. | £ | cwt. | £ | cwt. | £ | cwt. | £ | cwt. | £ | cwt. | £ | cwt. | £ | cwt. | tons | £ |
| 1857-58 | Bengal | 117 | 354 | 7 | 30 | .. | .. | 13 | 19 | .. | .. | .. | .. | 4 | 20 | 141 | 7 | 423 |
| | Madras | .. | 13 | .. | 17 | .. | .. | .. | .. | .. | .. | .. | .. | .. | 104 | .. | .. | 134 |
| | Bombay | .. | 21,755 | .. | 313 | .. | 170 | .. | 240 | .. | 3,297 | .. | .. | .. | 570 | 112 | .. | 26,457 |
| | ALL INDIA | .. | 22,122 | .. | 360 | .. | 170 | .. | 259 | .. | 3,297 | .. | .. | .. | 570 | .. | .. | 27,014 |
| 1858-59 | Bengal | .. | .. | 160 | 254 | .. | .. | 82 | 96 | .. | .. | .. | .. | 6 | 20 | 248 | 12 | 370 |
| | Madras | .. | .. | .. | .. | .. | .. | .. | .. | .. | .. | .. | .. | 28 | 65 | 28 | 1 | 65 |
| | Bombay | .. | 22,166 | 29 | 29 | .. | .. | .. | .. | 11,633 | 7,384 | .. | 956 | .. | 301 | .. | .. | 30,836 |
| | ALL INDIA | .. | 22,166 | 189 | 283 | .. | .. | 82 | 96 | 11,633 | 7,384 | .. | 956 | .. | 386 | .. | .. | 31,271 |
| 1859-60 | Bengal | .. | .. | 144 | 274 | .. | .. | .. | .. | 3 | 6 | .. | .. | 140 | 449 | 287 | 14 | 729 |
| | Madras | .. | 27 | .. | .. | .. | .. | .. | .. | .. | .. | .. | .. | .. | 132 | .. | .. | 159 |
| | Bombay | 9,878 | 10,065 | 36 | 27 | .. | .. | .. | .. | 6,278 | 5,019 | 432 | 901 | 125 | 214 | 16,749 | 837 | 16,226 |
| | ALL INDIA | .. | 10,092 | 180 | 301 | .. | .. | .. | .. | 6,281 | 5,025 | 432 | 901 | 265 | 795 | .. | .. | 17,114 |
| 1860-61 | Bengal | 37 | 88 | 40 | 224 | .. | .. | .. | .. | 82 | 229 | 113 | 147 | 10,028 | 8,345 | 10,300 | 515 | 9,033 |
| | Madras | .. | .. | .. | .. | .. | .. | .. | .. | .. | .. | .. | .. | .. | 102 | .. | .. | 102 |
| | Bombay | 9,671 | 16,307 | .. | .. | .. | .. | 75 | 75 | 4,250 | 3,288 | 129 | 245 | 267 | 373 | 14,392 | 720 | 20,288 |
| | ALL INDIA | 9,708 | 16,395 | 40 | 224 | .. | .. | 75 | 75 | 4,332 | 3,517 | 242 | 392 | .. | 8,820 | .. | .. | 29,423 |

2266. [2743] Acacia gum, Nuggur, Mysore.

2267. [2585] Do.

2268. [4898] *Acacia speciosa*, Chingleput. DR. SHORTT.

2269. [4896] *A. sundra*, Chingleput. DR. SHORTT.

2270. [6046] Kheir gum (*A. catechu*), Ahmedabad.

This tree yields the largest proportion of the Cutch, so much employed in tanning and dyeing, and which is obtained by boiling the wood and inspissating the liquor. The gum exudes freely from the bark when wounded, and is of the character of an ordinary gum arabic.

2271. [142*] *Vachellia farnesiana*, Bengal.

2272 [4001] *V. farnesiana*, Salem.

This is a useful gum arabic, and is exuded freely and in considerable quantities by the tree which produces it. The flowers, under the name of wattle flowers, are much employed in perfumery for their delicious fragrance.

2273. [6002] Dhowla gum, Ahmedabad.

2274. [6044] Dower goond, Do.

2275. [2860] Dhowrah tree gum, Raepore.

2276. [7184] Koonee gond (*Odina wodier*), Calcutta.

2277. [139*] Koonee gond (*O. wodier*), N. W. India.

This gum resembles the true gum arabic both in appearance and properties, and is often largely mixed up with the East India arabic of commerce, which often contains gum collected indiscriminately from a number of different trees, including several species of *Acacia, Odina wodier*, and *Feronia elephantum*.

2278. [4908] *Prosopis spicigera*, Chingleput. DR. SHORTT.

2279. [4911] Soap tree gum (*Sapindus emarginatus*), Madras. DR. SHORTT.

2280. [4919] *Wrightia tinctoria*, Chingleput. DR. SHORTT.

2281. [4903] *Erythrina indica*, Do. Do.

2282. [6432] Gum from jungle trees, Chota Nagpore.

2283. [6433] Do. Do.

2284. [6434] Do. Do.

2285. [4917] Shamgathalee gum, Chingleput. DR. SHORTT.

2286. [4918] Poongkalee gum, Chingleput. DR. SHORTT.

2287. [4901] Palmyra gum (*Borassus flabelliformis*), Do. Do.

2288. [73*] Margosa gum (*Melia azedarach*), Travancore.

2289. [37*] Do (*do.*), Palamcottah.

2290. [124*] Gum of *Buchanania latifolia.*

2291. [4904] Wood apple gum (*Feronia elephantum*), Chingleput. DR. SHORTT.

This gum is abundant. From its ready solubility without residue, it gives the best mucilage for making ink.

2292. [69*] Marking nut gum (*Semecarpus anacardium*), Travancore.

2293. [79*] Moringa (*Moringa pterygosperma*), Travancore.

2294. [4905] Do. Chingleput. DR. SHORTT.

The gum of the *Moringa* somewhat resembles tragacanth. It exudes freely whenever an incision is made in the bark. It is used by the natives in headache mixed with milk and rubbed on the temples, and is also employed as a local application for pains in the limbs.

2295. [4899] Gum of *Ailanthus excelsa*, Chingleput. DR. SHORTT.

2296. [6054] Kuda na gond (*Sterculia*, species), Ahmedabad.

2297. [132] Kuteera gum (*Sterculia urens*), Bengal.

2298. [6042] Do. Ahmedabad.

Kuteera gum was known as false tragacanth before the tree which produces it was correctly determined. Both this and the following gum resembles a coarse tragacanth, and is employed in India as a substitute for the genuine gum, which is the produce of the south of Europe.

2299. [38*] Gum of *Cochlospermum gossypium*, North-west India.

This gum is also sometimes called *Kuteera*. It exudes from every part of the tree when broken, and resembles an inferior tragacanth.

2300. [10154] Kotilla or Tragacanth, Calcutta.

2301. [4912] Gum of *Stereospermum suaveolens*, Chingleput. DR. SHORTT.

2302. [4913] Gum of *Terminalia alata*, Do. Do.

2303. [4895] Gum of *Calophyllum inophyllum*, Do. Do.

2304. [4897] Gum of *Cassia auriculata*, Chingleput. DR. SHORTT.

2305. [4906] Gum of *Poinciana alata*, Do. Do.

2306. [4759] Ta Mazie gum, Burmah. Messrs. HALLIDAY, FOX, & CO.

2307. [4760] Do. Do. Do.

2308. [6052] Daweej gum, Ahmedabad.

2309. [4916] Selembai, Chingleput.

## TINCTORIAL GUMS.

2310. [6382] Mangosteen gum (*Garcinia mangostana*), Malacca. A. A. DE WIND, Esq.

A few tear-like pieces of this gum is all that is exhibited. These resemble gamboge in appearance, but the sample is too small to venture a correct judgement.

2311. [189] Ardalla (*G. pedunculata*), Mangalore. V. P. COELHO, Esq.

This gum bears a great resemblance to the gamboge of the Wynaad; and is interesting as being the first sample sent to England derived from this source.

2312. [1*] Gamboge (*G. pictoria*), Wynaad.

A very superior kind of gamboge, to which the attention of the trade might be advantageously drawn.

2313. [1896] Gamboge, Bengal.

2314. [6010] Pipe gamboge, Bombay.

2315. [6663] Do. Siam. G. ANGUS, Esq.

2316. [6695] Gamboge (*G. cochinchinensis*), Cambodia. JOSE D'ALMEIDA, Esq.

This is the true pipe gamboge of Siam, and the ordinary gum gamboge of British commerce.

2317. [6383] Gutta Kandees, Malacca. A. A. DE WIND, Esq.

From the wild mangosteen.

2318. [5*] Gum from *Hebradendron cambogioides.*

This is the gamboge which is known commercially as Ceylon gamboge. It is of good quality, and is made up in cakes or irregular masses.

2319. [6016] Heradukhun (*Pterocarpus draco*), Bombay.

2320. [4063] Dragon's blood, Pinang.

2321. [11*] Do. Singapore.

2322. [6664] Do. Sumatra. G. ANGUS, Esq.

2323. [7136] Kino (*P. Wallichii*), Rangoon.

This is affirmed to be the produce of the above tree, which is one of the commonest forest trees in the locality whence it is derived; the ordinary commercial kino is obtained from *Pterocarpus marsupium.*

2324. [19*] Vangay kino (*P. dalbergioides*).

This kino differs but little in appearance and properties from the ordinary kino.

2325-6. [4909, 3356] Kino (*P. marsupium*), Bengal.

This is the common East India kino. It is employed medicinally as an astringent.

2327. [7127] Moochrus (*Bombax heptaphyllum*), Calcutta.

This astringent gum, which exudes from the bark rather freely, is employed for tanning leather as a substitute for kino; it is also used medicinally for its astringent property and its supposed strengthening virtue.

2328. [4907] Butea gum (*Butea frondosa*), Chingleput. DR. SHORTT.

2329. [6600] Dhak gum (*B. frondosa*), Lucknow.

This is sometimes called Pulas kino; it is very astringent, some specimens yielding 73 per cent. of tannin. In the North-west Provinces the natives employ it for precipitating their indigo and in tanning. It contains a durable colouring matter, which generally renders it objectionable for the latter purposes.

### MEDICINAL GUM RESINS.

2330. [6026] Assafœtida (*Narthex assafœtida*), Bombay.

2331. [1851] Assafœtida (*Narthex assafœtida*), Punjab.

The concrete juice of the root is obtained by slicing off the stem obliquely, collecting the liquid which exudes, and drying it. It is obtained chiefly from Bokhara. By the natives it is employed as a condiment to food, as a carminative in colic, and also as a stimulant tonic in paralysis, tremors, and epilepsy. Dose, gr. iv. Price, 2s. per lb.

2332. [31*] Ammoniacum (*Dorema ammoniacum*), Persia.

2333. [6033] Ooshak (*D. ammoniacum*), viâ Bombay.

Gum ammoniac is not an Indian product, but is obtained from Persia. It is a gum resin exclusively medicinal, and is employed as a stimulant expectorant and antispasmodic.

2334. [6035] Dikamali (*Gardenia lucida*), Bombay.

This resin is obtained chiefly in Canara and Mysore, and is stated to have been found useful in the hospitals in keeping away flies from sores.

2335. [1818] Sarcocolla (*Penea mucronata*), Kabul.

*Anzerut,* a gum resin obtained from the bark of the *Penea mucronata,* called sometimes Sarcocolla, and obtained chiefly from Kabul. It is used either in powder or infusion as a laxative, and as an alterative in cancer. Dose 1 drachm. Price 1s. per lb. Formerly used in European medicine, chiefly as an application for wounds, whence its name of Sarcocolla.

2336. [9924] Bdellium (*Balsamodendron* species), Calcutta.

2337. [7126] Googul (*B. agallocha*), Do.

2338. [51*] Do. (*B. myrrha*), Do.

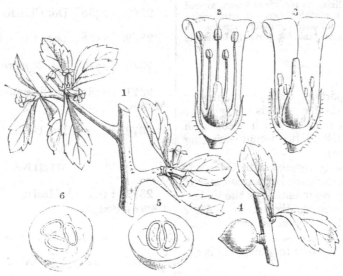

GOOGUL TREE (*Balsamodendron Mukul*).*

2339. [6051] Heera bol, Bombay.

2340. [6056] Googul (*B. Roxburghii*), Bombay.

The sources of these gums are given as sent from India; there is still some confusion in the botanical sources of the substance known in India as googul and of the bdellium and myrrh of commerce.

———

2341. [6572] Benzoin (*Styrax benzoin*), *viâ* Calcutta.

2342. [61*] Do. Sumatra.

2343. [6384] White benzoin, Malacca. A. A. DE WIND, Esq.

2344. [6674] Benzoin, Sumatra.

Benzoin has an agreeable fragrant odour, and a sweetish balsamic taste. It is obtained chiefly from Sumatra and Siam. A very interesting variety (No. 6384) is exhibited from Malacca. In Sumatra, benzoin is obtained by making incisions into the tree in its seventh year. The juice which first exudes is the purest and most fragrant; it hardens on exposure to the air, and becomes brittle and semi-transparent.

2345. [10147] Olibanum (*Boswellia thurifera*), Calcutta.

2346. [10152] Do. Do.

2347. [6048] Do. Bombay.

2348. [6219] Do. Calcutta.

This balsamic gum resin is a considerable article of export from Bombay and other ports of India. The best is found in pieces as large as a walnut, of a bright yellowish colour, sometimes inclining to reddish or brown, covered on the outside with a whitish bloom. It burns with a clear and steady light, diffusing a grateful fragrance. In taste it is slightly bitter, and not perfectly soluble in water or alcohol.

———

The most important resin belonging to the next group is lac. It is formed by the insect *Coccus lacca*, which feeds upon various trees in India. It is found encircling twigs and branches.

The broken twigs covered with these incrustations is called 'stick lac' in commerce, and is sufficiently represented, as the undernamed sources abundantly testify. After the colour has been extracted and farther purified, shell lac results. The Table (p. 109) indicates the extent to which lac is exported from India.

2349. [5396] Crude lac, Shahabad.

2350. [2*] Stick lac, Vizagapatam.

2351. [1949] Lac from *Acacia arabica*, Bengal.

2352. [10103] Shell lac, Do.

2353. [6429] Stick lac, Chota Nagpore.

2354. [7160] Do. Cuttack.

2355. [7162] Do. Do.

2356. [2839] Lac, Raepore.

2357. [1940] Shell lac, Umritsur.

2358. [1941] Do. Do.

2359. [6680] Do. Siam. G. ANGUS, Esq.

2360. [6005] Lac from *Zizyphus jujuba*, Admedabad.

2361. [7512] Do. from *Urostigma religiosa*, Ahmedabad.

2362. [6045] Chupra lac, Poona.

2363. [4002] Shell lac, Salem.

2364. [6606] Lac, Durrung, Assam. LIEUT. PHAIRE.

2365. [7161] Do. Do. Do.

2366. [7163] Stick lac, Burmah.

2367. [7164] Do. Do.

2368. [2586] Do. Chittledroog, Mysore.

2369. [2588] Lac, Do. Do.

2370. [2742] Do. Mysore.

2371. [2580] Sealing wax, Chittledroog, Mysore.

2372. [9170] Do. Cuttack.

### RESINS.

2373. [10152] Mastic (*Pistacia lentiscus*), Bengal.

2374. [6007] Do. Bombay.

2375. [1821] Do. Kabul.

A gum resin which exudes from the young branches and is scraped off. It forms white translucent tears,

TABLE SHOWING THE QUANTITIES (AS FAR AS CAN BE ASCERTAINED) AND THE VALUE OF LAC EXPORTED FROM INDIA AND EACH PRESIDENCY TO ALL PARTS OF THE WORLD FROM 1850-51 TO 1860-61.

COUNTRIES WHITHER EXPORTED

| Years | Whence Exported | United Kingdom — Quantity (cwt.) | United Kingdom — Value (£) | France — Quantity (cwt.) | France — Value (£) | Other Parts of Europe — Quantity (cwt.) | Other Parts of Europe — Value (£) | America — Quantity (cwt.) | America — Value (£) | China — Quantity (cwt.) | China — Value (£) | Arabian and Persian Gulfs — Quantity (cwt.) | Arabian and Persian Gulfs — Value (£) | Other Parts — Quantity (cwt.) | Other Parts — Value (£) | Total Exported to All Parts — Quantity (cwt.) | Total Exported to All Parts — Quantity (tons) | Total Exported to All Parts — Value (£) |
|---|---|---|---|---|---|---|---|---|---|---|---|---|---|---|---|---|---|---|
| 1850-51 | Bengal | 40,219 | 76,534 | 6,575 | 9,763 | 6,072 | 7,549 | 18,448 | 41,454 | .. | .. | 219 | 119 | 296 | 365 | 71,829 | 3,596 | 135,784 |
| | Madras | 973 | 898 | 762 | 515 | .. | .. | .. | .. | .. | .. | 54 | 107 | .. | .. | 1,789 | 89 | 1,520 |
| | Bombay | 848 | 1,014 | .. | .. | .. | .. | .. | .. | .. | .. | 586 | 858 | .. | 1 | 1,434 | 72 | 1,873 |
| | ALL INDIA | 42,040 | 78,446 | 7,337 | 10,278 | 6,072 | 7,549 | 18,448 | 41,454 | .. | .. | 859 | 1,084 | 296 | 366 | 75,052 | 3,757 | 139,177 |
| 1851-52 | Bengal | 30,449 | 58,041 | 2,076 | 3,687 | 5,268 | 6,457 | 18,641 | 36,780 | 4 | 6 | 9 | 5 | 18 | 20 | 56,465 | 2,823 | 104,996 |
| | Madras | .. | .. | 828 | 705 | .. | .. | .. | .. | .. | .. | 61 | 122 | .. | .. | 889 | 44 | 827 |
| | Bombay | .. | .. | .. | .. | .. | .. | .. | .. | .. | .. | .. | .. | .. | .. | .. | .. | .. |
| | ALL INDIA | 30,449 | 58,041 | 2,904 | 4,392 | 5,268 | 6,457 | 18,641 | 36,780 | 4 | 6 | 70 | 127 | 18 | 20 | 57,354 | 2,867 | 105,833 |
| 1852-53 | Bengal | 52,508 | 103,993 | 9,365 | 15,768 | 2,356 | 3,017 | 11,847 | 27,192 | 6 | 7 | 256 | 146 | 15 | 19 | 76,353 | 3,818 | 150,142 |
| | Madras | .. | .. | .. | .. | .. | .. | .. | .. | .. | .. | .. | .. | .. | .. | .. | .. | .. |
| | Bombay | 35 | 50 | .. | .. | .. | .. | .. | .. | .. | .. | 339 | 488 | .. | .. | 374 | 19 | 538 |
| | ALL INDIA | 52,543 | 104,043 | 9,365 | 15,768 | 2,356 | 3,017 | 11,847 | 27,192 | 6 | 7 | 595 | 634 | 15 | 19 | 76,727 | 3,837 | 150,680 |
| 1853-54 | Bengal | 41,246 | 64,374 | 7,413 | 13,818 | 2,973 | 3,839 | 16,987 | 28,919 | 5 | 7 | 27 | 18 | 1,594 | 1,376 | 70,245 | 3,512 | 102,351 |
| | Madras | 22 | 32 | 9 | 17 | .. | .. | .. | .. | .. | .. | 278 | 389 | 1 | 3 | 310 | 15 | 441 |
| | Bombay | .. | .. | .. | .. | .. | .. | .. | .. | .. | .. | .. | .. | .. | .. | .. | .. | .. |
| | ALL INDIA | 41,268 | 64,406 | 7,422 | 13,835 | 2,973 | 3,839 | 16,987 | 28,919 | 5 | 7 | 305 | 407 | 1,595 | 1,379 | 70,555 | 3,527 | 102,792 |
| 1854-55 | Bengal | 33,384 | 63,089 | 4,108 | 7,947 | 1,451 | 1,831 | 11,479 | 17,830 | 12 | 12 | .. | .. | 272 | 1,366 | 50,696 | 2,535 | 92,075 |
| | Madras | 22 | 31 | .. | .. | .. | .. | .. | .. | .. | .. | 102 | 126 | .. | .. | 134 | 7 | 157 |
| | Bombay | .. | .. | .. | .. | .. | .. | .. | .. | .. | .. | .. | .. | .. | .. | .. | .. | .. |
| | ALL INDIA | 33,406 | 63,120 | 4,108 | 7,947 | 1,451 | 1,831 | 11,479 | 17,830 | 12 | 12 | 102 | 126 | 272 | 1,366 | 50,830 | 2,542 | 92,232 |
| 1855-56 | Bengal | 26,425 | 59,404 | 6,296 | 10,443 | 2,386 | 3,008 | 21,710 | 52,582 | 438 | 578 | 8 | 11 | 472 | 884 | 57,735 | 2,887 | 126,411 |
| | Madras | .. | .. | .. | .. | .. | .. | .. | .. | .. | .. | 376 | 467 | .. | .. | 376 | 19 | 467 |
| | Bombay | .. | .. | .. | .. | .. | .. | .. | .. | .. | .. | .. | .. | .. | .. | .. | .. | .. |
| | ALL INDIA | 26,425 | 59,404 | 6,296 | 10,443 | 2,386 | 3,008 | 21,710 | 52,582 | 438 | 578 | 384 | 478 | 472 | 884 | 58,111 | 2,906 | 126,878 |
| 1856-57 | Bengal | 24,335 | 61,997 | 4,243 | 6,478 | 1,935 | 2,710 | 17,734 | 32,386 | 19 | 23 | .. | .. | 388 | 378 | 48,654 | 2,433 | 103,972 |
| | Madras | 308 | 369 | .. | .. | .. | .. | .. | .. | .. | .. | 802 | 1,234 | .. | .. | 1,110 | 56 | 1,603 |
| | Bombay | .. | .. | .. | .. | .. | .. | .. | .. | .. | .. | .. | .. | .. | .. | .. | .. | .. |
| | ALL INDIA | 24,643 | 62,366 | 4,243 | 6,478 | 1,935 | 2,710 | 17,734 | 32,386 | 19 | 23 | 802 | 1,234 | 388 | 378 | 49,764 | 2,489 | 105,575 |
| 1857-58 | Bengal | 33,540 | 76,755 | 8,614 | 13,156 | 1,760 | 2,060 | 7,703 | 14,521 | 4 | 6 | 9 | 11 | 379 | 465 | 52,009 | 2,600 | 105,974 |
| | Madras | 538 | 882 | .. | .. | .. | .. | .. | .. | .. | .. | .. | .. | 70 | 115 | 608 | .. | 997 |
| | Bombay | .. | .. | .. | .. | .. | .. | .. | .. | .. | .. | .. | 702 | 3 | 4 | 3 | .. | 706 |
| | ALL INDIA | 34,078 | 77,637 | 8,614 | 13,156 | 1,760 | 2,060 | 7,703 | 14,521 | 4 | 6 | 9 | 713 | 452 | 584 | 52,082 | 2,603 | 107,677 |
| 1858-59 | Bengal | 10,888 | 52,818 | 1,856 | 2,500 | .. | .. | 12,657 | 23,871 | .. | .. | .. | .. | 1,330 | 1,057 | 26,726 | 1,336 | 80,246 |
| | Madras | 264 | 181 | .. | .. | .. | .. | .. | .. | .. | .. | .. | .. | .. | .. | 264 | 13 | 181 |
| | Bombay | .. | .. | .. | .. | .. | .. | .. | .. | .. | .. | 100 | 140 | .. | .. | 100 | 5 | 140 |
| | ALL INDIA | 11,147 | 52,999 | 1,856 | 2,500 | .. | .. | 12,657 | 23,871 | .. | .. | 100 | 140 | 1,330 | 1,057 | 27,090 | 1,354 | 80,567 |
| 1859-60 | Bengal | 27,750 | 54,237 | 1,665 | 3,535 | 342 | 419 | 9,308 | 19,359 | .. | .. | .. | .. | 108 | 131 | 39,178 | 1,959 | 77,681 |
| | Madras | 37 | 52 | .. | .. | .. | .. | .. | .. | .. | .. | 215 | 304 | .. | 5 | 252 | 13 | 361 |
| | Bombay | .. | .. | .. | .. | .. | .. | .. | .. | .. | .. | .. | .. | .. | .. | .. | .. | .. |
| | ALL INDIA | 27,787 | 54,289 | 1,665 | 3,535 | 342 | 419 | 9,308 | 19,359 | .. | .. | 215 | 304 | 108 | 136 | 39,425 | 1,972 | 78,042 |
| 1860-61 | Bengal | 32,711 | 126,041 | 3,088 | 13,066 | 1,014 | 4,239 | 6,644 | 27,699 | 47 | 172 | .. | .. | 88 | 215 | 43,592 | 2,179 | 171,482 |
| | Madras | 20 | 52 | .. | .. | .. | .. | .. | .. | .. | .. | 116 | 162 | .. | .. | 136 | 7 | 214 |
| | Bombay | .. | .. | .. | .. | .. | .. | .. | .. | .. | .. | .. | .. | .. | .. | .. | .. | .. |
| | ALL INDIA | 32,731 | 126,093 | 3,088 | 13,066 | 1,014 | 4,239 | 6,644 | 27,699 | 47 | 172 | 116 | 162 | 88 | 215 | 43,728 | 2,186 | 171,696 |

obtained principally from Kabul. It is used by the natives as an astringent in diarrhœa, which property it owes to the volatile oil contained in it; 'also in all diseases depending on an undue accumulation of phlegm.' Price 1s. per lb.*

---

2376. [10699] Piney resin (*Vateria indica*), Mangalore. V. P. COELHO, Esq.

2377. [10699] Do. (refined) (*do.*) Do. Do.

2378. [88*] Do. Malabar.

2379. [147] Dhoop (*V. indica*).

2380. [77*] Do. Travancore.

2381. [90*] Do. Do.

As these specimens distinctly indicate, there are at least two distinct varieties of the piney resin. Some resemble amber both in lustre and fracture; others are of various shades of colour, from light green to yellow.

2382. [108*] Black dammar (*Canarium strictum*), Malacca.

Black dammar is of a shining black colour when viewed in mass, but on closer examination appears to be of a reddish hue and translucent. It occurs in stalactitic masses.

2383. [103*] Resin of *Canarium strictum*, Seebsaugor.

2384. [6686] Dammar Daging, or rose dammar, Singapore. G. ANGUS, Esq.

$4½ per picul.

---

2385. [6378] Dammar Batu (*Shorea robusta*), Singapore. G. ANGUS, Esq.

2386. [7135] Dhonah, Calcutta.

2387. [6426] Do. Chota Nagpore.

2388. [6427] Do. Do.

2389. [6700] Dammar Batu (*S. robusta*), Sumatra. T. NEUBRONNER, Esq.

$1½ per picul. Boiled with common oil and used for ships' bottoms.

2390. [117*] Saul resin (*S. robusta*), Midnapore.

2391. [6220] Sakhoo (*S. robusta*), Lucknow.

2392. [7129] Dhoona (*S. robusta*), Cuttack.

A resin, locally called '*Jhoona*' or '*Dhoona*.' It is obtained from the *Shorea robusta*, one of the most abundant forest trees in Indian jungles, and is procured by making incisions in the stem of the tree from which the resin exudes. The local retail price is about 3 annas, that is 4½d. per seer of 2½ pounds weight.*

2393. [4813] Dhoona (*S. robusta*), Burmah. Messrs. HALLIDAY, FOX, & Co.

---

2394. [6385] Dammar Keejee, Malacca. A. A. DE WIND, Esq.

Procurable in large quantities.

2395. [6379] Dammar Mata Kooching, Singapore. T. NEUBRONNER, Esq.

2396. [6678] Dammar Mata Kooching, Eastern Archipelago. G. ANGUS, Esq.

$8½ per picul. Procurable in large quantities. Boiled down for varnish.

2397. [6004] Dammar (*Dammara orientalis*), Bombay.

2398. [6707] Dammar, Malay Peninsula. G. ANGUS, Esq.
Largely exported from Singapore to Great Britain and other ports. It is a very clear transparent resin, much in request for the manufacture of varnishes.

---

2399. [7130] Dammar Pooy Nyat, Moulmein.

2400. [2518] African Animé. H.H. the SULTAN of ZANZIBAR.

2401. [6011] Animé, Bombay.

It is easily distinguished from copal by its softening in the mouth, and its ready solubility in alcohol.

---

2402. [110*] Gaup resin (*Embryopteris glutinifera*), Bhagulpore.

Gaup tree resin is dark-coloured in the mass, and approximates in appearance to the black dammar of Travancore, except that it is in smaller masses or tears, and is not so vitreous in its fracture.

2403. [105*] Resin (*Dipterocarpus*, species), Seebsaugor.

All dipterocarpous trees abound in a resinous juice, and from one of these this sample of resin has been obtained.

---

* Central Committee, *Lahore.*

* Local Committee, *Cuttack.*

2404. [2847] Surry tree resin (*Vatica*, a species of), Raepore.

2405. [4003] Resin, Salem.

2406. [7130] Ing Donay (*Vatica tambuggaia*), Moulmein.

2407. [4761] A resin, Burmah. Messrs. HALLIDAY, FOX, & CO.

2408. [6381] Gutta Runghus from red wood tree, Malacca. A. A. DE WIND, Esq.

Procurable in large quantities.

2409. [7128] Pine resin (*Pinus longifolia*), Calcutta.

2410. [6003] Pine resin, Bombay.

2411. [6465] Thus, Shahjehanpore.

2412. [6053] Suk Muniar, Bombay.

A resin coloured artificially and sold for scammony.

## OLEO-RESINS.

2413. [5184] Black varnish (*a*), (*Melanorrhœa usitata*), Burmah.

2414. [5185] Do. Do. (*b*)

2415. [5186] Do. Do. (*c*)

This is the celebrated Burmese black varnish, obtained from *Melanorrhœa usitata*. In varnishing an article, the Burmese first give it a coat of *a*; when this is dry, they lay a coat of *b* over it, and finish with a coat of *c* over all. Price 120R. for first quality, 80R. for second, and 60R. for third quality, for 365 lbs.*

2416. [7131] Tset-tsee, Moulmein.

The wood oils may be said to form the link between the oils and resins. They hold a balsamic resin in solution, and are obtained from various of the Order *Dipterocarpeœ*, some of which yield the dammars of commerce.

2417. [5711] Teak oil (*Tectona grandis*), Cuttack.

A kind of wood oil obtained from the teak tree.

2418. [5698] Emsleekee wood oil, Moulmein.

2419. [10700] Wood oil, Mangalore. V. P. COELHO, Esq.

2420. [5695] Capawawood oil, Moulmein.

2421. [5693] Do. Akyab.

2422. [7137] Eintsee, Arracan.

2423. [6337] Minia Kayu, Malay Peninsula.

2424. [1893] Gurjun (*Dipterocarpus turbinatus*), *viâ* Calcutta.

2425. [5702] Do. Chittagong.

This oil is obtained from a large tree, formerly common in the hills of this district. The oil is obtained by cutting a hole in the tree, about 3 feet from the ground; the cut being about 4 to 5 inches deep into the trunk of the tree. The base is hollowed out to retain the oil. The whole of the hollow is cleared with fire, without which no oil exudes; after it is cleared the oil exudes, and is collected in the hollow at the base, and removed at intervals. The oil is thus extracted year after year, and sometimes there are two or three holes in the same tree, while the tree does not die. The oil is allowed to settle, when the clear part separates from a thick portion, which is called the 'guad.' If a growing tree is cut down and cut to pieces, the oil exudes and concretes on the stem and ends of the pieces, very much resembling camphor, with an aromatic smell also. It is said that the tree yields from 3 to 5 maunds yearly, i. e. 240 to 400 lbs., and the same tree will yield oil for several years. It is a good balsamic medicine, and is very generally used as a substitute for copaiba; but it would be more valuable as a varnish: it is preservative to wood, to which it gives, with little trouble of application, a fine surface and polish; it becomes, however, white and milky if exposed to wet. It can be had at Chittagong in large quantities at 10R. per maund.*

2426. [5679] Choca oil, distilled from country resin.

2427. [4068] Barus camphor (*Dryobalanops camphora*), Sumatra.

The camphor is found secreted in cavities of the wood. It is of great value, and is only obtained from trees of a certain age. It sells at Padang at upwards of 3 dollars for a lb.

2428. [7132] Camphor (*Laurus camphora*), Calcutta.

2429. [6014] Do.

This is the produce of a Chinese and Japanese laurel, and is largely obtained from the island of Formosa. It is not so highly esteemed in the East as the Barus camphor, and consequently does not command such extravagant prices. All parts of the plant contain camphor diffused through them, but it is obtained chiefly from the wood of the stem branches, and root. These are cut in chips and boiled, and the camphor is obtained by distillation.

2430. [5130] Cup of camphor (*L. camphora*), Calcutta. BABOO GOPAUL CHUNDAR GOOPTA.

## ELASTIC AND PLASTIC GUMS.

2431. [6509] Caoutchouc, Durrung, Assam. LIEUT. W. PHAIRE.

2432. [5572] India rubber, Cossia Hills. J. N. MARTIN, Esq.

2433. [10229] Indian caoutchouc, Goruckpore. GEORGE OSBORNE, Esq.

'A new production collected and prepared from the various plants of the district of Goruckpore.'

2434. [6318] Gutta Babee caoutchouc, Malacca. T. NEUBRONNER, Esq.

2435. [6317] Gutta Gree grip Do., Do. Do.

2436. [6701] India rubber, Eastern Archipelago. G. ANGUS, Esq.

2437. [6380] Gutta Terbole, Malacca. A. A. DE WIND, Esq.

'Used to adulterate gutta percha.'

2438. [6324] Gutta percha (*Isonandra gutta*), Eastern Archipelago. G. ANGUS, Esq.

2439. [6693] Gutta percha as imported into Singapore. G. ANGUS, Esq.

2440. [6693] Do. Malay Peninsula. JOSE D'ALMEIDA, Esq.

2441. [6316] Do. Malacca. T. NEUBRONNER, Esq.

2442. [161*] Do. Singapore.

The gutta percha tree is a native of the Malayan Peninsula; but not the only one which yields it.

2443. [7083] Gutta percha cup. DR. RIDDELL.

2444. [6319] Pair of gutta percha buckets, Singapore. HON. COL. CAVANAGH.

2445. [164*] Pauchontee (*Isonandra acuminata*), Travancore.

The exudation from the trunk has some resemblance to the gutta percha of commerce. According to General Cullen, 'in 5 or 6 hours upwards of 1½ lbs. was collected from 4 or 5 incisions in one tree.'

2446. [154*] Cattimundoo (*Euphorbia Cattimundoo*), Madras. W. ELLIOT, Esq.

This gum is worthy of more attention than it has yet received.

2447. [4924] Gutta (*E. antiquorum*), Chingleput. DR. JOHN SHORTT.

The true Euphorbium is supposed to be obtained chiefly from *E. officinarum.*

2448. [4949] Gutta (*E. tortilis*), Chingleput. DR. JOHN SHORTT.

Very similar to the former in nature and properties.

2449. [4965] Mudar gutta (*Calotropis gigantea*), Chingleput. DR. JOHN SHORTT.

DR. RIDDELL calculates that 10 average sized plants will afford as much juice as will make 1 lb. of this gutta-percha-like substance.

## Subdivisions II. and III. — DYE STUFFS AND TANNING MATERIALS.

TABLE SHOWING THE QUANTITIES (AS FAR AS CAN BE ASCERTAINED) AND THE VALUE OF DYES (EXCLUSIVE OF INDIGO AND MUNJEET) EXPORTED FROM INDIA AND EACH PRESIDENCY TO ALL PARTS OF THE WORLD FROM 1857–58 TO 1860–61.

| YEARS | WHENCE EXPORTED | UNITED KINGDOM | | FRANCE | | OTHER PARTS OF EUROPE | | AMERICA | | CHINA | | ARABIAN AND PERSIAN GULFS | | OTHER PARTS | | TOTAL EXPORTED TO ALL PARTS | | |
|---|---|---|---|---|---|---|---|---|---|---|---|---|---|---|---|---|---|---|
| | | Quan. | Value | Quan. | Value | Quan. | Value | Quan. | Value | Quan. | Value | Quan. | Value | Quan. | Value | Quantity | | Value |
| | | cwt. | £ | cwt. | £ | cwt. | £ | cwt. | £ | cwt. | £ | cwt. | £ | cwt. | £ | cwt. | tons | £ |
| 1857-58 | Bengal | 11,035 | 26,017 | 23,755 | 6,785 | 1,722 | 505 | 2,938 | 1,074 | 394 | 293 | .. | .. | 482 | 848 | 40,326 | 2,016 | 35,522 |
| | Madras | 24,246 | 18,535 | 11,657 | 8,961 | 906 | 1,023 | .. | .. | .. | .. | 3,079 | 1,402 | 8,154 | 3,325 | 48,042 | 2,402 | 33,247 |
| | Bombay | .. | 5,604 | .. | 1,657 | .. | .. | .. | 443 | 135 | 2,740 | .. | 6,715 | .. | 1,148 | .. | .. | 18,307 |
| | ALL INDIA | .. | 50,156 | .. | 17,403 | 2,628 | 1,528 | .. | 1,517 | 529 | 3,033 | .. | 8,118 | .. | 5,321 | .. | .. | 87,076 |
| 1858-59 | Bengal | 60,042 | 57,102 | 21,367 | 16,930 | 9,097 | 3,806 | 17,281 | 8,624 | 21 | 604 | 111 | 29 | 2,437 | 1,332 | 110,555 | 5,528 | 88,421 |
| | Madras | 2,866 | 1,466 | 91 | 61 | .. | .. | .. | .. | .. | .. | 3,298 | 1,577 | 7,887 | 3,201 | 14,142 | 707 | 6,303 |
| | Bombay | 1,800 | 1,275 | 98 | 46 | .. | .. | .. | .. | 188 | 3,943 | 13,101 | 5,518 | 1,518 | 827 | 16,700 | 835 | 11,619 |
| | ALL INDIA | 64,708 | 59,843 | 21,545 | 17,037 | 9,097 | 3,806 | 17,281 | 8,624 | 209 | 4,547 | 16,510 | 7,116 | 11,842 | 5,360 | 141,397 | 7,070 | 106,343 |
| 1859-60 | Bengal | 64,078 | 47,168 | 26,970 | 14,184 | 13,087 | 5,054 | 12,889 | 6,440 | 9 | 183 | .. | .. | 21,828 | 12,087 | 138,861 | 6,943 | 85,116 |
| | Madras | 6,675 | 2,441 | 3,192 | 950 | 158 | 47 | .. | .. | 4 | 1 | 1,167 | 551 | 6,381 | 2,490 | 17,577 | 879 | 6,480 |
| | Bombay | 4,133 | 2,730 | 127 | 75 | .. | .. | 559 | 333 | 61 | 1,290 | 6,620 | 4,271 | 1,281 | 794 | 12,781 | 639 | 9,493 |
| | ALL INDIA | 74,886 | 52,339 | 30,289 | 15,209 | 13,245 | 5,101 | 13,448 | 6,773 | 74 | 1,474 | 7,787 | 4,822 | 29,490 | 15,371 | 169,219 | 8,461 | 101,089 |
| 1860-61 | Bengal | 50,452 | 91,706 | 7,231 | 10,541 | 916 | 861 | 3,322 | 3,421 | 53 | 18 | .. | .. | 1,261 | 1,158 | 63,235 | 6,662 | 107,705 |
| | Madras | 1,630 | 521 | .. | .. | .. | .. | .. | .. | .. | .. | 2,108 | 722 | 3,792 | 1,384 | 7,530 | 376 | 2,657 |
| | Bombay | 3,406 | 1,646 | 500 | 250 | .. | .. | 525 | 263 | 231 | 4,654 | 7,453 | 21,410 | 873 | 286 | 12,988 | 649 | 28,509 |
| | ALL INDIA | 55,488 | 93,873 | 7,731 | 10,791 | 916 | 861 | 3,847 | 3,684 | 284 | 4,672 | 9,561 | 22,132 | 5,926 | 2,828 | 83,753 | 7,687 | 138,871 |

The Table (p. 112) indicates the quantity and value of the dye stuffs, exclusive of Indigo and Munjeet (Indian Madder) exported from 1857 to 1861.

### INDIGO.

The importance and value of indigo as an article of Indian export will be gathered from the Table (p. 114).

But few samples of indigo have on this occasion been forwarded from Bengal, its chief seat of manufacture. This deficiency (as far as showing the public here the various qualities of this important dye imported into the London market is concerned) has been supplied by Messrs. AUERBACH & Co., of Mincing Lane.

2450. [10584] Indigo. Messrs. FISCHER & Co., *Salem.*

2451. [10585] Do. Do. Do.

2452. [106*] Do. Do.

2453. [92*] Do.

2454. [94*] Do.

2455. [10584] Kurpah indigo. W. R. ARBUTHNOT, Esq., *Madras.*

2456. [10623] Do. North Arcot. RAMA-SHESHIA.

2457. [10011] Do. Hansi. Messrs. JARDINE & Co.

2458. [5123] Do. Hooghly. H. STEEL, Esq.

2459. [5124] Do. Do. Do.

2460. [5125] Do. Do. Do.

2461. [5126] Do. Do. Do.

2462. [5127] Do. Do. Do.

2463. [5122] Do. Midnapore.

2464. [99*] Do. Jessore.

2465. [7109] Wild indigo, Cuttack.

Wild indigo is reported to be found in some parts of the forest jungles of this division, but the Cuttack Committee have been unable to verify the report. The specimen was procured from the state of Bunkee, where a small quantity of indigo is cultivated in a very imperfect manner: the plants being apparently allowed to grow so thickly that they shoot up into long slender reeds instead of developing the leaf. It was tried in a part of the Cuttack district, and proved an entire failure.*

2466. [7363] Indigo, Ulwar. H.H. the MAHARAJAH.

Indigo might be cultivated in Ulwar, but there is a great, though by no means insuperable, disinclination on the part of the people generally against its cultivation, as it is looked upon as 'napak,' unclean. The cost of production is about 2 seers for the rupee. When the railway is completed to Agra, that station will be the nearest connecting link towards Calcutta. A cart occupies five days in going from here to Agra, the cost of which is 8 annas, 12 annas, and 1 rupee a day marching, with half for return hire, according as there are two, three, or four bullocks to each.

2467. [6449] Indigo, Mooltan, Punjab.

2468. [6450] Do. Moozuffergurh, do.

Indigo has been manufactured from time immemorial in the districts of Moozuffergurh, Mooltan, and the country west of the River Indus, called the Derajat. It is exported, but not to any great extent, in the direction of Afghanistan. The article, as at present prepared in the Western Punjab, is quite unfitted for the European market, but under proper superintendence it might be produced of the finest quality, and to an almost unlimited extent.*

2469. [1048] Indigo, Jacobabad, Scinde.

2470. [1056] Do. Hyderabad, do.

2471. [820] Do. Shikarpoor, do.

2472. [4189] Do. Penang.

2473. [93*] Do.

2474. [4163] Do. Penang.

SAMPLES OF INDIGO EXHIBITED BY MESSRS. R. AUERBACH AND CO.

2475. [7198] H H; H & Co. Bengal Indigo Company, Khalbolia. Fine purple violet.

2476. [7199] R; J & R W. European make. Deep red violet.

2477. [7200] P & O; B. Do. Hatton Tirhoot. Fine red violet.

2478. [7201] C D & Co. Do. Sericole, Jessore. Good red violet.

2479. [7202] McA & Co.: R. Do. Meergunge, Jessore. Dull violet.

2480. [7203] Do. Do.

2481. [7204] G W C. Do. Bancore. Good violet and red.

2482. [7205] Moran, European make, Meerpore. Middling violet and red.

## CLASS IV.—*India.*

TABLE SHOWING THE QUANTITIES (AS FAR AS CAN BE ASCERTAINED) AND THE VALUE OF INDIGO EXPORTED FROM INDIA AND EACH PRESIDENCY TO ALL PARTS OF THE WORLD FROM 1850-51 TO 1860-61.

*Quantities in lbs. unless otherwise shown; Values in £. Total Exported Quantity also shown in tons (in parentheses).*

| Years | Whence Exported | United Kingdom Quantity | United Kingdom Value | France Quantity | France Value | Other Parts of Europe Quantity | Other Parts of Europe Value | America Quantity | America Value | China Quantity | China Value | Arabian and Persian Gulfs Quantity | Arabian and Persian Gulfs Value | Other Parts Quantity | Other Parts Value | Total Exported to All Parts Quantity | Total Exported to All Parts Value |
|---|---|---|---|---|---|---|---|---|---|---|---|---|---|---|---|---|---|
| 1850-51 | Bengal | 6,661,756 | 1,286,404 | 1,482,721 | 285,414 | 33,558 | 6,532 | 393,073 | 76,260 | .. | .. | 317,238 | 61,718 | 760 | 145 | 8,889,106 (3,968) | 1,716,474 |
|  | Madras | 2,022,192 | 235,925 | 111,384 | 14,876 | .. | .. | .. | .. | .. | .. | .. | .. | .. | .. | 2,133,576 (952) | 250,801 |
|  | Bombay | 39,395 | 3,632 | .. | .. | .. | .. | .. | .. | .. | .. | 71,681 | 9,954 | 614 | 35 | 111,590 (50) | 13,621 |
|  | ALL INDIA | 8,723,343 | 1,525,961 | 1,594,105 | 300,290 | 33,558 | 6,532 | 393,073 | 76,260 | .. | .. | 388,819 | 71,672 | 1,374 | 180 | 11,134,272 (4,970) | 1,980,896 |
| 1851-52 | Bengal | 6,635,848 | 1,271,813 | 2,039,142 | 396,816 | 138,426 | 27,801 | 308,602 | 59,730 | .. | .. | 484,288 | 63,119 | 493 | 95 | 9,606,799 (4,290) | 1,819,374 |
|  | Madras | 1,489,881 | 162,757 | 181,803 | 22,838 | .. | .. | .. | .. | .. | .. | 38 | 4 | 1,025 | 154 | 1,672,697 (747) | 185,753 |
|  | Bombay | 67,557 | 10,134 | 257 | 39 | .. | .. | .. | .. | .. | .. | 66,728 | 9,963 | 1,644 | 50 | 186,186 (61) | 20,186 |
|  | ALL INDIA | 8,193,236 | 1,444,704 | 2,221,202 | 419,693 | 138,426 | 27,801 | 308,602 | 59,730 | .. | .. | 551,954 | 73,086 | 3,162 | 299 | 11,415,682 (5,098) | 2,025,313 |
| 1852-53 | Bengal | 4,168,183 | 802,880 | 2,311,225 | 454,072 | 9,952 | 1,927 | 441,929 | 85,503 | .. | .. | 282,856 | 43,653 | 2,179 | 120 | 7,165,824 (3,199) | 1,388,155 |
|  | Madras | 2,575,701 | 329,616 | 372,983 | 47,715 | .. | .. | .. | .. | .. | .. | .. | .. | .. | .. | 2,948,684 (1,317) | 377,381 |
|  | Bombay | 29,276 | 4,391 | .. | .. | .. | .. | .. | .. | .. | .. | 267,093 | 39,545 | 2,041 | 263 | 298,410 (133) | 44,199 |
|  | ALL INDIA | 6,773,160 | 1,136,887 | 2,684,158 | 501,787 | 9,952 | 1,927 | 441,929 | 85,503 | .. | .. | 549,949 | 83,198 | 4,220 | 383 | 10,412,868 (4,649) | 1,809,685 |
| 1853-54 | Bengal | 5,669,328 | 1,102,602 | 2,119,376 | 398,712 | 85,840 | 7,012 | 530,880 | 77,627 | .. | .. | 475,328 | 97,368 | 108,864* | 20,729* | 8,989,616 (3,991) | 1,704,050 |
|  | Madras | 2,069,927 | 285,411 | 186,423 | 30,313 | .. | .. | 7,132 | 891 | .. | .. | 1,212 | 145 | .. | .. | 2,264,694 (1,011) | 316,760 |
|  | Bombay | 18,285 | 2,743 | 196 | 12 | .. | .. | .. | .. | .. | .. | 294,555 | 43,986 | 2,753 | 220 | 315,789 (141) | 46,961 |
|  | ALL INDIA | 7,757,540 | 1,390,756 | 2,305,995 | 429,037 | 85,840 | 7,012 | 538,012 | 78,518 | .. | .. | 771,095 | 141,499 | 111,617 | 20,949 | 11,520,099 (5,143) | 2,067,771 |
| 1854-55 | Bengal | 5,139,008 | 992,122 | 1,307,264 | 277,168 | .. | .. | 479,472 | 92,637 | 2,016 | 396 | 117,936 | 22,906 | 225,792* | 40,919* | 7,271,488 (3,246) | 1,426,149 |
|  | Madras | 1,320,320 | 209,936 | 78,966 | 13,159 | .. | .. | .. | .. | .. | .. | .. | .. | .. | .. | 1,399,286 (625) | 223,095 |
|  | Bombay | 15,344 | 2,302 | .. | .. | .. | .. | .. | .. | .. | .. | 353,537 | 49,879 | 4,721 | 400 | 373,602 (167) | 52,581 |
|  | ALL INDIA | 6,474,672 | 1,204,360 | 1,386,230 | 290,327 | .. | .. | 479,472 | 92,637 | 2,016 | 396 | 471,473 | 72,785 | 230,513 | 41,319 | 9,044,376 (4,038) | 1,701,825 |
| 1855-56 | Bengal | 6,003,088 | 1,127,298 | 2,835,392 | 548,882 | 360,080 | 70,947 | 762,720 | 191,475 | .. | .. | 140,448 | 27,339 | 5,040 | 922 | 10,106,768 (4,512) | 1,966,868 |
|  | Madras | 2,389,778 | 351,784 | 461,052 | 72,441 | .. | .. | 1,321 | 198 | .. | .. | .. | .. | 562 | 26 | 2,862,713 (1,273) | 424,449 |
|  | Bombay | 30,793 | 3,954 | .. | .. | 6,000 | .. | .. | .. | .. | .. | 235,695 | 28,769 | 4,672 | 297 | 273,160 (121) | 33,020 |
|  | ALL INDIA | 8,423,659 | 1,483,036 | 3,296,444 | 621,323 | 366,080 | 70,947 | 764,041 | 191,673 | .. | .. | 376,143 | 56,108 | 10,274 | 1,245 | 13,210,641 (5,906) | 2,424,332 |
| 1856-57 | Bengal | 4,703,440 | 863,407 | 2,112,656 | 388,419 | 82,992 | 15,145 | 589,512 | 106,214 | .. | .. | 164,192 | 29,914 | 322,112* | 58,757* | 7,967,904 (3,557) | 1,456,886 |
|  | Madras | 2,129,876 | 351,075 | 521,551 | 93,736 | 2,955 | 443 | .. | .. | .. | .. | .. | .. | 2,484 | 181 | 2,656,866 (1,186) | 445,435 |
|  | Bombay | 43,895 | 5,696 | .. | .. | .. | .. | .. | .. | .. | .. | 211,458 | 27,492 | 17,807* | 2,428* | 273,160 (122) | 35,616 |
|  | ALL INDIA | 6,877,211 | 1,220,178 | 2,634,207 | 477,155 | 85,947 | 15,588 | 582,512 | 105,214 | .. | .. | 375,850 | 57,406 | 342,403 | 61,366 | 10,897,930 (4,865) | 1,937,997 |
| 1857-58 | Bengal | 5,025,412 | 986,192 | 1,005,620 | 205,130 | 8,064 | 1,570 | 273,028 | 48,691 | .. | .. | 167,300 | 23,715 | 163,486* | 41,929* | 6,642,860 (2,966) | 1,307,227 |
|  | Madras | 1,488,986 | 270,166 | 488,346 | 95,877 | .. | .. | .. | .. | .. | .. | .. | .. | 275 | 44 | 1,977,607 (883) | 366,087 |
|  | Bombay | 77,815 | 11,102 | .. | .. | .. | .. | .. | .. | .. | .. | 422,871 | 47,511 | 16,702* | 2,411* | 517,388 (231) | 61,024 |
|  | ALL INDIA | 6,592,213 | 1,267,460 | 1,493,966 | 301,007 | 8,064 | 1,570 | 273,028 | 48,691 | .. | .. | 599,171 | 71,226 | 180,413 | 44,384 | 9,137,855 (4,080) | 1,734,338 |
| 1858-59 | Bengal | 4,128,336 | 1,038,909 | 1,465,660 | 387,174 | 59,556 | 12,584 | 619,920 | 172,956 | .. | .. | 310,576 | 61,863 | 217,308* | 41,522* | 6,796,856 (3,084) | 1,715,008 |
|  | Madras | 1,818,759 | 326,861 | 144,271 | 27,333 | .. | .. | .. | .. | .. | .. | .. | .. | 2,229 | 155 | 1,965,259 (877) | 354,349 |
|  | Bombay | 88,865 | 12,183 | .. | .. | .. | .. | .. | .. | .. | .. | 336,566 | 35,137 | 9,363* | 1,340* | 434,774 (194) | 48,660 |
|  | ALL INDIA | 6,035,960 | 1,377,953 | 1,609,931 | 414,507 | 59,556 | 12,584 | 619,920 | 172,956 | .. | .. | 647,142 | 97,000 | 228,890 | 43,017 | 9,196,389 (4,105) | 2,118,017 |
| 1859-60 | Bengal | 5,727,694 | 1,119,185 | 1,362,676 | 262,924 | 16,996 | 3,313 | 489,604 | 94,792 | .. | .. | 98,728 | 19,205 | 212,492* | 40,836* | 7,911,120 (3,532) | 1,540,255 |
|  | Madras | 2,180,233 | 387,075 | 339,063 | 66,520 | .. | .. | .. | .. | .. | .. | .. | .. | 1,600 | 72 | 2,520,896 (1,125) | 453,667 |
|  | Bombay | .. | .. | .. | .. | .. | .. | .. | .. | .. | .. | 285,031 | 27,238 | 1,804 | 129 | 286,835 (128) | 27,367 |
|  | ALL INDIA | 7,907,857 | 1,506,260 | 1,701,739 | 329,444 | 16,996 | 3,313 | 492,604 | 94,792 | .. | .. | 383,759 | 46,443 | 215,896 | 41,037 | 10,718,851 (4,785) | 2,021,289 |
| 1860-61 | Bengal | 5,851,328 | 1,136,652 | 1,731,184 | 339,395 | 31,136 | 6,048 | 242,256 | 46,676 | .. | .. | 106,736 | 20,802 | 64,400* | 12,936* | 8,027,040 (3,583) | 1,562,509 |
|  | Madras | 1,326,027 | 228,407 | 225,922 | 45,315 | .. | .. | .. | .. | .. | .. | .. | .. | 3,398 | 184 | 1,565,347 (694) | 273,906 |
|  | Bombay | 6,716 | 435 | 15,222 | 1,028 | .. | .. | .. | .. | .. | .. | 223,151 | 48,118 | 4,227 | 388 | 249,316 (111) | 49,961 |
|  | ALL INDIA | 7,184,071 | 1,365,494 | 1,972,328 | 385,738 | 31,136 | 6,048 | 242,256 | 46,676 | .. | .. | 329,887 | 68,920 | 72,025 | 13,508 | 9,831,703 (4,388) | 1,886,376 |

* The principal portion of this was exported to Suez.

2483. [7206] Up country, native make. Ordinary violet.

2484. [7207] R. & H. Up country. Copper red.

2485. [7208] Plant, Oude. Violet.

2486. [7209] <M> Oude, native make. Lean copper.

2487. [7210] I. C. European factory make, Hirzapore, Jessore. Bengal washing.

2488. [7211] A. & Co. Kurpah, European make, deep violet. Kurpah, native make, good red violet. Do. red. Do. coppery red.

2489. [7215] Pondicherry indigo (Kurpah).

2490. [7217] Madras dry leaf (native make).

2491. [7218] Do. Do.

2492. [7219] Penang indigo.

2493. [7220] Madras fig indigo.

2494. [7221] Spurious fig indigo, supposed to be made in England.

2495. [7223] H. & S. J. Bengal indigo, European make. Jeetnapore, Kishnaghur. Purple blue.

2496. [7224] Indigo. Do. Do.

2497. [7225] H. & S. J. Bengal indigo, European make. Jeetnapore, Kishnaghur. Deep blue.

————

2498. [9082] Indigo, P. & Co.; T.; Vellore. North Arcot. Messrs. PARRY & Co.

2499. [9083] Do. P. & Co.; W.; Vellore. Do. · Do.

2500. [9084] Do. P. & Co.; C.; Vellore. Do. Do.

2501. [9085] Do. P. & Co.; A.; Vellore. Do. Do.

2502. [9086] Do. (green leaf), T. D. South Arcot. Do.

2503. [9087] Dry leaf indigo, T. G. Do. Do.

————

2504. [7124] Room dye (*Ruellia* species), Muttock, Assam. W. G. WAGENTRIEBER, Esq.

2505. [7125] Do. Do.

2506-7. [6520-1] Cloth dyed therewith, Muttock, Assam. W. G. WAGENTRIEBER, Esq.

This dye, derived from the Assam Room plant, is prepared after the manner of indigo. Specimen 6520 has been steeped twenty-four hours, but not pressed for want of apparatus; 6521 has been prepared in a similar manner, but steeped forty-eight hours. Room grows wild in Assam. No attention is paid to the cultivation. This plant (or a species very nearly allied to it) is cultivated with the same object in Pegu and other parts of the Burmese Empire, a specimen of which is exhibited under the name of *Mai gyee* from Moulmein. It is believed that the Room contains indigo allied to that produced by species of Isatis and Wrightia. The source of this dye has been referred to *Ruellia comosa* Wall.

## MADDER.

The *Rubia munjista,* known in the market under the name of 'Munjeet,' furnishes the madder-root exported from India. The following Table (p. 116) shows the extent to which it is so.

2507a. [5277] Madder (*Rubia munjista*), Meeree and Duflah Hills. H. L. MICHEL, Esq.

Produced at the Hills occupied by the Meeree and Duflah tribes, North Luckimpore, Assam. Value 1½d. per lb.

2508. [5276] Do. Mishmee and Abor Hills. MAJOR H. S. BIVAR.

Produced at the Hills occupied by the Mishmee and Abor tribes, Suddyah Luckimpore, Upper Assam. Value 1½d. per lb.

2509. [9162] Do. Mishmee Hills. W. G. WAGENTRIEBER, Esq.

2510. [9163] Do. Durrung. LIEUT. W. PHAIRE.

2511. [9160] Do. Geipore. J. N. MARTIN, Esq.

2512. [1871] Do. Calcutta.

2513. [9161] Do. Nepal. H.H. SIR JUNG BAHADOOR, K.C.B.

2514. [1738] Do. (*Rubia munjista*), Punjab.

2515. [6031] Do. Bombay.

2516. [5221] Do. (*Rubia tinctoria*), Jullundur. COL. F. C. BURNETT.

Grown from seed imported two years ago from France.

————

2517. [5329] Chay-root (*Oldenlandia umbellata*), Cuttack.

CLASS IV.—*India.*

TABLE SHOWING THE QUANTITIES (AS FAR AS CAN BE ASCERTAINED) AND THE VALUE OF MUNJEET EXPORTED FROM INDIA AND EACH PRESIDENCY TO ALL PARTS OF THE WORLD FROM 1850–51 TO 1860–61.

| Years | Whence Exported | United Kingdom Quantity | United Kingdom Value | France Quan. | France Value | Other Parts of Europe Quan. | Other Parts of Europe Value | America Quan. | America Value | Arabian and Persian Gulfs Quan. | Arabian and Persian Gulfs Value | Other Parts Quan. | Other Parts Value | Total Quantity (lbs.) | Total Quantity (tons) | Total Value (£) |
|---|---|---|---|---|---|---|---|---|---|---|---|---|---|---|---|---|
| 1850-51 | Bengal | 32,917 | 120 | 412 | 1 | 4,362 | 16 | .. | .. | .. | .. | .. | .. | 37,691 | 17 | 137 |
| | Madras | .. | .. | .. | .. | .. | .. | .. | .. | .. | .. | .. | .. | .. | .. | .. |
| | Bombay | 1,370,320 | 13,324 | .. | .. | .. | .. | .. | .. | 896 | 10 | .. | .. | 1,371,216 | 612 | 13,334 |
| | ALL INDIA | 1,403,237 | 13,444 | 412 | 1 | 4,362 | 16 | .. | .. | 896 | 10 | .. | .. | 1,408,907 | 629 | 13,471 |
| 1851-52 | Bengal | 28,710 | 104 | .. | .. | .. | .. | 20,295 | 74 | .. | .. | .. | .. | 49,005 | 22 | 178 |
| | Madras | .. | .. | .. | .. | .. | .. | .. | .. | .. | .. | .. | .. | .. | .. | .. |
| | Bombay | 1,172,304 | 10,467 | .. | .. | .. | .. | .. | .. | 2,576 | 23 | 4,032 | 36 | 1,178,912 | 526 | 10,516 |
| | ALL INDIA | 1,201,014 | 10,571 | .. | .. | .. | .. | 20,295 | 74 | 2,576 | 23 | 4,032 | 36 | 1,227,917 | 548 | 10,694 |
| 1852-53 | Bengal | 128,205 | 311 | .. | .. | .. | .. | .. | .. | .. | .. | .. | .. | 128,205 | 57 | 311 |
| | Madras | .. | .. | .. | .. | .. | .. | .. | .. | .. | .. | .. | .. | .. | .. | .. |
| | Bombay | 3,216,192 | 28,717 | .. | .. | .. | .. | .. | .. | .. | .. | 6,832 | 62 | 3,223,024 | 1,439 | 28,779 |
| | ALL INDIA | 3,344,397 | 29,028 | .. | .. | .. | .. | .. | .. | .. | .. | 6,832 | 62 | 3,351,229 | 1,496 | 29,090 |
| 1853-54 | Bengal | 42,896 | 105 | .. | .. | .. | .. | 18,816 | 46 | .. | .. | .. | .. | 61,712 | 28 | 151 |
| | Madras | .. | .. | .. | .. | .. | .. | .. | .. | .. | .. | .. | .. | .. | .. | .. |
| | Bombay | 3,782,576 | 33,773 | .. | .. | .. | .. | .. | .. | .. | .. | 2,912 | 26 | 3,785,488 | 1,690 | 33,799 |
| | ALL INDIA | 3,825,472 | 33,878 | .. | .. | .. | .. | 18,816 | 46 | .. | .. | 2,912 | 26 | 3,847,200 | 1,718 | 33,950 |
| 1854-55 | Bengal | 122,304 | 344 | .. | .. | .. | .. | .. | .. | .. | .. | .. | .. | 122,304 | 55 | 344 |
| | Madras | .. | .. | .. | .. | .. | .. | .. | .. | .. | .. | .. | .. | .. | .. | .. |
| | Bombay | 1,621,088 | 15,094 | .. | .. | .. | .. | .. | .. | .. | .. | .. | .. | 1,621,088 | 724 | 15,094 |
| | ALL INDIA | 1,743,392 | 15,438 | .. | .. | .. | .. | .. | .. | .. | .. | .. | .. | 1,743,392 | 779 | 15,438 |
| 1855-56 | Bengal | 134,960 | 493 | .. | .. | .. | .. | .. | .. | .. | .. | 1,792 | 7 | 136,752 | 61 | 500 |
| | Madras | .. | .. | .. | .. | .. | .. | .. | .. | .. | .. | .. | .. | .. | .. | .. |
| | Bombay | 1,635,312 | 13,141 | .. | .. | .. | .. | .. | .. | .. | .. | 3,136 | 26 | 1,638,448 | 731 | 13,167 |
| | ALL INDIA | 1,770,272 | 13,634 | .. | .. | .. | .. | .. | .. | .. | .. | 4,928 | 33 | 1,775,200 | 792 | 13,667 |
| 1856-57 | Bengal | 187,712 | 685 | .. | .. | .. | .. | .. | .. | .. | .. | .. | .. | 187,712 | 84 | 685 |
| | Madras | .. | .. | .. | .. | .. | .. | .. | .. | .. | .. | .. | .. | .. | .. | .. |
| | Bombay | 4,161,136 | 33,429 | .. | .. | .. | .. | .. | .. | .. | .. | 224 | 2 | 4,161,360 | 1,858 | 33,431 |
| | ALL INDIA | 4,348,848 | 34,114 | .. | .. | .. | .. | .. | .. | .. | .. | 224 | 2 | 4,349,072 | 1,942 | 34,116 |
| 1857-58 | Bengal | 280,252 | 1,022 | .. | .. | .. | .. | .. | .. | .. | .. | .. | .. | 280,252 | 125 | 1,022 |
| | Madras | .. | .. | .. | .. | .. | .. | .. | .. | .. | .. | .. | .. | .. | .. | .. |
| | Bombay | 4,219,824 | 35,832 | 5,478 | 44 | .. | .. | 17,472 | 151 | .. | .. | 2,240 | 20 | 4,245,014 | 1,895 | 36,047 |
| | ALL INDIA | 4,500,076 | 36,854 | 5,478 | 44 | .. | .. | 17,472 | 151 | .. | .. | 2,240 | 20 | 4,525,266 | 2,020 | 37,069 |
| 1858-59 | Bengal | 186,144 | 679 | .. | .. | .. | .. | 39,984 | 146 | .. | .. | .. | .. | 226,128 | 101 | 825 |
| | Madras | .. | .. | .. | .. | .. | .. | .. | .. | .. | .. | .. | .. | .. | .. | .. |
| | Bombay | 1,672,832 | 14,936 | .. | .. | .. | .. | .. | .. | .. | .. | .. | .. | 1,672,832 | 747 | 14,936 |
| | ALL INDIA | 1,858,976 | 15,615 | .. | .. | .. | .. | 39,984 | 146 | .. | .. | .. | .. | 1,898,960 | 848 | 15,761 |
| 1859-60 | Bengal | 274,820 | 1,002 | .. | .. | .. | .. | .. | .. | .. | .. | .. | .. | 274,820 | 123 | 1,002 |
| | Madras | .. | .. | .. | .. | .. | .. | .. | .. | .. | .. | .. | .. | .. | .. | .. |
| | Bombay | 1,456,896 | 13,008 | 336 | 3 | .. | .. | 42,784 | 382 | .. | .. | 336 | 3 | 1,500,352 | 669 | 13,396 |
| | ALL INDIA | 1,731,716 | 14,010 | 336 | 3 | .. | .. | 42,784 | 382 | .. | .. | 336 | 3 | 1,775,172 | 792 | 14,398 |
| 1860-61 | Bengal | 271,824 | 1,973 | .. | .. | .. | .. | 27,328 | 186 | .. | .. | .. | .. | 299,152 | 134 | 2,159 |
| | Madras | .. | .. | .. | .. | .. | .. | .. | .. | .. | .. | .. | .. | .. | .. | .. |
| | Bombay | 2,353,792 | 31,032 | .. | .. | .. | .. | 131,936 | 1,178 | .. | .. | 1,127 | 10 | 2,486,855 | 1,110 | 32,220 |
| | ALL INDIA | 2,625,616 | 33,005 | .. | .. | .. | .. | 159,264 | 1,364 | .. | .. | 1,127 | 10 | 2,786,007 | 1,244 | 34,379 |

2518. [76*] Chay-root (*Oldenlandia umbellata*), Cuttack.

Is largely used by the Indian dyer in the south of India. It furnishes a red dye similar to Munjeet. Experiments in this country with the chay-root have hitherto failed, in consequence, it is supposed, of deterioration during the voyage. It is advisable in the case of this and of some other Indian dye stuffs, that the colouring matter should be extracted similarly to indigo before it is exported.

2519. [4013] Muddi chuckay (*Morinda umbellata*), Salem.

2520. [109*] Mungkudu (*M. umbellata*), Java.

2521. [75*] Mungkudu (*M. umbellata*), Malacca.

The tree producing this dye-stuff grows freely everywhere in India, and no particular care is required in gathering it. The best dye is procured from the bark of the roots of plants three years old. It is one of the commonest red dyes of India, though the colour is dull, yet it is considered faster than the brighter tints obtained from other substances.

2522. [6461] Al root (*Morinda citrifolia*), first quality, Banda District. H. W. DASHWOOD, Esq.

2523. [6462] Do. second do., Do. Do.

2524. [6463] Do. third do., Do. Do.

2525. [1950] Al root (*Morinda citri-folia*), Jhansee.

2526. [5800] Do. Burmah.

2527-8. [7501–2] Do. Dharwar.

The bark and root of this *Morinda* is used in the same manner as the last. Most of the Madras red turbans are dyed with this substance, which is very common in that presidency. It yields three different permanent shades — a bright red, a pink red, and a faint red.

MR. DASHWOOD adds the following note :—

The Al root is cultivated in the Banda Purgunnah of this district at the villages of Kupsa, Muttound, Khunna, &c.: the whole extent of cultivation is only about 400 beegahs. The Humeerpore district is the great producing country for al, probably from there being greater facilities for irrigation than in this district. The plants come to maturity in three years. The roots are then dug up and sorted into three kinds, according to the fineness of the fibres. The fibres are then cut and beaten down well, and then ground to powder, which latter is used for the dye. The proportion of prepared fibre is equal to the raw material. The uses to which it is applied are dyeing a red colour, as a remedy for tooth-ache, as well as for sprains of horses. The cost of the prepared fibre is 1 rupee per maund. It is transported on carts, horses, and bullocks to Banda and Rajapore for export to other parts of India, and the cost of transport is about 8 annas per maund.

2529. [9371] Mug dye, Chittagong.

This plant grows wild in the southern part of this district. The Mugs make yellow and red dye, by grinding the plant and its roots into powder, and boiling the same in water. The colours are dull, but seem to last for a long time.*

2530. [6021] Indian Alkanet, Havapiva (*Onosma* species), Bombay.

Several species of *Onosma* are employed as substitutes for the true alkanet, of which this appears to be the produce of one.

2531. [1806] Alkanet (*Anchusa tinctoria*), Punjab.

2532. [1807] Do. Do.

This is probably the true alkanet root.

2533. [4798] Turmeric (*Curcuma longa*), Burmah. Messrs. HALLIDAY, FOX, & Co.

2534. [7114] Do. Cuttack.

2535. [2852] Do. Raepore.

2536. [6024] Do. Bombay.

2537. [7504] Do. Dharwar.

The powder of the rhizome of this plant is much used in India for dyeing.

2538. [84*] Pupli chuckay (*Ventilago Maderaspatana*).

2539. [7506] Do. Dharwar.

2540. [7510] Do. powdered, Do.

The bark of the pupli root is used in Mysore and elsewhere, as yielding an orange dye. It is treated with alum, myrabolans, &c. This dye stuff is in very common use in India, and deserves a fair trial in this country. The pupli is seldom used alone, but generally as an adjunct, with chay root, to produce a rich chocolate colour, or, if with galls, a black.

2541. [6169] Jack wood (*Artocarpus integrifolia*), Akyab.

Used for dyeing phoongie (priests) clothes with yellow orange colour; 5,000 maunds procurable, but more can be had if required. Price 5 rupees per maund.*

2542. [81*] Kayu Lakah, Singapore.

This is a Malayan red wood. It is heavy and compact, somewhat resembling red sanders wood, but, when powdered, the colour is browner and not so brilliant.

2543. [82*] Kayu Laxka, Singapore.

This is a red dye wood, so closely resembling the *Kayu lakah* that it may be the produce of the same tree. Either of the above woods seem to be applicable to the same purposes as red sanders.

2544. [9372] Oak bark (*Quercus* species), Chittagong.

There are a great number of oak trees in the jungles in this district, both in the low lands and hills, particularly the latter. It is used for tanning as well as for firewood.

2545. [4015] Turwar (*Cassia auriculata*), Salem.

2546. [4953] Do. Madras. DR. SHORTT.

This bark is employed both in tanning and dyeing, and has been of late imported on two or three occasions into this country.

2547. [6170] Cassia fistula bark, Cuttack.

The bark of the *Cassia fistula*, locally *Soona Rea Chali*, is used in tanning leather. The tree is abundant in the jungles of the Tributary Mehals.†

2548. [77*] Kuephul (*Myrica sapida*), Rohilkund.

Used for dyeing yellow.

---

* Local Committee, *Chittagong.*

* Local Committee, *Akyab.*
† Local Committee, *Cuttack.*

2549. [23*] Samak or Sumach (*Cæsalpinia coriaria*) or Divi-divi bark, Singapore.

This bark is used both as a tanning material and as a dye-stuff.

2550. [7106] Lodh bark (*Symplocos racemosa*), Cuttack.

2551. [2844] Do. Raepore.

This bark is employed in dyeing yellow, and also sometimes as a mordant for other dyes.

2552. [15*] Garan-chal (*Ceriops Roxburghianus*).

The bark is used in India for dyeing, chiefly in the Presidency of Bengal.

2553. [21*] Sagah bark, Singapore.

This bark resembles mangrove bark in appearance, and is employed as a dyeing material.

2554. [24*] Mangrove bark (*Rhizophora mangle*), Singapore.

2555. [6708] Do. Do. G. ANGUS, Esq.

2556. [4101] Do. Penang.

It is used in Arracan as the source of a chocolate colour. This substance can be readily enough obtained if found valuable to the home dyer, as it is often imported for tanning.

2557. [4100] Tengah (*Rhizophora* species), Penang.

It is allied to the mangrove.

2558. [3809] Babool bark (*Acacia arabica*), Jhansee.

2559. [22] Do. Do.

A valuable tanning bark, sometimes also employed in dyeing.

2560. [20*] Jamoon (*Eugenia jambolana*).

This bark is chiefly employed for tanning.

2561. [40*] Trap bark (*Artocarpus* species).

2562. [78*] Thenaka bark, Tenasserim.

2563. [6023] Rusot extract (*Berberis lycium* and *Berberis aristata*), Bombay.

Furnishes a yellow dye. A similar dye is obtained in the Neilgherries from *Berberis asiatica*.

2564. [6009] Mara Manjil, Woniwol or Bangwellgetta (*Menispermum fenestratum*), Bombay.

Furnishes a yellow dye.

2565. [9373] Bukkum chips (*Cæsalpinia sappan*), Calcutta. KANNY LOLL DEY.

2566. [7104] Do. Cuttack.

2567. [31*] Do. Madras.

2568. [79*] Sapan wood (*C. sappan*), Singapore.

2569. [5270] Bukkum or Sapan, Assam.

2570. [5269] Do. Assam.

A large quantity of *Sappan* or *Bukkum* wood is grown in Malabar. The wood of the trunk, and also of the root, are rendered available.

The wood of *Cæsalpinia pulicata* is also employed in the East Indies as a dye stuff. There is every probability of its becoming an established article of British commerce.

2571. [80] Kayu Kudrang, Singapore.

A yellow dye wood; source undetermined.

2572. [86*] Woodunpar, Upper Assam.

2573. [6172] Kabine (dye tree bark), Akyab.

Used to colour fishing nets a red brown; 10,000 maunds could be obtained. Price 6 rupees per maund.*

2574. [6168] Rairo (Dye tree bark), Akyab.

This gives a yellow colour; with oil and plantain ash a red colour is made. Price 6 rupees per maund.†

2575. [1786] Ekalbir (*Datisca cannabina*), Punjab.

2576. [1787] Ukkulbeer,(*D. cannabina*), Do.

The bark and woody portions of the root have long been known and used as a yellow dye in Cashmere. It is in great esteem for dyeing silks.

2577. [6171] Thit-na-myeng, Akyab.

Thread is coloured yellow by it, and when oil and sealing-wax are added, a red colour is obtained: 1,000 maunds might be obtained. Price 8 rupees per maund.‡

2578. [6559] Gutteah, Chittagong.

Another tanning substance. It is a bush that grows on the sides of creeks and rivers, in low ground, which is inundated with the spring tide. It is cut for firewood, and the fishermen and shoemakers purchase it, and take the bark off to lay their fishing nets and leather, and afterwards sell the wood posts for firewood. It is very abundant at Chittagong.

2579. [18*] Bark of *Gmelina arborea*, Bengal.

* Local Committee, *Akyab.*
† Local Committee, *Akyab.*
‡ Local Committee, *Akyab.*

NEW DYE STUFFS EXHIBITED BY DR. R. F.
THOMPSON, MALDA.

2580. [2191] Vegetable green dye and cloth.

2581. [2194] Do. (1).

2582. [6645] Do. in a liquid state.

2583. [2153] Dyed cloth from the green dye, without mordants (2).

2584. [2152] Do. passed through a dilute sulphuric acid (3).

2585. [2158] Cake of green dye (4).

2586. [2156] Cloth dyed from the cake of green dye with acetate of copper (5).

2587. [2165] Dyed cloth from cake of green dye (6).

2588. [2157] Do. from leaves of do. (7).

2589. [2163] Leaves of green dye dried in the oven (8).

2590. [2159] Do. in powder (9).

2591. [2165] Dyed cloth from dried leaves of green dye (10).

2592. [2164] Cloth boiled in solution of leaves of the green dye (11).

———

2593. [2147] Yellow dye in cake.

2594. [2161] Do. in powder.

2595. [2164] Do.   do.

2596. [2154] Cloth dyed with yellow dye.

2597. [2168] Do.

DR. THOMPSON supplies the following note:—
This yellow dye is from the petals of a flower which gives out the dye largely, and which at once attaches itself to cloth permanently, without mordants. It can be had abundantly, and collected with greater facility than safflower.
Dissolve and boil in hot water, then dip the cloth or silk into it.
Nos. 2154 and 2168 have been dyed with it.
If these green and yellow dyes are approved of, India will be able to supply annually a very large quantity.

———

2598. [63*] Casan leaves (*Memecylon tinctorium*), Madras.

A small tree, common in the jungles of the Carnatic, the wood of which is used for firewood, and the leaves brought into the markets in large quantities for dyeing purposes. A cold infusion of the leaves yields a yellow dye. Crimson dye is also said to be obtained from them.

2599. [4017] Vattangee (*Cæsalpinia Sapan?*), Salem.

2600. [20*] Leaves of *Dicalyx tinctoria.*

The powdered leaves of *Dicalyx tinctoria* are employed at Mirzapore and elsewhere for dyeing red.

2601. [6027] Hunsraj (*Adiantum lunulatum*), Bombay.

2602. [7101] Mai Gyee (*Reuellia*, species), Moulmein.

The leaves of this plant are boiled in water, when the decoction gives a blue colour to cloth &c., inferior to indigo.*

2603. [4963] Henna (*Lawsonia alba*) Chingleput.  DR. SHORTT.

The leaves of this plant beaten up into a soft mass with conjee water are applied to the nails, finger ends, palms and soles of the feet over night: on being washed out the next morning these parts are found stained a deep red colour. The men use it to stain their moustaches and beards, and for dyeing the manes and tails of their horses. It is also used as an ordinary dye stuff. A decoction of the leaves is used in lepra &c. The flowers when distilled are used as a perfume.

2604. [49] Safflower (*Carthamus tinctorius*), Celebes.

2605. [7105] Do. Cuttack.

2606. [6602] Do. Hoogly.

2607. [6459] Safflower (*C. tinctorius*), Allahabad.

2608. [2848] Do. Raepore.

2609. [45] Do. Deccan.

2610. [6013] Do. Ahmedabad.

2611. [48] Do. Assam.

2612. [7907] Cake Do. Dacca. Messrs. AUERBACH & CO.

2613. [7908] Do.    Do.    Do.

2614. [7507] Do. Dharwar.

2615. [7508] Do.  Do.

This article is sent down from Sumbulpore. It yields a bright handsome colour. The wholesale price per maund of 100 lbs. English is about 8 rupees or 16*s*.†

2616. [58*] Godari flowers (*Grislea tomentosa*).

The red flowers and leaves are used for dyeing purposes. In the Northern Circars, where it is known under

* Local Committee, *Moulmein.*
† Local Committee, *Cuttack.*

the name of *Godari*, the leaves are employed in dyeing leather. Sheep-skins, steeped in an infusion of the dried leaves, become a fine red, of which native slippers are made. The dried flowers are employed in Northern India, under the name of *Dhauri*, in dyeing with Morinda bark; but perhaps more for their astringent than for their tinctorial properties. DR. GIBSON states that in Kandeish the flowers form a considerable article of commerce inland as a dye. It grows abundantly in the hilly tracts of the Northern Circars.

2617. [6206] Goolanna (*Punica granatum*), Calcutta.

2618. [7031] Saffron (*Crocus cashmerianus*), Cashmere.

This well-known substance, scarcely important as a dye-stuff, is produced in the vale of Cashmere, from indigenous species of *Crocus*.

––––––

2619. [65*] Siamese dye stuff.

2620. [66*] Burmese dye flowers.

––––––

2621. [6008] Tisso flowers (*Butea frondosa*), Cuttack.

2622. [7107] Do. Bombay.

These flowers, and probably also those of *Butea superba*, are used as a yellow dye in India. 16 seers per rupee.

––––––

2623. [7110] Hursinghar flowers (*Nyctanthes arbor tristis*), Calcutta.

2624. [4891] Do. Chingleput. DR. SHORTT.

These flowers are used for dyeing yellow or orange.

2625. [4958] Shoe flower (*Hibiscus rosa sinensis*), Chingleput. DR. SHORTT.

The flowers are of a rich scarlet colour, and yield a purple juice. They are employed for dyeing a lilac colour, and for blackening leather. DR. SHORTT adds, that he has frequently used an infusion of the shoe flower for preparing litmus, and found it to answer admirably.

2626. [7108] Kamala powder (*Rottlera tinctoria*), Cuttack.

2627. [1862] Do. Punjab.

2628. [51] Capila rung (*do.*)

2629. [83] Capela podi (*do.*)

2630. [4016] Kapila podi (*do.*), Salem.

2631. [7509] Kamala (*do.*), Dharwar.

Used for dyeing silk an orange colour. This powder consists of the stellate pubescence shaken from the surface of the fruit capsules of the *Rottlera tinctoria*, a large tree widely spread over many parts of the country. It is worthy of extended attention.

2632. [1246] Suringee (*Calysaccion longifolium*), Bombay.

The flower buds of this plant, a portion of which is figured below (*see opposite page*), are collected and employed for dyeing silk.

2633. [7111] Daleemka kola, or pomegranate rind (*Punica granatum*), Ahmedabad.

2634. [6022] Do.      Do.

2635. [3439] Mangosteen rind (*Garcinia mangostana*), Calcutta. KANNY LOLL DEY.

The coat or rind of the fruit of the mangosteen, and the bark of the Katapping or wild almond (*Terminalia catappa*), are used for dyeing black.

2636. [73*] Munjulde (*Terminalia chebula*), Assam.

Appears to be the fruit collected and dried before it is ripe. Under the name of Munjulde it is known in Assam, where it is employed in dyeing.

2637. [6041] Haridah (*T. chebula*), Ahmedabad.

2638. [6605] Do. Calcutta.

2639. [6019] Heerda (*T. chebula*), Bombay.

2640. [6030] Bherda (*T. chebula*), Do.

2641. [2842] Do. Raepore.

2642. [1817] Do. Punjab.

2643. [6486] Bohara (*T. belerica*).

2644. [2836] Bahara, Raepore.

2645. [9376] Humtokee.

2646. [6486] Haritoke, Calcutta.

2647. [2843] *Terminalia angustifolia*, Raepore.

These three kinds of Myrabolans yield, with alum, a good durable yellow, and with salts of iron, a black colour. They are in very common use in India, and have been so from time immemorial. It is not long since they were introduced into Great Britain for tanning purposes, and now a large quantity is annually imported.

2648. [9377] Aomla (*Phyllanthus emblica*).

2649. [2845] Owlah (*P. emblica*), Raepore.

2650. [7102] Do. Cuttack.

2651. [4945] Do. Madras.

––––––

2652. [6604] Bhellawa nuts (*Semecarpus anacardium*], Calcutta.

**2653.** [4019] Bhela nuts (*S. anacardium*), Madras.

The common marking nuts are thus designated. The juice is black, and is employed not only as a medium, but especially for marking all kinds of cotton cloth.

**2654.** [70*] Gaub fruits (*Diospyros glutinosa*).

**2655.** [6560] Tari (*Cæsalpinia* species), Chittagong.

It is the pod of a leguminous plant, which grows abundantly in the Hills, and is useful for tanning.

**2656.** [6571] Oom Tari, Chittagong.

Is the produce of a palm, which grows in this district, and is susceptible of the same as an ordinary tari.

**2657.** [6036] Amlee (*Tamarindus* species), Ahmedabad.

### LICHENS.

**2657a.** [36*] Burmese Orchil (*Roccella phycopsis*), Burmah.

Probably equal in value to the ordinary *R. tinctoria* of commerce.

**2658.** [1815] Chulcheleera, Punjab.

**2659.** [34*] Do.  Do.

A mixture of dye lichens employed for dyeing, contains *Parmelia Kamtschadalis, Parmelia perlata*, and its variety *sorediata, Usnea florida, Ramalina calicaris*, and fragments of *Physcia leucomela*. In this mixture the first-named species constitutes the greatest proportion.*

Suringee (*Calysaccion longifolium*).†

### TANNING SUBSTANCES.

Some of these have been already incidentally referred to, but as the following are distinct of their kind, they are better enumerated together :—

**2660.** [1812] Kakrasinghee galls on *Rhus kakrasinghee*, Kangra.

These galls are procured only in small quantities, and are not yet articles of commerce beyond the confines of India.

**2661.** [6015] Gool-i-pista, Bombay.

**2662.** [7507] Do. Dharwar.

These galls are produced on the Pistachio (*Pistacia vera*), and are brought down from the North.

**2663.** [7503] Galls of *Terminalia chebula*, Dharwar.

* M. C. Cooke, Esq.
† Received from *Pharmaceutical Journal.*

2664. [6032] Cadooca Poo (*Terminalia* species), Bombay.

2665. [6451] Mani dye, galls of *Tamarix furas*, Jhung District.

2666. [6006] Burree-mue, galls of *T. indica*, Bombay.

2667. [9378] Chotee, galls of *T. furas*, Bombay.

2668. [6020] Chotee-mue, Do.   Do.

The Tamarisk galls are exported in small quantities, and are occasionally found in the British market.

OAK GALLS.

2669. [6018] Maaphul, galls of *Quercus infectoria*, Bombay.

2670. [11*] Galls of *Q. infectoria*.

2671. [7103] Do. Calcutta.

These galls are exported, and are ordinarily to be met with in commerce.

2672. [15] Asaucum (*Terminalia tomentosa*).

An astringent extract somewhat of the nature of kino applicable for tanning and dyeing.

2673. [18] Dhak gond (*Butea frondosa*), North-west India.

The kino of the Pulas tree.

2674. [6601] Dhak gum (*B. frondosa*), Lucknow.

2675. [13] Pelachy extract.

The astringent extract of *Butea superba*, similar in properties and uses to kino.

2676. [16*] Kino (*Pterocarpus marsupium*), Malabar.

The concrete exudation of the *Pterocarpus marsupium*.

2677. [6017] Moocherus (*Bombax heptaphyllum*), Bombay.

The astringent gum of the silk cotton tree, which is employed for tanning.

2678. [6681] Gambir (*Nauclea gambir*), Singapore.  G. ANGUS, Esq.

2679. [6662] Do. Do.  JOSE D'ALMEIDA, Esq.

2680. [6704] Yellow gambir, Rhio.  G. ANGUS, Esq.

2681. [6665] Gambir, Do.  Do.

Gambir is the Malay name for *Terra japonica*.

2682. [6039] Cutch or Catechu (*Acacia catechu*), Bombay.

2683. [5341] Do. Cuttack.

2684. [5398] Do. Shahabad.

2685. [5397] Do. Do.

2686. [6211] Do. Calcutta.

2687. [6204] Do. Do.

2688. [5317] Do. Rangoon.

2689. [3332] Ava Cutch (*do.*)  Messrs. HALLIDAY, FOX, & Co.

2690. [4763] Cutch (*do.*), Burmah.  Do.

2691. [6028] Cutch (*Areca catechu*).

2692. [36*] Cake catechu (*do.*), Calcutta.

Subdivision IV. — FIBROUS SUBSTANCES, INCLUDING MATERIALS FOR CORDAGE AND CLOTHING.

## COTTON.

A very large and interesting collection has been forwarded.  The arrangement requires but little explanation.  The cottons are not placed according to the importance of the district, in relation to its exports up to the present time, but geographically, beginning with those from Bengal.  A distinct statement is made as to whether the sample is grown from native or foreign seed.

The cotton has been priced both by the Jury * and by the Commissioner † appointed by the COTTON SUPPLY ASSOCIATION.  Various details will be found condensed in the tabular synopsis commencing at p. 134, whilst the Table which now follows (p. 123) exhibits the exports of cotton from each Presidency for ten years.

### Cotton from Cuttack and Sumbulpore.

2693. [121] Kupass and bolls, Piplee. REV. G. TAYLOR.

N.B.—The native term 'Kupass' is applied to every sample of uncleaned cotton.

---

* For Jury, by THOMAS BAZLEY, Esq., M.P.
† W. WANKLYN, Esq.

## TABLE SHOWING THE QUANTITIES (AS FAR AS CAN BE ASCERTAINED) AND VALUE* OF COTTON EXPORTED FROM INDIA AND EACH PRESIDENCY TO ALL PARTS OF THE WORLD FROM 1850-51 TO 1860-61 INCLUSIVE.

COUNTRIES WHITHER EXPORTED

| YEARS | WHENCE EXPORTED | UNITED KINGDOM Quantity (lbs) | Value (£) | FRANCE Quantity (lbs) | Value (£) | OTHER PARTS OF EUROPE Quantity (lbs) | Value (£) | NORTH AMERICA Quantity (lbs) | Value (£) | CHINA Quantity (lbs) | Value (£) | ARABIAN AND PERSIAN GULFS Quantity (lbs) | Value (£) | ISLANDS AND SHORES OF THE INDIAN SEAS Quantity (lbs) | Value (£) | ALL OTHER PLACES Quantity (lbs) | Value (£) | TOTAL Tons | TOTAL Quantity (lbs) | TOTAL Value (£) |
|---|---|---|---|---|---|---|---|---|---|---|---|---|---|---|---|---|---|---|---|---|
| 1850-51 | Bengal | 985,096 | 12,010 | 2,497,850 | 30,369 | 60,042 | 730 | | | 18,248,478 | 221,865 | | | 1,290,009 | 15,684 | 49,761 | 605 | 10,396 | 23,131,166 | 281,263 |
| | Madras | 9,037,889 | 116,342 | 255,900 | 3,837 | | | | | 9,155,350 | 117,349 | | | 299,400 | 3,267 | 687,765 | 9,688 | 8,678 | 19,438,520 | 250,505 |
| | Bombay | 131,423,883 | 1,981,366 | 496,941 | 38,128 | | | | | 49,646,801 | 821,150 | 2,216 | 27 | 1,868,723 | 145,842 | 136,641 | 1,888 | 82,110 | 183,903,997 | 2,943,021 |
| | ALL INDIA | 141,446,798 | 2,059,718 | 3,250,691 | 72,334 | 60,042 | 730 | | | 77,050,629 | 1,160,364 | 2,216 | 27 | 3,458,132 | 164,793 | 874,167 | 12,126 | 101,104 | 226,473,683 | 3,474,789 |
| 1851-52 | Bengal | 642,637 | 7,812 | 47,952 | 583 | | | | | 38,151,251 | 463,840 | | | 1,909,269 | 23,214 | 8,143 | 88 | 18,196 | 40,759,152 | 495,537 |
| | Madras | 4,632,380 | 61,540 | 48,000 | 571 | | | | | 10,737,163 | 135,706 | 26,322 | 213 | 1,568,800 | 17,873 | 368,864 | 5,209 | 7,760 | 17,381,519 | 221,112 |
| | Bombay | 75,829,306 | 1,101,928 | | | | | | | 111,829,247 | 1,692,381 | 953,433 | 11,232 | 6,571,359 | 95,023 | 228,815 | 2,776 | 87,287 | 195,412,160 | 2,903,340 |
| | ALL INDIA | 81,104,423 | 1,171,280 | 95,952 | 1,154 | | | | | 160,717,651 | 2,291,927 | 979,755 | 11,445 | 10,049,428 | 136,110 | 605,822 | 8,073 | 113,193 | 253,552,831 | 3,619,989 |
| 1852-53 | Bengal | 6,853,728 | 83,328 | | | 131,928 | 1,604 | | | 24,848,383 | 302,108 | 14,969 | 182 | 1,635,130 | 19,880 | 1,269,827 | 17,789 | 14,948 | 33,484,138 | 407,102 |
| | Madras | 16,575,197 | 191,872 | | | | | | | 13,026,102 | 164,528 | 56,602 | 577 | 795,375 | 9,728 | 96,124 | 1,087 | 14,178 | 31,759,247 | 385,176 |
| | Bombay | 157,932,069 | 2,249,986 | | | | | 34,944 | 687 | 37,797,257 | 559,808 | 1,121,918 | 14,819 | 717,420 | 11,516 | | | 88,243 | 197,664,788 | 2,887,216 |
| | ALL INDIA | 181,360,994 | 2,525,186 | | | 131,928 | 1,604 | 34,944 | 687 | 75,671,742 | 1,026,444 | 1,193,489 | 15,578 | 3,149,125 | 41,119 | 1,365,951 | 18,876 | 117,369 | 262,908,173 | 3,679,494 |
| 1853-54 | Bengal | 2,065,056 | 25,153 | | | 350,448 | 4,257 | | | 11,663,904 | 140,036 | 66,428 | 631 | 146,100 | 1,746 | 16,800 | 204 | 6,293 | 14,096,208 | 169,650 |
| | Madras | 8,721,984 | 118,899 | 1,800 | 876 | 97,360 | 2,002 | | | 2,480,400 | 34,726 | | | 246,288 | 3,286 | 694,744 | 9,635 | 5,450 | 12,207,016 | 162,739 |
| | Bombay | 127,396,389 | 1,808,625 | 598,288 | 6,440 | | | | | 41,632,704 | 633,809 | 1,357,540 | 14,908 | 394,388 | 5,932 | 227,382 | 2,692 | 76,544 | 171,458,541 | 2,469,760 |
| | ALL INDIA | 138,183,429 | 1,947,777 | 600,088 | 9,584 | 447,808 | 6,259 | | | 55,777,008 | 808,571 | 1,423,968 | 15,539 | | | 938,876 | 12,531 | 88,287 | 197,761,765 | 2,802,149 |
| 1854-55 | Bengal | 59,186 | 730 | | | | | | | 7,436,128 | 90,489 | 25,439 | 229 | 134,512 | 1,636 | 1,456 | 18 | 3,407 | 7,631,232 | 92,873 |
| | Madras | 8,006,035 | 104,491 | 32,400 | 666 | | | | | 1,711,500 | 28,961 | 1,051,534 | 10,858 | 2,066,500 | 26,194 | 1,084,689 | 14,949 | 5,771 | 12,996,513 | 169,490 |
| | Bombay | 111,448,866 | 1,578,923 | 224,140 | 3,257 | 904,120 | 9,625 | | | 36,746,295 | 524,694 | | | 2,256,308 | 31,819 | 591,684 | 7,227 | 68,403 | 153,222,447 | 2,166,403 |
| | ALL INDIA | 119,513,537 | 1,684,144 | 256,540 | 3,923 | 904,120 | 9,625 | | | 45,893,923 | 639,144 | 1,076,973 | 11,087 | 4,457,320 | 58,649 | 1,677,779 | 22,194 | 77,581 | 173,780,192 | 2,428,766 |
| 1855-56 | Bengal | 598,192 | 7,270 | | | 896 | 10 | | | 12,372,080 | 150,357 | 1,148 | 12 | 1,052,032 | 12,204 | 42,280 | 477 | 5,809 | 13,013,470 | 158,115 |
| | Madras | 4,792,388 | 58,900 | 1,800 | 87 | | | | | 54,000 | 600 | 401,240 | 3,203 | | | 1,242,196 | 17,608 | 3,199 | 7,149,564 | 89,361 |
| | Bombay | 165,380,980 | 2,320,454 | 736,172 | 9,594 | 1,497,048 | 19,079 | | | 44,265,032 | 651,468 | | | 4,633,355 | 62,482 | 108,138 | 1,205 | 96,883 | 217,016,915 | 3,067,475 |
| | ALL INDIA | 170,771,510 | 2,386,624 | 737,972 | 9,621 | 1,497,944 | 19,089 | | | 56,691,112 | 802,425 | 402,388 | 3,215 | 5,685,409 | 74,687 | 1,393,614 | 19,290 | 105,884 | 237,179,949 | 3,314,951 |
| 1856-57 | Bengal | 3,484,928 | 41,744 | 63,728 | 775 | 508,592 | 6,158 | | | 12,610,864 | 153,260 | | | 605,696 | 7,361 | 3,024 | 36 | 7,690 | 17,226,832 | 209,334 |
| | Madras | 19,597,302 | 261,080 | 5,999 | 72 | 309,000 | 4,228 | | | 1,003,200 | 14,945 | 410 | 8 | 232,799 | 3,268 | 2,404,636 | 33,666 | 10,515 | 23,553,346 | 316,362 |
| | Bombay | 230,377,806 | 3,189,377 | 1,803,984 | 23,581 | 10,698,416 | 188,107 | | | 35,170,497 | 548,547 | 192,726 | 2,533 | 574,196 | 9,326 | 55,721 | 782 | 124,497 | 278,873,346 | 3,912,253 |
| | ALL INDIA | 253,410,036 | 3,492,201 | 1,873,711 | 24,428 | 11,516,008 | 198,488 | | | 48,784,561 | 715,852 | 193,136 | 2,541 | 1,412,691 | 19,955 | 2,463,381 | 34,484 | 142,702 | 319,653,524 | 4,437,949 |
| 1857-58 | Bengal | 164,948 | 1,950 | 29,988 | 365 | 1,527,500 | 21,306 | 13,888 | 405 | 635,488 | 7,723 | 537,211 | 7,842 | 106,944 | 1,291 | 2,996 | 36 | 420 | 940,364 | 11,365 |
| | Madras | 11,699,984 | 161,148 | 4,260,594 | 58,530 | | | | | 651,600 | 9,122 | | | 151,200 | 2,309 | 1,928,667 | 27,002 | 9,027 | 20,219,545 | 279,407 |
| | Bombay | 185,356,315 | 3,183,600 | 9,853,292 | 133,600 | 18,175,090 | 260,336 | | | 19,287,031 | 376,847 | | | 5,788,292 | 94,930 | 283,024 | 3,735 | 106,783 | 239,194,143 | 4,010,996 |
| | ALL INDIA | 197,421,247 | 3,296,698 | 14,143,874 | 194,386 | 19,702,590 | 281,642 | 13,888 | 405 | 20,574,119 | 393,492 | 537,211 | 7,842 | 5,996,436 | 98,530 | 2,214,687 | 30,772 | 116,230 | 260,354,052 | 4,301,708 |
| 1858-59 | Bengal | 296,386 | 3,602 | 2,386,500 | 88,411 | 8,945,480 | 53,945 | | | 30,268 | 500 | | | | | 25,550 | 22,893 | 157 | 352,204 | 4,242 |
| | Madras | 6,432,353 | 90,694 | 41,524 | 876 | | | | | 3,596,400 | 50,250 | 1,894 | 31 | | | 1,636,128 | 57,150 | 6,273 | 14,052,275 | 197,379 |
| | Bombay | 157,289,419 | 2,986,431 | | | | | | | 38,607,749 | 788,956 | 617,743 | 10,121 | | | 2,955,178 | 80,183 | 90,889 | 203,457,098 | 3,892,479 |
| | ALL INDIA | 164,018,158 | 3,080,727 | 2,428,024 | 89,287 | 8,945,480 | 53,945 | | | 42,234,417 | 839,706 | 619,637 | 10,152 | | | 4,615,856 | 160,226 | 97,259 | 217,861,577 | 4,094,100 |
| 1859-60 | Bengal | 223,468 | 2,444 | 1,548,044 | 21,413 | 785,509 | 11,004 | | | 5,448,716 | 59,596 | | | | | 376,860 | 4,285 | 2,700 | 6,049,044 | 66,275 |
| | Madras | 16,662,840 | 236,320 | 944,914 | 15,251 | 9,365,004 | 144,860 | 866,712 | 11,150 | 1,120,516 | 29,303 | | | | | 1,000,841 | 14,397 | 9,425 | 21,112,750 | 812,437 |
| | Bombay | 241,193,027 | 3,650,395 | | | | | | | 58,152,248 | 1,270,333 | 501,690 | 9,201 | | | 7,768,180 | 157,722 | 142,318 | 318,791,775 | 5,268,912 |
| | ALL INDIA | 258,079,335 | 3,889,159 | 2,487,958 | 36,664 | 10,150,513 | 155,864 | 866,712 | 11,150 | 64,721,480 | 1,359,232 | 501,690 | 9,201 | | | 9,145,281 | 176,354 | 154,443 | 345,953,569 | 5,637,624 |
| 1860-61 | Bengal† | 13,771,236 | 186,000 | 1,086,212 | 11,211 | 3,951,752 | 58,246 | | | 5,792,215 | 71,359 | | | | | 1,056,444 | 14,811 | 9,668 | 21,666,107 | 74,281 |
| | Madras | | | | | | | | | | | | | | | | | | | 292,906 |
| | Bombay | 279,271,662 | 5,380,477 | 1,429,288 | 28,357 | | | | | 66,144,785 | 1,443,873 | 588,056 | 9,866 | 3,602,256 | 71,906 | 3,602,256 | 71,906 | 158,477 | 854,987,799 | 6,972,725 |
| | ALL INDIA | | 5,546,532 | 2,515,500 | 39,558 | 3,951,752 | 58,325 | | 93 | 66,144,785 | 1,443,873 | 588,056 | 9,866 | | | | 89,362 | | | 7,339,862 |

* The Indian rupee is here converted at the exchange of two shillings.
† Returns of cotton for 1861 not yet to hand.

2694. [125] Clean cotton, Piplee. REV. G. TAYLOR.

Both this and the preceding sample are from New Orleans seed.

2695. [82] Cotton bolls, Cuttack.

2696. [83] Do.                Do.
> Both yellow and white upland.

2697. [102] Clean, Cuttack.
> Yellow lowland cotton.

2698. [122] Kupass, Cuttack.
> Yellow upland cotton.

2699. [11] Kupass, Cuttack.
> Yellow lowland cotton.

2700. [119] Clean cotton, Cuttack. GO-PUL PATUR.

> Yellow lowland cotton.

2701. [123] Kupass, Autgar, Cuttack.
> Yellow and white upland cotton.

2702. [120] Clean cotton, Dhenkanal.
> Yellow upland cotton.

2703. [126] Clean cotton, Dhenkanal.
> White upland cotton.

2704. [103] Clean cotton, Sumbulpore.

2705. [104] Do.                Do.

2706. [124] Kupass,            Do.
> Upland cotton.

The following remarks on the cottons of the above district are by MR. W. C. LACEY, Secretary to the CUTTACK COMMITTEE:—

The first sample of the raw material was presented to the Committee of Cuttack by the REV. GEO. TAYLOR, of Piplee, a missionary in connection with the General Baptist Mission Society, who has for some time persevered in an experiment on a small scale to grow, and thus encouraged others to grow, this useful staple. His plants are reared from American New Orleans seed, and yield, as calculated on his data, at the rate of 300 lbs. per English acre, at an outlay of about 21s. per acre. On a light sandy soil the plants require manure and irrigation only in the hot months, or from March to June. The others are specimens of the good cotton grown in Sumbulpore, from which district a quantity of cotton is annually brought down the Mahanuddy.

In settled parts, the crop is for the most part a stunted, weakly, annual one, frequently grown in the same beds with other crops, and therefore yielding a poor return. More cotton is raised in the Sumbulpore district comparatively than elsewhere in the province, and three-fourths of the produce is said to be exported to Cuttack and Calcutta.

The varieties from Dhenkanal may be described as two highland or upland varieties; the one called the Daloona—a name given probably because the plants throw out numerous branches and grow to a great height, some specimens being 12 feet;—the second kind of upland is what is called the yellow, from the colour of the flowers; —the flower of the Daloona being white. The third variety may be called the lowland, and is known locally as the 'Keda.' They would all appear to be indigenous.

The upland varieties, which are most extensively cultivated, are grown more or less all over the Hill States, and, in short, wherever the single condition of a virgin forest soil exists. They are grown generally in the Sumbulpore district and its dependencies, throughout the Tributary States, and share the cultivation with the overland variety in Dhenkanal and Khoordah. In different localities there are variations in the mode adopted in regard to the sowing of the seed and the after-tending of the plants. But, as a rule, the trouble taken with the crop is almost nominal, the vigour and richness of the new soil being looked to to compensate for the absence of these auxiliary efforts of the planter, which are absolutely necessary in the case of the lowland cotton on soil which every year is called on to yield one or two crops of one kind or another.

The ground selected for the cultivation of uplands is then, in the first instance, a patch of jungle. The site is elevated; the soil may be a dark-looking mould, a light clay, or a loose reddish gravelly one. The jungle is cut down, all the brushwood cleared, heaped, and burnt on the spot. But much trouble is not taken in the clearing, and the stems and roots of the larger trees are left in the ground. The ground thus cleared then receives a superficial ploughing and is weeded; at all events this is the case in Sumbulpore, in Khoordah, and Dhenkanal generally, so that the plants shoot and grow and arrive at maturity through the rainy months.

In Sumbulpore a selection is made from the following seeds, dwarf paddy, Sooa, *Panicum italicum, Eleusine corocana,* Bajra, a species of pulse, castor oil, melons, pumpkins, and others. Two, three, or more of these kinds of seed are, with the cotton seed, scattered broad-cast over the ground which has been prepared as before described. No artificial irrigation is of course applied. The edible seed crops spring up and mature rapidly, and in the third or fourth month are ripe, and gathered successively as they ripen. After they have been taken off the ground, say the castor-oil plants and the cotton remain. These then receive a little attention; the ground is weeded and turned about, and the luxuriant masses of light green foliage, which every stump in the clearance has meantime put forth, are all cut away. After this the two crops thrive rapidly together. In January and February the cotton plants yield the first picking, and a month after, the castor-oil seed is ripe, and the plants are plucked and removed, leaving the cotton alone. If the variety of cotton be the Daloona or tall cotton, the plants last for two or even three years, and yield three pickings annually, and reach a height of 9, 10, 11, or even 12 feet. With the yellow upland, it is not so generally the practice of sowing many other crops. In Khoordah, for example, it is principally cultivated to the exclusion, to a certain extent, of the taller white flower cotton. Its yield is considered more abundant, the plants average a height of 6 and 7 feet, but are treated as annuals, and except when cultivated along with the white, in the newest soil and in cool sheltered positions, die off, or are abandoned after the first year.

The yield of the upland so cultivated may be said, from the imperfect information afforded, to average about 50 lbs. of raw seed cotton per local beegah, and taking the Khoordah beegah, three of which are the equivalent of an English acre, 40 lbs. of clean good cotton (seedless) per English acre. Land of the character indicated is abundant in the province, the whole of the woodland estate of Ungool, which is Government property, being well adapted to the growing of upland cottons. A similar remark may be made of Khoordah. The extensive tract of country called the Gurjato, or Hill States, which, for the most part, are covered with luxuriant forest jungle, are also well suited to the same purpose. No opinion can here be ventured as to the quality of the staple produced in these tracts, of which specimens are submitted: it is probably short and poor, while the yield is undoubtedly small. But these are faults which may arise more from the rude and negligent treatment of the plant, than from any other cause.

The seed used throughout the district for lowland cotton is procured from Khoordah and Dhenkanal, it being alleged that none other will germinate in the lowland districts. It undergoes the following preparatory processes, before being placed in the ground. It is placed in a pot, and soaked in dung and water for a night, and then dried by exposure to the sun on the following day. It is afterwards laid on straw contained in an earthen vessel covered over with castor-oil leaves and placed near a fire. So soon as the seed splits and shoots it is carried away and planted; after two leaves have sprouted water is applied to the plant at intervals of two, three, and four days.

This kind is planted, for it cannot be said to be sown in ridges, the intervening channels being necessary for irrigation, which in this case is indispensable, and must be abundant and frequent. November and December are the usual months for the planting. The plants are annual, and attain a height of 4 to 5 or sometimes 6 feet. The cold weather showers falling occasionally in December, January, and February, favour the plants, and when plentiful, constitute a good season. The pickings are obtained continuously in April, May, and June; in the latter month all the bolls are picked off the plants, and open on exposure to the sun. The same plan is adopted with the uplands, when an anticipated down-pour is likely to interfere with the natural developement of the mature cotton pods on the plants at the time. After the month of June, the lowland cotton plants are plucked up and the land cleared for a pulse crop.

The yield of this variety may be said to be 200 lbs. of raw cotton per acre, giving in the usual proportions about 50 lbs. of clean cotton per acre.

The proportion of wool to seed and loss for all the varieties is very uniformly stated to be as 1 to 4, that is out of 4 lbs. of raw cotton there is obtained 1 lb. of good clean staple.

### Cotton from Chittagong.

2707. [26] Kupass, Chur Burrea, Backergunge. SHAIK UKBAR, cultivator.

2708. [44] Clean cotton. Do.     Do.

MR. C. H. HARVEY reports that this is produced from native seed. It is sown from the middle of October to the beginning of November, and is ready for picking in May. It requires no irrigation.

2709. [182] Kupass, Chittagong.

2710. [181] Do. yellow or Nankin variety, Chittagong.

2711. [246] Do. Bengal. MANCHESTER COTTON SUPPLY ASSOCIATION, *Liverpool Market.*

### Cotton from Burdwan.

2712. [24] Kupass, uncleaned or seed-cotton, Cutwa.

2713. [23] Clean, Do.

2714. [62] Kupass, Bood Bood.

2715. [63] Clean, Do.

2716. [70] Kupass, Poobthul.

2717. [71] Clean, Do.

2718. [72] Kupass, Do.

2719. [138] Kupass, Mungulkote.

2720. [137] Clean, Do.

2721. [53] Kupass, Salamabad.

2722. [54] Clean, Do.

2723. [76] Kupass, Burdwan.

2724. [74] Clean, Do.

2725. [52] Do. Do.

MR. TWEEDIE, Secretary of the Local Committee at *Burdwan,* supplies the following information:—

The seed employed is indigenous, and the soil itself light. The height of the plant is from four to seven feet, and the approximate yield of clean cotton per acre six maunds seed-cotton giving one fourth of clean.

The soil is ploughed four or five times before the seed is put into the ground. For three or four days previous to sowing, the seed is kept in water, and is taken out on the day before it has to be sown. It is then mixed with ashes and cow-dung, and in this state is scattered over the ground, which is again ploughed. Some cultivators, however, put four or five seeds in small holes at the interval of about 1½ cubit. No irrigation is employed just after the seed is sown. When in the months of January and February the plants rise half a cubit high, they are watered. This practice is not followed in the species of cotton called 'Nurma,' which is cultivated in June. No irrigation is required for 'Nurma' cotton, as it is a rainy season plant.

The picking of the first-mentioned variety of cotton, called in the native dialect, 'Wesbee,' is commenced in April and finished in June and July. It requires watering three or four times in the course of its growth. The pods of 'Nurma' cotton are picked in the months of November and December.

### Cotton from Bancoorah.

2726. [92] Cleaned,     Gungajulghatee. RAINE HAZAREE, cultivator.

2727. [90] Kupass, Bishenpore. NUSSER DEY, cultivator.

2728. [93] Cleaned,     Do.     Do.

2729. [94] Do. Sonamookhey. KENARAIN DUTT, cultivator.

The soil is sandy, but mixed with clay. The cotton is grown from native seed. Its average height is 3 feet, and produces about 1 maund 35 seers of clean cotton, and about 3 maunds and 30 seers of seed. MR. WELLS of *Bancoorah,* states:— 'After ploughing the land three times before planting, water is sprinkled on the seed, and it is mixed with cow-dung and covered with earth, and allowed to remain two or three days in this state, and then planted out. It is irrigated three or four times from the time it is planted till bolls show themselves. Picking generally commences in April and lasts to the middle of July.

2730. [91] Cleaned, Jail Garden, Bancoorah. J. ANDREWS, Esq.

2731. [3] Cotton seed, Do.     Do.

The seed in this case is Egyptian: the soil a damp argillaceous sand. Only 1½ acre of land was planted, and yielded 3 maunds * of cotton, worth about 9 maunds of seed. Mr. W. S. Wells adds: — 'The soil was irrigated but little, as the unusually heavy rains this year caused the trees to shoot up somewhat quickly; consequently the production has been small and untimely, and much of the strength of the plant wasted in height and leaves.'

The sample now sent is from the second picking; the bolls commenced ripening about December, and it was then picked by the prisoners.

### Cotton from Midnapore.

2732. [117] Kupass.

2733. [118] Cleaned, Midnapore. Joy- sing & Bagroy Sootol, of *Hateamohun;* Seedookotal, of *Goorma;* Doorgaram Sing, of *Ekoor;* and Pooroo Sing, of *Moorakatee,* cultivators.

The cotton is indigenous, and grows from 1 to 5 feet high. Mr. R. V. Cockerell states that the bolls are picked from October to middle of January. The sample was obtained from the second picking. As the rains were very heavy, no irrigation was employed.

2734. [101] Kupass, Midnapore.

### Cotton from Bogra.

2735. [66] Cleaned, village of Deogong. Doorgakanth Hore, cultivator.

The seed is indigenous. The height of the plant from 3 to 4½ feet. This sample, says Mr. T. P. Larkins, was cultivated in the beginning of April, and is called the *Chaugtah* cotton. The month of April is the proper time for the cultivation of such cotton, but there is another sort of cotton called the *Bureâttee*, which is cultivated in October. Before sowing the seed the land is repeatedly ploughed and watered.

The average number of bolls grown on both plants is 150 or upwards, according to the soil. The picking of the Chaugtah cotton commences in the beginning of September. The sample sent was obtained from the last picking, which took place in November last.

The approximate quantity of seed required to cultivate a beegah of land with cotton is 10 or 12 seers, and the average produce of cotton per beegah will be about 1½ maund, which, being cleaned, would yield 20 seers of clean cotton.

There being no demand for country cotton in the market, the cultivation is very indifferent at present. If proper encouragement was held out, a vast extent of land could be cultivated in this district.

### Cotton from Pubna.

2736. [51] Cleaned, Pubna.

### Cotton from Monghyr.

2737. [1] Cotton bolls, Mouzah Hussun- gunge. Shaik Yar Ally Ruhman Khan, cultivator.

2738. [34] Kupass.

The seed was obtained from Tirhoot. The plant is from 4½ to 6 feet high, and affords 1 maund 10 seers per

acre. Mr. W. H. Henderson writes: — 'The seed is sown in June, after the soil has been well ploughed, and is scattered broadcast with Indian corn. In April the first picking takes place, but little is obtained; in the second, more; and yet more after several pickings. Cotton is not generally cultivated in the district.'

2739. [35] Cotton bolls, Fort Garden. W. H. Henderson, Esq.

2740. [36] Kupass, Do. Do.

2741. [37] Do. Do. Do.

The seed is acclimatised New Orleans. The cotton was obtained from the first picking. The plants grew to the height of 5 feet, and afforded 1 maund 20 seers of clean cotton (to the acre?)

2742. [19] Kupass.

2743. [17] Do.

2744. [38] Do.

### Cotton from Behar.

2745. [56] Kupass, Mouzah Klinjoor, Nurbut. Bujjoo Sing, Teka, Mabtoo, Ghun- sham, Muktoo, and Robee Muktoo, cul- tivators.

2746. [55] Cleaned, Mouzah Klinjoor, Nurbut. Bujjoo Sing, Teka, Mabtoo, Ghun- sham, Muktoo, and Robee Muktoo, cul- tivators.

The seed was native and the soil sandy; the average height of the plant was 4 feet. Mr. J. S. Drummond, of *Behar,* remarks: — 'The field in which the cotton was grown was twice ploughed, once in the month of June and again immediately prior to the sowing. Previous to sowing, the seed is allowed to soak in water and saltpetre- earth for about a couple of hours.

'The date of picking could not be accurately ascertained, but the sample sent was gathered about the end of May, or the commencement of June. Judging from its appear- ance, it was probably gathered at a late picking, as no considerable deterioration in the article is said to take place up to the sixth picking: the sample sent is very poor.

'The seed is separated from the cotton by a primitive machine of the description of the model forwarded.

'Irrigation takes place four times; once about a fortnight after the sowing, and afterwards at intervals of twenty days or a month, as the land may require it, and the land is always manured with ashes of cowdung each time previous to the irrigation.'

### Cotton from Sarun.

2747. [73] Kupass 'Bhogla' cotton, Ben- doosaree, Barrah. Chutter Dharee, Mubto, and Ramunnograh Tewary, cultivators.

2748. [75] Cleaned, Do. Do.

2749. [132] Kupass 'Boochree' cotton, Dewreah, Burrye. Gudharee Dooby, culti- vator.

2750. [131] Cleaned, Do. Do.

The seed is indigenous. The Bhogla cotton, according

---

* The weight avoird. of this maund is not stated. At the usual rate of 82 lbs. the above would give 164 lbs. to the acre.

to MR. ROSE of *Sarum*, is superior in quality, and is grown in small quantities on all sorts of land, excepting the low land. The fields are ploughed three or four times according as the soil is soft or hard, and weeded after the plants have germinated. The date of picking is the beginning of May. The above remarks are also applicable to the Boochree cotton, but this cotton is inferior in quality.

### Cotton from Patna.

2751. [12] Kupass, Patna. BABOO GOOREE SHUNKER, cultivator.

2752. [18] Cleaned, Do. Do.

The seed is indigenous: the soil a black loam. Irrigation is employed. The sample is last year's. — E. L. LATOUR, Esq., Collector, *Patna.*

### Cotton from Shahabad.

2753. [95] Cotton bolls, Arrah. MR. MYLNE, on the estate of H. BURROWS, Esq.

2754. [97] Do. Do. Do.

2755. [114] Do. Do. Do.

2756. [96] Kupass, Do. Do.

2757. [98] Do. and cleaned, Do. Do.

2758. [115] Cleaned, Arrah. MR. MYLNE, on the estate of H. BURROWS, Esq.

The foregoing are from Egyptian seed.

2759. [99] Cotton bolls. Do. Do.

2760. [·100] Kupass, Do. Do.

2761. [116] Cleaned, Do. Do.

The above are from New Orleans seed.

2762. [13] Sample of Nankin cotton, kupass, with yarn and cloth of do., Chynepore. R. W. BINGHAM, Esq.

### Cotton from Tirhoot.

2763. [42] Kupass and bolls, Nanpore. CHOWDREE ROODERPERSHAD, cultivator.

2764. [43] Cleaned, Do. Do.

Native seed yielding 1 maund and 28½ seers per acre. According to MR. T. B. LANE, the cotton produced in Tirhoot is of the kinds called Bhojra, Bhogla, and Kooktee, but as the two former do not ripen till April or May, they have not been forwarded. The Kooktee ripened in September 1861. The fabric manufactured from this cotton is not white, but of darkish colour: the white is alone produced from the Bhojra and Bhogla kinds.

### Cotton from Assam, Cachar, Silhet, the Garrow Hills, and Darjeeling.

#### *From Assam.*

2765. [79] Kupass, Assam. CAPTAIN MORTON.

2766. [78] Kupass, Kamroop, Assam. CAPT. LAMB.

'The produce might be greatly increased, were there sufficient demand.'

2767. [81] Clean cotton, Mattock, Assam. W. G. WAGENTRIEBER, Esq.

2768. [80] Do. Assam. LIEUT. PHAIRE.

2769. [67] Kupass, Gowhatty, Assam. REV. R. BLAND.

Grown in Mr. Bland's garden.

2770. [77] Do. Luckimpore, Assam. MAJOR BIVAR.

2771. [16] Cotton bolls, Assam.

#### *From Cachar.*

2772. [49] Kupass, Kookee, on southern slopes of the Burail Mountain to north of the river Barak, Cachar. CAPT. R. STEWART.

2773. [50] Cleaned cotton, Kookee, Cachar. CAPT. R. STEWART.

2774. [68] Kupass, Naga, Do. Do.

2775. [69] Cleaned cotton, Do., Do. Do.

CAPT. R. STEWART, Superintendent, *Cachar*, remarks as follows upon the soil of Cachar and the cultivation of cotton:—

'The soil upon which the cotton plant in Cachar is grown consists of a rich red clay, considerably mixed with sand, which forms the soil of the principal hills in the district, and also of the small ranges of hillocks that run through it. The cultivation lies on the slopes of these hills and mountains, such lands being never inundated, although they are wonderfully retentive of moisture. The Hill tribes, to whom the cultivation is limited, raise all that is necessary for their subsistence on the same fields which produce cotton. The same hills and slopes are now in great request for the cultivation of the tea plant, the soil being peculiarly adapted for its growth; when left to nature the soil is such as to produce the most dense jungle, and in cultivation it is fertile enough for any purpose.

'The soil of the plains of Cachar is a rich alluvial one, formed from the washings of the mountains which surround it on three sides.

'The climate is a very temperate one for Bengal, the thermometer in the shade rarely, if ever, rising above 93° or 94° Fahrenheit in the hottest weather of July and August.

'The seed produced upon the land is the same that is again sown upon it. Several endeavours have been made to introduce Mauritius and Sea-Island cotton, and other varieties, but without avail. These varieties do not appear to suit the primitive mode of cultivation to which they are subjected.

'The mode of cultivation is a peculiar one, and requires description. The cotton-bearing tracts forming the slopes of the hills in the district are, in their virgin state, covered either with glorious timber forests or dense jungles of bamboo. The latter, being more easily cut, is considered better for the cultivation, but the former is also frequently cleared for the purpose. Early in the cold season large parties of the cultivators, the wild tribes of Kookees and Nagahs, proceed to the jungles in the vicinity of their village, and selecting good patches commence felling the forest. The bamboos and small trees are cut off about two feet from the ground, and the stumps allowed to remain in the soil. The larger trees are merely ringed and allowed to die away. When a sufficient space has been cleared, the felled bamboos and jungles are left to rot on the ground, and the effect of one or two showers at intervals, coupled with the continued dryness of the cold season, renders them by the months of March and April as inflammable as jin cotton. Towards the end of the cold season these fields of cut bamboos and jungles, sometimes embracing the whole of a hill, at other times stretching along the faces of long ridges and valleys, are set on fire in various places. Nothing can exceed the fierceness of the conflagration, or the glorious effect produced by such large masses of flame roaring and lapping the hills on all sides, and the enormous volumes of smoke that are emitted and hover like clouds in the air. The conflagration is over in a few hours, and leaves on the ground a coating of ashes about an inch or two in thickness, and this is the only manure necessary to make these hills yield fertile crops of almost every kind. By means of a small hoe, or kodalee, the soil lying below the ashes is mixed with them in the places between the stumps of the felled trees and bamboos, which are still left to cumber the ground. Nor is the immunity enjoyed by these stumps the effect of indolence, or a desire to save labour at the expense of the crop on the part of the cultivators, but, on the contrary, an established custom, which experience has forced them to adopt, for the roads and stumps serve in a great measure to prevent the loose soil being washed away from the faces of the hills, and furthermore facilitate the fresh growth of the jungle when cultivation on the spot is abandoned. The soil being thus prepared, the seeds are dropped in, nor is care taken to allot the different vegetables different compartments, but paddy, sugar-cane, tobacco, cotton, and cucumbers are found growing on the same beds. The harvest is reaped in September and October, sometimes as late as December, and the same ground is again made in many instances to yield for another year or two, after which it is abandoned and reclaimed by its pristine jungle. In seven years' time, if bamboo jungle, it is again fit for cultivation, but not for twenty or thirty if tree forests have to be cleared.

'The cotton seeds, together with others, are put in in March and April; they are planted irregularly, but never closer than 3 or 4 feet apart. The whole cultivation is weeded three or four times during the rains. The cotton flowers in July and August, the picking commences in September, and is continued till December. In such a system of cultivation of course it is impossible to obtain correct statistics as to the weight produced per acre, or the amount of labour required. Suffice to say, that although carried on on a very small scale, it is the most profitable cultivation practised by the Hill tribes, for not only does it supply them with their own scanty clothing, but it is almost their sole article of barter with the people on the plains for salt, iron, gunpowder, cattle, and ornaments.

'No manure, save the ashes of the jungle, is ever used. When the cotton seed is returned to the same soil, the stalks of the last crop are again burned on the surface. No irrigation is ever required.'

### From Silhet.

2776. [48] Kupass, Silhet. F. SMITH, Esq.

2777. [47] Cleaned cotton, Do. Do.

The seeds are of last year's plants. The cotton was grown on a range of small hills.

### From Garrow Hills.

2778. [29] Kupass, Goja and Dobunba, Garrow, Bhalukmara Hill, Pergunnah Soosoong, Garrow Hills. A. SMITH, Esq.

2779. [64] Kupass and bolls, Goja and Dobunba, Garrow, Bhalukmara Hill, Pergunnah Soosoong, Garrow Hills. A. SMITH, Esq.

The seed was originally brought from Gowalparsh, and the plant is from 3 to 6 feet high. MR. A. SMITH, Officiating Collector of *Mymensing*, reports:—

'New ground is selected every year about the end of February; the jungle is burnt down to the ground, which is covered with the ashes of the burnt jungle; holes are dug at sufficient intervals, and two or three seeds thrown into each hole along with the seeds of other vegetables, such as pumpkins, and covered.

'The picking goes on, as bolls get ripe, from November to January.

'The specimens sent are from the lower Hills, but a large quantity is grown beyond the limits, to which the natives of the plains dare not go, by the uncivilised Hill tribes, who bring it for sale to the hâts, on the British frontier; no irrigation is needed, there being pretty constant rain from May to October.'

### From Darjeeling.

2780. [15] Clean cotton, Darjeeling. CAPT. MAGSON.

From Egyptian seed.

### Cotton from Chota Nagpore.

2781. [40] Cotton bolls, Burbee and Bussureah, Hazareebaugh. C. E. BLECHYNDEN, Esq.

2782. [39] Seed cotton, Do. Do.

2783. [58] Clean cotton, Do. Do.

The above samples are from New Orleans seed.

2784. [57] Clean, Do. Do.

2785. [59] Do. Do. Do.

2786. [129] Do. Do. Do.

These samples are from Egyptian seed.

'The specimens from New Orleans seed,' writes MR. BLECHYNDEN, 'were received in June 1860 from the Manchester Cotton Supply Association, and were first grown at Chumparun, twelve miles farther west of this station: the seed there produced was sown here in 1861.

'The Egyptian seed was also received from the Manchester Supply Association in the beginning of July 1861.

'The New Orleans cotton plant was from 2 to 5 feet high—the Egyptian from 1 to 3. Both were grown upon a common black soil.

'Commenced picking New Orleans in November. Sample from first picking. Plants now covered with bolls and flowers. Egyptian from first picking. Plants with bolls and flowers still on.

'The season has been very much against the cultivation of cotton; in the latter part too much moisture, and cloudy weather, with occasional rain, every change of the moon. The cloudy weather has prevented the usual fall of dew, and deteriorated the quality of the cotton, the heat not being sufficient to bring the bolls to maturity, and make them burst. No irrigation has been given.

'The advanced period of the season in which the cultivation was commenced has also tended to make this season's results unfavourable, both as to quality and quantity. As the crop is only in the act of being gathered, the yield of clean cotton per acre has not been determined; the same cause has also prevented the sending of the quantity of sample of cleaned cotton as required.

'The other specimen of cotton was grown from indigenous seed; in this the second year of cultivation a marked improvement is observed. The plants are 7 to 8 feet high. It is worthy of remark that this cotton has changed its time of bearing, the cotton crop of the country having been gathered two months ago, whereas these plants are continuing to blossom and bear bolls, the same as the New Orleans and Egyptian.'

2787. [8] Clean cotton, Seetagurrah Plantation, Hazarebaugh. M. LEIBERT, Esq.

2788. [7] Bolls, Do. Do.

From New Orleans seed.

2789. [84] Clean, Seetagurrah Plantation, Hazarebaugh. M. LEIBERT, Esq.

From Sea Island cotton.

2790. [9] Clean, Seetagurrah Plantation, Hazarebaugh. M. LEIBERT, Esq.

From Egyptian seed.

2791. [88] Kupass, Sillee.

2792. [86] Kupass and bolls, Rahey.

2793. [87] Cleaned, Do.

2794. [111] Kupass, Boloamat.

2795. [106] Clean, Torepa?

2796. [108] Do. Peetoria?

2797. [109] Kupass, Leshgunge?

2798. [110] Clean, Munika.

2799. [112] Do. Chuttur.

2800. [113] Kupass, Lohardagga.

2801. [128] Clean, Burgurh.

2802. [130] Kupass, Tamar.

All these samples are grown from indigenous seed by various native cultivators, whose names have not been forwarded. CAPT. R. C. BIRCH supplies the following information. The plants are from 6 to 7 feet high, and the yield per acre 5 maunds, one third clean cotton and two-thirds seed.

The cotton is grown on the hills and high lands, where the soil is naturally dry, undulating, and well-drained.

The mode of planting is as follows: From the month of February the people burn their fields with wood procured from the jungles, and leave the ashes there until a shower or two falls in July, when the seeds are sown, and the soil is then slightly ploughed. On sprouting, in August, the fields are carefully weeded, which is continued till the plants are grown up. The shrubs flower in September, and the bolls are fully open in October. In some parts of Nagpore the field is tilled and manured with ashes and cow-dung before sowing. In Purgunnah Boondoo, besides the common kind, there are two other sorts of cotton, called 'Tureea' and 'Guteh.' The former is sown in October, and picked in April and May, the field being tilled ten or twelve times before sowing; the latter is sown in July, and grows in two years. Cotton is picked two or three times in April; the trees last from

three to four years, producing cotton every year, and they are 7½ feet high. This is grown by the poorest class on their own premises.

The time of picking, speaking generally, is the whole of November and December, excepting in Purgunnah Boondoo, where, as already stated above, the 'Tureea' and 'Guteh' are picked in the months of April and May.

In this country rain falls abundantly from the middle of June to the end of September, consequently no irrigation is required.

2803. [2] Cotton bolls, Maunbhoom. JADOO MANJEE and others, cultivators.

2804. [105] Clean cotton, Do. Do.

The cotton is grown in every part of the district on high lands from indigenous seed. The plants are from 4 to 5 feet high, and yield from 3 to 4 maunds per acre. CAPT. G. N. OAKES, who forwards the samples, says that this cotton is called 'Kherooa' in this district. The land is ploughed five or six times, and then the seed is sown. After the plants have grown a little, the land is kept carefully clean. The cotton is picked in October. It requires no irrigation. The samples were obtained from the third picking.

2805. [85] Cotton bolls, Jail Garden, Chybassa, Singbhoom. CAPT. R. C. BIRCH.

2806. [107] Kupass, Do. Do.

2807. [127] Clean cotton, Do. Do.

The cotton is indigenous to Singbhoom; it grows on gravelly soil to an average height of 3½ feet. MR. W. H. HAYES, Deputy Commissioner, states that the first picking commences in November. The plant is perennial, and bears well for three years; it requires little irrigation.

2808. [60] Kupass, Chota Nagpore.

Is the produce from the native crop, but the exact locality is not given.

2809. [89] Kupass, Chota Nagpore.

Grown from American seed. Grown in the German Lutheran Mission.

## Cotton from the North-Western Provinces, Bundelkhund, Gwalior, and Ulwar.

2810. [46] Kupass and cleaned cotton, Kool Puhar, in the district of Humeerpore.

2811. [22] Kupass, Poongree, Banda District. H. W. DASHWOOD, Esq.

From a long and elaborate note of MR. DASHWOOD's, we extract the following statement regarding the cultivation in Banda:—

'Cotton is a precarious crop; its success or failure depends entirely on the rain-fall. It is injured by drought, but its chief risk is from excess of rain. This year considerable damage has been done to the cotton crops by the severity of the wet season, and especially by the heavy fall of September. Cotton being entirely a "Khurreef" or rain crop in this district, it cannot be sown in the low rich lands, which bear the best spring crops, and which are generally flooded in the wet season. It is sown generally on high ground, on the sides of ravines for instance, or on elevated spots where the water cannot lodge. In its best season it is more remunerative than ordinary crops; but it is uncertain and precarious, and in the long run it is not sufficiently remunerative to induce ryots to cultivate more than a small portion of their land with it. If a man has 10 or 15 beegahs of land, he will almost certainly cultivate two or three with cotton, but he will not cultivate more. If the price were to rise, the cultivation would no doubt be extended: *and if it rose enough to make the cultivation of cotton considerably more profitable than that of other crops, it might be extended almost indefinitely.*

'In this district the same land is never cultivated with cotton in two consecutive years. After a field has borne cotton, two, and sometimes three, years are allowed to intervene before it is again sown with that crop. Entirely new lands are said to be more fruitful than any. Fresh cotton is always sown. The same plants are never allowed to remain standing for a second crop, as it is the common opinion of the people of the district that the plants produce nothing in the second year. The sowing takes place in the month of July, commonly after the first fall of rain. The ground is generally manured, if the ryot can afford it. It is ploughed only once, and never watered. As a general rule, it is weeded three or four times; occasionally even five and six times. The weeding takes place during August or September and the beginning of October. About the end of October the crop is generally matured, and the gathering takes place between this and the end of December.'

2812. [27] Clean cotton, Raepore.

2813. [28] Kupass (second quality), Do.

2814. [41] Clean cotton, Jaloun, Bundelkhund. CENTRAL COMMITTEE, *Allahabad.*

2815. [61] Do. and bolls, Do. Do.

2816. [30] Clean cotton, Sutwaree, in the native state of Chirkaree, Bundelkhund.

2817. [31] Do. Do.

2818. [32] Do. Do.

The last two are from American seed.

2819. [45] Clean cotton, Bundelkhund.

From Egyptian seed.

2820. [140] Kupass, Gwalior. H. H. the MAHARAJAH of GWALIOR.

2821. [141] Clean cotton, Do. Do.

2822. [14] Do. Ulwar. H. H. the MAHARAJAH of ULWAR.

## Cotton from the Punjab and Sind.

2823. [139] Clean cotton, Dehra Ismael Khan. LAHORE COMMITTEE.

Grown from New Orleans seed.

2824. [21] Clean cotton, first sort, Jung District. Do.

2825. [20] Do. second sort, Do. Do.

2826. [135] Kupass, Umballah. Jung District. LAHORE COMMITTEE.

2827. [136] Clean cotton, Do. Do.

2828. [25] Do. Hoshiarpore. Do.

2829. [134] Do. Mooltan. Do.

Value per maund of 80 lbs., 13R. in Mooltan. 14,764 maunds annually produced in the district.

2830. [10] Clean cotton, Leia District. Do.

The following note is by the CENTRAL COMMITTEE, *Lahore:*—

'Cotton is grown largely in the Punjab, chiefly for home consumption; but the soil is generally not so suited for cotton as the basaltic soils of Central India. Still there is no doubt, from experiments which have been made, that the Punjab is capable of producing cotton suitable for the English market. But efforts to secure it must not be confined to making speeches at Manchester; the only practical plan is to depute persons of skill and capital to direct the people in the best mode of preparing the cotton, and to buy up the produce on the spot. Specimens of cotton from the principal cotton-growing districts of the Punjab have been included in the collection, and also a specimen of cotton grown from American seed in the Dehra Ismael Khan district. The localities best suited for the growth of cotton are the submontane districts of Umballah, Hoshiarpore, Gujerat, and Peshawur; but with irrigation it might be produced almost anywhere. From official returns published in the *Punjab Gazette* of the 28th August 1861, it would appear that altogether about 467,513 acres are under cotton cultivation in the Punjab and its dependencies. The average produce per acre of cleaned cotton varies from 50 to 150 lbs.; its price varies from 2*d.* to 4⅓*d.* per lb.; and the whole cotton produced per annum amounted to 20,000 tons, of which not more than 3,500 tons were exported. The time of sowing varies from February in the south, to the middle of June in some of the northern districts. The flowering commences, according to locality, between August and December; the picking following about a month after the flowering, is continued at intervals for two months.'

2831. [133] Clean cotton, Punjab. Messrs. SMITH, FLEMING, & Co.

## Cotton from Sind.

2832. [142] Kupass, Shikarpoor. COLLECTOR of SHIKARPOOR.

2833. [143] Clean cotton, Jacobabad.

2834. [144] Kupass, Do.

2835. [147] Do. Do.

2836. [149] Clean cotton, Khyrpoor. H. H. MEER ALI MOORAD.

2837. [145] Partly cleaned cotton, Hyderabad.

2838. [238] Clean cotton, Do.

2839. [146] Do. Seebee.

2840. [148] Kupass, Do.

2841. [150] Clean cotton, Do.

## Cotton from Western India and Berar.

2842. [156] Nurma cotton, clean, Ahmedabad. BOMBAY GOVERNMENT.

2843. [159] Lalia cotton, do. Do.

2844. [154] Do. Kupass, Dundooka. Do.

2845. [155] Lalia cotton, clean, Do. Do.

2846. [160] Wagaria cotton, Kupass, Do. Do.

2847. [153] Do. clean, Do. Do.

2848. [151] Do. Kutch. H.H. the RAO of KUTCH.

2849. [152] Do. Kutch. Messrs. SMITH, FLEMING, & Co.

2850. [247] Do. Mangrole, Kattywar. MANCHESTER COTTON SUPPLY ASSOCIATION, *Liverpool Market.*

2851. [158] Dhollerah cotton, clean, Dhollerah, Guzerat. Messrs. SMITH, FLEMING, & Co.

2852. [243] Do. Dhollerah, Guzerat. MANCHESTER COTTON SUPPLY ASSOCIATION, *Liverpool Market.*

2853. [168] Do. Broach. Messrs. SMITH, FLEMING, & Co.

2854. [169] Do. Do. Do.

2855. [248] Do. Do. MANCHESTER COTTON SUPPLY ASSOCIATION, *Liverpool Market.*

2856. [240] Do. Kandeish. Do.

2857. [163] Cotton, cleaned, Poonah. MANCHESTER COTTON SUPPLY ASSOCIATION, *Liverpool Market.*

*From indigenous seed.*

2858. [162] Cotton, cleaned, Poonah. Do.

*From Egyptian seed.*

2859. [172] Cotton, cleaned, Belgaum. — HEARN, Esq.

*From Egyptian seed.*

2860. [157] Coompta cotton, clean, Dharwar. Messrs. SMITH, FLEMING, & Co.

2861. [170] Do. Coompta. Do.

2862. [167] Do., Kupass, Coompta. Do.

2863. [241] Coompta cotton, clean, Coompta. MANCHESTER COTTON SUPPLY ASSOCIATION, *Liverpool Market.*

2864. [244] Do. Dharwar. Do.

2865. [161] Kupass, Bombay. Messrs. SMITH, FLEMING, & Co.

2866. [165] Do. Do. Do.

2867. [166] Do. Do. Do.

2868. [180] Clean cotton, Do. Do.

2869. [171] Do. Do. Do.

2870. [173] Do. Do. Do.

2871. [174] Do. Do. Do.

2872. [175] Do. Do. Do.

2873. [176] Kupass, Do. Do.

2874. [178] Clean cotton, Do. Do.

2875. [179] Kupass, Do. Do.

2876. [177] Clean cotton, Do. Do.

The two last are from New Orleans seed.

## Cotton from Berar.

2877. [183] Clean cotton, Bolarum. DR. RIDDELL.

2878. [184] Do. Do. Do.

2879. [185] Carded cotton, Do. Do.

2880. [186] Clean cotton, Bolarum. DR. RIDDELL.

*From Mauritius seed.*

2881. [200] Kupass, Oomrawattee. Messrs. SMITH, FLEMING, & Co.

2882. [245] Do. MANCHESTER COTTON SUPPLY ASSOCIATION, *Liverpool Market.*

2883. [164] Clean cotton, Hingenghaut. Messrs. SMITH, FLEMING, & Co.

2884. [249] Do. MANCHESTER COTTON SUPPLY ASSOCIATION, *Liverpool Market.*

## Cotton from Madras.

2885. [207] Oopum cotton, Kupass, Salem. Messrs. FISHER & Co.

*From native seed.*

2886. [195] Bourbon cotton, first quality, cleaned, Salem. Messrs. FISHER & Co.

2887. [196] Do., second quality, cleaned, Do. Do.

2888. [194] Do. clean, Chingleput. DR. J. SHORTT.

*Uncultivated.*

2889. [201] Clean cotton, Do. Do.

*From Gossypium arboreum.*

2890. [203] Clean cotton, Chingleput. DR. J. SHORTT.

*From Egyptian seed.*

2891. [188] Clean cotton, Tinnevelly. Messrs. SMITH, FLEMING, & Co.

2892. [189] Do. Do. Do.

2893. [242] Do. Do. MANCHESTER COTTON SUPPLY ASSOCIATION, *Liverpool Market.*

2894. [191*a*] Kupass, Nellore. GOVERNMENT.

2895. [191*b*] Kupass and clean, Do. Do.

2896. [187] Clean cotton, Do. Messrs. SMITH, FLEMING, & Co.

2897. [197] Do. Madras. Messrs. SMITH, FLEMING, & Co.

This is from Bourbon seed.

2898. [206] Do. Western Madras. Do.

2899. [239] Do. Do. Do.

### Cotton from Mysore.

2900. [198] Clean cotton, Mysore. GOVERNMENT of MYSORE.

2901. [192] Cotton bolls, Do. Do.

*From New Orleans seed.*

2902. [193] Cotton bolls, Mysore. GOVERNMENT of MYSORE.

*From Egyptian seed.*

2903. [204] Kupass, Mysore. GOVERNMENT of MYSORE.

2904. [205] Clean cotton, Do. Do.

2905. [208] Kupass, Do. Do.

2906. [202] Clean cotton, Do. Do.

2907. [199] Clean cotton, Mysore. GOVERNMENT of MYSORE.

### Cotton from Arracan, Pegu, Tenasserim, and Straits Provinces.

2908. [214] Clean cotton, Arracan. Messrs. HALLIDAY, FOX, & Co.

*From Egyptian seed.*

2909. [216] Kupass, Akyab. LOCAL COMMITTEE.

2910. [215] Clean cotton, Do. Do.

2911. [217] Clean. Red or Nankin cotton, Do. Do.

Cotton is principally grown by the Hill tribes in this province; but little is brought down to Akyab or other markets. Price from 6 to 7 rupees per maund.

2912. [212] Kupass, Moulmein.

2913. [213] Do. Do.

2914. [210] Clean cotton, Pegu. LOCAL COMMITTEE, *Rangoon.*

The RANGOON COMMITTEE writes as follows:—
The area under cotton cultivation in Pegu in 1860–61 was 17,500 acres, and the estimated produce in cleaned cotton 2,116,300 lbs.
The general character of the native cotton is—fibre coarse, curly, harsh, and rather short, most tenaciously attached to the seed; but it is exceedingly strong, and in this respect lies its excellence.
Persevering efforts have been made to induce the Burmese to grow foreign cotton, but, as in the case of tobacco, without success. There are many millions of acres in Pegu now lying waste, where cotton can be grown of a quality far superior to any now known in the province. It is hoped that the recent offer by the Government of India of the fee-simple in waste lands, free of tax for ever, at the low rates of 5s. and 10s. an acre, will attract both European superintendence and capital to the mutual benefit of England and Pegu.

2915. [209] Clean cotton, Burmah. Messrs. HALLIDAY, FOX, & Co.

2916. [211] Do. Rangoon. Messrs. SMITH, FLEMING, & Co.

2917. [224] Do. Washington Estate, Province Wellesley, Penang. ALEXANDER HUTCHINSON, Esq.

This is raised from New Orleans seed.
This is a new estate, the proprietor and manager both being Americans. The latter is a cotton planter from the Southern States. The sample sent has been grown on the same principle of culture as cotton growers follow in the Southern States. It is from a yield the average of which gives 1425 lbs. per acre of Kupass or seed-cotton. It was planted on the 11th September 1861, came into bloom in the first week of December, and began to bear about the first week of January in the present year.
As there were at the time no gins on the estate, the cotton was separated by hand, and is consequently somewhat deteriorated by perspiration.

2918. [225] Pernambuco cotton, clean, Glugor Estate, Penang.

2919. [226] Sea-Island cotton, clean, Do.

2920. [221] Clean cotton, Singapore. COL. CAVANAGH.

2921. [223] Do.      Do.      Do.

2922. [218] Do.      Do.      Do.

2923. [219] Sea-Island cotton, clean, Do. G. H. BROWN, Esq.

2924. [220] Pernambuco cotton, clean, Do. Do.

2925. [222] Malacca cotton, clean, Do. J. BAUMGARTEN, Esq.

[*For Tabular Synopsis, see pages 134–7.*]

### Silk Cottons.

The seed pods of various genera of plants supply a material which, from its appearance, is called 'silk cotton.' It is deficient in strength, and difficult to spin, on account of the smoothness of the individual fibres. Some specimens of cloth manufactured from an admixture of cotton and the floss of the Ak (*Calotropis Hamiltonii*) are shown, and an interesting application of the material by itself is supplied by the rug (2942) exhibited and entered below.

There are two species of *Calotropis*—one the Mûdar (*Calotropis gigantea*), the other the Ak (*C. Hamiltonii*), which produce this floss in great abundance. One or other of these grow luxuriantly in all parts of the country; and should the material, as now expected *, prove of commercial value, it could be furnished at a cheap rate in large quantities. Hitherto its chief use has been for stuffing pillows &c. The sample (2926) is a portion from that referred to in the note below.

2926. [9386] Floss or 'silk cotton' from the Ak (*Calotropis Hamiltonii*), Agra. DR. W. WALKER.

---

* Attempts in this country to work this material by means of machinery have hitherto failed. At the suggestion, however, of Mr. Stuart Clark, Inspector-General of Prisons, North-Western Provinces, a considerable quantity of it was lately forwarded to my department by Dr. Walker of the Agra Jail, and a portion having been submitted to Messrs. Thresher & Glenny (who have for a long time been devoting attention to the subject), these gentlemen are at length enabled to report their ability to turn it to account, if obtainable here in a clean, good condition at 30*l.* per ton.—J. F. W.

DR. WALKER appends the following remarks:—
The charge of the down is merely that of the labour employed in collecting it, and the charges incurred in packing. It may be collected at about 1R. 8A. (3*s.*) per maund (82 lbs.). The plant is to be found in the greatest abundance everywhere, growing most luxuriantly in those dry sandy tracts where nothing else will flourish. The down ought to be collected in May and June, and its collection is spread at least over two months.

2927. [229] Floss or silk cotton from Simul tree (*Bombax malabaricum*), Chota Nagpore. GOVERNMENT.

2928. [235] Do. Mysore (Bangalore Division). GOVERNMENT of MYSORE.

2929. [231] Do. from *B. heptaphyllum*, Chingleput. DR. SHORTT.

2930. [233] Do. from *B. pentandrum*, Do. Do.

2931. [234] Do. from *Chynanchum*, Do. Do.

2932. [232] Do. from *Calotropis gigantea*, Do. Do.

2933. [236] Do. (*C. gigantea*), Madras. CAPT. J. PUCKLE.

2934. [230] Do. Sutwarree, Bundelkhund.

2935. [8137] Do.

2936. [228] Floss or silk cotton, uncleaned, Singapore. HON. COL. CAVANAGH.

2937. [237] Flowers of Lallang grass, Do. TAN KIM SING.

2938. [227] Floss or silk cotton, Do. Do.

---

The following manufactures, consisting chiefly of Mûdar floss, are, for convenience, exhibited in this class.

2939. [10304] Cloth, one part cotton and four parts Mûdar floss thread, Central Prison, Agra. DR. W. WALKER.

2940. [10304*a*] Do. one part cotton and one part Mûdar floss thread, Do. Do.

2941. [10304*b*] Cloth made entirely of Mûdar floss, Do. Do.

2942. [5191] Rug made of Mûdar floss, Shahpore Jail, Punjab.

# INTERNATIONAL EXHIBITION, 1862.

## INDIA COTTON: TABULAR SYNOPSIS.

| Number: Of Sample | Number: In Catalogue | Number: In Jury List and that of Cotton Supply Association | Place of Growth | Seed Employed | By whom Exhibited, Name of Cultivator, &c. | Length of Staple in Inches and Parts of an Inch: Min. (Decimal Fractions) | Max. (Decimal Fractions) | Mean: Decimal Fractions | Mean: Vulgar Fractions (approximate) | Price per Lb.: By Jury | Price per Lb.: By Commissioner, Cotton Supply Association | Remarks by Commissioner appointed by Cotton Supply Association to examine the various Cottons in the International Exhibition (1862) |
|---|---|---|---|---|---|---|---|---|---|---|---|---|
| | | | BENGAL | | | | | | | Pence | Pence | |
| 1 | [2697] | 119 | Cuttack . . . . . | Native or indigenous | Gov. of India, Gopul Patur, cultivator | ·80 | 1·10 | ·95 | 1 | 7¼ | 8½ | Short strong fibre |
| 2 | [2700] | 102 | Do. . . . . . . | Do. | Government . . . . . . | ·80 | 1·00 | ·90 | 9-10ths | .. | .. | Not worth growing |
| 3 | [2702] | 120 | Dhenkanal . . . . . | Do. | Do. . . . . | ·50 | ·85 | ·68 | 2-3rds | .. | 6 | Very red; worthless |
| 4 | [2703] | 126 | Do. . . . . . | Do. | Do. . . . . | ·65 | ·95 | ·80 | 4-5ths | 7¼ | 7; would have been worth 9 | Good; injured in gimming |
| 5 | [2704] | 103 | Sumbulpur . . . . | Do. | Do. . . . . | ·60 | ·90 | ·75 | 3-4ths | .. | .. | |
| 6 | [2705] | 104 | Do. . . . . . | Do. | Do. . . . . | ·65 | ·95 | ·80 | 4-5ths | 7¼ | 7 | Very white; short and tender |
| 7 | [2713] | 23 | Cutwa, Burdwan . . . | Do. | Do. . . . . | ·50 | ·80 | ·65 | 2-3rds | 7½ | 8 | Very clean; very short; very weak |
| 8 | [2715] | 63 | Bood Bood, Do. . . . | Do. | Do. . . . . | ·70 | ·90 | ·80 | 4-5ths | 7½ | .. | Short and coarse; spoiled; very badly ginned |
| 9 | [2717] | 71 | Poobthul, Do. . . . | Do. | Do. . . . . | ·70 | ·90 | ·80 | 4-5ths | 7¼ | 8 | Inferior; very short |
| 10 | [2720] | 137 | Mungulkote, Do. . . . | Do. | Do. . . . . | ·60 | ·90 | ·75 | 3-4ths | 7¾ | 8 to 9 | Very short, but very clean. |
| 11 | [2722] | 54 | Salamabad, Do. . . . | Do. | Do. . . . . | ·80 | 1·00 | ·90 | 9-10ths | 7¼ | 8 | Inferior; very short, but good colour |
| 12 | [2724] | 74 | Burdwan . . . . . | Do. | Do. . . . . | ·55 | ·85 | ·70 | 2-3rds | 7¾ | .. | |
| 13 | [2725] | 52 | Do. . . . . . | Do. | Do. . . . . | ·90 | 1·20 | 1·05 | 1 | 7½ | .. | Inferior and short |
| 14 | [2726] | 92 | Gungajulghatee, Bancoorah | Do. | Do. Raine Hazaree, cultivator . | ·70 | ·90 | ·80 | 4-5ths | 6½ | 7 to 8 | Very inferior, but clean |
| 15 | [2728] | 93 | Bishenpur Do. . | Do. | Do. Nusser Dey, Do. . . . | ·80 | 1·00 | ·90 | 9-10ths | 6¾ | 7 to 8 | Do. do. |
| 16 | [2729] | 94 | Sonamookhey Do. . | Do. | Do. Kenarain Dutt, Do. . . . | ·60 | ·90 | ·75 | 3-4ths | 7 | 7 to 8 | Do. do. |
| 17 | [2733] | 118 | Midnapur . . . . . | Do. | Do. Joysing and Bagroy Sootol, Seadookotal, Doorgaram Sing and Pooroo Sing, cultivators | ·65 | ·95 | ·80 | 4-5ths | 6½ | .. | |
| 18 | [2735] | 66 | Deogong, Bogra . . . | Do. | Do. . . . . | ·60 | ·90 | ·75 | 3-4ths | 8 | .. | Silky; spoiled in ginning |
| 19 | [2736] | 51 | Pubna . . . . . . | Do. | Do. . . . . | ·75 | ·95 | ·85 | 7-8ths | 7½ | 7½ to 8 | Inferior and short |
| 20 | [2746] | 55 | Mouzah Klinjoor, Nurbut, Behar | Do. | Do. Bujjoo Sing, Teka, Mabtoo Gunsham, Muktoo and Robee Muktoo, Do. | ·75 | ·90 | ·83 | 4-5ths | 8 | .. | Clean, but very short |
| 21 | [2748] | 75 | Bendoosaree, Barrah, Sarun | Do. | Do. Chatter Dharee, Mubto and Ramun nograh Tewary, Do. | ·70 | ·90 | ·80 | 4-5ths | 6¾ | .. | Ruined in cleaning; chopped to bits |
| 22 | [2750] | 131 | Dewreah, Burrye, Do. | Do. | Do. Gudharee Dooby | ·75 | 1·05 | ·90 | 9-10ths | .. | .. | Dirty and inferior |
| 23 | [2752] | 18 | Patna . . . . . . | Do. | Baboo Gooree Shunker . . . | ·50 | ·80 | ·65 | 2-3rds | 6½ | .. | Chopped to bits in ginning |
| 24 | [2764] | 43 | Nanpore, Tirhoot . . . | Do. | Chowdree Rooderpershad . . | ·50 | ·70 | ·60 | 3-5ths | 6¼ | .. | |
| 25 | [2768] | 80 | Assam . . . . . . | Do. | Lieut. Phaire . . . . . | ·60 | ·90 | ·75 | 3-4ths | 7¾ | 8 | Very short; good colour |
| 26 | [2767] | 81 | Mattock, Assam . . . | Do. | W. G. Wagentrieber, Esq. . . | ·60 | ·90 | ·75 | 3-4ths | 7 | 8 | Coarse, short, strong; bad colour |
| 27 | [2773] | 50 | Kookee, Cachar . . . | Do. | Capt. R. Stewart . . . . | ·60 | ·90 | ·75 | 3-4ths | 7½ | 8 to 9 | Good colour; naturally very short |
| 28 | [2775] | 69 | Naga, Do . . . | Do. | Do. . . . . . | ·70 | 1·00 | ·85 | 7-8ths | 8 | 8 to 9 | Very short, but white |
| 29 | [2777] | 47 | Silhet . . . . . . | Do. | F. Smith, Esq. . . . . . | ·50 | ·80 | ·65 | 2-3rds | 7¼ | 7 to 8 | Very short rough fibre; very white |
| 30 | [2793] | 87 | Rahey, Lohurdagga, Chota Nagpore | Do. | Native cultivators, through Capt. R. C. Birch | ·65 | ·85 | ·75 | 3-4ths | 6¾ | .. | White, but too short to be of any value |
| 31 | [2795] | 106 | Torepa (?), Do. | Do. | Do. Do. . . | ·70 | 1·00 | ·85 | 7-8ths | 6 | .. | |
| 32 | [2796] | 108 | Peetoria (?), Do. | Do. | Do. Do. . . | ·50 | ·85 | ·68 | 2-3rds | 7 | 6 or under | Poor and weak; dusty |
| 33 | [2798] | 110 | Munika, Do. | Do. | Do. Do. . . | ·75 | ·95 | ·85 | 7-8ths | 7¼ | 7 | White; clean; exceedingly short |
| 34 | [2799] | 112 | Chuttur, Do. | Do. | Do. Do. . . | ·60 | ·90 | ·75 | 3-4ths | 6¾ | 6 or less | Poor and weak; dusty |
| 35 | [2801] | 128 | Burgurh, Do. | Do. | Do. Do. . . | ·70 | 1·00 | ·85 | 7-8ths | .. | 5 to 6 | White; short; very weak |
| 36 | [2804] | 105 | Maunbhoom, Do. | Do. | Jadoo Manjee and others . . | ·80 | 1·10 | ·95 | 1 | 6 | .. | |
| 37 | [2807] | 127 | Jail Garden, Chybassa, Singbhoom, Do. | Do. | Capt. R. C. Birch . . . . | ·90 | 1·20 | 1·05 | 1 | 8 | 9 to 11 | Good white; somewhat over-ginned |
| | | | | | MEAN . . . . | ·66 | ·93 | ·80 | 4-5ths | | | |

| | | | | | | | | | | | | |
|---|---|---|---|---|---|---|---|---|---|---|---|---|
| | | | NORTH-WESTERN PROVINCES, &c. | | | | | | | | | |
| 38 | [2810] | 46 | Kool Puhar, Humeerpur | Do. | Government | ·65 | ·95 | ·80 | 4-5ths | 7½ | 7 to 8 | Well cleaned, but very short |
| 39 | [2812] | 27 | Jaepore | Do. | Do. | .. | .. | .. | .. | .. | 9; if well cleaned 11 | Silky staple; overgrown, yet tough; badly ginned: strong staple; bad colour |
| 40 | [2814] | 41 | Jaloun, Bundelkhund | Do. | Central Committee, Allahabad | ·80 | 1·10 | ·95 | 1 | 7 | .. | White; badly cut in ginning |
| 41 | [2815] | 61 | Do. Do. | Do. | Do. | ·60 | ·90 | ·75 | 3-4ths | 7¾ | .. | Short and coarse; spoilt by bad ginning |
| 42 | [2816] | 30 | Sutwaree, Chirkaree, Do. | Do. | . . . . . . | ·70 | ·95 | ·83 | 4-5ths | 7¾ | 8 | White; clean; short |
| 43 | [2818] | 32 | Do. Do. Do. | Do. | . . . . . . | ·50 | ·80 | ·65 | 2-3rds | 7 | .. | Clean; very tender; badly ginned. |
| 44 | [2821] | 141 | Gwalior | Do. | H. H. the Maharajah of Gwalior | .. | .. | .. | .. | .. | 9 | Silky; short; strong |
| 45 | [2822] | 14 | Ulwar | Do. | H. H. the Maharajah of Ulwar | ·75 | ·95 | ·85 | 7-8ths | 7½ | 6 | Poor; short; bad colour |
| | | | | | MEAN | ·66 | ·94 | ·80 | 4-5ths | | | |
| | | | PUNJAB AND SIND | | | | | | | | | |
| 46 | [2825] | 20 | Jung District, Punjab | Do. | Lahore Committee, Government | ·65 | ·95 | ·80 | 4-5ths | 6½ | 7 to 8 | Short, strong, and clean |
| 47 | [2824] | 21 | Do. Do. | Do. | Do. Do. | ·50 | ·80 | ·65 | 2-3rds | 6¾ | 8 to 9 | Do. do. |
| 48 | [2827] | 136 | Umballah, Do. | Do. | Do. Do. | ·70 | ·90 | ·80 | 4-5ths | 8 | .. | Chopped to bits in ginning |
| 49 | [2828] | 25 | Hoshiarpur, Do. | Do. | Do. Do. | ·70 | 1·00 | ·85 | 7-8ths | 7½ | 9 | Shell; short; strong; badly ginned |
| 50 | [2829] | 134 | Mooltan, Do. | Do. | Do. Do. | ·60 | ·90 | ·75 | 3-4ths | 8 | 8 | Very weak, but of good colour |
| 51 | [2830] | 10 | Leia District, Do. | Do. | Do. Do. | ·70 | ·90 | ·80 | 4-5ths | 8½ | 8 | White; clean |
| 52 | [2831] | 133 | Punjab | Do. | Messrs. Smith, Fleming, & Co. | ·90 | 1·20 | 1·05 | 1 | 8¾ | 8 | Very weak, but of good colour |
| 53 | [2833] | 143 | Jacobabad, Sind | Do. | . . . . . . | ·70 | 1·00 | ·85 | 7-8ths | 7½ | 5 to 6 | Very short; dusty; inferior, but white |
| 54 | [2836] | 149 | Khyrpur, Do. | Do. | H.H. Meer Ali Moorad | ·75 | 1·05 | ·90 | 9-10ths | 7 | 6 | Short; dusty; weak |
| 55 | [2837] | 145 | Hydrabad, Do. | Do. | . . . . . . | ·65 | ·95 | ·80 | 4-5ths | .. | .. | Dusty; very short staple; rotten fibre; naturally bad |
| 56 | [2838] | 238 | Do. Do. | Do. | . . . . . . | ·70 | ·95 | ·83 | 4-5ths | 8½ | .. | Shelly, weak; probably spoilt in ginning |
| 57 | [2839] | 146 | Seebee, Do. | Do. | . . . . . . | ·85 | 1·15 | 1·00 | 1 | 7½ | 8 | Shelly; white; weak |
| 58 | [2841] | 150 | Do. Do. | Do. | . . . . . . | ·90 | 1·20 | 1·05 | 1 | 9 | 11 to 12 | Good, strong, coarse, white, well cleaned; if indigenous, very good |
| | | | | | MEAN | ·71 | 1·00 | ·86 | 7-8ths | | | |
| | | | BOMBAY AND BERAR | | | | | | | | | |
| 59 | [2842] | 156 | Ahmedabad | Do. | Bombay Government | 1·00 | 1·30 | 1·15 | 1 1-7th | 10½ | 13 to 14 | Very good; well prepared. *Nurma cotton* |
| 60 | [2843] | 159 | Do. | Do. | Do. | ·90 | 1·10 | 1·00 | 1 | 9 | 9 | Badly ginned; short. *Lalia cotton* |
| 61 | [2845] | 155 | Do. | Do. | Do. | ·85 | 1·15 | 1·00 | 1 | 7½ | 9 | Silky staple; short; damaged in ginning. *Lalia cotton* |
| 62 | [2847] | 153 | Do. | Do. | Do. | 90 | 1·10 | 1·00 | 1 | 8½ | 8 to 9 | Spoiled in ginning. *Wagaria cotton* |
| 63 | [2848] | 151 | Kutch | Do. | H.H. the Rao of Kutch | ·95 | 1·15 | 1·05 | 1 | 7½ | 7 to 8 | Coarse, harsh, short fibre; weak; very clean |
| 64 | [2849] | 152 | Do. | Do. | Messrs. Smith, Fleming, & Co. | ·90 | 1·20 | 1·05 | 1 | 7½ | 9 to 10 | Good; not well cleaned |
| 65 | [2850] | 247 | Mangrol, Kattywar | Do. | Manchester Cotton Supply Association, Liverpool Market | ·90 | 1·10 | 1·00 | 1 | .. | 8 to 9 | Leafy; irregular; strong |
| 66 | [2851] | 158 | Dhollerah, Guzerat | Do. | Messrs. Smith, Fleming, & Co. | ·90 | 1·20 | 1·05 | 1 | 8½ | 9 | Badly ginned; short. *Dhollerat cotton* |
| 67 | [2852] | 243 | Do. Do. | Do. | Manchester Cotton Supply Association, Liverpool Market | ·80 | 1·10 | ·95 | 1 | .. | 8 | Coarse, strong, shelly. Do. |
| 68 | [2853] | 168 | Broach | Do. | Messrs. Smith, Fleming, & Co. | ·75 | 1·05 | ·90 | 9-10ths | 8 | 10 to 11 | White; good staple; regular and strong |
| 69 | [2854] | 169 | Do. | Do. | Do. | ·80 | 1·10 | ·95 | 1 | 7½ | 11½ to 12½ | Very superior; well prepared |
| 70 | [2855] | 248 | Do. | Do. | Manchester Cotton Supply Association, Liverpool market | 1·05 | 1·25 | 1·15 | 1 1-7th | .. | 8 | White; leafy; weak |
| 71 | [2856] | 240 | Kandeish | Do. | Do. | ·85 | 1·15 | 1·00 | 1 | .. | 6½ | Very dirty; inferior |
| 72 | [2857] | 163 | Poona | Do. | Do. | ·90 | 1·10 | 1·00 | 1 | 11½ | 10½ | Very good |
| 73 | [2860] | 157 | Dharwar | Do. | ssrs. Smith, Fleming, & Co. | ·95 | 1·25 | 1·10 | 1 1-10th | 8¾ | 9½ to 10 | Shelly and dusty; good staple; strong. *Coompta cotton* |
| 74 | [2861] | 170 | Coompta | Do. | Me Do. | ·90 | 1·20 | 1·05 | 1 | 7½ | 9 to 10 | Good staple; badly prepared; if properly ginned, 12*d.* to 13*d.* *Coompta cotton* |
| 75 | [2863] | 241 | Do. | Do. | Manchester Cotton Supply Association, Liverpool Market | ·75 | 1·05 | ·90 | 9-10ths | .. | 6½ | Very dirty; inferior |
| 76 | [2864] | 244 | Dharwar | Do. | Do. | ·90 | 1·20 | 1·05 | 1 | .. | 10 to 11 | Very good staple; yellowish |
| 77 | [2868] | 180 | Bombay | Do. | Messrs. Smith, Fleming, & Co. | ·70 | 1·00 | ·85 | 7-8ths | 7 | .. | Very inferior |
| 78 | [2869] | 171 | Do. | Do. | Do. | ·90 | 1·20 | 1·05 | 1 | 7½ | .. | Very inferior |
| 79 | [2870] | 173 | Do. | Do. | Do. | ·85 | 1·15 | 1·00 | 1 | 8 | 7 | Dirty; ill prepared; of low quality |
| 80 | [2871] | 174 | Do. | Do. | Do. | ·80 | 1·10 | ·95 | 1 | 7¾ | 7 | Do. do. do. |
| 81 | [2872] | 175 | Do. | Do. | Do. | ·80 | 1·10 | ·95 | 1 | 8 | .. | Short; dusty |
| 82 | [2874] | 178 | Do. | Do. | Do. | ·85 | 1·15 | 1·00 | 1 | .. | .. | |
| 83 | [2883] | 161 | Hingenghaut, Berar | Do. | Do. | ·90 | 1·20 | 1·05 | 1 | 9 | 10 | Good |
| 84 | [2884] | 219 | Do. | Do. | Manchester Cotton Supply Association, Liverpool Market | ·80 | 1·10 | ·95 | 1 | .. | 10 | White; silky; strong medium staple |
| | | | | | MEAN | ·86 | 1·14 | 1·00 | 1 | | | |

| Number: Of Sample | Number: In Catalogue | Number: In Jury List and that of Cotton Supply Association | Place of Growth | Seed Employed | By whom Exhibited, Name of Cultivator, &c. | Length of Staple in Inches and Parts of an Inch: Min. (Decimal Fractions) | Max. (Decimal Fractions) | Mean: Decimal Fractions | Mean: Vulgar Fractions (approximate) | Price per Lb.: By Jury | Price per Lb.: By Commissioner, Cotton Supply Association | Remarks by Commissioner appointed by Cotton Supply Association to examine the various Cottons in the International Exhibition (1862) |
|---|---|---|---|---|---|---|---|---|---|---|---|---|
| | | | MADRAS AND MYSORE | | | | | | | Pence | Pence | |
| 85 | [2889] | 201 | Chingleput (*Gossypium arboreum*) | Native or indigenous | Dr. J. Shortt | ·90 | 1·10 | 1·00 | 1 | 8½ | 9 | Chopped to bits in ginning; should have been 12*d.* *Gossypium arboreum* |
| 86 | [2893] | 242 | Tinnevelly | Do. | Manchester Cotton Supply Association, Liverpool Market | ·85 | 1·10 | ·98 | 1 | .. | 8 | Well cleaned; naturally very weak and short |
| 87 | [2891] | 188 | Do. | Do. | Messrs. Smith, Fleming, & Co. | ·85 | 1·15 | 1·00 | 1 | 7½ | 6 to 7 | Reddish; very weak |
| 88 | [2892] | 189 | Do. | Do. | Do. | ·90 | 1·20 | 1·05 | 1 | 11½ | 7 to 8 | Weak white; well cleaned |
| 89 | [2895] | 191*b* | Nellore | Do. | Government | ·75 | ·95 | ·85 | 7-8ths | 9½ | .. | Very short staple; damaged in ginning |
| 90 | [2896] | 187 | Do. | Do. | Messrs. Smith, Fleming, & Co. | 1·00 | 1·30 | 1·15 | 1 1-7th | 8 | 8½ | Good colour; very weak |
| 91 | [2898] | 206 | 'Western Madras' | Do. | Do. | 1·10 | 1·30 | 1·20 | 1 1-5th | 7¼ | 6 | Badly prepared; naturally weak and poor |
| 92 | [2899] | 239 | Do. | Do. | Do. | ·90 | 1·20 | 1·05 | 1 | .. | 6½ | Very dirty; inferior |
| 93 | [2900] | 198 | Mysore | Do. | Government of Mysore | ·80 | 1·10 | ·95 | 1 | 8¾ | 6 | Short, coarse, weak fibre; red colour; very nappy; over-ginned. Should have been 9*d.* |
| 94 | [2904] | 205 | Do. | Do. | Do. | ·80 | 1·25 | ·95 | 1 | 12½ | 8 | Very clean; somewhat tender fibre; short |
| 95 | [2907] | 199 | Do. | Do. | Do. | ·90 | 1·25 | 1·08 | 1 1-10th | 10½ | 10 | White; medium staple; rather shelly; slightly damaged in ginning |
| 96 | [2906] | 202 | Do. | Do. | Do. | ·90 | 1·20 | 1·05 | 1 | .. | .. | |
| | | | | | MEAN | ·86 | 1·17 | 1·05 | 1 | | | |
| | | | ARRACAN, BURMAH, TENASSERIM, AND STRAITS' PROVINCES | | | | | | | | | |
| 97 | [2910] | 215 | Akyab | Do. | Akyab Committee (Gov.) | ·75 | ·95 | ·85 | 7-8ths | 6¾ | 6 to 7 | Very short and fuzzy |
| 98 | [2911] | 217 | Do. | Do. | Do. Do. | ·65 | ·90 | ·78 | 4-5ths | 5 | .. | |
| 99 | [2914] | 210 | Pegu | Do. | Rangoon Committee Do. | ·75 | 1·05 | ·90 | 9-10ths | 7 | .. | Clean, but very inferior |
| 100 | [2915] | 209 | Burmah | Do. | Messrs. Halliday, Fox, & Co. | ·75 | ·95 | ·85 | 7-8ths | 7½ | 7 to 8 | Good white; strong fibre, but very short; probably cut in ginning |
| 101 | [2916] | 211 | Rangoon | Do. | Messrs. Smith, Fleming, & Co. | ·70 | 1·00 | ·85 | 7-8ths | 7¼ | .. | Good colour; very short staple; probably spoilt in ginning |
| 102 | [2922] | 218 | Singapore | Do. | Hon. Colonel Kavanagh | 1·20 | 1·40 | 1·30 | 1 3-10ths | 13 | 13 to 15 | Very good colour; staple strong |
| 103 | [2920] | 221 | Do. | Do. | Do. | 1·10 | 1·40 | 1·25 | 1 1-4th | 12½ | 12 | Good strong wiry cotton; slightly injured in ginning. |
| 104 | [2921] | 223 | Do. | Do. | Do. | 1·10 | 1·30 | 1·20 | 1 1-5th | 12½ | 10 | Very good cotton; slightly spoiled in ginning; should have been 13*d.* |
| 105 | [2925] | 222 | Malacca | Do. | J. Baumgarten, Esq. | 1·10 | 1·40 | 1·25 | 1 1-4th | 12½ | 13 to 15 | Very good white cotton; coarse fibre; well cleaned |
| | | | | | MEAN | ·90 | 1·15 | 1·03 | 1 | | | |
| | | | | | Do. TOTAL OF NATIVE | ·80 | 1·08 | ·95 | 9-10ths | | | |
| | | | BENGAL, &c. | | | | | | | | | |
| 106 | [2694] | 125 | Piplee, Cuttack | New Orleans | Rev. G. Taylor | ·80 | 1·10 | ·95 | 1 | 11 | 13 to 14 | Very good white; regular in fibre; very well prepared, though slightly injured in ginning |
| 107 | [2761] | 116 | Arrah, Shahabad | Do. | Mr. Mylne, on the estate of H. Burrows, Esq. | 1·00 | 1·30 | 1·15 | 1 1-7th | 12½ | 13 to 14 | Good white; fair Boweds character |
| 108 | [2783] | 58 | Burbee and Bussureah, Hazarebaugh, Chota, Nagpore | Do. | C. E. Blechynden, Esq. | ·90 | 1·10 | 1·00 | 1 | 12¼ | 12 to 12½ | White; very nearly equal to American in staple |
| 109 | [2787] | 8 | Seetagurrah Plantation, Hazarebaugh | Do. | M. Leibert, Esq. | 1·00 | 1·30 | 1·15 | 1 1-7th | .. | 10½ to 11 | Good colour, but short in staple; not a satisfactory sample of New Orleans, but very superior to indigenous |

| | | | | | | | | | | | | |
|---|---|---|---|---|---|---|---|---|---|---|---|---|
| 110 | [2817] | 31 | Sutwaree, Chirkaree, Bundelkhund | Do. | . . . . . . . . . . | ·80 | 1·10 | ·95 | 1 | 11 | 12 to 13 | Well prepared; good staple; tender |
| 111 | [2823] | 139 | Dehra Ismael Khan, Punjab | Do. | Lahore Committee (Government) | ·90 | 1·20 | 1·05 | 1 | .. | 12 | Good useful cotton; white; 14*d.* if properly picked and ginned |
| 112 | [2876] | 177 | Bombay . . . . | Do. | Messrs. Smith, Fleming, & Co. | ·80 | 1·10 | ·95 | 1 | 8¾ | .. | Short, weak, dusty, and shelly |
| 113 | [2877] | 183 | Bolarum, Berar . . . | Do. | Dr. Riddell . . . . . . | ·95 | 1·25 | 1·10 | 1 1-10th | 11 | 11 to 12 | Strong staple; very good; a little shorter than New Orleans |
| 114 | [2878] | 184 | Do. Do. . . . | Do. | Do. . . . . . . . | ·90 | 1·20 | 1·05 | 1 | 11 | 11 to 12 | Do. do. do. |
| 115 | [2879] | 185 | Do. Do. . . . | Do. | Do. . . . . . . . | 1·10 | 1·30 | 1·20 | 1 1-5th | 11½ | .. | |
| 116 | [2917] | 224 | Province Wellesley, Penang | Do. | A. Hutchinson, Esq. . . . | 1·10 | 1·40 | 1·25 | 1 1-4th | 14 | 14 | Rather high-coloured; good strong staple; superior to New Orleans |
| | | | | | MEAN (N. Orleans) | ·93 | 1·21 | 1·07 | 1 | | | |
| 117 | [2918] | 225 | Glugor Estate, Do. | Pernambuco | . . . . . . . . . | 1·20 | 1·40 | 1·30 | 1 3-10ths | 13 | 16 | Beautiful white cotton; very good |
| 118 | [2924] | 220 | Singapore . . . . | Do. | G. H. Brown, Esq. . . . . | 1·10 | 1·30 | 1·20 | 1 1-5th | 12½ | 12 to 13 | Good strong, coarse cotton; very clean; injured by ginning; would have been worth 18*d.* to 20*d.* |
| 119 | [2880] | 186 | Bolarum, Berar . . . | Mauritius | Dr. Riddell . . . . . . | 1·50 | 2·10 | 1·80 | 1 4-5ths | 14½ | .. | Long staple mixed with short * |
| 120 | [2888] | 194 | Chingleput, Madras . . | Bourbon | Dr. J. Shortt . . . . . | ·90 | 1·20 | 1·05 | 1 | 10½ | 13 to 14 | Very good silky cotton, especially commended |
| 121 | [2886] | 195 | Salem Do. . . | Do. | Messrs. Fisher & Co. . . . | 1·00 | 1·20 | 1·10 | 1 1-10th | 12½ | 14 to 15 | Beautiful colour; fully equal to fair New Orleans; would have been worth, if cleaned by Macarthy's gin, 15*d.* to 16*d.* |
| 122 | [2887] | 196 | Do. . . . . . . | Do. | Do. . . . . | 1·15 | 1·35 | 1·25 | 1 1-4th | 12 | 14 to 15 | Do. do. do. |
| 123 | [2897] | 197 | Madras . . . . . . | Do. | Messrs. Smith, Fleming, & Co. | 1·00 | 1·25 | 1·13 | 1 1-8th | 12½ | .. | Very good; equal to New Orleans; well cleaned, but somewhat overgrown |
| 124 | [2730] | 91 | Jail Garden, Bancoorah . | Egyptian | J. Andrews, Esq. . . . . . | 1·25 | 1·45 | 1·35 | 1 1-3rd | .. | 14 | Very white; good strong fibre; equal to New Orleans |
| 125 | [2757] | 98 | Arrah, Shahabad . . . | Do. | Mr. Mylne, on the estate of H. Burrows, Esq. | 1·20 | 1·50 | 1·35 | 1 1-3rd | | 14 or more | Do. do. resembling New Orleans |
| 126 | [2758] | 115 | Do. Do. . . . | Do. | Do. Do. . . | 1·20 | 1·45 | 1·33 | 1 1-3rd | 13½ | 15 to 16 | Good strong, serviceable cotton; longer than New Orleans, but not equal to Egyptian |
| 127 | [2786] | 129 | Burbee and Bassureah, Hazarebaugh | Do. | C. E. Blechynden, Esq. . . . | ·90 | 1·20 | 1·05 | 1 | .. | 10 to 11 | Very good colour, but not equal to New Orleans |
| 128 | [2784] | 57 | Do. Do. Do. | Do. | Do. . . . | ·80 | 1·10 | ·95 | 1 | 13 | 13 and more | Very good cotton; quite equal to middling American |
| 129 | [2785] | 59 | Do. Do. Do. | Do. | Do. . . . | 1·20 | 1·40 | 1·30 | 1 3-10ths | .. | 14 to 15 | Very good cotton |
| 130 | [2790] | 9 | Seetagurrah Plantation, Hazarebaugh | Do. | M. Leibert, Esq. . . . . | ·95 | 1·15 | 1·05 | 1 | .. | 9 | White; short; deteriorated |
| 131 | [2819] | 45 | Sutwaree, Chirkaree, Bundelkhund | Do. | . . . . . . . . . | 1·30 | 1·60 | 1·45 | 1½ | 17 | 16 to 18 | Very like true Egyptian |
| 132 | [2858] | 162 | Poona . . . . . | Do. | Manchester Cotton Supply Association, Liverpool Market | 1·25 | 1·55 | 1·40 | 1 2-5ths | 14 | 13 to 14 | Good colour; not so long as Egyptian |
| 133 | [2859] | 172 | Belgaum . . . . | Do. | Mr. Hearn . . . . . . | ·90 | 1·20 | 1·05 | 1 | 12 | .. | Very good; like Egyptian |
| 134 | [2890] | 203 | Chingleput . . . . | Do. | Dr. J. Shortt . . . . . | ·95 | 1·25 | 1·10 | 1 1-10th | 11¼ | 13 to 15 | Very good, desirable cotton; not so long as Egyptian |
| 135 | [2908] | 214 | Arracan . . . . . | Do. | Messrs. Halliday, Fox, & Co. | 1·10 | 1·30 | 1·20 | 1 1-5th | 13 | 12½ | Good useful cotton, but overgrown; reddish |
| | | | | | MEAN (Egyptian) | 1·09 | 1·34 | 1·22 | 1½ | | | |
| 136 | [2789] | 84 | Seetagurrah Plantation, Hazarebaugh | Sea Island | M. Leibert, Esq. . . . . | 1·55 | 1·85 | 1·70 | 1 2-3rds | 9½; if well cleaned 24 | 12 to 16 | Very short for Sea Island; probably spoilt in ginning; yellowish |
| 137 | [2919] | 226 | Glugor Estate, Penang . | Do. | . . . . . . . . . | 1·60 | 2·00 | 1·80 | 1 4-5ths | 24 | 36 | A very good sample of Sea Island cotton |
| 138 | [2923] | 219 | Singapore . . . . | Do. | G. H. Brown, Esq. . . . . | 1·35 | 1·65 | 1·50 | 1½ | 14 | 18 to 24 | Beautiful cotton. |
| | | | | | MEAN . . . . | 1·50 | 1·83 | 1·66 | | | | |

* But for an accidental admixture previously to examination, this cotton would have been valued at 2*s.* and upwards. In February 1860 it was valued at 16*d.* in Liverpool.

Length of the long staple kind only inserted.

PRINTED FOR HER MAJESTY'S COMMISSIONERS

BY

SPOTTISWOODE AND CO., NEW-STREET SQUARE, LONDON

DURENNE'S FOUNTAIN.

that of the Rhea. As soon as arrangements have been effected for its production, along with that of other species of nettle which abound in various parts of India, it is anticipated that fibres from this class of plants will eventually occupy a position second only in importance to that of cotton and flax. Late experiments have shown that the fibre of the Rhea can be turned to account for the manufacture of a variety of fabrics of a very valuable and useful description; and its extended cultivation in India is worthy of every attention and encouragement.* The great desideratum is an efficient machine for the separation of the fibre from its parent stem.

2949. [5266] Rhea fibre (*Bœhmeria nivea*), Gowhatty, Assam. H. BAINBRIDGE, Esq.

2950. [7783] Do. Debrooghur, Assam. REV. E. H. HIGGS.

2951. [5330] Do. Luckimpore, Assam. AMEER of LUCKIMPORE.

2952. [5331] Do. Do. H. L. MICHEL, Esq.

2953. [8113] Do. Do.

2954. [8114] Do. Do.

2955. [9379] Talee Ramee (*B. nivea*), Malacca. C. EVANS, Esq.

#### c. PUYA (*Bœhmeria puya*).

Although botanically a different species, the fibre of this plant is almost identical with that furnished by the Rhea. It flourishes at Darjeeling and other places in the north of India. Its commercial value is the same as that of Rhea.

2956. [8134] Puya stems (*Bœhmeria puya*), Deyra Dhoon.

2957. [8132] Do. fibre, Do., Darjeeling.

2958. [8155] Do. do., finally prepared by chemical action, Do.

2959. [8156] Do. do. do. Do.

#### d. NILGIRI NETTLE (*Urtica heterophylla*).

This nettle abounds in the Nilgiri Hills and also in some other parts of the country. It furnishes a fibre of such a nature that the term 'vegetable wool' has been applied to it. The samples exhibited have been valued at 70*l*. to 80*l*. per ton.

2960. [4850] Nilgiri nettle (*Urtica heterophylla*), from old wood. MR. W. G. McIVOR.

2961. [4850*a*] Do. from new wood. Do.

MR. McIVOR in forwarding these excellent specimens appends the following remarks:—

The plant grows wild all over the Nilgiris, and as it is well known to the natives, its cultivation might be readily extended. Its cultivation on the Hills would pay well as soon as the rail is finished to Coimbatore.

With reference to the preparation of the fibre, that from the bark of the old wood was steeped in cold water for about six days.

The bark of the young wood was steeped in hot water for about twenty-four hours, when the fibre was found to separate readily from the pulp. Neither of the specimens were exposed more than three weeks to bleach, and, consequently, the colour is not so good as would have been obtained had more time been occupied in the bleaching process.

#### e. MÛDAR or YERCUM (*Calotropis gigantea*), also AK (*C. Hamiltonii*)

Have already been referred to under the head of Silk Cottons. The stems of these two species furnish a valuable fibre, which is, however, very difficult of extraction.

The following samples have been forwarded for exhibition:—

2962. [8133] Yercum (*Calotropis gigantea*), Madras. DR. HUNTER.

2963. [4843] Do. Do. Do.

2964. [4854] Tow from Do. Do. Do.

2965. [4888] Do. or Mûdar (*do.*), Chingleput. DR. J. SHORTT.

Prepared as follows:—

The branches are gathered and dried in the sun for from twenty-four to thirty-six hours, when they are taken up, the bark peeled from the woody parts, and the fibres gathered. If placed out in the dew for a night, they lose their greenish tint, and become white.

2966. [4011] Yerooka (*C. gigantea*), Salem. H. A. BRETT, Esq.

2967. [9385] Yercum (*do.*), Bolarum. DR. RIDDELL.

2968. [8157] Do. Shapore, Punjab.

2969. [8115] Do. Do.

2970. [8136] Handkerchief made of Mûdar fibre. INDIA MUSEUM.

#### f. BEDOLEE SUTTA (*Pederia fœtida*).

This excellent fibre is the produce of a creeper which grows abundantly on grass allu-

---

* Some bales of Rhea fibre, lately put up for sale, realised at the rate of 80*l*. per ton.

vial deposits along some parts of the banks of the Brahmaputra in Assam. It is particularly worthy of attention, as its fibre is of a very valuable description. From its appearance it has been named 'vegetable silk.'

2971. [8161] Stems of the *Pederia fœtida*, Assam. INDIA MUSEUM.

2972. [8161*a*] Fibre from *P. fœtida*. Do. Do. Do.

II.—FIBRES SUITED FOR SPINNING AND MANUFACTURING PURPOSES, BUT OF INFERIOR DURABILITY TO THE PRECEDING.

### *a.* JUTE.

There are two species which afford this well-known article of commerce, viz. *Corchorus olitorius* and *C. capsularis*. Both are largely cultivated. The extent to which jute, in either its raw or manufactured form, is exported from India, will be gathered from the subjoined tables. The first (p. 142) shows the quantity of jute and jute-rope exported from India to all parts from 1850-51 to 1860-61, and the second (p. 143) the quantity of gunnies (pieces of gunny cloth) and gunny-bags (which are made almost entirely of jute) to all parts for the same period.

2973. [5793] Jute (*Corchorus olitorius*), Hooghly.

2974. [9168] Do. Midnapore.

2975. [3191] Do. Darjeeling. DR. CAMPBELL.

2976. [5924] Do. Assam. LIEUT. PHAIRE.

2977. [9384] Do. Cuttack.

The plant is to be found under cultivation all over the district. Every farmer requires rope and twine, and so grows a little jhote. The fibre is extracted as in the case of the 'sunn' hemp. In the bazaar jute sells at 10 lbs. per shilling, and the rope at from 5 to 7 lbs. weight for do. Gunnies are made of this fibre at Cuttack and elsewhere in the district, while a good proportion is taken up to Calcutta.*

2978. [5069] Jute, bleached for manufacture of paper &c. C. F. JEFFREY, Esq.

2979. [5070] Do. Do. Do.

In the above samples the difficulties attendant upon the bleaching of jute appear to have been, to a very considerable extent, overcome.

* Local Committee, *Cuttack.*

*b.* SUFET BARIALA (*Sida rhomboidea*).

This fibre is very similar to jute in appearance; but it is considered to be intrinsically so superior that it is worth from 5*l.* to 6*l.* more per ton, and it has accordingly been placed next to that fibre, in order to attract to it the attention which it deserves.

2980. [9387] Sufet bariala (*S. rhomboidea*). R. STURROCK, Esq.

*c.* AMBAREE or HEMP-LEAVED HIBISCUS (*Hibiscus cannabinus*).

This plant furnishes a portion of the so-called 'brown hemp,' exported from Bombay. It is readily cultivated, and with more attention to its preparation, is calculated to compete with jute.

2981. [8116] Patwa (*Hibiscus cannabinus*), Lucknow.

Every ryot sows a small quantity along the edges of his usual crops for his own use. It is not, but it might be, cultivated extensively all over Oude, and in all kinds of soil. It is sown in the beginning of the rains, and when it commences to flower, it is cut and treated exactly in the same way as 'sunn hemp' from *Crotalaria juncea*. The proportion of fibre is about half the weight of the plant. It is used for making rope, sackcloth, twine, paper, &c. The cost of the prepared fibre is from three to four rupees per maund, according to its strength, length, and cleanliness.*

2982. [4811] Palungoo (*H. cannabinus*), Chingleput. DR. SHORTT.

Prepared carelessly by the native rolling process, which accounts for its dark colour.

2983. [8117] Ambaree (*H. cannabinus*), Bolarum. DR. RIDDELL.

2984. [10060] Sunn Okra (*H. cannabinus*), Lahore.

This grows abundantly, but its fibre is inferior to that of the true hemp. Price at Lahore from 10*l.* to 14*l.* per ton.†

*d.* ROSELLE (*Hibiscus sabdariffa*).

Commonly cultivated in gardens for the sake of its leaves, which are eaten in salads. Worthy of extended cultivation on account of its fibre.

* Central Committee, *Lucknow.*
† Central Committee, *Lahore.*

## TABLE SHOWING THE QUANTITIES (AS FAR AS CAN BE ASCERTAINED) AND VALUE* OF JUTE AND JUTE ROPE EXPORTED FROM INDIA AND EACH PRESIDENCY TO ALL PARTS OF THE WORLD, FROM 1850-61 TO 1860-61 INCLUSIVE.

COUNTRIES WHITHER EXPORTED

| Years | Whence Exported | United Kingdom Quantity (lbs.) | United Kingdom Value (£) | France Quantity (lbs.) | France Value (£) | Other Parts of Europe Quantity (lbs.) | Other Parts of Europe Value (£) | North America Quantity (lbs.) | North America Value (£) | China Quantity (lbs.) | China Value (£) | Arabian and Persian Gulf Quantity (lbs.) | Arabian and Persian Gulf Value (£) | Islands and Shores of the Indian Seas Quantity (lbs.) | Islands and Shores of the Indian Seas Value (£) | All Other Places Quantity (lbs.) | All Other Places Value (£) | Total E to All Quantity (lbs.) | Total E to All Value (£) |
|---|---|---|---|---|---|---|---|---|---|---|---|---|---|---|---|---|---|---|---|
| 1850-51 | Bengal | 63,273,168 | 191,069 | 1,146,320 | 3,441 | 43,568 | 130 | 760,480 | 2,382 | … | … | … | … | … | … | 4,480 | 14 | 65,228,016 | 196,986 |
| | Madras | … | … | … | … | … | … | … | … | … | … | … | … | … | … | … | … | … | … |
| | Bombay | … | … | … | … | … | … | … | … | … | … | … | … | … | … | … | … | … | … |
| | ALL INDIA | 63,273,168 | 191,069 | 1,146,320 | 3,441 | 43,568 | 130 | 760,480 | 2,382 | … | … | … | … | … | … | 4,480 | 14 | 65,228,016 | 196,936 |
| 1851-52 | Bengal | 55,947,920 | 168,974 | 1,874,768 | 5,623 | … | … | 2,126,432 | 6,379 | … | … | … | … | … | … | … | … | 59,949,120 | 180,976 |
| | Madras | … | … | … | … | … | … | … | … | … | … | … | … | … | … | … | … | … | … |
| | Bombay | … | … | … | … | … | … | … | … | … | … | … | … | … | … | … | … | … | … |
| | ALL INDIA | 55,947,920 | 168,974 | 1,874,768 | 5,623 | … | … | 2,126,432 | 6,379 | … | … | … | … | … | … | … | … | 59,949,120 | 180,976 |
| 1852-53 | Bengal | 35,278,544 | 101,380 | 1,817,984 | 5,178 | 30,016 | 85 | 2,098,208 | 5,804 | … | … | … | … | 27,440 | 127 | 784 | 4 | 39,182,976 | 112,578 |
| | Madras | … | … | … | … | … | … | … | … | … | … | … | … | … | … | … | … | … | … |
| | Bombay | … | … | … | … | … | … | … | … | … | … | … | … | … | … | … | … | … | … |
| | ALL INDIA | 35,278,544 | 101,380 | 1,817,984 | 5,178 | 30,016 | 85 | 2,098,208 | 5,804 | … | … | … | … | 27,440 | 127 | 784 | 4 | 39,182,976 | 112,578 |
| 1853-54 | Bengal | 41,498,240 | 117,982 | 4,596,704 | 12,908 | … | … | 10,833,648 | 30,452 | … | … | 112 | 1 | 49,840 | 197 | 86,240 | 259 | 57,064,784 | 161,769 |
| | Madras | … | … | … | … | … | … | … | … | … | … | … | … | … | … | … | … | … | … |
| | Bombay | … | … | … | … | … | … | … | … | … | … | … | … | … | … | … | … | … | … |
| | ALL INDIA | 41,498,240 | 117,982 | 4,596,704 | 12,908 | … | … | 10,833,648 | 30,452 | … | … | 112 | 1 | 49,840 | 197 | 86,240 | 259 | 57,064,784 | 161,769 |
| 1854-55 | Bengal | 63,363,888 | 184,588 | 3,870,720 | 12,280 | … | … | 10,494,400 | 30,432 | … | … | … | … | 44,128 | 300 | 578,256 | 1,641 | 78,351,392 | 229,241 |
| | Madras | … | … | … | … | … | … | … | … | … | … | … | … | … | … | … | … | … | … |
| | Bombay | … | … | … | … | … | … | … | … | … | … | … | … | … | … | … | … | … | … |
| | ALL INDIA | 63,363,888 | 184,588 | 3,870,720 | 12,280 | … | … | 10,494,400 | 30,432 | … | … | … | … | 44,128 | 300 | 578,256 | 1,641 | 78,351,392 | 229,241 |
| 1855-56 | Bengal | 85,751,568 | 285,499 | 6,220,368 | 20,705 | 118,720 | 393 | 6,085,520 | 20,223 | 173,152 | 560 | … | … | 255,248 | 818 | 166,768 | 600 | 98,771,344 | 328,798 |
| | Madras | 92,736 | 278 | … | … | … | … | … | … | … | … | … | … | … | … | … | … | 92,736 | 278 |
| | Bombay | … | … | … | … | … | … | … | … | … | … | … | … | … | … | … | … | … | … |
| | ALL INDIA | 85,844,304 | 285,777 | 6,220,368 | 20,705 | 118,720 | 393 | 6,085,520 | 20,223 | 173,152 | 560 | … | … | 255,248 | 818 | 166,768 | 600 | 98,864,080 | 329,076 |
| 1856-57 | Bengal | 55,164,816 | 201,122 | 7,277,200 | 26,483 | 10,864 | 40 | 12,569,984 | 45,824 | 137,088 | 500 | … | … | 35,168 | 110 | 268,352 | 978 | 75,463,472 | 275,057 |
| | Madras | … | … | … | … | … | … | … | … | … | … | … | … | … | … | … | … | … | … |
| | Bombay | … | … | … | … | … | … | … | … | … | … | … | … | … | … | … | … | … | … |
| | ALL INDIA | 55,164,816 | 201,122 | 7,277,200 | 26,483 | 10,864 | 40 | 12,569,984 | 45,824 | 137,088 | 500 | … | … | 35,168 | 110 | 268,352 | 978 | 75,463,472 | 275,057 |
| 1857-58 | Bengal | 75,065,312 | 251,275 | 5,042,240 | 24,274 | 70,448 | 235 | 7,709,520 | 25,884 | … | … | … | … | 81,024 | 189 | 429,072 | 1,435 | 88,347,616 | 308,292 |
| | Madras | … | … | … | … | … | … | … | … | … | … | … | … | … | … | … | … | … | … |
| | Bombay | … | … | … | … | … | … | … | … | … | … | … | … | … | … | … | … | … | … |
| | ALL INDIA | 75,065,312 | 251,275 | 5,042,240 | 24,274 | 70,448 | 235 | 7,709,520 | 25,884 | … | … | … | … | 81,024 | 189 | 429,072 | 1,435 | 88,347,616 | 308,292 |
| 1858-59 | Bengal | 13,547,744 | 451,588 | 12,163,760 | 40,546 | 25,200 | 84 | 9,452,576 | 31,509 | … | … | … | … | … | … | 439,600 | 1,456 | 35,603,680 | 525,099 |
| | Madras | … | … | … | … | … | … | … | … | … | … | … | … | … | … | … | … | … | … |
| | Bombay | … | … | … | … | … | … | … | … | … | … | … | … | … | … | … | … | … | … |
| | ALL INDIA | 13,547,744 | 451,588 | 12,163,760 | 40,546 | 25,200 | 84 | 9,452,576 | 31,509 | … | … | … | … | … | … | 439,600 | 1,456 | 35,603,680 | 525,099 |
| 1859-60 | Bengal | 76,418,048 | 260,544 | 5,609,632 | 18,699 | … | … | 2,663,024 | 8,871 | … | … | 74,928 | 250 | … | … | 463,680 | 1,570 | 85,254,512 | 290,018 |
| | Madras | … | … | … | … | … | … | … | … | … | … | … | … | … | … | … | … | … | … |
| | Bombay | … | … | … | … | … | … | … | … | … | … | … | … | … | … | … | … | … | … |
| | ALL INDIA | 76,418,048 | 260,544 | 5,609,632 | 18,699 | … | … | 2,663,024 | 8,871 | … | … | 74,928 | 250 | … | … | 463,680 | 1,570 | 85,254,512 | 290,018 |
| 1860-61 | Bengal | 103,447,456 | 346,364 | 11,203,479 | 37,512 | … | … | 6,385,064 | 21,380 | 53,760 | 180 | 155,288 | 520 | … | … | 1,080,240 | 41,927 | 122,325,280 | 409,871 |
| | Madras | … | … | … | … | … | … | … | … | … | … | … | … | … | … | … | … | … | … |
| | Bombay | … | … | … | … | … | … | … | … | … | … | … | … | … | … | … | … | … | … |
| | ALL INDIA | 103,447,456 | 346,364 | 11,203,472 | 37,512 | … | … | 6,385,064 | 21,380 | 53,760 | 180 | 155,288 | 520 | … | … | 1,080,240 | 41,927 | 122,325,280 | 409,371 |

* The Indian rupee is here converted at the exchange of two shillings.

TABLE SHOWING THE NUMBER AND VALUE (AS FAR AS CAN BE ASCERTAINED) OF GUNNY BAGS AND OF GUNNIES (PIECES OF GUNNY CLOTH) EXPORTED FROM INDIA AND EACH PRESIDENCY TO ALL PARTS OF THE WORLD FROM 1850–51 TO 1860–61 INCLUSIVE.

| YEARS | WHENCE EXPORTED | UNITED KINGDOM Quantity | Value | FRANCE Quantity | Value | OTHER PARTS OF EUROPE Quantity | Value | NORTH AMERICA Quantity | Value | CHINA Quan. | Value | ARABIAN AND PERSIAN GULFS Quantity | Val. | ISLANDS AND SHORES OF THE INDIAN SEAS Quantity | Value | OTHER PARTS Quantity | Value | TOTAL EXPORTED TO ALL PARTS Quantity | Value |
|---|---|---|---|---|---|---|---|---|---|---|---|---|---|---|---|---|---|---|---|
| | | pieces | £ | pieces | £ | pieces | £ | pieces | £ | pieces | £ | pieces | £ | pieces | £ | pieces / bags | £ | pieces / bags | £ |
| 1850-51 | Bengal | 6,636 | 1,565 | ... | ... | 2,180 | 94 | 2,290,437 | 130,249 | ... | ... | 4,000 | 44 | 1,286,150 | 17,685 | 701,145 / 27,700 | 8,450 | 4,353,538 / 27,700 | 158,087 |
| | Madras | ... | ... | ... | ... | ... | ... | ... | ... | ... | ... | ... | ... | ... | ... | 1,150 | 292 | 1,700 | 292 |
| | Bombay | ... | ... | ... | ... | ... | ... | ... | ... | ... | ... | 550 | 8 | ... | ... | ... | 34 | ... | 42 |
| | ALL INDIA | 6,636 | 1,565 | ... | ... | 2,180 | 94 | 2,290,427 | 130,249 | ... | ... | 4,550 | 52 | 1,286,150 | 17,685 | 702,295 / 27,700 | 8,776 | 4,355,238 / 27,700 | 158,421 |
| 1851-52 | Bengal | 6,650 | 789 | ... | ... | 100 | 20 | 8,391,719 | 260,865 | 17,375 | 211 | ... | ... | 690,918 | 10,179 | 561,710 / 39,050 | 7,206 | 9,729,472 / 41,550 | 279,270 |
| | Madras | ... | ... | ... | ... | ... | ... | ... | ... | ... | ... | 154 | 7 | 2,500 | 25 | 445 | 399 | 699 | 424 |
| | Bombay | ... | ... | ... | ... | ... | ... | ... | ... | ... | ... | ... | ... | ... | ... | ... | 19 | ... | 26 |
| | ALL INDIA | 6,650 | 789 | ... | ... | 100 | 20 | 8,391,719 | 260,865 | 17,375 | 211 | 154 | 7 | 690,918 | 10,204 | 562,155 / 39,050 | 7,604 | 9,730,071 / 41,550 | 279,720 |
| 1852-53 | Bengal | 100 | 52 | 25,000 | 256 | 274 | 11 | 3,347,580 | 169,648 | 245,750 | 2,518 | ... | ... | 1,128,323 | 20,026 | 664,269 / 17,870 | 8,770 | 5,416,296 / 17,870 | 201,281 |
| | Madras | ... | ... | ... | ... | ... | ... | ... | ... | ... | ... | ... | ... | ... | ... | ... | 236 | ... | 236 |
| | Bombay | ... | ... | ... | ... | ... | ... | ... | ... | ... | ... | 6,000 | 150 | ... | ... | ... | ... | 6,000 | 150 |
| | ALL INDIA | 100 | 52 | 25,000 | 256 | 274 | 11 | 3,347,580 | 169,648 | 245,750 | 2,518 | 6,000 | 150 | 1,128,323 | 20,026 | 664,269 / 17,870 | 9,006 | 5,422,296 / 17,870 | 201,667 |
| 1853-54 | Bengal | 1,124 | 193 | ... | ... | 17,100 | 289 | 5,539,788 | 138,489 | 19,700 | 244 | 6,400 | 106 | 1,017,660 | 18,251 | 1,264,746 / 34,950 | 16,794 | 7,881,618 / 34,950 | 174,366 |
| | Madras | ... | ... | ... | ... | ... | ... | ... | ... | ... | ... | ... | ... | ... | ... | 6,010 | 379 | 10,380 | 379 |
| | Bombay | ... | ... | ... | ... | ... | ... | ... | ... | ... | ... | 4,370 | 90 | ... | ... | ... | 91 | ... | 181 |
| | ALL INDIA | 1,124 | 193 | ... | ... | 17,100 | 289 | 5,539,788 | 138,489 | 19,700 | 244 | 10,770 | 196 | 1,017,660 | 18,431 | 1,270,756 / 34,950 | 17,264 | 7,891,898 / 34,950 | 174,926 |
| 1854-55 | Bengal | 5,300 | 515 | 8,300 | 115 | ... | ... | 9,138,141 | 180,954 | 1,000 | 11 | 10,500 (bags) | 208 | 778,076 | 14,732 | 1,262,146 / 115,440 | 17,414 | 11,229,663 / 115,440 | 213,626 |
| | Madras | ... | ... | ... | ... | ... | ... | ... | ... | ... | ... | ... | ... | ... | ... | ... / 850 | 1,381 | ... / 11,350 | 1,381 |
| | Bombay | ... | ... | ... | ... | ... | ... | ... | ... | ... | ... | ... | ... | ... | ... | ... | 131 | ... | 339 |
| | ALL INDIA | 5,300 | 515 | 8,300 | 115 | ... | ... | 9,138,141 | 180,954 | 1,000 | 11 | 10,500 (bags) | 208 | 778,076 | 14,732 | 1,262,146 / 116,290 | 18,926 | 11,229,663 / 126,790 | 215,346 |
| 1855-56 | Bengal | 8,616 | 3,160 | 3,705 | 57 | ... | ... | 9,251,758 | 273,660 | 29,000 | 375 | 400 | 10 | 970,553 | 14,301 | 510,789 / 17,010 | 10,493 | 10,860,416 / 17,010 | 302,114 |
| | Madras | ... | ... | ... | ... | ... | ... | ... | ... | ... | ... | ... | ... | ... | ... | ... | ... | ... | ... |
| | Bombay | ... | ... | ... | ... | ... | ... | ... | ... | ... | ... | 364 (bags) | 4 | ... | ... | ... / 364 | 222 | ... / 364 | 4 |
| | ALL INDIA | 8,616 | 3,160 | 3,705 | 57 | ... | ... | 9,251,758 | 273,660 | 29,000 | 375 | 400 | 14 | 970,553 | 14,301 | 510,789 / 17,374 | 10,715 | 10,860,416 / 17,374 | 302,340 |
| 1856-57 | Bengal | 5,830 | 418 | 21,000 | 275 | 380 | 60 | 6,355,142 | 324,752 | 3,000 | 75 | 1,990 | 22 | 2,046,189 | 36,223 | 933,046 / 2,643 | 14,474 | 9,381,282 / 2,643 | 376,081 |
| | Madras | ... | ... | ... | ... | ... | ... | ... | ... | ... | ... | ... | ... | ... | ... | ... | ... | ... | 167 |
| | Bombay | ... | ... | ... | ... | ... | ... | ... | ... | ... | ... | 20 (bags) | 5 | ... | ... | ... / 20 | 167 | ... / 20 | 5 |
| | ALL INDIA | 5,830 | 418 | 21,000 | 275 | 380 | 60 | 6,355,142 | 324,752 | 3,000 | 75 | 1,990 | 27 | 2,046,189 | 36,223 | 933,046 / 2,663 | 14,641 | 9,381,282 / 2,663 | 376,253 |
| 1857-58 | Bengal | 2,000 | 1,139 | 225 | 2 | ... | ... | 4,155,890 | 167,094 | ... | ... | 5,900 | 83 | 574,273 | 31,563 | 1,063,016 / 8,683 | 16,423 | 5,892,079 / 8,683 | 216,577 |
| | Madras | ... | ... | ... | ... | ... | ... | ... | ... | ... | ... | ... | ... | ... | ... | ... / 2,172 | 469 | ... / 2,172 | 469 |
| | Bombay | ... | ... | ... | 50 | ... | ... | ... | ... | ... | ... | 10 (cwt) | 21 | ... | ... | ... | 25 | 2,000 | 46 |
| | ALL INDIA | 2,000 | 1,139 | 225 | 52 | ... | ... | 4,155,890 | 167,094 | ... | ... | 5,900 | 104 | 574,273 | 31,563 | 1,063,016 / 10,855 | 16,917 | 5,894,079 / 10,855 | 217,092 |
| 1858-59 | Bengal | 1,934 | 3,646 | 1,000 | 17 | ... | ... | 5,142,367 | 348,113 | 4,000 | 100 | ... | ... | ... | ... | 2,576,557 | 39,168 | 7,792,088 | 391,029 |
| | Madras | 3,424 | 286 | 300 | 5 | ... | ... | ... | ... | ... | ... | ... | ... | ... | ... | 15,858 | 709 | 30,282 | 995 |
| | Bombay | ... | ... | ... | ... | ... | ... | ... | ... | ... | ... | ... | 338 | ... | ... | ... | 12 | ... | 400 |
| | ALL INDIA | 5,358 | 3,932 | 1,300 | 22 | ... | ... | 5,142,367 | 348,113 | 4,000 | 100 | ... | 338 | ... | ... | 2,592,415 | 39,889 | 7,822,365 | 392,424 |
| 1859-60 | Bengal | 2,604 | 5,415 | ... | ... | ... | ... | 4,543,629 | 296,799 | 11,500 | 425 | 950 | 14 | ... | ... | 1,837,869 | 30,647 | 6,437,452 | 333,317 |
| | Madras | ... | ... | ... | ... | ... | ... | ... | ... | ... | ... | 5,475 | 147 | ... | ... | 6,074 | 864 | 6,074 | 369 |
| | Bombay | ... | ... | ... | ... | ... | ... | ... | ... | ... | ... | 6,425 | 161 | ... | ... | 10,000 | 144 | 15,475 | 291 |
| | ALL INDIA | 2,504 | 5,415 | ... | ... | ... | ... | 4,543,629 | 296,799 | 11,500 | 425 | 6,425 | 161 | ... | ... | 1,853,643 | 31,655 | 6,459,001 | 333,977 |
| 1860-61 | Bengal | 3,400 | 5,135 | ... | ... | 2,000 | 40 | 2,265,659 | 307,921 | 17,900 | 317 | 500 | 10 | ... | ... | 2,380,099 | 45,467 | 4,705,558 | 358,891 |
| | Madras | ... | ... | ... | ... | ... | ... | ... | ... | ... | ... | ... | ... | ... | ... | 2,434 | 66 | 2,424 | 66 |
| | Bombay | ... | ... | ... | ... | ... | ... | ... | ... | ... | ... | ... | ... | ... | ... | 4,000 | 86 | 4,000 | 86 |
| | ALL INDIA | 3,400 | 5,135 | ... | ... | 2,000 | 40 | 2,265,659 | 307,921 | 17,900 | 317 | 500 | 10 | ... | ... | 2,386,533 | 45,619 | 4,711,982 | 359,043 |

COUNTRIES WHITHER EXPORTED

2985. [8118] Roselle (*Hibiscus sabda-riffa*), Bolarum. DR. RIDDELL.

2986. [8120] Do.    Do.    Do.

2987. [8121] Do.    Do.    Do.

2988. [4008] Gango (*H. sabdariffa*), Salem. H. A. BRETT, Esq.

### e. OKRO (*Abelmoschus esculentus*).

DR. RIDDELL, from his experiments, strongly recommends this plant, as furnishing an excellent fibre for the manufacture of paper.

2989. [8122] Okro (*A. esculentus*), Bolarum. DR. RIDDELL.

2990. [9181] Benda (*A. esculentus*), Madras. H.H. the RAJAH of VIZIANAGRAM.

2991. [10723] Do. Mangalore. V. P. COELHO, Esq.

Other fibres from the same natural family (*Malvaceæ*) are likewise worthy of attention. They include the following:—

2992. [4880] Indian mallow (*Abutilon indicum*), Chingleput. DR. SHORTT.

The plants are gathered and freed of their leaves and twigs, and are put out to dry in the sun for a couple of days. They are then taken up, tied into bundles, and placed under water for about ten days, after which they are taken out, and the fibres are well washed to remove the bark and other foreign matter that may be adhering to them, and are placed in the sun to dry.

2993. [8131] Indian mallow (*A. indicum*), Madras. DR. HUNTER.

2994. [4853] Do.    Do.    Do.

2995. [10724] Jungle Bende (*Hibiscus* sp.), Mangalore. V. P. COELHO, Esq.

2996. [8119] Jungle mallow (*Hibiscus* sp.) DR. RIDDELL.

2997. [8125] Wild Bende (*Hibiscus* sp.) DR. RIDDELL.

2998. [9165] Paharea Jute (*Hibiscus* sp.), Cuttack.

2999. [5791] *Urena lobata*, Burmah.

This plant is the pest of Rangoon and its neighbourhood, springing up spontaneously wherever the jungle is cleaned, and rapidly forming a dense mass of luxuriant vegetation.

The specimen of fibre exhibited has been manufactured in the jail by simple maceration, and afterwards beating the stalks. Very good 'gunny' has been made from it, and it is believed that this fibre might, if treated with due care and skill, prove valuable.

Any quantities of the plant may be had for the mere trouble of gathering it.*

3000. [8142] Fibre of *Malope grandiflora*. DR. RIDDELL.

3001. [4852] *Cryptostegia grandiflora*, Madras. DR. HUNTER.

The plant is common in the south, and yields a fine silky fibre capable of being spun into fine yarn, and of employment for many of the purposes to which flax is applicable.

### III.—FIBRES CHIEFLY SUITED FOR THE MANUFACTURE OF CORDAGE, TWINE, ETC.

#### a. HEMP (*Cannabis sativa*).

It is cultivated in many parts of India for the sake of the 'Bhang' or intoxicating resin of its leaves, but as yet only occasionally for its fibre.

3002. [10245] Sunn Bhang (*C. sativa*), Kangra, Punjab.

It grows spontaneously and in abundance everywhere in the submontane tracts, but is cultivated for the fibre only in the eastern portions of the Kangra, and in the Simla Hills. In 1859 an experimental consignment of two tons of Himalayan hemp was valued in the English market at from 30*l.* to 32*l.* per ton, and during the past year another larger consignment of hemp has been despatched at Government expense, by request of the merchants of Dundee. The price at Lahore is about 15*l.* or 16*l.* per ton.†

3003. [8135] 'Himalayan hemp' (*C. sativa*), Kangra.

3004. [8123] Do. Do.

3005. [8124] Do. Do. R. STURROCK, Esq.

Opened mechanically in Dundee. Worth 40*l.* per ton.

---

* Local Committee, *Rangoon.*
† Central Committee, *Lahore.*

3006. [2731] Hemp (*C. sativa*), Bangalore.

3007. [5902] Do. Calcutta. Messrs. AHMUTY & Co.

### b. SUNN HEMP (*Crotalaria juncea*).

This plant furnishes the vast proportion of the so-called hemps exported from India. The table on the next page shows the quantity and value of the 'hemp' and also hemp rope, of this kind sent from India to all parts.

3008. [7772] Sunn (*Crotalaria juncea*), Hooghly.

3009. [2866] Do. Raepore.

3010. [9169] Do. Lucknow.

Cultivated near cities by hundreds of beegahs; but in the vicinity of villages only in small quantities, principally for the purpose of making fishing nets. Its cultivation can be extended all over Oude, and principally where a light soil exists. It is sown very thickly at the beginning of the rains, so that it may grow tall and thin. When it begins to flower, it is cut near the root, tied in large bundles and immersed in water, putting some weight on it (generally mud) to prevent its being carried away. After remaining immersed from four to eight days it is withdrawn from the water, taken by handfuls, beaten on a piece of wood or stone, and washed till quite clean, and the cuticle with the leaves completely removed from the other portion of the plant. Each handful is then piled musket fashion, and left to dry. When perfectly dry, the woody portion, which has been more or less broken, is separated from the fibre by farther beating and shaking. From 3 to 6 maunds of fibre are extracted from each beegah of plant. The fibre is used for making rope, sackcloth, nets, twine, and paper. The raw material on the field, as plant, costs from two to four rupees per beegah, according to quality; and the prepared fibre costs from four to ten rupees per maund, according to strength, length, and cleanliness of fibre.*

3011. [9411] Sunn (*Crotalaria juncea*), Cuttack.

Under its local synonymes, *Chuniput* and *Chumese*, this plant is grown in this district in sufficient quantities to supply its wants, and probably more. It requires comparatively but little tillage, and not much after-tending. The plants, when site and soil agree, attain to a height of 8 or 9 feet. The fibre is separated by threshing and beating, after the plant, which, at the time of cutting, is tied into convenient bundles for the purposes, has been kept immersed in water several days. The hemp is bought in the bazaar about 7 lbs. per shilling, and rope made of it at 5 lbs. weight for the shilling. The country paper is made from this article.†

* Central Committee, *Lucknow.*
† Local Committee, *Cuttack.*

3012. [10076] Sunn (*C. juncea*), Lahore.

It is extensively cultivated for its fibre, especially near rivers. Sunn prepared for the native market can be obtained at Lahore for 14*l.* per ton.

3013. [5899] Sunn (*C. juncea*), Nipal. Messrs. AHMUTY & Co.

3014. [5900] Do. Do. Do.

3015. [3810] Brown hemp (*C. juncea*), Bombay.

3016. [553] Sunn (*C. juncea*), Dharwar.

3017. [9412] Do. Bolarum. DR. RIDDELL.

3018. [4014] Janapan (*C. juncea*), Salem. H. A. BRETT, Esq.

### c. JUBBULPORE HEMP (*Crotalaria tenuifolia*).

3019. [1497] Jubbulpore hemp (*C. tenuifolia*), Jubbulpore.

In forwarding this sample, the JUBBULPORE COMMITTEE appends the following remarks:—

The cultivation of 'hemp' in the district has received considerable impulse of late. Several years ago, Mr. Williams having occasion to send to Calcutta samples of wax, oilseeds, and other materials, filled up the box with indigenous hemp to prevent breakage of the bottles. On arrival at Calcutta, the cleanness and brightness of the fibre struck the consignee, who had it immediately examined by one of the proprietors of the patent Rope-walks, who pronounced it equal to the best Russian hemp, and at once sent an order for 400 maunds of it. The trade has since gradually increased, and Mr. Williams now sends about 6,000 maunds of this fibre annually to Calcutta. The plant is regularly cultivated, but the cultivation is limited. About 10 per cent. of the fibre is lost in the process of heckling, and the cost varies according to the several places in the district and seasons of the year. The price of the prepared fibre is from 3R. 8A. to 4R. per maund. The present means of inland transport is by country carts to Mirzapore, which costs 1R. 8A. per maund, and from thence to Calcutta by boats, at a farther cost of 1R. 4A. to 1R. 8A., which, with other contingencies, such as covering for carts, or guards' hire, duty in native states, and agency charges at Mirzapore, brings up the cost of the material to 7 rupees per maund before it reaches Calcutta. Mirzapore is, at present, the nearest place of export. The great length of time in getting down bulky produce from Central India, and the enormous expense of transport, have hitherto prevented Mr. Williams sending his hemp to England ; but all this will be overcome the moment the railway line from Bombay to Jubbulpore opens, when the hemp can be landed at Liverpool in as many days as it now takes weeks by country carts and native boats to convey it from this station to Calcutta, and there is no doubt that in a very few years hemp, and also flax, will become large articles of export from the Saugor and Nerbudda territories.

# TABLE SHOWING THE QUANTITIES (AS FAR AS CAN BE ASCERTAINED) AND VALUE* OF HEMP† AND HEMP ROPE† EXPORTED FROM INDIA AND EACH PRESIDENCY TO ALL PARTS OF THE WORLD FROM 1850-51 TO 1860-61 INCLUSIVE.

Quantities in lbs.; values in £.

| Years | Whence Exported | United Kingdom Qty | U.K. Value | France Qty | France Value | Other Parts of Europe Qty | Other Europe Value | North America Qty | N. America Value | China Qty | China Value | Arabian and Persian Gulf Qty | Arab. & Pers. Value | Islands and Shores of the Indian Seas Qty | Islands Value | All Other Places Qty | All Other Value | Total Exported to All Parts Qty (lbs) | Total (tons) | Total Value |
|---|---|---|---|---|---|---|---|---|---|---|---|---|---|---|---|---|---|---|---|---|
| 1850-51 | Bengal | 475,440 | 1,225 | 3,024 | 7 | 6,608 | 16 | 286,160 | 1,970 | 4,144 | 35 | 8,400 | 38 | 71,456 | 701 | 248,752 | 2,641 | 1,092,660 | 488 | 6,588 |
| | Madras | 32,928 | 132 | | | | | | | | | | | | | | | 41,328 | 18 | 170 |
| | Bombay | 879,312 | 3,740 | | | | | | | | | 341,824 | 1,430 | | | 3,584 | 16 | 1,224,720 | 547 | 5,186 |
| | ALL INDIA | 1,387,680 | 5,097 | 3,024 | 7 | 6,608 | 16 | 286,160 | 1,970 | 4,144 | 35 | 350,224 | 1,468 | 71,456 | 701 | 252,336 | 2,657 | 2,358,608 | 1,053 | 11,944 |
| 1851-52 | Bengal | 117,376 | 285 | | | | | 996,016 | 6,592 | 5,824 | 50 | 1,344 | 6 | 103,264 | 1,024 | 262,304 | 2,273 | 1,484,784 | 663 | 10,224 |
| | Madras | 388,976 | 3,490 | | | | | | | | | 348,880 | 1,879 | | | 7,728 | 23 | 398,048 | 178 | 3,619 |
| | Bombay | 1,163,120 | 6,231 | | | | | | | | | | | | | | | 1,512,000 | 675 | 8,110 |
| | ALL INDIA | 1,669,472 | 10,006 | | | | | 996,016 | 6,592 | 5,824 | 50 | 350,224 | 1,885 | 103,264 | 1,024 | 270,032 | 2,296 | 3,394,832 | 1,516 | 21,853 |
| 1852-53 | Bengal | 711,312 | 1,786 | | | | | 1,192,464 | 6,291 | 11,088 | 95 | 8,288 | 10 | 268,576 | 2,479 | 256,144 | 2,481 | 2,450,896 | 1,094 | 13,149 |
| | Madras | 394,240 | 1,810 | | | | | 672 | | | | 2,016 | 9 | | | 448 | 7 | 397,376 | 177 | 1,835 |
| | Bombay | 2,614,640 | 14,007 | | | | | | | | | 310,128 | 1,661 | | | 5,600 | 29 | 2,930,368 | 1,308 | 15,697 |
| | ALL INDIA | 3,720,192 | 17,603 | | | | | 1,191,136 | 6,300 | 11,088 | 95 | 320,432 | 1,680 | 268,576 | 2,479 | 262,192 | 2,517 | 5,778,540 | 2,579 | 30,681 |
| 1853-54 | Bengal | 2,695,840 | 6,864 | 224 | 1 | 1,568 | 4 | 3,076,192 | 12,202 | 35,280 | 496 | 2,912 | 36 | 681,744 | 5,582 | 1,111,488 | 9,124 | 7,568,176 | 3,379 | 33,808 |
| | Madras | 347,312 | 1,202 | | | | | | | | | 112 | 1 | 112 | 1 | | | 347,760 | 155 | 1,205 |
| | Bombay | 7,309,568 | 39,238 | | | 94,080 | 840 | | | | | 476,244 | 1,956 | | | | | 7,785,792 | 3,476 | 41,194 |
| | ALL INDIA | 10,352,720 | 47,304 | 224 | 1 | 95,648 | 844 | 3,076,192 | 12,202 | 35,280 | 496 | 479,248 | 1,993 | 681,856 | 5,583 | 1,111,488 | 9,124 | 15,701,728 | 7,010 | 76,207 |
| 1854-55 | Bengal | 6,329,408 | 13,680 | 38,080 | 93 | 13,104 | 48 | 1,623,888 | 8,825 | 31,696 | 420 | 1,008 | 12 | 344,736 | 3,382 | 885,136 | 7,887 | 8,259,104 | 3,687 | 34,279 |
| | Madras | 877,184 | 3,058 | | | | | | | | | 112 | 1 | | | 1,344 | 9 | 878,640 | 392 | 3,068 |
| | Bombay | 6,230,000 | 37,767 | | | 189,056 | 1,688 | | | | | 300,608 | 2,536 | | | | | 6,624,688 | 2,957 | 41,143 |
| | ALL INDIA | 12,436,592 | 54,505 | 38,080 | 93 | 202,160 | 1,736 | 1,623,888 | 8,825 | 31,696 | 420 | 301,728 | 2,549 | 344,736 | 3,382 | 886,480 | 7,896 | 15,762,432 | 7,036 | 78,590 |
| 1855-56 | Bengal | 1,697,696 | 6,350 | 116,816 | 576 | 19,824 | 108 | 1,476,944 | 8,363 | 19,152 | 227 | 5,040 | 55 | 175,056 | 1,717 | 215,936 | 2,431 | 3,732,288 | 1,666 | 19,960 |
| | Madras | 528,640 | 2,446 | 672 | 3 | 2,464 | 11 | | | | | 7,616 | 36 | | | | | 636,928 | 289 | 2,485 |
| | Bombay | 3,463,936 | 30,928 | 15,344 | 137 | | | | | | | 423,136 | 3,777 | | | | | 4,091,472 | 1,827 | 36,530 |
| | ALL INDIA | 5,690,272 | 39,724 | 132,812 | 716 | 22,288 | 119 | 1,476,944 | 8,363 | 19,152 | 227 | 435,792 | 3,868 | 175,056 | 1,717 | 215,936 | 2,431 | 8,360,688 | 3,732 | 58,975 |
| 1856-57 | Bengal | 508,144 | 2,014 | 33,824 | 138 | | | 3,288,320 | 14,088 | 19,152 | 227 | 9,856 | 119 | 107,520 | 1,307 | 903,952 | 5,421 | 4,870,768 | 2,174 | 23,284 |
| | Madras | 271,152 | 1,047 | | | | | | | | | 336 | 1 | | | 224 | 1 | 308,000 | 138 | 1,198 |
| | Bombay | 2,676,576 | 23,851 | | | | | | | | | 363,552 | 3,246 | | | 59,136 | 528 | 3,099,264 | 1,384 | 27,625 |
| | ALL INDIA | 3,455,872 | 26,912 | 33,824 | 138 | | | 3,288,320 | 14,088 | 19,152 | 227 | 373,744 | 3,366 | 107,520 | 1,307 | 963,312 | 5,950 | 8,278,032 | 3,696 | 52,107 |
| 1857-58 | Bengal | 2,054,080 | 13,662 | 14,896 | 118 | | | 1,002,288 | 7,779 | | | 2,464 | 30 | 248,304 | 2,555 | 669,088 | 6,216 | 3,991,120 | 1,782 | 30,360 |
| | Madras | 637,840 | 2,237 | 3,024 | 10 | 6,272 | 26 | | | | | 21,280 | 95 | | | 29,904 | 533 | 698,320 | 311 | 2,901 |
| | Bombay | 2,894,528 | 18,094 | 57,232 | 255 | | | 2,464 | 11 | | | 448,784 | 2,071 | | | 1,792 | 10 | 3,404,800 | 1,520 | 20,441 |
| | ALL INDIA | 5,586,448 | 33,993 | 75,152 | 383 | 6,272 | 26 | 1,004,752 | 7,790 | | | 472,528 | 2,196 | 248,304 | 2,555 | 700,784 | 6,759 | 8,094,240 | 3,613 | 53,702 |
| 1858-59 | Bengal | 645,120 | 2,951 | 29,960 | 109 | | | 746,228 | 6,754 | | | 2,520 | 22 | 665,868 | 7,827 | | | 2,089,696 | 933 | 17,663 |
| | Madras | 117,040 | 378 | | | 2,240 | 10 | | | | | 36,064 | 161 | | | | | 155,344 | 69 | 549 |
| | Bombay | 3,753,344 | 16,757 | | | | | | | | | 719,824 | 3,212 | | | | | 4,473,168 | 1,997 | 19,969 |
| | ALL INDIA | 4,515,504 | 20,086 | 29,960 | 109 | 2,240 | 10 | 746,228 | 6,754 | | | 758,408 | 3,395 | 665,868 | 7,827 | | | 6,718,208 | 2,999 | 38,181 |
| 1859-60 | Bengal | 503,440 | 19,929 | 77,784 | 2,836 | | | 289,128 | 25,903 | | | 3,192 | 333 | | | 468,888 | 45,803 | 1,342,432 | 599 | 94,804 |
| | Madras | 350,112 | 1,109 | 21,280 | 58 | | | | | | | 3,808 | 20 | | | 1,232 | 3 | 376,432 | 168 | 1,190 |
| | Bombay | 3,044,608 | 13,592 | 87,808 | 392 | | | 29,456 | 131 | | | 283,696 | 1,267 | | | | | 3,445,568 | 1,538 | 15,382 |
| | ALL INDIA | 3,898,160 | 34,630 | 186,872 | 3,286 | | | 318,584 | 26,034 | | | 290,696 | 1,620 | | | 470,120 | 45,806 | 5,164,432 | 2,305 | 111,376 |
| 1860-61 | Bengal | 372,428 | 2,781 | 1,344 | 8 | | | 205,744 | 3,190 | 448 | 2 | | | | | 327,572 | 4,433 | 907,536 | 405 | 10,414 |
| | Madras | | | 3,808 | 43 | | | | | | | 1,344 | 10 | | | | | 5,152 | 2 | 53 |
| | Bombay | 1,710,576 | 7,670 | 3,248 | 15 | | | 20,720 | 92 | | | 413,728 | 2,162 | | | 14,448 | 65 | 2,162,720 | 965 | 10,004 |
| | ALL INDIA | 2,083,004 | 10,451 | 8,400 | 66 | | | 226,464 | 3,282 | 448 | 2 | 415,072 | 2,172 | | | 342,020 | 4,498 | 3,075,408 | 1,372 | 20,471 |

* The Indian rupee is here converted at the exchange of two shillings.

† This is the term used in the returns from India, but it is a misnomer. The 'hemp' here specified is almost entirely the produce of the leguminous plant called 'sunn' (*Crotalaria juncea*), not of the *Cannabis sativa*, or true hemp.

*d.* **DUNCHEE** (*Sesbania aculeata*).

For rope-making purposes, well worthy of attention.

3020. [5902] Dunchee (*S. aculeata*), Nepal? Messrs. AHMUTY & CO., *Calcutta.*

*e.* **JETEE** (*Marsdenia tenacissima*).

3021. [9410] Jetee (*M. tenacissima*). INDIA MUSEUM.

3022. [5954] Kumbhee (*Careya arborea*), Lucknow.

This is a large tree which abounds in mountainous districts. The fibre it affords is coarse, and only applicable for rough cordage. Matches for matchlocks are made from its bark.

3023. [10726] Kavane (*Sterculia urens*), Mangalore. V. P. COELHO, Esq.

3024. [5336] Do. Assam. H. L. MICHEL, Esq.

The bark of this, as well as almost every other species of *Sterculia*, affords a strong, but not very handsome, fibre, which may be employed for making ropes.

3025. [9402] Kodal, Cuttack.

This substance is the inner bark of a forest tree. It is reported to make the strongest and most durable rope, which is said not to be liable to deteriorate from wet, and hence it is made into boat cables. The specimen was procured from Autgurh, where the fibre is collected for sale on requisition by the Sahars. The tree, which the Committee have not been able to examine, may possibly be the *Sterculia villosa*, which in Assam is called the 'Oadal,' and the fibres are employed for making ropes with which to secure wild elephants.*

3026. [7757] Scalie, Cuttack.

This is the fibre of a gigantic twining plant, common throughout the forest jungles of the province. It is used in these parts for cordage, and is made into twine for mat-making and roofing purposes.†

IV.—FIBRES FOR MISCELLANEOUS PURPOSES, ADAPTED FOR TWINE, CORDAGE, AND PAPER, OCCASIONALLY CAPABLE OF MANUFACTURE INTO FABRICS SUITED FOR WOMEN'S DRESSES; IMITATION HORSEHAIR CLOTH, ETC.

This division embraces the fibres furnished by the leaves and stems of endogenous plants.

*a.* **PINE APPLE** (*Ananassa sativa*).

This plant supplies the only fibre of the group which is at all likely to be employed for spinning by machinery. Its fibres are fine and very divisible.

3027. [4886] Pine-apple fibre (*A. sativa*), Chingleput. DR. SHORTT.

The leaves are gathered in the same way as the aloe, and are placed on a piece of board and scraped with a blunt knife. The fibres that are loosened are drawn out, the leaves turned over, and from four to six inches of the stem end scraped as before, and as soon as the fibres are loosened by the removal of the pulp in that part of the leaf, the fibres are taken hold of by the fingers and drawn out. These fibres are again laid on the board, and any remaining portion of the pulp gently scraped out with the aid of water, when they are gathered and dried in the sun.

By another mode of treatment, the leaves are laid in the sun so as to dry up a portion of the sap, when, on being taken up and bruised by the hand, the fibres become loosened, and may be taken hold of, and drawn out. But a great loss of fibre results, so that this method cannot be recommended.

3028. [8138] Pine-apple fibre (*A. sativa*), Bolarum. DR. RIDDELL.

3029. [10720] Do. Mangalore. V. P. COELHO, Esq.

3030. [10721] Jungle pine-apple, Mangalore. V. P. COELHO, Esq.

3031. [6345] Do. Malacca. TAN KEIN SING.

3032. [6377] Do. Do.

*b.* **MOORVA, MAROOL, or BOW-STRING HEMP** (*Sanseviera zeylanica*).

This plant supplies a fibre in point of strength and other qualities well calculated, when properly prepared, to compete with the 'Manilla hemp' of the Philippine Islands.

3033. [7784] Moorga or Moorgavee (*S. zeylanica*), Cuttack.

This plant is both indigenous and common in the province, growing alike in low, marshy, shady spots along the coast, as in Balasore, on high gravelly grounds in the interior, and in the jungle in the tributary Mehals. The fibre it yields in this district is only used for the manufacture of bowstring.

The sample of rope (see Class XIX.) was made for the Local Committee by Captain Bond of Balasore; the leaves are to be had for the collecting. They are, when matured, about two feet long, and each leaf yields from thirty to forty threads. The same tedious process of detaching the thread from the cellular tissue is employed with this as

* Local Committee, *Cuttack.*
† Local Committee, *Cuttack.*

with the Agave leaf, and it would be necessary to introduce a mechanical method of doing so, ere the production could be made profitable on a large scale. The bowstring and sample of fibre were prepared to order in Cuttack: not being a marketable article, no attempt at a specification of price can be made.*

3034. [4009] Munjum (*S. zeylanica*), Salem. H. A. BRETT, Esq.

3035. [4887] Bowstring hemp (*S. zeylanica*), Chingleput. DR. SHORTT.

3036. [8140] Do. DR. RIDDELL.

3037. [4847] Munjee (*S. zeylanica*), Bangalore.

3038. [4844] Marool (*S. zeylanica*), Madras. DR. HUNTER.

3039. [4851] Tow of *Sanseviera cylindrica*, Do. Do.

c. **AGAVE**† or **ALOE FIBRES** (*Agave americana*, also *Agave vivipara* or *Fourcroya gigantea*).

Although neither of these plants is indigenous, both are now cultivated in many parts of the country. A variety of specimens have been forwarded for exhibition. After suitable preparation, the agave fibre is usually employed for the manufacture, amongst other things, of an imitation 'horsehair' cloth.

3040. [4884] Aloe (*A. americana*), Chingleput. DR. SHORTT.

3041. [4879] Great aloe (*F. gigantea*), Chingleput. DR. SHORTT.

The leaves, cut close to the stem, are placed on a piece of board, and beaten with a short stout stick. After being thus bruised, the pulpy portions are scraped out with a blunt knife, and the fibres subsequently washed in clean water and dried in the sun.

3042. [9405] 'Hatteecheeghar' (*F. gigantea*), Lucknow.

Planted in hedges, and grows luxuriantly without any farther cultivation. The extent of present cultivation is very limited, but it is capable of being extended all over Oude, and in any soil. The fibre has been prepared only on a small scale. In the Lucknow jail rope and sackcloth *have been made of it*.‡

3043. [4845] Agave (*Agave americana*), Madras. DR. HUNTER.

3044. [2725] Kathali, long aloe, Bangalore. DR. KIRKPATRICK.

3045. [2726] Chicca Kathali, short aloe, Do. Do.

3046. [10727] Wild aloe (*Aloe indica*), Mangalore.

3047. [8126] Aloe (*Agave americana*). DR. RIDDELL.

3048. [8140] Aloe fibre dressed. Do.

3049. [8127] Do. dyed in colours. Do.

3050. [8128] Do. do. Madras.

3051. [554] Agave, Dharwar.

3052. [6483] Aloe (*Agave americana*), Chota Nagpore.

3053. [2726] Do. Do.

3054. [9406] Do. Agra.

3055. [10306] Aloe fibre, dyed and undyed, Bareilly.

3056. [8144] Specimens of Agave fibre dyed and prepared. W. STAUFEN, Esq.

3057. [8144] Brush manufactured from do. Do.

By process originally patented in this country by Exhibitor.

d. **ADAM'S NEEDLE** (*Yucca gloriosa*).

This plant, although not yet cultivated for economic purposes, produces fibre of very considerable value when properly prepared.

3058. [4841] Adam's needle (*Y. gloriosa*), Chingleput.* DR. SHORTT.

3059. [9407] Adam's needle (*do.*), Chingleput. DR. RIDDELL.

3060. [2733] Howah Kathali (*do.*) Bangalore. DR. KIRKPATRICK.

3061. [8141] Adam's needle (*Y. angustifolia*), Madras. DR. HUNTER.

e. **PLANTAIN** (*Musa paradisiaca*).

Universally cultivated for its fruit. Its leaves afford a fibre suited for certain purposes. Ordinarily it is inferior to Manilla hemp (*Musa textilis*) in point of strength.

---

* Local Committee, *Cuttack*.
† Misnamed aloe, but having now become the 'trade' term, it is likely to be retained.
‡ Central Committee, *Lucknow*.

* Central Committee, *Lucknow*.

3062. [5901] Plantain (*M. paradisiaca*), Nepaul. Messrs. AHMUTY & Co.

3063. [4889] Do. Chingleput. DR. SHORTT.

These fibres were prepared from the inner footstalks of the plantain tree. These were taken of certain length, placed on a piece of board, and the pulpy mass scraped out with a blunt knife. Both sides of the stalks having been thus scraped, whilst clean water was poured on to wash away the remains of the pulp, the fibres were dried in the sun.

3064. [9408] Do. Bolarum. DR. RIDDELL.

3065. [4846] Manilla hemp (*M. textilis*), Madras. DR. HUNTER.

3066. [4842] Plantain (*M. paradisiaca*), Do. Do.

3067. [4849] Do. Bangalore.

3068. [10728] Red do. (*M. Cavendishii*), Mangalore. V. P. COELHO, Esq.

3069. [10725] Jungle plantain (*M. superba*), Do. Do.

3070. [10722] Plantain (*M. paradisiaca*), Do. Do.

3071. [9176] Do. Vizagapatam. H. H. the RAJAH.

3072. [8129] Do. Singapore.

*f.* SCREW-PINE (*Pandanus odoratissimus*).

The leaves of this plant furnish a fibre which can be turned to account for the manufacture of paper and some common purposes. It is, however, in every respect inferior to those in this group above entered.

3073. [4856] Screw pine (*P. odoratissimus*), Chingleput. DR. SHORTT.

3074. [8146] Screw pine (*P. odoratissimus*), Madras. DR. HUNTER.

Prepared like the pine-apple, but no water should be used, simple scraping, after which the fibres are put to dry.

V.—FIBRES SUITED FOR THE MANUFACTURE OF MATS, BRUSHES, COARSE CORDAGE, IMITATION HORSE HAIR FOR STUFFING PURPOSES, ETC.

*a.* COIR.

This well-known material is furnished by the fibrous envelope of the nut of the cocoa palm (*Cocos nucifera*). It is exported from India in considerable quantities, as will be gathered from the table (p. 150), which indicates the amount of coir and coir-rope supplied from India and each Presidency to all parts of the world from 1850-1 to 1860-1 inclusive.

3075. [4874] Coir (*Cocos nucifera*), Chingleput. DR. SHORTT.

3076. [453] Do. DR. MCPHERSON.

3077. [2728] Do., Bangalore. DR. KIRKPATRICK.

The outer rind of the nut is taken, bruised, and steeped in water for two or three days, when it is taken up and the fibres separated by the fingers and scraped gently with a blunt knife and dried in the sun. If steeped in water too long, they get dark-coloured.

*b.* GOMUTI (*Arenga saccharifera*).

This fibre is considered superior to all others yet made use of for the manufacture of artificial bristles for brushes, imitation horse-hair for stuffing, and such like purposes.

3078. [6349] Gomuti (*A. saccharifera*), Singapore. COL. COLLYER.

3079. [4091] Ejow (*A. saccharifera*), Province Wellesley.

3080. [4091] Do. Penang.

3081. [8143] Do. dyed and prepared for use as brush bristles, &c. (patented). W. STAUFEN, Esq.

3082. [8143] Hair-brush manufactured from do. Do.

*c.* MOONJ (*Saccharum Munja*).

This grass supplies a strong good fibre, which is beginning to attract attention in this country, and is now being exported from Kurachi in Sinde.

3083. [10242] Moonj (stems) (*Saccharum Munja*), Lahore.

3084. [10240] Do. in raw state. Do.

3085. [10241] Do. in that of fibre. Do.

Used for rope making. The outer rind of the grass called *Sirki* (*Saccharum Moonja*). It is largely used for well ropes, tow lines, and for attaching buckets to Persian wheels, and in all cases where the rope is exposed to the action of water. Price, at present, from 5*l.* to 8*l.* per ton, but it can ordinarily be obtained much cheaper.*

* Central Committee, *Lahore.*

## TABLE SHOWING THE QUANTITIES (AS FAR AS CAN BE ASCERTAINED) AND VALUE* OF COIR AND COIR ROPE EXPORTED FROM INDIA AND EACH PRESIDENCY TO ALL PARTS OF THE WORLD FROM 1850-51 TO 1860-61 INCLUSIVE.

| Years | Whence Exported | United Kingdom Qty (lbs) | Value (£) | France Qty (lbs) | Value (£) | Other Parts of Europe Qty (lbs) | Value (£) | North America Qty (lbs) | Value (£) | China Qty (lbs) | Value (£) | Arabian and Persian Gulf Qty (lbs) | Value (£) | Islands and Shores of the Indian Seas Qty (lbs) | Value (£) | All Other Places Qty (lbs) | Value (£) | Total Exported to All Parts Qty (lbs) | (tons) | Value (£) |
|---|---|---|---|---|---|---|---|---|---|---|---|---|---|---|---|---|---|---|---|---|
| 1850-51 | Bengal | 32,704 | 212 | 88,816 | 182 | | | 52,928 | 498 | 10,416 | 126 | 681,856 | 1,767 | 1,008 | 10 | 56,000 | 475 | 152,656 | 68 | 1,321 |
| | Madras | 1,745,072 | 3,258 | | | | | | | | | 247,968 | 735 | 10,304 | 19 | 589,280 | 1,179 | 3,065,328 | 1,369 | 6,405 |
| | Bombay | 4,196,304 | 12,751 | | | | | 97,552 | 166 | | | | | | | 86,912 | 266 | 4,628,736 | 2,066 | 13,918 |
| | ALL INDIA | 5,974,080 | 16,221 | 88,816 | 182 | | | 150,080 | 664 | 10,416 | 126 | 929,824 | 2,502 | 11,312 | 29 | 732,192 | 1,920 | 7,845,720 | 3,593 | 21,644 |
| 1851-52 | Bengal | 2,707,824 | 5,097 | | | | | 113,904 | 310 | 5,488 | 68 | 810,096 | 1,974 | | | 372,960 | 834 | 4,064,256 | 1,814 | 8,387 |
| | Madras | 896,896 | 3,610 | 59,472 | 122 | | | 124,880 | 558 | 6,944 | 31 | 291,424 | 1,290 | 18,480 | 83 | 94,192 | 894 | 1,452,816 | 640 | 5,966 |
| | Bombay | | | | | | | | | | | | | | | | | | | |
| | ALL INDIA | 3,604,720 | 8,707 | 59,472 | 122 | | | 238,784 | 868 | 12,412 | 99 | 1,101,520 | 3,264 | 18,480 | 83 | 503,104 | 1,556 | 5,538,512 | 2,472 | 14,699 |
| 1852-53 | Bengal | 11,200 | 68 | 9,408 | 23 | | | 5,264 | 25 | 5,264 | 26 | | | | | 896 | 7 | 22,694 | 10 | 126 |
| | Madras | 3,270,400 | 5,722 | | | | | 61,600 | 137 | 25,312 | 70 | 927,360 | 3,688 | 31,808 | 101 | 310,688 | 944 | 4,636,576 | 2,070 | 10,685 |
| | Bombay | 1,277,248 | 4,055 | | | | | | | 22,064 | 98 | 426,384 | 1,776 | 20,160 | 93 | 106,288 | 467 | 1,852,144 | 897 | 6,489 |
| | ALL INDIA | 4,558,848 | 9,845 | 9,408 | 23 | | | 66,864 | 162 | 52,640 | 194 | 1,353,744 | 5,464 | 51,968 | 194 | 417,872 | 1,418 | 6,511,324 | 2,977 | 17,300 |
| 1853-54 | Bengal | 3,808 | 48 | 14,672 | 27 | 141,668 | 247 | 107,408 | 691 | 26,656 | 360 | 841,232 | 2,000 | 21,728 | 229 | 18,776 | 159 | 173,376 | 77 | 1,487 |
| | Madras | 4,784,912 | 8,928 | 169,792 | 569 | | | | | 12,992 | 36 | 307,888 | 1,277 | 11,200 | 77 | 345,744 | 1,243 | 6,102,320 | 2,724 | 12,658 |
| | Bombay | 2,499,056 | 9,079 | | | | | | | | | | | | | 185,024 | 671 | 3,161,760 | 1,412 | 11,696 |
| | ALL INDIA | 7,287,776 | 18,055 | 184,464 | 596 | 141,568 | 247 | 107,408 | 691 | 39,648 | 396 | 1,149,120 | 3,277 | 32,928 | 306 | 544,544 | 2,073 | 9,437,456 | 4,213 | 25,641 |
| 1854-55 | Bengal | 119,840 | 1,524 | 10,752 | 20 | 79,744 | 132 | 74,704 | 460 | 14,784 | 90 | 660,016 | 1,787 | | | 248,864 | 2,107 | 458,192 | 205 | 4,181 |
| | Madras | 3,899,616 | 9,268 | 66,528 | 277 | | | 136,640 | 607 | 9,184 | 41 | 12,768 | 557 | | | 780,576 | 3,227 | 5,380,704 | 2,402 | 14,484 |
| | Bombay | 1,918,784 | 6,532 | | | | | | | | | | | | | 95,760 | 409 | 2,239,664 | 999 | 8,423 |
| | ALL INDIA | 5,938,240 | 17,324 | 77,280 | 297 | 79,744 | 132 | 211,344 | 1,067 | 23,968 | 131 | 672,784 | 2,344 | | | 1,075,200 | 5,743 | 8,078,560 | 3,606 | 27,938 |
| 1855-56 | Bengal | 9,072 | 66 | 21,280 | 39 | 142,016 | 359 | 34,720 | 321 | 28,560 | 267 | 336 | 4 | 71,120 | 713 | 75,376 | 618 | 219,184 | 98 | 1,989 |
| | Madras | 1,875,328 | 4,549 | | | | | 74,032 | 444 | 42,448 | 178 | 512,288 | 1,464 | 66,976 | 223 | 658,336 | 4,105 | 3,350,256 | 1,496 | 11,183 |
| | Bombay | 1,498,112 | 6,251 | | | | | | | | | 180,208 | 1,037 | 1,008 | 4 | 67,648 | 277 | 1,789,424 | 799 | 7,747 |
| | ALL INDIA | 3,382,512 | 10,866 | 21,280 | 39 | 142,016 | 359 | 108,752 | 765 | 71,008 | 445 | 692,832 | 2,505 | 139,104 | 940 | 801,360 | 5,000 | 5,358,864 | 2,393 | 20,919 |
| 1856-57 | Bengal | 70,672 | 601 | 55,664 | 474 | 195,776 | 494 | 1,568 | 13 | 11,984 | 102 | 474,320 | 1,168 | 38,304 | 327 | 172,928 | 1,470 | 351,120 | 157 | 2,987 |
| | Madras | 2,210,768 | 4,703 | 17,360 | 31 | | | 158,480 | 436 | 8,064 | 22 | | | | | 1,581,104 | 3,768 | 4,645,872 | 2,074 | 10,622 |
| | Bombay | 2,784,656 | 9,961 | | | | | | | 64,512 | 288 | 277,872 | 1,214 | 61,876 | 274 | 108,304 | 482 | 3,296,720 | 1,472 | 12,219 |
| | ALL INDIA | 5,066,096 | 15,265 | 73,024 | 505 | 195,776 | 494 | 160,048 | 449 | 84,560 | 412 | 752,192 | 2,382 | 99,680 | 601 | 1,862,336 | 5,720 | 8,293,711 | 3,703 | 25,828 |
| 1857-58 | Bengal | | | 7,056 | 76 | 169,456 | 415 | | | 2,016 | 17 | | | 35,056 | 283 | 18,928 | 257 | 226,576 | 101 | 1,849 |
| | Madras | 3,327,744 | 6,729 | 50,064 | 133 | 10,864 | 39 | 163,520 | 1,816 | | | 520,128 | 1,254 | | | 706,944 | 1,968 | 4,774,336 | 2,131 | 10,499 |
| | Bombay | 4,170,880 | 18,756 | | | | | 87,584 | 370 | | | 194,880 | 984 | 80,192 | 400 | 57,792 | 284 | 4,602,192 | 2,055 | 20,833 |
| | ALL INDIA | 7,498,624 | 25,485 | 57,120 | 209 | 180,320 | 454 | 251,104 | 1,586 | 2,016 | 17 | 715,008 | 2,238 | 115,248 | 683 | 783,664 | 2,509 | 9,603,104 | 4,287 | 33,181 |
| 1858-59 | Bengal | 100,408 | 1,220 | 3,528 | 30 | 70,000 | 157 | 212,110 | 1,943 | 448 | 11 | | | | | 186,050 | 1,669 | 502,544 | 224 | 4,873 |
| | Madras | 3,715,600 | 7,503 | 3,472 | 7 | | | 315,952 | 599 | | | 562,464 | 1,384 | | | 953,120 | 2,212 | 5,680,608 | 2,509 | 11,862 |
| | Bombay | 3,306,240 | 15,799 | | | | | 316,288 | 1,358 | | | 232,176 | 1,174 | | | 272,944 | 1,369 | 4,127,648 | 1,842 | 19,700 |
| | ALL INDIA | 7,122,248 | 24,522 | 7,000 | 37 | 70,000 | 157 | 844,350 | 3,900 | 448 | 11 | 794,640 | 2,558 | | | 1,412,114 | 5,250 | 10,250,800 | 4,575 | 36,435 |
| 1859-60 | Bengal | 2,268 | 253 | 3,920 | 6 | 89,200 | 222 | 1,876 | 164 | | | | | | | 108,500 | 8,859 | 112,644 | 50 | 8,776 |
| | Madras | 5,192,320 | 9,644 | | | | | 224,224 | 998 | 2,800 | 15 | 448,336 | 1,092 | | | 441,504 | 939 | 6,125,280 | 2,734 | 11,903 |
| | Bombay | 4,167,856 | 18,602 | | | | | | | | | 82,992 | 437 | | | 93,184 | 470 | 4,571,056 | 2,041 | 20,522 |
| | ALL INDIA | 9,362,444 | 28,499 | 3,920 | 6 | 89,200 | 222 | 226,100 | 1,162 | 2,800 | 15 | 531,328 | 1,529 | | | 643,188 | 9,758 | 10,808,980 | 4,825 | 41,201 |
| 1860-61 | Bengal | 1,036 | 9 | 54,992 | 212 | 106,400 | 396 | 255,584 | 1,333 | | | | | | | 49,000 | 478 | 50,026 | 22 | 487 |
| | Madras | 6,353,872 | 22,574 | | | | | | | 7,840 | 42 | 680,848 | 2,612 | | | 1,440,128 | 4,162 | 8,626,240 | 3,851 | 29,956 |
| | Bombay | 3,573,808 | 21,314 | | | | | | | | | 329,392 | 2,550 | | | 221,760 | 1,602 | 4,388,384 | 1,959 | 26,841 |
| | ALL INDIA | 9,928,716 | 43,897 | 54,992 | 212 | 106,400 | 396 | 255,584 | 1,333 | 7,840 | 42 | 1,010,240 | 5,162 | | | 1,710,888 | 6,242 | 13,064,660 | 5,832 | 57,284 |

* The Indian rupee is here converted at the exchange of two shillings.

3086-8. [5920-2] Moonj in state of fibre, Lucknow.

Grows wild all over Oude, and is planted in hedges. The moonj or fibre is prepared from the vagina of the leaf just when the stem begins to bear flower; ropes for towing boats on rivers, and twine for bottoms of charpoys (bedsteads), are made from this fibre. If it is not occasionally wetted, and allowed to become too dry, it easily *breaks when used.* The prepared fibre costs two rupees per maund.*

3089. [595] Bunkuss. Lucknow.

3089*a.* [10243] Dab grass. Lahore.

3089*b.* [4873] Palmyra (*Borassus flabelliformis*), Chingleput. DR. SHORTT.

3090. [4010] Do. Salem.

*Series of Mat-making Materials, exhibited by* DR. SHORTT, *Chingleput.*

3091. [4864] Screw-pine leaves (*Pandanus odoratissimus*).

Used also to cover umbrellas.

3092. [4858] Mat-grass (*Cyperus textilis*).

3093. [*] Bulrush (*Typha elephantina*).

3094. [4882] Palmyra leaves (*Borassus flabelliformis*).

Used also for native books.

3095. [4865] Date leaves (*Phœnix dactylifera*).

3096. [4860] Cocoa-nut leaves (*Cocos nucifera*).

3097. [5118] Vittivayr (*Andropogon muricatum*).

Used for thatching houses. Roots employed as cuscus tats (air coolers).

3098. [*] Mat-grass (*Typha angustifolia*).

*Materials for Baskets, Brooms, &c. exhibited by* Dr. SHORTT, *of Chingleput.*

3099. [*] Midrib of leaves of *Cocos nucifera.*

3100. [4867] Basket material (*Vitex negundo*).

3101. [4870] Basket material (*Terreola buxifolia*).

3102. [4868] Do. (*Elate sylvestris*).

3103. [4869] Cacha codie.

Stems of a creeper used for tying bundles and other purposes instead of twine.

3104-7. [4080-78-90-89] Mandrong rushes, for sugar and rice bags, Province Wellesley.

3108. [4086] Mangkwang (*Pandanus* sp.) for matting, Do.

3109. [6711] Glam tree bark (*Melaleuca viridiflora*), Malacca.

3110. [6347] Talee trap (*Artocarpus* sp.) for fishing nets. Do.

3111. [6347] Talee Taras, Singapore. JOSE D'ALMEIDA, Esq.

3112. [6348] Bark used as twine, Siam. Do.

3113. [3157] Reeds and grasses from Darjeeling. DR. CAMPBELL.

*Basts from Akyab and Burmah.*

3114. [9374] Heng-kyo Shaw, Akyab.

3115. [7788] Dam Shaw, Do.

3116. [7602] Thanot Shaw, Do.

3117. [7601] Wapreeloo Shaw, Do.

3118. [7789] Shaw Goung, Do.

The above are used in preparing cordage for boats, nets, &c.: wholesale market price, 2R. 8A. per maund. The inner bark of large trees.

3119. [9375] Shaw Nee, Akyab.

3120. [7788] Shaw Phru, Do.

3121. [7600] Thengnan Shaw, Do.

The above are used in preparing cordage for boats, nets, &c.: wholesale market price, 1R. 12A. per maund. More plentiful than those above.

3122. [7603] Guand-young Shaw, Akyab.

Used for cables and strong nets: wholesale market price, 3R. 4A. per maund.
The whole of these fibres are used much by the inhabitants of the province.*

---

* Central Committee, *Lucknow.*

* Local Committee, *Akyab.*

3123. [5788] Inner bark of *Sterculia,* Thatpootnet Shaw, Burmah.

3124. [5789] Do. Shaw Laybway, Do.

3125. [5790] Do. Shaw Nee, Do.

The above are extensively used for making ropes.
These fibres, or basts, appear to be the inner barks of various species of *Sterculia* and allied plants, which abound in the districts from whence they are forwarded. They are strong and enduring in their nature, and some of them have been tested with satisfactory results, but they are coarse and ill prepared, so that in their present condition they would scarcely command a price in the English markets. No information has been afforded of the plants from whence they have been obtained.

*Miscellaneous Fibres, exhibited with the native names only, and without definite information.*

3126. [2729] Dudi, Bangalore. GOVERNMENT of MYSORE.

3127. [2730] Pundi, Do. Do.

3128. [9177] Jillado, Do.

3129. [9178] Sagao, Do.

3130. [9179] Udda, Do.

3131. [9180] Mogally, Do.

3132. [9182] Urasa (*Ananassa* sp.), Do.

3133. [10707] Shreetaly, Mangalore. V. P. COELHO, Esq.

3134. [10708] Daddal, Mangalore. V. P. COELHO, Esq.

3135. [10709] Daddas, Do. Do.

3136. [10710] Kappace, Do. Do.

3137. [10711] Nareeta Bally, Do. Do.

3138. [10712] Haralley, Do. Do.

3139. [10714] Purelley, Do. Do.

3140. [10715] Vatta, Do. Do.

3141. [10716] Kalley, Do. Do.

3142. [10717] Basarry, Do. Do.

3143. [10719] Mut'alla, Do. Do.

3144. [10721] Jungle apple, Do. Do.

3145. [10713] Banyan (*Ficus*). Do. Do.

3146. [10718] Kinni golly (*do.*) Do. Do.

## RATTANS AND CANES.

The Table below shows the value of the rattans and canes exported from India during the last four years.

3146*a.* [5175] Specimen of cane used for thatching houses, Akyab. GOVERNMENT of INDIA.

3146*b.* [3157] Rattans, Darjeeling. DR. CAMPBELL.

TABLE SHOWING THE VALUE OF RATTANS AND CANES EXPORTED FROM INDIA AND EACH PRESIDENCY TO ALL PARTS OF THE WORLD FROM 1857-58 TO 1860-61.

| YEARS | WHENCE EXPORTED | COUNTRIES WHITHER EXPORTED | | | | | | | TOTAL EXPORTED TO ALL PARTS |
|---|---|---|---|---|---|---|---|---|---|
| | | UNITED KINGDOM | FRANCE | OTHER PARTS OF EUROPE | AMERICA | CHINA | ARABIAN AND PERSIAN GULFS | OTHER PARTS * | |
| | | Value | Value | Value | Value | Value | Value | Value | Value |
| 1857-58 | Bengal | £ 2,179 | £ 622 | £ 158 | £ 20 | £ .. | £ .. | £ 232 | £ 3,211 |
| | Madras | .. | .. | .. | .. | .. | .. | .. | .. |
| | Bombay | 164 | .. | .. | 76 | 266 | .. | 36 | 542 |
| | ALL INDIA | 2,343 | 622 | 158 | 96 | 266 | .. | 268 | 3,753 |
| 1858-59 | Bengal | 923 | 1,113 | 60 | 255 | 22 | .. | 270 | 2,643 |
| | Madras | .. | .. | .. | .. | .. | 4 | 7 | 11 |
| | Bombay | 923 | 729 | .. | 97 | 22 | 14 | 100 | 1,885 |
| | ALL INDIA | 1,846 | 1,842 | 60 | 352 | 44 | 18 | 377 | 4,539 |
| 1859-60 | *Bengal | 884 | 547 | 204 | 28 | .. | 151 | 431 | 2,245 |
| | Madras | .. | .. | .. | .. | .. | 17 | 11 | 28 |
| | Bombay | 51 | 490 | .. | .. | .. | .. | 14 | 555 |
| | ALL INDIA | 935 | 1,037 | 204 | 28 | .. | 168 | 456 | 2,828 |
| 1860-61 | Bengal | 701 | 91 | .. | .. | .. | .. | 1,062 | 1,854 |
| | Madras | .. | .. | .. | .. | .. | 24 | 6 | 30 |
| | Bombay | 98 | 431 | .. | 81 | .. | .. | 6 | 616 |
| | ALL INDIA | 799 | 522 | .. | 81 | .. | 24 | 1,074 | 2,500 |

\* The greater part exported to the Cape of Good Hope, Mauritius, and New South Wales.

Subdivision C.—v. CELLULAR SUBSTANCES.

This subdivision has only one representative.

3147. [10161] Sago palm tinder, Luckimpore, Assam. H. S. BIVAR, Esq.

3148. [9409] Palm tinder, Malacca.

Subdivision C.—vi. TIMBER AND FANCY WOODS USED FOR CONSTRUCTION AND ORNAMENT, AND PREPARED FOR DYEING.

The collection of woods is particularly valuable, on account of the accuracy with which the majority of the samples have been named. Although some progress has been made since 1851, the adaptation of many of these woods is still but imperfectly known. Experiments* and observations have, however, been commenced with the view of testing the strength and other qualities of those forwarded to the present Exhibition, which, when completed, will afford the means of judging of their relative value as compared with the woods of this and other countries, and so far assist in determining to what extent they are likely to prove worthy of attention for export.†

Various specimens of timbers, forming portions of the collections from Lucknow, Midnapore, Chota Nagpore, Umritsur, Cuttack, Jubbulpore, Akyab—from Dr. Campbell, Darjeeling; Dr. Graham, Moulmein; Dr. Brandis, British Burmah; Col. Reid and Lieut. Phaire, Assam; Drs. Hunter and Shortt, Madras—and from H. H. Inche Wan Aboo Bakar, Hon. Col. Cavanagh, G. Angus, Esq., Col. Collyer, C. Evans, Esq., and Hon. Major Man, Malayan Peninsula and Singapore—have, for want of space and other reasons, been deposited at the India Museum. The whole have, however, been carefully gone over, and the more promising ones selected for experiment &c.

## WOODS FROM CUTTACK.

Exhibited by T. W. ARMSTRONG, Superintending-Engineer of the Cuttack Division.

3149. [5600] Sissoo, black (*Dalbergia Sissoo*), ·875 sp. gr., 1s. per cubic ft.

3150. [5606] Do., red (*D. Sissoo*), 1·000 sp. gr., 1s. per cubic ft.

Used for every description of furniture, both by natives and Europeans. In grain and colour it somewhat resembles rosewood. The heart of this timber is generally unsound.

3151. [5610] Koozoom, 1·286 sp. gr., 7½d. per cubic ft.

Used for the handles of tools, and native cart axles, and might be applied to other purposes.

3152. [5599] Teak, 'Sagoon' (*Tectona grandis*), ·875 sp. gr., 1s. 3d. per cubic ft.

To what extent this valuable timber exists in the Sumbulpore district and its dependencies, and some of the tributary Mehals of Cuttack, has never been certainly ascertained, but is a question well worthy of careful enquiry.

3153. [5608] Koozoom, ·714 sp. gr., 6d. per cubic ft.

Used for ordinary purposes, such as packing-cases, common doors, &c.

3154. [5597] Guringa, ·714 sp. gr., 9d. per cubic ft.

A light wood, principally used by the turners of Cuttack, and for palankeen poles, &c.—purposes where lightness is a necessity.

3155. [5598] Sâl (*Shorea robusta*), 1·000 sp. gr., 1s. per cubic ft.

Common in our jungles; large quantities are floated down the river Mahanuddy, and sold at Cuttack. By the natives it is used for almost every purpose to which wood can be applied—young trees being cut down even for fuel and palings. A good supply used, some years ago, to be obtained from the jungles skirting the principal waterways of the district, for the Government gun manufactory in the Madras Presidency. In mature trees the heart is always unsound. Temporary bridges, gun-carriages, boats, beams, door-frames, trusses, &c., are generally made of this wood.

3156. [5605] Jack, 'Punsee' (*Artocarpus integrifolia*), ·750 sp. gr., 1s. per cubic ft.

Wood of which the native oil-mill or 'ghana' is made. It is also a handsome wood for furniture purposes, having a neat fresh appearance, which darkens with age. As a timber tree, however, it cannot be said to be plentiful, as it is not a forest tree, and the fruit it yields renders it more valuable than if it were simply cultivated for the sake of the timber.

3157. [5607] Peasal (*Buchanania latifolia*), ·875 sp. gr., 9d. per cubic ft.

This useful wood is worked up generally into furniture, house doors and windows, presses, tables, &c. It requires to be polished, otherwise it stains a burnt sienna colour any cloth brought into contact with it.

3158. [5601] Burdur, 1·000 sp. gr., 1s. per cubic ft.

Excellent wood for carriage poles, shafts, and wheels, and in all coach-builder's work.

3159. [5609] Keehar, 1·250 sp. gr., 9d. per cubic ft.

A hard useful wood for mallets, pounders, rammers, and such like articles, and would, perhaps, make up strong furniture.

3160. [5604] Gumbaree.

The trees which furnish these timbers are found more or less plentifully throughout the forest jungles of the Sumbulpore district, and on the banks of the Mahanuddy, Brahming, and Byturg rivers. The main difficulty attending timber transactions is the, at present, almost insurmountable one of conveying the timber from the spot where it is felled (which, of course, is for the most part in the interior of the dense forest) to the nearest spot whence water-carriage is available. Teams of buffaloes are employed for this purpose, in the present mode of operating; but if the distance to be traversed is at all considerable, it may easily be conceived that this method becomes so expensive and dilatory that much cannot be undertaken. The only means of modifying or evading this difficulty, if it were desired to embark in any extensive transactions, would apparently be to have the timber sawn and cut up on the spot, and roughly shaped for the purposes for which it might be required. Planks and sleepers, for example, or the various constituent portions of a gun-carriage, might perhaps, with advantage, be roughly shaped and cut on the spot. For the transport of timber there are, however, some facilities, which it may be proper to notice. The timber country of this division is traversed by three large streams — the Mahanuddy, Brahming, and Byturnee; so that when the united difficulty of getting the wood from the forest to the water-side has been overcome, it is a matter of ease, at the proper season, to float it down in rafts to any depôt which might be established for the purpose at the mouths of the rivers, or others connected with them. From Cuttack to False Point harbour is a distance of 65 miles, viâ the Mahanuddy. This harbour affords a safe anchorage for vessels at all seasons, and the route is available for about five months in the year, namely, from the middle of June generally to the middle of November; while from a point on the route, 42 miles from Cuttack, at Tadânda, namely, on to the sea, the river is navigable all the year round. Another route from Cuttack is viâ the Beeroopa and Brahming to the Dhamree harbour, the distance being 96 miles, and the Dhamree port safe for sailing vessels from November to February. Small steamers could ply to the harbour all the year round, and vessels of 150 to 200 tons can clear the bar. But this route from Cuttack is an uncertain one, and not open for more than three and a half months in the height of the monsoon, that is, it may be said, from July to September.*

3161. [5603] Assân (*Terminalia tomentosa*), 9*d.* per cubic ft.

3162. [5602] Abloos or Kândoo (*Diospyros melanoxylon*), 1s. 6*d.* per cubic ft.

A very handsome fancy wood.

## WOODS FROM JUBBULPORE.

3163. [4657] Seba Sagoon Teak (*Tectona grandis*).

This is called by the natives 'Oil Teak,' or 'Seba Sagoon,' and is found on the Bindhyers, north of the Nerbudda, almost exclusively, and is the best in these provinces.

3164. [4658] Putteereea Sagoon. Do.

Called by the natives 'Putteereea Sagoon,' or 'Stony Teak;' is shorter and more knotty than the last; is found in the more hilly tracts.

3165. [4659] Doodheea Sagoon. Do.

Called by the natives 'Doodheea Sagoon,' or 'Milky Teak;' is the softest timber of the three, and is found chiefly south of the Nerbudda, on the Satpoora. The only difference in the above three woods is the soil they grow on.

* Local Committee, *Cuttack.*

3166. [4660] Surrye (*Shorea robusta*).

Found chiefly in large forests in the south Mundlah, and one forest near the Puchmurries.

3167. [4661] Jiomrassee (botanical name not known).

Is a beautiful close-grained wood, the leaf oblong, and serrated edge; it is found in the more hilly tracts, but does not attain any great size.

3168. [4662] Dhengun (*Cordia Macleoda*).

A remarkably beautiful wood, found in Mundlah, Seonee.

3169. [4663] Saj (*Terminalia arguna*).

Very useful for beams and rafters; grows abundantly in all the districts to a great size, 40 to 50 feet long, and 2 to 3 feet broad; will not last if exposed to the weather.

3170. [4664] Beejah (*Pterocarpus* sp.)

An excellent wood, easily worked, grows to a large size, is found in all parts, but not very abundant.

3171. [4665] Kowah (*Terminalia arguna*).

Grows to a large size along the banks of rivers, all over the district; is an excellent lasting timber, somewhat similar in quality to ash.

3172. [4666] Ghattoo (*Zizyphus zylopyxa* or *glabra*).

It grows to a fine large tree, but is a scarce wood, and close-grained and excellent.

3173. [4667] Trosum (botanical name not known).

Good timber, but does not exist in any quantity.

3174. [4668] Dhowrah (*Conocarpus latifolius*).

A tough, knotty wood, hard to work, grows abundantly everywhere: used much for cart axles.

3175. [4669] Serlee (*Boswellia thurifera*).

Very abundant, but is soft, and has a bad character for lasting.

3176. [4670] Bher (*Zizyphus jujuba*).

Is abundant, but not often found of large size. Timber inferior as to transverse, but otherwise good.

3177. [4671] Baubul (*Acacia arabica*).

A close-grained, hard, and tough wood, but does not attain any great size; very valuable for the spokes and felloes of wheels.

3178. [4672] Khumee (botanical name not known).

Is a light, strong, and easily-worked wood, much in request by natives.

3179. [3947] Gunjah (botanical name not known).

The same as the foregoing specimens of wood.

3180. [3948] Siris (*Acacia sirisa*).

A splendid timber, but now very scarce in these parts.

3181. [3949] Hurdoo (*Nauclea cordifolia*).

Abundant, and much in request; is light and easily worked. Its strength is not great, but it is lasting, if not exposed to the weather.

3182. [3950] Kaim (*N. parvifolia*).

Somewhat similar to Hurdoo, but is a stronger, better timber.

3183. [3951] Pindra (*N. orientalis*).

Not abundant. A good joiner's wood.

3184. [3952] Jymungul.

A large tree, not of much use.

3185. [3953] Rohnee (*Acacia leucoploca?*)

An excellent and tough wood, but does not work smoothly. Abundant in the Deinwah valley and Hossingabad.

3186. [3954] Londya.

A common wood, suited for poles.

3187. [3955] Kardahee (*Conocarpus mystifolium*).

A tough wood, but difficult to work; tolerably abundant (similar to Dowrah); grows along the banks of the Nerbudda.

3188. [3956] Taman (*Eugenia jambolana*).

A coarse-grained wood, used for well steps, and in other wet places, where it is almost indestructible.

3189. [3957] Tine or Sisso (*Dalbergia Sissoo*).

A splendid timber, but not abundant; small in this part of India.

3190. [3958] Pandur.

A coarse wood, common, and is a good, strong, and lasting timber.

3191. [3959] Kumbee (*Careya arborea*).

The wood is not much used; the bark is made into slow matches for matchlocks.

3192. [3960] Hurrah.

Is abundant in the hilly tracts, but attains no great size.

3193. [3961] Mowah (*Bassia longifolia*).

This tree is so valuable for its fruit, out of which arrack is made, that it is seldom felled, except when barren; but its wood is excellent.

3194. [3962] Tendoo (*Diospyros ebenum*).

The heart-wood of the tendoo; it is found to a large size in the Seonee district, but generally small elsewhere.

3195. [1214] Doodhee (*Asclepias rosea*).

An inferior timber of no transverse strength.

3196. [1215] Karee (*Uvaria*).

Used by natives for making toys.

3197. [1216] Damin (*Grewia tiliafolia*).

Not abundant, and now very difficult to procure of any size.

3198. [1217] Sissoo (*Dalbergia latifolia*).

A very strong and useful timber.

3199. [1218] Gurraree (*Acacia procera*).

3200. [1219] Toon (*Cedrela Toona*).

3201. [1220] Unjun (*Hardwickia binata*).

Of the woods from Assam forwarded by COL. REID the following are exhibited [*]:—

3202. [5273] Nahori (*Mesua ferrea*), Assam.

3203. [9001] Ajar or Jarool (*Lagerstrœmia regina*).

3204. [9002] Saum (*Artocarpus*).

3205. [9003] Kantal (*A. integrifolia*).

3206. [9004] Poma (*Cedrela Toona*).

3207. [9005] Gomari (*Gmelia arborea*).

3208. [9009] Gondhosoroi (*Laurus Sassafras*).

3209. [9010] Uriam (*Andrachne trifoliata*).

3210. [9011] Bheh (*Salix tetraspermum*).

3211. [9012] Reghu (*Nauclea cadamba*).

3212. [9013] Hilikha (*Terminalia citrina*).

3213. [9015] Bual (*Ehretica serrata*).

3214. [9016] Aum (*Mangifera indica*).

3215. [9017] Teham (*Artocarpus*).

[*] The remainder of COL. REID's collection, along with some specimens presented by LIEUT. W. PHAIRE, for which space could not be found, and to which the native names only have been attached, have been deposited at the India Museum for farther reference.

3216. [9018] Joba Hingoru (*Quercus*).

3217. [5272] Sissoo (*Dalbergia*).

3218. [9019] Koroi (*Acacia*).

## WOODS FROM DARJEELING.

3219. [7235] Horcul timber. DR. CAMP-BELL.

3220. [7276] Kerhoola. Do.

3221. [7237] Roobees. Do.

3222. [1779] Cospie. Do.

3223. [1769] Kuttoos (chesnut). Do.

3224. [1770] Tacar (*Chelonia*). Do.

3225. [1771] Toon (*Cedrela Toona*). Do.

3226. [1772] Chump (*Magnolia*). Do.

3227. [1752] Keranee. Do.

3228. [1751] Boheeleear. Do.

## WOODS FROM CHOTA NAGPORE.

A numerous collection, embracing some good specimens, has been forwarded. The following are exhibited:—

3229. [7576] Dhan Dhauta.

Hard white timber.

3230. [7575] Dhela Kata.

Hard yellow timber.

3231. [7594] Siris (*Mimosa Serisa*).

Hard light-brown timber.

3232. [10217] Sisa (*Dalbergia Sisoo*).

Hard brown timber.

3233. [10216] Sal Sakhna (*Shorea robusta*).

Hard brown timber.

3234. [7541] Belunnan.

Hard brown timber.

3235. [7582] Dhaman (*Grewia* sp.)

3236. [7535] Asân (*Terminalia alatatomentosa*).

Hard brown timber.

3237. [7577] Pindar (*Grewia nudiflora*).

Hard white timber.

3238. [10219] Gora.

3239. [10244] Tun (*Cedrela Toona*).

3240. [10221] *Nauclea cordifolia*, Philibeet. CENTRAL COMMITTEE, *Allahabad.*

3241. [10222] Usyna, Do. Do.

3242. [10223] Kame, Do. Do.

3243. [10224] Toon (*Cedrela Toona*), Do. Do.

3244. [10225] Saul (*Shorea robusta*), Do. Do.

3245. [10226] Sissoo (*Dalbergia Sisoo*), Do. Do.

3246. [10236] Kutha (*Acacia Catechu*), Shahjehanpore.

3247. [1221] Indian rosewood (*Dalbergia* sp.), Jhansee.

3248. [1222] Thurdai, Do.

3249. [7239] Mahogany (*Swietenia Mahagoni*) grown in Calcutta. C. LAZARUS, Esq.

3250. [748] Bamboo (*Bambusa arundinacea*).

3251. [749] Do. Do.

The collection of woods from Oude, amongst others *, comprises the following:—

3252. [7527] Neem (*Melia azedirachta*), Lucknow.

Plentifully in Oude.

3253. [7523] Peepul (*Ficus religiosa*), Lucknow.

In various parts of Oude.

3254. [7530] Mulseree (*Mimusops elengi*), Lucknow.

In various parts of Oude : not extensively.

3255. [7517] Toon (*Cedrela Toona*), Lucknow.

Spontaneously in the northern parts of Oude. It is considered the best for furniture of a high polish.

* Deposited at the India Museum, on account of want of space.

3256. [7526] Bahera (*Terminalia bellerica*), Lucknow.

Grows spontaneously in the Taraee. Not a very large tree: used for all purposes. From its fruit blacking is made.

3257. [7516] Jamun (*Eugenia jambolana*), Lucknow.

All over Oude. This is a large-sized tree, and bears a black astringent small fruit, about the size of a large olive, which the natives eat. Used for various kinds of woodwork.

3258. [7529] Asna or Asan (*Terminalia tomentosa*), Lucknow.

Spontaneously in the Taraee jungles. Considered durable and elastic for many purposes ; preferable to sâl.

3259. [7525] Aum (*Mangifera indica*), Lucknow.

Cultivated extensively in Oude.

3260. [7524] Kaitha (*Feronia elephantum*), Lucknow.

In various parts of Oude, not extensively. The fruit of this tree is used for the same purposes with that of *Ægle marmelos*; but the latter is preferable.

3261. [7528] Bael (*Ægle marmelos*), Lucknow.

All parts of Oude, extensively. The fruit of this tree is extensively used dry in powder, and also for making shurbut for bowel complaints.

3262. [7518] Bair (*Zizyphus Jujuba*), Lucknow and various parts of Oude.

Used principally for making native *clogs.*

3263. [7522] Arar (*Ailanthus excelsa*), Lucknow.

Extensively all over Oude. The principal use made of this wood is for sword scabbards.

3264. [7513] Gooler (*Ficus glomerata*), Lucknow.

All over Oude. For furniture. From this tree some of the lac of commerce is gathered: the fruit is eaten by natives.

3265. [7519] Mhowah (*Bassia latifolia*), Lucknow.

All over Oude in groves. From the seeds oil is extracted, and from the fruits or flowers spirits are distilled.

3266. [7532] Saul (*Shorea robusta*), Midnapore.

3267. [7514] Sakhoo (*do.*), Lucknow.

Spontaneously and extensively in the Taraee. This is the timber generally used for building purposes, bridges, &c. ; is durable, and is considered the best for such works.

Of the collection from the Punjab* the following specimens are exhibited :—

3268. [3729] Walnut wood, Mehra Forest, near Abbottabad, Hazara.

3269. [3730] Toon wood (*Cedrela Toona*), Do.

3270. [3740] Buroongi, Do.

An evergreen oak bearing acorns, leaves of young plant like those of the holly; a variety of oak, *Quercus* (*Quercus Ilex*).

3271. [3741] Umloke, Mehra Forest, near Abbottabad, Hazara.

3272. [3742] Mulberry, Do.

3273. [3743] Loon, Do.

Apparently a species of wild pear.

3274. [3744] Kungur or Kukker, Mehra Forest, near Abbottabad, Hazara.

A species of toon.

3275. [3745] Deodar (*Cedrus Deodara*), Mehra Forest, near Abbottabad, Hazara.

3276. [3746] Do. Do.

---

From Chittagong, as under :—

3277. [563] Toon (*Cedrela Toona*).

3278. [564] Chuckwah.

3279. [565] Tazeboil.

3280. [566] Loehah.

3281. [567] Chuckrassee (*Chickrassia tabularis*).

3282. [568] Gootgooteah.

3283. [569] Kandeb.

3284. [570] Jarrool (*Lagerstrœmia regina*).

3285. [571] Gamar.

3286. [572] Chaplass.

Forwarded by the CENTRAL COMMITTEE, *Chittagong.*

---

* A series of small specimens of timber in the form of a round table, forwarded from Umritsur, has lost its value in consequence of the labels on the individual pieces having in many instances become so damaged as to be illegible.

The following are representatives from the collection from Arracan, forwarded through the LOCAL COMMITTEE at *Akyab* :—

**3287. [9118] Kashy (*Erythina indica*), Akyab.**

A strong wood, used as floor and wall-planking. It grows to a large size, and is procurable in the Sandoway district.

**3288. [9119] Thykadah (*Erythina*), Akyab.**

Used for making banghies, also for boxes. This tree grows to a large size, and is procurable throughout the province.

**3289. [9158] Tahoot, Akyab.**

For making banghies, and other fine work. It grows to a moderate size, but is not very plentiful.

**3290. [9159] Toung-gangan, Akyab.**

Sometimes used for planks. It grows to a large size, and is not very plentiful.

**3291. [9144] Ka-moung, Akyab.**

Used for planks, posts, &c. Grows to a large size, and is plentiful.

**3292 [9145] Pya, ironwood (*Inga xylocarpa?*), Akyab.**

Used in making rice mills. Grows to a moderate size, and is plentiful in Sandoway and Ramree districts.

**3293. [9157] Thenganet (Tilsa), Akyab.**

A very good wood, used for work of all kinds. Grows to a large size, and is very plentiful in the Akyab and Ramree districts.

**3294. [9121] Phathan (*Bignonia stipulata*), Akyab.**

Used by natives for bows &c. It is a moderate-sized tree, very plentiful in the province.

**3295. [9122] Bamaw, Akyab.**

**3296. [9123] Khoongho (*Dipterocarpus* sp.), Akyab.**

Used for making oars for boats, and sometimes in house-building. It grows to a large size, and is plentiful in the Sandoway district.

---

Specimens of timber furnished by the Superintendent of the Gun-carriage Manufactory, *Madras*. COL. J. MAITLAND.

**3297. [2501] Pegu teak (*Tectona grandis*).**

Colour light brown. Grain straight and open. Free from knots.
*Uses.*—For all parts of light field carriages (except the beams) ; waggons and their limbers (except poles and splinter bars), as well as heavy field and garrison carriages, garrison traversing platforms, and gun and mortar platforms, and all parts of heavy and light mortar carts ; store carts (with the exception of poles and splinter bars) ; platform line and water carts ; gins and wheel work ; heavy and light field ammunition boxes ; transport carriages and limbers, and furniture work.

**3298. [2502] Saul (*Shorea robusta*).**

Strong coarse-grained timber. Colour both white and dirty whitish-brown. The white is the better of the two. The grain is often short and cross, and contains little or no fibre ; is liable to expand and contract ; contains a good deal of acid.
*Uses.*—For beams of gun and howitzer carriages ; light field axle-cases of all kinds ; all parts of carts ; transport carriage cheeks ; handspikes of all sorts ; perches of waggons, poles, short perches, braces, framing and splinter-bars of limbers ; gun and waggon, and framing of all carts.

**3299. [2503] Peddowk (*Pterocarpus dalbergioides*).**

A good strong wood ; colour deep, and pale red. The pale is the lighter of the two, but the red is the stronger ; it has an aromatic smell, and is slightly pungent to the taste, and when steeped in water imparts a deep indigo tinge to it.
*Uses.*—For light field beams, cheeks, axle-cases, perches, poles, limber-framing, waggon-perches, and framing ; heavy field-cheeks, transoms, axle-cases, handspikes, poles, braces, framing, &c. All parts of garrison carriages, garrison traversing platforms, as well as gun and mortar platforms ; transport carriages and limbers, and cart work of all sorts ; wheels, heavy and light field.

**3300. [2504] Peemah (*Lagerstroemia regina*).**

A light tough straight-grained wood. Colour pale red.
*Uses.*—For light field-cheeks, felloes, and cart naves ; framing and boards of waggons, limbers, and platform-carts, and ammunition box-boards, and heavy field-cheeks.

**3301. [2505] Trincomallie (*Berrya ammonilla*).**

A good strong close-grained reddish-coloured wood.
*Uses.*—For light field splinter bars, poles, pole rests, waggon handspikes, wheel-props, &c. ; heavy field-poles, handspikes, splinter-bars, pole-rests, garrison carriage handspikes, poles of carts, sick cart framing, spokes, yokes, &c. ; and shafts.

**3302. [2506] Satin (*Chloroxylon Swietenia*).**

A strong curly-grained wood ; its colour is whitish or yellow.
*Use.*—For naves.

**3303. [2507] Rose (*Dalbergia latifolia*).**

A strong fibrous and close-grained wood ; its colour is good purple, mottled with whitish veins. This wood is also called black wood.
*Uses.*—For light field-beams, cheeks, axle-cases, braces, perches, poles, splinter-bars, waggon-perches and framing, spokes and felloes.

**3304. [2508] Peengandoo (*Inga xylocarpa*).**

A very heavy hard close-grained wood ; it is of a reddish-brown colour, and of a brittle nature. It is also called iron wood.

*Uses.*—Poles, axle-cases, and braces for transport limbers; poles and yokes for water-carts; cheeks, axle-cases for transport carriages; light mortar-carts.

## 3305. [2509] Chittagong (*Chickrassia tabularis*).

A pale red-coloured light wood; it is rather soft, and some of it beautifully veined, somewhat resembling mahogany, and is susceptible of a high polish.
*Uses.*—Plane-tables and furniture work.

## 3306. [2510] Model or Putchavettoo (*Nauclea cordifolia*).

This wood is close-grained and soft, resembling the box in colour and texture; lighter, and easily worked; not durable; will not stand the alternations of dryness and moisture.
*Uses.*—A good wood for model work.

## 3307. [2511] Paula (*Mimusops hexandra*).

A very fine close-grained heavy wood, hard and very brittle, colour chocolate.
*Uses.*—For rulers, knobs, handles of tools, such as chisels, &c., and other articles of turnery.

## 3308. [2512] Thumbagum (*Vatica tambuggaia*).

A very strong, close-grained, splintery heavy wood. The tree yields dammer resin.
*Uses.*—It was tried in the arsenal, Fort St. George, for fuzes, some time ago.

## 3309. [2513] Malavemboo (*Melia azedarach*).

A light wood, colour reddish brown.
*Use.*—Will answer for boxes.

## 3310. [2514] Ebony (*Diospyros ebenaster*).

Hard, brittle, and heavy, and takes a high polish.
*Use.*—For ornamental work.

## 3311. [2515] Congo.

Colour brown, grain straight and not fine.
*Use.*—For fuzes.

### TABLE SHOWING THE RESULTS OF EXPERIMENTS BY·COLONEL MAITLAND ON THE WOODS UNDERNOTED.

| NAMES OF TIMBERS | BREAKING WEIGHT IN TESTING EXPERIMENTAL PIECES, EACH 3 FT. LONG AND 1½ IN. SQUARE | AVERAGE DEFLECTION CALCULATED FROM A SERIES OF EXPERIMENTS | PERIOD OF SEASONING IN LOG | PERIOD OF SEASONING IN HALF-WROUGHTS | SPECIFIC GRAVITY OBTAINED BY APPROXIMATION | SOURCES FROM WHENCE OBTAINED |
|---|---|---|---|---|---|---|
| | lbs. | inch | yrs. | | lbs. | |
| Pegu Teak . . | 702 to 1000 | ·4 to 1·3 | 5 to 6 | 18 months for the larger component parts, 12 months for the smaller | 41 | Burmah |
| Saul . . . . | 627 to 1187 | ·4 to 1·8 | 6 | 3 years for larger component parts, 2 years for the smaller | 56 | Cuttack, Bengal Provinces, and Burmah |
| Peddowk . . | 739 to 1336 | ·5 to 1·5 | 5 to 6 | 2 to 3 years for the larger component parts, 18 months for the smaller | 56·6 | Burmah |
| Peemah . . . | 664 to 888 | ·5 to 1·5 | 2 | 1 to 2 years . . | 36·25 | Burmah |
| Rose . . . . | 888 to 1187 | ·4 to 1·5 | 3 | 18 months . . . | 50·5 | Annamullay forests |
| Trincomallee . | 739 to 1224 | ·5 to 1·5 | 2 | 12 do. . . . | 50 | Ceylon |
| Satin . . . . | 739 to 1262 | ·5 to 1·7 | 3 | 6 do . . . | 60·75 | Ceylon and Southern India |
| Thumbagum . | 1075 to 1187 | ·4 to 1·4 | | | 67·75 | Southern India |
| Peengaudoo . | 702 to 1064 | ·3 to 1·8 | 5 | 1 to 2 years . . | 58·6 | Burmah |
| Chittagong . . | | | 2 | 6 to 12 months . . | 31 | Annamullay forests |
| Model or Putchavettoo . | | | 1 | 6 months . . . | 42 | Annamullay forests and Malabar coast |
| Ebony . . . | | | 3 | 12 months . . . | 75 | Ceylon and Southern India |
| Malavemboo . | | | | | | Annamullay forests |
| Congo . . . | | | 1½ | 6 months . . . | 59 | Malabar coast |
| Paula . . . | | | 2 | 12 do. . . . | 72 | Southern India |

### WOODS OF SOUTH CANARA.

#### EXHIBITED BY V. P. COELHO, ESQ.

## 3312. [140] Sandal wood (*Santalum album*).

## 3313. [141] Savagani or Teak (*Tectona grandis*).

## 3314. [142] Halsu or Jack (*Artocarpus integrifolia*).

## 3315. [143] Kebalsu or Wild Jack (*Artocarpus incisa*).

## 3316. [144] Bengha.

3317. [145] Bou.

Strong wood used for building purposes.

3318. [146] Bannapoo (*Guettarda speciosa*).

A strong wood employed for building.

3319. [147] Terruvah.

Strong and useful for building purposes.

3320. [148] Marava (*Terminalia alata*).

Strong and useful building timber.

3321. [149] Jembu Nerlu.

3322. [150] Votte Kully.

3323. [151] Tamarind (*Tamarindus indica*).

3324. [152] Uru Sampige (*Michelia champaca*).

3325. [153] Kaddi Sampige.

3326. [154] Daddalu.

3327. [155] Torenha or Pumbilo.

3328. [156] Kalu boghe.

3329. [157] Pattu bage.

3330. [158] Shere Kane.

Used in boat-building and for spars.

3331. [159] Uru Hone.

Used for making boats.

3332. [160] Mango (*Mangifera indica*).

3333. [161] Jarrige.

3334. [162] Nanne.

3335. [163] Andippu naru.

3336. [164] Cadippilan.

Used in the preparation of a dye.

3337. [165] Manjutty.

The bark is employed for medicinal purposes.

3338. [166] Purrally.

3339. [167] Nalikai (*Emblica officinalis*).

For making frameworks for wells. Does not rot in water.

3340. [168] Santamarry.

Makes good gunstocks.

3341. [169] Renje.

3342. [170] Page or Gargass.

Leaves used instead of sandpaper for polishing wood.

3343. [171] Ardalla or gamboge tree (*Garcinia pedunculata*).

3344. [172] Cinnamon (*Laurus* sp.)

3345. [173] Mannadike.

3346. [174] Jungle Geru Kai (*Semecarpus anacardium*).

The juice is vesicatory. Nuts employed for markin cloth.

3347. [175] Cashew (*Anacardium occidentale*).

3348. [176] Halley.

3349. [177] Tally.

A very strong wood for building purposes.

3350. [178] Cocoa nut (*Cocos nucifera*).

3351. [179] Kunttal.

3352. [180] Karmara.

3353. [181] Dhûppa (*Vateria indica*).

3354. [182] Loukatty.

3355. [183] Tarrolly.

3356. [184] Areca nut (*Areca catechu*).

3357. [185] Blackwood (*Dalbergia frondosa*).

3358. [186] Ebony (*Diospyros melanoxylon.*)

3359. [187] Jummikai.

3360. [188] Takote Kai.

3361. [7436] Pith of *Æschynomene.*

## WOODS FROM MYSORE.

(*See Tables, pp.* 162-166).

3362. [2680] Teak (*Tectona grandis*), Nuggur Division, Mysore.

3363. [2706] Sandalwood (*Santalum album*), Do.

3364. [2707] Do.     Do.

## COLLECTION OF WOODS FROM RANGOON.

By Messrs. HALLIDAY, FOX, & Co.

3365. [4754] Ironwood (*Inga xylocarpa*).

3366. [2345] Tenasserim mahogany.

3367. [2339] Muniahban.

3368. [2337] Thengan (*Hopea odorata*).

3369. [2335] Teak (*Tectona grandis*).

3370. [2340] Mango (*Mangifera indica*).

3371. [2336] Ting-daik-nits.

3372. [2338] Ger-doma.

The residue of this series, being the same as shown by other exhibitors, and space being limited, are removed to the India Museum, Whitehall.

3373. [9395] Two young Teak trees (*Tectona grandis*), Pegu. D. BRANDIS, Esq.

## WOODS FROM MOULMEIN.

Forwarded by DR. GRAHAM.

3374. [7618] Thin Gan (*Hopea odorata*).

A very strong durable wood; used for making canoes.

3375. [10459] Pyen-ka-doe (*Inga xylocarpa*).

Wood extremely hard; used for house-posts.

3376. [7619] Ah Nan (*Xylocarpus granatum*).

A very strong wood; used for making gun-stocks and scabbards.

3377. [10460] Toun Phain (*Artocarpus echinatus*).

Used for making boats and carts.

3378. [7620] Bun Boay (*Careya arborea*).

A strong durable wood; used for house-posts.

3379. [7621] Gyeo Tha (*Mellicocca trijuga*).

This wood is used for bows, being tough and elastic.

3380. [7622] Oak An.

This wood is made into canoes.

3381. [7623] Toung Pain Nai (*Artocarpus echinatus*).

Fruit edible. Used in house-building.

3382. [7624] Mya-ya-gyee (*Grewia floribunda*).

Made into any common house-building material.

3383. [7625] Fonk-sha-gyee (*Vitex arborea*).

Fruit eaten. The wood is used for any common purpose.

3384. [10461] Khan Tha.

This wood is made into any house-building material.

3385. [10462] Myouk Ngo, Moulmein lance wood.

Ditto.

3386. [10463] Tsouk Yo (*Dalbergia ovata*).

A tough wood; much used for tool-handles.

3387. [7626] Mohmagah (*Galex* sp.).

Used in common purposes of building.

3388. [10496] Thau Thet Ngai (*Bignonia* sp.)

Ditto.

3389. [7691] Goay-pin-gyee (*Adenanthera pavonina*).

Ditto; and also its seed for weight in weighing gold.

3390. [7692] Balawa (*Garcinia speciosa*).

Ditto.

3391. [7627] Yee Pyee (*Phyllanthus emblica*).

Ditto.

3392. [10464] Setphan.

Ditto.

3393. [7628] Goay Tha.

Ditto.

3394. [7629] Bom Mai Za (*Inga* sp.)

Wood hard. Used for making musical instruments.

3395. [7630] Thet Ya (*Gordonia floribunda*).

This wood is made use of for ordinary house-building purposes.

3396. [10465] Dedoap Tha.

Ditto.

3397. [7631] Ka Nat Tha.

Ditto.

SPECIMENS AND STATEMENT OF RESULTS OF EXPERIMENTS ON THE TRANSVERSE STRENGTH OF SEASONED WOODS FROM THE FORESTS OF WESTERN MYSORE.—CAPTAIN J. PUCKLE.*

| Number | No. in List | Names of Woods | No. of Specimens | Deflection in Inches and Tenths 126 lbs. | 238 lbs. | 350 lbs. | 462 lbs. | Breaking weight in lbs. | Ultimate deflection | Specific gravity | Description | Uses to which applied, &c. | Whether likely to be attacked by insects or not | Kind of fracture | Average size of rough squared log procurable, and if abundant or not | REMARKS |
|---|---|---|---|---|---|---|---|---|---|---|---|---|---|---|---|---|
| 3398 | I | *Can.* Boghy / *Lat.* Acacia (probably) | 1 / 2 | 0·10 / 0·10 | 0·20 / 0·20 | 0·40 / 0·30 | 0·80 / 0·60 | 462 / 686 | 0·90 / 0·90 | 915 | Dull brown, close grain | For furniture, is strong, and tough | No | Short fibrous | 12 × 1 × ¾. Yes | No. 2 cracked at 630 lbs. |
| 3399 | II | *Can.* Soojbel / *Lat.* Acacia (probably) | 1 / 2 / 3 / 4 / 5 | 0·17 / 0·10 / 0·25 / 0·20 / 0·20 | 0·37 / 0·25 / 0·40 / 0·40 / 0·40 | 0·58 / 0·40 / 0·60 / 0·60 / 0·63 | … / 0·60 / 1·00 / 0·80 / … | … / 658 / 574 / 462 / 400 | 1·70 / 1·60 / 0·90 / 0·75 | 973 | Bright brown, long grain, rather open | Has great resilience—useful for all purposes | No | Long fibrous | 8 × ¾ × ¾. Yes | No. 1 deflected 0·68 with 380 lbs.; 0·72 with 400 lbs.; 0·98 with 446 lbs.; 1·02 with 456 lbs. It was then unloaded, and recovered its shape all but 0·100 of an inch |
| 3400 | III | *Can.* Baughy / *Tam.* Vaghy / *Lat.* Acacia speciosa | 1 / 2 / 3 / 4 | 0·05 / 0·10 / 0·20 / 0·10 | 0·25 / 0·30 / 0·30 / 0·20 | 0·40 / 0·50 / 0·50 / 0·40 | 0·55 / 1·00 / 0·60 / 0·60 | 630 / 480 / 582 / 639 | 1·55 / 1·40 / 1·10 / 1·20 | 904 | Light reddish brown, and with minute cavities, filled with secretion | For carriages, and house building | No | Long fibrous | 12 × 1 × ¾. Yes | Cracked at 460 lbs. / No. 4 Did not break quite through |
| 3401 | IV | *Can.* Hoonsay / *Tam.* Poolia Marum / *Lat.* Tamarindus Indica / *Tel.* Chinta Chettoo | 1 / 2 / 3 / 4 | 0·12 / 0·10 / 0·20 / 0·10 | 0·38 / 0·20 / 0·30 / 0·22 | … / 0·45 / 0·50 / 0·30 | … / 0·75 / 0·70 / … | 294 / 602 / 486 / 345 | 0·60 / 0·95 / 0·80 / 0·45 | 1323 | Heartwood is red and black, streaked; close-grained, knotty | For naves of wheels, oil mills, mallets, rice pounders, &c., excellent for brick and tile burning | No | Short fibrous | 10 × 1¼ × 1. Yes | No. 1 broke in two after half a minute's suspension; specimen bad / No. 2 was not a fair specimen, it had a longitudinal crack; but it sustained the weight 40 seconds. Yields the best charcoal for gunpowder, the stones of the fruit pounded, and boiled with thin glue, make best wood cement |
| 3402 | V | *Can.* Biti / *Tam.* Yeti Marum / *Lat.* Dalbergia latifolia. / *Tel.* Eroopootoo | 1 / 2 / 3 / 4 | 0·10 / 0·07 / 0·15 / 0·10 | 0·20 / 0·15 / 0·30 / 0·20 | 0·40 / 0·30 / 0·45 / 0·40 | 0·60 / 0·50 / 0·60 / 0·60 | 574 / 602 / 522 / 597 | 0·70 / 0·95 / 0·80 / 1·10 | 818 | Black, and streaked red, like rosewood, rather open grain | Furniture of every description | No | Short fibrous | 18 × 1½ × 1½. Yes, very | No. 1 broke after sustaining weight 20 seconds / No. 2 cracked at 518 lbs., but sustained the last weight for one minute / Much larger timber is procurable in the forests of Malabar, where it grows to a stately and handsome tree; but in the largest specimens, earth is often found embedded. A fixed oil is procured from the seeds, and the root is medicinal / No. 4 Did not quite break through |

| No. | | Name | | | | Sp. gr. | Grain and colour | Uses | | Fibre | Dimensions | | Remarks |
|---|---|---|---|---|---|---|---|---|---|---|---|---|
| 3403 [9093] | VI | Can. Honagul — Tam. — Lat. Terminalia | 1<br>2<br>3<br>4 | 0·10 / 0·37 / 0·70 / ·· / 450 / 1·20<br>0·10 / 0·25 / 0·40 / 0·70 / 480 / 0·85<br>0·10 / 0·20 / 0·35 / 0·55 / 576 / 0·60<br>0·15 / 0·25 / 0·35 / 0·60 / 569 / 1·00 | | 913 | Dark straw colour, compact | Furniture and house building | No | Long fibrous | 17×1¾. | Yes | No. 1 cracked after sustaining 448 lbs. for one minute, but broke through with an additional two lbs. No. 2 cracked and broke through, *to neutral axis only*. No. 3 broke straight through. This is perhaps the most abundant tree in the forests |
| 3404 [9094] | VII | Can. Wulla Honay — Lat. Pterocarpus | 1<br>2<br>3<br>4<br>5 | 0·25 / 0·45 / 0·80 / ·· / 385 / 1·25<br>0·10 / 0·30 / 0·60 / 1·10 / 546 / 1·10<br>0·15 / 0·33 / 0·65 / 1·30 / 574 / 1·30<br>0·25 / 0·50 / 0·80 / ·· / 406 / 0·90<br>0·20 / 0·40 / 0·70 / ·· / 429 / ·· | | 1020 | Light-coloured, open grain, an excellent timber | Furniture and house building | No | Rather long, and fibrous | 17×1½×1. | Yes | A variety of the *Pterocarpus Marsupium.* This wood is lighter, but apparently as good, and does not stain yellow, as the Rugta Honay does. No. 1 cracked in the centre, and broke (but not quite through) after sustaining the weight for a few seconds |
| 3405 [9095] | VIII | Can. Nellee — Tam. Nellee — Lat. Emblica officinalis — Tel. Assereki | 1<br>2 | 0·20 / 0·30 / 0·48 / 1·00 / 490 / 1·20<br>0·10 / 0·25 / 0·40 / 0·80 / 574 / 1·30 | | 1080 | Dark flesh colour, smooth, very close grain, compact and tough | For veneering; good for well-rings, does not decay under water, well adapted for turning | Yes | 8 inches long and fibrous | 8×1×¾. | Yes | No. 1 broke in two pieces, after sustaining the weight half a minute. No. 2 scarcely broke beyond neutral axis. The fruit (country gooseberry) is pickled and preserved: the bark is astringent and used in tanning—the young branches are often put in wells to purify the water |
| 3406 [9096] | IX | Can. Nundee — Tam. Benteak — Lat. Lagerstroemia microcarpa | 1<br>2<br>3<br>4<br>5 | 0·15 / 0·30 / 0·60 / 1·10 / 467 / 1·20<br>0·10 / 0·20 / 0·40 / 0·50 / 574 / 0·70<br>0·10 / 0·25 / 0·50 / 1·20 / 462 / 1·20<br>0·15 / 0·30 / 0·40 / 0·60 / 518 / 0·90<br>0·20 / 0·40 / 0·57 / 0·70 / 541 / 1·00 | | 660 | Light brown, rather open grain | Useful for a variety of purposes, has great 'stiffness;' wooden bridges have been built of this | Yes | Long fibrous | 18×1½×1. | Yes | The fracture of No. 1 was 6" long. This wood is much used in the dockyard, but does *not* answer for sleepers, rotting quickly under ground |
| 3407 [9097] | X | Can. Billawar — Lat. Acacia odoratissima | 1<br>2<br>3<br>4<br>5 | 0·10 / 0·30 / 0·45 / 0·65 / 546 / 1·30<br>0·10 / 0·30 / 0·45 / 0·65 / 504 / 2·20<br>0·15 / 0·25 / 0·45 / 0·55 / 518 / 0·80<br>0·10 / 0·25 / 0·40 / 0·60 / 578 / 0·80<br>0·10 / 0·20 / 0·30 / 0·50 / 711 / 1·20 | | 730 | Handsome grain, red and brown streaked, rather open | Has great toughness or elasticity, makes handsome furniture, resembling walnut, and much used in carriage building, for the frame-work, felloes, and spokes | No | Short fibrous | 12×1½×1. | Yes | No. 1 cracked at 496 lbs, but sustained 546 lbs. for a second; at the moment of fracture, it deflected two inches; it broke very slowly, and not quite through. No. 2 cracked after sustaining 476 lbs.; broke through at 504 lbs. |
| 3408 [9098] | XI | Can. Rugta Honay — Tam. Vengay — Lat. Pterocarpus marsupium — Tel. Yegasa | 1<br>2<br>3<br>4 | 0·12 / 0·30 / 0·50 / 1·20 / ·· / ··<br>0·20 / 0·40 / 0·60 / 0·70 / 518 / 0·70<br>0·15 / 0·30 / 0·50 / ·· / 578 / 0·70<br> | | 820 | Do. Do. but closer grain, and a little darker | Makes handsome furniture, and resembles fine mahogany, but must be well seasoned, or it stains yellow | No | Short, and rather splintery | 17×1×1. | Yes | No. 1 cracked after bearing 475 lbs., for a few seconds; the weights were then taken off, and the specimen remained whole. This tree, which grows to a large size, yields the gum 'Kino,' and is abundant in all the forests of southern India |

* The weight was suspended from the centre of each specimen, which was 24 inches long and 1 one inch square, and supported at both ends for one inch—so that the bearing was just 22 inches. *The experiments were perhaps carried on too quickly to give the best results.*

| Number | No. in List | NAMES OF WOOD | No. of Specimens | Defl. 1 (126 lbs.) | Defl. 2 (238 lbs.) | Defl. 3 (50 lbs.) | Defl. 4 (462 lbs.) | Breaking weight in lbs. | Ultimate deflection | Specific gravity | Description | Uses to which applied, &c. &c. | Whether likely to be attacked by insects or not | Kind of fracture | Average size of rough squared log procurable, and if abundant or not | REMARKS |
|---|---|---|---|---|---|---|---|---|---|---|---|---|---|---|---|---|
| 3409 [9099] | XII | Can. Nowladdi | 1 | 0·05 | 0·20 | 0·40 | 0·60 | 518 | 0·70 | 907 | Greenish brown, dull, close grain | Polishes well, is used for house building, and furniture, &c. | No | Long, and rather splintery | 11 × 1 × ¾. Yes | This is one of the woods used as sleepers in the South Western Railway |
| 3410 [9100] | XIII | Can. Haudiga | 1 | 0·20 | 0·40 | 0·70 | .. | 364 | 1·00 | 657 | Light, mottled brown; long regular grain | Furniture; polishes and turns well, useful for the cabinet maker; and would do for veneering | No | Splintery | 9 × ¾ × ¾. Not very | No. 1 bad specimen No. 2 cracked at 462 lbs., and broke after sustaining the weight one minute |
| | | | 2 | 0·15 | 0·30 | 0·50 | 1·10 | 462 | 1·20 | | | | | | | |
| 3411 [9101] | XIV | Can. Jálari . Tam. Talura . Lat. Vatica laccifera . | 1 | 0·20 | 0·40 | 0·60 | 0·80 | 462 | 1·00 | 689 | Light yellow; long fibrous grain | Strong useful wood for a variety of purposes | No | Short | 10 × 1 × 1. Yes | A large tree, which the lac insect attacks; the shellac of commerce is procured from it |
| | | | 2 | 0·20 | 0·35 | 0·55 | .. | 494 | 0·70 | | | | | | | |
| 3412 [9102] | XV | Can. Kurraymutti . Tam. Kurray Maradah. Lat. Terminalia tomentosa Tel. Muddie . | 1 | 0·05 | 0·20 | 0·40 | 1·00 | 462 | .. | 892 | Dark brown; open grain | House building; bears a good transverse strain; a wood much esteemed for all railway purposes | No | Rather splintery | 18 × 1¼ × 1. Yes, very | No. 1 cracked at 406 lbs.: very abundant everywhere |
| | | | 2 | 0·10 | 0·30 | 0·50 | 1·00 | 462 | .. | | | | | | | |
| | | | 3 | 0·20 | 0·30 | 0·45 | 0·60 | 602 | 1·10 | | | | | | | |
| | | | 4 | 0·10 | 0·20 | 0·40 | 0·50 | 589 | 0·10 | | | | | | | |
| 3413 [9103] | XVI | Com. Jambay . Tam. Erool . Lat. Inga xylocarpa Tel. Eezoovaloo . | 1 | 0·15 | 0·35 | 0·60 | .. | 350 | 0·70 | 934 | Dull dark chocolate brown, with orange tint; close grain | Furniture, shafts, plough heads and knees, and crooked timbers in ship building; and railway sleepers | No | Long | 14 × 1 × 3¾. Yes | No. 1 cracked at 550 lbs., and snapped with that weight after half a minute A large tree, the timber is hard, durable, and in great demand The bark is astringent, and used for dyeing black |
| | | | 2 | 0·10 | 0·25 | 0·65 | .. | 406 | 0·70 | | | | | | | |
| 3414 [9104] | XVII | Can. Sagvan . Tam. Theke Marums Lat. Tectona grandis Tel. Teka . | 1 | 0·20 | 0·40 | 0·70 | .. | 400 | .. | 684 | Light brown, open grain | Ship building, house ditto, furniture, &c. &c. | No | Splintery | 30 × 1¼ × 1¼. Yes, very | No. 1 broke in two suddenly No. 2 snapped short in two, after sustaining weight for a few seconds No. 3 Do. N.B.—The Nuggur teak appears not to be so good as that of the Mysore and Malabar forests. Sometimes timber of much larger scantling is procurable, but seldom longer |
| | | | 2 | 0·20 | 0·48 | 0·80 | .. | 350 | 0·80 | | | | | | | |
| | | | 3 | 0·20 | 0·40 | 0·60 | .. | 406 | 0·70 | | | | | | | |
| | | | 4 | 0·25 | 0·55 | .. | .. | 322 | 0·50 | | | | | | | |
| | | | 5 | 0·20 | 0·40 | .. | .. | 387 | | | | | | | | |

| | | | | | | | | | | | | | | | | | |
|---|---|---|---|---|---|---|---|---|---|---|---|---|---|---|---|---|---|
| 3415 | [9105] | XVIII | *Can.* Dinduga | 1 | 0·15 | 0·35 | 0·70 | .. | 406 | 0·80 | 1037 | Close compact grain, yellowish striped, and something like satin wood. Very handsome | House building, shafts and yokes, and general use for railway purposes; but makes very good cabinet furniture | No | Splintery | 12×¾×¾. | Yes, very | Snapped like the preceding; but is much used at Bangalore |
| | | | *Tam.* Vella Nága | 2 | 0·15 | 0·30 | 0·50 | .. | 406 | 0·80 | | | | | | | | |
| | | | *Lat.* Conocarpus latifolia | 3 | 0·25 | 0·40 | 0·65 | 0·90 | 510 | 1·25 | | | | | | | | |
| | | | *Tel.* Síri Maun | 4 | 0·10 | 0·30 | 0·50 | 0·70 | 513 | 0·80 | | | | | | | | |
| 3416 | [9106] | XIX | *Can.* Kuddavailoo | 1 | 0·10 | 0·35 | 0·80 | .. | 378 | 0·90 | 681 | Brownish yellow, close compact grain, tough | For various kinds of furniture | Yes | Splintery | 12×1½×¾. | Yes | Snapped like the preceding. A large and ornamental tree which yields extensive shade |
| | | | *Tam.* Vella Cadambay | 2 | | | | | | | | | | | | | | |
| | | | *Lat.* Nauclea Cadamba | 3 | 0·15 | 0·60 | .. | .. | 294 | 0·60 | | | | | | | | |
| | | | *Tel.* Rudrakshakamba | | | | | | | | | | | | | | | |
| 3417 | [9107] | XX | *Can.* Sumpaghy | 1 | 0·20 | 0·50 | 0·90 | .. | 350 | 0·90 | 671 | Rich brown, rather close grain | Very handsome furniture, and polishes well; grows to a very large size; has a yellow, sweet-scented flower | Yes | Splintery | 15×3½×2½. | Very | No. 1 snapped at 350 lbs.; but there being a knot in the middle, it was not a fair specimen. This tree is highly venerated by the Hindoos, &c., dedicated to Vishnu; the bark is bitter, and very acid |
| | | | *Tam.* Chumpaca | | | | | | | | | | | | | | | |
| | | | *Lat.* Michelia Champaca | 2 | 0·15 | 0·40 | 0·90 | .. | 350 | 0·90 | | | | | | | | |
| 3418 | [9108] | XXI | *Can.* Mauvena | 1 | 0·10 | 0·30 | .. | .. | 294 | 1·00 | 597 | Light greyish straw, open grain, and rather soft | For solid wheels of the country carts, and rough furniture | Yes | 11 inches long | 9×½×1½. | Very | Snapped suddenly. The wood is sacred, and used by the Hindoos for burning corpses. A reddish brown gum resin, hardening by age, and resembling bdellium, is procured from this tree |
| | | | *Tam.* Maah Marum | 2 | 0·20 | 0·30 | .. | .. | 317 | 0·45 | | | | | | | | |
| | | | *Lat.* Mangifera Indica | 3 | 0·10 | 0·30 | 0·70 | .. | 350 | 0·70 | | | | | | | | |
| | | | *Tel.* Mámada Chettoo | 4 | 0·27 | 0·50 | 0·60 | .. | 850 | 0·60 | | | | | | | | |
| 3419 | [9109] | XXII | *Can.* Godda | 1 | 0·40 | .. | .. | .. | 182 | .. | 639 | Light flesh, close even grain | Polishes well, and is good for turning | Yes | Splintery, brittle | 9×¾×¾. | Very | No. 1 deflected much, and broke in two, almost immediately at 182 lbs. No. 2 snapped in three pieces very suddenly |
| | | | *Lat.* Cedrela | 2 | 0·20 | 0·40 | 0·80 | .. | 350 | 0·85 | | | | | | | | |
| 3420 | [9110] | XXIII | *Can.* Bayvena | 1 | 0·30 | 0·70 | 1·00 | .. | 294 | 1·00 | 783 | Light flesh, open grain, sweet scented | Common furniture, but it warps and splits | No | Long splintery | 15×1½×1¼. | Very | No. 1 cracked at 224 lbs., and broke after sustaining the last weight for one minute. No. 2 broke near one end suddenly. The heart wood is good, but the branches are very apt to break and snap off in high wind. Margosa oil is extracted from the yellowish green seeds, which are about the size of small gooseberries |
| | | | *Tam.* Vepa Marum | | | | | | | | | | | | | | | |
| | | | *Lat.* Melia Azadirach | 2 | 0·20 | 0·60 | .. | .. | 294 | 1·70 | | | | | | | | |
| | | | *Tel.* Taruka vépa | | | | | | | | | | | | | | | |
| 3421 | [9111] | XXIV | *Can.* Mussee | 1 | 0·30 | 0·65 | .. | .. | 280 | .. | 839 | Very light brown, open grain | In general demand | No | Shortish and rather splintery | 10×¾×¾. | Yes | Broke near one end suddenly, near a knot |
| | | | | 2 | 0·20 | 0·50 | .. | .. | 261 | 0·60 | | | | | | | | |
| | | | *Lat.* Lauracea. | 3 | 0·20 | 0·50 | .. | .. | 238 | 0·50 | | | | | | | | |
| | | | | 4 | 0·40 | 0·57 | .. | .. | 238 | 0·60 | | | | | | | | |
| 3422 | [9112] | XXV | *Can.* Halasoo | 1 | .. | .. | .. | .. | 112 | 0·30 | 676 | Yellow at first, but on exposure it assumes a darker tint, approaching to dull mahogany | Furniture, chairs, tables, &c., picture frames, &c.; but must be well seasoned, or it will warp and crack | Yes | Splintery | 12×1×1½. | Yes | No. 1 bad specimen. Broke suddenly, without sustaining the weight a single moment. Broke suddenly. Produces the large 'Jack' fruit. Birdlime is manufactured from the juice of the bark, and the leaves are greedily eaten by cattle; the roasted seeds are much used by the poorer people |
| | | | *Tam.* Peelah Marum | 2 | 0·20 | 0·54 | 0·80 | .. | 485 | 0·90 | | | | | | | | |
| | | | *Lat.* Artocarpus integrifolia | 3 | 0·15 | 0·50 | .. | .. | 280 | 0·70 | | | | | | | | |
| | | | *Tel.* Panasa | 4 | 0·30 | 0·60 | 1·00 | .. | 350 | 1·20 | | | | | | | | |

| Number | No. in List | NAMES OF WOOD | No. of Specimens | Deflection in Inches and Tenths: 1 (126 lbs.) | 2 (238 lbs.) | 3 (350 lbs.) | 4 (462 lbs.) | Breaking weight in lbs. | Ultimate deflection | Specific gravity | Description | Uses to which applied, &c. &c. | Whether likely to be attacked by insects or not | Kind of fracture | Average size of rough squared log procurable, and if abundant, or not | REMARKS |
|---|---|---|---|---|---|---|---|---|---|---|---|---|---|---|---|---|
| 3423 [9113] | XXVI | *Can.* Yettáyga<br>*Tam.* Munja Cadumbay<br>*Lat.* Nauclea cordifolia<br>*Tel.* Daduga | 1<br>2<br>3<br>4 | 0·45<br>0·40<br>0·40<br>0·45 | 0·90<br>0·90<br>0·75<br>0·70 | ..<br>..<br>..<br>.. | ..<br>..<br>..<br>.. | 252<br>245<br>238<br>277 | 0·90<br>1·00<br>0·75<br>0·95 | 581 | Dull yellow, close short grain | Polishes well, resembles box wood, and is good for turning; cracks and warps; is light and durable if kept from wet | Yes | Very splintery and brittle | $14 \times 1\frac{1}{4} \times 1\frac{1}{4}$. Very | No. 2 snapped in two like a carrot<br>No. 3 do. after one minute's suspension<br>No. 4 the last two lbs. caused it to snap like the others |
| 3424 [9114] | XXVII | *Can.* Thengana<br>*Tam.* Thenga Marum<br>*Lat.* Cocos nucifera<br>*Tel.* Tenkaia | 1<br>2<br>3<br>4 | 0·20<br>0·20<br>0·10<br>0·15 | 0·50<br>0·30<br>0·30<br>0·40 | ..<br>0·47<br>..<br>.. | ..<br>..<br>..<br>.. | 238<br>457<br>238<br>294 | 0·50<br>0·50<br>0·55<br>0·60 | 747 | Light speckled, something like a nutmeg in the section; pithy | Ridge poles for temporary roofs, aqueducts, &c. &c. | Yes | Very fibrous | $28 \times \frac{3}{4} \times \frac{1}{2}$. Yes | With 182 lbs. No. 1 deflected much, but it broke nearly through after sustaining the weight (238 lbs.), and remained so, kept together by its stringy fibres, with a deflection of three inches. This tree produces the cocoa nut oil and fibre of commerce; the leaves are used for thatching houses; Toddy is also extracted |
| 3425 [9115] | XXVIII | *Can.* Somy<br>. . . . . | 1<br>2 | 0·20<br>0·30 | 0·30<br>0·60 | ..<br>.. | ..<br>.. | ..<br>294 | <br>0·80 | | Handsome red, rather open grain, but it warps and cracks | Furniture | No | Splintery | Yes | Neither specimen good |
| 3426 [9116] | XXIX | *Can.* Baulay<br>*Tam.* Toombra<br>*Lat.* Diospyros melanoxylon<br>*Tel.* Toomida | 1 | | | | | | | 1200 | | | No | | Yes | |
| 3427 [9117] | XXX | *Can.* Thadsal<br>. . . . . | 1<br>2 | 0·10<br>0·12 | 0·20<br>0·30 | 0·30<br>0·40 | 0·60<br>0·70 | 558<br>533 | 0·90<br>0·80 | | Light, close even grain | | Yes | Straight, short, fibrous | $16 \times 1\frac{1}{4} \times 1$. Yes | Apparently a very excellent wood |

N.B.—The size of the logs is under, rather than over stated.

## EXTRA EXPERIMENTS WITH WOODS OF LARGE SCANTLING.

| | Scantling | Weight Suspended in lbs. | Deflection in inches | REMARKS |
|---|---|---|---|---|
| Hoonagul | 15′—3″—5″ | 1058<br>1658<br>1858<br>2258<br>2558<br>2782<br>3006<br>3230 | 2<br>3<br>$3\frac{1}{4}$<br>4<br>$4\frac{1}{2}$<br>5<br>$5\frac{1}{2}$<br>$7\frac{1}{2}$ | On previous occasion this piece of wood was suffered to remain five weeks with a weight of 1,600 lbs. suspended from the centre, and exposed to alternate rain and sun. The deflection was three inches, but when the weight was removed the wood regained its original position<br>Commenced to creak and broke after six minutes' suspension at a small knot a foot from the centre |
| Kuraymutti | 15′—3″—5″ | 1058<br>2258 | $1\frac{1}{2}$<br>3 | Commenced to creak and broke at $1\frac{1}{2}$ feet from the centre after two minutes' suspension. Not a good specimen |

## WOODS FROM BURMAH.

The following numbers from 3428 to 3540, embracing 112 samples of wood, are the excellent collection of DR. D. BRANDIS, Superintendent of Forests in the *Pegu, Tenasserim,* and *Martaban* provinces.

| Number | BURMESE NAME | SYSTEMATIC NAME OF TREE | Weight of 1 cubic ft. in lbs. | Average size of full-grown trees on good soil | | REMARKS * |
|---|---|---|---|---|---|---|
| | | | | Girth measured at 6 ft. from ground | Length of trunk to first branch | |
| 3428. [10341] | Zimbjoon . . | Dillenia aurea, Sm. . | 48 | 9 | 20 | Abundant in the plains and on the hills. Wood occasionally used in house-building, but mostly for fire-wood. Br. weight 198 lbs. |
| 3429. [10342] | Bjooben . . | Dillenia pentagyna, Roxb. | 69 | 6 | 20 | Abundant in the Eng forest (forest of *Dipterocarpus grandiflora*). Wood hard and strong, used for rice mills |
| 3430. [10343] | Thabyoo . . | Dillenia speciosa, Thunb. | 41 | 5 | 15 | On the banks of the mountain streams. Wood not used |
| 3431. [10344] | Thabootkyee . | Meliusa velutina, Hf. & Th. | 42 | 5 | 15 | All over the plains. Wood used for the poles of carts and harrows, yokes, spear-shafts, oars, &c. |
| 3432. [10345] | Lepan . . . | Bombax malabaricum, Dc. | 28 | 15 | 60 | The cotton tree, abundant in the plains. Wood light and loose-grained, used for coffins. The cotton used for stuffing pillows |
| 3433. [10346] | Let-Khop . . | Sterculia fœtida, L. . | 33 | 10 | 50 | Common in the plains and on the hills. Wood not used |
| 3434. [10347] | Pinlay Kana-zoe | Heritiera sp. . . . | 66 | 6 | 30 | Common in the Delta of the Irrawaddy. Wood used for house posts and rafters, and for firewood for the manufacture of salt. The tree is nearly related to the 'Soon-dree' of Bengal |
| 3435. [10348] | Petwoon . . | Berrya mollis, Wall.. | 56 to 62 | 7 | 50 | Found on elevated ground. Wood red, much prized for axles, the poles of carts and ploughs; also used for spear handles |
| 3436. [10349] | Dwa-Nee . . | Eriolœna sp. . . . | 47 | 7 | 50 | Trees not uncommon, but not very large. Wood of a beautiful brick-red colour, tough and elastic, used for gun-stocks, paddles, and rice-pounders. A wood well worth attention, the weight being moderate |
| 3437. [10350] | Mya-ya . . | Grewia microcos, L. . | 51 | 4 | 10 | Found on elevated ground. Wood not used |
| 3438. [10351] | Ka-nyin . . | Dipterocarpus alata, Wall. | 38 | 25 | 100 | The wood oil tree of Burmah. The wood decays very fast: used for canoes, which last only from three to four years |
| 3439. [10352] | Eng . . . . | Dipterocarpus grandiflora, Wall. | 55 | 10 | 60 | This tree forms, in company with a few other kinds, extensive forests which cover upwards of 2,000 square miles in the Province of Pegu. Wood somewhat more durable than that of 'Kanyin' (No. 3438): used for canoes, house-posts, planking, &c. |
| 3440. [10353] | Kyau-thoo . | Dipterocarpus sp. . | 43 | 20 | 80 | A large tree found in the hills. Wood used for canoes and cart-wheels |

* The figures marked 'Br. weight' denote the weight required to break a piece 4 ft. long, 1 in. square, laid on supports 36 inches apart. These results were obtained by a few preliminary experiments, and are subject to corrections.

| Number | BURMESE NAME | SYSTEMATIC NAME OF TREE | Weight of 1 cubic ft. in lbs. | Average size of full-grown trees on good soil | | REMARKS |
|---|---|---|---|---|---|---|
| | | | | Girth measured at 6 ft. from ground | Length of trunk to first branch | |
| 3441. [10354] | Thingan . . | Hopea odorata, Roxb. | 46 | 12 | 80 | One of the finest timber trees of the country. Found near mountain streams and in evergreen forests. Large specimens of this valuable tree are common east of the Sittang river, but rather scarce in the greater part of Pegu. Wood much prized for canoes and cart-wheels. Boats made of this wood are said to last for more than twenty years |
| 3442. [10355] | Thingadoe | Hopea sp. . . . . | 52 | 20 | 100 | Large trees abound in the same localities as the foregoing, but the wood is not equally valued |
| 3443. [10356] | Engyin . . . | Hopea suava, Wall. . | 55 | 7 | 60 | This valuable tree is found in the Eng forest. Large trees not common in Pegu. Wood tough and hard, but heavy, used in house-building, for bows, and a variety of other purposes: said to be as durable as teak |
| 3444. [10357] | Theya . . . | Shorea obtusa, Wall . | 57 | 7 | 50 | In the Eng forest and on the brow of hills in Pegu. Wood valued equally with Engyin |
| 3445. [10358] | Gangau . . | Mesua ferrea, L. . . | 69 | 5 | 20 | Cultivated in Pegu on account of the beauty and fragrance of its flowers, but wild in Tenasserim. Wood said to be used for furniture |
| 3446. [10359] | Toung-tha-lay | Garcinia cowa, Roxb. | 42 | 6 | 20 | Scattered over the hills. Wood not used |
| 3447. [10360] | Tha-ra-phee . | Calophyllum sp. . . | 57 | 4 | 20 | Wood used for carving images, occasionally for canoes |
| 3448. [10361] | Poonyet . . | Calophyllum sp. . . | 39 | 12 | 60 | Firewood |
| 3449. [10362] | Gyo . . . | Schleichera trijuga, Willd. | 70 | 12 | 25 | One of the heaviest woods known in Burmah, common in the plains as well as on the hills: used for cart-wheels, the teeth of harrows, the pestles of oil mills, &c. |
| 3450. [10363] | Tsheik-khyee. | Sapindus sp. . . . | 66 | 6 | 40 | Found on the hills and in the forests skirting them. Wood prized for house-posts, ploughs, &c. Colour grey, with a beautifully mottled grain |
| 3451. [10364] | Pinlay-oong . | Xylocarpus granatum, Koen. | 47 | 7 | 20 | In the forests of the Delta. Wood used for house-posts and musket-stocks |
| 3452. [10365] | Thit-kadoe . | Cedrela toona, Roxb. | 28 | 8 | 40 | On the hills and in the plains, plentiful in some districts. If not identical with the Toon of Bengal, certainly nearly related to it |
| 3453. [10366] | Yimma . . | Chickrassia tabularis, Juss. ? | 24 | 8 | 80 | Scattered throughout the forests on elevated ground, large trees scarce. Either identical with 'Chittagong wood,' or nearly related to it |
| 3454. [10367] | Boomayza . . | Albizzia stipulata, Boiv. | 66 | 9 | 30 | Common throughout the forests on elevated ground; heartwood brown, beautifully streaked, but rather small, the sap wood being very large. Much prized for cart-wheels, also used for the bells of cattle |

| Number | BURMESE NAME | SYSTEMATIC NAME OF TREE | Weight of 1 cubic ft in lbs. | Average size of full-grown trees on good soil | | REMARKS |
|---|---|---|---|---|---|---|
| | | | | Girth measured at 6 ft. from ground | Length of trunk to first branch | |
| 3455. [10368] | Seet . . . | Albizzia alata . . . | 42 to 55 | 10 | 40 | Abundant throughout the country in the plains, particularly near the banks of rivers. This wood may at a future time become an important article of trade. The heartwood is strong and durable, and less heavy than that of most trees of the same family. The only drawback is, that the proportion of sapwood is large. Used by the Burmans for bridges and house-posts. Br. weight 250 lbs. |
| 3456. [10369] | Sha . . . . | Acacia catechu, L. var. a. | 56 | 6 | 20 | Common all over the plains and scattered over the hills. Immense |
| 3457. [10370] | Sha . . . . | Acacia catechu, L. var. b. | 70 | 6 | 20 | numbers of these trees are annually cut down and made use of for the extraction of cutch. The wood is considered more durable than teak, and is used for house-posts, spear and sword handles, bows, &c. There are several varieties, differing in shade, specific weight, and yield of cutch |
| 3458. [10371] | Boay-gyin . . | Bauhinia malabarica, Roxb. | 42 | 4 | 15 | Common in the plains. Wood used for the cross pieces of harrows, house posts, &c. |
| 3459. [10372] | Hpa-lan . . | Bauhinia racemosa, Lam. | 44 | 3 | 10 | Resembles No. 3458 |
| 3460. [10373] | Gnoo-shwoay . | Cathartocarpus fistula, L. | 66 | 4 | 15 | Common in the plains and on the hills. Wood used for bows, axles of carts, &c. |
| 3461. [10374] | Gnoo-gyee . | Cassia sp. . . . . | 57 | 4 | 15 | Same as No. 3460 |
| 3462. [10375] | May-za-lee . | Cassia florida . . . | 58 | 6 | 15 | Cultivated. Heartwood almost black: used for helves, walking-sticks, mallets, &c. |
| 3463. [10376] | Yin-dike . . | Dalbergia sp. . . . | 64 | 9 | 35 | Common in the plains and on the hills. A kind of black wood well worth notice. The sapwood of this tree decays rapidly, but the heartwood is extremely durable; it is black, sometimes with white and red streaks, elastic, but full of natural cracks. Used for ploughs, bows, handles of dahs and spears. There are probably two kinds in the country |
| 3464. [10377] | Pynkado . . | Inga xylocarpa, L. | 60 to 66 | 9 | 50 | A magnificent tree, abundant throughout the forests on and near the hills. The *Ironwood* of Pegu. The sapwood is attacked by white ants, and decays easily, but is very small in large trees. The heartwood of full-grown trees is said to last as long as teak. This wood would be invaluable if it were not for its weight. Used for house and bridge posts, ploughs, boat-anchors, in the construction of carts, and for other purposes |
| 3465. [10378] | Thitpouk . . | Leguminosæ . . . | 35 | 4 | 20 | A light wood, not much used |

| Number | BURMESE NAME | SYSTEMATIC NAME OF TREE | Weight of 1 cubic foot in lbs. | Girth measured at 6 ft. from ground | Length of trunk to first branch | REMARKS |
|---|---|---|---|---|---|---|
| | | | | Average size of full-grown trees on good soil | | |
| 3466. [10379] | Padouk . . . | Pterocarpus dalbergi-oides | 60 | 9 | 35 | Trees of the largest size, of this strong and beautiful timber, abound in the forests east of the Sitang river, also in the valley of the Salween river, and its tributaries, the Thoungyeen, Yoonzalen, Hlineboay, Houndraw, and Attaran. Much less frequent in Pegu, and entirely wanting in some districts. Wood prized beyond all others for cart-wheels. The trees are felled green, and are split up into short planks, 3 ft. 6 in. long, 2 ft. wide, and 9 in. thick. Three of these pieces make one wheel, and a pair is sold on the spot, in the forests of the Prome district, at from 12 to 25 Rs. The wood is extensively used in the gun-carriage manufactories in India |
| 3467. [10380] | Kokoh . . . | Albizzia sp. . . . | 48 | 12 | 60 | In the northern districts of Pegu, on and near the hills. The wood is valued by the natives as much as Padouk (No. 3466), or even more so. It is used for cart-wheels, oil-presses, and canoes. In the Prome district, a special tax was levied on the felling of 'Kokoh' and 'Padouk' under the Burmese rule. Large trees are becoming very scarce in the Irrawaddy valley, but are not uncommon in the Toungoo district |
| 3468. [10381] | Thinwin . . | Pongamia sp. . . . | 60 | 6 | 20 | Not uncommon in the dry forest, in the plains, and on the hills. The heartwood, which is black and tough, but rather small, is used for the cross pieces of Burmese harrows, the teeth being made of Sha (No. 3456), Myoukkhyau (No. 3485), and Gjo (3449) |
| 3469. [10382] | Poukthenma-myek-kyouk | Leguminosæ . . . | 58 | 5 | 15 | A light-coloured, close-grained wood, much prized by Burmans |
| 3470. [10383] | Tounkatseet . | Leguminosæ . . . | 45 | 10 | 50 | Not uncommon on the hills. Wood used for canoes |
| 3471. [10384] | Thitsee . . . | Melanorhœa usitatissima, Wall. . . . | 54 | 9 | 30 | The varnish-tree of Burmah. Rare in the Irrawaddy valley, common in the forests east of the Sitang river, particularly south-east of Sitang Town. Wood dark red, hard and close-grained; used by the Burmese for the stocks of their wooden anchors, tool helves, &c. |
| 3472. [10385] | Khyong-yook . | Garuga pinnata, Roxb. | 52 | 9 | 40 | Tree rather common in plains, and on the hills. Wood not much used |
| 3473. [10386] | Nabhay . . | Odina wodier . . . | 65 | 12 | 50 | Tree rather common on the hills. Heartwood red, used for sheaths of swords, spear-handles, oil-presses, and rice-pounders |
| 3474. [10387] | Titseim . . | Terminalia bellerica, Roxb. | 40 | 12 | 80 | Common throughout Pegu. Wood not used |
| 3475. [10388] | Pangah . . | Terminalia chebula, Retz. | 53 | 12 | 80 | Common on the hills. A valuable wood, used for yokes and canoes; heartwood yellowish brown |

| Number | BURMESE NAME | SYSTEMATIC NAME OF TREE | Weight of 1 cubic foot in lbs. | Average size of full-grown trees on good soil | | REMARKS |
|---|---|---|---|---|---|---|
| | | | | Girth measured at 6 ft. from ground | Length of trunk to first branch | |
| 3476. [10389] | Lein . . . | Terminalia bialata, Roxb. | 39 | 12 | 80 | Common. Wood not used |
| 3477. [10390] | Htoukgyan . | Terminalia macrocarpa | 58 | 12 | 80 | One of the largest trees in Pegu, very common, and the stems of very regular shape. Heartwood dark brown. Used for house-posts and planking |
| 3478. [10391] | Yoong . . . | Conocarpus acuminatus | 50 to 57 | 12 | 80 | Almost equal to the preceding, in size and the regular growth of its stem. Wood reddish brown, hard and strong. Br. weight 226 lbs. NOTE.—If it were not for their weight, Nos. 3473, 3475, 3477, and 3478 would be most valuable for furniture |
| 3479. [10392] | Bambouay . . | Careya arborea, Roxb. var. a. (dark) | 55 | 9 | 20 | Common throughout the country. Wood used for gun-stocks, house-posts, planks, &c. |
| 3480. [10393] | Bambouay . | Careya arborea, Roxb. var. b. (light) | 55 | 9 | 20 | Same as foregoing |
| 3481. [10394] | Thabyehgjo . | Eugenia obtusifolia, Roxb. | 48 | 9 | 20 ⎞ | The different kinds of Thabyeh have a hard, red-coloured wood, |
| 3482. [10395] | Thabyehgyin . | Eugenia cerasoides, Roxb. | 51 | 9 | 40 | but not straight-grained, and supposed to be brittle. The stems are occasionally used for canoes, espe- |
| 3483. [10396] | Thabyehthapan | Eugenia. sp. . . . | 50 | 9 | 30 | cially those of Thabyehgah. Br. |
| 3484. [10397] | Thabyehgah . | Eugenia caryophyllæfolia, Roxb. | 56 | 6 | 20 ⎠ | weight of the Thabyehgah, 254 lbs. |
| 3485. [10398] | Myouk-kyau . | Blackwellia tomentosa, Vent. | 56 | 6 | 70 | Wood tough, of a light yellow colour, used for the teeth of harrows |
| 3486. [10399] | Laizah . . . | Lagerstrœmia pubescens, Wall. | 53 | 12 | 100 | A very large tree, stem not always perfectly round, inclined to form buttresses. Timber valued for bows and spear handles, also used for canoes and cart-wheels |
| 3487. [10400] | Thitpyoo . . | Lagerstrœmia sp. | 30 to 38 | 12 | 80 | A light but comparatively strong wood, colour white and pinkish, probably a valuable wood for furniture. Used for planking. Br. weight 153 to 179 lbs. |
| 3488. [10401] | Pyimma . . | Do. Regina, Roxb. var. a. wood light red | 37 | 12 | 30 | A splendid tree, abundant throughout the country. Wood |
| 3489. [10402] | Pyimma . . | Do. Regina, var. b. wood dark red | 44 | 12 | 30 | used more extensively than any other, except teak: used generally for the fittings of boats, sometimes for the hulls of canoes, for house-posts, planking, beams, scantling for roofs, carts, and a variety of other purposes. Large quantities are now employed for ordnance purposes. The wood of the light-coloured variety is less heavy, and is said to be less durable |
| 3490. [10403] | Tsambelay . | Do. parviflora, Wall. | 40 | 5 | 15 | Wood not much used |
| 3491. [10404] | Myoukgnau . | Duabanga grandiflora, Wall. | 30 | 12 | 80 | Wood used in house-building |
| 3492. [10405] | Hnsu . . . | Nauclea cordifolia, Roxb. | 42 | 10 | 80 | Trees large, of regular growth, but not very common. Wood yellow, rather close-grained, used to make combs — may be expected to prove valuable for furniture |
| 3493. [10406] | Bingah . . | Nauclea diversifolia, Wall. | 45 | 7½ | 60 | Wood of a light yellow colour, not much used, but may be recommended for furniture |

| Number | BURMESE NAME | SYSTEMATIC NAME OF TREE | Weight of 1 cubic ft. in lbs. | Average size of full-grown trees on good soil | | REMARKS |
|---|---|---|---|---|---|---|
| | | | | Girth measured at 6 ft. from ground | Length of trunk to first branch | |
| 3494. [10407] | Maookadoon . | Nauclea cadamba, Wall. | 37 | 15 | 70 | Wood of a deep yellow colour, but loose-grained; recommended for furniture |
| 3495. [10408] | Ma-oo lettan . | Nauclea undulata, Wall. | 23 to 34 | 15 | 100 | A soft useless wood, decays in less than a year. Br. weight 80 to 120 lbs. |
| 3496. [10409] | Htein . . . | Nauclea parviflora, Roxb. | 43 | 6 | 30 | Used for planking |
| 3497. [10410] | Hteingalah . | Nauclea sp. . . . | 43 to 56 | 6 | 40 | Wood of a light chestnut colour, recommended for furniture. Br. weight 208 lbs. |
| 3498. [10411] | Hteinthay . | Do. . . . . | 35 | 6 | 30 | Wood not used. Br. weight 170 lbs. |
| 3499. [10412] | Tsaythambyah | Gardenia lucida, Roxb. | 49 | 3 | 15 | A white close-grained wood, apparently well adapted for turning. This wood, like that of several other species of Gardenia and Randia, is used for making combs |
| 3500. [10413] | Ouk-khyin-za | Diospyros sp. . . . | 41 | 9 | 30 | A beautifully white and black mottled wood, used for house-posts |
| 3501. [10414] | Gjoot . . . | Do. sp. . . . . | 49 | 3 | 15 | Wood similar to that of the foregoing, but a much smaller tree. Small quantities of black heartwood (ebony) are occasionally found near the centre of very old trees of this and another kind nearly related to it (Taybeu) |
| 3502. [10415] | Khaboung . | Strychnos nux vomica, L. | 52 | 3 | 15 | Trees small, but common. Wood close-grained and hard |
| 3503. [10416] | Toung-za-lat . | Wrightia sp. . . . | 55 | 5 | 40 | A beautiful wood |
| 3504. [10417] | Paet-than . . | Spathodea stipulata, Wall. | 48 | 4 | 20 | Used for bows and spear handles, also for paddles and oars |
| 3505. [10418] | Thit-lin-da . | Spathodea sp. . . | 63 | 6 | 50 | A white wood, not much used |
| 3506. [10419] | Tha-khoot-ma | Spathodea Rheedii, Spreng | 35 | 7 | 30 | Wood used for yokes and cart poles |
| 3507. [10420] | Than-day . . | Bignonia sp. . . . | 33 to 36 | 7 | 30 | A light loose-grained wood, not much used. Br. weight 125 lbs. |
| 3508. [10421] | Kyoun-douk . | Do. . . . . | 23 | 2 | 15 | Wood not used |
| 3509. [10422] | Thanat . . . | Cordia myxa, L. . . | 33 | 4 | 15 | Wood soft, not used. Leaves collected extensively, sold for cover leaves for cigars |
| 3510. [10423] | Kjeyoh . . | Vitex sp. . . . | 45 | 3 | 15 | Wood used for tool handles, much prized, but rather scarce |
| 3511. [10424] | Htouk-sha . | Vitex leucoxylon, Roxb. | 42 | 12 | 30 | A large tree very common in the plains. Wood grey, deserves attention for furniture: used for cartwheels. Br. weight 142 lbs. |
| 3512. [10425] | Kyoon-na-lin . | Premna pyramidata, Wall. | 52 | 5 | 30 | Wood strong, used for weaving shuttles. Trees small |
| 3513. [10426] | Kuyon—Teak | Tectona grandis, L. . | 40 to 51 | 18 | 90 | The best teak forests in British Burmah are on the hills between the Sitang and Irrawaddy rivers, and in the Thoungyen valley; but even these forests are poor compared with the extensive tracts covered with teak-producing forests to the north of the British boundary, especially on the feeders of the Sitang and Salween rivers, and some of the tributaries of the Meinam, or Bankok river. The trees also are, as a rule, much larger, and the shape of the stem more regu- |

| Number | BURMESE NAME | SYSTEMATIC NAME OF TREE | Weight of 1 cubic ft. in lbs. | Average size of full-grown trees on good soil | | REMARKS |
|---|---|---|---|---|---|---|
| | | | | Girth measured at 6 ft. from ground | Length of trunk to first branch | |
| | | | | | | lar, in the forests of the Burmese empire, the Siamese kingdom, and the Karennee country. The tallest teak tree measured in Pegu was 106 ft. high to the first branch. The strength and density of teak timber vary exceedingly, according to the locality where the tree is grown. The extremes observed in preliminary experiments were 40 and 50 lbs. per cubic foot, and 190 lbs. to 289 lbs. breaking weight. Teak, when young, grows very rapidly. The two stems sent were dug out by me in July 1858, at the Thinganenoung nursery in the Attaran forests. The seed had been sown in March and April 1856. The plants, therefore, were two years and three months old. The largest seedlings had a girth of 13 in. measured one foot from the ground, and of 8 in. at 6 ft. from the ground. They were 32 ft. high, but this is an instance of uncommonly rapid growth. Trees ten years old have usually a girth of 18 in., measured at 6 ft. from the ground; with 22 years a girth of 3 ft. is attained: but full-grown trees of 9 ft. in girth cannot be supposed to be less than 160 years old |
| 3514. [10427] | Yemaneh . . | Gmelina arborea, Roxb. | 35 | 12 | 50 | A large tree with white light wood, used for house-posts, planks, and for carving images. Recommended for planking and furniture |
| 3515. [10428] | Thit-kya . . | Quercus semiserrata, Roxb. | 48 | 4 | 20 | Used for plugs or pins to join together the three pieces which compose the body of a Burmese cart-wheel |
| 3516. [10429] | Momakha . . | Salix tetrasperma, Roxb. | 37 | 3 | 10 | Wood not used |
| 3517. [10430] | Tounbein . . | Artocarpus mollis, Wall. | 30 | 12 | 80 | Immense trees, wood used for canoes and cart-wheels. On the hills, large trees rather scarce |
| 3518. [10431] | Toun-pain-nai | Artocarpus sp. . . | 39 | 12 | 80 | Wood yellow, used like the preceding |
| 3519. [10432] | Myouklouk . | Artocarpus lacoocha, Roxb. | 40 | 6 | 30 | Used for canoes |
| 3520. [10433] | Thaphon . . | Ficus lanceolata, Roxb. | 27 | 12 | 25 | Wood soft, useless |
| 3521. [10434] | Theetmin . . | Podocarpus neriifolia | 50 | 6 | 20 | The meaning of the Burmese name is, ' the prince of trees.' Large trees with stems not very regularly shaped are found on the higher hills between the Sitang and Salween rivers, and on the range which skirts the coast of the Tenasserim provinces. The wood is close-grained, and may prove a substitute for boxwood |
| 3522. [10435] | Tinyooben . | Pinus Massoniana, Lamb. | | 6 | 50 | The pines of British Burmah. Pinus Massoniana is a moderate-sized tree, found in the forest of Dipterocarpus grandiflora (Eng |
| 3523. [10436] | Tinyooben . | Pinus Khasyana . . | | 9 | 80 | |

| Number | Burmese Name | Systematic Name of Tree | Weight of 1 cubic ft. in lbs. | Average size of full-grown trees on good soil | | Remarks |
|---|---|---|---|---|---|---|
| | | | | Girth measured at 6 ft. from ground | Length of trunk to first branch | |
| | | | | | | forest), east of the Salween river. Spars of this species have occasionally been brought down to Maulmain P. Khasyana is found on the hills between the Sitang and Salween rivers, at an elevation exceeding 3,000 ft. It is a stately tree, sometimes as high as 200 ft. to the top; but owing to the difficulties of transport from these hills, no timber of this species has as yet been brought to Maulmain. The wood of both kinds is very rich in resin |
| 3524. [10437] | Kanazoe . . | Pierardia sapida . . | 61 | 4 | 15 | A small tree, wood not used |
| 3525. [10438] | Nasha . . . | Phyllanthus sp. . . . | 35 | 6 | 30 | A light-coloured wood, exhibiting a natural shine or polish when planed |
| 3526. [10439] | Yagine . . | Rottlera sp. . . . | 35 | 6 | 30 | A moderate-sized tree, common on the low ground near streams. Br. weight from 153 to 170 lbs. |
| 3527. [10440] | Bamau . . . | Unknown . . . . | 52 | 6 | 30 | Close-grained, possibly a substitute for boxwood, prized by Karens for bows |
| 3528. [10441] | Palawah . . | Do. . . . . | 52 | 6 | 45 | A beautiful red, but heavy wood |
| 3529. [10442] | Nattamin . . | Do. . . . . | 33 | 6 | 60 | Wood loose-grained, reddish grey, recommended for cigar boxes. Br. weight 129 lbs. |
| 3530. [10443] | Moondein . . | Do. . . . . | 33 to 38 | 10 | 50 | Wood fine-grained, light, recommended for furniture. Br. weight 121 lbs. |
| 3531. [10444] | Koothan . . | Do. . . . . | 28 | 6 | 40 | A loose-grained light wood, recommended for packing cases, used for blackboards in Burmese schools. Br. weight 114 lbs. |
| 3532. [10445] | Thakooppo . | Stereospermum chelonioides | | 5 | 30 | Wood used in house-building |
| 3533. [10446] | Maneioga . . | Carallia integerrima, Dc. | 60 | 10 | 50 | A large tree, common north of Rangoon and throughout Pegu. Wood of a peculiar structure, thick medullar rays going through from the centre to the circumference; colour red; may possibly be found useful for cigar boxes. Used for planks and rice-pounders |
| 3534. [10447] | Thitnee . . | Unknown . . . . | 80 | 8 | 50 | A beautifully red, but heavy wood |
| 3535. [10448] | Lumbo . . . | Buchanania latifolia . | 36 | 6 | 30 | A soft light wood, not used |
| 3536. [10449] | Chloctni . . | Erioloena sp.? . . | | | | A red wood, used like Dwanee (No. 3436) |
| 3537. [10450] | Thitpagan . . | Pongamia sp. . . | | 9 | 40 | A soft wood, said to be useless |
| 3538. [10451] | Kaungmhoo . | Dipterocarpus sp. . | | 12 | 100 | Trees of an immense size, used for canoes |
| 3539. [10452] | Katsitka . . | Unknown . . . . | | 6 | 30 | A red wood, abundant in the forests north of Rangoon, used for boats; said to last from 5 to 6 years |
| 3540. [10453] | Anambo . . | Henslowia paniculata, Migu. | | 9 | 50 | A reddish-coloured wood, not straight-grained; used occasionally for cart-wheels, mostly for firewood |

**WOODS FROM MOULMEIN.**—*Continued.*

3541. [7632] Young Zalai (*Garcinia mangostana*).

This wood is made use of for ordinary house-building purposes. Fruit edible.

3542. [7633] Dain Tha (*Moringa pterygosperma*).

Flowers, bark, and root used medicinally. Wood made into dolls.

3543. [10466] Mya Ya Ngai (*Grewia micrococos*).

This wood is used for ordinary house-building purposes.

3544. [10467] Pyen Ma Nee, or Jarue of Chittagong (*Lagerstræmia regina*).

Wood used for boats and carts, also for flooring houses.

3545. [10468] Ein Gyin (*Dipterocarpus vatica*).

A very strong durable wood, as strong as Pyengado; when kept long in water it is said to become petrified.

3546. [10469] Phangah (*Terminalia chebula*).

Is very hard and heavy. Used to make rice pounders, furniture, &c.

3547. [7634] Kamala (*Sonneratia apetala*).

An inferior wood for boats, which last but two or three years.

3548. [10470] La Moo (*Sonneratia acida*).

The fruit is an article of food.

3549. [7635] Ya Tha Nat.

Ditto ditto.

3550. [10471] Ka Na Zo (*Pierardia sapida*).

A very hard wood. Used for wheel axles.

3551. [7636] Kya Nan.

Red wood; used generally by carpenters.

3552. [7637] Tha-ran (*Grewia* sp.).

A wood used to make dancing dolls.

3553. [7638] Ya tha pyo

The fruit is edible. Used for house-building purposes.

3554. [7639] Oan Naih.

Ditto.

3555. [7640] Bon Sone.

Ditto.

3556. [7641] Thin-win.

The root is used medicinally.

3557. [10472] Phàt Than.

Used for chisel handles.

3558. [7642] Koung Mhoo (*Vatica* sp.)

Used for making carts and boats.

3559. [7643] Pyen Ma Phoo (*Lagerstræmia* sp.)

Used for making oars and for rough house-building.

3560. [7644] Konk Koe (*Acacia* sp.)

This wood is made into boats, carts, and other ordinary house-building material.

3561. [10473] Myouk Shaw (*Dalbergia* sp.)

This wood is used in ordinary house-building.

3562. [10474] Kyee (*Cassia sumatrana*).

Ditto.

3563. [7645] Mazalee (*Cassia florida*).

Ditto.

3564. [7969] Tha Khoot.

Ditto.

3565. [7646] Zinpyun Gyee.

Ditto.

3566. [7647] Ya Ka Ngine.

Ditto.

3567. [7648] Ouk Kyine.

Ditto.

3568. [7649] Yamani.

Ditto.

3569. [7650] Thapya (*Water Dalbergia*).

Ditto.

3570. [7651] Koan Tha Nath (*Cordia myxa*).

The leaf is made into cigar-wrappers.

3571. [7652] Tha Yat (*Mangifera indica*).

Fruit is eaten.

3572. [7653] Koun Soay-dan.

Ditto.

3573. [7654] Thet Kon Nyen.

Fruit is eaten.

3574. [7655] Tsan-saypen.

Used for ordinary house-building purposes. Leaf is eaten boiled as greens.

3575. [7656] Ahline Ngai.

Ditto.

3576. [10475] Manee Auka.

Bark is used medicinally.

3577. [10476] Ngoo Tha (*Cassia* sp.)

Made into house-posts. Fruit and bark used medicinally.

3578. [7657] Youg Tha Ngai (*Arbus?*)

Used in ordinary building materials.

3579. [7658] Tha-man-tha.

Ditto.

3580. [7659] Kaboung (*Strychnos Nux vomica*).

Fruit used as medicine.

3581. [7660] Mai Kin.

Used in ordinary building material. Fruit used as medicine.

3582. [7661] Yamana (*Gmelina arborea*).

Ditto.

3583. [10477] Kay Yoob.

Ditto.

3584. [10478] Nat Gyee.

Used for posts and knife handles.

3585. [10479] Anan Pho (*Gordonia* sp.)

A strong wood, good for building purposes.

3586. [10480] Yin Yo.

Ditto.

3587. [7662] Sha Bya Gyin (*Eugenia* sp.)

Wood soft; used as an ordinary building material.

3588. [10481] Pyen-ma-zoat Gyee (*Lagerstræmia* sp.)

Ditto.

3589. [10482] Pune Tha.

Ditto.

3590. [7663] Paran Tha.

Wood soft; used as an ordinary building material.

3591. [10483] Ma-oo-tha (*Nauclea ca-camba*).

Used for building purposes.

3592. [7664] Tsat Tha.

Ditto.

3593. [7665] Dhane Eha (*Moringa pterygosperma*).

Ditto.

3594. [7234] Yin-gat? (*Gardinia coronaria*).

Ditto. Fruit edible.

3595. [10484] Mayan (*Mangifera oppositifolia*).

Ditto.

3596. [10485] Padouk (*Pterocarpus Dalbergioides*).

A very strong wood, admirable for furniture, used by the Burmese to make their musical instruments.

3597. [7666] Kya Zo.

Used for building material.

3598. [7667] Na Bai (*Odina Wodier*).

A red wood. Bark used medicinally.

3599. [7668] Nyoung Lan.

Used for building material.

3600. [10486] Tha Bya Nee (Red Jambo).

Ditto.

3601. [10497] Nyoay Sha.

Ditto.

3602. [7974] Monk Kyan (*Homalium tomentosum*).

A strong wood for any ordinary purpose.

3603. [7669] Nga Thingyee (*Ficus cordifolia?*).

Ditto.

3604. [10487] Nyoung Tha.

Ditto.

3605. [10488] Woot Tha.

Ditto.

3606. [10489] Kya Ya (*Mimusops elengi*).

Ditto. The flower is used medicinally and for scent.

3607. [7670] That Pan (*Bombax* sp.)

A strong wood for any ordinary purpose. The flower is used medicinally and for scent.

3608. [7671] Tha Bya (*Eugenia* sp.)

Ditto.

3609. [7672] Zin Pyun Ngan (*Dillenia speciosa*).

Ditto. Fruit edible.

3610. [7673] Ma Shoay (*Bignonia stipulata*).

A strong wood for any ordinary purpose. Fruit edible.

3611. [7674] Tonk Tsa (*Vitex arborea*).

Ditto.

3612. [7675] Ah See Eha.

Wood hard. Used for making musical instruments.

3613. [10490] Than-that-gyee.

Used for building materials.

3614. [7676] That Yat (*Mangifera indica*).

Ditto. Fruit edible.

3615. [7677] Tseek Tha (*Acacia sirissa*).

Wood reddish colour. Used for furniture.

3616. [7678] Ein Win.

Used for all ordinary purposes of building.

3617. [7679] Thet Lendah.

Ditto.

3618. [10495] Wiha Oung.

Ditto.

3619. [7680] Kha Gyee (*Strychnos Nux vomica*).

Ditto. Fruit used medicinally.

3620. [10491] Zangyeeoat-doup (Oak-leaved Polypod).

Ditto.

3621. [7681] Ah Nan (*Cyrtophyllum fragrans*).

A strong wood, good for building purposes.

3622. [7682] Kyan-pho.

Ditto.

3623. [7683] Za Padrup.

Ditto.

3624. [7684] Yendike (*Dalbergia* sp.)

A hard heavy black wood, useful for furniture.

3625. [7685] Bha Woon (*Grewia* sp.)

Converted into planks for building.

3626. [10492] Tngtha (*Dipterocarpus grandis*).

Ditto.

3627. [10493] Tsouk Yoa (*Dalbergia alata*).

Used for tool handles.

3628. [7686] Toung Ma Yoa.

Wood smooth. Used generally for Burmese slate or writing boards.

3629. [10494] Thit Nee.

Converted into boxes, tables, &c. &c.

3630. [7687] Thit Nya (*Castanea martabanica*).

The fruit eaten exactly like chestnuts.

3631. [7688] Pani Nai, vulgo 'Jack tree' (*Artocarpus integrifolius*).

Fruit eaten. Wood yellow. Used to dye the yellow Pongyee (Burmese Priest) cloths.

## SPECIMENS OF WOODS FROM MALACCA.

Forwarded by C. EVANS, Esq.

3632. [2465] Marabow.

3633. [2466] Billian Wangee.

3634. [2467] Madang Katana.

3635. [2468] Pannaga.

These four species are the very best description of timber procurable in Madras, and command a market at very high prices. They are strong, solid, and very durable, being principally used for girders, rafters, joists, door and window posts, and timber for bridges, standing the sudden changes of the climate remarkably well. The Marabow is also used for furniture. Not subject to dry rot, and when well seasoned is known to last nearly half a century.

3636. [2469] Patalin.

3637. [2470] Klat Mera.

3638. [2471] Kasso.

These hold a second position in the art of house-building, but are much more commonly used, being more abundant and easily procurable. Nos. 3636 and 3637 are commonly used for door and window frames, but No. 3637 is apt to split in the sun, consequently is always used within doors in the Straits.

3639. [2472] Tumboosoo.

3640. [2473] Giam.

3641. [2474] Brombong.

Best and most durable species of timber, known to resist the effects of a damp soil; invariably used for foundation piles, palisading and supporting piles for bridges; Nos. 3639 and 3670 sawn into planks are the very best description of timber that can be used for the platform of a timber bridge supporting a gravel road.

### 3642. [2475] Traling.

### 3643. [2476] Marsawa.

### 3644. [2477] Pasal Antoo.

These three species grow to an enormous height and girth, with huge buttresses, which are eagerly sought for solid cart-wheels, in common use among the Malays, and the trunks are converted into jaloors or river boats, consisting of one solid block scooped out in the shape of a canoe.

### 3645. [2478] Bintangore Batoo.

### 3646. [2479] Do. Akar.

### 3647. [2480] Do. Boonoot.

Tough, hard, crooked-grained, fibrous wood. In general use for masts and spars of vessels. No. 3645 is often used for purlins among the natives.

### 3648. [2481] Marpoyan.

### 3649. [2482] Marbatoo.

### 3650. [2483] Marpadang.

Used for fishing stakes, piles, and is the best description of fuel for steamers. Makes very good charcoal for a blacksmith's forge.

### 3651. [2484] Madang Kuniet.

### 3652. [2485] Do. Pao.

### 3653. [2486] Do. Klade.

### 3654. [2847] Do. Lawang.

### 3655. [2488] Do. Saraya Batoo.

### 3656. [2489] Marantee.

### 3657. [2490] Niatoo.

### 3658. [2491] Doorian Doorian.

In general use for planks, tile laths, except No. 3651, which being soft and cohesive is used by the Chinese for carvings. No. 3656 is very inferior in quality, and is only used by the poorer classes, great quantities being brought down to market as floats for heavier descriptions of timber.

### 3659. [2492] Kampas.

Used as charcoal, which is of excellent quality and much used by the tin miners.

### 3660. [2493] Klaydang.

Used principally by the Chinese for coffins, and planking vessels.

### 3661. [2494] Minia Jantan.

The wood oil of commerce is tapped from this tree, and it yields tolerably good planks for bridges.

### 3662. [2495] Kranjee.

A good, heavy, valuable timber, somewhat like iron wood. Used for machinery, mortar and pestle, &c.

### 3663. [2496] Alban.

Used for ribs of vessels and boats.

### 3664. [2497] Ensanna.

Close-grained, mottled, and valuable wood for furniture; it takes a high polish, and when well seasoned does not warp.

### 3665. [2498] Karantey.

Employed for gun stocks. It is a white soft wood, close and compact, and fit for turning purposes.

### 3666. [2499] Gelotong.

Light, pithy wood, coarse-grained and porous. Used for sandals, stoppers for bottles, and covers for cooking utensils.

### 3667. [2500] Neepis Koolit.

Light and pliant. Used for oars and buggy shafts.

### 3668. [2500a] Pangarawan.

A very valuable tree; the bark is used in lieu of planks by the poorer classes of natives. The trunk yields excellent planks for shipbuilding; and the valuable gum known in commerce as Damar Matakooching, or Gum Copal, is procured from this tree.

### 3669. [7986] Rambey Dahoon.

Good for planks, and the tree yields Damar Batoo, a coarse resin, much used in manufacturing torches.

### 3670. [7987] Rangas.

Red wood. Much used for furniture.

### 3671. [7987a] Kamooning.

This tree is not indigenous to the place, but thrives well on private grounds. It yields superior planks for manufacturing small boxes, and the roots, not unlike Kayoo bookoo, are made into handles for krisses.

### 3672. [7988] Glam Tambaga.

Used for piles and posts under water: the paper-like bark is much used by the Malays in caulking the seams of vessels.

## SPECIMENS OF WOODS FROM THE MALAYAN PENINSULA.

### 3673. [79] Garro, Malay Peninsula.

Used for burning as incense. TAN KIM SENG.

### 3674. [2464] Kaimooning, Do.

Scarce. Hard and close-grained, resembling box. H. H. INCHE WAN ABOO BAKAR.

### 3675. [2445] Moodang Tandoo (Sipoo?) Do.

Used for furniture, and doors and windows. Do

3676. [2448] Maraboo, Malay Peninsula.

Wheels, buggy shafts, and junks' masts. H. H. INCHE WAN ABOO BAKAR.

3677. [2453] Kranjee, Do.

Junks' masts. Do.

3678. [2449] Mursawah, Do.

Do. Do.

3679. [2447] Julatong, Do.

Used for making coffins and boxes. Do.

3680. [2458] Jamah, Do.

Common house work, doors, &c. Do.

3681. [2451] Pawang, Do.

Boat building. Do.

3682. [2454] Rengas, Do.

Furniture. Do.

3683. [2455] Serayah, Do.

Doors, windows, and ordinary floors. Do.

3684. [2446] Mudang, Do.

House fittings. Do.

3685. [2450] Kuning, Do.

Boat purposes. Do.

3686. [2457] Samarang, Do.

House fittings. Do.

3687. [7984] Casuarina (*Casuarina*), Singapore grown.

Felloes for wheels and spokes. COL. COLLYER.

3688. [7983] Mangrove (*Rhizophora mangle*), Salt swamp, Do.

Used for piles. Do.

3689. [2459] Pinagah, Do.

Used for boats' knees. Do.

3690. [7985] Kompas, Do.

House building. Do.

3691. [7982] Teak (*Tectona grandis*), Do.

Building. Do.

3692. [7981] Daroo, Do.

Do. Do.

3693. [2460] Malow, Do.

Junks, masts, &c. HON. COL. CAVANAGH.

3694. [2461] Tampenis, Malay Peninsula.

House building. HON. COL. CAVANAGH.

3695. [2462] Balow, Do.

Piles and junks' masts. Do.

3696. [2463] Changal, Do.

Ships' lower masts. G. ANGUS, Esq.

3697. [2452] Trabang, Do.

Do. Do.

3698. [7238] Cladang, Do.

Doors and windows. Do.

3699. [2456] Do.

Used for carvings. Do.

### SPECIMENS OF WOODS FROM THE FOREST ON PINANG HILL.

HON. MAJOR MAN.

3700. [9418] Pulai.

3701. [7070] Bahkoh.

3702. [7061] Champada.

3703. [9419] Bayor.

3704. [9239] Bayang Bada.

3705. [7064] Jurai.

3706. [7068] Nangka.

3707. [9240] Brangan.

3708. [7065] Gaham Bada.

3709. [7069] Tumusu.

3710. [7067] Bia-babi.

3711. [7080] Mungkudu.

3712. [7091] Ahtow.

3713. [7078] Kampas.

3714. [7074] Tumpang.

3715. [7066] Rungas.

3716. [7073] Juntang-malah.

3717. [7061] Champada Ayer.

3718. [7062] Tampineh.

3719. [9241] Nangka pipet.

2720. [7072] Klat.

3721. [7076] Jelutong.

3722. [7063] Jong-purlis.

3723. [7075] Jermalang.

3724. [7090] Kumpas.

3725. [7088] Jelatoh.

3726. [7086] Dammer-laut.

3727. [7093] Gading-gading.

3728. [7087] Koolin.

3729. [7079] Pisang-pisang.

3730. [7071] Murbow.

3731. [7089] Bintaling.

3732. [7092] Madang-Serai.

3733. [7077] Sittola.

Subdivision C.—vi. (3) PREPARED WOODS.

Of this there is only one representative.

Specimens of wood prepared with CULLEN's patent composition. MR. ADAMS, *Bow, London.*

3735. [9397] A piece of Deodar wood, or Himalaya pine, not prepared with the composition.

3736. [9396] A piece of wood cut from the same log as No. 3735, and coated with the composition.

These two specimens were buried in a white ants' nest, at Meean Meer, for the space of five months, from March to July, 1860. At the end of this period, No. 3735, which had not been previously coated, was nearly destroyed by the ants; and No. 3736, which had been prepared with the composition, was in a perfect state.

SECTION VII.—MISCELLANEOUS SUBSTANCES.

1. *Substances used as Soap.*

Of substances belonging to this division, the Indian soap berry is worthy of more attention than it has yet received. It is admirably adapted for cleaning silks and other fabrics, the colours of which are likely to be damaged by the use of ordinary soap. It besides readily forms a lather with water, however hard.

3737. [9925] Soap berries from *Sapindus emarginatus*, Calcutta.

The natives use them as soap for washing the hair, silk, &c. Agitated with hot water they form seeds. Price 4 annas per lb.

3738. [4793] Kinmon (*Mimosa abstergens* or *Acacia concinna*), Burmah.

3739. [6047] Sikakai (*do.*), Ahmedabad.

A considerable trade is carried on in some parts of India in the pods of this plant, which resemble the soap-nut, and like it are used for washing the head.

2. *Substances used mechanically.*

3740. [4978] Aerial roots of Banyan (*Ficus indica*), Madras.

Used as a tooth brush.

3. *Seeds employed for Bracelets and other Ornamental Purposes.*

3741. [7157] Kooneh (*Abrus precatorius*) (white), Calcutta.

3742. [6222] Do. (red), Do.

3743. [7013] Do., Bombay.

Of the seeds of this creeper there are several varieties, white, scarlet, and black. Those of a bright scarlet colour, with a black speck at the tops, are used not only for ornament, but also as weights by jewellers and druggists.

3744. [4923] Mimosa seeds (*Desmanthus virgatus*), Chingleput. DR. SHORTT.

Amongst miscellaneous substances are also placed the following:—

Charcoal.

3745. [7233] Charcoal fire balls, Vizagapatam.

3746. [4893] Mudar root charcoal (*Calotropis gigantea*), Chingleput. DR. SHORTT.

3747. [7056] Charcoal, Ahmedabad.

3748. [7059] Charcoal of *Euphorbia antiquorum*, Do.

3749. [7057] Do. (*Bassia latifolia*), Do.

3750. [7058] Babool charcoal (*Acacia arabica*), Bombay.

3751. [6431] Buglar tree bark, Chota Nagpore.

3752. [6437] Buglar tree bark powdered, Chota Nagpore.

This is used for uniting wood, as a substitute for glue.

## X.—GENERAL MANUFACTURES FROM WOOD
### (*not being Furniture*).

#### I.—TURNERY.—*Plain and lacquered.*

3753-4. [5845-6] Two betel stands, Lahore. *Lahore Committee.* GOVERNMENT of INDIA.

3755-8. [5841-4] Four large circular boxes, Do. Do.

3759-66. [5833-40] Eight small do. Do. Do.

3767-9. [5830-2] Three flower vases, Do. Do.

3770. [5829] Pair candlesticks, Do. Do.

3771-5. [5851-5] Five cigar cases, Do. Do.

3776-7. [5825-6] Pen cases, Do.

3778-82. [5856-60] Five plates, Do. Do.

3783. [5847] Glass-shaped vessel with cover, Do. Do.

3784. [5848] Do. without cover, Do. Do.

3785. [5850] Ten toys, Do. Do.

These articles have been manufactured by the turners of Påk Puttan in the Googaria District.

3786. [5433] Table, lacquered, Lahore. R. TAYLOR, Esq.

3787. [5434] Candlestick, do. Do. Do.

3788. [1509] Vases, black and silver, ornamented in chemical amalgam, Umritsur. *Lahore Committee.* GOVERNMENT of INDIA.

3789. [1510] Do. Do. Do.

3790-1. [1537-8] Two wooden cups, Umritsur. Do. Do.

3792. [1639] Eighty-two specimens of coloured turnery (toys, &c.) Do. Do.

3793. [7998] Set of black japanned articles.

3794. [5752] Pen-tray, lacquered, Meerut. SYUD HOSSEIN ALLEE KHAN.

3795. [5753] Workbox, do. Do. Do.

3796. [8108] Boxes, vases, and other articles made of lac at Beerbhoom. GOVERNMENT of INDIA.

3797. [3164] Drinking cup of Lamas, Darjeeling. DR. CAMPBELL.

3798. [3165] Do. Do. Do.

3799. [3166] Two wooden spice boxes, Do. Do.

3800. [3169] Wooden bowls, Do. Do.

3801. [3162] Thibetan cup, Do. Do.

3802. [3170] Four meal boxes, Do. Do.

3803. [3185] Wooden bowl, Do. Do.

3804. [3179] Wooden cup, Do. Do.

3805. [1070] Circular box, Hyderabad. SINDE LOCAL COMMITTEE.

3806. [1071] Do. Do. Do.

3807. [1074] Cigar case, Do. Do.

3808. [1075] Do. Do. Do.

3809. [1064] Wooden boxes, Do. Do.

3810. [1065] Nest of boxes, Do. Do.

3811. [1066] Do. Do. Do.

3812-3. [1067-8] Two boxes melon-shaped, Do. Do.

3814. [1069] Nest of seven boxes, Do. Do.

3815. [1072] Flower vase with stand, Do. Do.

3816. [1073] Do. Do. Do.

3817. [2516] Painted box, Sinde. BOMBAY GOVERNMENT.

3818. [4715] Pen-tray, Sawunt Warree. Do.

3819. [4716] Pen cases, Do. Do.

3820. [4721] Churka model, lacquered, Do. Do.

3821. [4722] Two pots, Do. Do.

3822. [4724] Egg cups, Do. Do.

3823. [4725] Polpat and Satnee, Do. Do.

3824. [4726] Mortar and pestle, Do. Do.

3825. [4728] Two lacquered toys, Do. Do.

3826. [10508] Two trays, lacquered ware, Bangapully, Kurnool, Madras. GOVERNMENT of MADRAS.

3827. [10509] Two boxes, do. Bangapully, Kurnool, Madras. GOVERNMENT of MADRAS.

3828. [2686] Toys, lacquered, various, Bangalore, Mysore. GOVERNMENT of MYSORE.

3829. [4482] Box, Burmese, Rangoon. Messrs. HALLIDAY, FOX, & Co.

3830. [4494] Do.          Do.          Do.

3831. [4495] Nest of nineteen boxes, Do.          Do.

3832. [4499] Nests of boxes, Do.          Do.

3833-4. [4477-8] Two water cups, Do. Do.

3835. [469] Knitting box, Burmah, Canara. DR. M'PHERSON, *Madras.*

3836. [4832] Bowls, Do.  Do.

3837. [6322] Wooden boxes, turned, supposed to be tamarind, Trinjanu. Turned and presented by H. H. the RAJAH.

3838. [6323] Lacquered tray, Singapore. COL. COLLYER.

3839. [6311] Box and cup for betel nut, Singapore. HON. COL. CAVANAGH.

3840. [6749] Box lacquered in imitation of Mangosteen, Do.          Do.

3841. [6321] Box and cup for betel nut, Singapore. H. H. INCHE WAN ABOO BAKAR.

3842. [6327] Circular box, Palembang, Sumatra. Do.          Do.

## II.—CARVINGS IN SANDALWOOD ETC.

3843. [3438] Spoon, Umritsur. RAJAH of PUTTIALA.

3844. [1501] Box, Puttiala.          Do.

3845. [968] Walking-stick, Do.          Do.

3846. [3041] Large box, with pictures on lid, Delhi. ISHMAIL KHAN.

3847. [10048] Five spoons, Budaon. GOVERNMENT of INDIA.

3848. [10049] Five do.,          Do.          Do.

3849. [5155] Two ebony necklaces and bracelets, Moonghyr. Do.

3850. [10030] Pen-box, Bijnour. ABDOOLA of NUGEENA.

3851. [5356] Ebony combs, Bijnour. SAADUT ALLEE.

3852. [5357] Do.          Do.          Do.

3853. [5359] Do.          Do.          Do.

3854. [5360] Do. Do.  KURREEM BUX.

3855. [5361] Do.          Do.          Do.

3856. [5362] Do.          Do.          Do.

3857. [5364] Do.          Do.          Do.

3858. [5365] Do.          Do.          Do.

3859. [5352] Carved ebony casket, Do.

3860. [5353] Carved ebony writing-case, Do.

3861. [5354] Carved ebony casket, Do.

3862. [355] Glove-box, Bombay. BOMBAY GOVERNMENT, *per* J. MACFARLANE, Esq.

3863. [343] Portfolio,          Do.          Do.

3864. [306] Book-stand,  Do.          Do.

3865. [313] Portfolio,          Do.          Do.

3866. [354] Glove-box,          Do.          Do.

3867. [524] Card-basket, Do.          Do.

3868. [286] Card-case,  Do.          Do.

3869. [285] Do.          Do.          Do.

3870. [346] Card-basket, Do.          Do.

3871. [270] Writing-desk, Do.          Do.

3872. [338] Do.          Do.          Do.

3873. [523] Glove-box,          Do.          Do.

3874. [316] Cabinet, Coompta Malabar. JAMSETJEE HEERJEE PARSEE.

3875. [1243] Knitting-box, Do.          Do.

3876. [320] Do.          Do.          Do.

3877. [1240] Do.          Do.          Do.

3878. [1241] Do.          Do.          Do.

3879. [1239] Do.          Do.          Do.

3880. [317] Envelope-case, Do.          Do.

3881. [318] Do.          Do.          Do.

3882. [1333] Large box,          Do.          Do.

3883. [1242] Knitting-box, Do.          Do.

3884. [319] Do.          Do.          Do.

3885. [334] Writing-desk, Do.          Do.

3886. [1244] Twenty-four paper-knives, Coompta Malabar. JAMSETJEE HEERJEE PARSEE.

3887. [321] Knitting-box, Surat, Bombay.

3888. [322] Do.          Do.          Do.

3889. [323] Do.          Do.          Do.

3890. [324] Do.          Do.          Do.

3891. [325] Do.          Do.          Do.

3892. [326] Do.          Do.          Do.

3893. [327] Do.          Do.          Do.

3894. [328] Do.          Do.          Do.

3895. [329] Do.          Do.          Do.

3896. [307] Writing-desk, Do.          Do.

3897. [308] Do.          Do.          Do.

3898. [311] Card-case,    Do.          Do.

3899. [312] Do.          Do.          Do.

3900. [310] Do.          Do.          Do.

3901. [332] Card-basket,  Do.          Do.

3902. [330] Do.          Do.          Do.

3903. [331] Do.          Do.          Do.

3904. [303] Work-box,     Do.          Do.

3905. [305] Inkstand,     Do.          Do.

3906. [309] Chessboard,   Do.          Do.

3907. [304] Glove-box,    Do.          Do.

3908. [648] Carved writing-case, Kutch. H. H. the RAO of KUTCH.

3909. [4731] Pen-case, Madras.

3910. [4733] Tray, Do.

3911. [2521] Chess-table, North Canara. GOVERNMENT of MADRAS.

3912. [2522] Paper-knives, Do.          Do.

3913. [10750] Large box,   Do.          Do.

3914. [443] Box,     Kurnool.     DOCTOR M'PHERSON.

3915. [444] Do.          Do.          Do.

3916. [423] Four card-cases, Vizagapatam. Do.

3917. [2714] Pen-rack, Bangalore, Mysore. J. LACEY, Esq.

3918. [2662] Pen-case,     Do.          Do.

3919. [2675] Card-case, Bangalore, Mysore. J. LACEY, Esq.

3920. [2711] Work-box,     Do.          Do.

3921. [2668] Walking-stick, Do.          Do.

3922. [2669] Do.          Do.          Do.

3923. [2663] Fly fan,      Do.          Do.

3924. [2664] Do.          Do.          Do.

3925. [2674] Paper-knife,  Do.          Do.

3926. [2666] Album-cover,  Do.          Do.

3927. [2667] Do.          Do.          Do.

3928. [2665] Book-cover,   Do.          Do.

3929. [2673] Paper-knife,  Do.          Do.

3930. [2677] Portfolio,    Do.          Do.

3931. [2671] Bracelet,     Do.          Do.

3932. [2672] Do.          Do.          Do.

3933. [2712] Box,          Do.          Do.

3934. [2713] Envelope-case, Do.          Do.

3935. [569] Box, Do.   CAPT. PUCKLE.

3936. [434] Two picture frames, Malacca. DR. M'PHERSON.

3937. [471] Fan, British Burmah.   Do.

3938. [472] Paper-knife, Do.       Do.

3939. [469] Bracelet,     Do.       Do.

## MODELS CARVED IN WOOD.

3940. [8158] Model of Benares Chuttaree, in logwood, Benares.

3941. [4698] Do. Hindoo temple, sandalwood, Do.   RAJAH DOONARUM SING.

3942. [8159] Do. Rulianishwar temple.

3943. [7080] Do. Hindoo temple, sandalwood.

3944. [2579] Do. Minar, in blackwood, Ahmedabad.

3945. [449] Do. Malacca temple, British Burmah.   DR. M'PHERSON.

3946. [5896] Do. Tasoung or pavilion, Pegu.   COL. PHAYRE.

3947. [6743] Do. Malay house, Malacca. HON. CAPT. BURN.

3848. [6744] Model Pondoh Eery (Hut-seim), Malacca. A. A. De Wind, Esq.

---

XII. — Manufactures from Straw, Grass, and other similar Materials.

3949. [5174] Reeds (*Saccharum* sp.) used as a substitute for quills, Calcutta.

3950. [5167] Hand fan, Do. A. M. Dowleans, Esq.

3951. [5168] Do. Do. Do.

3952. [5169] Do. Do. Do.

3953. [5170] Do. Do. Do.

3954. [5171] Do. Do. Do.

3955. [3155] Palm-leaf fans, Hooghly.

3956. [3156] Do. Do.

3957. [2901] Basket of straw, Moonghyr. Government of India.

3958. [2902] Do. Do. Do.

3959. [2903] Do. Do. Do.

3960. [2904] Do. Do. Do.

3961. [5163] Set of table mats, bamboo, Do. Do.

3962. [5164] Do. Do. Do.

3963. [5165] Do. Do. Do.

3964. [5166] Fan with blue edges, Do. Do.

3965. [6482] Leaf cloak worn by natives in wet weather, Chota Nagpore.

The Chookul or Choput is in general use amongst the natives in Manbhoom during wet weather.*

3966. [2688] Fancy worked basket, Darjeeling. Dr. Campbell.

3967. [3159] Basket for infusing Murwa, Do. Do.

3968. [3176] Three baskets, Do. Do.

3969. [3171] Basket, Do. Do.

3970. [4422] Do. Do. Do.

3971. [4428] Do. (curious manufacture), Do. Do.

3972. [4429] Do. (do.) Do. Do.

3973. [4752] Basket, Darjeeling. Dr. Campbell.

3974-8. [4423-27] Five bamboo tubs, Do. Do.

3979. [3178] Large straw hats, Do. Do.

3980. [4269] Straw hat and waterproof cover, Do. Do.

3981. [4431] Halter of bamboo, Do. Do.

3982-5. [5908-11] Grass basket, Nepal. H. H. Sir Jung Bahadoor, K.C.B.

3986-7. [5539-40] Kuskus fans, Ulwar. H. H. the Maharajah.

3988. [4730] Nest of four baskets, Sawunt Warree. Government of Bombay.

3989. [2888] Four Kuskus baskets, Poonah. Do.

3990. [4866] Basket made from the leaves of the *Cocos nucifera*, Chingleput. Dr. Shortt.

3991. [4868] Basket made of wild date leaves, Do. Do.

3992. [600] Two leaf caps, Salem, Madras.

3993-6. [9349-52] Four do., S. Canara, Do. V. P. Coelho, Esq.

3997. [2710] Nest of bamboo boxes, Bangalore. J. Lacey, Esq.

3998. [3735] Betel boxes, Burmah. *Calcutta Committee.* Government of India.

3999. [3737] Do. Do. Do.

The framework of these boxes is formed of thin strips of bamboo plaited into the shape of a box; the basket-work foundation is then coated with 'Theetsee,' painted and varnished. Every Burman has one or more of these shaped boxes to hold his betel, cigars, money, &c.; whilst the women, in addition to the above purposes, use them as jewel and dressing-cases. The specimens exhibited are the finest procurable, and come from Pagan, in Burmah, celebrated for the manufacture of these boxes. Inferior sorts are made all over Pegu and in the Shan States. The higher classes of Burmese use boxes of silver, whilst the nobles of the Court of Ava use gold.*

4000. [4455] Straw hat, Rangoon. Messrs. Halliday, Fox, & Co.

4001. [6709] Mat of Pandong rush, Singapore. Col. Collyer.

4002. [6320] Cocoa-nut colander, Do. G. Angus, Esq.

---

* Local Committee, *Chota Nagpore.*

* Local Committee, *Rangoon.*

4003. [4102] Two water buckets of the spathe of the Nibong (*Nipa fruticans*), Penang.  Messrs. HALLIDAY, FOX, & Co.

4004. [4752] One do.    Do.    Do.

4005. [4092] Work-basket of *Pandanus odoratissimus*, Province Wellesley. PENANG COMMITTEE.

4006-7. [4104-6] Three plate covers of do. Do.  Do.

4008. [4110] Bird cage,    Do.    Do.

4009. [4088] Mat bags of Mandrong rush, Do.  Do.

4010. [6364] Tapeesan, Malacca.  T. NEUBRONNER, Esq.

4011. [6368] Kokoosan or bucket, Do. Do.

4012. [6369] Eery sieve, Do.  Do.

4013. [6370] Neeroo, Do.  Do.

4014. [6372] Sarenda or Malay fisherman's hat, Do.  HON. CAPT. BURN.

4015. [6363] Sikole, Do.  CHEE YAM CHUAN.

4016. [6365] Tapeesan, Do.    Do.

4017. [6366] Gayong, Do.    Do.

## SECTION D. — PERFUMERY ETC.

The value of the Perfumery of all kinds exported from India is shown by the following Table.

**TABLE SHOWING THE VALUE OF PERFUMERY EXPORTED FROM INDIA AND EACH PRESIDENCY TO ALL PARTS OF THE WORLD FROM 1857-58 TO 1860-61.**

| YEARS | WHENCE EXPORTED | COUNTRIES WHITHER EXPORTED | | | | | | TOTAL EXPORTED TO ALL PARTS |
|---|---|---|---|---|---|---|---|---|
| | | UNITED KINGDOM | FRANCE | OTHER PARTS OF EUROPE | CHINA | ARABIAN AND PERSIAN GULFS | OTHER PARTS | |
| | | Value | Value | Value | Value | Value | Value | Value |
| | | £ | £ | £ | £ | £ | £ | £ |
| 1857-58 | Bengal | 16 | 1 | .. | .. | 9 | 33 | 59 |
| | Madras | .. | .. | .. | .. | .. | .. | .. |
| | Bombay | 2 | .. | .. | 1,460 | 122 | 1,253 | 2,837 |
| | ALL INDIA | 18 | 1 | .. | 1,460 | 131 | 1,286* | 2,896 |
| 1858-59 | Bengal | 28 | .. | .. | 4 | 29 | 18 | 79 |
| | Madras | 15 | .. | .. | .. | 25 | 22 | 62 |
| | Bombay | 12 | .. | .. | 4,315 | 37 | 1,484 | 5,848 |
| | ALL INDIA | 55 | .. | .. | 4,319 | 91 | 1,524* | 5,989 |
| 1859-60 | Bengal | 283 | 7 | .. | 9 | 20 | 36 | 355 |
| | Madras | .. | .. | .. | .. | .. | .. | .. |
| | Bombay | 7 | .. | .. | 9,326 | 30 | 688 | 10,051 |
| | ALL INDIA | 290 | 7 | .. | 9,335 | 50 | 724* | 10,406 |
| 1860-61 | Bengal | 156 | .. | .. | 84 | 80 | 87 | 407 |
| | Madras | .. | .. | .. | .. | 7 | 83 | 90 |
| | Bombay | .. | .. | .. | 6,936 | 53 | 972 | 7,961 |
| | ALL INDIA | 156 | .. | .. | 7,020 | 140 | 1,142* | 8,458 |

\* Exported chiefly to Ceylon, Straits, Aden, and Africa.

### I.—PERFUMES OF ANIMAL ORIGIN.

Under this division only two specimens are exhibited.  Of these musk is the one of chief importance.  The Table on next page indicates the value of that exported from India to all parts for the last four years.

4018. [1*] Musk (*Cervulus moschatus*), Nepal.

TABLE SHOWING THE VALUE OF MUSK EXPORTED FROM INDIA AND EACH PRESIDENCY
TO ALL PARTS OF THE WORLD FROM 1857-58 TO 1860-61.

| YEARS | WHENCE EXPORTED | COUNTRIES WHITHER EXPORTED | | | | | | TOTAL EXPORTED TO ALL PARTS |
|---|---|---|---|---|---|---|---|---|
| | | UNITED KINGDOM | FRANCE | OTHER PARTS OF EUROPE | CHINA | ARABIAN AND PERSIAN GULFS | OTHER PARTS | |
| | | Value | Value | Value | Value | Value | Value | Value |
| | | £ | £ | £ | £ | £ | £ | £ |
| 1857-58 | Bengal | 1,142 | 42 | .. | .. | .. | 700 | 1,884 |
| | Madras | .. | .. | .. | .. | .. | .. | .. |
| | Bombay | .. | .. | .. | .. | .. | 72 | 72 |
| | ALL INDIA | 1,142 | 42 | .. | .. | .. | 772* | 1,956 |
| 1858-59 | Bengal | 1,347 | 142 | 19 | .. | 102 | 1,586 | 3,196 |
| | Madras | .. | .. | .. | .. | .. | .. | .. |
| | Bombay | .. | .. | .. | .. | .. | 40 | 40 |
| | ALL INDIA | 1,347 | 142 | 19 | .. | 102 | 1,626* | 3,236 |
| 1859-60 | Bengal | 1,304 | .. | .. | .. | 160 | 328 | 1,792 |
| | Madras | .. | .. | .. | .. | .. | .. | .. |
| | Bombay | .. | .. | .. | .. | .. | .. | .. |
| | ALL INDIA | 1,304 | .. | .. | .. | 160 | 328† | 1,792 |
| 1860-61 | Bengal | 2,343 | .. | .. | .. | 63 | 1,255 | 3,651 |
| | Madras | .. | .. | .. | .. | .. | .. | .. |
| | Bombay | .. | .. | .. | 20 | .. | .. | 20 |
| | ALL INDIA | 2,343 | .. | .. | 20 | 63 | 1,255‡ | 3,671 |

* The whole to Suez.        † To Suez and Mauritius.        ‡ Chiefly to Suez.

4019. [2*] Civet (*Viverra zibetha*), Madras.

II.—PERFUMES DERIVED FROM PLANTS ETC.

4020. [6421] Garoo wood (*Aquilaria agallocha*), the produce of Pahang.

This wood has a great reputation in the East, and yields one of the kinds of Lign aloes.

4021. [2891] Agarbuttees, or stick pastiles. Messrs. DOSS & DEY.

4022. [4430] Do., another kind.

4023. [2256] Citronelle oil. DR. CLEGHORN.

4024. [6446] Lemon grass oil.

4025. [4081] Lemon grass from whence the oil is obtained, Penang.

4026. [4087] Citronelle grass, Do. Do.

4027. [5055] Roosa grass oil (*Andropogon nardoides*). DR. R. RIDDELL.

4028. [2128] Do. Do., Jubbulpore.

This is also known under the names of grass oil and ginger grass oil. It has an odour distinct from that of lemon grass and citronele.

The specimen sent from Jubbulpore is not supposed to be the pure produce, as every endeavour to obtain unadulterated oil has failed. The best is said to be *pressed* at Ajmere. A miserable imitation of this oil is occasionally manufactured at Saugor. Twenty seers of the grass, which grows wild over the station and district, are mixed with two seers of common teel oil, and then slowly distilled. The oil thus becomes highly impregnated with the peculiar roosa flavour, and is sold as such at 4R. a seer. Grass oil is never taken internally by natives, but they have a great faith in it as a stimulant to the functions of the several organs, when rubbed on externally. They also use it as a liniment in chronic rheumatism and neuralgic pains, and though they place great reliance on its virtues, its expense prevents its being used generally. It has a fragrant aromatic smell, persistent, and very agreeable at first, but after a time the odour becomes unpleasant, and gives many people a feeling of sickness with headache. The natives use it for slight colds also, to excite perspiration, by rubbing in a couple of drachms on the chest before the fire or in the heat of the sun. From information collected, it appears that the pure unadulterated oil has been used by many European officers with most wonderful effect in cases of severe rheumatism; and indeed such appears to have been the effect of its application, that two good rubbings of the pure oil on the part affected produced such severe burning as to render a third application almost impracticable. In the cases brought to notice, the second application was found sufficient to insure perfect cure.§

4029. [2254] Essential oil of orange. BAULOO MODELIER.

4030. [2255] Do. DR. CLEGHORN.

4031. [10082] Essence of Bahoor. Messrs. DOSS & DEY.

4032. [10081] Belgachia villa bouquet. Do.

4033. [10084] Hair dye. Do.

4034. [10083] Do. Do.

4035. [4199] Patchouli oil (*Pogostemon patchouli*), Penang. GEORGE SCOTT, Esq.

4036. [4198] Nutmeg oil, Do. Do.

§ Local Committee, *Jubbulpore.*

4037. [6683] Nutmeg oil, Singapore. H. G. BROWN, Esq.

4038. [10686] Sandal wood oil (*Santalum album*), Mangalore. V. P. COELHO, Esq.

4039. [2749] Do. Do.

4040. [6477] Khus ka attur (*Andropogon muricatum*), Lucknow.

Essential oil, extracted from the roots of *A. muricatum.* Bazaar price 2 rupees per tola. Grows spontaneously and plentifully in all the jungles of Oude. The roots are also used for making tatties, and leaves for thatching.

4041. [6478] Keoula ka attur (*Pandamus odoratissimus*), Lucknow.

Essential oil, extracted from the male flowers of *P. odoratissimus.* Bazaar price 2 rupees per tola (about a rupee's weight). Cultivated in very small quantities in Lucknow.

4042. [6475] Chamelee ka attur (*Jasminum grandiflorum*), Lucknow.

Essential oil, extracted from the petals of *J. grandiflorum.* Bazaar price 2 rupees per tola. This plant is extensively cultivated in gardens in Lucknow for the sake of its flowers.

4043. [6474] Motiah or Belak ka attur (*Jasminum sambac*), Lucknow.

Essential oil, extracted from the petals of *J. sambac.* Bazaar price 2 rupees per tola. Cultivated extensively in gardens in Lucknow for the sake of its flowers. Coloured red by means of dragon's blood.

4044. [6476] Hina or Mehudee ka attur (*Lawsonia inermis*), Lucknow.

Essential oil of the petals of *Lawsonia inermis.* Bazaar price 2 rupees per tola. Forms extensive hedges in all native gardens all over Oude. Coloured red by means of dragon's blood.

4045. [6479] Golab ka attur (*Rosa damascena*), Lucknow.

Essential oil of the petals of *Rosa damascena.* Bazaar price 2 rupees per tola. Cultivated in gardens in Lucknow for the sake of its flowers.

4046. [3254B] Champa ka attur (*Michelia champaca*), Calcutta. KANNY LOLL DEY.

4047. [3256B] Kawrah ka attur (*Pandanus odoratissimus*), Do. Do.

4048. [3257B] Donna ka attur (*Artemisia indica*), Do. Do.

4049. [3258B] Nagkusur ka attur (*Mesua ferrea*), Do. Do.

4050. [3259B] Motia ka attur (*Jasminum hirsutum*), Do. Do.

4051. [3260B] Pucha put ka attur (*Pogostemon patchouli*), Calcutta. KANNY LOLL DEY.

4052. [3261B] Kurna ka attur (*Phœnix dactylifera*), Do. Do.

4053. [3262B] Tore ka attur (*Jasminum grandiflorum*), Do. Do.

4054. [3263B] Bookool ka attur (*Mimusops elengi*), Do. Do.

4055. [3255B] Sohag ka attur.

4056. [10085] Mattagussa oil, Calcutta. Messrs. DOSS & DEY.

4057. [2891-1] Attar of Champa (*Michelia champaca*), Poona, Bombay.

4058. [2] Attar of Patch (*Pogostemon patchouli*), Do.

4059. [3] Attar of Kewda (*Pandanus odoratissimus*), Do.

4060. [4] Attar of Goolab (*Rosa damascena*), Do.

4061. [5] Essence of Chumbellee (*Jasminum grandiflorum*), Do.

4062. [6] Attar of Chundun, Do.

4063. [7] Essence of Mogra, Do.

4064. [8] Attar of Mussala, Do.

4065. [9] Essence of Motia (*Jasminum sambac*), Do.

4066. [10] Attar of Downa (*Artemisia indica*), Do.

4067. [11] Attar of Buckoola (*Mimusops elengi*), Do.

4068. [13] Essence of Ood, Do.

4069. [14] Essence of Dalchenee, Do.

4070. [15] Essence of Lorung, Do.

4071. [16] Essence of Joyphul, Do.

4072. [17] Essence of Joyputtree, Do.

4073. [18] Essence of Elldorah, Do.

The following, also from Poona, are in

the form of ointments for anointing the person:—

4074. [19] Kessur ke golee.

4075. [20] Limbolee arguja.

4076. [21] Arguja.

4077. [22] Kewda ka arguja.

4078. [23] Keshri arguja.

4079. [24] Kustori ka arguja.

4080. [25] Arguja.

4081. [26] Patch ka arguja.

4082. [27] Keshri mussalaka arguja.

4083. [28] Ootna.

4084. [29] Badam ka ootna.

---

4085. [2891] Kathgolie, Poona.

Small pills to be chewed with beera.

4086. [5720] Rose-water (*Rosa damascena*).

4087. [5721] Do.

4088. [5722] Do.

4089. [5723] Keora-water (*Pandanus odoratissimus*).

4090. [5724] Do.

4091. [5725] Do.

4092. [10086] Mataghussa (a kind of *Pot pourri*).

4093. [1913] Putcha pat (*Marrubium odoratissimum*).

4094. [7050] Patchouli (*Pogostemon patchouli*).

4095. [5119] Khus khus (*Anatherum muricatum*).

4096. [5120] Do.

4097. [1884] Do.

4098. [10046] Kharee (scented grass), Jhung District.

4099. [7024] Putchuk (*Costus arabicus*).

4100. [942] Screw-pine flowers (*Pandanus odoratissimus*).

4101. [1784] Ustakhudas (*Lavandula stœchas*).

4102. [1785] Do.

4103. [1870] Do.

4104. [1809] Indian Nard (*Nardostachys Jatamansi*).

4105. [7043] Goolab ke phul (*Rosa damascena*), Bombay.

4106. [1877] Nagkesur flowers (*Mesua ferrea*).

4107. [6486] Southernwood (*Artemisia indica*).

---

4108. [78*] Sweet Fennel-seed oil (*Pimpinella anisum*), Madras.

4109. [75*] Oil of Fenugrec (*Trigonella fœnum grœcum*), Do.

4110. [73*] Pepper oil (*Piper nigrum*), Do.

4111. [71*] Bishop's weed (*Anethum sowa*), Do.

4112. [69*] Cuscus (*Andropogon muricatum*), Do.

---

The following miscellaneous collection of articles of perfumery, &c. is from the INDIA MUSEUM:—

4113. [8*] Three cakes of scented paste or ointment from Malwa.

4114. [9*] Arguza, Beejapoor.

Pigment for the body. 1R. 14A. per tola.

4115. [10*] Indooree Booka, Satara.

Scented powder used in offerings.

4116. [11*] Vijarpoorchey Till Chumalee, Do.

Scented sesamum seed. 2A.

4117. [12*] Ashtagundh Tupkeree, Do.

Snuff-coloured perfume of eight ingredients. 5R. per tola.

4118. [13*] Ashtagundh Keshree, Do.

Saffron-coloured perfume of eight ingredients. 2R. 10A.

4119. [14*] Beejapoor Booka, Do.

Scented powder used in offerings. 6A.

4120. [19*] Clarified Gum Benjamin (*Styrax benzoin*), Madras.

4121. [20*] Cinnamon sugar-candy.

4122. [22*] Scented sticks, Sattara.

For fumigating apartments.

4123. [23*] Scented sticks, Sattara.

4124. [24*] Do. Do.

4125. [25*] Do. Do.

4126. [284*] Oil of Keora (*Pandanus odoratissimus*), Do.

4127. [285*] Oil of Paudhree Matee (or white earth), Do.

4128. [286*] Oil of Bukolee (*Mimusops elengi*), Do.

4129. [287*] Oil of Murra or hyssop, Do.

4130. [288*] Oil of Duvna (*Artemisia indica*), Do.

4131. [289*] Oil of Goolab or rose (*Rosa damascena*), Do.

4132. [290*] Oil of Son Chapa (*Michelia Champaca*), Do.

4133. [292*] Oil of Gaee (*Jasminum sp.*), Do.

4134. [293*] Attar of Kheora (*Pandanus odoratissimus*), Do.

4135. [294*] Attar of Josee, Do.

4136. [295*] Volatile attar of *Lavandula carnosa*, Do.

4137. [304*] Attar Keora (*Pandanus odoratissimus*), Calcutta.

4138. [305*] Attar 'Santal' (*Santalum album*), Do.

4139. [306*] Nag Kesur (*Mesua ferrea*), Do.

4140. [307*] Attar of Ughur, Do.

4141. [308*] Attar (Bela) (*Jasminum sambac*), Do.

4142. [309*] Attar of Jasmine (single), (*do.*) Madras.

4143. [310*] Do. (double), Do.

4144. [311*] Do. Do.

4145. [312*] Do. (single), Do.

4146. [313*] Do. Patchouli (*Pogostemon Patchouli*), Do.

4147. [314*] Do. Do.

4148. [315*] Attar of Rose (*Rosa centifolia*), Madras.

4149. [316*] Do. Do.

4150 [317*] Do. Do.

4151. [318*] Do. of mixed roots and flowers, Do.

4152. [319*] Do. Do.

4153. [320*] Do. Lemon Grass (*Andropogon schœnanthus*), Do.

4154. [321*] Do. Do.

4155. [322*] Do. Fragrant Percularia, (*Percularia odoratissima*), Do.

4156. [323*] Do. Do.

4157. [324*] Do. Do.

4158. [326] Do. mixed flowers, Do.

4159. [327*] Do. Aloes Wood, Do.

4160. [328*] Do. Do.

4161. [329*] Do. Southernwood (*Artemisia indica*), Do.

4162. [330*] Do. Fragrant screw pin (*Pandanus odoratissimus*), Do.

4163. [331*] Do. Pointed-leaved Mimusops (*Mimusops elengi*), Do.

4164. [333*] Do. (*Pandanus odoratissimus*), Do.

4165. [334*] Do. Gulzar, Do.

4166. [335*] Do. (*Andropogon muricatus*), Do.

4167. [336*] Do. (*Jasminum auriculatum*), Do.

4168. [337*] Do. Nuttee, Do.

4169. [338*] Do. Muhak Pari, Do.

4170. [339*] Do. *Jasminum grandiflorum*.

4171. [340*] Do. Do.

4172. [344*] Grass Oil (*Andropogon Iwarancusa*), Calcutta.

4173. [345*] Attar of Moteya (*Jasminum hirsutum*), Benares.

4174. [346*] Do. Sohag. Do.

4175. [347*] Do. Chumpa (*Michelia Champaca*), Do.

4176. [348*] Attar of Khus-khus (*Andropogon muricatus*), Benares.

4177. [349*] Do. Uggu (Aloes wood). Do.

4178. [350*] Do. Wuroos. Do.

4179. [351*] Do. Sogundra. Do.

4180. [352*] Oil of Beyla (*Jasminum sambac*), Benares.

4181. [353*] Oil of Beyla (*J. sambac*), Benares.

4182. [354*] Attar of Goolzur. Do.

4183. [355*] Do. Raehrookh. Do.

4184. [356*] Sireh oil.

*Note.*—In Classes III. and IV. Mr. M. C. Cook has, from his botanical knowledge, been enabled to render valuable assistance.

# Class V.—RAILWAY PLANT, INCLUDING LOCOMOTIVE ENGINES AND CARRIAGES.

### I.—RAILWAY MACHINERY AND PERMANENT WAY.

4185. [10200] Cast-iron railway chair, Kumaon.

Manufactured by the Kumaon Iron Works.

4186. [10199] Three specimens of cast-iron tramplates, Kumaon.

Invented by Mr. HARDY WELLS, chief engineer, Rohilcund Tramway.

4187. [2425] Iron rails, Kumaon.

# Class VI. —CARRIAGES NOT CONNECTED WITH RAIL OR TRAMROADS.

Of carts and wagons of all kinds, except those distinctly agricultural, there are but two.

4188. [5815] Sumbulpore cart (model), drawn by bullocks, Cuttack.

4189. [3326] Burmese cart (model), Rangoon. Messrs. HALLIDAY, FOX, & Co.

# Class VII. —MANUFACTURING MACHINES AND TOOLS.

Section A. — MACHINERY EMPLOYED IN SPINNING AND WEAVING, OR IN MAKING FELT AND LAID FABRICS.

*For the Manufacture of all Spun, Woven, Felted, or Laid Fabrics.*

4190. [5813] Cotton spinning machine, Rangoon.

4191. [5811] Weaving machine, Rangoon.

4192. [5819] Cotton-carder's bow, Cuttack.

4193. [561] 'Pingara,' or cotton beater, Dharwas. C. W. ANDERSON, Esq.

After being separated from the seed, it is beaten in this manner to open out the fibre and fit it for spinning.

4194. [544] 'Rahat,' or spinning-wheel (½ real size). C. W. ANDERSON, Esq.

4195. [545] 'Tanwul,' or rack, (½ real size), on which the thread is wound to form into hanks for sale.

4196. [2744] Silk spinning wheel, Mysore.

4197. [8096] Loom, Rangoon. Messrs. HALLIDAY, FOX, & Co.

4198. [3915] Instrument used in spinning, Durrung, Assam. LIEUT. PHAIRE.

4199. [5918] Loom, Assam. Do.

4200. [5919] Do. Do.

4201. [5811] Weaving apparatus, Cuttack.

4202. [1649] Shawl manufacturer's loom, with goat's wool, &c., Umritsur.

4202a. [3830] Carpet loom, Jhansee.

———

Although not strictly coming under the above heading in the 'Jury Directory,' the cotton-cleaning gins or churkas are inserted here.

4203. [5820] Churka, Patna.

4204. [3827] Cotton gin.

4205. [2831] Do. Behar.

4206. [2681] Do. Mysore.

4207. [5818] Churka, Ulwar.

4208. [2873] Cotton gin, Cuttack.

Used in the Sumbulpore district only.

4209. [5812] Cotton gin, Cuttack.

4210. [562] The foot roller, for cleaning cotton.

The iron is worked with two feet on a stone by a woman sitting, or rather balancing herself on a low stool. The seeds are rolled out in front, and the cotton drawn away as fast as it is freed from the seed, and piled up behind under the stool. A small model of a woman cleaning cotton with this instrument is sent.

4211. [537] 'Ratee,' or roller (½ the real size) is sometimes used for separating the seed from the cotton.

———

Section B.—MACHINERY USED IN THE MANUFACTURE OF VEGETABLE SUBSTANCES.

4212. [5822] Oil mill, Shahabad.

4213. [5816] Do. Cuttack.

4214. [5821] Oil mill, Patna.

4215. [8097] Kailhoo or oil machine.

4216. [2722] Oil mill, Mysore.

4217. [3330] Burmese pestle, Rangoon. Messrs. HALLIDAY, FOX, & Co.

4218. [3323] Husking machine, Do. Do.

4219. [3324] Do. Do. Do.

4220. [5809] Model of rice-cleaning machine, Burmah. Do.

4221. [5810] Model of rice-husking machine, Do. Do.

4222. [3061] Model of mill for hulling and dressing rice, Rangoon. ORIENTAL RICE COMPANY.

The following remarks accompany the above:—

The mill was invented by THOMAS SUTHERLAND, merchant and mill-owner, Melbourne, Australia, at which place the first mill was erected, and is now working.

A great many merchants having tested the invention at once formed themselves into a company, named as above, and sent the inventor and patentee to India to erect a mill. Rangoon was selected. MR. SUTHERLAND arrived there on the 13th of November 1860; and in a little over three months had these extensive and substantial buildings finished, the walls being all brick, and covered with corrugated galvanised iron. The main building is 175 ft. long, 40 ft. wide, and 35 ft. high, three stories. The engine-room is 70 ft. long, 20 ft. wide, and 18 ft. high; the boiler-house, 50 by 60, and 18 ft. high; work-shop, 20 by 20; the smithy and moulding-shop, 35 by 16; European's cottage, 40 by 24; office, 30 by 20 — two stories high.

The mill contains, as shown by the model, 40 run of stones, 20 being conical and 20 circular; besides these are fifty other machines, viz. polishing-machines, elevators, fans, screens, &c.

These are driven by two large engines, each engine driving a half of the mill; but one engine is capable of driving all, if necessary, in case of a break-down. A clutch is placed on the main shaft, by which it is joined or separated at pleasure, and one engine then drives the entire mill. The steam is supplied by four large boilers.

The mill has now been several months at work, and has surpassed the best expectations of the owners, both as to the quantity and quality of the rice produced. Three hundred and fifty tons can be turned out in the twenty-four hours, with a mere trifle as to manual labour, as nearly all the work is done by machinery. Ships load at the company's wharf, in front of the mill — the sacks of rice being sent from the mill to the ship's side in trucks running on a tramway. This alone saves immense labour. Such is the success of the mill that several others are now ordered for this and other ports.

The value of rice produced by the Company's mills is already valued at 1s. a cwt. over native-cleaned rice; but it is believed, when the trade once know the quality, it will fetch 2s. a cwt. over any hand-cleaned rice.

Messrs. JAMES WYLLIE & Co., of No. 2 King William Street, London, have full power from the patentee to license parties to erect and work this patent. They further also possess working plans, &c., and will give all necessary information to those who may desire to erect such mills in Europe or elsewhere.

4223. [2721] Sugar mill, Mysore.

4224. [5814] Sugar-cane mill, Cuttack.

4225. [538] 'Gana,' Dharwar. W. C. ANDERSON, Esq.

Mill for extracting the juice of the sugar-cane ($\frac{1}{10}$ real size), worked by two bullocks, or, by putting an extra bearer, by four bullocks.

## CLASS VIII. — MACHINERY IN GENERAL.

Under the division of this class devoted to 'Hydraulic Machines, Pile Drivers, &c.,' a particularly interesting model of a pile-driving machine, invented by DR. FORBES, is exhibited, along with a photograph of a bridge constructed in accordance with the arrangement shown in the model. The object of the machine is the construction of substantial wooden bridges at a small cost, in 'black soil' districts, where building materials are not procurable.

4226. [269] Pile-driving machine, Dharwar. DR. FORBES.

4227. [9420] Photograph of bridge constructed by DR. FORBES.

Under the head of 'Measuring and Registering Machines for Commercial, and not for Philosophical Purposes,' is placed Mr. Wood's ingenious and useful 'East Indian Wages Calculating Machine.'

4228. [8099] Wages calculating machine for the East Indies. C. WOOD, Esq.

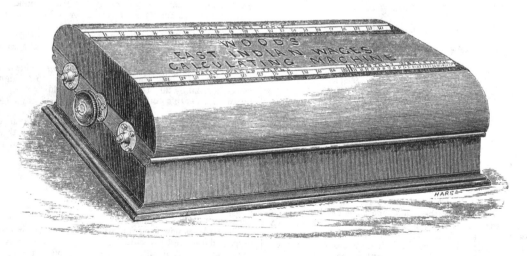

The inventor having long experienced the labour and liability to error in making up the weekly wages account, in works where a number of hands are employed, recommends the above machine to the notice of railway contractors, builders, managers of iron works, coffee plantations, cutcherries, &c. as a valuable adjunct to their accountant's offices.

The CALCULATING MACHINE is arranged upon a plan so simple, that persons of the most ordinary capacity may be made familiar with its use in a few minutes, and the inventor believes that it will be found to repay the purchaser its original cost in the space of a few weeks, viz. by reducing the number of clerks required under the present system of calculating wages.

This machine will calculate wages in rupees, annas, and pice, at rates varying from 6 pice per day to 12 annas per day, and from $\frac{1}{4}$ day to 35 days; and also for monthly wages varying from 1 rupee per month up to 20 rupees per month, or for any number of days during that time.

*Directions for Use.*—Required the amount of wages due to a workman for 18¾ days, at the rate of 7 annas and 9 pice per day:—

Turn the rollers round until you find 7/9 on the first row of large figures (on the left-hand side), then find 18¾ in the fixed *Index* above the opening under weekly wages table, immediately under which number you will see 9/1/3 upon the roller, being the amount, 9 rupees, 1 anna, and 3 pice.

Again, let a man's wages be 9 rupees *per month*, required the amount due to him for 7 days' work :—
Look for the division marked monthly wages table. Turn the rollers round until you find 7, in the last row of large figures on the left-hand side (of this division), then find 9 in the fixed *Index* above the opening, and immediately under which number you will see 2/1/3 upon the roller — 2 rupees, 1 anna, and 7 pice.

Upon the roller of each machine will be found farther instructions. Price in England, packed ready for India, 120 rupees.
These machines may be obtained of Messrs. Ransomes & Sims, Ipswich, or of the inventor, Mr. Charles Wood, Bramford, near Ipswich.

# Class IX. — AGRICULTURAL AND HORTICULTURAL MACHINES AND IMPLEMENTS.

Although the ordinary agricultural implements, viz. those used in simple tillage, are often of the very rudest description, and although a wide field is opened for improvement, it must nevertheless be borne in mind that the soils of India are of a very different character to those of this country, and do not require either the same amount or kind of mechanical treatment in order to produce the requisite effect. The description of the implements used in Dharwar agriculture given with the models forwarded by Mr. Anderson, will be read with interest.

## AGRICULTURAL IMPLEMENTS USED IN THE DHARWAR COLLECTORATE.

Collected by C. W. Anderson, Esq.

**4229. [536] Large plough ($\frac{2}{5}$ the real size).**

On ground being brought into cultivation for the first time, it is ploughed with this, lengthways and crossways. If the land is heavy eight bullocks are used ; if light four are sufficient. It is used in cotton and also in grain cultivation.

**4230. [533] Small plough ($\frac{1}{4}$ real size).**

Used in black soil at intervals of from six to ten years, and worked with two or four bullocks according to the depth of ploughing and stiffness of the soil. Used in cotton and also in grain cultivation; in red soils it is used every year.

**4231. [528] The ' Kooloo ' ($\frac{1}{3}$ the real size).**

Used with two bullocks after ploughing for farther breaking up the soil, and also used without previous ploughing in the years when the black cotton soil is not ploughed. After the seed, whether cotton or grain, is sown with the drill, the iron and wooden supports are removed from this instrument, and the soil smoothed over the seed with the upper wood alone drawn by two bullocks, and kept by the foot of the driver.

**4232. [531] The ' Teephun ' or drill ($\frac{2}{5}$ the real size).**

Used for sowing cotton, drawn by two bullocks ; the two seed tubes are fed by a woman each.

**4233. [543] The ' Koolpee ' ($\frac{1}{4}$ the real size).**

Drawn by two bullocks between the rows of cotton, to eradicate weeds by this means: also the soil about the roots of the cotton plants is loosened and piled up — a rough substitute for hoeing.

**4234. [532] Koorie, or drill ($\frac{1}{4}$ real size).**

Used in sowing grain, worked with two bullocks, which one man drives, and this man feeds the receptacle for the seed communicating to the four tubes, and a third man works the extra tubes at the side, with which another description of seed or oil seed is very commonly sown in every fifth row. A model of this implement in work is sent.

**4235. [530] Kolpa ($\frac{1}{3}$ real size).**

Drawn by two bullocks, used for rooting up the weeds between the rows of grain ; the row of grain is left untouched in the interval in the middle; the earth is also by the same operation loosened around the roots of the grain. Two of these are frequently worked together with one pair of bullocks and two men.

**4236. [555] ' Eela,' or reaping-hook ($\frac{1}{6}$ real size).**

**4237. [556] Pickaxe ($\frac{1}{8}$ real size), two descriptions, and axe ($\frac{1}{8}$ real size.)**

**4238. [535] ' Hullee Bundee ' ($\frac{1}{10}$ real size).**

Not seen much of large size in the Deccan, but very common in the S. M. C., drawn by eight bullocks. The tires are of much heavier iron than appears from the model. They are commonly six inches deep. A pair of wheels cost up to 120 rupees ; they last 50 or even 100 years, and are handed down as heirlooms in families.

## RICE CULTIVATION.

**4239.** [550] 'Nangur' plough ($\frac{1}{4}$ real size).

Used for rice cultivation, worked with two bullocks. Rice land is ploughed with this two or three times every year.

**4240.** [549] 'Dou,' or clod-crusher ($\frac{1}{10}$ real size).

Drawn with two bullocks: the driver stands on the implement when working it.

**4241.** [528] 'Kooloo' ($\frac{1}{5}$ real size).

Used after the clod crusher for farther evening the ground. With the scarifier removed, it is used for covering in the seed after it is drilled in.

**4242.** [534] Manure-cart ($\frac{1}{14}$ real size).

Worked with two bullocks.

**4243.** [534] A 'Koree,' or drill used in rice cultivation ($\frac{1}{6}$ real size).

Is similar to the drill used for the other grain, except that there are six tubes, and no extra tube for other grain is used, rice being sown alone: worked by two bullocks.

**4244.** [542] 'Koolpa,' or weed-extractor ($\frac{1}{4}$ real size).

The loose handle is used to keep it true in its line. Worked by two bullocks.

**4245.** [559] 'Khora,' or hoe ($\frac{1}{5}$ real size).

**4246.** [551] 'Khorpee,' or weeder ($\frac{1}{4}$ real size).

For cleaning away any weeds which may have escaped the khoolpa or weeder drawn by bullocks.

**4247.** [558] 'Eela,' or rice reaping-hook ($\frac{1}{5}$ real size).

**4248.** [560] 'Akree,' or hook, for collecting the grain in straw together ($\frac{1}{8}$ real size).

**4249.** [547] 'Phowra' ($\frac{1}{8}$ real size).

May be called a solid rake for heaping together grain.

**4250.** [541] 'Dantala,' or rake ($\frac{1}{3}$ real size).

**4251.** [548] 'Fewuttee,' or stool ($\frac{1}{10}$ real size).

For standing on when winnowing. It is six or seven feet high.

**4252.** [540] 'Bheerut, or mill ($\frac{1}{8}$ real size).

Used for removing the husk off rice.

## SUGAR-CANE CULTIVATION.

**4253.** [557] 'Ghoorda' ($\frac{1}{6}$ real size).

Used for raising water three or four feet; worked by men holding the ropes at the corners and surging it backwards and forwards.

---

**4254.** [2715] Model of plough, Mysore.

**4255.** [2716] Harrow,       Do.

**4256.** [2717] Sowing-machine,    Do.

**4257.** [2718] Weeding-machine,   Do.

**4258.** [2719] Levelling-machine,   Do.

**4259.** [2720] Harrow,       Do.

**4260.** [3288] Common plough, Jhansee.*

**4261.** [5817] Plough, Cuttack.

**4262.** [3829] Plough (*Bukhur*), Jhansee.*

---

* Forwarded by the Central Committee, *Allahabad.*

# CLASS X.—CIVIL ENGINEERING, ARCHITECTURAL AND BUILDING CONTRIVANCES.

**Section A.** — CIVIL ENGINEERING AND BUILDING.

Under the fifth division of this section MR. ELTON TEMPLEMORE exhibits a model of a diving-bell and apparatus in successful application in connection with the operations now being carried on in the Madras roads.

**4263.** [375] Model of diving-bell and apparatus, Madras. ELTON TEMPLEMORE, Esq.

---

**Section C.** — OBJECTS OF ARCHITECTURAL BEAUTY OR DECORATION.

iv. — *Manufactures in Stone, Alabaster, &c. for useful and ornamental purposes.*

**4264.** [5596] Cup and cover, gilt edges, Patna. GOVERNMENT OF INDIA.

**4265.** [5219] Stone dish, Do.     Do.

**4266.** [5799] Small stone goblet, Do. Do.

4267. [5587] Brahmin bull, Gyah. GOVERNMENT of INDIA.

4268. [5588] Do. cow, Do. Do.

4269. [5589] Buffalo, Do. Do.

4270. [5590] Buffalo cow, Do. Do.

4271. [5584] Rhinoceros, Do. Do.

4272. [5586] Camel, Do. Do.

4273. [5591] Figure of Runjeet Sing, Do. Do.

4274. [5583] Bear, Do. Do.

4275. [5592] Bull, Do. Do.

4276. [5593] Stag, Do. Do.

4277. [5594] Religious mendicant, Do. Do.

4278. [5595] A dog, Do. Do.

4279. [5570] Stone cup, Nilgiri, Cuttack, Do.

4280. [5571] Pair drinking cups, Do. Do.

4281. [5220] Turned stone plate, Do. Do.

4282. [1207] Square marble chessboard, Agra, Do.

4283. [1333] Octagonal, Do. Do.

4284. [1506] Marble inkstand, Do. Do.

4285. [1344] Do. Do. Do.

4286. [1530] Tray, Do. Do.

4287-90. [1533-6] Four paper weights, Do. Do.

4291. [3817] Inkstand carved in soapstone, Do. Do.

4292. [3815] Do. Do. Do.

4293. [3818] Large tray, Do. Do. Do.

4294. [3819] Do. Do. Do. Do.

4295. [3821] Square tray, Do. Do. Do.

4296. [3822] Small tray, Do. Do. Do.

4297. [3823] Small tray, carved in soapstone, Agra. GOVERNMENT of INDIA.

4298-4300. [3824-6] Small plates, Do. Do. Do.

4301. [3816] Octagonal box, Do. Do. Do.

4302. [1638] Model of Hyderabad Temple, Do. Do. Do.

4303-5. [1633-5] Vessels in soapstone, Jhansee. KHEREE RAMBUX.

4306. [5874] Ruler of pink tinted marble, Ulwar. H. H. the MAHARAJAH.

4307. [5877] Paper weight, Do. Do. Do.

4308. [5878] Do. Do. Do.

4309. [5879] Do. white marble, Do. Do.

4310. [5880] Do. Do. Do.

4311. [5881] Cup of pink tinted marble, Do. Do.

4312. [5882] Spoon, Do. Do. Do.

4313. [4980] Plate, white marble, Jeypore.

4314. [9380] Elephants fighting, Do.

4315. [9381] Cow and calf, Do.

4316. [3740] Flower vases, Bellary, Madras. GOVERNMENT of MADRAS.

4317. [3741] Two paper weights, Do. Do.

4318. [3742] Inkstand, Do. Do.

4319. [3743] Water jug, Do. Do.

4320-1. [3774-5] Butter-pot and stand, Do. Do.

4322. [10624] Goglet, N. Arcot. THATHNA LINGACHARRI, manufacturer. Do. Do.

4323. [10625] Vase, Do. Do. Do.

4324. [10626] Do. Do. Do. Do.

4325. [5888] Model of temple in Burmese alabaster, Pegu. COL. PHAYRE.

4325a. [4832] Burmese idol, Rangoon. Messrs. HALLIDAY, FOX, & Co.

# CLASS XI. — MILITARY ENGINEERING.

## Section C. — ARMS ETC.

## Division III. — SMALL ARMS.

The arms exhibited on the present occasion are not nearly equal in number to those shown in the Exhibitions of 1851 and 1855; nevertheless some of them are exceedingly choice, and of exquisite workmanship.

4326. [5764] Rifle, Enfield pattern, Nepaul. H. H. SIR JUNG BAHADOOR, K.C.B.

4327. [5765] Musket and bayonet, Do. Do.

4328. [5763] Breech-loading carbine, Do. Do.

4329. [5766] Revolver, six-barrelled, Do.

Manufactured in SIR JUNG BAHADOOR's Arsenal at Nepaul.

4330. [10249] Matchlock flint, with apparatus, &c., Lahore. H. H. the NAWAB of BHAWULPORE.

---

4331. [5761] Matchlock, Ulwar. H. H. the MAHARAJAH of ULWAR.

4332. [5759] Musket, enclosing a second, Do. Do.

. Exhibited as an excellent specimen of native ingenuity. It was made by SETARAM, head-smith of the RAJAH of ULWAR.

DIRECTIONS FOR TAKING OUT THE INNER GUN.

*a.* Sight-screw to be taken off. This will separate the length of the outer gun and discover the smaller one inside.
The box at the end of the butt of the latter contains:—
*b.* Screw, to be fixed on the left side of the outer barrel after the moveable breech has been fixed in.—*c.* Moveable breech, to be fixed to outer barrel.—*d.* Breech screwer.—*e.* Small gun.—*f.* The knife securing gun-caps at its handle pushes down the trigger through the recess, and the cap is fixed on the down-pointed nipple by the knife's handle.

---

4333. [788] Rifle, twisted and engraved barrel, Khyrpoor. H. H. MEER ALI MORAD.

4334. [802] Matchlock, Arabian pattern, with accoutrements, Kutch. H. H. the RAO of KUTCH.

4335. [803] Do., and accoutrements, Do. Do.

4336. [5762] Matchlock, with punch and powder-horn, Cuttack. GOVERNMENT of INDIA.

Made at Bedhanpore, in the Hill State of Dhenkanal. Time has not allowed of its being finished in the best style, but it is a fair piece of workmanship for this country, and the maker is noted among native Shekarees (hunters), his guns carrying a small charge, yet shooting very hard and well. The match is made of the Koombhee (*Careya arborea*). Price of this piece 30R. or 3*l*.*

## SWORDS, DAGGERS, ETC.

4337. [5771] Dagger (Kuttar), gold-mounted, Ulwar. H. H. the MAHARAJAH of ULWAR.

4338. [5768] Sword (Tulwar), blade of two kinds of steel, Do. Do.

4339. [5769] Do. small (Zuffur Tukia), Do. Do.

4340. [5770] Knife (Choora), Do. Do.

---

4341. [971] Dagger, ivory handle, gold-mounted, Puttiala. H. H. the MAHARAJAH of PUTTIALA.

4342. [972] Do., silver handle, ivory scabbard, Do. Do.

4343. [973] Sword, gold-mounted, Do. Do.

4344. [974] Do. Do. Do. Do.

---

4345. [5776] Dagger, pearl handle, Lahore. H. H. the NAWAB of BHAWULPORE.

4346. [5777] Do., ivory handle, green velvet scabbard, Lahore. CAPT. HARRISON, *H. M. 79th Highlanders.*

4347. [8175] Sword and scabbard, Calcutta. JOHN MARTIN, Esq.

4348. [3051] Dagger, fine steel, gold-inlaid handle, green velvet scabbard, Lucknow. Do.

---

* Local Committee, *Cuttack*.

4349. [2186] Sword, presented by Calcutta Volunteers to MAJOR DAVIES, Calcutta. Messrs. ALLEN & HAYES, *Jewellers, Calcutta.*

Manufactured by native workmen.

4350. [3052] Dagger, Damascene blade, massive ivory handle, in a gold-mounted velvet scabbard, Calcutta. Messrs. ALLEN & HAYES, *Jewellers, Calcutta.*

4351. [4252] Do., with scabbard, with two smaller knives in handle, Rampore. H. H. the NAWAB of RAMPORE.

4352. [4253] Do. Do.    Do.

4353. [5772] Hunting-knife, with scabbard, Cuttack. GOVERNMENT of INDIA.

Price 10*s.* The native blacksmiths of Cuttack turn out excellent specimens of cutlery.

4354. [5778] War knife, Peshawur. MAHOMED ZUMA.

4355. [6495] Dagger or Chura, Bokhara, Peshawur. H. H. NAWAB KHARROOLA KHAN.

4356. [6496] Do.    Do.    Do.

4357. [4254] Knife (Kookee), silver-mounted, Darjeeling. DR. CAMPBELL.

4358. [4255] Do.  do.  Do.   Do.

4359. [4256] Do., plain,  Do.   Do.

4360. [4257] Do. (Bootanese), Do.  Do

4361. [4258] Dagger-shaped instrument, wooden, Do. Do.

Found in altars, in Lamooneers.

4362. [4268] Sword (Bootanese), in crimson scabbard, Darjeeling. DR. CAMPBELL.

4363. [1381] Dagger, jade green buffalo-head handle, jade scabbard mount. COL. GUTHRIE, *Bengal Engineers.*

4364. [1382] Do., jade ram's-head handle, leather scabbard. Do.

4365. [1383] Do., white jade handle, jewelled. Do.

4366. [1384] Do. do., gold-inlaid. Do.

4367. [1385] Do., plain jade handle, two jade mountings. Do.

4368. [1386] Dagger, rock-crystal leopard's head handle, topaz eyes, coral tongue, plain velvet scabbard. COL. GUTHRIE, *Bengal Engineers.*

4369. [1387] Do., rock-crystal handle, blue velvet scabbard. Do.

4370. [1388] Do. do., red velvet scabbard. Do.

4371. [789] Do., with horn handle, silver-mounted scabbard, Khyrpoor. H. H. MEER ALI MORAD.

4372. [786] Sword, gold-mounted, Do. Do.

4373. [792] Sword or Talwa, inlaid with gold and silver, Kutch. H. H. the RAO of KUTCH.

4374. [795] Dagger, inlaid with gold and silver, Do. Do.

4375. [796] Do. do. Do.  Do.

4376. [799] Do. or Bichwa, inlaid, Do. Do.

4377. [800] Do., with red glass handle, Do. Do.

4378. [801] Do., inlaid with gold, Do. Do.

4379. [1076] Knife, ivory handle and scabbard, Hyderabad. GOVERNMENT of INDIA.

4380. [1077] Do., wooden scabbard, Do. Do.

4381. [4719] Hunting knife or Shikaree, with accoutrements, Sawunt Warree.  Do.

4382. [452] Two Malay knives, Malabar. DR. M'PHERSON, *Madras.*

4383. [10590] Two do.   Do.   Do.

4384. [6360] Six Krisses, viz. 'Panjang,' 'Sepucal,' 'Chinankas,' 'Toomboo Ladah,' 'Bladohe,' 'Badeh,' Singapore. HON. COL. CAVANAGH.

These Krisses are those commonly in use and have their names according to their uses or qualities, as *Panjang,* long — *Sepucal,* straight, &c.

4385. [9422] Two metal daggers, small.

4386. [9423] Two long swords, with gauntlet handles. INDIA MUSEUM.

## BATTLE-AXES, SPEARS, ARROWS, &c.

**4387.** [5779] Weapon, Bysakhee, Khyrabad, Oude. C. LINDSAY, Esq., *Deputy Commissioner of Hurdin.*

A powerful weapon of offence and defence; also used as a rest for the arm by fakeers. An article of very great antiquity.

**4388.** [5767] Weapon, Nepaulese 'Korah.' H. H. SIR JUNG BAHADOOR, K.C.B.

Silver handle and gold mountings, Nepaul.

**4389.** [10235] Bow and arrows, Tilhur, Shahjehanpore. GOVERNMENT of INDIA.

**4390.** [5773-5] Battle-axes, Cuttack. GOVERNMENT of INDIA.

**4391-2.** [7857] 2 spears, highly ornamented. Jhend.

**4393.** [8172] Shield of basket-work, Darjeeling. DR. CAMPBELL.

**4394.** [7816] Horn priming flask, Jeypore. H. H. the RAJAH.

**4395.** [5660] Shield or Dhal, Ulwar. H. H. the MAHARAJAH.

**4396.** [646] Shield of rhinoceros skin, Kutch. H. H. the RAO.

**4397.** [651] Battle-axe or Koolung, Do. Do.

Inlaid with gold, &c.

**4398.** [797] Do. or Tubbur, Do. Do.

Inlaid with gold, &c.

**4399.** [798] Do. Do. Do.

Inlaid with gold, &c.

**4400.** [976] Do. Da. J. Do. Do.

Inlaid with gold, &c.

**4401.** [794] Lance, Kutch, H. H. the RAO.

Gold mounted (in two parts).

**4402.** [8174] Chain armour, Do. Do.

Complete suit of five pieces.

**4403.** [787] Shield, gold mounted, Khyrpoor. H. H. MEER ALI MORAD.

**4404.** [8173] Chain armour, viz., one coat of mail; one head-piece; one legging; two gauntlet-gloves, Do. Do.

**4405.** [920] Steel·bow and twenty-six arrows, Rewa Kanta. DR. NICHOLSON, *Bombay Army.*

**4406.** [451] Battle-axe, British Burmah. DR. MCPHERSON, *Madras.*

**4407.** [466] Quiver of Upas poisoned arrows for the simpoon or blow-pipe gun, with spare Upas poison in case attached, Do. Do.

**4408.** [6343] Six spears, with royal gold ferules. H. H. the RAJAH of TRINGANU.

These spears are borne as marks of royalty before the Rajah of Tringanu.

**4409.** [9424] Four shields on outside of arm's case. INDIA MUSEUM.

**4410.** [9425] Eight spears, various. Do.

**4411-2.** [2443-4] Two Hankas, or Elephant drivers' spears. KENNETH MCLEOD, Esq., *Grishinish, Isle of Skye.*

**4413.** [9426] Two very fine Mahout spears, of the best cut steel. E. W. WINGROVE, Esq., *Twickenham.*

# CLASS XII.—NAVAL ARCHITECTURE AND SHIPS' TACKLE.

### Sections A. AND B.

The models of boats, &c. exhibited on the present occasion are but few in number. In order to form a correct notion of the forms employed, the India Museum, which contains the admirable collection of models belonging to this class exhibited in 1851, and again in Paris in 1855, must be referred to.

**4414.** [5806] Ferry-boat, Kishnaghur.

**4415.** [3315] Burmese pulling-boat, Rangoon. Messrs. HALLIDAY, FOX, & Co.

4416. [3316] Burmese royal boat, Rangoon. Messrs. HALLIDAY, FOX, & Co.

4417. [3317] Do. racing-boat, Do. Do.

4418. [3318] Do. cargo-boat, Do. Do.

4419. [3319] Do. Do. Do.

4420. [3320] Do. long-boat, Do. Do.

4421. [3322] Do. dauk, Do. Do.

4422. [3325] Chinese junk, Do. Do.

## Section C. — SHIPS' TACKLE ETC.

Under this section is exhibited an interesting appliance, the invention of MR. ELTON TEMPLEMORE, of *Madras*, the object of which is to prevent the loss and facilitate the recovery of ships' moorings, and of submerged property in general.

4423. [7996] Submarine recovery buoys (patent). ELTON TEMPLEMORE, Esq.

4424. [7996] Tank, with buoys, models of boats, &c., showing application of the above. Do.

TEMPLEMORE'S SUBMARINE RECOVERY BUOY.

This invention is stated to supersede the ordinary and doubtful security afforded by the present plan of buoying ships' anchors and cables off with a rope and surface buoy, when slipped at the anchorage ground; in which case the rope rots, or is broken, the buoy disappears, and, finally, the moorings are lost. The submarine buoy is for the purpose of being attached to property intended to be submerged; and whilst being chiefly applicable to ships' moorings, it may be used with advantage in all cases of submerged property, where the use of a buoy is necessary.

The following advantages are claimed for this invention:—

1st. It preserves ships' moorings from all risk of loss, at whatever description of anchorage ground.

2nd. The moorings are preserved submerged, in a position to be recovered after any lapse of time.

3rd. It affords additional facilities for ships to slip their cables, when it is known they are perfectly safe with the submarine recovery buoy attached.

4th. It is economical, as it saves the destruction of good rope, which must otherwise be required for the purpose of buoying the cables.

The submarine recovery buoy can be used in lieu of, or in addition to, the ordinary surface float, in all cases of property to be submerged. The inventor, being engaged by the Madras Government in recovering and raising ships' moorings from the Madras roads, has thus been enabled, from practical experience, to mature and submit a plan which will, for the future, prevent the loss of submerged property. The Madras roadstead, consisting as it does of a soft yielding sand, presents greater difficulties than are to be met with elsewhere; it allows articles to sink beneath the surface of the sand, by which they are *irrecoverably lost,* by being buried out of sight of the divers.

Patentee's agent W. H. PAYN, Esq., Solicitor, Dover.

# CLASS XIII. — PHILOSOPHICAL INSTRUMENTS.

4425. [5058] Troughton's improved land level and prismatic compass, Roorkee, N.W. Provinces.  SIR PROBY CAUTLEY, K.C.B.

These excellent specimens of native workmanship are copies of instruments furnished by Messrs. TROUGHTON & SIMMS to the government workshops at Roorkee.

With the exception of the glass, which was procured from England, every part of these instruments was made by native workmen under the supervision of CAPTAIN AUGUSTINE ALLEN, the Superintendent, and MR. MASON, the head of the Mathematical Instrument Department.

4426. [10050] Trisector, Agra.  MOONSHEE SADA SUKH LALL.

Invented by the exhibitor.

# CLASS XIV. — PHOTOGRAPHIC APPARATUS AND PHOTOGRAPHY.

Division II. — PHOTOGRAPHIC IMPRESSIONS EXHIBITED AS PHOTOGRAPHIC PICTURES.

The following photographs reached in time for exhibition. Since then a large collection, embracing portraits from almost the whole of the tribes and classes of India, taken under the auspices of the Indian Government, have been received and are now being arranged for publication as lithographs, along with copious descriptive notes of the various subjects delineated. Amongst these are included the tribes of Central India, photographed by the Rev. E. Godfrey and Lieut. Waterhouse; those of Bhurtpore by Messrs. Shepherd and Robinson; of Nagpore, Sikhim, and Bootan (specimens of which are exhibited) by Dr. Simpson; of the North-Western Provinces by Dr. Tressider; of Oude by Captain Fitzmaurice and Lieut. R. H. De Montmorency; of Hazara by T. T. Davies, Esq.; of Bombay and Sinde by Captain Houghton and Lieut. Tanner; and also of the tribes or inhabitants from the places undernoted which have come to hand unaccompanied with the artists' names, viz., from Moradabad, Benares, Allyguib, Goruckpore, Dehra Dhoon, and the Punjab; from the various collectorates of Madras; from Koorg, and, lastly, from Burmah and Penang.

A. — *Landscapes, Architectural Subjects, Natural Objects.*

4427. [8018] View of the town and temples of Hurdwar and head of Ganges Canal. SIR PROBY CAUTLEY, K.C.B.

4428. [8006] 1. Great Temple; 2. The Singh Durwajah, or Lion's Gate, of the Great Temple; 3. A curious tank to the east of the Great Temple, surrounded by seventy or eighty small temples; 4. Group of temples to the west of the Great Temple; 5. The Boital Temple; 6. The Mookteswar Temple, and idol's swing; 7. A grave near the Mookteswar Temple, and ruins; 8. The Temple Annuntoo Basa Davey, on the large tank Bindoo Sagur — evening; 9. The Temple Pursuram Iswar. Bobeneswar, Orissa.

In one frame. Photographed by CAPTAIN HENRY DIXON, 22nd M.N.I.

4429. [8007] 12. The Palace of Rajah Lalet Indra Kesari, or Ranee Goompah, rock cut caves (right view); 13. The Palace of Rajah Lalat Indra Kesari, or Ranee Goompah, rock-cut caves (left view); 14. Do. Do. Do.; 16. The Ganesa Cave, cut in the rock, said to be the most ancient, and dating about 200 years B.C.; 17. The Hill of Khundaghirri, with modern Jain temple. 19. Group of rock-cut caves. Oodyaghirri, Orissa.— 10. The Temple Sideswar; 11. Group of temples. Bobeneswar, Orissa.—22. View of the Great Temple of Juggernaut, showing the lion's gate and beautiful monolith; 23. The Street leading to the Great Temple. Pooree. CAPTAIN HENRY DIXON.

In one frame.

———

4430. [8008] Idols carved out of rock, in the Badamee Caves; Carved Juggernauth Car; Stone ditto; Stone Idol called Rockus (the Evil Spirit); Brahminee Bull. Bombay.

Photographed by CAPTAIN SELLON, Bombay Engineers.

4431. [8009] Elephant Stables at Beejanuggur; Temples at ditto; Lesser Fort at Badamee; Linguite Temple and Tank at Bunshunkeree. Do.

Photographed by CAPTAIN SELLON, Bombay Engineers.

4432. [8010] Wooden Bridge over the Sangum River at Peona; Falls of Gokak; Cocoa-nut Trees; Encampment at Ironey; Banian Tree at Sutguttee. Do.

Photographed by CAPTAIN SELLON, Bombay Engineers.

4433. [8011] Fort at Copal; Large Fort at Badamee; Fort at Ironey, on the Toongabudra; Lesser Fort at Badamee; Hill and Fort of Munshur and Munsuntash. Do.

Photographed by CAPTAIN SELLON, Bombay Engineers.

4434. [8012] Hill and Fort of Nurgoond; Town of ditto; Cotton Carts. Cotton District of S.W. India.

Photographed by CAPTAIN SELLON, Bombay Engineers.

4435. [8013] Mahomedan Temple at Badamee; Temples at Sassoor, West Poona; Temples, Sacred Tanks, Idols, &c., at Mahacote; Grand Hindoo Temple (150 ft. high), at Beejanugger; Linguite Temple at Badamee. Do.

Photographed by CAPTAIN SELLON, Bombay Engineers.

4436. [2207] Calcutta drainage works at Kotrung, steam brick-making shops. PETER NICHOLL, Esq.

4437. [2208] Do. machine-shops, &c. Do.

4438. [2209] Do. view from the works across the river Hooghly. Do.

4439. [2206] Do. engineer's bungalow Do.

4440. [2205] Do. superintendent's bungalow. Do.

4441. [2204] Do. work-people. Do.

4442. [8014] 1. Dog; 2. Brahmin Priest; 3. An Officer of Irregular Cavalry; 4. A Dancing Beggar; 5. A Gooroo, or Seikh Priest; 6. A Woman selling Indian Corn; 7. A Tailor; 8. A Madras Bandy; 9. A Parsee; 10. A Barber; 11. A Hindoo Woman; 12. Jewellers. CAPTAIN ALLEN N. SCOTT.

In one frame.

4443. [8015] 13. Hindoo Temple, Secunderabad; 14. Tomb at Golconda; 15. Fruit and Flowers; 16. Tomb of a Mahomedan Saint; 17. Tomb and Banian Tree; 18. Hill Fort at Golconda; 19. A Native Cart; 20. Women grinding Wheat; 21. A Bullock Cart; 22. Camels; 23. Dead Tiger; 24. Dead Cheetah or Leopard. CAPTAIN ALLEN N. SCOTT.

In one frame.

———

4444. [8094] Stereographs of Trichinopoly, Tanjore, Madura, &c., 1855. 2 vols. CAPTAIN TRIPE.

N.B.—Several other volumes of photographs, forming part of this collection, are in the India Museum, Fife House, Whitehall Yard, S.W.

4445. [9427] Copy of paintings of the Adjunta Caves.

Photographed by MAJOR GILL.

B. — *Portraits, Single or in Groups.*

4446. [1650] H. H. Mehundur Singh, son of the Maharajah of Puttiala, and heir-apparent to the throne.

4447. [1651] Sirdar Narain Sing, Sahib of Sealbah, son-in-law to the Maharajah.

4448. [1652] Interior of the principal reception-room in the palace of the Maharajah of Puttiala.

4449. [8016] Portraits of natives (15 pictures in one frame). Printed at Industrial School of Arts, Madras.

Photographed by DR. SCOTT.

4450. [8017] Do. (12 pictures in one frame). Do.

Photographed by DR. SCOTT.

4451. [2210] Lagoatna Brahmin.
4452. [2211] Do. Do.
4453. [2212] Sanoorna Do.
4454. [2213] Bhugela Thakoor.
4455. [2214] A Marwaree.
4456. [2215] Goundhés, or bricklayers (men and women).
4457. [2216] Do. Do. Do.

**Photographic Likenesses of Natives of various Parts of India.**

4458. [8019] Cheeboo Lama, Dowan of Sikkim. DR. SIMPSON.
4459. [8020] Bhootia (Bhotan).
4460. [8021] Do. (Thibet).
4461. [8022] Sauwar (Nepaul).
4462. [8023] Bhootia (Sikkim).
4463. [8024] Thibetan.
4464. [8025] Limbo female (Nepaul).
4465. [8026] Lepcha female (Sikkim).
4466. [8027] Sauwar female (Nepaul).
4467. [8028] Bhootia (Thibet).

4468. [8029] Magar (Nepaul).
4469. [8030] Kus (Gookha Nepaul).
4470. [8031] Bhootia (Bhotan), interpreter at the Darjeeling Court.
4471. [8032] Lama Pernianchi Sikkim.
4472. [8033] Lepcha (male), Do.
4473. [8034] Bhootia (female), Lassa Thibet.
4474. [8035] Muddick Group.
4475. [8036] Newar (Nepaul).
4476. [8037] Moormi, Nepaul.
4477. [8038] Limbo, Do.
4478. [8039] Goorung, Do.
4479. [8040] Mech, Terai.
4480. [8041] Lepcha, Sikkim.
4481. [8042] Kamee, Nepaul.
4482. [8043] Bhootia female, Sikkim.
4483. [8044] Oraon Cole female, Chota Nagpore.
4484. [8045] Korwah, Do.
4485. [8046] Rajah Chutroo Gimadbit, Maumboom, Rajpoot.
4486. [8047] Moonda female, Chota Nagpore.
4487. [8048] Cole Christian, Do.
4488. [8049] Rajpoot, Singbhoom.
4489. [8050] Cheeroo, Palamow.
4490. [8051] Bhogta, Chota Nagpore.
4491. [8052] Oran, Do.
4492. [8053] Moonda male, Do.
4493. [8054] Rajwar, Behar.
4494. [8055] Aheer, Shahabad.
4495. [8056] Dosadh, Behar.
4496. [8057] Mullick, Do.
4497. [8058] Musahar, Do.
4498. [8059] Pohariah, Bhaugulpore Hills.
4499. [8060] Garrow.

4500. [8061] Maram Nagar, Muneepore.

4501. [8062] Cacharee, Assam.

4502. [8063] Rajbunsee, Koch Behar.

4503. [8064] Burmese.

4504. [8065] Mugh, Akyab.

4505. [8066] Kyang, Burmah.

4506. [8067] Rajpoot Christian, Bhojpore.

4507. [8068] Ghilzie, Kandahar.

4508. [8069] Dooranee, Cabul.

4509. [8070] Persian, Calcutta.

4510. [8071] Lurka Cole, Singbhoom.

4511. [8072] Pathan.

4512. [8073] Tartar horse-dealers, Thibet.

4513. [8074] Bhootan group, Bhootan.

4514. [8075] Goorung group, Nepaul.

4515. [8076] Sunwar family group, Do.

4516. [8077] Lepcha peasants, Sikkim.

4517. [8078] Newar group, Nepaul.

4518. [8079] Kus group, Do.

4519. [8080] Limbo group, Nepaul.

4520. [8081] Mech group, Terai.

4521. [8082] Bhootia group, Bhootan.

4522. [8083] Moormi group, Nepaul.

4523. [8084] Do. Do.

4524. [8085] Lepcha water-carrier.

4525. [8086] Rajbunsee group, Koch Behar.

4526. [8087] Dosadh group, Behar.

4527. [8088] Affghan group, Cabul.

4528. [8089] Korwah group, Chota Nagpore.

4529. [8090] National dance, Coles.

4530. [8091] Do. Do.

4531. [8092] Do. Chota Nagpore.

4532. [8093] Santhal group, Bhaugulpore.

4533. [8095] Santhal, Do.

4534. [2218] A Kuttik Thakoor, Etah district, Agra division.

4535. [2219] A Mahomedan priest, Do.

4536. [2220] A Kamboo Mahomedan, Do.

4537. [2217] A Khateeh, Do.

# Class XV. — HOROLOGICAL INSTRUMENTS.

4538. [3129] Silver watch (with chain), Ulwar. Presented by H. H. the MAHARAJAH of Ulwar.

This watch is stated to have been made by the watchmaker to H. H. the MAHARAJAH.

# Class XVI. — MUSICAL INSTRUMENTS.

These are few in number, but an almost complete collection is to be found at the India Museum.

4539. [3173] Double-headed drum, Darjeeling. Dr. Campbell.

4540. [4462] Burmese gong, Rangoon. Messrs. Halliday, Fox, & Co.

4541. [3312] Burmese 'putalah,' Do. Do.

4542. [4459] Small gong, Do. Do.

4543. [4458] Gong tolled after prayer, Do. Do.

4544. [1764] Trumpet made of a human thigh bone, Darjeeling. Dr. Campbell.

4545. [4480] Bell hung to the pinnacle of Burmese pagodas, Rangoon. Messrs. Halliday, Fox, & Co.

4546. [1766] Horn bells, Darjeeling. Dr. Campbell.

4547. [4464] Burmese harp, Rangoon. Messrs. Halliday, Fox, & Co.

4548. [1245] Burmese gong. H. S. Baily, Esq.

4549. [8100] Guitar.

4550. [9428] Do.

4551. [9429] Flageoletto.

4552. [9430] Do.

4553. [9431] Fiddle.

4554. [9432] Do.

4555. [9433] Pair of cymbals.

# Class XVIII. — COTTON.

The Table (p. 205) shows the value of the cotton goods, including twist and yarn, exported from India to all parts of the world from 1850–51 to 1860–61, and indicates the persistency which has so far attended the Indian export trade in manufactured cotton to various countries. Of the large collection of samples forwarded for exhibition on the present occasion, a selection only has been shown, the remainder, as in other similar instances, having been deposited at the India Museum.

Although not manufactured in India, space was readily accorded to the Cotton Supply Association for the exhibition of a complete and important series of specimens illustrating the adaptability of the native or indigenous cotton of India to the manufacture in this country of goods, in every respect, whether dyed or bleached, of first-class quality, and such as, until now, it had been deemed impracticable to produce from 'Indian cotton.' *

---

* Much credit is due to Mr. R. Burn, of Manchester, for his exertions in this matter.

## I.—Cotton Yarn and Thread.

The collection from India under this head is of very small extent. The samples of fine yarn exhibited by the Cotton Supply Association are, however, worthy of particular attention, as showing the capability of Indian cotton for the manufacture of yarns of high numbers—a fact which is still farther illustrated by the numerous specimens of manufactured cotton, chintzes, &c. shown in Case 30.

4556. [5338] Cotton yarn, Jumbulpore, Cuttack.

4557. [5129] Do. Piplee, Cuttack.

4558. [5485] Do. Assam. Lieut. Phaire.

4559. [5486] Do. Do.

4560. [7880] Do. Poona.

4561. [1261] Cotton yarn, spun by females with mills termed 'Bhettia,' Broach.

4562. [1265] Do. No. 8. Do.

TABLE SHOWING THE VALUE OF COTTON GOODS (INCLUDING TWIST AND YARN) EXPORTED FROM INDIA AND EACH PRESIDENCY TO ALL PARTS OF THE WORLD FROM 1850-51 TO 1860-61.

| Years | Whence Exported | COUNTRIES WHITHER EXPORTED | | | | | | | VALUE OF TOTAL EXPORTED TO ALL PARTS |
| | | United Kingdom | France | Other Parts of Europe | America | China | Arabian and Persian Gulfs | Other Parts | |
| | | Value | Value | Value | Value | Value | Value | Value | Value |
| | | £ | £ | £ | £ | £ | £ | £ | £ |
| 1850-51 | Bengal | 917 | 4 | .. | 942 | 1,289 | 17,364 | 93,403 | 113,919 |
| | Madras | 46,054 | 246 | .. | 54 | 12 | 3,261 | 125,106 | 174,733 |
| | Bombay | 1,830 | .. | .. | .. | 1,442 | 267,992 | 77,735 | 348,999 |
| | ALL INDIA | 48,801 | 250 | .. | 996 | 2,743 | 288,617 | 296,244 | 637,651 |
| 1851-52 | Bengal | 794 | .. | .. | 861 | 929 | 11,908 | 103,046 | 117,538 |
| | Madras | 70,163 | .. | .. | .. | 2 | 3,087 | 132,562 | 205,814 |
| | Bombay | 11 | .. | .. | 1 | 774 | 305,862 | 118,493 | 425,141 |
| | ALL INDIA | 70,968 | .. | .. | 862 | 1,705 | 320,857 | 354,101 | 748,493 |
| 1852-53 | Bengal | 461 | .. | 40 | 1,572 | 8,310 | 13,990 | 99,948 | 124,321 |
| | Madras | 136,095 | 113 | .. | 1 | 56 | 1,521 | 123,981 | 261,767 |
| | Bombay | 66 | 112 | .. | .. | 1,749 | 411,663 | 130,379 | 543,969 |
| | ALL INDIA | 136,622 | 225 | 40 | 1,573 | 10,115 | 427,174 | 354,308 | 930,057 |
| 1853-54 | Bengal | 4,681 | 35 | .. | 2,775 | 474 | 16,246 | 16,125 | 40,336 |
| | Madras | 102,607 | .. | .. | 1 | .. | 2,077 | 105,262 | 209,946 |
| | Bombay | 818 | .. | .. | .. | 353 | 363,460 | 154,226 | 518,857 |
| | ALL INDIA | 108,106 | 35 | .. | 2,776 | 827 | 381,783 | 275,613 | 769,139 |
| 1854-55 | Bengal | 15,018 | .. | .. | 270 | 758 | 20,125 | 16,722 | 52,893 |
| | Madras | 68,050 | .. | .. | .. | .. | 700 | 114,405 | 183,155 |
| | Bombay | 512 | .. | .. | .. | 3,592 | 392,965 | 183,986 | 581,055 |
| | ALL INDIA | 83,580 | .. | .. | 270 | 4,350 | 413,790 | 315,113 | 817,103 |
| 1855-56 | Bengal | 1,440 | 814 | 423 | 89 | 1,006 | 11,288 | 16,124 | 31,184 |
| | Madras | 35,410 | 39 | .. | 253 | .. | 648 | 97,532 | 133,882 |
| | Bombay | 288 | .. | .. | .. | 553 | 431,386 | 181,944 | 614,171 |
| | ALL INDIA | 37,138 | 853 | 423 | 342 | 1,559 | 443,322 | 295,600 | 779,237 |
| 1856-57 | Bengal | 5,352 | 224 | 185 | 104 | 328 | 33,117 | 13,226 | 52,536 |
| | Madras | 39,952 | 42 | .. | 40 | 2 | 1,086 | 84,599 | 125,721 |
| | Bombay | 30 | .. | .. | .. | 878 | 493,443 | 208,216 | 702,567 |
| | ALL INDIA | 45,334 | 266 | 185 | 144 | 1,208 | 527,646 | 306,041 | 880,824 |
| 1857-58 | Bengal | 1,504 | 10 | 350 | 345 | 29,210 | 24,939 | 25,315 | 81,673 |
| | Madras | 33,515 | 98 | .. | .. | .. | 1,246 | 94,197 | 129,056 |
| | Bombay | 78 | .. | .. | .. | 2,425 | 428,534 | 167,417 | 598,454 |
| | ALL INDIA | 35,097 | 108 | 350 | 345 | 31,635 | 454,719 | 286,929 | 809,183 |
| 1858-59 | Bengal | 3,686 | 238 | .. | 339 | 229 | 15,677 | 13,651 | 33,820 |
| | Madras | 22,594 | 78 | .. | .. | .. | 987 | 89,460 | 113,119 |
| | Bombay | 17 | .. | .. | .. | 1,198 | 504,836 | 160,614 | 666,665 |
| | ALL INDIA | 26,297 | 316 | .. | 339 | 1,427 | 521,500 | 263,725 | 813,604 |
| 1859-60 | Bengal | 1,510 | 167 | 35 | 518 | 4,816 | 10,128 | 9,112 | 26,286 |
| | Madras | 27,284 | 41 | 11 | 1 | .. | 801 | 83,609 | 111,747 |
| | Bombay | 310 | .. | .. | .. | 7,558 | 387,813 | 229,872 | 625,553 |
| | ALL INDIA | 29,104 | 208 | 46 | 519 | 12,374 | 398,742 | 322,593 | 763,586 |
| 1860-61 | Bengal | 293 | 5 | .. | 246 | 1,185 | 3,885 | 16,762 | 22,376 |
| | Madras | 39,617 | 822 | .. | .. | 5 | 224 | 82,419 | 123,087 |
| | Bombay | 17 | .. | .. | .. | 9,944 | 433,095 | 190,596 | 633,652 |
| | ALL INDIA | 39,927 | 827 | .. | 246 | 11,134 | 437,204 | 289,777 | 779,115 |

4563. [1267] Cotton yarn, No. 16. Broach.

4564. [1264] Do. No. 20. Do.

4565. [1266] Do. No. 22. Do.

4566. [1263] Do. No. 30. Do.

4567. [850] Do. Shikarpoor. GOVERNMENT of INDIA.

4568. [9434] Yarn No. 70s double. Spun in Manchester from Broach cotton. COTTON SUPPLY ASSOCIATION.

4568a. [9434a] Yarns Nos. 60s and 80s spun from exotic cotton grown near Madras. COTTON SUPPLY ASSOCIATION.

## II. — MUSLINS.

Some excellent specimens of Dacca manufacture are exhibited. The piece referred to in the following note by MR. H. HOULDSWORTH, is superior to any shown in 1851, the best specimen on that occasion proving to be of No. 357s yarn instead of 380s as in the present instance. It will be seen, however, that, as then, some machine-made muslin in the present Exhibition is superior in point of fineness, according to the mode of computation adopted.

'MULMUL KHAS' MUSLIN PIECE.

The number of warp threads in one
  inch appeared to be   .   .   .   104
And of weft    .    .    .    .   100
The width of piece .    .    .    3 feet
The length    .    .    .    .   31 „
The number of square feet   .   .   93 „
The weight of the piece  .   .   1565 grs.

The formula by which the numbers of the yarn is deduced from the above is warp threads + weft threads in 1 sq. in. × 144 (in. in 1 sq. ft.) × 7000 grains × sq. ft. in the piece, and ÷ by 36 in. in a yard × 840 yards in a hank × weight of piece in grains = Nos. of the yarn.

This, if the constants are inverted, resolves itself into

$$\frac{\text{threads} \times \text{threads} \times \text{sq. ft.} \times 333}{1565 \text{ grs.}}$$

Thus $\dfrac{204 \times 93 \times 333}{1565} = 380$,— the Nos. of the yarn according to our English mode of computation, which is as follows. A hank of cotton yarn contains 7 ropes or leys, each of 80 threads, each thread being 54 in. long (which is the circumference of the cotton reel). Thus there are 560 threads in one hank (7 × 80) each 54 in. or $1\frac{1}{2}$ yards in length, and 560 × $1\frac{1}{2}$ = 840 yards, the length of 1 hank of cotton yarn, and the fineness is indicated by the number of hanks in one pound of 7000 grains.

In the French department, in the case of MONS. THIVEL MICHON, of Tavare, there are several pieces of fine muslin or tarletan, woven in France, the yarn of which is No. 440, spun by THOMAS HOULDSWORTH & Co., of Manchester; and there are a few yards of muslin woven of No. 700 yarn: but this last is too imperfect for any purpose, except to point the limits of fineness at which cotton yarn can be woven at all. The pieces of No. 440 are, however, very perfect, and a great advance on any muslin exhibited in 1851, chiefly owing to the introduction since then of NEILMAN'S combing machine for cotton, by which the quality of fine yarn has been vastly improved, and made nearly as perfect as the fibre will admit. A comparison, however, of this muslin with the Dacca piece, as tested by the eye and feel, would lead to the opinion that the Indian piece was the finer. This arises from the difference in the finishing or getting up of the two muslins — the French pieces being got up hard and wiry by means of starch, which coats the threads and makes them appear coarser than they are; while the Dacca muslin is soft, and appears perfectly free from all starch or other dressing. It may also be that the India threads, spun by hand, are more condensed in their substance by the compression of the fingers in the act of spinning than the machine-spun 440 of the Manchester yarn.

The fact, however, still is, that by our mode of computing the fineness of cotton yarn, the French muslin is 15 per cent. finer than that of the Dacca muslin.

It may be of interest to add, that the fineness of the finest select seed, Sea Island cotton fibre, which my firm has ever spun, appears to be from No. 8,000 to 10,000, supposing the fibres to be laid end to end in a continuous line, and assuming each fibre to average $1\frac{1}{2}$ to $1\frac{1}{4}$ in length; thus if we assume the latter length, 1 pound of such fibres would extend nearly 4,770 miles, and 1 grain weight of such fibres would extend about 1,200 yards.*

### MUSLINS ETC.

4569. [3396] 'Mulmul Khas,' Muslin, Dacca. HURMOHUN ROY.

4570. [3371] 'Buddon Khas' muslin, Dacca. Do.

4571. [9435] Checked muslin, Do. Do.

---

* H. HOULDSWORTH, Esq.

4572. [3378] Charkana muslin, Dacca.

4573. [3366] 'Junglekhassa' muslin. Do. Do.

4574. [3401] Striped Dooreah, Do. Do.

4575. [3401] Do. Dooreah, Do. Do.

4576. [3400] Spotted Bootee, Do. Do.

4577. [1707] Striped muslin, Dacca. JUGGUT CHUNDER DOSS.

4578. [1703] Allabully muslin, Do. JUGGUT CHUNDER DOSS.

4579. [1709] Red spotted Bootee, Do. CHUNDER DOSS.

4580. [3393] Spotted muslin, Do. KHAJEH ABDOOL GUNNY.

4581. [3372] 'Jamdanee' do., Do. Do.

4582. [3365] 'Abrowan' do., Do.

4583. [3367] 'Sircar Ali,' Do.

4584. [3364] 'Mulmul Khass,' Do.

4585. [3364] 'Mulmul Khas,' Do.

4586. [3370] 'Nyansook, Do.'

4587. [3376] 'Figured muslin,' Do.

4588. [3394] Jamdanee muslin, Do. KHAJEH ABDOOL GUNNY.

4589. [3400] Spotted Bootee, Do.

4590. [4744] Striped muslin, Assam. CAPT. LLOYD.

4591. [8266] Chunderkoora Mulmul, Hooghly.

4592. [10514] Kurnool muslin, 1st quality, Kurnool.

4593. [10515] Do., 2nd do., Do.

4594. [10516] Do., 3rd do., Do.

4595. [10517] Do., 4th do., Do.

4596. [10518] Do., 5th do., Do.

4597. [10627] 'Maderpak' muslin, North Arcot.

4598. [90] Maderpak muslin, from Manellore, Maderpak division, Do. PETTIANGERI CHETTI and PONNERI JANGAM CHETTI.

The thread used in weaving muslins of this kind is spun from a peculiar kind of cotton, known by the vernacular name of 'Pu Parthi.' The weavers state that they purchase this thread ready spun from a colony of pariahs who have settled in the neighbourhood, and who have the speciality of its preparation. It is not stated wherein this peculiarity of manufacture chiefly consists, as all that the local report says on this head is, that after the cotton is carefully cleaned and picked it is laid by in cloth bundles for two or three years, when it is rolled in plantain bark and then wound off.

## CALICOES ETC.

4599, 4603. [1661-5] Calicoes glazed, Puttiala. H. H. the RAJAH.

4604. [10035] Do., Jullundurh. GOVERNMENT of INDIA.

4605. [3805] Gazereen, Umritsur. Do.

4606. [5214] Gambroon (twilled cotton lining), Khanjurh. Do.

4607. [8272] Do., Sealkote Jail. Do.

4608. [8274] Do., Loodianah do., Do.

4609. [866] Doputtas, Ahmedabad. Do.

4610. [7800] Do., Sauganeer. Do.

4611. [7876] Do., yellow, Poonah. Do.

4612. [7869] 'Chudders' (2 pairs), Santepore. Do.

4613. [8269] Do., Hooghly. Do.

4614. [5198] Handkerchiefs, Goojerat Jail. Do.

4615. [7801] Do., Sauganeer. Do,

## TABLE CLOTHS.

4616. [615] Table cloth, Salem, Madras. GOVERNMENT of INDIA.

4617. [2049] Do. 8 ft. square, 5¼R., Madras. Do.

4618. [2051] Do. (2) damask, 6 ft. square, 3R. each, Madras. Do.

4619. [2642] Do., bleached, Chittledroog, Madras. Do.

4620. [1492] Do., coloured, Benares Jail. Do.

4621. [1973] Do. 36 ft. × 7, 12R. each, Broach. Do.

4622. [1974] Do. 36 ft. × 7, 10R. 8A. each, Do. Do.

4623. [1975] Table cloth, 36 ft. × 7, 11R. each, Broach. GOVERNMENT of INDIA.

4624. [1979] Do. 7 ft. × 6, 1R. 15A. Do. Do.

4625. [1996] Do., Allahabad. Do.

4626. [3268] Do. Jubbulpore School of Industry. Do.

4627. [3269] Do. Do. Do.

4628. [5199] Do., Lahore Central Jail. Do.

4629. [8254] Do., coloured, Patna. Do.

4630. [8255] Do. Do. Do.

4631. [8256] Do. (2), 3 yards square, Beerbhoom. Do.

4632. [8259] Do. (2), Do. Do.

4633-5. [8263-5] Do. (3), Do. Do.

4636. [8281] Do. (6), various patterns, Dinapore. Do.

## NAPKINS AND DOYLEYS.

4637. [417] Napkins, one dozen, 6R. 8A., Madras. GOVERNMENT of INDIA.

4638. [1976] 'Doyleys,' 1R. per doz., Broach. Do.

4639. [2066] Do., checked (6) 22 inches square, 1R. 5A. Madras. Do.

4640. [2072] Do., coloured check, 3¼ yards square, 1R. 8A. Do. Do.

4641. [2074] Do., coloured, 12 inches square, 1R., Do. Do.

4642. [2075] (6) Napkins, checked, 2R. 8A., Do. Do.

4643. [3270] Do. Jubbulpore School of Industry. Do.

4644. [5195] Do. Lahore Central Jail. Do.

4645. [5196] Do. Do. Do.

4646. [5200] Do. Punjaub Jails. Do.

4647. [5202a] Do. Punjaub Jails. Do.

4648. [5204] Do. Goojerat Jail. Do.

4649. [5205] Napkins, Goojerat Jail. GOVERNMENT of INDIA.

4650. [5224] Do., coloured, Dinapore.

4651. [6325] Do., bleached and unbleached, Singapore. COL. COLLYER.

4652-3. [8257-8] 6 large, 6 small, Beerbhoom. GOVERNMENT of INDIA.

4654. [8280] Do. 1R. 8A., Broach. BOMBAY GOVERNMENT.

4655. [8282] Do. 5 dozen, Dinapore. GOVERNMENT of INDIA.

## TOWELLINGS, DUSTERS, SHEETINGS, ETC.

4656. [1495] Ribbed, 5R. 6P., Benares Jail. DR. N. H. CHEKE, Superintendent.

4657. [1496] Bathing, 12A., Do. Do.

4658. [1969] 3R., Broach. BOMBAY GOVERNMENT.

4659. [1980] Bathing, 2R., Broach. Do.

4660. [1991] Baden Baden, Allahabad Jail. GOVERNMENT of INDIA.

4661. [1993] Bathing, Allahabad Jail. Do.

4662. [2061] Ribbed, 1¼ yards × 1 yard, Madras. GOVERNMENT of MADRAS.

4663. [2062] Do. 8A. 8P., Do. Do.

4664. [5192] Do. Shahpore Jail. GOVERNMENT of INDIA.

4665. [5197] 'Huckaback,' Jhelum Jail. Do.

4666. [5207] 'Huckaback,' Goojerat Jail. Do.

4667. [5208] Dhera Ghazee Khan Jail. Do.

4668. [5209] Punjaub Jails. Do.

4669. [5210] Lahore Central Jail. Do

4670. [5211] Huckaback, Ferozepore Jail. Do.

4671. [5212] Do. Do.

4672. [5568] Bathing, 12R. per doz. Meerut Jail. Do.

4673. [8252] 1 dozen, Patna. Do.

4674. [8253] 1 do. Do. Do.

4675. [8260] (6) Washing, Beerbhoom. BOMBAY GOVERNMENT.

4676. [8251] Dusters, 1 dozen, Patna. Do.

4677. [766] Sheeting, Khyrpoor. H. H. MEER ALI MORAD.

4678. [769] Do. Do.

4679. [1138] Moorgode, Belgaum. BOMBAY GOVERNMENT.

4680. [2626] Fine Dungaree, Astagram. GOVERNMENT of MYSORE.

4681. [724] Pillow case, Kutch. H. H. the RAO.

## CANVAS.

Canvas for tents, sails, &c. is in India almost entirely made of cotton. The list given below of the tents, &c. manufactured for sale at the Jubbulpore School of Industry, will afford an idea of the variety, sizes, and prices, of a class of appliances in the manufacture of which a large quantity of cotton is yearly consumed.

4682. [3279] Canvas. JUBBULPORE SCHOOL OF INDUSTRY.

LIST OF TENTS AND OTHER ARTICLES MANUFACTURED IN THE JUBBULPORE SCHOOL OF INDUSTRY.

DOUBLE-POLED REGULATION TENTS, of four cloths throughout, with four feet verandah, complete, with carpets, bamboo chicks, purdahs, gunny bags and ropes—

| | |
|---|---|
| 28 feet by 16 feet . . . . . | 670R. |
| 26 do. „ 16 do. . . . . | 645 |
| 24 do. „ 16 do. . . . . | 600 |
| 23 do. „ 14 do. . . . . | 520 |

SINGLE-POLED REGULATION TENTS, of four cloths throughout, with four feet verandah, complete as above—

| | |
|---|---|
| 16 feet square . . . . . | 450R. |
| 15 „ „ . . . . . | 430 |
| 14 „ „ . . . . . | 390 |
| 13 „ „ . . . . . | 350 |
| 12 „ „ . . . . . | 310 |

HILL TENTS, with four folds of cloth throughout, complete as above—

| | |
|---|---|
| 14 feet square, with two sybans . | 265R. |
| 13 „ „ . . . . . | 240 |
| 12 „ „ . . . . . | 215 |
| 10 „ „ . . . . . | 200 |

SWISS COTTAGE TENTS, double fly, with sybans, complete as above—

| | |
|---|---|
| 16 feet square . . . . . | 390R. |
| 14 „ „ . . . . . | 340 |
| 13 „ „ . . . . . | 320 |
| 12 „ „ . . . . . | 300 |

Rowties, five cloths to the fly, and four to the kunauts, complete as above—

| | | |
|---|---|---|
| 16 feet by 14 feet | . . . . . | 210R. |
| 16 do. „ 12 do. | . . . . | 200 |
| 14 do. „ 12 do. | . . . . | 185 |
| 13 do. „ 12 do. | . . . . | 175 |

### SHAMEEANAHS.

| | | |
|---|---|---|
| 40 feet square, and 12 poles | . . | 400R. |
| 30 „ „ 12 „ | . . | 300 |
| 24 „ „ 12 „ | . . | 225 |
| 20 „ „ 8 „ | . . | 150 |
| 18 „ „ 8 „ | . . | 125 |
| 15 „ „ 4 „ | . . | 90 |

Carpets and kunauts extra.

Bechobah Tents, five cloths in fly, and four in kunauts, as above—

| | | |
|---|---|---|
| 12 feet square | . . . . | 135R. |
| 10 „ „ | . . . | 120 |
| 9 „ „ | . . . | 100 |
| 8 „ „ | . . . | 90 |

Sleeping Pals, four folds of cloth, including chintz, with lacquered poles, and carpet, with four feet kunauts—

| | | |
|---|---|---|
| 12 feet square | . . . . | 110R. |
| 10 „ „ | . . . | 100 |

Without carpet, 15R. less.

### SERVANTS' PALS.

| | | |
|---|---|---|
| Madras pattern, 12 feet square | . | 66R. |
| Bengal „ 12 „ „ | . | 55 |
| Necessary tents | . . . | 27 |

Stable Tents, and every other description of tent, made to order.

| | | |
|---|---|---|
| Division kunauts to large tents | . | 35R. |
| „ „ to small do. | . | 30 |
| Bhoots, 3 feet, for increasing the inside of the tent, each | . . . | 26 |
| „ 2½ feet | . . . | 20 |
| „ 2 „ | . . . | 18 |
| Tent Thermantidote, carried by one cooly | . | 25 |
| Zenanah kunauts, per running yard | . | 3½ |
| Bath rooms, according to size ordered | . | 0 |

#### SUNDRY ARTICLES MANUFACTURED.

| | | |
|---|---|---|
| Turkey carpets, fine texture, per square yard | . . . . . | 4R. |
| Scotch carpeting, per yard, 33 inches wide | . . . | 1½ |
| Suttrinjees, per square yard | . . | 1⅛ |
| Kidderminster ditto, one yard wide, per yard | . . . . | 1⅛ |
| Plaids of (9) yards English thread | . | 5 |
| Table cloths, of all sizes, per running foot | 11A. |
| Table napkins, per dozen | . . | 3½R. |
| Stamped floor chintz on Dosootee, per piece of 12 yards | . . . | 4 |
| Towels, per dozen | . . . | 4 |
| Huckaback towels, per dozen | . . | 6½ |
| Bathing towels, per dozen | . . | 7½ |
| Horse clothing, per thân of 12 yards | . | 4 |
| Horse rollers, to match the above | . | 2 |

The tents are made of the best materials and of new cloth. Brass eyelets are used in lieu of leather ones; the ropes are all manufactured with English machinery, and will be found far superior to, and more durable than, ropes made with country machines.

4683. [3274] 'Dosootee,' Jubbulpore School of Industry. GOVERNMENT of INDIA.

4684. [967] Canvas, eight samples of various texture, Rewa Kanta. GOVERNMENT of BOMBAY.

4685. [1288] 'Purum.' Narrow strips, Dharwar. W. C. ANDERSON, Esq.

These strips are sewn together, and used as curtains for the front of shops and houses, and also as tents to protect goods on transit. This is made of the waste warp-ends of pieces of cotton cloth joined together.

4686. [1289] Do. Another description used more for making bags, &c. Halfpiece, 9A. Do.

4687. [1290] 'Kadee.' Common white cotton cloth, used in great quantities for all purposes, 2R. 8A. Do. Do.

4688. [1293] Tape. Principally used for lacing of cot frames for sleeping upon, and also in making horse trappings, 1R. 6A.

4689. [1298] 'Gint,' two Thread cloth, used by the working classes for making breeches, &c., 1R. 12A., Do. Do.

4690. [1981] (2) for Tents, 18 yards, 4R. per piece, Broach. GOVERNMENT of BOMBAY.

4691. [1995] for Tents, Allahabad Jail. GOVERNMENT of INDIA.

4692. [5289] 'Dosootee,' for Tents, two Thread, Meerut Jail. Do.

4693-4. [5291-2] 'Teensootee,' for Tents, Do. Do.

4695. [10579] for Tents, North Arcot, Madras. GOVERNMENT of MADRAS.

## COTTON PIECE GOODS (COLOURED), CHINTZES, ETC. ETC.

4696. [705] Striped 'Neelo Kakamia, Kutch. H. H. the RAO.

4697. [711] Coloured stripe, Do. Do.

4698. [719] 'Phaleu,' coloured stripe, Do. Do.

4699. [725] Checked, Do. Do.

4700. [727] 'Tanjeri,' Do. Do.

4701. [730] 'Puncheree,' Do. Do.

4702. [806] Coarse, Do. Do.

4703. [869] 'Chowtara,' 32½ ft. × 2 ft. 2 in., 2R. 4A., Ahmedabad. BOMBAY GOVERNMENT.

4704. [870] 'Chowtara,' 16½ ft. × 2 ft. 6 in., 2R. 3A. 2P., Do. Do.

4705. [853] 'Elacha,' 12 ft. × 2 ft. 9 in., 15A., Ahmedabad. BOMBAY GOVERNMENT.

4706. [883] 'Kholia,' Do. Do.

4707. [931] Do., 3 ft. 6 in. × 1 ft. 6 in., 1R. 8A., Do. NEKNAMDAR SHETANEE HURKOOVURBHAEE.

4708. [1087] Checked, Broach. BOMBAY GOVERNMENT.

4709. [1251] Checked. 26 yards, 5R. 14A., Broach, Bombay. Do.

4710. [1253] Do., 6¾ do., 2R., Do. Do. Do.

4711. [1254] Do., 25 do. × 2 ft., Do. Do.

4712. [1257] Do., 26 yards, 6R. 8A., Do. Do. Do.

4713. [1985] Duck, 10 yards, 5R., Do. Do.

4714. [1986] Checked, 10½ yards, 2R. 10A., Do. Do.

4715. [2077] Blue drill, 3 yards, 10A. per yard, Madras. GOVERNMENT of MADRAS.

4716. [2078] Checked do., 3 do. × 28 in. 10A. do., Do. Do.

4717. [2079] Blue stripe, 3 do., 9A. do., Do. Do.

4718. [2080] Do. ticken, 3 do., 9A. do. Do. Do.

4719. [2081] Do., 3 do., 9A. do., Do. Do.

4720. [2082] Horse 'Jhool,' 4 ft. wide. 1¼R. do, Do. Do.

4721. [2083] Do., 1 yard broad, 9A. do., Do. Do.

4722. [2628] Thick four thread, Astagram. GOVERNMENT of MYSORE.

4723. [3193] Striped, Darjeeling. DR. CAMPBELL.

4724. [5033] Coloured, Dharwar. BOMBAY GOVERNMENT.

4725. [5201] 'Guzzee,' second quality, Sirsah Jail. GOVERNMENT of INDIA.

4726. [5202b] Punjaub Jail. Do.

4727. [5203] 'Docootee,' Goojerat Jail. Do.

4728. [5215] Printed, Seetapore, Oude. GOVERNMENT of INDIA.

4729. [5238] 'Lahanza,' green, Lucknow. Do.

4730. [5239] Do. blue, Do. Do.

4731. [5240] 'Urhnee,' Do. Do.

4732. [9436] Two pieces checked cotton, Hyderabad.

4733. [5241] 'Mughie,' Chittagong. GOVERNMENT of INDIA.

4734. [5242] Do. Do. Do.

4735. [5243] Do. Do. Do.

4736. [5293] Imitation of Duck, Meerut Jail. Do.

4737. [5561] Common quality, Cuttack. Do.

4738. [5562] Medium quality, 3s., Do.

4739. [7875] 'Joth,' 2R. 4A., Poonah. GOVERNMENT of BOMBAY.

4740. [8262] 'Dimity,' Beerbhoom. GOVERNMENT of INDIA.

4741. [8270] 'Nawar,' or band, Sirsah Jail. Do.

4742. [8271] Punjaub Jails. Do.

4743. [8273] 'Konia Bor Kapoo,' Durrang, Assam. LIEUT. W. PHAIRE.

4744. [8275] Checked, Bhaugulpore. GOVERNMENT of INDIA.

4745. [8276] Do. Do. Do.

4746. [8277] 'Selleng,' Durrang, Assam. LIEUT. W. PHAIRE.

4747. [8278] Checked, Bhaugulpore. GOVERNMENT of INDIA.

4748. [8279] Do. Do. Do.

4749-51. [8289-91] Do. Lucknow. Do.

4752. [9211] Striped, Bhaugulpore. Do.

4753. [751] Chintz, 3R. 4A., Khyrpoor. H. H. MEER ALI MORAD.

4754. [753] Do., 3R. 4A., Do. Do.

4755. [755] Do., 2R. 8A., Do. Do.

4756. [757] Do., 2R. 4A., Do. Do.

4757. [758] Do., 2R. 4A., Do. Do.

4758. [759] Chintz (2), 2R. 8A., Khyrpoor. H. H. MEER ALI MORAD.

4759. [760] Do., 1R. 10A., Do. Do.

4760. [1035] Do., 3R. 13A., Hyderabad. BOMBAY GOVERNMENT.

4761-3. [7796-8] Printed, Sauganeer. Do.

4764-72. [10536-44] Chintz. North Arcot, Madras. VENKATA RAO and BAPANA RAO.

4773. [10545] Do. A. RAB RAO.

4774. [10546] Do. Do. BASETH RAO.

4775. [9437] Four pieces of Chintz, Hyderabad.

———

4776. [8294] Cotton printed fabrics (various) in Case 30. Manufactured exclusively from East India Cotton, exhibited by the COTTON SUPPLY ASSOCIATION, *Manchester.*

These fabrics, manufactured by Mr. J. Cheetham; bleached by Messrs. Eden & Thwaites; dyed by Messrs. Ainsworth, Son, & Co., F. Steiner & Co., and Greenwood & Lieber; printed by Messrs. McNaughtan & Thom, Thomas Hoyle & Sons, Salis Schwabe & Co., Daniel Lee & Co., Walter Crum & Co., Daglish & Co., Inglis & Wakefield, Thomas Clarkson & Co., Grimshaw, Gibson & Co., Edmund Potter & Co., and F. & W. Grafton & Co., prove the capability of *Indian Cotton* for the production of goods of *first class quality* in every respect.

4777. [9438] Specimens of Furniture Chintz, manufactured exclusively of Indian Cotton. Printed by THOMAS CLARKSON & Co. (Case 24.) COTTON SUPPLY ASSOCIATION.

———

The under-noted specimens of cotton manufactures, for which space could not be found, are deposited for reference at the Indian Museum; viz., towellings from Beerbhoom; Jails in the Punjaub; Benares Jail; Jubbulpore School of Industry; Dinapore; Patna; Allahabad; Mysore; Madras, and Broach.

Table-cloths from Beerbhoom and Benares Jail; Mysore; Madras; Broach.

Sheetings from Mysore; Madras; Dharwar; Broach.

Gambroon and chintzes, Sealkote Jail; Khyrpoor, H. H. MEER ALI MORAD; Hyderabad and Chundawarra.

Striped, checked, and other cottons, from Allahabad; Bhandere; Darjeeling, DR. CAMPBELL; Hooghly; Jaloun; Jhansee; Jubbulpore; Lucknow; Meerut; Jails in the Punjaub, &c.; Madras; Ahmedabad; Broach; Belgaum; Dharwar; Shikarpoor, and Poonah.

# CLASS XIX. — MANUFACTURES IN FLAX, HEMP, AND OTHER FIBRES.

Of manufactures belonging to this class exported from India, those furnished by Jute occupy an important position. The amount of these may be again seen by reference to the Table (p. 142). The Table (p. 146), likewise indicates that a certain proportion of 'Hemp,' in the manufactured form of rope, is also exported.

4778. [5962] Twilled cloth and sacks, No. 1, Baranagore. BORNEO COMPANY (LIMITED).

4779. [5960] Twilled cloth and sacks, No. 2. Do.

4780. [5961] Twilled cloth and sacks, No. 3. BORNEO COMPANY (LIMITED).

4781. [5964] Plain power loom bag, C. Do.

4782. [5963] Twilled jail bag. Do.

4783. [5965] American gunny cloth. Do.

The above specimens were manufactured at the steam works of the Borneo Company, Limited,' at Baranagore, near Calcutta. The Company work up about 400 tons of jute monthly, and employ 2,500 natives. Contributed by THOS. DUFF, Esq., Manager and Agent.

4784. [5290] One piece of canvas made of hemp, Meerut.

Made by the prisoners in the jail at Meerut.

4785. [5289] One piece of Dosootee (made of two threads) for tent-making, Do.

Made by the prisoners in the jail at Meerut.

4786. [5291] One piece of Teensootee (made of three threads) for tent-making, Do.

Made by the prisoners in the jail at Meerut.

4787. [5292] One piece of broad Dosootee (two threads) for tent-making, Do.

Made by the prisoners in the jail at Meerut.

4788. [9237] Bag-a-Chut, Peshawur.

4789. [10312] Sacking of aloe fibre, Chota Nagpore.

Made in the Hazareebaugh jail by prisoners.

4790. [5197] Newar cloth, Jhelum Jail. CENTRAL COMMITTEE, *Lahore*.

4791. [8270] Do., Sirsa Jail. Do.

4792. [3192] Four specimens of jute cloth, Darjeeling. DR. A. CAMPBELL.

4793. [9071] *Crotalaria juncea*, Gunny cloth or rope for sackcloth, Lucknow.

This is sackcloth manufactured from the fibre of the above, used for making bags and sacks of various sorts.

4794-5. [9331-2] Two pieces of cloth made from the fibre of the stinging-nettle, Nepaul. H. H. SIR JUNG BAHADOOR, K.C.B.

These specimens are very inferior, but in the cold season the Hill people bring into Katmandoo small quantities of a very superior quality, which very much resembles thick canvas.

4796. [5267] Net of Rhea fibre, Assam.

Presented by H. BAINBRIDGE, Esq.

*Fabrics manufactured from Aloe Fibre.*

4797. [10306] Raw aloe fibre, Central Prison, Bareilly.

4798. [10306] Dyed aloe fibre, Do. Do.

4799. [8139] Cloth made of aloe fibre dyed. INDIA MUSEUM.

4800. [8145] Do. Do.

4801. [8148] Do. DR. R. RIDDELL.

4802. [8149] Cloth made of aloe, double fibre. DR. R. RIDDELL.

4803. [8150] Do., and cotton. Do.

4804. [8151] Do., and silk. Do.

4805. [8152] Do., and Tusser silk. Do.

4806. [8153] Do., aloe fibre dyed and silk. Do.

4807. [8154] Cloth made of coloured aloe fibre. Do.

VI. *Cordage of all Kinds.*

4808. [5946] Rope Manilla, of plantain fibre, Shal-i-mar Ropery, Seebpore. Messrs. AHMUTY & CO.

Running rigging for ships, or tackling for land purposes.

4809. [5940] Rope, West Indian hemp (*Cannabis sativa*). Do.

Standing rigging for ships, or for use in water.

4810. [5942] Rope, West Indian hemp.

4811. [5937] Do. of *Sesbania aculeata*.

Running rigging for ships, and tackling for land purposes.

4812. [5939] Do. of *Sesbania aculeata*.

4813. [5938] Do., Coir (cocoa-nut fibre).

Boats' and ships' running gear and hawsers for cables, also for all tackling purposes when exposed to wet, especially salt water.

4814. [5944] Do., of *Crotalaria juncea*.

Tackling in dry places, without exposure to wet. These specimens will show the great improvement in the manufacture of the various kinds of cordage in India since 1851.

The foregoing were manufactured at the Shal-i-mar Ropery, Seebpore, near Calcutta, by the Exhibitors, Messrs. AHMUTY & CO.

4815. [10074] Rope of a fibrous plant called Chuyan, Lahore.

4816. [10062] Do., Sunn Okra, Do.

4817. [10061] Do., common hemp (*Cannabis sativa*), Do.

4818. [10064] Do., fibre called 'Bugar,' Do.

4819-20. [10044-5] Do., two samples, Lahore.

Made in the Shahpore Jail. CENTRAL COMMITTEE, *Lahore*.

4821. [10059] Do., made of palm leaves, Do. CENTRAL COMMITTEE, *Lahore*.

4822. [10069] Rope made of Dah grass, Lahore. CENTRAL COMMITTEE, *Lahore.*

4823. [10065] Do., of plantain leaves, Do. Do.

———

4824. [5898] Rope of Bail grass, Assam. LIEUT. PHAIRE.

4825. [5330] Do. (*Urtica tenacissima*), Luckimpore. JADOORAN BOROOAH.

4826. [5336] Do. (*Sesbania aculeata*), Do. Do. H. L. MICHEL, Esq.

4827. [5678] Do. (four samples), Tezpore, Do. TEZPORE JAIL.

4828-9. [5264-5] String of twisted Rhea fibre, Assam. H. BAINBRIDGE, Esq.

4830. [7754] Bowstring of Moorgave fibre (*Sanseviera zeylanica*), Do.

4831. [5566] Rope of cotton, Meerut Jail. GOVERNMENT of INDIA.

4832. [10313] Do. aloe fibre, Chota Nagpore. Do.

4833. [10315] Twine, Do., Do. Do.

4834. [5953] Rope of *Sacharum Moonja*, Lucknow. Do.

Twine made from the fibre of the leaf sheath: a kind little thicker than the present specimen is used for towing boats; when dry, it does not possess much strength, when wet it is strong and durable. The moonj is used for ropes, thatching, &c.

4835. [5957] Twine of *Crotalaria juncea*, Lucknow. Do.

4836. [7761] Rope of Agave fibre, Balasore. Do.

4837. [7762] Do. Moorgave fibre, Do. Do.

4838. [7763] Do. Curved Agave fibre, Do. Do.

4839. [7757] Do. Sealee fibre, Do. Do.

4840. [7760] Do. (*Sterculia ramosa*), Cuttack. Do.

4841. [5949] Twine of *Crotalaria juncea*, Hooghly. GOVERNMENT of INDIA.

4842. [7755] Cord and fibre of Red Sterculia bast, Moulmein. Do.

4843. [7756] Do. do. Do. Do.

4844. [4094] Rope of rattan (two samples), Penang, Malay Peninsula. SINGAPORE COMMITTEE.

Used for drawing water, and as halters for cattle.

4845. [4751] Do. bast, Burmah. Messrs. HALLIDAY, FOX, & CO.

4846. [2889] Do. 'Mole,' Poonah. BOMBAY GOVERNMENT.

4847. [4873] Do. Palmyra fibre (*Borassus flabelliformis*), Chingleput, Madras. DR. SHORTT.

4848. [2727] Do. Plantain fibre (*Musa paradisiaca*), Mysore. DR. KIRKPATRICK.

4849. [2728] Do. Cocoa-nut fibre (*Cocos nucifera*), Do. Do.

4850. [2729] Do. Dudi, Do. Do.

4851. [2730] Do. Pundi, Do. Do.

4852. [2732] Do. Sereum (*Calotropis gigantea*), Do. Do.

4853. [2733] Do. (*Yucca gloriosa*), Do. Do.

4854. [2683] Do. Cotton, Do. LOCAL COMMITTEE.

4855. [2684] Do. do. Do. Do.

The undernamed specimens belonging to this class, forming a portion of the consignments from India to the Exhibition, are for the present deposited in the INDIA MUSEUM.

Samples from Lucknow; Hooghly; Darjeeling; Lahore; Guzerat; Rajgolee; North Arcot; Broach; Poona; Rewa Kanta, &c.; and Bolts of Canvas, by the BORNEO COMPANY, LIMITED.

And of cordage and twine as under—

4856. [5950] Rope of *Gossypium herbaceum*, Lucknow. GOVERNMENT of INDIA.

Produced all over Oude. This kind of rope is made of cotton thread, and is used for hanging and pulling punkahs, and the ropes of all tents are made of it; it is considerably strong.

4857. [5951] Rope Koombhee, Lucknow. GOVERNMENT of INDIA.

4858. [5956] Do. of *Crotalaria juncea*, Do. Do.

4859. [5954] Do. Bunkuss, Do. Do.

4860. [5955] Do. *Hibiscus cannabinus*, Do. Do.

4861. [9172] Do. Sunn, Ulwar. H. H. the MAHARAJAH.

4862. [9173] Rope Moonj, Ulwar. H. H. the MAHARAJAH.

4863. [9174] Do. Cotton, Do. Do.

4864. [2932] Do. of aloe plant for tent-makers, Meerut. GOVERNMENT of INDIA.

4865. [10066] Do. Bark of the Phalsa tree (*Grewia asiatica*), Lahore. Do.

4866. [10068] Do. grass called 'Dib.' Do. Do.

4867. [10072] Do. Madar plant, Do. Do.

4868. [10067] String of the Putta plant, Do. Do.

4869. [5263] Rope (ten samples), Calcutta. Messrs. AHMUTY & Co.

4870. [5958] Rope Hooghly. GOVERNMENT of INDIA.

4871. [5959] Do. Do. Do.

4872. [2731] Do. (*Cannabis sativa*), Mysore. DR. KIRKPATRICK.

4873. [779] Twine of hemp (two lbs.), Khyrpoor. H. H. MEER ALI MORAD.

4874. [7889] Do. aloe fibre, Poonah. BOMBAY GOVERNMENT.

4875. [2886] Rope of hemp, Do. Do.

4876. [7890] Do. flax, Do. Do.

4877. [7894] Do. 'Ghayal,' Do. Do.

4878. [6346] Do. Manilla plantain fibre, Manilla. HON. COL. CAVANAGH.

# CLASS XX.— SILK AND VELVET.

The quantity, in pieces, and value of the silk goods exported from India and each Presidency to all parts of the world will be seen by referring to the Table (p. 215).

## I.—SILK YARNS.

With the exception of the specimens under-noted, these have already been entered under Section B, Class IV. pp. 102, 103.

4879. [775] Thirteen specimens of dyed silk yarn, Khyrpore. H. H. MEER ALI MORAD.

4880. [8227] Silk thread, Assam. LIEUT. W. PHAIRE.

4881. [8228] Do. (Moonga.) Do. Do.

4882. [8229] Do. Do. Do. Do.

4883. [5471] Ornamental fastening of red silk thread, Lahore.

4884. [5469] "Fly protectors," Do. Do.

## II.—PLAIN AND FANCY SILKS.

4885. [3147] Fifteen silk pieces of colours, Berhampore. GOVERNMENT of INDIA.

4886. [3150] Silk piece, Do. Do.

4887. [6630] Silk piece, white, Do. Do.

4888. [1744] Silk piece, Benares. GOOLBADAN THAN. GOVERNMENT of INDIA.

4889. [1745] Do. Do. Do.

4890. [5075] Silk piece, white gauze, Nimtollah, Midnapore. RAM DOSS, manufacturer.

4891. [8202] Do. orange-check, Dasspore. RAM GUI. GOVERNMENT of INDIA.

4892. [8201] Do. white. Do. MADHUB PORRAMANICK. Do.

4893. [8200] Do. red gauze, Do. KRISTO PORRAMANICK, manufacturer. Do.

4894. [8204] Do. white. Do. Do. Do.

4895. [8199] Do. figured, Puddumpore, Midnapore. KARTIC NUNDEE, manufacturer. Do.

4896. [8203] Silk piece, Bellaghat, Midnapore. NOBIN HYTHE, manufacturer. GOVERNMENT of INDIA.

4897. [9439] Do. Russickgunge. BACHARAM PERA, manufacturer. Do.

4898. [5085] Silk piece, white, Burdwan, BIPRODOSS DUTT, manufacturer. GOVERNMENT of INDIA.

TABLE SHOWING THE QUANTITIES (AS FAR AS CAN BE ASCERTAINED) AND THE VALUE OF SILK GOODS (EXCLUSIVE OF CHUSSUMS) EXPORTED FROM INDIA AND EACH PRESIDENCY TO ALL PARTS OF THE WORLD FROM 1850-51 TO 1860-61.

| Years | Whence Exported | United Kingdom Quantity | United Kingdom Value £ | France Quantity | France Value £ | Other Parts of Europe Quantity | Other Parts of Europe Value £ | America Quantity | America Value £ | China Quantity | China Value £ | Arabian and Persian Gulfs Quantity | Arabian and Persian Gulfs Value £ | Aden and Suez Quantity | Aden and Suez Value £ | Straits Settlements Quantity | Straits Settlements Value £ | Other Parts Quantity | Other Parts Value £ | Total Exported to All Parts Quantity | Total Exported to All Parts Value £ |
|---|---|---|---|---|---|---|---|---|---|---|---|---|---|---|---|---|---|---|---|---|---|
| 1850-51 | Bengal | 559,892 | 309,419 | 1,957 | 1,289 | 1,032 | 720 | 7,543 | 4,330 | 559 | 730 | 908 | 275 | ... | ... | 23,154 | 12,955 | 4,559 | 2,974 | 599,604 | 332,657 |
| | Madras | 21 | 55 | ... | ... | ... | ... | ... | ... | ... | ... | 7 | 11 | ... | ... | 3,781 | 1,805 | 34 | 38 | 3,843 | 1,909 |
| | Bombay | 571 | 1,737 | ... | ... | ... | ... | ... | ... | 983 | 304 | 19,337 | 11,138 | 855 | 517 | 6,237 | 4,267 | 2,594 | 2,694 | 30,577 | 20,657 |
| | All India | 560,484 | 311,211 | 1,957 | 1,459 | 1,032 | 720 | 7,543 | 4,330 | 1,542 | 1,034 | 20,252 | 11,424 | 855 | 517 | 33,172 | 19,027 | 7,187 | 5,706 | 634,024 | 355,223 |
| 1851-52 | Bengal | 408,256 | 224,406 | 405 | 266 | 294 | 182 | 2,089 | 1,058 | 76 | 42 | 259 | 75 | ... | ... | 1,006 | 417 | 27,280 | 13,288 | 439,615 | 239,734 |
| | Madras | 36 | 96 | ... | ... | 4 | 4 | 15 | 6 | ... | ... | 203 | 67 | ... | ... | 129 | 68 | 472 | 228 | 637 | 392 |
| | Bombay | 12 | 33 | ... | ... | ... | ... | ... | ... | 9,050 | 955 | 10,498 | 7,867 | 935 | 834 | 15,185 | 7,661 | 3,256 | 2,739 | 38,955 | 20,099 |
| | All India | 408,304 | 224,535 | 495 | 266 | 298 | 186 | 2,104 | 1,064 | 9,126 | 997 | 10,757 | 7,942 | 935 | 834 | 16,320 | 8,146 | 30,958 | 16,255 | 479,207 | 260,225 |
| 1852-53 | Bengal | 502,960 | 267,803 | 423 | 230 | 1,879 | 1,010 | 2,608 | 1,455 | ... | 276 | 2,407 | 897 | ... | ... | ... | 777 | 5,977 | 14,725 | ... | 287,173 |
| | Madras | 25 | 51 | ... | ... | ... | ... | ... | ... | ... | ... | 203 | 67 | ... | ... | 67 | 42 | ... | 621 | 6,272 | 781 |
| | Bombay | ... | 2 | ... | ... | ... | ... | ... | ... | ... | 211 | ... | 15,892 | ... | 1,723 | 11,460 | 7,035 | ... | 2,378 | ... | 27,241 |
| | All India | 502,985 | 267,856 | 423 | 230 | 1,879 | 1,010 | 2,608 | 1,455 | ... | 487 | ... | 16,856 | ... | 1,723 | 11,460 | 7,854 | ... | 17,724 | ... | 315,195 |
| 1853-54 | Bengal | 522,883 | 276,747 | 2,128 | 1,068 | 1,004 | 502 | 12,577 | 7,431 | 262 | 215 | 429 | 105 | 9,113 | 4,802 | 2,055 | 997 | 19,312 | 9,804 | 569,763 | 301,671 |
| | Madras | 17 | 29 | ... | ... | ... | ... | ... | ... | ... | ... | ... | ... | ... | ... | 79 | 42 | 1,692 | 790 | 1,788 | 861 |
| | Bombay | ... | 53 | ... | ... | ... | ... | ... | ... | ... | 150 | ... | 5,149 | ... | 1,382 | ... | 4,322 | ... | 2,743 | ... | 13,799 |
| | All India | 522,900 | 276,829 | 2,128 | 1,068 | 1,004 | 502 | 12,577 | 7,431 | ... | 365 | ... | 5,254 | 9,113 | 6,184 | ... | 5,361 | ... | 13,337 | 513,970 | 316,331 |
| 1854-55 | Bengal | 443,187 | 230,214 | ... | ... | 6 | 3 | 5,244 | 3,470 | ... | ... | 1,461 | 436 | 11,830 | 6,022 | 1,890 | 578 | 16,273 | 6,901 | 479,891 | 248,577 |
| | Madras | 26 | 51 | ... | ... | ... | ... | ... | ... | ... | ... | ... | ... | ... | ... | 1,349 | 863 | 846 | 720 | 2,221 | 1,634 |
| | Bombay | 4 | 19 | 35 | 72 | ... | ... | ... | ... | 1,277 | 172 | 19,061 | 6,874 | 5,135 | 2,936 | 4,372 | 1,788 | 2,009 | 1,453 | 31,858 | 13,242 |
| | All India | 443,217 | 230,284 | ... | 943 | 6 | 3 | 5,244 | 3,470 | 1,277 | 202 | 20,522 | 7,310 | 16,965 | 8,958 | 7,611 | 3,229 | 19,128 | 9,074 | 513,970 | 261,453 |
| 1855-56 | Bengal | 569,104 | 317,188 | 297 | 211 | 606 | 318 | 3,339 | 1,940 | 36 | 23 | 5,572 | 1,334 | 3,773 | 2,589 | 1,215 | 377 | 2,027 | 1,033 | 585,969 | 325,008 |
| | Madras | 45 | 243 | ... | ... | ... | ... | ... | ... | ... | ... | ... | ... | ... | ... | 2,428 | 1,327 | 1,205 | 880 | 3,678 | 2,450 |
| | Bombay | 140 | 128 | ... | ... | ... | ... | ... | ... | 79 | 172 | 8,604 | 6,227 | 6,684 | 4,494 | 9,207 | 1,349 | 2,322 | 1,257 | 27,036 | 13,577 |
| | All India | 569,289 | 317,554 | 297 | 211 | 606 | 318 | 3,339 | 1,940 | 115 | 195 | 14,176 | 7,561 | 10,457 | 7,023 | 12,850 | 3,053 | 5,554 | 3,170 | 616,683 | 341,035 |
| 1856-57 | Bengal | 437,860 | 241,230 | 3,108 | 1,618 | 153 | 109 | 2,187 | 1,243 | ... | 71 | 3,928 | 959 | 3,587 | 2,016 | 749 | 620 | 856 | 853 | 452,431 | 248,719 |
| | Madras | 40 | 119 | ... | ... | ... | ... | ... | ... | ... | ... | ... | ... | ... | ... | ... | 482 | 1,645 | 685 | 1,645 | 1,286 |
| | Bombay | ... | 54 | ... | ... | ... | ... | ... | ... | 2,158 | 624 | ... | 14,300 | ... | 9,320 | 13,668 | 4,560 | ... | 2,587 | ... | 31,445 |
| | All India | ... | 241,403 | 3,108 | 1,618 | 153 | 109 | 2,187 | 1,243 | 2,158 | 695 | ... | 15,259 | 3,587 | 11,336 | 13,668 | 5,662 | 1,286 | 4,125 | ... | 281,450 |
| 1857-58 | Bengal | 154,510 | 92,682 | ... | 570 | 455 | 250 | 823 | 354 | 14 | 7 | 219 | 120 | 3,526 | 3,183 | ... | 4,081 | 2,051 | 2,014 | ... | 103,261 |
| | Madras | 15 | 10 | ... | ... | ... | ... | ... | ... | 9 | 3 | ... | ... | ... | ... | ... | 26 | 998 | 948 | 2,119 | 987 |
| | Bombay | ... | ... | ... | ... | ... | ... | ... | ... | 33,720 | 1,368 | ... | 27,130 | 19,995 | 16,742 | 19,078 | 5,146 | ... | 3,590 | ... | 53,976 |
| | All India | 154,525 | 92,692 | ... | 759 | 455 | 250 | 823 | 354 | 33,734 | 1,375 | 13,350 | 27,253 | ... | 19,925 | 19,078 | 9,253 | ... | 6,552 | ... | 158,224 |
| 1858-59 | Bengal | 275,880 | 157,374 | 786 | 449 | 447 | 242 | 686 | 995 | 23 | 12 | 61 | 31 | 2,629 | 1,459 | 1,266 | 896 | 4,442 | 2,499 | 286,220 | 162,956 |
| | Madras | ... | 221 | ... | ... | ... | ... | 1 | 1 | ... | ... | ... | ... | ... | ... | 2,604 | 1,058 | 1,915 | 1,396 | 4,519 | 2,137 |
| | Bombay | 2,800 | 843 | 35 | 72 | ... | ... | ... | ... | ... | 1,275 | ... | 20,154 | ... | 8,220 | ... | 2,649 | ... | 3,917 | ... | 48,015 |
| | All India | 278,680 | 158,438 | 786 | 831 | 447 | 242 | 687 | 996 | ... | 1,287 | ... | 20,634 | ... | 9,557 | ... | 6,253 | ... | 7,812 | ... | 213,108 |
| 1859-60 | Bengal | ... | 146,212 | ... | 916 | ... | ... | ... | 1,071 | ... | 30 | ... | 31 | ... | 1,459 | ... | 896 | ... | 2,499 | ... | 152,795 |
| | Madras | ... | 6 | ... | ... | ... | ... | ... | 1 | ... | ... | ... | ... | ... | ... | ... | 1,058 | ... | 1,396 | ... | 2,461 |
| | Bombay | ... | 120 | ... | ... | ... | ... | ... | ... | ... | 1,071 | ... | 18,748 | ... | 8,098 | ... | 4,299 | ... | 3,917 | ... | 36,253 |
| | All India | ... | 146,338 | ... | 919 | ... | ... | ... | 1,083 | ... | 1,083 | ... | 18,779 | ... | 9,557 | ... | 6,253 | ... | 7,812 | ... | 191,509 |
| 1860-61 | Bengal | 167,262 | 96,493 | 1,513 | 916 | ... | 14 | 823 | 1,322 | 25 | 30 | 261 | 121 | 832 | 430 | 115 | 1,616 | 2,098 | 901 | 171,847 | 101,843 |
| | Madras | 27 | 237 | ... | ... | ... | ... | ... | ... | ... | ... | ... | ... | ... | ... | ... | 53 | 2,240 | 1,609 | 2,240 | 1,689 |
| | Bombay | ... | 237 | 2 | 3 | ... | ... | ... | ... | 7,219 | 789 | 13,350 | 8,647 | 6,056 | 3,698 | 9,085 | 4,390 | 1,376 | 1,691 | 87,455 | 19,255 |
| | All India | 167,656 | 96,757 | 1,515 | 919 | ... | 14 | 823 | 1,322 | 7,244 | 819 | 13,611 | 8,658 | 6,888 | 4,028 | 9,085 | 6,059 | ... | 4,201 | 211,542 | 122,787 |

4899. [8206] Silk piece, white, Burdwan. BHOOBUN DUTT. GOVERNMENT of INDIA.

4900. [8207] Do., Do. MADHUB DUTT. Do.

4901. [8208] Do., Do. NOBOKISSEN DUTT. Do.

4902. [8215] Do., Do. NOBOCOMAR NUNDEE. Do.

4903. [8216] Silk piece, white, Malda. GOVERNMENT of INDIA.

4904. [8217] Do., Do., Do. Do.

4905. [8218] Do., grey, Do. Do.

4906. [5092] Do., Do., Do. Do.

4907. [8221] Do., figured, Malda. BABOO HANS GEER GOSSAIN.

4908. [8222] Do., Do., Do. DR. R. F. THOMPSON.

4909. [5083] Do., Do., Do. T. M. LEWIS, Esq.

4910. [5086] Silk piece, white, Gonatea, Beerbhoom. — RAIT, Esq.

4911. [5084] Silk piece, bordered, Assam. H. BAINBRIDGE, Esq.

4912. [8232] Do., Bhotan red, Do. Do.

4913. [5088] Do., yellow, Do. LIEUT. W. PHAIRE.

4914. [1195] Silk piece, crimson, Puttiala. H. H. the MAHARAJAH.

4915. [1198] Do., white, Do. Do.

4916. [1199] Do., crimson, Do. Do.

4917. [3777] Do., striped, Do. Do.

4918. [3778] Do., Do., Do. Do.

4919. [3780] Do., Do., Do. Do.

4920. [3781] Do., Do., Do. Do.

4921. [3751] Silk piece, red, Punjaub. GOVERNMENT of INDIA.

4922. [3752] Do., green, Do. Do.

4923. [3753] Do., Do., crimson stripe, Do. Do.

4924. [3754] Silk piece, rose colour, Punjaub. GOVERNMENT of INDIA.

4925. [3755] Do., yellow, Do. Do.

4926. [3756] Do., scarlet, Do. Do.

4927. [3757] Do., yellow, Do. Do.

4928. [3758] Do., green, Do. Do.

4929. [3759] Do., crimson, Do. Do.

4930. [3760] Do., purple, Do. Do.

4931. [8241] Do., pink, Do. Do.

4932. [5465] Silk piece, shot, Peshawur. GOVERNMENT of INDIA.

4933. [5466] Do. Do. Do.

4934. [8247] Figured and gold embroidered, Do. Bhawulpore. H. H. the NAWAB.

4935. [8248] Do. Do. Do.

The principal places of silk manufacture are the cities of Peshawur, Lahore, Umritsur, Mooltan, and the capital of the neighbouring state of Bhawulpore.—The silks of the latter place are considered the best, and the next those of Mooltan.*

4936. [6497] Silk pieces, Bokhara. H. H. the NAWAB PHAIROOLLA KHAN.

4937. [6498] Do. shot, Do. Do.

4938. [6499] Do. red, Do. Do.

4939. [838] Two pieces do. red, Shikarpoor, Sindh. Do.

**BOMBAY AND SOUTHERN INDIA, &c.**

4940. [891] Silk pieces, striped, 12R. 4A. Ahmedabad. GOVERNMENT of INDIA.

4941. [894] Do. 'Gagrapat,' Do. Do.

4942. [909] Two pieces Do. Do. SHETT JAYSINGBHAEE HUTTEESING.

4943. [910] Two pieces 'Gujeeam,' Do. Do.

4944. [911] Do. Do. Do.

4945. [937] Do. striped Panchputta, Do. BAHADOOR SHETT MUGGUNBHAEE KRAMCHUND.

4946. [915] Figured Do. with stamped figure, 2R. 4A. Ahmedabad. SHETT JAYSINGBHAEE HUTTEESING.

4947. [916] Do. black, Do. Do.

* Central Committee, *Lahore.*

4948. [917] Figured silk, crimson, Ahmedabad. SHETT JAYSINGBHAEE HUTTEESING.

4949. [2606] Do. checked, Mysore. GOVERNMENT of MYSORE.

### III.—HANDKERCHIEFS.

4950. [8231] Silk piece for handkerchiefs (Bhootan), Assam. H. BAINBRIDGE, Esq.

4951. [7870] Do. Santipore. GOVERNMENT of INDIA.

4952. [2369] Do. red check, Mysore. GOVERNMENT of MYSORE.

4953. [2613] Do. Do. Do. Do.

### 'MUSHROO' OR SATINS WITH COTTON BACK.

A satin, the back or warp of which consists of cotton.

4954. [1746] Mushroo, Benares. GOVERNMENT of INDIA.

4955. [1747] Do. Do. Do.

4956. [933] Six pieces do., 2R. 4A. per yd., Ahmedabad. BAHADOOR SHETT MUGGUNBHAEE KRAMCHUND.

4957. [934] Do., 3R. 8A. per yd., Do. Do.

4958. [935] Do., 4R. per yd., Do. Do.

4959. [936] 'Hemroo' do., 2R. per yd., Do. Do.

4960. [944] 'Panchputta' do., 11A. 6P. per yd., Do. NUGGER SHETT PREMABHAEE HEMABHAEE.

4961. [945] 'Chumkhee' do. (red), 1R. per yd., Do. Do.

4962. [946] Do. (blue and white stripe), 1R. per yd., Do. Do.

4963. [947] Do. (white and red stripe), 1R. per yd., Do. Do.

4964. [948] Do. (Goolkhar), 1R. per yd., Do. Do.

4965. [949] Do. (Goolkhar Choondry), 1R. 6A. per yd., Do. Do.

4966. [697] 'Chumkhee, 1R. 6A. 10P. per yd., Kutch. H. H. the RAO of KUTCH.

4967. [698] Do., 2R. 7A. 1P. per yd., Do. Do.

4968. [699] Do., 2R. 7A. 1P. per yd., Do. Do.

4969. [702] 'Chumkhee,' 2R. 7A. 1P. per yd., Kutch. H. H. the RAO of KUTCH.

4970. [715] Do., 1R. 2A. 8P. per yd., Do. Do.

4971. [716] Do., 1R. 6A. 4P. per yd., Do. Do.

### TUSSUR SILK.

4972. [5081] Tussur silk piece plain, Beerbhoom. GOVERNMENT of INDIA.

4973. [5082] Do. Bhaugulpore. Do.

4974. [5093] Do. Do. Do.

4975. [5080] Do. shot, Do. Do.

4976. [8214] Do. Do. Do.

4977. [5089] Do. checked, Do. Do.

4978. [5090] Do. Do. Do.

4979. [8210] Do. Do. Do.

4980. [8211] Do. Do. Do.

4981. [8212] Do. Do. Do.

4982. [8213] Do. Do. Do.

4983. [5079] Do. coloured border and ends, Burdwan. Do.

4984. [8025] Do. Do. Do.

4985. [8209] Do. Do. Do.

4986. [9189] Do. Do. Do.

4987. [9190] Do. Do. Do.

4988. [9200] Do. plain, Cuttack. Do.

### MOONGA SILK.

4989. [7741] Moonga silk piece, Luckimpore. BABOO KISSERAM BOROOAH PESHKAR.

4990. [7743] Do. Do. BABOO MOHUN CHUNDER.

*Used as a mosquito curtain.*

4991. [8225] Do. Assam. LIEUT. W. PHAIRE.

### ERIA SILK.

4992. [5078] Eria silk piece, Assam. LIEUT.-COL. VETCH.

4993. [7742] Do. Do. BABOO JEWRAM DEKA.

4994-5. [8223-4] Two pieces, Do. LIEUT. W. PHAIRE.

4996. [8230] Two pieces, Assam. H. BAINBRIDGE, Esq.

**MEZANKOOREE SILK.**

4997. [5076] Mezankooree silk piece, bordered, Assam. LIEUT.-COL. VETCH.

4998. [8220] Mezankooree silk piece, bordered, Assam. GOVERNMENT of INDIA.

4999. [8233] Do. Do. H. BAINBRIDGE, Esq.

# CLASS XXI. — WOOLLEN AND WORSTED FABRICS, INCLUDING MIXED FABRICS GENERALLY.

## I.—YARNS.

The specimens of woollen yarns — chiefly those from the wool of the Kashmere goat — exhibited, have already been entered in Section B, Class IV. pp. 97–99.

## II.—WOOLLEN FABRICS, BLANKETINGS, ETC.

Under this head are placed 'cumblies,' a kind of small blanket or rug, generally of a coarse material, used as a covering at night during the cold season, or worn over the head or round the shoulders during the day.

The under-noted articles embrace a selection from the common woollens forwarded for exhibition on the present occasion. The remainder, for which space could not be found, have been deposited at the INDIA MUSEUM.

5000. [2638] Cumblies, white, 26R. Chittledroog. GOVERNMENT of MYSORE.

5001. [2629] Do. 4R. 6A. 4P. Astagram. Do.

5002. [2956] Black cumblies, check, Chittledroog. Do.

5003. [2658] Do. plain, Do. Do.

5004. [2648] Do. Do. Bangalore. Do.

5005. [1153] Do. Do. Do. Belgaum. GOVERNMENT of BOMBAY.

5006. [1160] Cumbly, black, 3R. 4A. Belgaum. Do.

5007-11. [7832-6] Five blankets, Jeypore.

5012. [8306] Felt blankets, Peshawur.

5013. [819] Felt cloth, 10R., Shikarpoor.

5014. [1656] Woollen horse cloth 'Taroo,' 3R. 12A., Umritsur.

5015. [1272] Cumbly or blanket, 1R. 8A. Dharwar. W. C. ANDERSON, Esq.

Commonly worn by all classes of natives as a protection from cold, heat, or wet.

5016. [1273] Do. striped, 1R. 12A. Do. Do.

Worn by Lingayut priests.

5017. [1274] 'Numbda,' 8R. Do. Do.

Felt used for sleeping on, and as a pad under saddles.

## III.—KASHMERE SHAWLS, ETC.

A large collection of Kashmere shawls has been forwarded. Those exhibited are the result of a careful selection from the very numerous examples sent by the different exhibitors.

The importance of this manufacture will be seen by reference to the Table (p. 219), which shows the value and, as far as can be ascertained, the quantity, in pieces, of the Kashmere shawls exported from India for the past eleven years.

The subjoined remarks are by the CENTRAL COMMITTEE, *Lahore.*

This is now by far the most important manufacture in the Punjaub: but thirty years ago it was almost entirely confined to Kashmere. At the period alluded to, a terrible famine visited Kashmere; and, in consequence, numbers of the shawl-weavers emigrated to the Punjaub, and settled in Umritsur, Nurpûr, Dinangar, Tilaknath, Jelalpûr, and Loodianah, in all of which places the manufacture continues to flourish. The best shawls of Punjaub manufacture are manufactured at Umritsur, which is also an emporium of the shawl trade. But none of the shawls made in the Punjaub can compete with the best shawls made in Kashmere itself; first, because the Punjaub manufacturers are unable to obtain the finest species of wool; and, secondly, by reason of the inferiority of the dyeing, the excellence of which in Kashmere is attributed to some chemical peculiarity in the water there. On receipt of the raw pashum or shawl wool, the first operation is that of cleaning it: this is done generally by women; the best kind is cleaned with lime and water, but ordinarily the wool is cleaned by being shaken up with flour. The next operation is that of separating the hair from the pashum; this is a tedious operation, and

TABLE SHOWING THE QUANTITIES (AS FAR AS CAN BE ASCERTAINED) AND THE VALUE OF KASHMERE SHAWLS EXPORTED FROM INDIA AND EACH PRESIDENCY TO ALL PARTS OF THE WORLD FROM 1850-51 TO 1860-61.

COUNTRIES WHITHER EXPORTED

| Years | Whence Exported | United Kingdom Quantity | United Kingdom Value | France Quantity | France Value | Other Parts of Europe Quantity | Other Parts of Europe Value | America Quantity | America Value | China Quantity | China Value | Arabian and Persian Gulfs Quantity | Arabian and Persian Gulfs Value | Suez Quantity | Suez Value | Other Parts Quantity | Other Parts Value | Total Exported to All Parts Quantity | Total Exported to All Parts Value |
|---|---|---|---|---|---|---|---|---|---|---|---|---|---|---|---|---|---|---|---|
| | | Pieces | £ | Pieces | £ | Pieces | £ | Pieces | £ | Pieces | £ | Pieces | £ | Pieces | £ | Pieces | £ | Pieces | £ |
| 1850-51 | Bengal | 1,258 | 15,339 | 3 | 9 | .. | .. | 168 | 1,316 | .. | .. | 3 | 14 | .. | .. | 26 | 221 | 1,458 | 17,499 |
| | Madras | 42 | 238 | .. | .. | .. | .. | .. | .. | .. | .. | 1 | 7 | .. | .. | .. | .. | 43 | 245 |
| | Bombay | 8,675 | 118,561 | .. | .. | .. | .. | .. | .. | 2 | 105 | 1,330 | 16,760 | 1,830 | 18,022 | 67 | 517 | 11,904 | 153,965 |
| | ALL INDIA | 9,975 | 134,738 | 3 | 9 | .. | .. | 168 | 1,316 | 2 | 105 | 1,334 | 16,781 | 1,830 | 18,022 | 93 | 738 | 13,495 | 171,799 |
| 1851-52 | Bengal | 359 | 4,758 | .. | .. | .. | .. | 119 | 1,017 | .. | .. | .. | .. | .. | .. | 19 | 154 | 497 | 5,929 |
| | Madras | .. | .. | .. | .. | 185 | 3,582 | .. | .. | .. | .. | .. | .. | .. | .. | .. | .. | .. | .. |
| | Bombay | 7,137 | 109,090 | 1 | 35 | .. | .. | .. | .. | 2 | 18 | 1,684 | 21,079 | 972 | 5,725 | 116 | 812 | 10,097 | 140,341 |
| | ALL INDIA | 7,496 | 113,848 | 1 | 35 | 185 | 3,582 | 119 | 1,017 | 2 | 18 | 1,684 | 21,079 | 972 | 5,725 | 135 | 966 | 10,594 | 146,270 |
| 1852-53 | Bengal | 1,048 | 13,606 | 1 | 57 | .. | .. | 249 | 601 | .. | .. | .. | .. | .. | .. | 2 | 43 | 1,300 | 14,307 |
| | Madras | 12 | 307 | .. | .. | .. | .. | .. | .. | .. | .. | .. | .. | .. | .. | 1 | 30 | 13 | 337 |
| | Bombay | 10,151 | 165,498 | .. | .. | 197 | 2,906 | .. | .. | 14 | 445 | 1,554 | 21,954 | 1,365 | 9,446 | 117 | 766 | 13,398 | 201,015 |
| | ALL INDIA | 11,211 | 179,441 | 1 | 57 | 197 | 2,906 | 249 | 601 | 14 | 445 | 1,554 | 21,954 | 1,365 | 9,446 | 120 | 839 | 16,711 | 215,659 |
| 1853-54 | Bengal | .. | 5,039 | .. | 3,512 | .. | 30 | .. | 2,732 | .. | 50 | .. | .. | .. | 10,377 | .. | 420 | .. | 22,110 |
| | Madras | .. | 236 | .. | .. | .. | .. | .. | .. | .. | .. | .. | .. | .. | .. | .. | .. | .. | 236 |
| | Bombay | 8,000 | 122,659 | 2 | 100 | .. | .. | .. | .. | 15 | 188 | 912 | 11,777 | 1,229 | 11,110 | 469 | 1,973 | 10,627 | 147,807 |
| | ALL INDIA | .. | 127,934 | 2 | 3,612 | .. | 30 | .. | 2,732 | 15 | 188 | 912 | 11,777 | 1,229 | 21,487 | 469 | 2,393 | 10,627 | 170,153 |
| 1854-55 | Bengal | 6,941 | 14,730 | 39 | 4,368 | .. | 70 | .. | 7,923 | .. | .. | .. | .. | .. | 12,652 | .. | 631 | .. | 40,494 |
| | Madras | .. | 348 | .. | .. | .. | .. | .. | .. | .. | .. | .. | .. | .. | .. | .. | .. | .. | 348 |
| | Bombay | .. | 134,567 | .. | 1,141 | .. | .. | .. | .. | 7 | 124 | 536 | 17,020 | 447 | 3,081 | 1,258 | 185 | 8,228 | 157,118 |
| | ALL INDIA | .. | 149,645 | .. | 5,509 | .. | 70 | .. | 7,923 | .. | 174 | 536 | 17,020 | 447 | 15,733 | 1,258 | 816 | 8,228 | 197,890 |
| 1855-56 | Bengal | .. | 14,332 | .. | 16,018 | .. | 10 | .. | 2,021 | .. | 77 | .. | .. | .. | 18,749 | .. | 152 | .. | 51,349 |
| | Madras | .. | 716 | .. | .. | .. | .. | .. | .. | .. | .. | .. | .. | .. | .. | .. | 26 | .. | 742 |
| | Bombay | 5,745 | 137,129 | 54 | 1,355 | .. | .. | 9 | 37 | 15 | 469 | 578 | 5,619 | 1,783 | 12,159 | 57 | 420 | 8,241 | 157,188 |
| | ALL INDIA | 5,745 | 152,167 | 54 | 17,373 | .. | 10 | 9 | 2,058 | 15 | 546 | 578 | 5,619 | 1,783 | 30,908 | 57 | 598 | 8,241 | 209,279 |
| 1856-57 | Bengal | 21 | 10,793 | 41 | 5,969 | .. | 495 | .. | 4,239 | .. | 185 | .. | 2 | .. | 11,851 | .. | 339 | .. | 33,873 |
| | Madras | .. | 139 | .. | .. | .. | .. | .. | .. | .. | .. | .. | .. | .. | .. | .. | .. | 21 | 139 |
| | Bombay | 11,479 | 216,975 | .. | 1,338 | .. | .. | .. | .. | 4 | 68 | 850 | 9,228 | 4,041 | 28,342 | 266 | 677 | 16,681 | 256,628 |
| | ALL INDIA | .. | 227,907 | .. | 7,307 | .. | 495 | .. | 4,239 | 4 | 253 | 850 | 9,230 | 4,041 | 40,193 | 266 | 1,016 | 16,681 | 290,640 |
| 1857-58 | Bengal | 25 | 2,246 | .. | 6,219 | .. | .. | .. | 2,640 | .. | 321 | .. | .. | .. | 392 | .. | 39 | .. | 11,857 |
| | Madras | .. | 245 | .. | .. | .. | .. | .. | .. | .. | .. | .. | .. | .. | .. | .. | 245 | .. | 245 |
| | Bombay | 6,788 | 169,038 | 384 | 18,796 | .. | .. | .. | .. | 27 | 149 | 1,102 | 6,407 | 1,957 | 19,173 | 554 | 1,953 | 10,812 | 215,516 |
| | ALL INDIA | .. | 171,529 | .. | 25,015 | .. | .. | .. | 2,640 | 27 | 470 | 1,102 | 6,407 | 1,957 | 19,565 | 554 | 1,992 | 10,812 | 227,618 |
| 1858-59 | Bengal | 43 | 7,785 | .. | 12,576 | .. | 280 | .. | 1,306 | .. | 8 | .. | .. | .. | 884 | .. | 135 | .. | 22,974 |
| | Madras | .. | 255 | .. | .. | .. | .. | .. | .. | .. | .. | .. | .. | .. | .. | 2 | 5 | .. | 260 |
| | Bombay | 10,265 | 220,772 | 1,145 | 40,325 | .. | .. | .. | .. | 4 | 115 | 2,529 | 17,808 | 944 | 7,180 | 177 | 593 | 15,064 | 286,793 |
| | ALL INDIA | .. | 228,812 | .. | 52,901 | .. | 280 | .. | 1,306 | 4 | 123 | 2,529 | 17,808 | 944 | 8,064 | 177 | 733 | 15,064 | 310,027 |
| 1859-60 | Bengal | 63 | 18,534 | .. | 12,513 | .. | 80 | .. | 1,501 | .. | 500 | .. | .. | .. | 1,972 | .. | 1,672 | 45 | 36,772 |
| | Madras | .. | 526 | 1 | 4 | .. | .. | .. | .. | .. | .. | .. | .. | .. | .. | 4 | 13 | .. | 543 |
| | Bombay | 8,339 | 182,038 | 245 | 4,296 | 9 | 230 | .. | .. | 4 | 155 | 1,749 | 22,984 | 963 | 4,858 | 186 | 952 | 11,495 | 215,513 |
| | ALL INDIA | .. | 201,098 | .. | 16,813 | .. | 310 | .. | 1,501 | 4 | 655 | 1,749 | 22,984 | 963 | 6,830 | 186 | 2,617 | 68 | 252,828 |
| 1860-61 | Bengal | 28 | 23,520 | .. | 10,748 | .. | 60 | .. | 2,731 | .. | 652 | .. | .. | .. | 1,542 | .. | 180 | .. | 39,433 |
| | Madras | .. | 300 | .. | .. | .. | .. | .. | .. | .. | .. | .. | .. | .. | .. | .. | .. | 28 | 300 |
| | Bombay | 13,598 | 266,944 | 1,133 | 19,144 | .. | .. | .. | .. | 14 | 437 | 1,984 | 16,505 | 451 | 7,665 | 109 | 665 | 17,289 | 311,360 |
| | ALL INDIA | .. | 290,764 | .. | 29,892 | .. | 60 | .. | 2,731 | 14 | 1,089 | 1,984 | 16,505 | 451 | 9,207 | 109 | 845 | .. | 351,093 |

the value of the cloth subsequently manufactured varies with the amount of care bestowed upon it. The wool thus cleaned and sorted is spun into thread with the common 'churka' or native spinning-machine. This is also an operation requiring great care. White pashumeea thread of the finest quality will sometimes cost as much as 2*l*. 10*s*. a lb. The thread is next dyed, and is then ready for the loom, a model of which is included in the collection. The shawls are divided into two great classes —1. Woven shawls, called 'Teliwalah;' 2. Worked shawls.

Shawls of the former class are woven into separate pieces, which are, when required, sewn together with such precision that the sewing is imperceptible. These are the most highly prized of the two. In worked shawls, the pattern is worked with the needle upon a piece of plain pashumeea or shawl cloth.

A woven shawl made at Kashmere of the best materials, and weighing 7 lbs., will cost in Kashmere as much as 300*l*.; of this amount, the cost of the material, including thread, is 30*l*., the wages of labour 100*l*., miscellaneous expenses 50*l*., duty 70*l*.

Besides shawls, various other articles of dress, such as chogas, or outer robes, ladies' opera-cloaks, smoking-caps, gloves, &c. are made of pashumeea.

Latterly great complaints have been made by European firms of the adulteration of the texture of Kashmere shawls; and there is no doubt that such adulteration is practised, especially by mixing up Kirmanee wool with real pashum. In order to provide some guarantee against this, it has been proposed that a guild or company of respectable traders should be formed, who should be empowered to affix on all genuine shawls a trade-mark, which should be a guarantee to the public that the material of the shawl is genuine pashum, especially as the Indian Penal Code provides a punishment for those who counterfeit or falsify trade-marks, or knowingly sell goods marked with counterfeit or false trade-marks.

At Delhi shawls are made up of pashumeea, worked with silk and embroidered with gold lace. A very delicate shawl is made of the wool of a sheep found in the neighbourhood of Ladak and Kûlu: the best wool is procurable in a village near Rampûr, on the Sutlej; hence the fabric is called 'Rampûr chudder.' Other woollen manufactures in the Punjaub are Peshawur chogas, made of the wool of the Damba sheep, and of camel's hair, and chogas made of Patti, or the hair of the Cabul goat.

5018. [4576] Shawl 'Chudder,' green, Umritsur. DAVEE SAHAI & CHUMBA MULL.

5019. [4575] Do., blue, Do. Do.

5020. [4574] Do., slate-colour, Do. Do.

5021. [4587] Do., crimson, Do. Do.

5022. [4590] Do. do., Do. Do.

5023. [4591] Do., green, Do. Do.

5024. [3676] Do. Kashmere, Do. Do.

5025. [3677] Do. do., Do. Do.

5026. [3678] Do. do., Do. Do.

5027. [3498] Shawl, Kashmere, Umritsur. DAVEE SAHAI & CHUMBA MULL.

5028. [3548] Shawl, Kashmere, Umritsur. DAVEE SAHAI & CHUMBA MULL.

5029. [3682] Choga, embroidered, Do. Do.

5030. [3684] Do., Do. Do.

5031. [3687] Do., Do.

5032. [3453] Cape, blue and white, Do. Do.

5033. [3466] Do., white, Do. Do.

5034. [3476] Do., Do. Do.

5035. [3469] Do., Do. Do.

5036. [3480] Do., Do. Do.

5037. [3691] Cap, Kashmere, Do. Do.

5038. [3693] Do. do., Do. Do.

5039. [3697] Do. do., Do. Do.

5040. [3698] Do. do., Do. Do.

5041. [3699] Do. do., Do. Do.

5042. [3700] Do. do., Do. Do.

5043. [4641] Bordering, do., Do. Do.

5044. [4642] Do. do., Do. Do.

5045. [4322] Shawl, Jamewar. ROY LALL CHUND, *Bahadoor*.

5046. [4314] Do. Kashmere, Do.

5047. [4312] Do. do. Do.

5048. [4308] Do. do. Do.

5049. [4311] Do. do. Do.

5050. [3546] Shawl, Kashmere, Umritsur. GOVERNMENT.

5051. [3547] Do. Do. Do. Do.

5052. [3441] Do. Do. BABOO MOHUN LALL.

5053. [3440] Do. Do. Do.

5054. [3853] Do., Jamewar. RAI HIVE, *Dyab of Umritsur*.

5055. [3668] Do., Kashmere. BAHAI KOLYAN SINGH.

5056. [3664] Do., Do. Do.

5057. [4506] Do., Do. MAHOMED SHAH SAPIDEEN.

5058. [4520] Shawl, Jamewar. MAHOMED SHAH SAPIDEEN.

5059. [4501] Do., Kashmere. Do.

5060. [4549] Do., Jamewar. MUNSEE RAM & KUNSHEE, *Loodianah.*

5061. [4556] Do., Do., Do. Do.

5062. [5278] Do., Kashmere. NARAIN DOSS.

5063. [5646] Do., Do. GOVERNMENT.

5064. [1683] Do. Scarf, Dacca. HURMOHUN ROY.

5065. [1685] Do. Do. Do.

5066. [1682] Do. Do. Do.

5067. [5281] Do. Loodianah. FUTTEH BUTT.

5068. [5282] Do. Do. Do.

5069. [5280] Shawl, Do. Do.

5070. [5279] Scarf. SOOBHAM JOO.

5071. [5647] Do., Mooltan.

5072. [5648] Do., Do.

5073. [8303] Chudder, Rampore, Serinugger.

5074. [8304] Shawl, Do., 'Kangra,' Do.

5075. [8305] Do., Do., 'Tusha,' Do.

5076. [9413] Do., Kashmere. Imported by Messrs. FARMER & ROGERS, *Regent Street.*

### III.—SILK SHAWLS.

5077. [2603] Silk shawl. GOVERNMENT of MYSORE.

5078. [2604] Do. Do.

5079. [2605] Do. Do.

### IV.—MIXED FABRICS, SOOSEES, ETC.

5080. [3776] Soosee, brown, 15¼ yds, Umritsur.

5081. [3768] Do., green and white stripe, Do.

5082. [3772] Do., dark purple stripe, Buttala, Umritsur.

5083. [3773] Do., yellow, do., Do.

5084. [3771] Do., do., Do.

5085. [3769] Do., dark green, do., Do.

5086. [3774] Do., purple, do., Do.

5087. [3775] Do., white, red stripe, Do.

5088. [3770] Do., Do., green stripe, Do.

5089. [1033] Do., Hyderabad.

5090. [1034] Do., Do.

5091. [855] Elacha, cotton and silk piece, Ahmedabad.

5092. [5547] Moonga, silk and cotton piece, Assam. W. A. O. BECKET, Esq.

# CLASS XXII. — CARPETS, RUGS, AND MATS.

The manufactures embraced under this class are likely to eventually occupy an important position amongst articles of export from India to this and other countries.

The substance of the following interesting remarks has been furnished by A. M. DOWLEANS, Esq.

## I.—CARPETS.

The chief places in which carpets are manufactured, are Lahore, Meerut, Bareilly, Jubbulpore, Gorruckpore, Mirzapore, Rungpore, and Benares, in the Presidencies of Bengal, the North-West Provinces and the Punjaub, and at Masulipatam in the Madras Presidency.

At Lahore, Meerut, and Bareilly, the manufacture is solely carried on by prisoners in the jails; but as it has only been commenced since the introduction of the new prison discipline, the annual production is limited; so far, however, as quality is concerned, the carpets are excellent. They can be made there of any size and pattern, the average price being from seven to nine shillings per square yard. The great drawback to the exportation of carpets from the above places is the heavy expense of inland transport, which, however, will remedy itself as soon as the three great lines of railway, now in the course of construction, have been completed.

At Jubbulpore, the manufacture of carpets, rugs, and suttringees (cotton carpets), has been regularly carried on for years; chiefly in jail, where Thugs and other prisoners are extensively employed upon them. The Jubbulpore carpets are considered of extremely good texture, and are remarkable for their cheapness. The annual consumption, though large, is limited to a comparatively small area. The nearest place to which they are at present conveyed is Mirzapore, on the Ganges, whence they are forwarded by river boats to Calcutta. The expense of transport from Jubbulpore to Calcutta, including duty and agency charges, &c., is very nearly 90 per cent. on the original cost of the articles at the place of production.

When the railway from Jubbulpore to Bombay is completed, the cost of bringing these carpets to Bombay will be reduced to about 20 per cent. on the cost price, so that a carpet costing at Jubbulpore 10*l.* will be capable of being landed in London for 14*l.* at the outside. The Jubbulpore School of Industry, as it is called, receives direct orders for any amount of carpets, and their fixed prices are as follows:—

| | | |
|---|---|---|
| Turkey carpets . | . 4R. or | 8*s.* 0*d.* per sq. yd. |
| Scotch carpeting | . 1R. 8A. or 3*s.* 0*d.* per yd. 33 feet wide. | |
| Suttrinjees . | . 1R. 2A. or 2*s.* 3*d.* per sq. yd. | |
| Kidderminster | . 1R. 2A. or 2*s.* 3*d.* per yd. 1 yard wide. | |

At present, however, the places which supply the greater portion of India, as well as the export demand, are Mirzapore and Benares. There is no specific price per yard, as carpets, both at Mirzapore and Benares, are generally sold at so much a piece. The Mirzapore carpets are noted for excellent staple and durability of wear, but are dearer than those from Jubbulpore, though for purposes of export they are cheaper, as the place is situated on the Ganges, and has, therefore, the advantage of easy transport to Calcutta. When the railways come into full operation, the carpets of Mirzapore and Benares will be, in all probability, superseded by those of Meerut, Bareilly, Lahore, and Jubbulpore. The manufacture of carpets

is also carried on at Gorruckpore; they are, therefore, more expensive than those of the neighbouring districts of Mirzapore and Benares. In the Madras Presidency, Masulipatam is the chief seat of the manufacture. The trade is carried on to a considerable extent, and entirely by natives, who, as in Bengal, combine it with agricultural undertakings adapted to the season of the year.

The above remarks apply exclusively to carpets not less than 10 feet square.

## CARPETS EXHIBITED.

5093. [6487] Carpet made by the Thugs, Lahore. MAJOR MCANDREW.

This carpet is made from the common wool of the Punjaub, obtained from sheep kept in the waste pasture lands of the Lahore district (called the Bah); the thread was spun, dyed, and made up into this carpet at the School of Industry attached to the Thuggee Department at Lahore; the spinning was done by the wives of Thug approvers, and the weaving of the carpet by eight boys, sons of approvers, the eldest of whom is only fourteen years of age.

Approvers are men who have been tried and convicted as having belonged to a band of Thug murderers, but who, having made a full confession of their crimes (in some individual cases amounting to the murders of as many as eighty persons) and denounced their associates, have received a conditional pardon.*

5094. [3296] Large Persian carpet, Lahore.

Made in Mooltan Jail.
Ornamental carpets of thread, with a woollen and sometimes with a silken pile, are made up in Mooltan, Peshawur, Umritsur, Bhawulpore, and Kashmere. Those of Mooltan are perhaps most celebrated.†

5095. [2896] Carpet, Kashmere, Lahore. M. DALLAS, Esq.

5096. [5824] Do. large woollen, Meerut Jail. GOVERNMENT of INDIA.

5097. [5981] Do. Do. Do.

5098. [5345] Do. 15 feet square, Jubbulpore School of Industry. Do.

Made by Thug prisoners.

5099. [5346] Do. Do. Do.

5100. [10198] Do. Bokhara. NAZIR BHARRULL, *Khan of Peshawur.*

5101-2. [9414-5] Do. Masulipatam. Messrs. WATSON, BONTOR, & Co., *London.*

5103. [9416] Do. Indian manufacture. Messrs. ROBINSON & Co., *London.*

---

* Central Committee, *Lahore.*
† Central Committee, *Lahore.*

5104. [1493] Carpet, imitation of Kidderminster, Benares Jail. Dr. N. H. Cheke, *Superintendent.*

## II.—Rugs.

The manufacture of rugs is very extensive and comprises many localities. At Peshawur, Bareilly, Shahpore, Sealcote, and Sirsa, the manufacture is entirely confined to the jails. The places, however, where a regular manufacture and trade are carried on, are, Benares, Mirzapore, Allahabad, and Gorruckpore in Bengal; North Arcot, Tanjore, Ellore, and Malabar in the Madras Presidency; and also at Mysore, as well as at Shikarpore, Kyrpore, and Hyderabad in Sinde. Those of Bengal commend themselves by extraordinary cheapness; they are extensively used throughout India, and also somewhat largely exported. In point of texture and workmanship, however, the rugs from Ellore, Tanjore, and Mysore, though they are comparatively much dearer, are greatly preferred.

The employment of rugs throughout India is most extensive, as every native who can afford to purchase one uses it to sit upon and smoke his hookah. It is impossible to form an estimate of the annual value of this manufacture, as only the small portion exported is entered in the official records, and as no steps have hitherto been taken to ascertain the local trade. The rugs made in Bengal vary in length from 3 to 3½ feet; their average width being 1¾ feet, and their value from 1*l.* to 1*l.* 10*s.* The rugs from Ellore, Tanjore, and Mysore are made of various sizes, and are valued from 2*l.* to 4*l.* each; those from Shikarpore and Kyrpore as well as from Hyderabad (Sinde) are of a lighter texture, but excellent workmanship; their width is generally uniform, but in length and consequent cost they vary from 2*l.* to 5*l.* each.

The finest articles of this description, however, are the silk rugs from Tanjore and Mysore, the blending of colours and workmanship being excellent. They are made of all sizes, up to even in squares of 10 feet; but being too costly for general adoption, this manufacture is very limited.*

5105. [5985] Rug, small square, Agra. Government of India.

5106. [5284] Do. Patna. Do.

5107. [5987] Do. for Buggy (a kind of gig), Sealcote Jail. Do.

5108. [3255] Do. Allahabad Central Jail. Do.

5109-10. [852-3] Three rugs, Persian, Shikarpoor, Sinde. Do.

5111. [5986] Rug, large. Do. Do.

5112. [2523] Do. small, Ahmedabad. Do.

5113-4. [10527-8] Two rugs, North Arcot, Madras. Subbi, Chetti, & Co.

5115. [2659] Rug, large, Bangalore, Mysore. Government of India.

5116-7. [10667-8] Two rugs, Ellore, Madras. Do.

5118. [5984] Two rugs. Messrs. Robinson & Co., *London.*

5119. [5285] Rug of silk, Bhawulpore. H. H. the Nawab.

The work of Sadige and Ola Buksh, of Khairpore.

5120. [5983] Do. do., Madras. India Museum.

5121. [587] Do. do. Tanjore. Saccarim Sahib, son-in-law of the late Rajah of Tanjore.

5122. [3256] Rug of aloe fibre, Allahabad Central Prison. Government of India.

5123. [10337] Do. dyed do., Bareilly. Dr. J. M. Cuningham.

## SUTTRINGEES, OR COTTON CARPETS AND RUGS.

These fabrics, which are entirely made of cotton, may be considered a cheap substitute for woollen carpets. They are used by every one, European or native, throughout India, and the annual manufacture is consequently very considerable, especially in Bengal, where they form a large and important branch of inland trade. They are of all sizes, from that of the largest carpet to the smallest rug, but generally of one and the same pattern throughout India, the only difference being the colour. Blue and white, and red and white, stripes constitute the prevalent patterns, but in some one colour of darker and lighter hues is employed. In Meerut, Bareilly, and Patna, new patterns have of late been tried with considerable success, but though preferred by the Europeans, are not by natives, who like the striped patterns because they wear better in daily use, and do not lose the freshness of colour by washing. The principal localities where suttringees are manufactured are Agra, Bareilly, Patna, Shahabad, Beerbhoom, and Burdwan. Those manufactured at Agra are considered the best, and the value of its annual production is about 10,000*l.* In Shahabad, the quantity manufactured last year was nearly 7,000*l.*; and the same may be assumed to have been produced in the other places above-mentioned. Suttringees vary in price according to size and quality. The small ones are valued from 3*s.* to 15*s.*, and the larger ones (carpet size) from 1*l.* 10*s.* to 4*l.*, the price in many cases being regulated by weight.*

5124. [5564] Suttringee carpet of cotton, Patna. Government of India.

5125. [10001] Do. or Dhurree, Bhawulpore. H. H. the Nawab.

5126. [1490] Do., cotton, Benares Jail. Dr. N. H. Cheke, *Superintendent.*

5127. [1491] Do., English thread, Do.

5128. [10197] Do., large, of dyed cotton, Bareilly Jail. Dr. J. M. Cuningham, *Superintendent.*

5129. [10338] Do. or Dhurree, Do. Do.

5130. [5567] Suttringee carpet, 'hemp cotton,' Meerut Jail.  DR. J. N. CUNINGHAM.

5131. [5565] Do., do., Do.  Do.

5132. [8293] Do., do., Luckimpore.  BABOO KISSERAM, *Darojah*.

5133. [3297] Do., common quality, Lahore.  Chief of the LOGHASEE TRIBE.

5134. [5988] Do., blue stripe, Patna.  GOVERNMENT of INDIA.

5135. [5990] Do., white and blue border, Do.  Do.

5136. [10563] Do., do., N. Arcot, Madras.  SUBBI, CHETTI, & CO.

5137. [10587] Do., do., Do.  Do.

5138. [5982] Do.

———

5139. [1291] Suttringee carpet, blue, Dharwar, 4R. 6A.  W. C. ANDERSON, Esq.

Used for sitting and sleeping on.  Made of all sizes.

5140. [2954] Do., Shahabad.  R. W. BINGHAM, Esq.

5141. [2930] Striped rug.  Do.

5142. [2931] Do. Do.  Do.

These are made wholly of cotton, and almost invariably striped.  From being made of cotton they are cool and pleasant, and are in invariable use by the better class of natives and by all Europeans.  The smaller kinds are used as quilts for beds, and of late the Government has given them to its European soldiers for that purpose. The manufacturers, called in this district Kalleeun Bap, are almost invariably Mussulmen of the weaver class, who will make carpets of any size and pattern given in stripes. The two local seats of manufacture in Shahabad are Bubbooah and Sasseeram.  In the former place, from 10,000 to 12,000 rupees worth are yearly manufactured and sold, and in the latter from 30,000 to 40,000 rupees. These dhurrees or carpets are sold readily in all the bazaars around, and at all the neighbouring fairs, particularly at Berhampore, and Hurrier Chutter, or Sonapore; probably two-thirds of the whole quantity made are exported from the district, while the annual expenditure in the district will vary from 20,000 to 25,000 rupees worth per annum.

The dhurrees or carpets generally made for sale are the following:—

I.—6 yards long and 2 yards broad, thick and strong, of any colour, sold at from 6R. to 6R. 8A.

II.—A small kind used as quilts, or to spread in lieu of any other bedding on the ground.  They weigh from 2 to 3 lbs. each, and are 1¼ to 1½ yards broad, by about 2 yards long: they sell at from 14A. to 1R. 8A. each, according to thickness and quality.  (The specimen accompanying is 14A. only.)

III.—*Hauzhassica.*—This is the better kind of carpet, and often displays much taste in the arrangement of the striped colours.  It is made of any size to fit any room, and is always sold by weight.  The price varies according to quality from 1R. 4A. to 1R. 12A. and sometimes as high as 2R. 4A. per seer.  It is sold in all the fairs and in all the large cities around, such as Patna, Ghazeepore, Daodnuggur, Gyah, &c.  No merchant's or banker's

shop, or rich native's reception room, is complete without these being spread.

This is the kind generally used by Europeans for their drawing and public rooms.

IV.—Is a small kind of carpet made for use in zemindarree and other small cutcherries, and much used from its portability.  It is from 3 to 4 yards long, and from 1½ to 2 yards broad, and sells at from 3 to 4R. each carpet.  It is generally made from five colours, from which cause it obtains the name of Dhurree Panch Rungha.

Any other description wanted are made, but these are the principal in use.  The supply of these articles is only limited by the local demand.  MR. BINGHAM thinks that Manchester might manufacture them with great advantage, and by copying and improving on the native patterns command a very large sale indeed.  If Manchester would make these articles in long webs and in all widths and patterns, she would be certain to drive the native manufacture from the market to other trades, and command a valuable trade all over India for herself; while the superior stiffness, thickness, and quality of Manchester goods would, as in the matter of her calicoes and cotton, surely but slowly supersede the native manufacture altogether: but to do so, it is important to work from native patterns.  The natives are a people of routine even in their carpets, and would not patronise sudden changes in the patterns and colours to which they had been used from childhood.  But there yet exists abundant ground for superior work, and for Manchester to improve on the established native patterns, as Europeans (and they would probably consume one-third of the supply) would be glad to have some other than the monotonous stripes of native manufacture; and as other patterns come into use among them, they would slowly but surely find their way among the native population.

*Gulleecha*, or carpets.—These are only manufactured in Sasseeram, and are almost always woollen, of florid but neat patterns, in imitation of the Persian carpet.  They are used to a considerable extent by the rich natives in their zenanas and by Europeans also.  The size usually manufactured is 2 yards long by 1 yard broad, and they sell at from 2R. to 4R. 8A. per carpet.  Any other sizes and patterns can be made according to order, and some of the patterns are extremely pretty.

The European carpet manufacturer could not compete with these as to price and actual value, as the wool costs but little in this country, and the native dyes answer admirably for the purpose, while also the coarse local wools, which would not pay for exportation, answer for carpet work.  The colours are harmonious, and I have but little doubt that it would pay any enterprising merchant to export these to Europe.  The annual manufacture at present in Sasseeram is about 10,000 to 12,000 R.

V.—Another kind, in imitation of the above, but wholly of cotton, is also made: prices nearly the same. The patterns are pretty, but they rapidly become spoiled by dirt and dust.  They are invariably made of only two colours, blue and white.*

———

## Division II.—MATTING OF HEMP, COCOA NUT FIBRE, STRAW REEDS AND GRASSES, FOR FLOORS OR WALLS.

In point of value the export of mats from India to this and other countries, as will be seen from the following Table, is not yet of great importance.

———

* R. W. BINGHAM, Esq.

TABLE SHOWING THE VALUE OF MATS EXPORTED FROM INDIA AND EACH PRESIDENCY
TO ALL PARTS OF THE WORLD FROM 1857-58 TO 1860-61.

| YEARS | WHENCE EXPORTED | COUNTRIES WHITHER EXPORTED | | | | | | TOTAL EXPORTED TO ALL PARTS |
|---|---|---|---|---|---|---|---|---|
| | | UNITED KINGDOM | FRANCE | AMERICA | CHINA | ARABIAN AND PERSIAN GULFS | OTHER PARTS | |
| | | Value | Value | Value | Value | Value | Value | Value |
| 1857-58 | Bengal. . . . | £ 115 | £ 47 | £ .. | £ 39 | £ 26 | £ 474 | £ 701 |
| | Madras . . . | 7 | .. | .. | .. | 3 | 707 | 717 |
| | Bombay . . . | .. | 5 | .. | .. | 23 | 130 | 158 |
| | ALL INDIA . . . | 122 | 52 | .. | 39 | 52 | 1,311 | 1,576 |
| 1858-59 | Bengal. . . . | 21 | 101 | 1,058 | 164 | 2 | 317 | 1,663 |
| | Madras . . . | 4 | .. | .. | .. | 25 | 697 | 726 |
| | Bombay . . . | .. | .. | .. | .. | 10 | 106 | 116 |
| | ALL INDIA . . . | 25 | 101 | 1,058 | 164 | 37 | 1,120 | 2,505 |
| 1859-60 | Bengal. . . . | 16 | 21 | 912 | 54 | 11 | 332 | 1,346 |
| | Madras . . . | 23 | .. | .. | .. | 3 | 568 | 594 |
| | Bombay . . . | 5 | .. | .. | .. | 18 | 416 | 439 |
| | ALL INDIA . . . | 44 | 21 | 912 | 54 | 32 | 1,316 | 2,379 |
| 1860-61 | Bengal. . . . | 65 | 178 | 695 | 46 | .. | 506 | 1,492 |
| | Madras . . . | 11 | .. | .. | .. | .. | 524 | 535 |
| | Bombay . . . | 25 | .. | .. | .. | 18 | 24 | 67 |
| | ALL INDIA . . . | 101 | 178 | 695 | 46 | 18 | 1,054 | 2,094 |

The internal trade in mats is, however, very extensive, as they are in universal use by both Europeans and natives, and are therefore made of kinds and varieties to suit everybody's taste and means. Europeans use only the better kinds of mats, and almost exclusively for the covering of floors in their houses, but natives employ them for a variety of other purposes, such as to sleep upon, smoke, &c. Every Mahomedan, however poor, after having performed the prescribed ablutions, spreads a small mat before him, while saying his prayers. The Hindoo uses it as a sort of table-cloth; in many a poor hut it constitutes the only piece of furniture perceptible.

Though mats are made in almost every part of India, the finest kinds are manufactured at Midnapore, near Calcutta. These are only manufactured to special order, but can be made of any size required. The piece varies according to the size of the border, which is coloured either red or black, and the large mat, No. 5143, 25 feet square, cost at Midnapore 30*l*. Smaller mats may be valued in proportion. But besides these extremely fine mats, a description is manufactured, of which considerable numbers are exported to Madras, Bombay, Mauritius, and South Australia; these are much cheaper, and a good strong mat, about 20 feet square, may be had for 4*l*. if plain, and 5*l*. 10*s*. with a black or red border. The mats next in point of fineness are those from Jessore, also in the vicinity of Calcutta, and called Sittulputtee; these, however, are never made, if Indian, of the size of an entire room-floor, but only in the shape of rugs, and have invariably a red border, sometimes also a red-flowered centre. They are generally made about 4 to 5 feet long, and 2 broad, and cost from 2*l*. to 3*l*. each. At Hooghly, near Calcutta, an inferior kind of small mat is made, of which very large quantities are exported by the emigrants to Mauritius and Demerara, and lately several shipments have been made to New South Wales. The largest variety of small mats is, however, made in the Madras Presidency; North Arcot, and the whole of the Malabar coast, are celebrated for these handsome fabrics. There are at least 200 varieties of design and colouring, the price varying from 3*s*. to 3*l*. per mat, according to quality and length. All mats in India are made by a special caste, who devote themselves exclusively to that description of manufacture. There are no statistical records to show the number and value of mats annually manufactured, but if it be considered that everybody, high or low, rich or poor, uses some kind of mat, it can easily be imagined that a very large number

of people must be employed in making mats to supply the demand, not only of the immense local population, but also that for export.

### MATS EXHIBITED.

5143. [2925] Large mat, Midnapore. H. B. COCKERELL, Esq.

This mat is made of the rushes exhibited under reeds and grasses, specimens of which have been exposed to the sun for three days; when about to be used they are soaked in water for an hour, and then split into thin strips, as shown in the specimen. It is made more or less fine, according to the quality of the mat required. If the border of the mat is to be coloured, the rushes are dipped into a red dye to the necessary depth. This process of manufacture consists in plaiting the rushes thus prepared on threads highly strung between two bamboos, a sley being used, as in weaving, for compressing them tightly together. The finest kind of mats take from one to six months to manufacture, both the time occupied and the cost depending on the size required. Small mats are much used by the natives, but of an inferior quality, and of much cheaper descriptions than those sent. Mats of this kind are exported largely to Calcutta. They can be made of any pattern. Price, 300R.*

5144-8. [5415-19] Mushnud mats, Midnapore. GOVERNMENT OF INDIA.

5149. [5420] Seetalputtee mat, Calcutta. R. D. TURNBULL, Esq.

5150. [5421] Do., Jessore. BABOO RAMDHONE GHOSE.

5151-3. [5422-4] Mats, Hooghly. GOVERNMENT OF INDIA.

Made from the Katee reed, at Mundul Ghât.

5154. [5897] Grass mat, 'Punch Kungee Mandra,' Nepaul. H. H. SIR JUNG BAHADOOR, K.C.B.

* H. B. COCKERELL, Esq.

5155-6 . [8311-7] Mats and matting, Malabar.

5162-88. [8319-45] Do. Do.

Some of these mats (contributed by the GOVERNMENT of MADRAS) were manufactured at Pulghaut, Malabar; the others are of a similar character, but are contributions from the INDIA MUSEUM.

5189-95. [10529-35] Mats, North Arcot.

There are 63 mat weavers in Wandawash, six of whom are reported to be skilful workmen. The manufacture of this article is not carried on so largely as it used to be, owing to a series of unfavourable seasons. The reeds or grass of which these mats are made grow in Kasba Wandawash, on a kani of land, which is situated in the vicinity of a tank. They are also largely cultivated in Palle-Konda, Pondicherry, and Cuddalore in the South Arcot District, on river-banks or river Poramboke, &c. At Wandawash a kani of land would yield a produce of two bandy-loads of grass, if the season be favourable. The price of a bandy-load at Pallikouda is 30R., exclusive of the bandy hire thence to Wandawash, viz., 7R. or thereabouts. A superior kind of mat grass is to be had at Véláni, Tanjore District, but the charges are so heavy that they prohibit import.

5196. [4864] Screw pine matting, Madras. DR. J. SHORTT.

5197. [4861] Rush mat, Madras. Do.

5198. [10323] Mats of wild date leaf, Chota Nagpore.

Made all over the district, and in universal use among natives.*

5199. [6709] Coir mats, Singapore. COL. COLLYER.

5200. [6710] Rattan mat, Do. Do.

These mats were manufactured by prisoners at Singapore.

5201. [6335] Mats made of the leaf of the Pandong rush, Lubeck, on the Coast of Java. H. H. INCHE WAN ABOO BAKAR.

5202. [6361] Mats, Malacca. CHEE YAM CHUAN.

5203. [6414] Do. Do.

Made by Malay women, from palm leaves.

5204-5. [9382-3] Large mats. Messrs. ROBINSON & Co., *London.*

A curious application of ivory is exemplified by the mat next entered:—

5206. [5760] Mat made of ivory strips, Chittagong. H. H. the RAJAH of TIPPERAH.

* Local Committee, *Chota Nagpore.*

# CLASS XXIII. — WOVEN, SPUN, FELTED OR LAID FABRICS, WHEN SHOWN AS SPECIMENS OF PRINTING OR DYEING.

The various specimens exhibited under this head are not numerous, and, as a whole, unimportant.

5207. [5103] Cloth dyed with Koosum flowers, Cuttack.

5208. [10331] Do. Goolabee, dyed with Koosoom, Allahabad.

5209. [10328] Soorkhee, do. indigo. Do.

5210. [10329] Peazee, do. Do. Do.

5211. [10336] Do. do. Koosoom, Do.

5212. [10327] Kasnee do. Koosoom and indigo, Do.

5213. [10334] Narungee, do. mixture of Koosoom, Do.

5214. [10335] Baijanee, do. Koosoom and indigo, Do.

5215. [10330] Budawee, do. Koosum and Hursinghar, Do.

5216. [10325] Khuskhus colour, do. Koosum and a little indigo, Do.

5217. [10326] Sonlay, dyed with Koosum and Hursinghar, Allahabad.

5218. [10332] Goolanar, do. mixture of Koosum, Do.

5219. [10333] Chumpyee, do. Do. Do.

5220-2. [5557-9] Roomals, shown for dyeing, Ulwar. H. H. the RAJAH.

5223-4. [5245-6] Do. do. Do.

5225. [6520] Cloth dyed with Roam dye, Assam. W. E. WAGENTRIEBER, Esq.

5226. [5622] Specimens of dyeing produced in Nepaul. H. H. SIR JUNG BAHADOOR, K.C.B.

A paper-book, containing patterns of cloth, showing the various dyes produced in Nepaul.

About the authenticity of this production there exists some doubt. The only person who professes to be able to produce the colours is a dyer in the employ of MAHARAJAH SIR JUNG BAHADOOR, K.C.B., who states that the book (which contains many colours that he cannot reproduce) was the result of repeated mixtures in various proportions of two or more of the following dyes:—

1. Bukkum, or sappanwood (*Cæsalpinia sappan*).

2. Al (*Morinda citrifolia*).
3. Lac Dye.
4. Buhera (*Terminalia belerica*).
5. Hurra, Myrobalan nut (*Terminalia chebula*).
6. Koossoom, safflower (*Carthamus tinctorius*).
7. Huldee or turmeric (*Curcuma longa*).
8. Hursinghar, Weeping Nyctanthes (*Nyctanthes arbortristis*).
9. Nil, indigo.
10. Potash (*Butea frondosa*).
11. Khyr (*Mimosa catechu*).
Also blue vitriol; iron'; Nepaul madder; lime; yellow ochres earth; and a preparation of the vetch called *Oord* or *Dolichos pilosus*.

5227. [10561] Dyed cloth, North Arcot, VENKATA RAO & BAPANA RAO.

5228. [10552] Do. blue, Do.      Do.

5229-33. [10547, 49, 51, 59, 62] Do. red, Do. Do.

5234. [10558] Do. green, Do.      Do.

5235. [10555] Do. red, Do.      Do.

5236-7. [10556-7] Do. do., Do. GOVERNMENT of INDIA.

5238-42. [10550-54] Dyed cloth, red, North Arcot. VENKATA RAO & BAPANA RAO.

5243. [10560] Do. blue, Do.      Do.

5244. [5544] Thrown silk, seven specimens of dyeing, Burmah.

5245. [4006] Kapila silk thread, Salem.

5246. [4007] Kiruvunga do., Do.

5247. [807] Red cotton cloth, Kutch. H. H. the RAO.

5248. [10548] Dyed cloth, North Arcot.

5249. [5105] Cloth dyed with *Cæsalpinia sappan*, Cuttack.

5250. [5106] Do. Kamba goonda, Do.

5251. [709] Do. yellow with pomegranate bark, Kutch.

5252. [978] Cloth dyed with indigo, Jhansee. BALMOK CHUND.

5253. [9417] Series of cotton thread and cloth dyed of various colours, Madras.

5254. [9266] Patterns of English Madderpauts, dyed at Peethapoor, in the Mahee Kanta, expressly for the use of the Siamese.

The dyers are supplied with cloth by the Bombay merchants. Cost of dyeing, 1*d.* per yard.

5255. [9268] Patterns of Madderpaut prints, stamped at Wasna Zillee Baweesee, in the Mahee Kanta.

The export of these to Siam is increasing, as is proved by the additional number of dyers employed within the last few years at Peethapoor and Wasna.

5256. [979] Cloth dyed with Al, Jhansee. SHUMSHERE, dyer.

5257. [10305] Four specimens of Saloo cloth, Banda. H. H. DASHWOOD, Esq.

This cloth is dyed with the Al-root, with a mixture of castor oil in the proportion of one *pâo* to every piece of cloth, each piece of cloth being eight yards. Besides castor oil, 'Russee,' a kind of earth, is also mixed, and goats' dung and alum. The cloth is first rubbed for ten days in the castor oil, 'Russee,' and goat's dung, and then dried in the sun. After ten days it is well washed and dried, and then steeped in the oil for five days; afterwards washed and dried in the sun, and after a third application of soap and water the cloth is ready for sale. The cost of dyeing different kinds of cloth is as follows:—

> Dyeing Long cloth, 1¼A. per yard.
> Ditto Nynsook cloth, 1A. per yard.
> Ditto Mulmul cloth, ¾A. per yard.
> Ditto Pugrees, ½A. per yard.

It is not easy to ascertain the extent and value of the quantity of saloo cloth annually manufactured. It is not confined to local consumption, but exported to other parts of India, and its use is general, and not limited to particular castes. The wholesale market value is about 1R. 6P. 6P. per piece, according to the quality of the cloth dyed.*

* H. H. DASHWOOD, Esq.

# CLASS XXIV.—TAPESTRY, LACE, AND EMBROIDERY.

In the following arrangement of this important class, the Jury Directory has, as far as practicable, been complied with; the great variety of Indian articles of embroidery has, however, led to the adoption of the following classification or grouping :—

I.—TAPESTRY, COUNTERPANES, QUILTS, ETC.

5258. [1666] Counterpanes, embroidered, Puttiala. H. H. the RAJAH.

5259. [3803] Do.      Do.      Do.

5260. [5108] Counterpanes, Cuttack. GOVERNMENT of INDIA.

II.—ORNAMENTAL TAPESTRY OF SILK, WOOL, MOHAIR, LINEN, COTTON, AND OF THESE MATERIALS MINGLED TOGETHER, OR WITH METAL WIRES, WHETHER WOVEN IN THE LOOM OR OF ANY KIND OF NEEDLEWORK.

The only Indian manufactures answering to the above are the 'Kincobs,' or loom-made fabrics of silk and with gold and silver wire, which, although only employed in India as articles for personal wear, might be used in this country for covering chairs, couches, &c.

5261. [3309] Kincob, light blue, gold and silver, Benares. MOHUN LALL.

5262. [3308] Kincob, blue and gold, Benares. MOHUN LALL & CHITOO LALL.

5263. [3307] Do., white and gold, Do. SILHUT CHUMDRABHAN.

5264. [3306] Do., dark blue, gold and silver, Do. Do.

5265. [3310] Do., gold and silver, Do. DABEE PERSHAUD.

5266. [951] Do.. Ahmedabad. NUGGER SHET PREMABHAEE HEMABHAEE.

5267. [899] Do., green and gold, Do.

5268-70. [896-8] Do., red and gold, Do.

5271-2. [895, 950] Do., Do.

5273. [952] Do., green and gold, Do. NUGGER SHET PREMABHAEE HEMABHAEE.

5274. [1235] Do., Surat.

5275. [1229] Luppoo Rosperi, silver tissue, Do.

5276. [1228] Luppoo Sooneri, tissue, red and gold, Do.

5277. [1237] Kincob, red and gold, Do.

5278. [1236] Do., green and gold, Do.

5279. [1233] Do. Do.

5280. [1234] Do. Do.

### III.—LACE.

5281. [10601] Lace, black, No. 130, 10 yds. *Per* COLLECTOR of TINNEVELLY.

Manufactured at Mission Station, Edayangudi. Directress, MRS. CALDWELL.

5282. [10602] Do. do., No. 132, 6 yds. Do.

5283. [10603] Do. do., veil border. Do.

5284-7. [10604-7] Four do. do., lappet. Do.

5288. [10608] Do., white, veil border. Do.

5289. [10609] Do. do., collar and sleeves. Do.

5290. [10610] Do. do., do. and gauntlets. Do.

5291. [10611] Do. do., Berthé No. 1. Do.

5292. [10612] Do. do., Do. „ 2. Do.

5293. [10613] Lace, white, No. 5, 1R. per yard. *Per* COLLECTOR of TINNEVELLY.

5294. [10614] Do. do., No. 10, 4R. per yard. Do.

5295-7. [10615-7] Three do. collars, Do.

5298. [10618] Do., insertion, at 14A. per yard, 16 yards. Do.

5299-5302. [10619-22] Four do., lappet, Do.

5303. [3995] Lace made by children in Missionary School, Nagracole, South Travancore. MRS. MACKINNON.

### IV.—PLAIN EMBROIDERY, OR 'CHICKUN WORK.'

#### *a.* On Tusser Silk.

5304. [3558] Dresses, double skirt, 120R., Calcutta. SHAIK GOLAB. GOVERNMENT of INDIA.

5305. [3559] Do., 80R., Do. Do. Do.

#### *b.* On Net.

5306. [493] Dresses, tamboured, Madras. Do. Do.

#### *c.* On Muslin.

5307. [416] Two dresses, do., Do. Do. Do.

5308. [488] Do., 12R. 8A. each, do. Do. Do.

5309. [773] Do., Khyrpoor. H. H. MEER ALI MORAD.

5310. [3553] Do., 100R., Calcutta. SHAIK GOLAB. GOVERNMENT of INDIA.

5311. [3554] Do., two flounces, 80R., Do. Do. Do.

5312. [3571] Do. Do. Do.

5313. [3572] Babe's robe, 20R., Do. Do.

5314. [3574] Dress bodies, 1R. 8A., Do. Do.

5315. [3583] Bernous, 40R., Do. Do. Do.

5316. [3584] Do., 30R., Do. Do. Do.

5317-20. [3579-82] Scarfs, 40R., Do. Do.

5321. [3556] Skirts, 8R., Do. Do.

5322-9. [3563-70] Petticoats, 8R., Do. Do.

5330-4. [3585-9] Collars and sleeves, 2R., Do. Do. Do.

5335-6. [3643-4] Collars and cuffs, 3R. each, Calcutta. SHAIK GOLAB.

5337-48. [3594-3605] Handkerchiefs, 12R., Do. Do.

5349-59. [3606-16] Do., 1R. 8A., Do. Do.

5360-1. [3618-9] Do., 8A., Do. Do.

5362. [3622] Do., 4A., Do. Do.

5363. [1712] Do., on plain muslin, 20 yards, 50R., Dacca. JUGGET CHUNDER DOSS.

5364. [3376] Two do., 30R., Do. GOVERNMENT of INDIA.

5365. [3378] Do., 'Charkana,' 8R., Do.

5366. [3398] Do., plain, 30R., Do. HURMOHUN ROY.

5367. [3899] Two do., 25R., Do. Do.

### V.—GOLD AND SILVER EMBROIDERY.

#### *a.* Gold on Silk.

5368. [2988] Shawls, 200R., Delhi. MANAK CHUND, Exhibitor and manufacturer.

5369. [1725] Palungposhe, 153R., Agra. GOVERNMENT of INDIA.

5370. [3800] Choput or chesscloth, Umritsur. LALL CHUMBER MULL, exhibitor and manufacturer.

#### *b.* Gold on Cashmere, Merino, etc.

5371. [1694] Shawl, 175R., Dacca. JUGGET CHUNDER DOSS.

5372. [2983] Do., Delhi, MANAK CHUND.

5373. [2987] Do. Do. Do.

5374. [2999] Do., 200R., Do. Do.

5375. [3000] Do., Do. GOVERNMENT of INDIA.

5376. [3304] Do., Benares. MOHUN LALL & CHITTOO LALL.

5377. [6252] Do., 110R., Delhi. MANAK CHUND.

5070. [1991] Oupt, Doi Doi

5379. [3303] Doputtas, Benares. MOHUN LALL & CHITTOO LALL.

5380. [3305] Do. Do. Do.

5381. [1667] Scarfs, 100R., Dacca. HURMOHUN ROY.

5382. [1726] Scarfs, 95R., Benares. DABEE PERSHAUD.

5383. [1727] Do., 75R., Do. Do.

5384. [1728] Do. Do. Do.

5385-8. [1731-4] Do., 10R., Do. Do.

5389. [1738] Do., 46R., Do. Do.

5390. [1739] Do., 48R., Do. Do.

5391. [1748] Do. Do. Do.

5392. [2989] Do., Delhi. MANAK CHUND.

5393. [3311] Do., Benares. MOHUN LALL & CHITTOO LALL.

5394. [3387] Three do., 90R. each, Delhi. MANAK CHUND.

5395. [6251] Do., Delhi. Do.

5396. [6254] Do., 65R., Do. Do.

5397. [1735] Roomals, 175R., Benares. DABEE PERSHAUD.

5398. [1737] Do., 90R., Do. Do.

5399. [9413] Shawl. Messrs. FARMER & ROGERS, *London.*

#### *c.* Gold on Muslin.

5400. [486] Dresses, with beetle wings, 200R., Madras. GOVERNMENT of INDIA.

5401. [487] Do. black, 100R., Do. Do.

5402-3. [2207-8] Two bottles of beetles' wings.

Used for ornamental work and embroidery.

5404. [1714] Doputtas, 123R. 8A., Agra. Do.

5405. [1719] Do., 67R. 8A., Do. Do.

5406. [1720] Do., 123R. 8A., Do. Do.

5407. [1721] Do., 136R., Do. Do.

5408. [3802] Roomals, Puttiala. H. H. the MAHARAJAH.

5409. [631] Scarfs, Vizagapatam. H. H. the RAJAH of VIZIANAGRAM.

5410. [1695] Do., 60R. each, Dacca. JUGGET CHUNDER DOSS.

5410a. [1710] Do., 60R. each, Do. Do.

5410b. [3389] Five do., 60R. each, Do. GOVERNMENT of INDIA.

5411. [3395] 'Luchuck Kusseeda.' KHAJEH ABDOOL GUNNY.

5412. [1499] Handkerchief, Benares Jail. DR. N. H. CHEKE.

#### d. Gold on Cotton.

5413. [1999] Roomals (diaper pattern), Rampore. H. H. the NAWAB.

5414. [2000] Do. (do.), Do. Do.

#### e. Gold on Net.

5415. [485] Two dresses, with beetle wings, Madras. GOVERNMENT of INDIA.

5416. [3403] Two shawls, 100R. each, Dacca. HURMOHUN ROY.

5417. [1715] Doputtas, 92R., Agra. GOVERNMENT of INDIA.

5418. [1718] Do., 72R., Do. Do.

5419. [1698] Scarf, 55R., Dacca. JUGGET CHUNDER DOSS.

5420. [3385] Three do., 60R. each, Do. GOVERNMENT of INDIA.

5421. [6255] Scarf, 30R., Delhi. MANAK CHUND.

#### f. Gold and Silver on Muslin and Net.

5422. [1717] Doputtas, 52R. 8A., Agra. GOVERNMENT of INDIA.

5423. [1722] Do., 107R., Do. Do.

5424. [1723] Do., 153R., Do. Do.

#### g. Gold Lace.

5425. [632] 'Danasary' border, Vizagapatam. H. H. the RAJAH of VIZIANAGRAM.

5426. [633] 'Banjeebund' do., Do. Do.

5427. [1230] Lace for trousers, Surat. GOVERNMENT of INDIA.

5428. [1231] Do. Do. Do.

5429. [1226] Gold lace edging, Surat.

5430-2. [954-6] Narrow gold lace, Do. NUGGER SHET PREMABHAEE HEMABHAEE.

5433. [2196] Gold lace, ¾-inch, Patna.

5434. [1224] Do., 2-inch, Surat.

5435-6. [5454-5] Gold lace edging, Do.

5437. [5472] Gold lace, Lahore.

#### h. Gold and Silver Lace.

5438. [1100] Lace, Bombay. GOVERNMENT of INDIA.

5439. [1101] Do. on red silk ground, Do. Do.

#### i. Silver on Cashmere or Merino.

5440. [1736] Shawl, Benares. DABEE PERSHAUD.

5441. [1729] Scarfs, 43R., Do. Do.

5442. [3388] Three scarfs, 70R. each, Dacca. GOVERNMENT of INDIA.

#### k. Silver on Muslin.

5443. [1668] Dresses, 95R., Dacca. HURMOHUN ROY.

5444. [1716] Doputtas, 59R., Agra. GOVERNMENT of INDIA.

5445. [3390] Seven scarfs, 40R. each, Dacca. Do.

#### l. Silver on Net.

5446. [3404] Three shawls, 80R. each, Dacca. GOVERNMENT of INDIA.

5447. [3386] Three scarfs, 40R. each, Do. Do.

5448. [1669] Head-dress, 4R., Do. HURMOHUN ROY.

### VI.—SILK EMBROIDERY.

#### a. Silk on Silk.

5449. [704] Sachet, in coloured silks, Kutch. H. H. the RAO.

5450. [6332] Pillow-ends, Singapore. TAN KIM SING.

#### b. Silk on Cashmere or Merino.

5451. [1091] Shawls, 60R., Bombay. GOVERNMENT of INDIA.

5452. [1092] Do., 60R., Do. Do.

5453. [2984] Do., 50R., Delhi. MANAK CHUND.

5454. [2990] Do., 50R., Do. Do.

5455. [6268] Do., 30R., Do. Do.

5456. [2968] Capes, Do. Do.

5457-9. [2969-71, 2996] Do. Do. GOVERNMENT of INDIA.

5460-1. [1678-9] Do., 7R., Dacca. HURMOHUN ROY.

5462. [1681] Capes, 7R., Dacca. HUR-MOYUN ROY.

5463. [1689] Do., 3R., Do. Do.

5464. [1690] Do., 3R., Do. Do.

5465-7. [2974-6] Do., Delhi. MANAK CHUND.

5468. [2979] Do. Do. Do.

5469-70. [2982, 2998] Do., Do. GOVERNMENT of INDIA.

5471. [3380] Two capes, 40R. each, Dacca. GOVERNMENT of INDIA.

5472. [3381] Two do., 35R. each. Do. Do.

5473. [6256] Do., 30R., Delhi. MANAK CHUND.

5474. [6257] Do. 30R., Do. Do.

5475-9. [6258-62] Capes, 10R., Do. Do.

5480. [6269] Do., 30R., Do. Do.

*c.* **Silk on Cloth (not Cashmere).**

5481. [1002] Table cover, scarlet, embroidered in silver, 230R. 1A. 4P., Hyderabad.

5482. [1007] Do., green, do. silk, 71R. 15A. 1P., Do.

5483. [1001] Do., scarlet, do. gold, 230R. 1A. 4P., Do.

5484. [1020] Chair cushion, green, do. and silver, 29R. 3A. 9P., Do.

5485. [1014] For cap, velvet, scarlet, do. gold, 11R. 13A., Do.

5486. [1013] Two do., cloth, black, do. and silver, 14R. 9A. 6P., Do.

5487. [1003] Table cover, scarlet, do. silk, 77R. 1A. 4P., Do.

5488. [1028] Apron, velvet, black, do. gold, 35R. 8A., Do.

5489. [1018] Slippers, cloth, black, do. and silver, 5R. 12A. 9P., Do.

5490. [1029] Apron, velvet, black, do. gold, 35R. 8A., Do.

5491. [1016] Two caps, green, do. and silver, 13R. 0A. 6P., Do.

5492. [1008] Table cover, scarlet, do. and silk, 41R. 12A., Do.

5493. [1017] Slippers, cloth, black, embroidered in gold and silver, 8R. 12A. 9P., Hyderabad.

5494. [1030] Apron, velvet, black, do., 17R. 8A., Do.

5495. [1031] Bottle stands, scarlet, do. and silks, 10R. 7A. 6P., Do.

5496. [1031a] Do., scarlet, do. and silver, 10R. 7A. P., Do.

5497. [1006] Table cover, green, embroidered in silver and silk, 172R. 5A. 4P., Do.

5498. [1012] Cushion, velvet, green, do., 15R. 1A. 2P., Do.

5499. [1027] Book cover, velvet, blue, do., 19R. 1A. 4P., Do.

5500. [1015] Cap piece, velvet, green, do. and silver, 13R. 0A. 6P., Do.

5501. [1026] Chess cloth, velvet, scarlet, do., 31R. 9A., Do.

5502. [1032] Bottle stands, velvet, green, do. and silver, 3R. 4A. 4P., Do.

5503. [1031b] Do., velvet, scarlet, do. and silver, 10R. 7A. 6P., Do.

5504. [1009] Cushion, cloth, black, do. and silver, 15R. 4A. 9P., Do.

5505. [700] Apron, silk, do. silk, 36R. 15A. 2P. RAO of KUTCH.

5506. [5067] Table cover, do. silk, Sinde. BURZORJEE, SONS, & CO.

5507. [5068] Do., do. silk, Do. Do.

5508. [1010] Cushion, green, do. gold and silk, 16R. 0A. 7P., Hyderabad.

5509. [701] Apron, silk, do. silk, 36R. 15A. 2P. RAO of KUTCH.

5510. [1005] Table cover, black, do. silk, 100R. 2A. 7P., Hyderabad.

5511. [710] Apron, do. silk, 179R. 10A. RAO of KUTCH.

5512. [703] Do., do. silk, 36R. 15A. 2P. Do.

5513. [1011] Cushion, velvet, violet, do. gold and silver, 17R. 7A. 1P., Hyderabad.

5514. [1023] Do., velvet, scarlet, do. and silk, 32R., Do.

5515. [1019] Do., cloth, black, do. and silver, 28R. 5A. 4P., Do.

#### d. Silk on Muslin.

5516. [3391] Five shawls, 60R. each, Dacca. GOVERNMENT of INDIA.

5517. [1713] Scarfs, 'Jamdanee,' Do. HURMOHUN ROY.

5518. [3392] Two do. 25R. each, Do. GOVERNMENT of INDIA.

#### e. Silk on Net.

5519. [1711] Three dresses in ten pieces, 60R., Dacca. JUGGET CHUNDER DOSS.

5520. [2985] Shawls, 3*l.*, Delhi. GOVERNMENT of INDIA.

5521. [2986] Do. 30R., Do. Do.

5522. [1672] Do. 50R., Dacca. HURMOHUN ROY.

5523. [1692] Do. 4R., Do. Do.

5524. [3392] Two do., 25R. each, Do. GOVERNMENT of INDIA.

5525. [6270] Shawls, 12R., Delhi. MANAK CHUND.

5526. [6271] Do. Do. Do.

5527. [6272] Two do., Do. Do.

5528. [713] Scarfs, 36R. 15A. 8P., Kutch. H. H. the RAO.

5529-33. [6263-7] Three do. (out of five), Delhi. MANAK CHUND.

5534. [6270] Scarfs, 12R., Do. Do.

5535. [6273] Do., 35R., Do. Do.

5536. [2991] Mantles, Do. Do.

#### Beadwork.

5537. [3801] Choput or chess cloth, Umritsur. LALL CHUMBER MULL.

5538. [923] 'Churee,' bead-worked sceptre, Ahmedabad.

5539. [1505] Chess cloth, with beads, H.H. RAJAH of PUTTIALA.

5540. [924] 'Bajat,' bead ornament, Ahmedabad.

#### MISCELLANEOUS ARTICLES OF EMBROIDERY NOT INCLUDED AMONGST THE PRECEDING.

5541. [1189] Gold embroidered mat, green, Benares. DABEE PERSHAUD.

5542. [1167] Gold embroidered bag, green, Benares. DABEE PERSHAUD.

5543. [1170] Silver do. do., blue, Do.

5544. [1183] Gold do. table mat, Benares.

5545. [1192] Silver do. do., Do.

5546. [10053] 3816. Lacquered hand screen, Bareilly.

5547. [10054] Do. Do.

5548. [9443] Pair of Chimdas or tops of a cap, Surat.

5549. [2919] Ivory fan, Chittagong. RAJAH of TIPPERAH.

5550. [5496] Two pouches, Assam. LIEUT. PHAIRE.

5551. [1191] Gold embroidered mat, black, Benares.

5552. [1187] Silver do. do., Do.

5553. [5432] State parasol, embroidered, in gold and silver handle, Moram, Oude. H. H. the RAJAH GAREE SHUNKER.

5554. [1194] Gold embroidered mat, blue, Benares.

5555. [1178] Do., do., Do.

5556. [1186] Do., do., Do.

5557. [1169] Do. bag, blue, Do.

5558. [1161] Do. do., Do.

5559. [1162] Do. do., Do.

5560. [5507] Gold embroidered purse, Lucknow. NAWAB SHURFOOD DOWLAH.

5561. [5508] Do. Do. Do.

5562. [5509] Do. Do. Do.

5563. [1175] Do. table mat, Benares.

5564. [2361] Goolductal or golden tree, Do. LALLA BUNERAIN.

5565 [4419] Do. Do. Do.

5566. [2907] Necklace of honour, Calcutta.

5567. [7829] Do., Jeypore.

5568. [1172] Silver embroidered bag, Benares.

5569. [1166] Silver embroidered bag, Benares.

5570. [919] Chukur gold and silver, embroidered round table cover, Ahmedabad. NEKNAMDAR SUKAWUTTEE SHETANEE HURKOOVURBHAEE.

5571-3. [1179-81] Gold embroidered table mats, Benares.

5574. [1173] Do. Do.

5575. [1185] Do. Do.

5576. [1165] Do. bag, Do.

5577. [1174] Silver embroidered table mat, Do.

5578. [1193] Gold embroidered table mat, Do.

5579. [1177] Gold do., Do.

5580. [1164] Do. bag, Do.

5581. [1184] Do. table mat, Do.

5582. [1182] Do. Do.

5583. [1163] Do. bag, Do.

5584. [1190] Do. mat, Do.

5585. [976] Small embroidered umbrella, Umritsur.

5586. [1168] Silver embroidered bag, Benares.

5587. [1171] Gold do., Do.

5588. [1176] Do. mat, Do

5589. [5537] Do. belt, set with pearls, emeralds and rubies, Nepaul. H. H. SIR JUNG BAHADOOR, K.C.B.

5590. [975] Large gold embroidered umbrella, Umritsur.

5591. [5450] Eight gold and crimson silk tassels, Lahore.

5592. [903] Gold lace edging, Kenaree, Ahmedabad.

5593. [928] Kothlee Kusbee embroidered purse, Ahmenabad. NEKNAMDAR SUKAWUTTEE SHETANEE HURKOOVURBHAEE.

5594. [929] Do. Do.

5595. [2195] Gold thread, Patna.

5596. [4712] Saddle cloth, green velvet, embroidered in gold. SAWUNT WARREE.

5597. [902] Gotta or silver lace, Ahmedabad.

5598. [904] Silver lace, Keenaree, Ahmedabad.

5599. [5071] Embroidered blue velvet mat, Benares. DEONARAIN SING.

5600. [1997] Velvet rug, gold embroidered, Benares.

5601. [2197] Silver thread, Patna.

5602. [925] 'Wutwa,' or purse, embroidered with pearls, Ahmedabad.

5603. [926] Kothlee, do. NUGGER SHETTAINE PREMABHAEE HEMABHAEE.

5604. [922] Wutwa, do. Do.

5605-8. [940-3] Keenarees of gold and silver lace, Ahmedabad. NEKNAMDAR SHETANEE SUKAWUTTEE HURKOOVURBHAEE.

5609. [938] Do. Do.

5610. [3767] Belts, gold embroidered, Umritsur. CHUMBA MULL.

5611. [1225] Silver lace, Surat.

5612-14. [5451-3] Silk and gold lace bands, Lahore.

5615. [1223] Silver lace, Surat.

5616-7. [5467-8] Gold thread, 'Kulla Buttoo,' Peshawur.

5618. [7830] Do., or wire, Jeypore.

5619. [7831] Silver thread, Do.

5620. [5473] Silver lace, Do.

# CLASS XXV. — SKINS, FUR, FEATHERS, AND HAIR.

The annual value of the feathers and hair exported from India, and each Presidency, will be seen by reference to the tables which follow:—

TABLE SHOWING THE QUANTITIES (AS FAR AS CAN BE ASCERTAINED) AND THE VALUE OF FEATHERS EXPORTED FROM INDIA AND EACH PRESIDENCY TO ALL PARTS OF THE WORLD FROM 1857–58 ·TO 1860–61.

| YEARS | WHENCE EXPORTED | UNITED KINGDOM | FRANCE | OTHER PARTS OF EUROPE | CHINA | | OTHER PARTS | | TOTAL EXPORTED TO ALL PARTS | |
|---|---|---|---|---|---|---|---|---|---|---|
| | | Value | Value | Value | Quantity | Value | Quantity | Value | Quan. | Value |
| | | £ | £ | £ | Nos. | £ | Nos. | £ | Nos. | £ |
| 1857-58 | Bengal | 233 | 327 | .. | .. | 84 | .. | 1,188 | .. | 1,832 |
| | Madras | 9 | .. | .. | .. | .. | .. | 3,712 | .. | 3,721 |
| | Bombay | .. | .. | .. | .. | .. | .. | .. | .. | .. |
| | ALL INDIA | 242 | 327 | .. | .. | 84 | .. | 4,900* | .. | 5,553 |
| 1858-59 | Bengal | 164 | 121 | 6 | .. | 295 | .. | 949 | .. | 1,535 |
| | Madras | .. | .. | .. | .. | .. | 200,226 | 2,663 | 200,226 | 2,663 |
| | Bombay | .. | .. | .. | .. | .. | .. | .. | .. | .. |
| | ALL INDIA | 164 | 121 | 6 | .. | 295 | .. | 3,612* | .. | 4,198 |
| 1859-60 | Bengal | 290 | 629 | 7 | .. | 958 | .. | 2,190 | .. | 4,074 |
| | Madras | .. | .. | .. | 31,231 | 814 | 187,952 | 4,715 | 219,183 | 5,529 |
| | Bombay | .. | .. | .. | .. | .. | .. | .. | .. | .. |
| | ALL INDIA | 290 | 629 | 7 | .. | 1,772 | .. | 6,905* | .. | 9,603 |
| 1860-61 | Bengal | 163 | 880 | 1 | .. | 2,517 | .. | 2,451 | .. | 5,012 |
| | Madras | .. | .. | .. | 13,357 | 293 | 134,675 | 2,911 | 148,032 | 3,204 |
| | Bombay | .. | .. | .. | .. | .. | .. | .. | .. | .. |
| | ALL INDIA | 163 | 880 | 1 | .. | 2,810 | .. | 5,362* | .. | 8,216 |

\* The greater part of this was exported to Ceylon and the Straits Settlements.

TABLE SHOWING THE QUANTITIES (AS FAR AS CAN BE ASCERTAINED) AND THE VALUE OF HAIR EXPORTED FROM INDIA AND EACH PRESIDENCY TO ALL PARTS OF THE WORLD FROM 1857–58 TO 1860–61.

| YEARS | WHENCE EXPORTED | UNITED KINGDOM | | FRANCE | | AMERICA | | ARABIAN AND PERSIAN GULFS | | OTHER PARTS | | TOTAL EXPORTED TO ALL PARTS | | |
|---|---|---|---|---|---|---|---|---|---|---|---|---|---|---|
| | | Quan. | Value | Quan. | Value | Quan. | Value | Quan. | Value | Quan. | Value | Quantity | | Value |
| | | cwt. | £ | cwt. | £ | cwt. | £ | cwt. | £ | cwt. | £ | cwt. | tons | £ |
| 1857-58 | Bengal | .. | .. | .. | .. | .. | .. | .. | .. | .. | .. | .. | .. | .. |
| | Madras | .. | .. | .. | .. | .. | .. | .. | .. | .. | .. | .. | .. | .. |
| | Bombay | 5,402 | 2,701 | 86 | 43 | .. | .. | .. | .. | .. | .. | 5,488 | 274 | 2,744 |
| | ALL INDIA | 5,402 | 2,701 | 86 | 43 | .. | .. | .. | .. | .. | .. | 5,488 | 274 | 2,744 |
| 1858-59 | Bengal | .. | .. | .. | .. | .. | .. | .. | .. | .. | .. | .. | .. | .. |
| | Madras | .. | .. | .. | .. | .. | .. | .. | .. | .. | .. | .. | .. | .. |
| | Bombay | 4,298 | 2,149 | .. | .. | .. | .. | 159 | 79 | 52 | 26 | 4,509 | 225 | 2,254 |
| | ALL INDIA | 4,298 | 2,149 | .. | .. | .. | .. | 159 | 79 | 52 | 26 | 4,509 | 225 | 2,254 |
| 1859-60 | Bengal | .. | .. | .. | .. | .. | .. | .. | .. | .. | .. | .. | .. | .. |
| | Madras | .. | .. | .. | .. | .. | .. | .. | .. | .. | .. | .. | .. | .. |
| | Bombay | 4,566 | 2,398 | .. | .. | 78 | 39 | 31 | 15 | 271 | 136 | 4,946 | 247 | 2,588 |
| | ALL INDIA | 4,566 | 2,398 | .. | .. | 78 | 39 | 31 | 15 | 271 | 136 | 4,946 | 247 | 2,588 |
| 1860-61 | Bengal | .. | .. | .. | .. | .. | .. | .. | .. | .. | .. | .. | .. | .. |
| | Madras | .. | .. | .. | .. | .. | .. | .. | .. | .. | .. | .. | .. | .. |
| | Bombay | 3,409 | 1,705 | .. | .. | .. | .. | 169 | 84 | 238 | 89 | 3,816 | 191 | 1,878 |
| | ALL INDIA | 3,409 | 1,705 | .. | .. | .. | .. | 169 | 84 | 238 | 89 | 3,816 | 191 | 1,878 |

## Section A.—SKINS AND FURS.

5621-2. [8171, 8171*a*] Two Bengal tigers, mounted in glazed cases, shot in the Deyra Doon, March 1860. COL. CHARLES REID, C.B.

5623. [2348*] Skin of the domestic yâk, Darjeeling. DR. A. CAMPBELL.

5624. [2346*] Do. of the 'Dung,' or wild yâk, Do. Do.

Of manufactures from skins from which the hair has not been removed, a few specimens of the coats and cloaks in common use in Upper India and Sinde during the cold season have been forwarded from Lahore and Shikarpore, and also from Darjeeling, by DR. CAMPBELL.

## Section B.—FEATHERS AND FEATHER WORK.

5625. [3334] Feather plumes, Calcutta. GOVERNMENT of INDIA.

5626-7. [3335-6] Muffs and boas, Do. Do.

5628-9. [3337-8] Do. and victorines, Do. Do.

5630. [3339] Do. and boa, Do. Do.

5631. [3340] Do., boa and victorine, Do. Do.

5632. [3341] Six boas, Do. Do.

5633. [3342] Do. Do. Do.

5634. [3343] Six victorines, children's, Calcutta. GOVERNMENT of INDIA.

5635. [3344] Six cuffs, Do. Do.

5636. [3345] Six powder puffs, Do. Do.

5637-55. [3346-64] Feather plumes, Do. Do.

5656. [627] Do., Bulbul feathers, Madras. RAJAH of VIZIANAGARUM.

5657. [5662] Pair peacocks' feather fans, Calcutta. A. M. DOWLEANS, Esq.

5658. [5661] Three peacocks' feather fans, Do. Do.

5659. [3154] Two do, Hoogly. Do.

5660. [4713] Pair fans, bordered with peacocks' feathers. SAWUNT WARREE. GOVERNMENT of BOMBAY.

5661. [2925] Peacocks' feather fan, silver handle, Nepaul. SIR JUNG BAHADOOR, K.C.B.

5662-3. [5929-30] Peacocks' feather baskets, Do. Do.

5664. [5429] Peacocks' feather umbrella, Do. Do.

5665-6. [5430-31] Do. fly flapper, Do. Do.

5667. [5785] Do. plume, Assam. COL. VETCH.

# CLASS XXVI.—LEATHER, INCLUDING SADDLERY AND HARNESS.

As chiefly connected with this class a Table, indicating the extent to which Hides and Skins now form an article of export from India, is here introduced. (P. 236.)

## Section A.—LEATHER AND ARTICLES CHIEFLY MADE FROM LEATHER.

5668. [5933] Chamois leather, Nepaul. H. H. SIR JUNG BAHADOOR, K.C.B.

5669. [5934] Coloured leather, Do. Do.

5670. [2961] 'Kimmookht,' Bareilly.

A kind of prepared leather which looks like 'shagreen.' Is used for making native shoes, and also for sword scabbards. About 2,000R. worth is annually exported from Bareilly to Delhi, and other places.

5671. [8101] Red leather sheepskin, cost 50R. per 100, Cawnpore.

5672. [8102] Buffalo leather, cost 10R. per maund of 40 lbs., Do.

5673. [8103] Cowhide, do., Do.

5674. [8104] Red goat leather, cost 62R. 8A. per 100 lbs., Do.

## Section B.—SADDLERY, HARNESS, ETC.

5675. [6488] Set of buggy harness, Cawnpore. LUCHMEE PERSHAD. GOVERNMENT.

5676. [6489] Set of artillery harness for one horse, Do. Do.

The staple manufacture of Cawnpore is leather. There

# CLASS XXVI.—*India.*

TABLE SHOWING THE QUANTITIES (AS FAR AS CAN BE ASCERTAINED) AND THE VALUE OF HIDES AND SKINS EXPORTED FROM INDIA AND EACH PRESIDENCY TO ALL PARTS OF THE WORLD FROM 1851-52 TO 1860-61.

| Years | Whence Exported | United Kingdom Quantity | United Kingdom Value £ | France Quantity | France Value £ | Other Parts of Europe Quantity | Other Parts of Europe Value £ | America Quantity | America Value £ | China Quan. No. | China Value £ | Arabian and Persian Gulfs Quan. No. | Arabian and Persian Gulfs Value £ | Other Places Quantity | Other Places Value £ | Total Exported to all Parts Quantity | Total Exported to all Parts Value £ |
|---|---|---|---|---|---|---|---|---|---|---|---|---|---|---|---|---|---|
| 1851-52 | Bengal | Pcs. 1,869,936 | 165,184 | | | | | | | | | | | Pcs. 966,629 | 111,071 | Pcs. 2,836,565 | 276,206 |
| | Madras | No. 762,734 | 15,009 | | | | | | | | | | | No. 279,956 | 6,814 | No. 1,042,690 | 21,323 |
| | Bombay | | 5,550 | | | | | | | | | | | | 11 | | 5,561 |
| | ALL INDIA | | 185,593 | | | | | | | | | | | | 117,396 | | 303,089 |
| 1852-53 | Bengal | Pcs. 2,050,232 | 196,613 | Pcs. 202,724 | 14,956 | Pcs. 182,353 | 17,387 | Pcs. 1,431,562 | 92,939 | | | 4,817 | 260 | Pcs. 170 | 37 | Pcs. 3,867,041 | 321,889 |
| | Madras | No. 439,064 | 7,951 | No. 5,960 | 68 | | | | | 3,500 | 39 | | 108 | No. 490,543 | 1,871 | No. 943,884 | 9,689 |
| | Bombay | | 6,153 | | | | | | | | | | 365 | | 22 | | 6,278 |
| | ALL INDIA | | 210,717 | | 15,024 | | 17,337 | | 94,939 | | 39 | | 365 | | 1,430 | | 337,849 |
| 1853-54 | Bengal | No. 740,594 | 215,128 | No. 3,648 | 9,492 | No. 615 | 8,217 | | 141,418 | | | | | No. 78,631 | 1,303 | No. 835,947 | 375,558 |
| | Madras | | 17,080 | | 86 | | 6 | | | | | 12,459 | 740 | | 687 | | 18,599 |
| | Bombay | | 8,045 | | | | | | | | | | 92 | | 71 | | 8,208 |
| | ALL INDIA | | 240,253 | | 9,578 | | 8,223 | | 141,418 | | | | 832 | | 2,061 | | 402,365 |
| 1854-55 | Bengal | No. 781,995 | 173,639 | No. 44,961 | 15,446 | | 7,577 | | 156,754 | | 1,585 | 30,192 | 54 | No. 181,769 | 766 | No. 1,038,917 | 365,821 |
| | Madras | | 29,282 | | 1,663 | | 2,238 | | | | | | 1,848 | | 2,300 | | 85,091 |
| | Bombay | | 8,698 | | | | | | | | | | 504 | | 34 | | 11,474 |
| | ALL INDIA | | 211,619 | | 17,109 | | 9,815 | | 155,754 | | 1,585 | | 2,406 | | 3,100 | | 402,386 |
| 1855-56 | Bengal | No. 1,170,390 | 220,273 | No. 47,808 | 26,959 | No. 6,720 | 12,468 | No. 56,500 | 111,286 | | 167 | 8,388 | 482 | No. 232,205 | 1,095 | No. 1,451,961 | 372,248 |
| | Madras | | 40,806 | | 2,132 | | 67 | | 1,646 | | | | 94 | | 2,364 | | 47,487 |
| | Bombay | | 10,512 | | | | 1,254 | | | | | | 576 | | 134 | | 11,994 |
| | ALL INDIA | | 271,591 | | 29,091 | | 13,789 | | 112,932 | | 167 | | | | 3,583 | | 431,729 |
| 1856-57 | Bengal | No. 1,320,218 | 252,742 | No. 59,061 | 44,506 | No. 30,982 | 82,737 | No. 3,000 | 167,038 | | 388 | 21 | 1 | No. 24,402 | 524 | No. 1,437,629 | 497,935 |
| | Madras | | 56,612 | | 3,197 | | 1,901 | | 105 | | | | 85 | | 602 | | 62,418 |
| | Bombay | | 11,899 | | | | 193 | | | | | | 86 | | | | 12,177 |
| | ALL INDIA | | 321,253 | | 47,703 | | 84,831 | | 167,143 | | 388 | | | | 1,126 | | 573,530 |
| 1857-58 | Bengal | No. 2,084,824 | 235,879 | No. 267,155 | 44,918 | No. 46,550 | 28,552 | | 138,733 | | 155 | 2,236 | 134 | No. 173,250 | 953 | No. 2,574,015 | 449,190 |
| | Madras | | 113,760 | | 21,452 | | 1,954 | | 676 | | | | 57 | | 4,164 | | 141,464 |
| | Bombay | | 41,521 | | 3,912 | | 2,794 | | | | | | 191 | | 88 | | 49,048 |
| | ALL INDIA | | 391,160 | | 70,282 | | 33,300 | | 139,409 | | 155 | | 191 | | 5,205 | | 639,702 |
| 1858-59 | Bengal | No. 2,558,537 | 206,419 | No. 68,380 | 32,314 | | | | 145,448 | | 16 | 2,005 | 120 | No. 240,493 | 1,700 | No. 2,870,915 | 385,919 |
| | Madras | | 96,373 | | 3,476 | | 36 | | | 1,500 | | | 246 | | 3,678 | | 103,663 |
| | Bombay | | 53,085 | | 940 | | | | 689 | | 2 | | 366 | | 138 | | 55,098 |
| | ALL INDIA | | 355,877 | | 36,730 | | 36 | | 146,137 | | 18 | | | | 5,516 | | 544,680 |
| 1859-60 | Bengal | No. 3,008,617 | 199,067 | No. 21,900 | 7,955 | | | | 86,138 | 640 | 66 | 675 | 841 | No. 63,207 | 3,280 | No. 3,094,399 | 309,806 |
| | Madras | | 100,592 | | 1,551 | | 12,574 | | 1,685 | | 18 | | 89 | | 1,189 | | 103,371 |
| | Bombay | | 29,091 | | 441 | | | | | | | | 64 | | 82 | | 31,361 |
| | ALL INDIA | | 328,750 | | 9,947 | | 12,574 | | 87,823 | | 79 | | 944 | | 4,501 | | 444,537 |
| 1860-61 | Bengal | No. 2,824,303 | 307,257 | No. 81,968 | 30,347 | | | | 141,060 | | | 6,781 | 403 | No. 101,485 | 27,084 | No. 2,965,177 | 505,814 |
| | Madras | | 118,257 | | 2,148 | | | | 5,444 | | | | 1,506 | | 2,511 | | 128,382 |
| | Bombay | | 20,320 | | 24 | | | | | | | | | | 189 | | 22,433 |
| | ALL INDIA | | 445,834 | | 32,519 | | | | 146,504 | | | | 1,909 | | 29,784 | | 656,629 |

are about fifty tanneries, each of which turns out on an average 1,200 hides a year, or 60,000 hides yearly on the whole. The average value of a hide tanned by the native process being about 4R., the total annual value of the out-turn of tanned hides is 240,000R. Besides the hides tanned in Cawnpore itself, about 180,000 hides are imported yearly, which are tanned either at Meerut or in the adjacent villages of the Cawnpore district. Those from Meerut are said to be of superior quality: the village hides, on the other hand, are inferior to those tanned in Cawnpore, and are mostly taken from cattle which have died of disease. The total number of hides used yearly at Cawnpore is thus about 240,000, and their value 960,000R. In addition to the tanning trade there is an extensive business done at Cawnpore in the manufacture of saddlery, harness, boots, shoes, and other leathern articles: there are 52 saddlers and 55 shoemakers' firms, which make up goods for the European market, and about 200 shops which supply the native market—the latter deal chiefly in shoes made in the native fashion. The value of the shoes made for the native market is about 40,000R. annually. Leathern articles made at Cawnpore are exported to Meerut, Benares, Central India, and Rohilkund, whence they find their way extensively to other parts of India. The places above-named carry on a direct trade with Cawnpore. There is no export trade to Europe or America. There can be no doubt that the quality of the leather produced by the native process is decidedly inferior. The native tanner does not leave his hides to soak in the pits containing his bark infusion, but, having sewed up a quantity of bark in the skin (made into a kind of bag) he exposes it to a constant stream of water, which forces the astringent matter into the pores of the hide very rapidly: but to make the process still more rapid the hide is taken out, wrung, and refilled every four or five days. A hide can be ready in this way in about a month, but the leather is less strong, durable, and pliable than English leather. The currying process also is often entirely omitted, and when performed it is generally on a minute scale, and very inefficiently. The only attempt hitherto made to introduce the English process is being conducted on the part of Government by CAPT. STEWART, Commissary of Ordnance.

The English system of tanning is more costly than the native. This circumstance would be a bar to the introduction of English tanned leather into the native market, where cheapness is the first requisite. But for the supply of Government contracts, and for the articles used by the European community in India, there is no question that Cawnpore could be made to afford ample quantities of leather and leathern articles at comparatively moderate prices. As it is, the articles supplied to Government and to private purchasers are hardly inferior to those manufactured in England, except in the quality of the leather and other materials. The workmanship is very little worse than that of an average English artisan. I should think, therefore, that improvement in the quality of Cawnpore leather might be expected to lead to a considerable increase in its consumption in India. As regards the European market, it is true that Indian hides cannot compete in quality with the hides of well bred and fed English cattle; but then the hides of English cattle do not supply the whole of the English market. There is already a considerable trade in raw hides between India and England, and it seems not unlikely that if a better process of tanning were established, tanned hides might be exported to England with economy. It is to be remarked that Cawnpore does not appear to be in a position of exceptional natural advantage as regards the leather trade. The circumstances which made it an emporium of this trade appear to be—first, that a great part of the population of the old town of Cawnpore were chumars (shoemakers); next, that under our Government Cawnpore became a considerable military and civil station.

Babool bark is almost exclusively used by the natives for tanning purposes: it is the only bark that can be procured in large quantities and cheap. Price from ½ to 2 rupees per maund.

In the Government experiments now being promoted, various barks are being used. That of the *Cassia fistula*, or native Amultas, is found to contain a good quantity of tannin, but it is rare in these provinces.

The *Acacia obtusifolia*, or native Chakoor, is more plentiful—but still not sufficiently so for manufacture on anything but a small scale.

The leaves and small shoots of the native 'Aura,' supposed to be one of the *Terminalia* species, have also been used. It has properties somewhat similar to the American sumach. The leaves of this 'Aura' have been known to be useful in tanning for some years; but the natives have not used it, owing to the difficulty of finding it in large quantities. It gives to leather a light colour.*

5677. [1459] 'Charjaweh' saddle, bridle, &c., embroidered in gold, Puttiala. H. H. the MAHARAJAH.

5678. [1460] 'Kathee' saddle, bridle, &c., Do. Do.

5679. [736] Sinde saddle, silver mounted, Khyrpoor. H. H. MEER ALI MOORAD.

5680. [734] Embroidered saddle cloth, Do.

5681. [740] Camel saddle, complete, Do. H. H. MEER ALI MOORAD.

5682. [4261] Saddle cloth, Thibet. DR. CAMPBELL.

5683. [10057] Whip, Futtehpore.

5684. [10058] Do. Do.

* Local Committee, *Cawnpore.*

# CLASS XXVII.—ARTICLES OF CLOTHING.

The vast proportion of what may be called articles of clothing in India being in the form of *loom-made* scarfs of various descriptions, the following arrangement and classification have been adopted with the view of indicating not only the nature of the materials so employed, but also the manner in which variety of pattern is secured by the introduction of silk or of gold and silver thread into the borders and ends of even ordinary cotton fabrics.* By this means an attractive garment is produced, and it is only by attention to such points that the manufacturer will be enabled to fully suit the tastes of the people, and thus extend the market for his goods.

Case 26 (see plan) contains one out of twenty sets of pattern books which have, with the view of affording facilities for this purpose, been prepared for presentation to the chief seats of commerce in this country. Each set is identical with the one exhibited, and comprehends upwards of 600 'working samples' illustrative of not only the articles of Indian clothing in the scarf-form made in the loom, but also specimens of the cotton and silk piece goods, woollens, and other textile manufactures of the country.†

The same case likewise contains a book forwarded by the LAHORE CENTRAL COMMITTEE, showing samples of the cotton cloths commonly worn in the Punjab.

Section I.—ARTICLES OF MALE ATTIRE (NATIVE).

### A. COVERINGS FOR HEAD.

#### I.—TURBAN PIECES.

##### *a.* Cotton.

5685. [873] Ahmedabad. GOVERNMENT of BOMBAY.

5686. [875] Ahmedabad. GOVERNMENT of BOMBAY.

5687. [1159] Red, Belgaum. Do.

5688. [1294] Moondasa (kind worn by the poorer classes), 1R. 4A., Dharwar. W. C. ANDERSON, Esq.

5689. [1295] Do., 12A., Do. Do.

5690. [1285] Roomal (for cultivators' wear), 13A., Do. Do.

Wrapped round the head, when in the fields.

##### *b.* Cotton with Silver thread in ends.

5691. [1296] Moondasa, 3R. 2A., Dharwar. W. C. ANDERSON, Esq.

Worn by poorer classes on festival days.

##### *c.* Cotton with Gold thread in ends.

5692. [1269] Rewa Kanta. GOVERNMENT.

##### *d.* Cotton and Gold.

5693. [707] 'Mundel,' 8R. 4A. 11P. Kutch. H. H. the RAO.

5694. [901] Ahmedabad. GOVERNMENT of INDIA.

5695-6. [1740–1] Black muslin, Benares. Do.

5697. [1742] Red and gold, 40R. Do. Do.

##### *e.* Cotton and Silk—Gold printed.

5698-5709. [7817–28] (12) Jeypore. H. H. the RAJAH.

##### *f.* Silk.

5710. [1200] Puttiala. H. H. the MAHARAJAH.

5711-13. [1201-3] Do. Do.

5714. [2885] (Child's), Poonah. GOVERNMENT of BOMBAY.

5715. [3545] Crimson, Umritsur. Do.

5716. [5463] From Peshawur. FUZL AHMUD.

* Some of the fabrics which, for the above-named reason, have been introduced into this class, are included in the Official Industrial Catalogue' under the head of embroideries, and as such were submitted by me to the Jury of Class XXIV. The arrangement here adopted is not, therefore, intended as a guide to the classification of such goods of the same description as may on a future occasion be contributed from India—the position of these having to be determined with reference to the kind of work or manufacture which a particular article may most aptly illustrate—a point which has to be carefully decided before exposition.

† It is expected that these pattern books will be ready for distribution early in January. It is intended at the same time to publish a paper on the 'Textile Manufactures of India,' with illustrations showing the different ways in which the various fabrics are worn as well as made up by the Indian consumer.

### g. Silk with Gold thread in ends.

5717. [1204] Yellow, Puttiala. H. H. the MAHARAJAH.

5718. [1205] (2) Buff, Do.   Do.

5719. [1206] Shot,   Do.   Do.

5720-2. [5443-5] Grey, Lahore. GOVERNMENT of INDIA.

5723. [5446] White,   Do.   Do.

5724. [5447] Crimson,  Do.   Do.

5725. [5448] Grey,   Do.   Do.

5726. [5449] Shot,   Do.   Do.

5727. [5464] Do. Peshawur.   Do.

### h. Silk, Gold embroidered throughout.

5728. [3543] Yellow, Umritsur. Government of INDIA.

5729. [3544] Red,   Do.   Do.

## II. — HATS, CAPS, ETC.

### Topees or Native Hats, Embroidered.

5730. [1097] Topee, Bombay. GOVERNMENT of INDIA.

5731. [5510] Do., Lucknow, Oude. RAJAH TAJ KISHEN.

5732. [5511]   Do.   Do.

5733. [5512] Do., Maraon, Do. H. H. RAJAH GOREE SHUNKER.

5734. [5516] Do., Lucknow, Do. NAWAB SHURFOOD DOWLAH.

5735. [5517] Do., Ulwar. H. H. the MAHARAJAH.

### Caps, Embroidered.

5736-39. [5513-5, 5518] Cap, Maraon, Oude. H. H. RAJAH GOREE SHUNKER.

5740. [3782] Do., Peshawur. GOVERNMENT of INDIA.

5741-44. [5476-9] Four do., Peshawur. Do.

5745-7. [5480-2] Three do. Do. Do.

5748. [3783] Cap, Benares. Do.

5749-50. [5474-5] Do., Lahore. Do.

5751. [9330] Do., Cuttack. Do.

5752. [1095] Do., Bombay. Do.

5753-4. [783-4] Two Sindee caps, Khyrpoor. H. H. MEER ALI MOORAD.

5755. [831] Cap, Shikarpoor. GOVERNMENT of INDIA.

5756. [842] Two do. Do. Do.

5757. [844] Three do. Do. Do.

5758. [1757] Woollen 'felt,' Darjeeling. DR. CAMPBELL.

5759-61. [1760-62] Four do. Do. Do.

5762. [706] Silk cap, Kutch. H. H. the RAO.

5763. [1758] Cap, Darjeeling. DR. CAMPBELL.

5764. [5460] Parandas, or hair ornaments of silk, Lahore. GOVERNMENT of INDIA.

5765. [5461]   Do.   Do.

5766. [5462]   Do.   Do.

## B.—CLOTHING FOR UPPER PORTION OF BODY, SHOULDERS, ETC.
### (Loom made.)

### I.—LOONGEES OR SCARFS, WORN BY MALES.

#### a. Cotton.

5767. [722] Loongee, Kutch. H. H. the RAO.

5768. [728] Do. Do.

5769. [729] 'Dhawood khani,' 1R. 7A., Do. Do.

5770. [605] Loongee, Madras. GOVERNMENT of INDIA.

5771. [726] Do., Kutch. H.H. the RAO.

5772. [732] Do. Do.

5773. [5244] 'Patso,' Burmah. GOVERNMENT of INDIA.

5774. [5554] Do., Do. Do.

These are the ordinary dresses of the poorer classes of Burmah and Pegu.*

5775. [1297] Selya, sheet or body covering, Dharwar. W. C. ANDERSON, Esq.

Used by the lower classes; always presented to the bridegroom by relations of the bride, together with a turban. Price 1R. 12A.

5776. [1280] Do.   Do.   Do.

Worn commonly by cultivators and labourers, wrapped round their shoulders and body, when employed in the fields. Price 1R. 4A.

* Local Committee, *Rangoon.*

5777. [1281] Selya, Dharwar. W. C. ANDERSON, Esq.

Another description of the same kind of cloth. Price 1R. 4A.

### b. Cotton with Silk borders and ends.

5778. [884] Loongee, Ahmedabad, 1R. 6A. GOVERNMENT.

5779. [8244] Do., Bhawulpore. H. H. the NAWAB.

5780. [8245] Do.        Do.

5781-2. [8283-4] Do., Peshawur. GOVERNMENT.

5783. [10004] 'Phalkaree,' Lahore. NAWAB of FAREEDKOTE.

5784. [10005] Do.        Do.        Do.

5785. [10007] Do.        Do.        Do.

5786-8. [607-9] Loongee, Madras. GOVERNMENT.

5789. [1137] Do., Belgaum.        Do.

5790-1. [2611-2] Do., Mysore.        Do.

5792. [2618] Do.        Do.

5793. [1317] Scarf, Dharwar. W. C. ANDERSON, Esq.

Used by cultivators on festivals, wrapped around the shoulders, and worn by Brahmins around the waist and loins. Piece of two, 11R. 2A.

5794. [1324] 'Buchkane,' Do.        Do.

Worn round the shoulders and body by male children of cultivators on holidays, also wrapped around the waist of images of gods. Piece containing two, 5R. 2A.

### c. Silk and Cotton mixed.

5795. [717] Silk and cotton Loongee, Kutch. H. H. the RAO.

5796. [5553] 'Potsan Patso,' Akyab. GOVERNMENT of INDIA.

Worn by the Mugs of the province. 1 piece, 24R.; made in Akyab; more of the same description of cloth is manufactured at Cox Bazaar, Chittagong.*

### d. Silk and Cotton with Silk borders and ends.

5797. [8249] From Bhawulpore. H. H. the NAWAB.

5798. [8246] Do.        Do.

The workmanship of SADIGE and OLA BUKSH of *Khanpore*.

---

* Local Committee, *Akyab.*

5799-5802. [8285-88] Four Loongees Peshawur. GOVERNMENT.

5803. [8292] Do.        Do.

5804. [8297] Do.        Do.

### e. Cotton with Silk and Gold in borders and ends.

5805. [2598] Loongee, Mysore. GOVERNMENT.

### f. Cotton, Gold embroidered.

5806. [5193] Loongee, Goojerat Jail. GOVERNMENT.

### g. Silk.

5807. [2619] Loongee, Mysore. GOVERNMENT.

5808. [8238] Do., Luckimpore. H. L. MICHEL, Esq.

### h. Silk, with Gold in borders and ends.

5809. [1743] Green and red double woven Loongee, Benares. GOVERNMENT.

5810. [3761] Do., Umritsur. Do.

5811-3. [3762-4] Crimson checked, Do. Do.

5814. [5286] Figured silk, Bhawulpore. H. H. the RAJAH.

Worked by NATTOO PATOLI.

5815. [5287] Do.        Do.        Do.

The workmanship of SADIGE and OLA BUKSH of *Khanpore.*

5816. [6500] Purple, Fareedkote.        Do.

5817. [10002] Green, Do.        Do.

5818. [3765] 'Khess,' Umritsur. GOVERNMENT of INDIA.

5819. [8242] Do. scarlet, Mooltan. Do.

5820. [8243] Do. green,        Do.        Do.

5821. [3766] Do. crimson, Do.        Do.

5822-3. [8239-40] Loongee,        Assam. MAJOR H. S. BIVAR.

### C.—UPPER CLOTHING MADE INTO COATS, JACKETS, ETC.

### I.—CHOGAS, NATIVE COATS, JACKETS, WAISTCOATS, ETC.

### a. Cotton.

5824. [635] Woven Choga, Vizagapatam. H. H. the RAJAH of VIZIANAGARUM.

### b. Cotton, embroidered with Silk.

5825. [5563] Piplee work, Cuttack. GOVERNMENT of INDIA.

### c. Cotton and Silk.

5826. [4471] 'Tindyne,' worn by Karens, Burmah. Messrs. HALLIDAY, FOX, & Co.

### d. Woollen, Gold embroidered.

5827. [847] Choga, Shikarpoor. GOVERNMENT.

### e. Silk, embroidered.

5828. [712] 'Jama,' Kutch. H. H. the RAO.

5829. [5757] Choga, Meerut. KOOER WUZEER ALI KHAN.

### f. Leather, embroidered with Silk.

5830. [10329] Jacket, Peshawur. GOVERNMENT.

---

5831. [828] Two gold embroidered woollen 'Neemchas,' scarlet, Khorassan, Shikarpoor. GOVERNMENT.

---

5832. [829] Two gold embroidered woollen waistcoats, red, Shikarpoor. Do.

## D.—COVERINGS FOR LOINS AND LOWER PORTIONS OF BODY
### (Loom made).

### I.—DHOTEES.

Waist and loin cloths: occasionally worn so as to fall over and cover the greater portion of the lower limbs.

### a. Cotton.

5833. [1282] Dhotee, Dharwar. W. C. ANDERSON, Esq.

Commonly worn by cultivators and labourers in the field. Piece contains four; price, 7s. 6d. each or 1R. 14A.

5834. [1283] Buchkhanee. Do.

Commonly worn as a waist cloth by children of respectable people; also worn by adults of the same class, while sleeping. Price, 1R. 2A.

5835. [1284] Punjee. Do.

Cloth used by well-to-do people, to dry themselves after bathing, and also worn as a waist cloth by poor people. Piece contains four. Price, 1R.

### b. Cotton, plain centre with coloured borders and ends.

5836. [720] Dhotee, Kutch, 1R. 10A. 1P. H. H. the RAO.

5837-8. [857, 860] Dhotees, Ahmedabad. GOVERNMENT of INDIA.

5839-40. [8267-8] Do., Hooghly. Do.

### c. Cotton, with silk borders and ends.

5841. [932] 'Chunderee,' Ahmedabad. RAO BAHADOOR SHETT MUGUNBHAEE KRAMCHUND.

5842. [1317] Dhotee, Dharwar, 11R. 2A. W. C. ANDERSON, Esq.

5843. [1998] Do., Rampore. H. H. the NAWAB.

5844. [861] Silk and cotton, Ahmedabad. GOVERNMENT of INDIA.

5845. [1316] Dhotees, Dharwar. W. C. ANDERSON, Esq.

This description is used by rich people. The piece contains two dhotees. 15R.

5846. [1318] Do., 4R. 6A. Do.

5847. [1319] Do. Do.

Common wear of the middle classes. Piece of two dhotees. 5R. 6A.

5848. [1320] Do., 3R. 11A. Do.

5849. [1321] Do., 3R. 4A. Do.

5850. [1322] Do. Do.

Common wear of male children of the rich and middle classes on feast days. Piece of two dhotees, 12R.

5851. [1323] Do. Do.

Common wear of male children of middle classes, and holiday wear of poorer classes. 9R.

### d. Cotton, with silk and gold in borders and ends.

5852. [863] Dhotee, Ahmedabad, 11R. GOVERNMENT of INDIA.

5853. [864] Do., 21R. Do.

### e. Silk.

5854. [1328] 'Peetamber,' red, Dharwar. W. C. ANDERSON, Esq.

Worn by middle classes at festivals, 18R.

5855. [1329] Do., yellow, Do. Do.

Worn by rich people daily at meals, and by middle classes at meals on festivals. 12R. 11A.

5856. [892] Do., 5R. 4A., Ahmedabad. GOVERNMENT.

5857. [893] Do. Do. Do.

5858. [5548] 'Peetamber,' white, Hooghly. GOVERNMENT of INDIA.

5859. [5549] Do., bordered, Do. Do.

5860. [8198] Do., red, do. Do. Do.

### *f.* Silk, with gold borders and ends.

5861-2. [1749-50] (2) Pink, Benares. GOVERNMENT.

### E.—WAISTBANDS, AS EZARBUNDS AND CUMMERBUNDS.

#### *a.* Cotton.

5863. [1292] Nurkuttoo or girdle, Dharwar. W. C. ANDERSON, Esq.

Worn by all classes when working or travelling. Money is ordinarily carried in it. 1R. 8A.

5864. [1799] Sauganeer. GOVERNMENT of INDIA.

#### *b.* Silk.

5865-70. [3537-42] Ezarbund, Umritsur. Silk reared by MR. H. COPE; reeled by JAFFER, of Goodarpoor; dyed, &c. under CHUMBER MULL.

5871-3. [5456-8] Do., Lahore. GOVERNMENT of INDIA.

5874. [5470] Two, Do. Do.

---

### II.—ARTICLES OF FEMALE ATTIRE.

### A.—SAREE.

Upper garment, in the form of scarf, for enveloping the person: one end usually brought over head as a covering (loom made).

#### *a.* Cotton.

5875. [1277] Saree, Dharwar. W. C. ANDERSON, Esq.

This description is such as the poorer classes wear. Each woman generally has a new one once a year. 2R. 8A. each.

5876. [1278] Do. Do.

Similar to the preceding. 2R. 2A.

5877. [1279] Do. Do.

Same use as the preceding, but of small size, for young girls. 2R. 8A.

5878. [1286] 'Putta,' Do. Do.

Female children of the poorer classes, between the ages of seven and twelve, wear this at weddings. 1R.

5879. [1287] 'Putta,' Dharwar. W. C. ANDERSON, Esq.

Another description, worn at weddings by female children under seven years old, and also used to dress up images of goddesses. 7R.

5880. [731] Saree, Kutch. 1R. 7A. 8P. H. H. the RAO.

#### *b.* Cotton with silk borders.

5881. [8295] Saree, with inscriptions on borders, Santipore. GOVERNMENT of INDIA.

#### *c.* Cotton with silk borders and ends.

5882. [879] Saree, Ahmedabad. GOVERNMENT of BOMBAY.

5883-4. [885-6] Do. Do.

5885. [2633] Do., Mysore. GOVERNMENT of MYSORE.

5886. [1309] Do., Dharwar. Do.

Another description, which is the common wear of the middle classes. 6R. 14A.

5887. [1310] Do. 5R. 14A. Do.

5888. [1311] Do. Do.

Common wear of the wives of farmers. 3R. 10A.

5889. [1312] Do. Do.

Worn by farmers' wives on festivals and at weddings. 12R. 3A.

5890. [1313] Do. 8R. 1A. Do.

5891. [1314] Do. Do.

For young girls of from eight to twelve years old. 7R. 12A.

5892. [1315] Do. Do.

Common wear of girls of poorer classes from eight to twelve years old. 1R. 8A.

#### *d.* Cotton, Gold embroidered.

5893-5. [7744-6] Saree, Runpore, Cuttack. GOVERNMENT of INDIA.

#### *e.* Cotton, with Silk and Gold embroidered ends.

5896. [887] Saree, Ahmedabad. BOMBAY GOVERNMENT.

5897. [2615] Do., Mysore. GOVERNMENT of MYSORE.

5898. [2594] Striped muslin. Do.

### f. Silk and Cotton mixed, and with Silk and Gold borders and ends.

5899. [880] Saree, Ahmedabad. GOVERNMENT of INDIA.

5900. [1158] Do. Belgaum. Do.

5901. [2617] Do. Mysore. Do.

5902-3. [2609-10] (2) Do. GOVERNMENT of MYSORE.

5904. [1307] Saree, Dharwar. W. C. ANDERSON, Esq.

This description used by middle classes at weddings and festivals (one saree in the piece). 28R.

5905. [1308] Do. Do.

Another description, worn by rich people daily and by middle classes at festivals. 20R.

5906. [1325] Putta, Do. Do.

Worn by female children of the richer classes at weddings. 23R.

5907. [2616] Saree, Mysore. GOVERNMENT of MYSORE.

### g. Silk.

5908. [930] Sarees, striped, 11R. 4A., Ahmedabad. NEKNAMDAR SUKAWATTEE SHETANEE HURKOOVURBHAEE.

5909. [2576] Do., Ahmedabad. GOVERNMENT of BOMBAY.

5910. [2577] Do., 12R. 8A., Ahmedabad. Do.

5911-2. [3148-9] (2) Do., Berhampore. GOVERNMENT of INDIA.

5913. [3151] Diagonal stripes, Do. Do.

5914. [5550] Crimson, Hooghly. Do.

5915. [5551] 'Tamieng,' Burmah. Do.

5916. [1327] Saree, Dharwar. W. C. ANDERSON, Esq.

Worn by Brahmins' wives while eating at festivals. 34R.

5917. [1330] Sulleedar, Do. Do.

Worn by women on festivals.

### h. Silk, Silk embroidered.

5918-19. [9440-1] (2) Green, worn by Parsee ladies, Bombay. GOVERNMENT.

### j. Silk, with Gold embroidered ends.

5920. [5552] Saree, Pegu. GOVERNMENT of INDIA.

### k. Silk, Gold borders and ends.

5921. [588] Figured, 115R., Tanjore. SACCARIM SAHIB, son-in-law to late RAJAH.

5922. [2614] Checked, Mysore. GOVERNMENT of INDIA.

5923. [2608] Do. Do. Do.

5924. [2592] Crimson and gold stripes, Bangalore. Do.

5925. [8234] Saree, Gowhatty, Assam. MAHINEE DEVYA, widow of AMINDARAM PHOOKEM.

## B.—SCARFS USED AS SKIRTS OR PETTICOATS.

### a. Cotton.

5926. [6336] 'Bugis,' Singapore. TAN KIM SING.

5927-35. [6393-401] 'Sarongs,' nine, Do. H. H. INCHE WAN ABOO BAKAR.

### b. Silk.

5936-40. [6731-5] 'Sarongs,' Singapore. H. H. INCHE WAN ABOO BAKAR.

### c. Silk embroidered.

5941. [827] Red scarf, Shikarpoor. GOVERNMENT of INDIA.

5942-7. [6736-7-9, 6740-2] 'Sarongs,' Singapore. H. H. INCHE WAN ABOO BAKAR.

## C.—KERCHIEFS, USUALLY WORN ON HEAD.

### a. Cotton.

5948-51. [6410-3] Kerchiefs, Singapore. H. H. INCHE WAN ABOO BAKAR.

### b. Silk.

5952. [4468] Worn by Kareen women, Burmah. Messrs. HALLIDAY, Fox, & Co.

5953-6. [6406-9] Do. Malay do., Singapore. H. H. INCHE WAN ABOO BAKAR.

5957. [7802] 'Fareens,' Sauganeer. GOVERNMENT.

## D.—MADE UP ARTICLES OF FEMALE ATTIRE.

### 1. Skirts.

### a. Cotton.

5958. [4474] Worn by Kareens, Burmah. Messrs. HALLIDAY, Fox, & Co.

### b. Silk.

5959. [4107] Skirt, native dyed, Penang. GOVERNMENT.

5960. [5552] Skirt, Burmah. Do.

This wrapped around the body forms the dress of a Burmese female. The quality of the material varies with the station of the wearer.*

### c. Silk, embroidered with Gold and Silver.

5961. [1096] 'Zubloo,' child's shirt, Surat. GOVERNMENT.

5962. [1098] Do. Do. Do.

5963. [1231] Do. Do. Do.

### 2. *Dresses, Gold embroidered.*

5964. [5109] Muslin, as worn by Princess of Ulwar. H. H. the RAJAH of ULWAR.

5965. [5538] Do. Do. Do.

5966. [10307] 'Khillat,' or dress of honour, Benares. GOVERNMENT of INDIA.

### 3. *Trowsers.*

#### a. Silk, Gold embroidered.

5967. [1099] 'Child's red,' Surat. GOVERNMENT of BOMBAY.

5968. [6334] From Singapore. TAN KIM SING.

#### b. Woollen.

5969. [5504] 'Churnee,' Kutch. H. H. the RAO.

### E.—MATERIAL FOR MAKING INTO CHOLEES OR BODICES.

Although more strictly coming under piece goods, the under-noted fabrics, which are used for the manufacture of women's bodices or under jackets, are here inserted.

#### a. Cotton.

5970. [1275] 'Khuns' or cholee pieces, Dharwar. W. C. ANDERSON, Esq.

Contains 4 cholees. This description is used by women working in the fields. 3A. for each cholee, or 12A.

5971. [1276] Do. Do. Do.

Another description, used by the same classes as the preceding. 2A. 9P. for each cholee, or 1R. 11A. 6P.

* Local Committee, *Rangoon.*

### b. Silk and Cotton mixed.

5972. [1299] Punchrungee, Dharwar. W. C. ANDERSON, Esq.

Woman's bodice cloth of five colours; worn by upper classes, or poorer classes on great days; 1 khun (or piece). Price, 1R. 1A.

5973. [1300] Do. Do. Do.

Contains 6 khuns, 6R.

5974. [1301] Do. Do. Do.

Another description used by the same classes as the preceding. Contains 5¼ khuns, 5R. 14A. 6P.

5975. [1302] Do. Do. Do.

Worn by middle classes; ordinary wear. Contains 4 khuns, 2R. 4A.

5976. [1303] Do. Do. Do.

Another description with the same uses as the preceding. This piece contains 7 khuns, 4R. 6A.

5977. [1304] Do. Do. Do.

Common wear of labouring women. Piece contains 1 khuns, 2R. 15A. 6P.

5978. [1305] Do. Do. Do.

Another piece containing 10 khuns, 1R. 14A.

5979. [1306] Do. Do. Do.

Warp of silk and weft of cotton; worn ordinarily by dancing women, not considered fit for respectable women, 1 khun, 1R. 12A.

#### c. Silk.

5980. [7881] 'Phudkee,' Poonah. GOVERNMENT.

5981. [7882] 'Cholee,' Do. Do.

5982. [1326] 'Khun,' or woman's bodice cloth, Dharwar. W. C. ANDERSON, Esq.

Worn by wives of Brahmins while eating at festivals. 2R.

### MISCELLANEOUS ARTICLES OF APPAREL.

#### Silk Scarfs.

5983. [5107] Two Tusser scarfs, Keonghur, Cuttack. GOVERNMENT of INDIA.

5984. [8219] Figured, 'Balisoree,' Gowhatty, Assam. Do.

#### Silk Scarfs with border.

5985. [2592] Of silk gauze, Assam. BABOO GORUCKCHUNDER, *Boreeah of Seebsaugor.*

5986. [8226] Of Eria silk, Do. LIEUT. W. PHAIRE.

### Cotton Scarfs.

5987. [1038] 'Khess,' chintz, 8R., Hyderabad.  Government of India.

5988. [3252] Scarf shawls of fine diaper material, Rampore.  H. H. the Nawab.

### Cotton, Gold Borders and Ends.

5989. [634] Muslin, Vizagapatam.  H. H. the Rajah of Vizianagarum.

5990. [708] Do., and gold woven, 19R. 4A. 7P., Kutch.  H. H. the Rao.

---

5991. [772] Chanduse, cotton, coloured border and ends, Khyrpoor.  H. H. Meer Ali Morad.

---

5992. [6738] Salimote, silk, Singapore. H. H. Inche Wan Aboo Bakar.

---

5993. [6404] Salendongs, silk, Singapore. Do.

*Worn only around loins.*

5994. [6405] Silk and cotton, Do.    Do.

5995-6. [6402-3] Cotton, Do.    Do.

---

5997. [8299] One pair stockings, Serinugger.

5998. [8300] Do. gloves, Do.

5999, 6000. [8301-2] Do. socks, Do.

---

### Umbrellas, Walking-Sticks, etc.

6001-5. [5780-4] Walking-sticks, bamboo, Ulwar.  H. H. the Maharajah.

6006. [2906] Do., of sago palm, Seebsagur.  Moonshee Koofaitoola.

6007. [968] Do., 'Churree,' Puttiala. H. H. the Maharajah.

6008. [969] Seven do., with ivory handles, Do.  Do.

6009. [970] Walking-sticks, do., Lahore. Government of India.

6010. [10248] Do.    Do.    Do.

6011. [5179] Do., silver mounted, Philibheet.  Sheik Budroodeen.

6012. [636] Walking-sticks of cocoa-nut wood, Vizagapatam.  H. H. the Rajah.

6013. [637] Do. of Palmyra wood, Do. Do.

6014. [638] Do. of betel palm, Do.  Do.

6015. [640] Do. of Kumba wood, Do. Do.

6016. [641] Do. of Chittunkoodoo wood, Do.  Do.

6017. [6342] Do., Singapore.  H.H. Inche Wan Aboo Bakar.

### Boots and Shoes.

6018-21. [5526-9] Four pairs shoes, various, 'Joota deesee,' worn by men, Ulwar. H. H. the Maharajah.

6022. [5520] Slippers, 'Jootee,' Lucknow.  Nawab Shurfood Dowlah.

6023. [5519] Shoes, 'boot,' Do.    Do.

6024. [5524] Slippers, 'Jootee,' embroidered in silver, worn by rich men, Lucknow. Rajah Goree Shunker, *Moraon, Oude.*

6025. [5525] Shoes, 'Jootee,' embroidered in silver, Do.  Do.

6026-8. [5521-3] Three pairs shoes, embroidered in gold, Do.  Do.

6029. [5756] Pair slippers, ornamented in gold tinsel, Meerut.  Koeer Wuzeer Ally Khan, *Dep. Magistrate of Meerut.*

6030-1. [2959-60] Two pairs clogs, Bareilly.  Government of India.

6032. [1759] Boots, Darjeeling.  Dr. Campbell.

6033. [5623] Green shoes, Cabul.  Government of India.

6034. [5625] Shoes, silver worked, Do. Do.

6035. [5624] Slippers, worked in chenille, Do.  Do.

6036. [1531] Shoes worn by Mezaree men, Lahore.  Do.

6037. [3812] Pair Bundela shoes, Jhansee. Do.

6038. [4720] Pair of sandals, Sawunt Warree.  Government of Bombay.

6039. [846] Four pairs shoes, Shikarpoor. GOVERNMENT of BOMBAY, *per* the COLLECTOR.

6040. [780] Pair riding boots, embroidered, Khyrpoor, Sinde. H. H. MEER ALI MOORAD.

6041-2. [781-2] Two pairs of shoes, Do.

6043-5. [621-3] Three do., Vizagapatam. H. H. the RAJAH.

6046-7. [624-5] Two pairs wooden sandals, Do. Do.

6048-9. [4466, 4496] Two pairs men's sandals (Mendoon Cutch), Rangoon. Messrs. HALLIDAY, FOX, & Co.

6050. [4497] Pair women's do., Do. Do.

6051-2. [6330-1] Two pairs women's slippers, embroidered, Singapore. TAN KIM SING.

6053. [4108] Pair women's sandals, Penang. PENANG LOCAL COMMITTEE.

6054. [4109] Do. men's do., Do. Do.

The articles of clothing for which space could not be afforded for exhibition are deposited at the India Museum. These include specimens from Ahmedabad, Belgaum, Darjeeling, Dharwar, Goruckpore, Guzerat, Jaloun, Jhansee, Jubbulpore, Mysore, Poona, Punaghur, Shikarpoor, &c. Selections from these collections are, however, on view.

# CLASS XXVIII. — PAPER, STATIONERY, PRINTING, AND BOOKBINDING.

## Section A. — PAPER, CARDBOARD, AND MILLBOARD.

### Paper in the Raw State as it leaves the Mill.

Although not strictly coming under the above heading, the subjoined samples of 'half stuff,' manufactured from jute, are here submitted.

6055. [5059] Red jute, prepared as 'paper stuff.' G. F. JEFFREYS, Esq.

6056. [5063] Do. do. Do.

6057-9. [5060–2] White jute, do. Do.

6060. [5364] Jute, bleached, do. Do.

### a. Printing and Writing Paper, etc.

6061. [2225] Mudar fibre paper (*Calotropis gigantea*), Bengal.

6062. [7764] Arsenical paper, Hooghly. — PALMER, Esq.

To preserve its contents from insects.

6063. [7765] Specimens of 'country paper,' Mahanad, Hooghly.

6064. [7766] Country paper, Pandooah, Do.

6065. [7767] Satgong paper, Do.

6066. [7768] Sha Bazar paper, Do.

6067. [7769] Doomurpore paper, Do.

6068. [5372] Paper made of hemp, Meerut. SUPERINTENDENT of JAIL.

Made by the prisoners in the Central Jail.

6069. [5375] Paper of plantain-leaf fibre. Do.

6070. [5374] Do. of aloe-leaf fibre, Do.

6071. [5371] Paper made of rags, Meerut. SUPERINTENDENT of JAIL.

6072. [5376] Do. made of old records, Do.

6073. [5373] Bibulous paper, Do.

6074. [5377] Herbarium paper (kyanised), Do.

---

6075. [10302] Paper made at Agra Central Jail. DR. W. WALKER.

Made of old ropes and gunny bags, bleached by means of carbonate of soda and lime. Such paper can be produced at 5R. or 10s. per ream.

---

6076. [3814] Kalpee paper, Kalpee, Jhansee.

---

6077. [7770] Paper of fibre of *Hibiscus cannabina*, Lucknow.

6078. [7248] Paper dyed with *Carthamus tinctorius*, Do.

---

6079. [6540] Stems of *Daphne cannabina*, and bark of Do., Nepaul. H. H. SIR JUNG BAHADOOR, K.C.B.

6080. [6541] · Paper brick of Do., Do. Do.

6081. [5904] Nepaul paper, 1st quality, Do.

6082. [5905] Do., 2nd quality. Do.

6083. [5906] Do., 3rd „ Do.

6084. [5907] Do., thinnest quality, Do.

Made from the *Daphne cannabina*. COLONEL RAMSAY, Resident at the Court of Nepaul, forwards the following remarks:—

The Daphne is a small evergreen perennial shrub, somewhat like a laurel, which bears poisonous berries. There are several species of it in Nepaul, from all of which, I am told, that paper is made. In some kinds the flowers are pure white, in others dirty white, tinged with pink or purple; and in my rambles last spring, at the back of a high mountain, north of the residency, I found two or three varieties of it, the flowers of which were bright yellow, very much like the large yellow jessamine, for which, at a little distance, I mistook it. These shrubs grow to a height of ten or twelve feet. I believe that the Nepaul paper has been sent to England in quantities, in all stages of preparation; but no notice has ever been taken of it, that I am aware of. Of the prepared pulp I sent two maunds to Messrs. Mackey & Co., of Calcutta, in November 1855. It cost 22R. 2A. per maund, delivered at Dinapore, and was intended for some Fibre Company in England; but as I have not had a line from those gentlemen since regarding it, I presume it was

not found to answer. There is an impression in the Plains (the *Friend of India* not long ago repeated it) that the Nepaul paper is prepared with arsenic. This is quite a mistake, for arsenic is not allowed to be sold here, nor any other virulent poison, under a heavy penalty. The whole tribe of plants bearing the name of Daphne are more or less poisonous; but the Daphne paper cannot retain the poisonous quality of the plant, as rats and insects often eat it with apparent avidity. In my opinion, this unsightly paper is much overrated. It is certainly tough when kept dry, and can be used like cloth, for wrapping up dry substances in; and it has one other good quality, which renders it superior in *that* respect to the ordinary country paper—it can be used after having been saturated with water, provided it be carefully dried within a reasonable time after it has been wetted.

6085. [10231] Nepaul paper, Goruckpore. W. OSBORNE, Esq.

6086. [1866] Stem, bark, and leaf of *Daphne cannabina*, Lahore.

Used for making Nepaul paper.

6087. [2223] 'Gundhera' paper (*D. cannabina*), Kangra.

Said never to be destroyed by insects.

---

6088. [6358] Nepaul paper, Behar.

This substance is manufactured almost exclusively in Nepaul from the bamboo, an arborescent grass. After being cut, it is beaten in wooden mortars until reduced to a pulpy mass, then thrown into a vat of water, the impurities separated, and when of a proper consistence, it is spread on linen to be dried; the surface is rendered smooth by friction, and with a pebble on boards. Its structure is very tough, and cannot be torn rectilineally: and it is most serviceable for filtration, as the fibres do not separate readily when saturated with moisture, and will resist in a moist condition considerable rough handling.

6089. [5368] Bamboo paper, Bullooah.

6090. [2224] Paper made from the bamboo, Lahore.

6091. [5368] Bamboo paper, Pegu.

---

6092. [6522] Paper 'Desee kaguz,' Ulwar. H. H. the MAHARAJAH.

---

6093. [7240] Tow of *Hibiscus esculentus*. DR. R. RIDDELL.

6094. [7241] Paper made therefrom. Do.

As already mentioned, DR. RIDDELL strongly recommends the *Hibiscus esculentus* for this purpose.

---

6095. [418] Paper of six kinds, Madras. RAMANJOOLO NAIDOO, manufacturer.

6096. [599] Paper of two sorts, Salem.

## Section B. — STATIONERY.

No representatives of this section.

## Section C. — PLATE, LETTERPRESS, AND OTHER MODES OF PRINTING.

**6097.** [10303] Specimen of lithographic printing, Agra Jail. DR. W. WALKER.

Taken, without selection or preparation, from the stones in use at the time, so as to give a fair idea of the style of work done in the prison.

**6098.** [5370] Specimen of lithographic printing, Meerut Jail. SUPERINTENDENT.

**6099.** [6643] Specimens of printing. BENGAL PRINTING COMPANY.

**6100.** [601] Native plants, nature-printed at Mangalore. V. P. COELHO, Esq.

**6101.** [8105] Two vols. nature-printed plants, Bengal.

**6102.** [532] Wood-cut block of sandal-wood, with impressions therefrom, showing the applicability of this wood to the purposes of the engraver. M. C. COOKE, Esq., *London.*

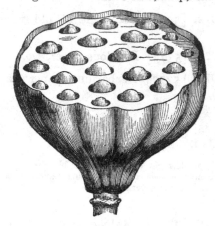

IMPRESSION OF SANDALWOOD BLOCK.

**6103.** [4265] Specimen of block printing (two kinds). DR. CAMPBELL.

**6104-6.** [4262–4] Printing blocks, Darjeeling. Do.

# CLASS XXIX. — EDUCATIONAL WORKS AND APPLIANCES.

## Section A.—BOOKS, GENERAL APPLIANCES FOR TEACHING, ETC.

**6107.** [9442] Specimen of Works published by the Christian Vernacular Education Society for India:—First and Second Reading Books in Punjabi; The Lanka Nidhana; Tamil Geographical Primer; Desopakari—an Illustrated Tamil Magazine, 2 Nos. and 2 Maps.

**6108.** [2221] Goldsmith's Traveller and Deserted Village, paraphrased for Schools and Colleges in India, Calcutta. Messrs. G. P. ROY & Co.

Paraphrased by BABOOS W. C. and G. C. MITTER.

**6109.** [2517] Sinde Educational Literature, 8 vols., bound with gold and silver, embroidered covers, Sinde.

These volumes are published by the Government Educational Department in Sinde. Director, MAJOR GOLDSMID.

**6110.** [3198] Memoirs of the Geological Survey of India, 2 vols. 8vo. PROF. OLDHAM.

**6111.** [3199] Palæontologia Indica, 1 vol. 4to. Do.

## NATIVE WORKS, WRITINGS, AND SPECIMENS OF CALIGRAPHY, ETC.

**6112.** [1768] 'Boom' (1 vol.), a Thibetan work in 12 vols., containing Tracts of the Eloopka Section, Darjeeling. DR. CAMPBELL.

The remaining 11 vols. are at the India Museum.

**6113.** [4300] Llama's Corona, Do. Do.

**6114.** [2198] Copy of the Treaty between the Government of Bengal and the Rajah of Sikkim. Do.

**6115.** [5406] Three Nagree writings, Ulwar.

**6116.** [5408] Nagree writing, Do.

**6117.** [5407] Do. Do.

**6118.** [5404] Four Persian writings. Do.

**6119.** [5401] Specimen of caligraphy, Do.

**6120.** [5403] Nastaley writing, Do.

**6121.** [5402] Do. Do.

6122-4. [1653-5] Specimen of caligraphy, Puttiala. H. H. the MAHARAJAH.

6125. [5410] Do., Loodianah. BAVA DAS-SOUN DHA SINGH.

6126. [5409] Cut writings, Do. Do.

6127. [5411] Do., executed with the thumb nail, Do.

6128. [5412] Three specimens of Persian writing, Lucknow.

6129. [5412] Specimen of caligraphy, Do.

6130. [5413] Portfolio.

6131. [2296] MSS. on Palmyra leaf.

Section B.—APPARATUS, ETC. EMPLOYED FOR EDUCATION, SUCH AS PLANS, SEC-TIONS, ETC.

6132. [8001] Series of drawings used in the School of Industrial Arts, Madras. DR. HUNTER.

6133. [4871] Eleven plates of geome-trical drawings, Do. Do.

Section C.—APPLIANCES FOR PHYSICAL EDUCATION, INCLUDING TOYS AND GAMES.

The specimens of ivory and horn-work coming under the head of 'Games,' are here inserted, along with the Table showing the value of the ivory and ivoryware exported from India (p. 250.)

6134. [3132] Set of chessmen, elabo-rately carved in ivory by BAWUL of *Berham-pore.* GOVERNMENT of INDIA.

6135. [3153] Pair dice boxes and 2 dice, Do. Do.

6136-7. [3146, 3152] Puzzles, Do. Do.

6138. [1658] Set of chessmen and dice, Do. Do.

6139. [5439] Do. and box, Lahore.

6140. [7250] Do., wooden, Cuttack. Do.

6141. [7997] Do., ivory, Luckimpore, Assam. BABOO LOTHONDHUR PHOOKIN.

6142-3. [1045, 8000] Two set of chess-men and dice, Puttiala. H. H. the MAHA-RAJAH.

6144. [10647] Chessboard, ivory and buffalo horn, Vizagapatam. GOVERNMENT of MADRAS.

6145. [10501] Do., satinwood and buf-falo horn, and two sets of men, Do. H. H. the RAJAH of VIZIANAGRAM.

6146. [10502] Chess table, do., Do. Do.

6147. [7810] Playing cards, two boxes, Jeypore. H. H. the RAJAH.

6148. [10324] Do., pack, Maunbhoom. GOVERNMENT of INDIA.

6149. [4717] Do., two boxes, Sawunt Warree. GOVERNMENT of BOMBAY.

Section D.—ILLUSTRATIONS AND SPECIMENS OF NATURAL HISTORY AND PHYSICAL SCIENCE.

A series of specimens of Fossil Cephalo-poda of the cretaceous rocks of Southern India. PROF. OLDHAM, *Superintendent of the Geological Survey of India,* viz. :—

6150. [a] *Belemnites fibula* Forbes, Ootatoor, Trichinopoly, 4 specimens.

6151. [b] Do.          Do.          9 do.

6152. [c] Do.          Do.          4 do.

6153. [d] Do.          Do.          9 do.

6154. [e] Do.          Do.          3 do.

6155. [f] *B. stilus* Blandford, Do. 3 do.

6156. [g] Do.          6 do.

6157. [h] Do.          4 do.

6158. [i] *B. seclusus* Blandford     9 do.

6159. [k] Do.          Do.          1 do.

6160. [l] Septum of Phragmocone of *Belemnites,* Odium, Trichinopoly.

6161. [1] *Nautilus Bouchardianus,* Shutanure, Do.

6162. [2] *Do.,* Olapandy, Do., pl. IV.,* fig. 1.

6163. [3] *Do.,* Pondicherry, Do., pl. V., fig. 2.

* These figures refer to the Illustrations in the Palæon-tologia Indica.

TABLE SHOWING THE QUANTITIES (AS FAR AS CAN BE ASCERTAINED) AND THE VALUE OF IVORY AND IVORYWARE EXPORTED FROM INDIA AND EACH PRESIDENCY TO ALL PARTS OF THE WORLD FROM 1850-51 TO 1860-61.

| Years | Whence Exported | United Kingdom Qty (cwt) | United Kingdom Value (£) | France Qty (cwt) | France Value (£) | Other Parts of Europe Value (£) | America Qty (cwt) | America Value (£) | China Qty (cwt) | China Value (£) | Arabian and Persian Gulfs Qty (cwt) | Arabian and Persian Gulfs Value (£) | Other Parts Qty (cwt) | Other Parts Value (£) | Total Qty (cwt) | Total Qty (tons) | Total Value (£) |
|---|---|---|---|---|---|---|---|---|---|---|---|---|---|---|---|---|---|
| 1850-51 | Bengal | | | | | | | | | | | | | | | | |
| | Madras | 1 | 26 | | | | | | | | | | | | 1 | | 26 |
| | Bombay | 2,641 | 40,927 | 11 | 203 | | 9 | 227 | 143 | 1,670 | 4 | 33 | | | 2,808 | 140 | 43,060 |
| | ALL INDIA | 2,642 | 40,953 | 11 | 203 | | 9 | 227 | 143 | 1,670 | 4 | 33 | | | 2,809 | 140 | 43,086 |
| 1851-52 | Bengal | | | | | | | | | | | | | | | | |
| | Madras | 7 | 140 | | | | | | | 3 | | | | | 7 | | 143 |
| | Bombay | 4,365 | 85,809 | 2 | 33 | | 90 | 2,110 | 370 | 2,034 | 2 | 10 | | | 4,829 | 241 | 89,996 |
| | ALL INDIA | 4,372 | 85,949 | 2 | 33 | | 90 | 2,110 | 370 | 2,037 | 2 | 10 | | | 4,836 | 242 | 90,139 |
| 1852-53 | Bengal | | | | | | | | | | | | | | | | |
| | Madras | | 243 | | | | | | | | | | | 1 | | | 244 |
| | Bombay | | 50,975 | | 15 | 31 | | 100 | | 4,379 | | 32 | | 249 | | | 55,781 |
| | ALL INDIA | | 51,218 | | 15 | 31 | | 100 | | 4,379 | | 32 | | 250 | | | 56,025 |
| 1853-54 | Bengal | 26 | 613 | | | | | | | | | | 8 | 143 | 34 | | 756 |
| | Madras | | 182 | | | | | | | | | | | 3 | | | 185 |
| | Bombay | | 77,957 | | 391 | 20 | | | | 1,068 | | 18 | | 520 | | 2 | 79,954 |
| | ALL INDIA | | 78,752 | | 391 | 20 | | | | 1,068 | | 18 | | 666 | | | 80,895 |
| 1854-55 | Bengal | | 613 | | | | | | | | | | | 82 | | | 695 |
| | Madras | | 359 | | | | | | | | | | | | | | 359 |
| | Bombay | | 64,601 | | 366 | 36 | | | | 460 | | 36 | | 484 | | | 65,867 |
| | ALL INDIA | | 65,473 | | 366 | 36 | | | | 460 | | 36 | | 566 | | | 66,921 |
| 1855-56 | Bengal | | 582 | | 39 | | | | | | | | | 349 | | | 970 |
| | Madras | | 187 | | | | | | | | | | | 4 | | | 191 |
| | Bombay | | 79,009 | | 51 | 38 | | | 204 | 1,921 | | 27 | | 179 | | | 81,223 |
| | ALL INDIA | | 79,778 | | 90 | | | | 204 | 1,921 | | 27 | | 532 | | | 82,384 |
| 1856-57 | Bengal | | 597 | | 44 | | | | | 11 | | | | 44 | | | 934 |
| | Madras | | 258 | | | | | | | | | | | 30 | | | 288 |
| | Bombay | | 123,298 | | 25 | 1,483 | | | | 1,010 | | | | 1,058 | | | 126,874 |
| | ALL INDIA | | 124,153 | | 69 | 1,521 | | | | 1,021 | | | | 1,132 | | | 128,096 |
| 1857-58 | Bengal | | 1,428 | | 38 | | | | | | | | | 444 | | | 1,916 |
| | Madras | | 408 | | | | | | | | | | | | | | 408 |
| | Bombay | | 14,880 | | 289 | 102 | | 6 | | 261 | | 13 | | 1,986 | | | 17,481 |
| | ALL INDIA | | 16,666 | | 327 | 102 | | 6 | | 261 | | 13 | | 2,430 | | | 19,805 |
| 1858-59 | Bengal | | 407 | | 136 | 25 | | | | 1,330 | | | | 946 | | | 2,844 |
| | Madras | | 345 | | | | | | | | | | | 10 | | | 355 |
| | Bombay | | 84,344 | | | | | | | 9,284 | | 283 | | 1,047 | | | 94,958 |
| | ALL INDIA | | 85,096 | | 136 | 25 | | | | 10,614 | | 283 | | 2,003 | | | 98,157 |
| 1859-60 | Bengal | | 341 | | | | | | 10 | 100 | | | | 445 | | | 886 |
| | Madras | | 583 | | | | | | | | | | | | | | 583 |
| | Bombay | | 85,098 | | 17 | | | 26 | | 9,599 | | 134 | | 783 | | | 95,657 |
| | ALL INDIA | | 86,022 | | 17 | | | 26 | | 9,599 | | 134 | | 1,228 | | | 97,126 |
| 1860-61 | Bengal | | 234 | | | | | | | 310 | | | | 33 | | | 577 |
| | Madras | | 484 | | | | | | | | | | | 31 | | | 515 |
| | Bombay | | 25,735 | | 12 | 6 | | 71 | | 3,688 | | 104 | | 1,430 | | | 31,046 |
| | ALL INDIA | | 26,453 | | 12 | 6 | | 71 | | 3,998 | | 104 | | 1,494 | | | 32,138 |

6164. [4] *Nautilus Bouchardianus* Arrialoor, Trichinopoly, pl. IV., fig. 6.

6165. [5] *Do.*, Shillagoody, Do., pl. V., fig. 3.

6166. [6] *Do.*, Arrialoor.

6167. [7.] *Do.*, young specimen, Arrialoor, Do.

6168. [8] *N. Clementinus*, Karapandy, Do., pl. VI., fig. 1.

6169. [9] *Do.*, Olapandy, Do., pl. VI.,fig. 2.

6170. [10] *N. Bouchardianus*, Arrialoor, Do., pl. VI. fig. 1.

6171. [11] *Do.*, Do., Arrialoor.

6172. [12] *Do.*, Koloture, Do., fig. 5.

6173. [13] *N. Huxleyanus*, Moonglepandy, Do., pl. IX., fig. 1.

6174. [14] *Do.*, Serdamungalun, Do., pl. VII. fig. 3.

6175. [15] *Do.* Andoor, Do.

6176. [16] *Do.* Shutanure, Do., pl. IX., fig. 3.

6177-8. [17–8] *Do.* Moonglepandy, Do.

6179. [19] *Do.* Andoor, Do.

6180. [20] *Do.* Coonum, Do., pl. VII., fig. 4.

6181. [21] *N. Danicus*, Sainthoray, Do.

6182. [22] *Do.* Ninnyoor, Do.

6183. [23] *N. Huxleyanus*, Andoor, Do.

6184-5. [24–5] *N. justus*, Odium, Do., pl. X., fig. 3.

6186-7. [26–7] *N. elegans*, Shutanure, Do., pl. VIII., fig. 4.

6188. [28] *N. splendens*, Odium, Do., pl. X., fig. 1.

6189. [29] *N. elegans*, Annapandy, Do., pl. XVI., fig. 2.

6190. [30] *N. formosus*, Karapandy, Do., pl. XIV., fig. 4.

6191. [31] *Do.* Andoor, Do., pl. XV., fig. 1.

6192. [32] *N. Kayeanus*, Ootatoor, Do., pl. XVIII., fig. 2.

6193. [33] *N. augustus*, Odium, Trichinopoly, pl. XIV., fig. 1.

6194. [34] *Nautilus clementinus*, Coothoor, Trichinopoly.

6195. [35] *N. justus*, Odium, Do., pl. X., fig. 2.

6196. [36] *N. Clementinus*, Ootacoil, Do., pl. VII., fig. 2.

6197. [37] *N. elegans*, Kunnanore, Do., pl. XVI., fig. 1.

6198. [38] *Do.* Andoor, Do., pl. XVI., fig. 4.

6199. [39] *N. Kayeanus*, Ootatoor, Do., pl. XVI., fig. 5.

6200. [40] *Do.* Do., pl. XVII., fig. 2.

6201. [41] *N. pseudo-elegans*, Odium, Do., pl. XVII., fig. 3.

6202. [42] *N. serpentinus*, Rayapoothapakkan, Do., pl. XII., fig. 1.

6203. [43] *N. splendens*, Odium, Do., pl. IX., fig. 5.

6204. [44] *N. Forbesianus*, Moraviatoor, Do.

6205. [45] *Do.* Do., pl. XIII., figs. 2, 3.

6206. [46] *N. Kayeanus*, Ootatoor, Do., pl. XVII., fig. 1.

6207. [47] *N. Forbesianus*, Odium, Do.

6208. [48] *N. Kayeanus*, Purawoy, Do.

6209. [49] *N. elegans*, Annapandy, Do., pl. XVI., fig. 3.

6210. [50] *N. formosus*, Kurribeem, Do., pl. XIV., fig. 3.

6211. [51] *N. Forbesianus*, Odium, Do.

6212. [52] *N. Kayeanus*, Ootatoor, Do., pl. XVIII., fig. 1.

6213. [53] *N. Negama*, Sirgumpore, Do., pl. XX., fig. 2.

6214. [54] *N. crebricostatus*, Ootatoor, Do., pl. XXII., fig. 1.

6215. [55] *N. pseudo-elegans*, Odium, Do., pl. XVIII., fig. 3.

6216. [56] *Do.* Do., pl. XX., fig. 1.

6217. [57] *N. Trichinopolitensis*, Arrialoor, Do.

6218. [58] *N. rota*, Mulloor, Do.

6219. [59] *N. crebricostatus*, Ootatoor, Trichinopoly, pl. XXI., fig. 3.

6220. [60] *Nautilus Kayeanus*, Trichinopoly, pl. XXI., fig. 2.

6221. [61] *N. pseudo-elegans*, Odium, Do., pl. XIX., fig. 1.

6222. [62] *N. Trichinopolitensis*, Arrialoor, Do., pl. XXIII., fig. 1.

6223. [63] *Nautilus pseudo-elegans*, Odium, Trichinopoly.

6224. [64] *N. rota*, Arrialoor, Do., pl. XXV., fig. 3.

6225. [65] *Do.* Karapandy, Do., pl. XXIV., fig. 3.

# Class XXX.—FURNITURE AND UPHOLSTERY, INCLUDING PAPER HANGINGS AND GENERAL DECORATIONS.

Section A.—FURNITURE AND UPHOLSTERY.

### Division I.—*Furniture (Carved, etc.)*

6226. [258] Large carved screen, with two side-tables, Bombay. BHRIMJEE BYRAMJEE, manufacturer. GOVERNMENT of BOMBAY.

6227. [5885] Carved cheffonier or sideboard. Do. Do.

6228. [5886] Carved sofa. Do.

6229. [5887] Do. Do.

6230. [251] Carved easy-chair. Do.

6231. [257] Do. Do.

6232. [256] Pair carved drawing-room chairs. Do.

6233. [266] Do. Do.

6234. [300] Do. Do.

6235-41. [252, 263-8] Carved flower-stands. Do.

6242. [8170] Do. flower-stands. Do.

6243-4. [259-60] Carved pier-tables. Do.

6245. [253] Carved teapoy or round table. Do.

6246. [255] Pair do. Do.

6247. [2883] Carved card-table. Do.

6248. [2884] Do. Do.

6249. [262] Carved oblong table. Do.

6250. [254] Carved oblong table, Bombay. J. GIBBS, Esq.

6251. [905] Carved flower vase, Ahmedabad. GOVERNMENT of BOMBAY, *per* AHMEDABAD COMMITTEE.

6252. [8168] Do. Do. R. ASHBURNER, Esq.

6253. [372] Carved frame for chimney-glass, Madras. J. DESCHAMPS, manufacturer.

6254. [373] Pair carved oval frames, Do. Do.

6255. [374] Carved Davenport or lady's writing-table, Do. Do.

6256. [371] Carved drawing-room sofa, Do. Do.

6257. [4451] Carved arm-chair (rosewood), Do. Do.

6258. [4452] Do. Do. Do.

6259. [4453] Pair carved small chairs, Do. Do.

6260. [4454] Do. Do. Do.

6261. [8169] Pianoforte in carved case, Madras. KIRKMAN & SON, *London.*

Exhibited as a specimen of carving by the native artisans of Madras, to whom the design was furnished by Messrs. KIRKMAN & SON, of London.

6262-4. [564-6] Flower-stands of stag horn, Vizagapatam. MADRAS GOVERNMENT.

Division II.—*Papier Maché, Japanned Goods (Inlaid Work), etc.*

### a. Papier Maché.

6265-6. [5638-9] Blotting case, Seringugger, Cashmere. CENTRAL COMMITTEE, *Lahore.*

6267. [5640] Tea caddy,　　Do.　Do.

6268-71. [5641-4] Glove boxes, Do. Do.

6272-3. [5636-7] Pin boxes,　Do. Do.

6274-9. [5626-31] Cigar cases, Do. Do.

6280-1. [5632-3] Card trays, Do. Do.

6282. [5635] Pipe case,　　Do. Do.

6283. [3245] Pen cases, Umritsur.

6284-6. [10008-10] Three inkstands, Budaon.

6287. [4445] Plate of grapes and pomegranates, papier maché, Madras. MR. W. KOMARECK.

6288. [4446] Plate of mangoes, pomegranates and oranges, Do.　Do.

6289. [4447] Plate with loaf of bread, Do. Do.

6290. [4448] Plate of roast turkey, Do. Do.

6291. [4449] Plate of pine apple, Do. Do.

6292. [4450] Pomegranates, Do. Do.

### b. Japanned Work.

6293-4. [1458, 3405] Box and stand, Bareilly. ALLAHABAD COMMITTEE.

6295-6. [6493-4] Two drawing room chairs, green velvet cushions, embroidered, Do. Do.

6297. [2955] Teapoys, Do.　　Do.

6298-9. [2957-8] Two footstools, Do. Do.

6300-1. [10031, 10051] Knitting boxes, Do. Do.

6302. [5355] Bookholder,　Do.　Do.

6303. [5442] Set of charpoy or bedstead legs, Meerut. Do.

6304. [3501] Folding bedstead, Moradabad. GOVERNMENT OF INDIA.

---

Divisions III. and IV.—*Indian Inlaid Work.*

The following note on the 'Inlaid Work' of Bombay is by DR. BIRDWOOD :—

This kind of work is stated to have been carried on in Bombay for about 60 years, having been originally introduced from Hyderabad, Sinde. It is said to have been introduced into Sinde about twenty years previously from Persia; its native seat is supposed to be Shiraz. From Bombay the work has been carried to Surat.

The following are the materials used in the work :—
A mineral *green dye* for dyeing the stag's horn.
*Tin Wire* (Kylacenotur) used in the ornamental veneering.
*Sandalwood, ebony* and *sappanwood* used in the framework, and sometimes entering into the ornamental veneer.
*Ivory,* Do.
*Stag's Horn,* Do. Dyed green with mineral dye.
*Glue,* for binding. Ahmedabad glue being esteemed far above all other kinds, including English.
The tools employed are a *wheel* for drawing the tin wire into different shapes for the preparation of the ornamental patterns;
*Saws* of different kinds, files, chisels, drills, planes, and a T square.

### PROCESS OF MANUFACTURE.

The only 'mystery' is in the portion of the work which appears inlaid, but which is not inlaid in the first sense of the term. The patterns are veneered on, and may be applied to any flat or gently rounded surface.

The ornamental veneer is prepared by binding the rods of ivory, tin, sappan, ebony, and green dyed stag's horn of different shapes together.

These rods are usually three-sided, cylindrical, and obliquely four-sided.

They are arranged so as when cut across to exhibit definite patterns, and in the mass present either the appearance of rods or of thin boards, the latter being to be sliced down into borders. The primary rods are sometimes bound together before being sliced, so as to form more complex patterns.

The patterns commonly found in Bombay ready prepared for use are :—

1st. *Chukra* (*i.e.* wheel), the smaller being of the diameter of a fourpenny bit, and the larger of a shilling.
2nd. *Kutkee,* or hexagonal, being composed of obliquely four-sided rods, of ivory, ebony or sandalwood, and of ebony, tin wire puttung, and green dyed stag's horn mixed.
3rd. *Trenkoonia gool* (*i.e.* three-sided flower), a three-sided pattern composed of tin wire, ebony, ivory, puttung, and green dyed stag's horn.
4th. *Gool* (flower), obliquely four-sided, and compounded as last. These are all for the central veneer.
The border patterns are :—
5th. *Teekee,* round and varying in size from a twopenny bit to a large pin's head, and used for the central patterns as well as for bordering.
6th. *Gundeerio* (plumb, full), composed of all the materials used in this work.
7th. *Ek dana* (one grain), having the appearance of a single row of tin beads set in ebony.
8th, 9th, and 10th. *Poree lehur,* '*Sanksoohansio*' and '*Porohansio,*' varieties of border ornaments not easy to distinguish from one another by mere description.
*Remarks.* Fifty manufacturers are established in Bombay, the majority having commenced business during the last few years; six, however, have been settled here from periods varying from twenty-five to forty-six years. A few employ workmen, but the majority work for themselves with the aid, in many cases, of a brother or son. The inlaid work resembles Tunbridge ware.

6305. [525] Materials, tools, and patterns referred to above, and exhibited in centre of Case 10.

## SPECIMENS EXHIBITED.

6306. [1507] Box, inlaid with ivory, Delhi. GOVERNMENT of INDIA.

6307. [302] Chessboard, Surat. GOVERNMENT of BOMBAY.

6308. [314] Jewel box, Do. Do.

6309. [271] Chessboard, Bombay. Do.

6310. [272] Workbox, small, Do. Do.

6311. [273] Inkstand, Do. Do.

6312. [274] Bookstand, Do. Do.

6313-4. [275–8] Watchstands, Do. Do.

6315-6. [279–80] Cribbageboard; double, Do. Do.

6316-7. [281–2] Do., single, Do. Do.

6318-9. [283–4] Card basket, Do. Do.

6320. [287] Two card cases, Do. Do.

6321. [288] Four card cases, Do. Do.

6322. [2751] Six needle cases, Do. Do.

6323. [2752] Do. Do. Do.

6324. [290] Square table, Do. Do.

6325. [291] Working and writing case, Do. Do.

6326. [292] Glove box, Do. Do.

6327. [293] Envelope case, diamond pattern, Do. Do.

6328-9. [294–5] Do., round, Do. Do.

6330. [296] Portfolio, Do. Do.

6331. [297] Card basket, Do. Do.

6332. [298] Glove box, Do. Do.

6333. [299] Wafer box, Do. Do.

6334. [335] Round table, with two drawers, Do. Do.

6335. [336] Writing desk, large, Do. Do.

6336. [337] Do., small, Do. Do.

6337. [339] Solitaire game, Do. Do.

6338. [340] Inkstand, Do. Do.

6339-40. [341–2] Card basket, Do. Do.

6341. [344] Portfolio, diamond pattern, Bombay. GOVERNMENT of BOMBAY.

6342. [345] Do., black, Do. Do.

6343-5. [347–9] Envelope cases, Do. Do.

6346-9. [350–3] Four glove boxes, various patterns, Do. Do.

6350-4. [356–60] Five do., Do. Do.

6355. [361] Bookstand, Do. Do.

6356. [362] Paper stand, Do. Do.

6357. [363] Pin cushion, Do. Do.

6358. [364] Round box, Do. Do.

6359-62. [365–8] Four wafer boxes, Do. Do.

6363-4. [369–70] Two round boxes, Do. Do.

6365. [502] Portfolio, diamond pattern, yellow, Do. Do.

6366. [503] Bookstand, Do. Do.

6367. [504] Solitaire game, board and 33 marbles, Do. Do.

6368. [505] 'Panjra,' or paper holder, Do. Do.

6369-71. [506–8] Paper holders, Do. Do.

6372. [509] Writing desk, Do. J. MACFARLANE, Esq.

6373. [510a] Chessboard, Do. Do.

6374. [510b] Two paper cutters, Do. Do.

6375. [511] Workbox, Do. Do.

6376. [512] Inkstand, Do. Do.

6377. [513] Bookstand, Do. Do.

6378. [514] Watchstand, Do. Do.

6379. [515] Card basket, Do. Do.

6380. [516] Card basket, Do. Do.

6381. [517] Glove box, round top, Do. Do.

6382-3. [518–9] Do., diamond pattern, Do. Do.

6384-5. [520–1] Portfolio, Do. Do.

6386. [522] Cribbage board, Do. Do.

6387-91. [10650–4] Knitting boxes, Vizagapatam. H. H. the RAJAH.

6392. [583] Card basket, Vizagapatam. COLLECTOR of VIZAGAPATAM. MADRAS GOVERNMENT.

6393. [584] Writing box, Do. Do. Do.

6394. [585] Book case, Do. Do. Do.

6395. [419] Envelope case, Do. DR. McPHERSON, Madras.

6396. [420] Tea caddy, Do. Do. Do.

6397. [421] Jewel case, Do. Do. Do.

6398. [429] Tortoise of ivory and tortoise shell, Do. Do. Do.

6399. [2676] Workbox, Bangalore, Mysore. J. LACEY, Esq.

6400. [2678] Card basket, Do. Do.

6401. [2679] Jewel case, Do. Do.

6402. [2405] Two paper cases. J. GLADDING, *London.*

6403-4. [2403-4] Workbox. Do.

6405. [2402] Writing desk. J. GLADDING, *London.*

6406. [2406] Inkstand. Do.

6407-9. [2407-8-10] Glove boxes. Do.

6410. [2442] Walking stick. Do.

6411. [2437] Workbox. J. NEAL, *London.*

6412. [2434] Writing desk. Do.

6413. [2436] Paper case. Do.

6414. [2432] Dressing case. Do.

6415. [2440] 'Fox and geese' board. Do.

6416. [2435] Sandalwood workbox. Do.

6417. [2438] Do. inkstand. Do.

6418. [2433] Envelope case. Do.

6419. [2441] Set of ebony chessmen. Do.

6420. [7751] Sandalwood envelope case. Messrs. FARMER & ROGERS, *London.*

6421. [7752] Portfolio. Do.

6422. [7753] Knitting box. Do.

# CLASS XXXI.—IRON AND GENERAL HARDWARE.

## HARDWARE, METAL WORK, ETC.

6423. [3299] Hookah bottom, brass, plated and engraved, large, Moradabad.

6424-5. [3300-1] Do. do., small, Do.

6426-7. [1527-8] Two tumblers, do. do., Do.

6428. [1517] Tumbler, large, do. do., Do.

6429. [1518] Do. small, do. do., Do.

6430-1. [1525-6] Do., do. do., with covers and platters, Do.

6432-3. [1512-13] Two plates, small, do. do., for betel leaf, Do.

6434. [1523] Spittoon, large, do. do., Do.

6435. [1542] Do. small, do. do., Do.

6436-7. [1521-2] Two cups, do. do., Do.

6438-9. [1519-20] Two brass wine goblets, plated and engraved, Moradabad.

6440. [1516] Plate, large, do. do., Do.

6441-2. [1514-5] Two do., small, do. do., Do.

The above articles are remarkable for the beauty of the patterns engraved on them, also for their shapes and the superior way in which they are tin-plated. Contributed by MAHOMED SYUT REHMUT ALEE, KADIR BUX, MAHOMED HAFIZ, and MAHOMED HYNE through the CENTRAL COMMITTEE, *Allahabad.*

6443. [8178] 'Ghurra,' brass, for fetching water, Hooghly. GOVERNMENT of INDIA.

6444. [8179] 'Garoo,' water vessel for cleaning purposes, Do. Do.

6445. [8183] 'Dabaree,' brass, vessel for steeping betel leaf, Do. Do.

6446. [8189] 'Bhogona,' do., for cooking curries, &c., Do. Do.

6447. [8187] 'Ghotee,' do., for drinking, Do. Do.

6448. [8188] 'Ghotee,' brass, engraved, for drinking, Hooghly. Government of India.

6449. [8180] 'Picdan,' do., a spittoon, Do. Do.

6450. [8190] 'Recaybee,' plate for tiffin or luncheon, Do. Do.

6451. [8185] 'Hookah Bytuck,' stand for hookah, Do. Do.

6452. [8181] 'Kassa,' or 'Thalla,' of mixed metal, plate for food, Do. Do.

6453-4. [8191-2] 'Batta,' do., do. with cover, on which prepared betel leaf is kept, Do. Do.

6455. [8182] 'Ghuttee,' of mixed metal, Do. Do.

For drinking water.

6456. [8184] 'Battee,' do., Do. Do.

For keeping curries.

6457-64. [8186-93] Mugs, do., Do. Do.

A recent invention, in imitation of European drinking glasses.

———

6465. [5253] Spittoon, brass, Sewan. Shaik Fukeer Haissain.

6466-8. [5252-4] Hookah, do., Do. Do.

6469. [5251] 'Catorah,' do., large cup, Do. Do.

6470. [5258] Do., do., small cup, Do. Do.

6471. [5257] Mug, do., Do. Do.

6472-3. [5255-6] Do., do., small, Do. Do.

———

6474. [5259] Dish, of brass, Luckimpore, Assam. Baboo Kessubram Borooah.

———

6475. 5573] Brass cups, Cuttack. Government of India.

6476. [5574] Brass bangles, &c., Do. Do.

———

6477. [5926] Sacrificial lamp, on elephant stand, Nepaul. H. H. Sir Jung Bahadoor, K.C.B.

6478. [5930] Plate, small, brass, Do. Do.

6479. [5927] Vessel, copper, used at festivals, Do. Do.

6480. [5925] Water pot, or 'Jharee,' Do. Do.

6481. [5928] Pen and ink case, brass, worn in the belt, Nepaul. H. H. Sir Jung Bahadoor, K.C.B.

6482-3. [5261-2] Two locks, used by native bankers, &c., Do. Do.

6484. [5260] Iron padlock, Patna. Government of India.

6485. [1532] Iron umbrella, comprising inkstand, dagger, spear, scissors, &c.; it may also be converted into a lanthorn, Sarun. Patna. Baboo Bindas Huree Pershad

6486. [8197] Water engine and pump, brass, Do. Do.

6487. [5366] 'Sarota,' for cutting betel nuts, silver, Shahjehanpore. Kunhye Loll Tehseldar, *Jehanabad*.

6488. [5367] Do., steel, Do. Mindaee Lohar.

6489-90. [1765, 8195] Bells, Darjeeling. Dr. Campbell.

6491. [3182] Copper vessel, Do. Do.

6492. [3163] 'Mane,' or praying cylinder, Do. Do.

6493. [4273] Model of Llamas' tomb, Do. Do.

6494. [8194] Three Dorjé sceptres, Do. Do.

6495-6. [3167-8] Two pairs Llamas' cymbals, Do. Do.

9497. [4266] Iron spoon, with Dorjé head, Do. Do.

6498. [8196] 'Mane,' or praying wheel, Thibet. India Museum.

———

6499-6500. [1582-3] Two castings in brass, Maunbhoom, Chota Nagpore. Prof. Thos. Oldham.

These specimens are interesting, not from the size or beauty and high finish of the workmanship, but from the ingenuity displayed in the mode of casting articles of this kind in hollow net work, &c. This is accomplished as follows:—

A core is made of plastic clay, all carefully shaped to the internal form of the fish or other object to be imitated. This core is then baked and indurated. On this, the pattern designed to be represented is formed with wax (ordinary clean bees' wax is used). This done, and the wax having cooled, it becomes tolerably hard. Soft clay is moulded over all. The whole is then baked, the heat indurating the outer coating of clay, but softening the wax, which all runs out of the mould, leaving empty the space occupied by it. The mould being sufficiently dried, the molten brass is then poured into the empty space, and, when cool, the clay is broken away, when the result is as exhibited. These are untouched after the casting, excepting on the smooth and flat surfaces, which are roughly filed.

These sell in the district at sixpence to one shilling each, and are used and hung from the neck or waist as boxes for tobacco, pán, &c.

6501. [996] Eleven pairs metal bangles, Jhansi. GOVERNMENT of INDIA.

6502. [2297] Two Burmese spoons, Rangoon. Messrs. HALLIDAY, FOX, & Co.

6503. [2299] Tattooing implement, Do. Do.

6504. [4476] Areca nut crushers, do. Do.

6505. [4479] Tweezers, Do. Do.

6506. [6358] Tin cup and stand, and goglet and stand, Malacca. HON. CAPTAIN BURN.

These are illustrative of the tin ore series exhibited by CAPT. BURN, in Class I.

6507. [10568] Cup, small, engraved, 'Jambu,' North Arcot.* SWARAMMUDU, VENGUBATHUDU, and KALAPA, manufacturers. GOVERNMENT of INDIA.

6508. [10569] Do., large, Do. Do. Do.

6509. [10570] Do., small, Do. Do. Do.

6510. [10571] Jug, Do. Do. Do.

6511. [590] Goglet, metal, Tanjore. W. M. CADELL, Esq., *Collector of Tanjore.*

6512. [591] Wine-glass, do., and cover, Do. Do.

6513. [592] Tumbler, do., Do. Do.

6514. [7998] Fishing reel, made by BOODRAJ, a native smith, at Aurungabad, Deccan. DR. RIDDELL.

6515. [1059] Rose-water sprinkler, Hyderabad. GOVERNMENT of INDIA.

* A variety of metal idols from North Arcot have been deposited at the India Museum.

# CLASS XXXII.—STEEL, CUTLERY, AND EDGE TOOLS.

6516. [1078] Pen-knife, Hyderabad.

6517. [10591] Three hunting knives and leather cases, Salem. BABOO ARNACHELLUM.

6518. [10593] Bread knife. Do.

6519. [10595] Two table knives. Do.

6520. [10592] Two spear heads and sheaths. Do.

6521-4 . [10596-9] Garden scizzors BABOO ARNACHELLUM.

6525. [10600] Pocket knife. Do.

6526. [10594] Turnscrew. Do.

6527. [6514] Four knives, Assam. LIEUT. W. PHAIRE.

# CLASS XXXIII. — WORKING IN PRECIOUS METALS

## AND IN THEIR IMITATIONS; JEWELLERY, AND ALL ARTICLES OF VERTU AND LUXURY NOT INCLUDED IN THE OTHER LISTS.

The value of the articles of this class exported from India up to the present time will be gathered from the Table below.

TABLE SHOWING THE VALUE OF JEWELLERY EXPORTED FROM INDIA AND EACH PRESIDENCY TO ALL PARTS OF THE WORLD FROM 1857–58 TO 1860–61.

| Years | Whence Exported | United Kingdom | France | Other Parts of Europe | America | China | Arabian and Persian Gulfs | Other Parts | Total Exported to All Parts |
|---|---|---|---|---|---|---|---|---|---|
| | | Value | Value | Value | Value | Value | Value | Value | Value |
| | | £ | £ | £ | £ | £ | £ | £ | £ |
| 1857-58 | Bengal | 1,433 | 334 | .. | .. | 567 | .. | 803 | 3,137 |
| | Madras | 1,299 | 9 | .. | .. | 34 | .. | 377 | 1,719 |
| | Bombay | 40 | 78 | .. | .. | 43 | 858 | 1,459 | 2,478 |
| | ALL INDIA | 2,772 | 421 | .. | .. | 644 | 858 | 2,639 | 7,334 |
| 1858-59 | Bengal | 2,583 | 300 | .. | .. | 17 | .. | 2,089 | 4,989 |
| | Madras | 1,221 | .. | .. | .. | .. | .. | 286 | 1,507 |
| | Bombay | 102 | 20 | .. | 10 | 182 | 1,241 | 1,162 | 2,717 |
| | ALL INDIA | 3,906 | 320 | .. | 10 | 199 | 1,241 | 3,537 | 9,213 |
| 1859-60 | Bengal | 356 | .. | .. | 13 | .. | .. | 1,495 | 1,864 |
| | Madras | 1,538 | .. | .. | .. | .. | .. | 377 | 1,915 |
| | Bombay | 139 | 20 | .. | .. | 44 | 256 | 2,366 | 2,825 |
| | ALL INDIA | 2,033 | 20 | .. | 13 | 44 | 256 | 4,238 | 6,604 |
| 1860-61 | Bengal | 656 | 5 | 214 | .. | 53 | .. | 3,658 | 4,586 |
| | Madras | 1,893 | 200 | .. | .. | .. | .. | 329 | 2,422 |
| | Bombay | 287 | 12 | .. | .. | 615 | 889 | 1,871 | 3,674 |
| | ALL INDIA | 2,836 | 217 | 214 | .. | 668 | 889 | 5,858 | 10,682 |

## ARTICLES OF GOLD AND SILVER, ETC.

### *a.* Works in Gold.

6528-9. [2257-8] Two gold bracelets, Madras. T. CHOCALINGUM, jeweller.

6530. [5255] Do. HENCKELL, DU BUISSON & Co., for MOOTIANASSARY, jeweller, *Trichinopoly.*

6531. [2264] Do. and three tassels, Trichinopoly. ROSALINGUM ASSARY, jeweller.

6532. [2265] Bangle, lion's head, gold. Do.

6533. [2266] Do., snake's head, do. Do.

6534. [2267] Bracelet ball, do. Do.

6535. [2268] Do., rose snake, do. Do.

6536. [2269] Filigree snake bracelet. Do.

6537. [2270] Bracelet, smooth snake, gold. Do.

6538. [2271] Chain rose, gold. ROSALINGUM ASSARY, jeweller.

6539. [2272] Do. bobbin, do. Do.

6540. [2273] Do. tape, do. Do.

6541. [2274] Brooch Sawmi, do. Do.

6542. [2275] Studs and sleeve links, do. Do.

6543. [2276] Brooch, elephant, do. Do.

6544. [2277] Bracelet, do. Do.

6545. [2278] Do., elephant, do. Do.

6546. [2279] Do., with three tassels, do. Do.

6547. [2280] Two puzzle rings, do. Do.

6548. [2281] Brooch, snake, do. Do.

6549. [2282] Do., with painting of Trichinopoly Rock. Do.

6550. [2287] Bracelet, gold, and beads, do. Do.

6551-5. [6776-80] Four thimbles, gold, Kutch. H. H. the RAO.

6556-7. [1062-63] Two scent bottles, Hyderabad. GOVERNMENT of INDIA.

———

6558. [3096] Necklace.

6559. [3097] Pair bracelets, gold.

6560 [3098] Nose ring.

6561. [3099] Pair earrings.

6562. [3100] Head button, gold.

Set of jewels worn by females, Ulwar. H. H. the MAHARAJAH.

———

6563. [2763] Head ornament, gold, set with rubies. SIR R. HAMILTON, BART.

Worn by Mahratta ladies.

6564. [2764] Pair earrings, gold, emeralds and pearls, Bhopaul. SIR R. HAMILTON, BART.

Worn by Mahomedans.

6565. [2765] Pair earrings, gold, emeralds and rubies. SIR R. HAMILTON, BART.

Worn by Mahomedans.

6566. [2766] Pair earrings, gold, pearls, emeralds, and rubies, Oossein. SIR R. HAMILTON, BART.

Worn by Hindoos.

6567. [2767] Pair earrings, enamel and pearls, Oossein. SIR R. HAMILTON, BART.

Worn by Mahratta ladies.

6568. [2768] Pair earrings, gold, Indore. SIR R. HAMILTON, BART.

Worn by Hindoos.

6569-70. [2769-70] Two pair earrings, enamel and pearls. SIR R. HAMILTON, BART.

Worn by Mahratta ladies.

6571. [2771] Pair earrings, pearls, emeralds, and rubies. SIR R. HAMILTON, BART.

Worn by Mahratta ladies.

6572. [2772] Pair earrings, pendant, emeralds and pearls, Delhi. SIR R. HAMILTON, BART.

6573. [2773] Nose ornament, gold and pearls, Indore. Do.

6574. [2774] Nose ornament, gold, pearls, and diamonds, Delhi. SIR R. HAMILTON, BART.

6575. [2775] Head ornament, gold, enamel, pearls, and emeralds, Madras. Do.

6576. [2776] Necklace, gold enamelled, 64 pearls, 29 rubies, 4 emeralds, Indore. Do.

6577-8. [2780-1] Two bracelets, massive gold, Mahratta. Do.

6579. [2782] Bangle, gold, Hindoo. Do.

6580. [2783] Rings for the toes, gold, Do. Do.

6581. [2784] Spice box, pendant, gold, rubies, and emeralds, Bhopaul. Do.

6582. 2785] Betel-cutter and box for lime, gold, Hindoo. Do.

6583-4. [2786-7] Two gold armlets, Do. Do.

6585. [2788] Two curious gold rings, Umferah. Do.

———

6586. [3065] Bracelet, gold, Babool work, six pieces, Delhi. HURRUCK CHUND, jeweller.

6587. [3066] Do., do., do., light ornamental, Do. Do.

6588. [3067] Brooch, do., and large amethyst, Do. Do.

6589. [3068] Do., do., circular, with pearls, Do. Do.

6590. [3069] Do., do., do., and painting on ivory of Kotoob Minoar, Do. Do.

6591. [3070] Do., do., and painting on ivory of Sufdar Jhung's tomb, Do. Do.

6592. [3071] Do., do., light gold work, rubies, and pearls, Do. Do.

6593. [3072] Do., do., topaz, and carbuncle, Do. Do.

6594. [3073] Do., do., filigree, turquoise, and pearl, Do. Do.

6595. [3074] Do., sword-shaped, Do. Do.

6596. [3075] Do., light gold, crescent-shaped, turquoise, Do. Do.

6597. [3080] Pair of bracelets, carbuncle and turquoise, nine pieces each, Do. Do.

6598. [3086] Brooch, gold, and five large carbuncles, Do. Do.

6599. [3087] Brooch, gold, and pendants, Delhi. HURRUCK CHUND, jeweller.

6600. [3088] Do., small square, four pearls and turquoise, Do. Do.

6601. [3089] Cornelian cross, Do. Do.

6602. [3090] Gold breast pin, Do. Do.

6603. [3113] 'Surpech,' 175R., Benares. CHOONEE LALL.

6604. [3105] Necklace or garland of pearls, 350R., Do. Do.

6605. [3114] 'Dusta' or bracelet, 100R., Do. BUKHTA DUR SINGH.

6606. [3076] Pair gold bracelets, Baboolwork, Delhi. JOWALLIE SHAW.

6607. [3077] Bracelet, gold and amethyst, Do. Do.

6608. [3078] Do., gold, with paintings on ivory, Do. Do.

6609. [3079] Do., gold, turquoise, Do. Do.

6610. [3081] Do., gold, Babool work, Do. Do.

6611. [3082] Four brooches, gold, crescent shaped, Do. Do.

6612. [3083] Brooch, gold, Babool pattern, Do. Do.

6613. [3084] Earrings (one pair), Do., Do. Do.

6614. [3054] Bracelet, gold filigree, with onyx, five pieces, Delhi. BHYARO DOSS.

6615. [3055] Do., gold and turquoise, Do. Do.

6616-7. [3056-7] Two do., gold filigree, topaz, and small stones, Do. Do.

6618. [3059] Brooch, gold and turquoise, Do. Do.

6619. [3060] Do., gold, double crescent, with turquoise, Do. Do.

6620. [3062] Small turquoise, with locket, Do. Do.

6621. [3064] Bracelets, gold and turquoise, Do. Do.

6622. [3091] Brooch, do., Do. Do.

6623-4. [3093-4] Six small turquoise crosses, Delhi. BHYARO DOSS.

6625. [3095] Brooch, horse-shoe pattern, Do. Do.

6626. [3107] Pair of Bazoo, 85R., Benares. BABOO FUKEER CHAND.

6627. [3108] 'Chumpa Kullee' necklace, 175R., Do. Do.

6628. [3109] Sattara garland of pearls, 325R., Do. Do.

6629. [3110] Pair of earrings, 120R., Do. Do.

6630. [3112] A Bundee, 175R., Do. Do.

6631. [3111] Pair of bracelets, 65R., Do. Do.

6632. [3115] Necklace of 122 pearls and emeralds, with diamond and topaz enamelled locket, Lucknow. JNO. MARTIN, Esq., *Calcutta.*

6633. [3116] Finger ring, gold, with an emerald, with Persian inscription, 'Badsha Jhazie Nusserwoollah Hysler Sultan,' Do. Do.

6634. [3117] Finger ring, gold, with 20 diamonds and Persian inscription, 'Ghazie ool Hyderabad Nussera Sultan,' Do. Do.

6635. [3118] Curious pearl with Persian inscription, 'Namee Shah Alum Gheer Bahadur Shah,' Do. Do.

5636. [3119] Two very curious shaped pearls, plain, Do. Do.

6637. [3120] Large pearl, gold mounted and set with rubies and other stones, Do. Do.

6638. [3996] Pearl ornament, taken at the capture of Seringapatam. DR. RIDDELL.

6639. [8307] Brooch, gold, aquamarine stone, set in pearls, Do.

6640. [10290] Pair tiger-claw bracelets set in gold, Calcutta. GOVERNMENT.

6641. [10291] Do. and turquoise, Do. Do.

6642. [10292] Four pairs earrings, gold, Do. Do.

6643. [10294] Three pairs sleeve links, gold, Do. Do.

6644. [10295] Three brooches, Calcutta. GOVERNMENT.

6645. [10296] Do., with turquoises, Do. Do.

6646. [10297] Two breast pins, Do. Do.

6647. [10299] Two vinaigrettes, Do. Do.

Nos. 6640 to 6647 were manufactured by native artists under the superintendence of Messrs. ALLEN & HAYES, Government jewellers, *Calcutta.*

6648. [626] Gold 'Chikoo' ring, Vizagapatam. RAJAH of VIZIANAGRAM.

6649. [8310] Three gold bracelets. J. NEAL, *Edgware Road.*

6650. [1396] Gold schist or archer's thumb ring, inlaid with rubies. COLONEL GUTHRIE.

6651. [1399] Gold enamelled and diamond schist or archer's thumb ring. Do.

6652. [4278] Gold locket. DR. CAMPBELL.

6653. [4289] Gilt image set in turquoise, Do.

6654. [4285] Gilt and malachite locket, Do.

6655. [4492] Sapphire ring, Rangoon. Messrs. HALLIDAY, FOX, & CO.

6656. [3101] Sapphire ring, Burmah Proper.

6657. [3102] Ruby, Do. Do.

The ruby and sapphire mines are to the north of Mandalay, the present capital of Burmah. All attempts of Europeans to visit them are frustrated by various impediments thrown in their way by the Burmese authorities, who are very jealous of too close a scrutiny into the source of wealth on which they so much pride themselves. Great numbers of these gems are brought down to Rangoon for sale, but a heavy price is always demanded for them, and it requires an experienced eye to purchase them with a view to profit.

Topazes are also found in the vicinity of the rubies and sapphires, but they are scarce, and fetch a higher price in Burmah than they would realise in England.*

*b.* **Works in Silver.**—Filigree Work, Plate, &c.

6658. [2411] Epergne, double branch, filigree, Cuttack, 599R. 1A. JUGGER NATH DOSS, maker.

6659. [2412] Do. do. Do. 377R. 8A. SEEBOO, maker.

6660. [2413] Do. do. Do. 285R. 5A. CHOTASEEBOO, maker.

6661. [2414] Jewel casket, filigree, Cuttack, 165R. KUNNYE, maker.

6662. [2415] Basket, do. Do. 81R. 4A. SAHOO SONAR RAMCHUNDER, maker.

6663. [2416] Pair of bracelets, do. Do. 15R. 15A.

6664. [2417] Do. Do. 24R. 3A. 6P.

6665. [2418] Brooch, in form of a cross, do. Do. 6R. 6A.

6666. [2419] Bouquet holder, do. Do. 15R. 15A.

6667. [2420] Ring stand, do. Do. 14R. 1A.

The native silversmiths of Cuttack have long been noted for the fineness, neatness, and lightness of their filigree work. This kind of work is executed, for the most part, under supervision, by mere boys, whose nimbler fingers and keener eyesight are supposed to enable them to bring out and put together the minute patterns with more distinctness and accuracy than their elders can; comparative cheapness is, perhaps, another reason for their employment. The ruling rates for this filigree work are from two to two and a half rupees, that is to say, taking the first rate, two rupees or four shillings is charged for every rupee weight of finished silver work, namely, one rupee for workmanship, and one rupee as the price of the silver. This branch of industry is, however, declining from want of sufficient demand. These articles are all of the purest silver. The filigree work in gold seems almost as good as that of Delhi.*

6668. [746] Calcutta trades' plate for 1860-61. G. PLOWDEN, Esq. *Calcutta.*

Manufactured by native artists under the superintendence of Messrs. ALLEN & HAYES, *Calcutta.*

6669-70. [5071-2] Silver vase. RAJAH DENORAIN SINGH BAHADOOR, *Benares.*

This vase, manufactured by native artists under the superintendence of Messrs. ALLEN & HAYES, Government jewellers, *Calcutta*, bears an inscription stating it to have been 'presented to Rajah Deonarain Singh Bahadoor, by His Excellency the Right Honourable the Governor-General and Viceroy of India, for his loyalty and devotion to the British government during the rebellion of 1857.'

6671. [4251] Fountain in massive silver. Do. Do.

6672. [5073] 'The Governor-General's plate.'

6673. [5074] 'The trades' plate,' 1862, Do. Do.

6674-7. [2421-4] Silver salver, claret jug and two cups, presented by Lodge 'True Friendship' to the worshipful Master. Do. Do.

Manufactured by native artists, under the superintendence of Messrs. ALLEN & HAYES, *Calcutta.*

6678-9. [644-5] Bouquet holder, silver, Bhooj, Kutch.  H. H. the RAO of KUTCH.

6680. [647] Salt cellar, do., Do.   Do

6681. [649] Box of stone, mounted in silver. Do. Do.

The stone of which this box is made was procured from the Hubba Hills, and polished at Bhooj. It is also used as a substitute for marble in the decoration of temples.

6682. [650] Mug, silver, Do.   Do.

6683. [655] Bottle labels, wild boars' tusks, silver mounted, Do.  Do.

6684-6. [656-8] Salt cellars, Do.   Do.

6687-9. [659-61] Muffineers, Do.   Do.

6690-1. [663-4] Card cases, Do.   Do.

6692-3. [665-6] Two cigar cases, Do. Do.

6694-7. [667-70] Four penholders, Do. Do.

6698-9. [671-2] Two paper-knives, Do. Do.

6700-3. [673-6] Four thimbles, silver, Do.  Do.

6704-5. [681-2] Goolabdanees or rose-water sprinklers and trays, Do.  Do.

6706-8. [683-5] Toasting forks, of horn and silver, Do.  Do.

6709. [687] Silver model of Mahomedan temple, Do.  Do.

This model is of a building erected over the tomb of SHEEAH MAHOMEDAN, at Bhooj.*

6710. [10032] Utterdan, silver, in the form of a fish.  PHILIBEET BUDROODEN. Aoula.

6711-2. [3122-3] Betel boxes, filigree work, Nepaul.  SIR JUNG BAHADOOR, K.C.B.

* A large number of articles in gold and silver are annually made at Bhooj, principally for Europeans. The Goolabdanas, or rose-water sprinklers, are, however, manufactured for native use. The silver and gold used is very nearly pure. The principal artizans are VISHRAM GOLDSMITH, JEWRAM SHAMJEE, and HEERJEE NAGJEE. The charge is at the rate of 8A. per tola weight.

6713-4. [5226-7] Snuff boxes, Trichinopoly. Messrs. HENCKELL, DU BUISSON, & Co., for MOOTIANASSARY, jeweller, *Trichinopoly.*

6715-7. [5228-30] Cigar cases, Do.  Do.

6718-23. [5231-6] Six cigar lighters, Do. Do.

6724. [2291] Cannon faleta, silver, 24R. 6A., Do.  ROSALINGUM ASSARY.

6725. [2283] Hair pins, silver filigree, 12R. 7A., Do.  Do.

6726. [2284] Bracelet, filigree ball, 13R. 14A., Do.  Do.

6727. [2285] Do. do., flat, 18R. 8A., Do. Do.

6728. [2286] Do. do., ball, 10R. 10A., Do. Do.

6729. [2290] Silver purse, 38R. 8A., Do. Do.

6730. [2292] Cheroot case, 70R., Do. Do.

6731. [2288] Card case, 25R., Do.  Do.

6732. [10662] Casket, filigree, 170R., Vizagapatam.  GAJAPATI RAO.

6733-4. [10663-4] Two pair of bracelets, do., 32R. each, Do.  Do.

6735. [10665] Necklace, beads and silver, do., 30R., Do.  Do.

These articles, No. 6732-5, are intended as presents to H. R. H. the Princess Alice, at the close of the Exhibition.

6736-7. [2259, 2293] Six dessert spoons, silver, 200R., Madras.  T. CHOCALINGUM.

6738. [2260] Six table do., do., 135R., Do.  Do.

6739-40. [2261, 2294] Six tea do., do., 100R. Do.  Do.

6741. [2262] Two ladles do., do., 30R., Do.  Do.

6742-3. [2263, 2289] Bracelet, do., 80R., Do.  Do.

6744. [1503] Tea pot and stand, do., Puttiala.  H. H. the MAHARAJAH.

6745. [4277] Butter do.  do., Do.  Do.

6746. [1502] Mug and lid, do., Do.  Do.

6747. [3048] Mango tree, do., Futtehpore.  RAI HALL, *Bahadoor of Jenanabad.*

6748. [5865] Box, massive silver, gilt inside, and enamelled lid, Lahore. GOVERNMENT.

6749. [2898] Spice box or 'Pandan,' Lucknow. SHA MAKHUM LALL, *Mahajaun of Lucknow.*

6750. [2897] Do. or 'Ilaeecheedan,' Do. Do.

6751. [2899] A chased silver box, Burmah. MAJOR T. P. SPARKS, *Rangoon.*

6752. [3434] Two antimony boxes, silver, Umritsur. CHUMBER MULL.

6753. [3435] One do., do., gilt inside, Do. Do.

6754. [3439] Three spindles for do., do., Do. Do.

6755. [3436] Two silver gilt hookah mouthpieces, Do. Do.

6756. [3049] Pandan box for holding betel leaf, Futtehpore. LALLA THAKOOR PERSHAD.

6757. [3050] Plate for do. do., Do. Do.

6758. [3130] Four silver Politas, Calcutta. GOVERNMENT.

6759. [3131] Two do., Do. Do.

6760. [3437] Five wild boars' tusks, mounted in silver as bottle labels, Umritsur. A. M. DOWLEANS, Esq.

6761. [3121] An antique vase of silver, exhibiting the twelve signs of the Zodiac, in basso relievo, and supported by the shield of a Burmese warrior, Ava. JOHN MARTIN, Esq., *Calcutta.*

6762. [5754] Ink and pen tray, glass, silver mounted, Meerut. KOOER WUZEER ALLY KHAN, *Meerut.*

6763. [8298] Basket, filigree work. Imported by J. NEAL, *Edgware Road, London.*

6764. [8298] Bracelets, do. Do.

6765-8. [7811-4] Four silver bottles for attar of roses, Jeypore. H. H. the MAHARAJAH's distillery.

6769. [1061] Ten enamelled buttons.

6770-1. [1062-3] Two scent bottles.

6772. [7850] 'Goolabdanee,' or rosewater sprinkler. JHEND.

6773. [7851] Peacock. Do.

6774-5. [7852-3] 'Goolabdanee,' or sprinkler and tray. JHEND.

6776. [7854] Pair bracelets. Do.

6777. [4279] Locket, silver, Darjeeling. DR. CAMPBELL.

6778. [4282] Bootanese bangles, silver, Do. Do.

6779. [4284] Silver locket necklace, Do. Do.

6780. [4283] Do. chain cloak clasp, Do. Do.

6781. [4287] Do. earrings, as worn by the Tumboo tribe, Do. Do.

6782. [4288] Do. locket, Do. Do.

6783. [3106] Pair bracelets, silver filigree, Banda. SETH OODEY KURI.

6784-6. [2793-5] Ornaments for the feet, silver, Hindoo. SIR ROBT. HAMILTON, BART.

6787. [2792] Three bangles, do., Do. Do.

6788. [2791] Anklet, do., Do. Do.

6789. [2790] Armlet, do., Do. Do.

6790. [2789] Pair bracelets, do., Do. Do.

6791. [4280] Cloak clasp, silver and malachite, Darjeeling. DR. CAMPBELL.

6792. [4281] Silver locket and coral necklace, Do. Do.

6793. [4290] Earrings, malachite, Do. Do.

6794. [4291] Earring, do., silver and coral, Do. Do.

6795. [4292] Earrings, silver and coral, worn by the Tumboos, Do. Do.

6796-7. [4294, 4305] Gentleman's earring, Do. Do.

6798-9. [3103-4] Two pairs agate bracelets, mounted in silver, Banda. GOVERNMENT.

6800. [8309] Knife handle.

## COINS.

6801. [906] Large gold coin, weight 4 tolas and 21 wals of acbars, and 2 silver, both of Jahageer, Ahmedabad. NUGGER SHETTE PREMABHAEE HEMABHAEE.

6802. [907] Square gold coin, period of
MAHOMED AHUNSHAR, Ahmedabad. NEKNAM-
DAR SUKAWATTEE SHETANEE HURKOOVUR-
BHAEE.

6803. [908] Three silver and one copper
coin, Ahmedabad. RAO BAHADOOH, SHET
MUGGUNBHAEE KRAMCHAND.

6804. [8250] Cashmere rupee, coined by
the late MAHARAJAH PUTAB SINGH, bearing
a Christian cross, and the letters, ' J. H. S.'

These rupees were first coined in 1849, shortly after
the annexation of the Punjaub, when the Maharajah was
very anxious to show his loyalty in a way which he supposed likely to be most gratifying to a Christian Government. Exhibited by SIR J. LOGAN.

6805. [3126] Silver bullion ' Daing,'
Burmah.

6806. [3127] Do. ' Ban,' Do.

6807. [3128] Do. 'Govetnee,' Do.

The currency of Burmah Proper, and formerly of Pegu,
until the British conquest, when the Indian coinage supplanted it.*

6808. [4299] Five Thibetan coins, Darjeeling. DR. CAMPBELL.

*c.* **Enamelling and Damascene Work, etc.**

*Enamelling on Gold and Silver.*

6809. [10286] Gold enamelled pân box,
Jeypore. H. H. the RAJAH.

6810. [10287] Do. do. vase and cover,
Do. Do.

6811. [10288] Pair of enamelled and
diamond bracelets, do.

6812. [10289.] Jewel-enamelled ankus or
elephant spear, do.

6813. [1059] Goolabdanee or rose-water
sprinkler, enamelled, Hyderabad. GOVERN-
MENT.

6814. [1060] Covered dish, do., Do. Do.

The following note on the enamelled work of India
is by A. M. DOWLEANS, ESQ :—
The finest enamelled work of India is produced in the
independent Rajpootana state of Jeypore, and considered
of great artistic merit. The enamellers came originally
from Lahore. The enamel is a kind of glass made in
earthen vessels, and when fused the colouring matters are
added; the whole is then allowed to cool, and in this state
is kept for use. Only pure silver or gold articles are
enamelled. From the silver, the enamel may come off
in course of time; but it never does from the gold.
All good enamel is consequently only applied to gold,

which must be free from alloy, or otherwise it would tar-
nish by contact with the enamel in the great heat to which
it is subsequently exposed. The gold is first carved of
the required pattern : the enamel, having been ground to
an impalpable powder, and made into a paste with water,
is then placed on the exact spot required by the pattern.
The article is then strongly heated, much skill being
required to take it out at the precise moment when the
enamel is thoroughly fused, but before the colours begin
to run into one another. As soon as removed, the workmen then exert the full power of their lungs in blowing
upon it as quickly and as violently as possible. The
hardest colours are first placed in the furnace and fused,
and then those which melt more easily. Afterwards, the
whole is ground and polished. The enamelled work of
Jeypore is very highly valued, and can only be procured
through H. H. the RAO of JEYPORE himself, by whom the
workmen are employed. The artisans themselves form a
small family, and the real process of enamelling is kept
by them as a secret, which descends from father to son
like an heirloom.
Enamelling, as applied to jewellery, consists of an extremely fine pencilling of flowers and fancy designs in a
variety of colours, the prevailing ones being white, red, and
blue, and is invariably applied to the inner sides of bracelets, armlets, anklets, necklaces, earrings, surpezes, tiaras,
and all that description of native jewellery, the value
depending upon the fineness of the work, and often exceeding that of the precious stones themselves. In general
the cost is moderate, as the finest specimens are only made
to order. The best come from Benares, Delhi, and the
Rajpootana states.
The manufacture of enamels on articles of domestic use
like the above is almost entirely restricted to Hyderabad:
It presents no varieties, but in general consists of a blue
coating interlined with white on a surface of silver, and is
applied to rose-water sprinklers, spice boxes, basins, and
such like articles. The merit of the manufacture lies in the
simplicity of the enamel itself, and in the lightness of the
silver article to which it is applied. Though pleasing, it
is the coarsest enamel produced in India.

*Bidri Ware, Composition Metal inlaid with
Silver.*

6815. [5218] ' Chillumchee,' or spittoon,
and cover, inlaid Bidri ware, Purneeah.
GOVERNMENT of INDIA.

6816. [5795] Water bottle, do., Do. Do.

6817–8. [5796-7] Hookah bottom, do.,
Do. Do.

Bidri, or Biddery ware, derives its name from Bider, a
city situated about 60 miles to the N.W. of Hyderabad.
It is a species of inlaid ware of excellent form and
graceful pattern. The stages of the manufacture are as
follows :—
A mass of finely-powdered and sifted old laterite dust
mixed with cow-dung is put upon a rude lathe, and when
dry, carefully turned to the correct shape. The model
having been smoothed with a chisel, is next covered with
a mixture of wax and oil boiled together; when dry, the
whole mass is carefully smoothed and turned. Over this
coating is plastered a second layer of laterite dust, moist-
ened with water alone; this coat is rough, and not subsequently smoothed down. The next stage consists in
boring two openings in the composite mould, and placing
it in the fire, the effect of which is to melt the interme-
diate layer of wax, and thus to leave a vacant space for
the reception of the alloy. Into this space is poured the
alloy, consisting of 1 part of copper and 4 parts of pewter.*

---

The vessel has now a dull leaden look; it is hard, but easily cut. This shell, as it may be called, is carefully turned, and upon its smooth surface the pattern is traced by hand. This tracing is done rapidly. The workman next takes a small chisel and hammer, and, following the lines of the pattern, cuts it deeply and expeditiously, scooping out the tracings of the little leaves, &c., and leaving an indented, but rough surface. This rough surface is next smoothed down by hammering gently with a blunt-pointed chisel, and the space is then ready for the process of inlaying. Thin plates of very pure silver are then taken, and the little leaves (or other patterns) are cut out with a small hammer and chisel; each little leaf is then raised separately by the chisel and finger tip, and hammered gently but carefully into the depression intended for it. This part of the process is tedious. In the more durable kinds of Bidri ware, silver wire is substituted for silver leaf. The vessel in this state is rough, and requires smoothing; this is done with a common file and a curved scraper of a rude and clumsy form. The hole in the bottom of the vessel is filled up with lead, and smoothed down. Finally, the vase is gently heated, and, whilst warm, is blackened by the application of a powder (supposed to consist of chalk and sal-ammoniac—chloride of ammonium). This imparts a brilliant black polish to the shell, and careful hand rubbing brings out the polish of the silver.*

### *Glass inlaid with Gold.*

6819. [2777] Pair of bracelets, gold enamelled, Pertaubghur. SIR R. HAMILTON, BART.

6820. [2778] Necklace, do., Do. Do.

6821. [2779] Five gold enamelled tablets for bracelet, Do. Do.

The manufacture is peculiar to Indore, in Central India, but it does not constitute a regular trade. It is invariably applied to articles of personal decoration, such as necklaces, armlets, brooches, earrings, &c., which are set by native jewellers according to the taste of the purchaser. These subjects generally consist in a representation of the avatars, or pictures of the metamorphoses of Indian deities; and the work is so perfect that it will stand, not only the influence of climate, but even rough handling.

The specimens of this kind of work have no fixed market value, and the price is, therefore, entirely dependent upon the number of competitors that may be in the field when any of them are offered for sale. A set of these ornaments, consisting of a necklace, earrings, two armlets and a brooch, in plain gold, contributed to the Exhibition of 1851, was valued at 1,700R., or 170*l.* A duplicate, forwarded to the Paris Exhibition in 1855, was purchased for 600R., or 60*l.*

### *Koftgari Work or Steel inlaid with Gold.*

6822-4. [5861-3] Caskets of Koftgari work, or steel inlaid with gold, Lahore.

6825. [5864] 'Kalamdan,' or pen case, inlaid with gold, Lahore.

6826-7. [5866-7] Pen trays, do. Do.

6828-31. [5868-71] Paper weights, do. Do.

6832. [5872] Paper knife, do. Do.

6833. [5873] Sword hilt, do. Do.

Koftgari work, or steel inlaid with gold, has, in former days, been carried on to a considerable extent in various parts of India. It was chiefly used for decorating armour; and among the collections exhibited on the present occasion, are some very fine specimens of guns, coats of mail, helmets, swords, and sword handles, to which the process of koftgari has been successfully applied. These specimens, however, are not the manufacture of the present day. Since the late rebellion in India, the manufacture of arms has been generally discouraged, and Koftgari work is, consequently, now chiefly applied to ornamenting a variety of fancy articles, such as jewel caskets, pen and card trays, paper weights, paper knives, inkstands, &c. The process is exactly the same as that pursued in Europe, and the workmen can copy any particular pattern required. The work is of high finish, and remarkable for its cheapness.*

Koftgari is chiefly carried on in Goojerat and Kotli, in the Sealkote district. It was formerly much in vogue for decorating armour and the blades and hilts of swords, but the artisans now confine themselves chiefly to the manufacture of ornamental paper knives, &c. The specimens above mentioned have been contributed by the Kotli artisans.†

Several admirable specimens of inlaid metal work by the native artizans of Bhooj will likewise be found in the collection of arms contributed by H. H. the RAO of KUTCH.

---

ARTICLES OF VERTU NOT INCLUDED IN THE PREVIOUS ENUMERATION.

#### *a.* Jade and Rock Crystal.

6834. [1342] Jade long box and top, both gold inlaid, Calcutta. COL. GUTHRIE.

6835. [1418] Do. leaf-shaped box, inlaid with gold, Do. Do.

6836. [1417] Do. carved octagon box and top, jewelled, Do. Do.

6837. [1375] Do. straight octagon box, jewelled, Do. Do.

6838. [1369] Do. pen box and top, jewelled, with six fittings all jewelled, viz. 2 ink bottles, 1 penknife, 1 pen rubber, 1 pencil, and 1 spoon, Do. Do.

6839. [1370] Do. small trefoil-shaped box and top, jewelled, Do. Do.

zinc 12,360 grains, copper 460 grains, and lead 414 grains, melted together; a mixture of resin and brick was being introduced into the crucible to prevent calcination. DR. HAYNE states that it is composed of copper 16 oz., lead 4 oz.. tin 2 oz., and that to every 3 oz. of this alloy, when melted for use, have to be added 16 oz. of zinc.

* The foregoing description of this interesting manufacture is chiefly taken from an article by DR. GEORGE SMITH in the *Madras Journal of Literature and Science* for October, 1856.

* A. M. DOWLEANS. Esq.
† Central Committee, *Lahore.*

6840. [1338] Jade, leaf-shaped box and top, carved in relief, Calcutta. Col. Guthrie.

6841. [1335] Do. large pen box and top, carved in relief, Do. Do.

6842. [1371] Do. white jade cup, 498 jewels, 5⅝ diameter, exclusive of handles, Do. Do.

6843. [1372] Green jade bowl, with handles, Calcutta. Col. Guthrie.

6844. [1373] White jade bowl, with handles, Do. Do.

6845. [1411] Pure white jade abcorah, with handles, Do. Do.

ARTICLES IN JADE, FORMING PART OF THE COLLECTION EXHIBITED BY COL. GUTHRIE.*

6846. [1340] Jewelled jade abcorah, with handles, Calcutta. Col. Guthrie.

6847. [1376] Grey jade abcorah and top, with handles, Do. Do.

6848. [1412] White jade jug, with handles and gold rim, Do. Do.

6849-50. [1353, 1428] White jade bowls, very thin, carved in relievo, Do. Do.

6851. [1427] Very remarkable thin cup, jade, carved all over, Do. Do.

6852. [1429] Thin green jade bowl, carved all over, Do. Do.

6853. [1377] Small oval greenish jade cup, with duck's head handles, jewelled eyes, Do. Do.

6854. [1424] Small green jade cup, with handles, Do. Do.

6855. [1430] Jade cup, turned up, Do. Do.

6856. [1431] Do., very thin, do., Do. Do.

6857. [1354] Jade cup, Calcutta, Col. Guthrie.

6858. [1355] Do. saucer, Do. Do.

6859. [1356] Small plain jade cup, Do. Do.

6860. [1374] Large coarse jade bowl, with handles, Do. Do.

6861. [1426] Green jade dwat, or ink bottle, carved in relief, Do. Do.

6862. [1432] Very thin jade saucer, Do. Do.

6863. [1341] Jade jug, with handle and top, Do. Do.

6864. [1416] Do., with handles, Do. Do.

6865. [1380] Yak chowrie, with jade handle, Do. Do.

6866. [1346] Green jade hilt, prepared for inlaying, Do. Do.

6867. [1413] White jade handle, in two parts, Do. Do.

6868. [1347] Handle, carved in relief, Do. Do.

* From the Mechanics' Journal record of the Exhibition, 1862.

6869. [1348] Large blueish jade handle, carved in relief, Calcutta. Col. Guthrie.

6870. [1414] Very fine white jade hilt, two onyx girdles, Do. Do.

6871. [1415] Hilt and two scabbards black, inlaid with white jade and rubies, Do. Do.

6872. [1420] Hilt, jewelled, greenish, Do. Do.

6873. [1396] Gold schist, or archer's thumb ring, inlaid with rubies, Do. Do.

6874. [1397] Jade white schist, inlaid with rubies and emeralds, Do. Do.

6875. [1398] One jade schist, prepared for inlaying, Do. Do.

6876–9. [1357–1360] Four plain schists, or archer's thumb rings, Do. Do.

6880. [1399] Gold enamelled diamond schist, or archer's thumb ring, Do. Do.

6881. [1400] White jade schist, jewelled, 30 rubies, 9 emeralds, Do. Do.

6882. [1401] White jade schist, jewelled, 13 rubies, 7 emeralds, Do. Do.

6883–4. [1402–3] Two schists, jade, prepared for inlaying, Do. Do.

6885. [1404] Silver enamelled schist, or archer's thumb ring, Do. Do.

6886. [1405] White jade schist, jewelled, 12 emeralds, 1 diamond, Do. Do.

6887. [1409] Five jade finger rings, Do. Do.

6888. [1366] White jade top to stick, jewelled, Do. Do.

6889. [1350] Green jade do., carved in relief, Do. Do.

6890. [1351] Black stone do., jewelled, Do. Do.

6891. [1425] Small white jade cup, jewelled, 12 jewels, Do. Do.

6892. [1419] Jade inlaid hookah bottom, Do. Do.

6893. [1406] White jade schist, 20 rubies, 2 emeralds, Calcutta. Col. Guthrie.

6894. [1407] Do., for inlaying, Do. Do.

6895. [1408] Do., do., Do. Do.

6896. [1336] Pair white jade bangles, Do. Do.

6897. [1364] Jade hookah mouthpiece, jewelled, Do. Do.

6898. [1365] White jade mouthpiece, Do. Do.

6899. [1389] Green jade pierced-work mirror, Do. Do.

6900. [1361] Small jewelled jade round mirror, Do. Do.

6901. [1362] Jade jewelled, 6 diamonds, 30 rubies, 1 emerald, Do. Do.

6902. [1363] Jade jewelled charm, 25 rubies, 6 emeralds, Do. Do.

6903. [3244] Jade box, set in gold, with rubies, Umritsur. Rai Nursing Doss.

A valuable relic of one of the rulers of Cabul.

6904. [1422] A jade charm, plain, Lahore. Government.

6905. [1423] Do., set in gold, Do. Do.

6906. [3124] A pair of bangles, a fine specimen of jade. Government of India, *per* Local Committee, *Rangoon.*

These bangles are made of jade from Mogoung, in the north of Burmah. The bright green tint seen in these specimens is the characteristic peculiarity of the Burmese jade, or precious serpentine. The Chinese have a perfect mania for it, using it for Mandarins' buttons, pipemouth pieces, and various articles of personal ornament and luxury. They estimate it according to the purity of the white and brightness of the green tints.

These bangles, though of good quality (they cost 125R. or 12*l.* 10*s.*, and were obtained from the owner with difficulty even at that price), are by no means of the finest description.

The Chinaman who sold the bangles showed the Committee a specimen which he assured them would fetch in China sixty times its weight in silver, and that the really first-rate is sold for as much as forty times its weight in gold; this appears incredible, but all enquiry tends to show that the Chinese will give almost anything for fine jade.*

6907. [1337] Rock crystal water pot and top, spout out of the same piece, Calcutta. Col. Guthrie.

---

* Local Committee, *Rangoon.*

ARTICLES IN ROCK CRYSTAL, FORMING PART OF THE SERIES EXHIBITED BY COL. GUTHRIE.*

6908-9. [1391-2] Two pairs rock crystal spoon bowls, one ribbed, one leaf-shaped, with handles, Calcutta. COL. GUTHRIE.

6910-1. [1378-9] Do., carved in relief, Do. Do.

6912. [1352] Rock crystal melon-shaped bowl and top, Do. Do.

6913. [1393] Rock crystal hilt, diamond cut, Do. Do.

6914. [1421] Do., carved in relief, Do. Do.

6915. [1394] Do., inlaid, Do. Do.

6916. [1395] Do., carved in relief, Do. Do.

6917. [1410] Rock crystal, carved in relief, Do. Do.

6918. [1390] Large plain bowl, with handles, Do. Do.

6919. [1367] Bowl, fluted octagon, Do. Do.

6920. [1368] Small mounted vase, with handles, Do. Do.

6921. [1339] Do., without handles, Do. Do.

6922-3. [1433-4] Two pairs crystal handles, Do. Do.

### b. Cornelian, Agate, etc.

6924. [5483] A cup of stone, called Zahmora, Cashmere. GOVERNMENT of INDIA, *per* LAHORE COMMITTEE.

6925. [2138] Four necklaces of cornelian, Lahore. SOBORHAM, maker. GOVERNMENT OF INDIA, *per* LAHORE COMMITTEE.

6926-7. [989-90] Two specimens of fortification agate, Jubbulpore. GOVERNMENT of INDIA.

6928. [991] Five brooch stones, moss agates, Do. Do.

6929. [992] Three specimens of moss agate pebbles, Do. Do.

6930. [993] Three specimens of bloodstone, Do. Do.

6931. [994] Two specimens of cornelians, Do. Do.

6932. [10273] Specimen of jasper (worked), Banda. Do.

6933. [1436] One dozen knife-handles, jasper, Do. Do.

6934. [1451] Do., do., Do. Do.

6935-6. [1448-9] Two paper weights, of jasper and Goodurrea stones, Do. Do.

6937. [1455] Jasper brooch-stone, Do. Do.

6938. [1435] One dozen knife-handles, agate, Do. Do.

6939. [1437] Grass agate, Do. Do.

6940. [1440] Fourteen stones for agate bracelets, Do. Do.

* Lent by J. H. JOHNSON, Esq.

6941. [1441] Sixteen stones for agate bracelets, Banda. GOVERNMENT of INDIA.

6942. [1442] Fourteen do., Do. Do.

6943. [1444] Forty-one do., Do. Do.

6944. [4445] Paper-knife, agate, Do. Do.

6945. [1446] Brooch-stone, do., Do. Do.

6946. [1447] Paper-knife, do., Do. Do.

6947. [1456] Six sets shirt-studs, do., Do. Do.

6948. [1457] Twenty stones for sleeve-links (five pairs), do., Do. Do.

6949–51. [10274-6] Agates, Do. Do.

6952. [1453] Brooch-stone, bloodstone, Do. Do.

6953. [1452] Four do., Lapis Lazuli, Do. Do.

6954. [10267] Map-stone, rough and polished specimens, from the Kane River, Do. Do.

6955. [4303] An agate schist, or archer's thumb-guard, Darjeeling. DR. CAMPBELL.

6956. [995] Specimen of agate, Guzerat. GOVERNMENT of INDIA, *per* CALCUTTA COMMITTEE.

6957. [1102] Twelve crochet needles, of various stones, Cambay, Guzerat. GOVERNMENT of BOMBAY.

6958. [1103] Ten paper-knives, do., Do. Do.

6959–60. [1104–5] Two cups and saucers, agate, Do. Do.

6961 [1106] Seven sets waistcoat buttons, do., set in silver, Do. Do.

6962. [1107] Three sets do., do., set in gold, Do. Do.

6963. [1108] Ten pairs sleeve-links, do., do., Do. Do.

6964. [1109] Twenty-four brooch-stones, do., Do. Do.

6965. [1110] Six doz. knife-handles, do., Do. Do.

6966. [1111] Ten crystal stones, do., Do. Do.

6967–8. [1112–3] Agate slabs, Do. Do.

6969. [1114] Do., with two side pieces, Do. Do.

6970. [1115] Three pairs do., Do. Do.

6971. [1116] One do., Do. Do.

6972. [1117] Thirty-four agate stones for bracelets, Cambay, Guzerat. GOVERNMENT of BOMBAY.

6973. [1118] Three pairs round beads, agate, for bracelets, Do. Do.

6974. [1119] One bracelet, black stones, Do. Do.

6975. [1120] Five pairs bracelet-stones, various colours, Do. Do.

6976. [1121] Ten cigar-holders, agate, Do. Do.

6977. [1122] Cup and saucer, do., Do. Do.

6978. [1123] Six rulers, various colours, do., Do. Do.

6979. [1124] Twelve sets shirt-studs, various colours, Do. Do.

6980. [1126] Two sets bracelet-stones, Do. Do.

6981. [4489] Pair serpentine earrings, Rangoon. Messrs. HALLIDAY, FOX, & Co.

6982. [4450] Bracelet, Do. Do.

6983. [4493] Stone earrings, set in horn, Do. Do.

*c.* Personal Ornaments in Tinsel, Ivory, Horn, Hair, etc.

6984. [2908] Tinsel box, Lucknow.

6985. [2909] Bracelets of tinsel, Do. SHEIK HASSAIN KHAN.

6986. [2910] Bracelet or necklet of do., Do. Do.

6987. [2911] Earrings of do., Do. Do.

6988. [2912] Ticklees or face ornaments, as used by native females, Patna. GOVERNMENT.

6989. [5755] Bangles of hair, Meerut. Do.

6990. [10047] Do. porcelain, Moonghyr. DR. SUTHERLAND.

6991. [7885] Do. glass, Poonah. GOVERNMENT.

6992. [7886] Rings, do., Do. Do.

6993. [5153, 8109] Necklaces of horn, Moonghyr. GOVERNMENT of INDIA.

6994-5. [5154, 8110] Bracelets, horn, Moonghyr. GOVERNMENT OF INDIA.

6996. [7999] Bangles for the wrists and ankles, do., Do. Do.

6997. [3432] Four and half doz. buttons, ivory, Umritsur. GOVERNMENT OF INDIA.

6998. [3433] Three doz. do. and studs, do., Do. MISSUR GEMA CHUND.

# CLASS XXXV. — POTTERY.

Although some of the specimens of pottery are very creditable, especially in the accuracy of their shape, the value and extent of the collection has been seriously diminished by the breakage of the greater portion during transit.

6999. [3295] Specimens of pottery, Amroha. SYUD AHMED KHAN, *principal Sudder Ameen.*

Thirty-three of these specimens, remarkable for the superior nature of the clay of which they were constructed, were sent to England for exhibition, but, unfortunately, two samples only came to hand uninjured.

Regarding these specimens the officiating magistrate reports, that the Umroha pottery work is only capable of improvement under European energy and capital, and gives the following information noted under the head of 'Manufactures.'

About 300,000 pieces are annually manufactured at an average of 1R. per hundred.

Articles made of this clay are much sought after, and are used by the natives from all parts of the N.W. Provinces.

The clay is not exported from this country.

Articles made from this clay are intended more for ornament than for use, but cups, saucers, and such like things, are used at festivals and feasts by people of rank. Mahomedans more than Hindoos use these articles.

The pottery articles can be made of all patterns and shapes, but the old oriental designs are those chiefly approved of.

The wholesale market price would be, on an average, 1R. 87A. per hundred.

The clay of which this pottery is composed is evidently of a very superior quality; it shows great strength and is used pure and unmixed in all articles composed of it. It is peculiar to Amroha, where it can be had in large quantities.

7000. [5347] Specimens of painted pottery, Allahabad. GOVERNMENT OF INDIA.

Fourteen specimens of this ornamented pottery were sent from India, but from their extremely brittle nature were, on receipt, all found to be more or less damaged.

7001. [5440] Specimen of Puttan pottery, Lahore. LAHORE CENTRAL COMMITTEE. GOVERNMENT OF INDIA.

7002. [2895] Specimens of glazed pottery, Lucknow. LUCKNOW COMMITTEE. Do.

7003. [7839] Figure of a buffalo, Jeypore.

7004. [3177] Specimens of pottery, Darjeeling. DR. CAMPBELL.

7005. [2893] Do., Hossengunge. SHAIK FAKEER HOSSAIN.

7006. [2894] Butter cooler, Sarun, Patna. PATNA COMMITTEE. GOVERNMENT OF INDIA.

Of sixteen specimens forwarded, this sample only came to hand uninjured.

7007. [1081] Six specimens 'Halla,' glazed tiles, Hyderabad. HYDERABAD COMMITTEE.

7008. [381] Five goblets, Madras Industrial School of Arts. DR. HUNTER.

7009. [398] Jars, Do. Do.

7010. [379] Salt-cellar, Do. Do.

7011. [378] A pedestal, Do. Do.

7012. [376] Tiles, plain, Do. Do.

7013. [377] Do., glazed, Do. Do.

7014. [384] Garden labels, plain, Do. Do.

7015. [406] Do., glazed, Do. Do.

7016. [397] Flower vase, Do. Do.

7017. [394] Red tile, Do. Do.

7018. [388] Four jars, Do. Do.

7019. [387] Porous cells, Do. Do.

7020-2. [389-91] Chemical utensils, four sets, Do. Do.

7023. [392] Flower basket, Do. Do.

7024. [393] Tray, Do. Do.

7025–6. [382-3] Raised-pie dish, Madras Industrial School of Arts. DR. HUNTER.

7027. [385] Ornamented tiles, glazed, Do. Do.

7028. [390] Jug, glazed, Do. Do.

7029. [396] Vase, do., Do. Do.

7030. [399] Two jars, do., Do. Do.

7031. [411] Models of cooking utensils, Do. Do.

7032. [402] Basket, Do. Do.

7033. [405] Butter cooler, Do. Do.

7034. [401] Goblet, Do. Do.

7035. [413] Three chemical lamps, Do. Do.

7036. [400] Two goblets, Do. Do.

7037. [407] Two water-pipes, Do. Do.

7038. [444] Ventilating brick, Do. Do.

7039. [409] Garden border tiles, Do. Do.

7040. [404] Cooking utensils, Do. Do.

7041. [408] Rail pattern balustrade, Do. Do.

7042. [410] A 'chattee,' glazed, Do. Do.

7043. [415] Vase, glazed, Do. Do.

7044. [4436] Tea pot, Madras. Do.

7045. [4435] Pair of jars, Do. Do.

7046. [4434] Jar and goblet, Do. Do.

7047. [4435] Pair of jars, Do. Do.

7048. [4436] Tea pot, Do. Do.

7049. [4437] Circular ornamental rim, Madras Industrial School of Arts. Dr. HUNTER.

7050. [4438] Pan, Do. Do.

7051. [4439] Chemical vessels, Do. Do.

7052–3. [594-5] Earthen cups, Vizagapatam. H. H. the RAJAH of VIZIANAGRAM.

7054. [10519] Goglet and cover, North Arcot, Madras. ARUMGA UDAYAR, maker. GOVERNMENT of MADRAS.

7055–6. [10520–1] Vase and cover, Do.

7057. [10522] Do. for wine, Do. Do.

7058. [10523] Do. for flowers, Do. Do.

7059. [572] Figure of a lion, Bangalore Convict School, Mysore. CAPT. J. PUCKLE.

7060. [573] An owl, Do. Do.

7061. [576] Drain tile, Do. Do.

7062. [577] Lid of butter cooler, Do., Do.

The last four specimens are made of potters' clay, buff, and felspar grit in the following proportions :— 2 buff, 1 brown potters', and 1 grit. This mixture is found very strong and serviceable.

7063. [575] Purple goglet or water bottle, Do. Do.

There is evidently much manganese in this clay, which hardens by hard firing ; the buff is found almost pure, and requires only washing to separate the clay from the grit. This mixture is found very strong and serviceable.

7064. [578] Fire brick, Do. Do.

Two parts buff clay and one part grit.

These specimens (Nos. 7059-64) are illustrative of the clays sent by the same exhibitor, and shown in Class I., No. 425.

# CLASS XXXVI.— DRESSING CASES, DESPATCH BOXES, AND TRAVELLING CASES.

Division I.— WRITING CASES AND FITTINGS.

7065–6. [3415-16] Paper-knives, Umritsur. MISSUR GEMA CHUND.

7067. [3419] Do., Do. GOVERNMENT of INDIA.

7068. [5751] Pen case and tray in ivory, Meerut. KOOER WUZEER ALLY KHAN, *Dep. Magistrate, Meerut.*

7069. [10635] Writing desk in ivory, Vizagapatam. H. H. the RAJAH of VIZIANAGRAM.

7070. [10636] Envelope case, ivory, Vizagapatam.  H. H. the RAJAH of VIZIANAGRAM.

7071. [10652] Inkstand in buffalo horn, Do.  GOVERNMENT of MADRAS.

7072. [10653] Envelope box, do., Do. Do.

7073. [10657] Do. in elk horn, Do.  Do.

7074. [10658] Inkstand in porcupine quills, Do.  Do.

7075. [10660] Pen tray, Do.  Do.

7076. [10643] Writing desk in buffalo horn, Do.  H. H. the RAJAH of VIZIANAGRAM.

7077–8. [10645-6] Paper weights, porcupine quills, Do.  Do.

7079. [567] Inkstand of horn, Do.  E. G. R. FANE, Esq.

7080. [568] Do., grotesque, Do.  G. N. TAYLOR, Esq., *Inam Commissioner.*

7081. [422] Do. elk and buffalo horn, Do.  DR. MCPHERSON.

Division II. — DRESSING CASES AND FITTINGS, TOILET ARTICLES, ETC.

7082–4. [3406-14-7] Combs of ivory, Umritsur.  MISSUR GEMA CHUND.

7085–7. [3418-20-3] Do., Do. GOVERNMENT of INDIA.

7088. [2913] Do., Seebsagur, Assam. BABOO POORMANUND BOOROOAH PESHKAR.

7089. [2916] Do., Do.  LIEUT. PHAIRE.

7090. [2921] Do., Luckimpore.  BABOO GOBIND RAM SHURMAH.

7091. [3425] Pincushion of ivory, Umritsur.  GOVERNMENT of INDIA.

7092. [3429] Thimbles, do. Do.  Do.

7093. [2914] Ear probes, do.  Seebsagur, Assam.  BABOO POORMANUND BOROOAH PESHKAR.

7094. [2917] Back scratcher, do., Do. GOVERNMENT of INDIA.

7095. [2920] Do. do., Gowhatty, Assam. BABOO PURSOORAM BOROOAH.

7096. [2919] Fan, do. Chittagong.  H. H. the RAJAH of TIPPERAH.

7097–8. [10503, 10649] Work baskets, octagonal, ivory, Vizagapatam.  GOVERNMENT of MADRAS.

7099. [10651] Jewel case, do., Do.  Do.

7100–1. [10639-40] Two do., octagonal, do., Do.  H. H. the RAJAH of VIZIANAGRAM.

7102. [10641] Workbox, do., Do.  Do.

7103. [10642] Glove box, do., Do. Do.

7104–6. [10637-8, 10666] Watch stands, do., Do.  Do.

7107. [2230] Two pairs glove stretchers, Travancore.  H. H. the RAJAH.

7108. [10648] Watch stand, ivory and buffalo horn, Vizagapatam.  GOVERNMENT of MADRAS.

7109. [10655] Tea caddy of elk horn, Do.  Do.

7110. [10656] Knitting box of do., Do. Do.

7111. [10659] Do. of porcupine quills, Do.  Do.

7112. [431] Shell tea ladle and horn handle, Malacca.  DR. MCPHERSON, *Madras.*

# FINE ARTS DIVISION.

## CLASS XXXVIII.—PAINTINGS IN OIL AND WATER COLOURS AND DRAWINGS.

### I.—PAINTINGS ON IVORY.

#### a. Architectural Subjects and Views.

These paintings are very beautiful, and exhibit a marked improvement in shade and perspective, whilst their minuteness of detail and brilliancy of colouring give evidence of great skill and accuracy of touch.

7113–4. [3002-3] Lahore palace gate at Delhi, inside Delhi. GOVERNMENT of INDIA.

7115. [3004] Taj at Agra. Do.

7116. [3005] Jummul Musjeed, Delhi. Do.

7117. [3006] Dewan Khas at Delhi. Do.

7118. [3007] Temple at Umritsur. Do.

7119. [3008] La Martiniere, Lucknow. Do.

7120. [3009] Mosque Koogat. Do.

7121. [3010] Emambarrah, Lucknow. Do.

7122. [3011] Peacock throne in the palace of Delhi. Do.

7123. [3012] Inside of the Taj at Agra. Do.

7124. [3013] Nurad Shaw Mosque, Delhi. Do.

7125. [3014] The Taj from the river side, Agra Do.

7126. [3015] Umritsur temple. Do.

7127. [3016] Interior of Dewan Khas, Delhi. Do.

7128. [3017] Twelve views of Agra, Delhi, and Lucknow. Do.

7129. [3018] Lahore palace gate. Do.

7130. [3019] Dewan Khas gate. Do.

7131. [3020] King's house, Delhi. Do.

7132. [3021] Kootub pillar, Delhi. GOVERNMENT of INDIA.

7133. [3022] Sufdur Tuness' tomb. Do.

7134. [3023] Hoomayoon's tomb, Delhi. Do.

7135. [3025] Kootub pillar, Delhi. Do.

7136. [3026] Nine views of Delhi. Do.

7137. [3027] Eight do., for shirt studs. Do.

7138. [3029] Feenut Mehal, Delhi. Do.

7139. [3035] Four views of Lucknow and Delhi. ISMAIL KHAN.

7140. [3036] Do. Delhi. Do.

7141. [3037] Do. Lucknow. Do.

7142. [3038] Do. Agra. Do.

7143. [3039] Do. Agra, Lucknow, and Delhi. Do.

7144. [3040] Do. Delhi. Do.

7145. [5657] View of Kaiser Bagh, Lucknow. SAH MAKKHUN LALL.

7146. [5658] State Procession of the late King of Lucknow. Do.

7147. [8003] Teroomaul Nuik's tank, Madura.

7148. [8004] Madura pagoda.

7149. [8005] Teroomaul Naik and ten rajahs inside of Poodoo Munabrun.

7150. [3333] Burmese mythological painting, Burmah. Messrs. HALLIDAY, FOX, & Co.

#### b. Portraits on Ivory.

7151. [3028] Ex-king of Delhi. ISMAIL KHAN.

7152. [3033] Three portraits of wives of the ex-king of Delhi. GOVERNMENT of INDIA.

7153. [3034] Portraits of Golab Singh, Runjeet Singh, and Bulad Shaw. GOVERNMENT OF INDIA.

7154. [3024] Eight portraits of kings and queens, Delhi. Do.

7155-6. [3030-1] Portraits of kings and queens, Delhi, for sleeve links. Do.

7157. [3032] Six portraits of wives of the ex-king of Delhi. Do.

7158. [5115] Portrait of Nawab Shurfood Dowlah, Shureef ol Mulk Gholam Rajah Khan, Sherafat Jung of Lucknow. Painted on ivory by the son of Mahomed Alee, a Lucknow artist. Lucknow. H. H. the NAWAB SHURFOOD DOWLAH.

## II. — PAINTINGS ON MICA, IN WATER COLOURS, ETC.

7159-60. [2199, 2201] Paintings on mica, small size, Patna. GOVERNMENT OF INDIA.

7161-2. [2202-3] Do., large do., Do. Do.

Most of these paintings, by native artists, represent the various domestic occupations and religious ceremonies of the Hindoos.

7163. [1540] Portrait of Rajah Dheean Singh, Prime Minister of H. H. the Maharajah Runjeet Singh, by native artists. CENTRAL COMMITTEE, *Lahore*. GOVERNMENT of INDIA.

7164. [1541] Ranee Surdan, wife of do. Do. Do.

7165. [1542] H. H. the Maharajah Runjeet Singh and Court. Do. Do.

7166-7. [1543-4] Sirdar Shere Singh of Utaree (Kanghur), Umritsur. Do. Do.

7168-70. [1545-7] H. H. Maharajah Runjeet Singh and Court, Do. Do. Do.

7171. [1551] Rajah Heera Singh, son of Rajah Dheean Singh, Do. Do.

7172. [1552] Ranee Sirdan, wife of H. H. Runjeet Singh, with her suite, Do. Do.

7173. [1553] The Maharajah Dhuleep Singh, son of H. H. Runjeet Singh, Do. Do.

7174. [1554] The Maharajah Shere Singh, son of Runjeet Singh, Umritsur. GOVERNMENT of INDIA.

7175-6. [1548-9] The Ameer Dost Mahomed Khan, Cabul, Do. Do.

7177. [1550] H. H. the Maharajah Golab Singh, Cashmere, Do. Do.

7178. [5111] Maharajah Bukhtawar Singh, grandfather of the present chief of Ulwar. Water-colour drawing by a native artist of Ulwar. H. H. the MAHARAJAH.

7179. [5110] The late Maharajah Bunee Singh, father of the present chief of Ulwar. Do. Do.

7180. [5116] H. H. the Maharajah Sheodan Singh Bahadoor, the present chief of Ulwar, and a liberal contributor to the International Exhibition of 1862. Age 16 years. Do. Do.

7181. [8002] View of the western portion of the city of Ulwar. Do. Do.

7182. [5117] Elephant carriage of the Maharajah of Ulwar. Do. Do.

7183. [5114] 'Umbapershad,' the favourite elephant of the Maharajah of Ulwar. Do. Do.

7184. [5112] 'Asphoor,' the favourite horse of the Maharajah of Ulwar. Do. Do.

7185. [5113] 'Kaisir,' mare, in jumping position, with Ahmedjan Khan. Do. Do.

## III.—OIL PAINTINGS.

7186. [9398] Portrait of the last Great Mogul, king of Delhi. Painted by August Schoefft.

7187. [9399] Portrait of the late Maharajah Shire Sing, king of the Punjaub. Do.

7188. [9400] The Thugs (men stranglers). Do.

This picture represents the moment when the Thugs expect from their chieftain's wife the signal for strangling unsuspecting travellers.

7189. [9401] The Court of Lahore, and other paintings of various interesting Indian subjects, by the same artist, can be seen in the principal dining-saloon in the Exhibition Building.

# Class XXXIX.—SCULPTURE, MODELS, ETC.

## I.—Statuettes, and other elaborate Carvings in Ivory.

7190. [3133] Royal yacht, carved by Bawul of Berhampore. Government of India.

7191. [3135] Palanquin, Do. Do.

7192. [3136] Juggernath car, Do. Do.

7193. [3137] Travelling cart, Do. Do.

7194. [3138] Eckha, Do. Do.

7195. [3139] Set of draughtsmen, red and white, Do. Do.

7196-7. [3140-1] Statuettes, Do. Do.

7198-9. [3142-3] Country boats, Do. Do.

7200-1. [3144-5] Paper knives, Do. Do.

7202. [3424] Boxes for antimony, elaborately carved in ivory, Umritsur. Government of India.

7203. [3426] Boxes with images, do., Do. Do.

7204. [3427] Chunkuna, do., Do. Do.

7205. [3428] Salt cellars, do., Do. Do.

7206. [3430] Box, with gold edges, do., Do. Do.

7207. [2431] Two cups, do., Do. Do.

7208. [5882] A cow and calf, do., Ulwar. H. H. the Maharajah.

7209. [1508] Table ornament, do., Puttiala. Do.

7210. [2228] A lion (a paper weight), do., Travancore. Do.

7211. [2229] A cow, do., do., Do. Do.

7312. [2231] Paper knife, crocodile handle, do., Do. Do.

7213. [2232] Do., serpent handle, do., Do. Do.

7214-7. [2233-6] Do., various designs, do., Do. Do.

7218-21. [2238-41] Parasol handles, do., Do. Do.

7222. [2242] Pair of birds, ivory, Travancore. H. H. the Rajah.

7223. [2246] Twenty-four various animals, small, do., Do. Do.

7224-5. [2247-8] Paper weights, serpents, do., Do. Do.

7226-7. [2249-50] Do., the human hand, do., Do. Do.

7228. [445] Three knife handles, large, Burmese. Dr. McPherson.

7229. [8106] Two dozen do., table size, Do. Do.

7230. [8107] One dozen do., dessert size, Do. Do.

7231. [10644] 'Tonjan,' or palanquin, carved in buffalo horn, Vizagapatam. H. H. the Rajah of Vizianagram.

7232. [10504] Two palanquins, do., Do. Do.

## II.—Plastic Models and Figures.

7233. [2939] A Rajpoot, Lucknow. Lucknow Committee. Government of India.

7234. [2940] A Mahratta, Do. Do.

7235. [2941] A Cabul fruitseller, Do. Do.

7236. [2942] A sweetmeat seller, Do. Do.

7237. [2943] A Sikh, Do. Do.

7238. [2944] A Mussulman fakeer, Do. Do.

7239. [2945] A moulvie, or Mussulman priest, Do. Do.

7240. [2946] A Brahmin, Do. Do.

7241. [2947] A Bengalee baboo, Do. Do.

7242. [2948] A Hindoo fakeer, Do Do.

7243. [2949] A tailor, Do. Do.

7244. [2950] A cloth merchant, Do. Do.

7245. [5801] Model of a bazaar, made by a native artist at Kishnaghur. Government of India.

7246. [5803] Do. of a plough, Do. Do.

7247. [5804] Model of hackery and two bullocks, made by a native artist at Kishnaghur. GOVERNMENT OF INDIA.

7248. [5805] Do. carriage and two bullocks, Do. Do.

7249. [5807] Do. palanquin and bearers, Do. Do.

7250. [5808] Various plastic figures, illustrating the different native trades and professions, Do. Do.

These include the following :—

7251. (1) Bheestie, or water carrier.

7252. (2) Up-country woman,

7253. (3) Brahmin praying.

7254. (4) Women collecting ghosee.

7255. (5) Boistom.

7256. (6) Songoter boistom.

7257. (7) Breadman.

7258. (8) Moyrah sweetmeat seller.

7259. (9) Santhal.

7260. (10) Mahomedan fakir.

7261. (11) Rowanee bearer.

7262. (12) Spearman.

7263. (13) Boonah.

7264. (14) Boistom with tomtom.

7265. (15) Toobriwalah.

7266. (16) Ploughman.

7267. (17) Carpenter.

7268. (18) Dhoonerie, cotton dresser.

7269. (19) Woman dressing cotton.

7270. (20) Spinning woman.

7271. (21) Dwijee (tailor).

7272. (22) Ooriah bearer.

7273. (23) Up-country shoemaker.

7274. (24) Shenkaree.

7275. (25) Hooka hurdar.

7276. (26) Up-country cloth seller.

7277. (27) Durwan.

7278. (28) Cook.

7279. (29) Sweeper.

7280. (30) Kejmutgar.

7281. (31) Chuprassee.

7282. (32) Native Shikaree.

7283. (33) Up-country sepoy.

7284. (34) Syee (groom).

7285. (35) Ayah (nurse).

7286. (36) Pundit.

7287. (37) Khansama.

7288. (38) Sheristadar.

7289. (39) Dhare.

7290. (40) Musalchee.

7291. (41) Goldsmith.

7292. (42) Bangy hurdar.

7293. (43) Washerman.

7294. (44) Plate maker.

7295. (45) Snake charmer.

7296. (46) Earthen pot maker.

7297. (47) Two women grinding oats.

7298. (48) Dhenkrewala rice grinding.

7299-7301. [2922-4] Kesah Kanie, Burmese figure, Assam. GOVERNMENT OF INDIA.

7302. [7874] Four figures, Poona. GOVERNMENT OF BOMBAY.

7303. [8164] Plastic ' Condapully,' figures illustrating the various native classes, trades and professions, Kistna, Madras. MADRAS GOVERNMENT.

These models include the following : —

7304. (49) Hermit.

7305. (50) Native drummers.

7306. (51) Singers.

7307. (52) Mussulman servant armed.

7308. (53) Servant of the zemindar.

7309. (54) Valama caste men.

7310. (55) Komali or Banian carrier.

7311. (56) Toddy drawer.

7312. (57) Shepherdess churning

7313. (58) Tappal runners or postmen.

7314. (59) Torchman.

7315. (60) Water carrier.

7316. (61) Mussulman peon.

7317. (62) Snake charmers.

7318. (63) Fowler.

7319. (64) Shepherd.

7320. (65) Woman of basket makers' caste.

7321. (66) Washerwoman.

7322. (67) Tank digger woman.

7323. (68) Woman of fowlers' caste.

7324. (69) Barber.

7325. (70) Woman of potters' caste.

7326. (71) Toddy drawer.

7327. (72) Carpenter.

7328. (73) Butler.

7329. (74) Satani beggar.

7330. (75) Sangam beggar.

7331. (76) Rohilla.

7332. (77) Beggar.

7333. (78) Raju.

7334. (79) Brahmin woman.

7335. (80) Meat seller.

7336. (81) Goldsmith at work.

7337. (82) Shroff.

7338. (83) Woman working at a mill.

7339. (84) Dasari beggar.

7340. (85) Horseman or groom.

7341. (86) Cooly man.

7342. (87) Zemindar's slave.

———

7343. [589] Elephant and figures of Shola pith, Tanjore. MADRAS GOVERNMENT. W. M. CADELL, Esq.

7344. [5344] Sample of pith (*Æschynomene aspera*) from which the above-named model was manufactured, Calcutta. GOVERNMENT of INDIA.

# INTERNATIONAL EXHIBITION, 1862.

## INDIA.

### MEDALS AND HONOURABLE MENTIONS AWARDED BY THE INTERNATIONAL JURIES.

| Presidency or Locality | To whom awarded | Medal | Hon. Mention (Certificate) | Class | Objects awarded and reasons for the award |
|---|---|---|---|---|---|
| **BENGAL** Calcutta . . . | Central Committee . . . | | 1 | 3B | For illustrative series of sugars and spices produced in India |
| | Do. | 1 | | 4C | For collection of useful vegetable products, including dye stuffs, &c. &c. |
| | Oldham, Professor T. . . | 1 | | 1 | For specimens with the analysis of a series of coals from many localities in India, and for the elaborate work of the Geological Survey conducted by him |
| | Surveyor General of India | 1 | | 1 | For the admirably executed maps of a part of the Himalayas, by the Topographical Survey now in progress |
| | Ahmuty & Co. . . . . | 1 | | 4C | For fibre prepared from *Œschynomene cannabina* |
| | Allen & Hayes . . . . | 1 | | 33 | For excellent workmanship and general merit of a silver vase, presented by the Government of India to Rajah Deonarain Singh, of Benares |
| | Borradaile, John, & Co. . | | 1 | 3B | For goodness of quality of tea |
| | George & Co. . . . . | 1 | | 3B | For the excellent quality of their arrowroot |
| | Guthrie, Col. . . . . . | | 1 | 1 | For the exhibition of his very beautiful series of works of art in jade and rock crystal |
| | Do. | | 1 | 11C | For his interesting collection of arms |
| | Kooney Lall Dey . . . | 1 | | 2B | For a large collection of East India drugs |
| | Do. | 1 | | 4D | For excellent quality of fragrant oils |
| | Lyall, Rennie, & Co. . . | | 1 | 20 | For raw silk from their mills at Calcutta and Berhampore |
| | Martin, John . . . . | | 1 | 11C | For his interesting collection of arms |
| | Sainte, Brothers . . . | | 1 | 4A | For goodness of manufacture of stearic candles |
| | Shaik Golab . . . . . | | 1 | 24 | For embroidery on muslin and tusser silk, deserving credit |
| | Simpson, Dr. . . . . . | 1 | | 14 | For a valuable series of portraits of the native tribes |
| | Steel, H. . . . . . | 1 | | 4C | For samples of indigo from Hooghly |
| Assam . . . . | Bivar, Major H. S. . . | 1 | | 4C | For madder from Assam |
| | Dhatooram Jemadar . . | 1 | | 3B | For excellence of manufacture, strength and flavour of his tea |
| | Martin, I. N.* . . . . | 1 | | 4C | For caoutchouc from the Cossia Hills |
| | Morgan, C. H. . . . . | 1 | | 3B | For excellence of manufacture, strength and flavour of his teas |
| | Phaire, Lieut. W. . . . | 1 | | 3B | For excellency of his peppers |
| | Do. | | 1 | 11C | For his interesting collection of arms |
| | Wagentrieber, W. G. . . | 1 | | 4C | For specimens of 'room' dye, an indigo made from a species of ruellia |
| Bhaugulpore . . | Sandys, T. . . . . . | | 1 | 10A† | For mortar, cement, and concrete of good quality |
| Cachar . . . | Cachar Tea Company . . | 1 | | 3B | For excellence of manufacture, strength and flavour of their tea |
| | Tydd, Forbes, & Co. . . | | 1 | 3B | For goodness of quality of their tea |
| Chota Nagpore . | Blechynden, C. E. . . . | 1 | | 4C | For cleaned cotton from New Orleans seed, @ 12½d. per lb. |

\* Initials erroneously given in the Jury List of awards as W. C.
† Entered in Catalogue in Class I.

| Presidency or Locality | To whom awarded | Medal | Hon. Mention (Certificate) | Class | Objects awarded and reasons for the award |
|---|---|---|---|---|---|
| BENGAL—*cont.* Chota Nagpore . | Blechynden, E. C. . . . | | 1 | 4c | For his Egyptian cleaned cotton, valued @ 13*d.* per lb. |
| | Leibert, M. T. . . . . | | 1 | 3B | For being the first producer of teas at Hazareebaugh |
| | Do. | | 1 | 4c | For Sea Island cotton, uncleaned, from Seetagurrah, value 9½*d.*, if cleaned 24*d.* * |
| Cuttack . . . | Local Committee . . . | | 1 | 3c | For leaf tobacco, sound and well-grown |
| | Juggernauth Doss . . . | 1 | | 33 | For general merit and good work of double branch epergné and jewel casket |
| | Taylor, Rev. George . . | 1 | | 4c | For his cotton from Piplee, valued at 11*d.* |
| Dacca . . . . | Hurmohun Roy . . . . | | 1 | 24 | For lace embroidered scarfs in black, silver, and gold |
| | Juggut Chunder Doss . . | | 1 | 24 | For a large collection of rich embroidered gold and silver scarfs |
| | Do. | | 1 | 24 | For India muslin scarfs, embroidered in gold, also some rich gold embroidered Cashmere shawls, small |
| Darjeeling . . | Brine, F., for Hope Town Tea Association | | 1 | 3B | For goodness of quality of tea |
| | Campbell, Dr. A. . . . | | 1 | 9 | For his models of agricultural implements |
| | Do. | | 1 | 11c | For his interesting collection of arms |
| | Do. | | 1 | 29 | For his collection of lichens and reeds |
| | Samler, Major . . . . | 1 | | 3B | For excellence of manufacture, strength and flavour of tea |
| | Scanlan, P. H. . . . . | | 1 | 3B | For goodness of quality of his tea |
| Malda . . . . | Thompson, Dr. R. F. . . | 1 | | 4c | For specimens of a new green vegetable dye, with illustrations of its application |
| | Do. | | 1 | 3c | For a specimen of mango spirit |
| Shahabad . . . | Mylne, Mr. . . . . . | 1 | | 4c | For cleaned cotton from Shahabad, Arrah, and for New Orleans seed, valued at 13½*d.* |
| | Government of India . . | | 1 | 1 | For a complete and instructive series of specimens, illustrating the dressing and smelting of tin ores from a new locality † |
| | Do. | 1 | | 2A | For a collection of chemicals, manufactured in India |
| | Do. | 1 | | 3A | For excellence of quality of paddy and rice, &c. |
| | Do. | | 1 | 4D | For collection of oils of excellent quality |
| | Do. | | 1 | 11c | For an interesting collection of arms |
| | Do. | 1 | | 18 | For superior manufactured cotton goods |
| | Do. | 1 | | 22 | For straw mats of excellence of design, colour, and manufacture |
| | Do. | 1 | | 23 | For a good collection of plain dyed cotton fabrics |
| | Do. | | 1 | 26B | For saddles, trappings, &c. creditable as native productions ‡ |
| | Do. | | 1 | 27D | For well-made native shoes, clogs, &c. § |
| | Do. | 1 | | 30AB | For a collection of sandalwood writing desks, jewel boxes, &c. of great artistic excellence ‖ |

* In Jury List awarded to the 'Collector Nagpore.'
† Chief contributor of these José D'Almeida, Esq., of Singapore.
‡ Chiefly manufactured at Cawnpore.
§ From Bareilly, Jhansi, Lahore, and Umritsur.
‖ From Delhi, &c.

| Presidency or Locality | To whom awarded | Medal | Hon. Mention (Certificate) | Class | Objects awarded and reasons for the award |
|---|---|---|---|---|---|
| BENGAL—*cont.* | Government of India . . | 1 | | 33 | For the general excellence of the gold and filigree work, brooches, bracelets and jewellery * |
| NORTH WEST PROVINCES AND OUDE | | | | | |
| Agra . . . . . | Central Committee . . | 1 | | 1 | For interesting collection of works executed in soapstone † |
| Benares . . . | Dabee Pershaud . . . | 1 | | 24 | For violet and gold, and mazarine and silk damask, very rich and beautiful Kincaub, well executed |
| | Do. | | 1 | 24 | Embroidered table mats and bags |
| | Mohun Lall and Chittoo Lall | 1 | | 24 | For two magnificent black and gold Kincaub scarfs; blue and gold Kincaub pieces, and other articles, well adapted for covering couches |
| | Do. | | 1 | 24 | Embroidered table mats and bags |
| | Silhut Chundrabhun . . | 1 | | 24 | For Kincaub pieces, blue and gold, and white and gold, of great merit |
| | Do. | | 1 | 24 | Embroidered table mats and bags |
| Dehra Doon . . | Dehra Doon Tea Company | | 1 | 3B | For goodness of quality of teas |
| Delhi . . . . | Manak Chund . . . . | 1 | | 24 | For black and gold, and green and gold (Cashmere cloth) embroidered shawls; scarlet cape, embroidered white, and a large assortment of similar articles of great beauty |
| | Do. | | 1 | 24 | For a net shawl, embroidered with floss silk, cardinal scarf, in good taste |
| Goruckpore . . | Osborne George . . . . | 1 | | 4C | For caoutchouc from Goruckpore |
| Gurwhal . . | Warrand, T. . . . . . | | 1 | 3B | For goodness of quality of tea |
| Jubbulpore . . | Superintendent of the School of Industry | 1 | | 22 | For the excellence of design, colour, and manufacture of carpets |
| Kumaon . . . | M'Ivor, K., of the Konsamire Plantation | 1 | | 3B | Tea, excellence of manufacture, strength, and flavour |
| | Troup, C. R., of the Megree Plantation | | 1 | 3B | For goodness of quality of teas |
| Meerut . . . | Superintendent of Jail. . | 1 | | 22 | For the excellence of design, colour, and manufacture of carpets |
| | Kooer Wuzeer Ali Khan, Deputy-Magistrate | | 1 | 27D | For good quality of shoes |
| Shahjehanpore . | Carew & Co. . . . . . | 1 | | 3B | For excellence of manufacture of sugar |
| | Do. | 1 | | 3C | For general excellence of their cane-juice rum |
| Sutwarree Bundelkhund | Dashwood, W., Collector of Banda | 1 | | 4C | For cleaned cotton from Sutwarree, unknown seed, 11*d.* and 17*d.* per lb. |
| OUDE | | | | | |
| Khyrabad . . | Lindsay, C., Deputy-Commissioner of Hurdin | | 1 | 11C | For his interesting collection of arms |
| Lucknow . . . | Local Committee . . . | 1 | | 4C | For a series of the vegetable fibres of Lucknow, and other vegetable products |
| PUNJAUB | | | | | |
| Hansi . . . . | Jardine & Co. . . . . | 1 | | 4C | For samples of their indigo |
| Lahore. . . . | Central Committee . . . | 1 | | 4C | For a collection of lacquered turnery |
| | Do. | | 1 | 4C | For cleaned cotton from the Leia district, value 8½*d.* per lb.‡ |
| | Do., for manufacturers of Bokhara carpets | 1 | | 22 | For excellence of design, colour, and manufacture of carpets |
| | Harrison, Captain, 79th Highlanders | | 1 | 11C | For a beautiful dagger |

* From Delhi and Benares.
† Awarded to Calcutta Committee.
‡ Awarded in error to Captain Mitchell, Madras.

| Presidency or Locality | To whom awarded | Medal | Hon. Mention (Certificate) | Class | Objects awarded and reasons for the award |
|---|---|---|---|---|---|
| PUNJAUB—*cont.* | | | | | |
| Lahore . . . | Superintendent of Central Jail | 1 | | 22 | For the excellence of design, colour, and manufacture of carpets |
| Peshawur. . . | Mahomed Zúma . . . | | 1 | 11c | For a collection of arms |
| Shahpore . . . | Superintendent of Jail . | 1 | | 4c | For a new application of Mádar floss to mat making: a very interesting experiment |
| Sealkote . . . | Imaum, Ad-deen. . . . | 1 | | 33 | For excellent workmanship of steel inlaid with gold |
| | Indian Flax Company (Belfast) | 1 | | 4c | For various samples of flax grown in the Punjaub |
| Umritsur . . . | Local Committee . . . | 1 | | 4c | For a collection of vases in black lacquered work, with ornamentation in chemical amalgam |
| NATIVE DIGNITARIES | | | | | |
| Benares . . . | Rajah Deonarain Sing. . | 1 | | 4c | For a model of a Hindoo temple in sandalwood |
| Bhawulpore . . | H. H. the Nawab . . . | | 1 | 11c | For his interesting collection of arms |
| Jeypore . . . | H. H. the Rajah. . . . | 1 | | 4d | For excellence of quality of fragrant oils |
| | Do. | | 1 | 11c | For his interesting collection of arms |
| | Do. | 1 | | 33 | For beautiful workmanship of enamelled gold spice box and cup |
| Oude, Lucknow . | Nawab Shurf-ood-Dowlah | 1 | | 27d | For excellence of work |
| | Do. | | 1 | 27d | For gold embroidered dress, turban, and slippers |
| Do., Moraon . . | Rajah Goree Shunker . . | 1 | | 27d | For finely embroidered shoes and slippers |
| Nepaul . . . | H. H. Sir Jung Bahadoor, K.C.B. | 1 | | 4c | For Nepaul madder |
| | Do. | | 1 | 11c | For his interesting collection of arms |
| | Do. | | 1 | 24 | For a great variety of gold embroidered articles of dress |
| Peshawur. . . | H. H. the Nawab . . . | | 1 | 11c | For a collection of arms |
| Puttiala . . . | H. H. the Rajah . . . | | 1 | 11c | For his interesting collection of arms |
| Rampore . . . | The Nawab . . . . . | | 1 | 11c | For his interesting collection of arms |
| Ulwar . . . . | H. H. the Rajah . . . | | 1 | 11c | For his interesting collection of arms |
| | Do. | | 1 | 24 | For a great variety of gold embroidered articles of dress |
| | Do. | | 1 | 27d | For well-made native shoes |
| BOMBAY AND SINDE | | | | | |
| Bombay . . . | Birdwood, Dr. . . . . . | 1 | | 4c | For an extensive collection of Indian vegetable products of economic value |
| | Do. | | 1 | 4d | For a collection of oils and their goodness of quality |
| | Bhrimjee Byramjee. . . | 1 | | 30AB | For good design and workmanship of carved furniture |
| | Jamsetjee Heerjee . . . | 1 | | 4c | For a large number of admirable carvings in sandalwood |
| | Sellon, Captain . . . . | | 1 | 14 | For a series of photographic views in India |
| | Smith, Fleming, & Co. . | 1 | | 4c | For cotton from Bourbon seed, cleaned, valued at 1½d. per lb., and various others |
| Ahmedabad . . | The Collector . . . . | 1 | | 4c | For Nurma native cotton from Ahmedabad, value 10½d. per lb. |
| | Do., for manufacturers of Kincaubs (Ahmedabad) | | 1 | 24 | For Kincaub handkerchiefs, and other specimens of gold and silver (loom) embroidery on silk |
| Belgaum . . . | Hearn, Mr. . . . . . | 1 | | 4c | For cleaned cotton, valued at 12d. per lb. |

| Presidency or Locality | To whom awarded | Medal | Hon. Mention (Certificate) | Class | Objects awarded and reasons for the award. |
|---|---|---|---|---|---|
| **BOMBAY** AND **SINDE**—*cont.* | | | | | |
| Dharwar . . . | The Collector . . . . | | 1 | 9 | For models of agricultural implements |
| Poonah . . . | Collector of . . . . . | 1 | | 4c | For cotton from Poonah, native seed |
| | Manufacturers of . . . | | 1 | 24 | For Kincaub handkerchiefs, and other specimens of gold and silver (loom) embroidery in silk cloth |
| **SINDE** | | | | | |
| Hyderabad . . | The Committee at . . . | 1 | | 4c | For a collection of lacquered turnery |
| | Do., for manufacturers . | | 1 | 24 | A variety of table covers, cloth, embroidered with silk, of great merit |
| Seebee. . . . | The Collector . . . . | | 1 | 4c | For cotton from Seebee, Sinde |
| Shikarpoor . . | Manufacturers of carpets. | 1 | | 22 | For the excellence of design, colour, and manufacture of their carpets |
| Bombay . . . | Government . . . . . | | 1 | 4c | For Lalia native cotton, uncleaned, from Ahmedabad. Value, 9*d.* |
| | Do. | 1 | | 4c | For sandalwood carvings |
| **NATIVE DIGNITARIES** | | | | | |
| Kutch . . . . | H. H. the Rao . . . . | | 1 | 11c | For his interesting collection of arms |
| | Do. | | 1 | 24 | For mats and table covers of various designs |
| | Do. | | 1 | 33 | For general merit in articles of silver |
| | Do. | 1 | | 4c | For a model of a Hindoo temple in sandalwood |
| | Do. | | 1 | 4c | For cleaned cotton. Value, 8¾*d.* per lb. |
| Khyrpoor. . . | H. H. Meer Ali Morad . | | 1 | 24 | For a great variety of gold embroidered articles of dress, turbans, slippers, &c. |
| | Do. | | 1 | 11c | For his interesting collection of arms |
| | Do. | 1 | | 27D | For a good display of native shoemakers' work |
| Zanzibar . . . | H. H. the Sultan . . . | 1 | | 4c | For specimens of copal |
| **MADRAS** | | | | | |
| Madras . . . | Central Committee. . . | | 1 | 11c | For an interesting collection of arms |
| | Cleghorn, Dr. . . . . | | 1 | 4D | For goodness of quality of fragrant oils |
| | Deschamps, J. . . . . | 1 | | 30AB | For good design and workmanship of rosewood arm-chairs |
| | Hunter, Dr. . . . . . | 1 | | 1 | For a carefully collected series of pottery clays and their manufactured products |
| | Do. | 1 | | 4c | For a collection of the fibrous materials of Madras, admirably prepared |
| | McPherson, Dr.. . . . | 1 | | 4c | For a collection of wood carvings and other manufactures and vegetable products |
| | Maitland, Col. . . . . | 1 | | 4c | For fifteen excellent slabs illustrating the woods now in use in Madras |
| | Mitchell, Capt. . . . | | 1 | 1 | For an instructive series of specimens of magnetic iron sand, employed in some parts of India by native iron smelters |
| Arcot, North . | Rámashishaia . . . . | 1 | | 4c | For indigo from N. Arcot. |
| | Venkata Raó and Bapana Raó | 1 | | 23 | For their good dyed cloth and printed chintzes |
| Beypore . . . | East Indian Iron Co. . . | 1 | | 1 | For an interesting and instructive collection of specimens illustrating the production of iron and steel in the Madras Presidency |
| Canara, South . | Coelho, V. Pedro . . . | | 1 | 3B | For goodness of quality of peppers |
| | Do. | 1 | | 4c | For a series of vegetable fibres and for gamboge from South Canara |
| | Do. | 1 | | 3c | For excellence of Palmyra arrack |
| | Do. | | 1 | 4c | For thirty-six specimens of wood |
| Chingleput . . | Shortt, Dr. John . . . | | 1 | 2B | For a collection of East India drugs |

| Presidency or Locality | To whom awarded | Medal | Hon. Mention (Certificate) | Class | Objects awarded and reasons for the award. |
|---|---|---|---|---|---|
| MADRAS—*cont.* | | | | | |
| Chingleput .. | Shortt, Dr. John ... | 1 | | 4c | For clean cotton from Chingleput, Egyptian seed. Value, 11d. per lb.* |
| | Do. | 1 | | 4c | For a series of mat-making materials and vegetable fibres for general purposes, and for other vegetable products of economic value |
| Coimbatore .. | McIvor, W. G., Superintendent of Gardens, Ootacamund | 1 | | 4c | For admirable specimens of fibres prepared from the Neilgherry nettle |
| Ellore.. ... | Carpet manufacturers .. | 1 | | 22 | Collective medal for excellence of design, colour, and manufacture of carpets † |
| Nellore ... | Collector of .... | | 1 | 4c | For cleaned cotton from Nellore. Value, 9½d. |
| Salem .... | Collector of .... | 1 | | 4c | For Oopum native cotton from Salem, valued at 11d.‡ |
| | Chetumbara Pillay... | 1 | | 4d | For excellence of fragrant oils |
| | Fisher & Co. .... | 1 | | 2a | For excellent quality of their saltpetre |
| | Do. | | 1 | 3b | For goodness of quality of coffee |
| | Do. | 1 | | 4c | For samples of indigo, and for cleaned cotton from Bourbon seed, valued at 12¼d. per lb. |
| Tanjore ... | Carpet manufacturer .. | 1 | | 22 | Medal for the excellence of design, colour, and manufacture of carpet (rug)§ |
| Tinnevelly .. | Edaiyangudi Missionary School, Mrs. Caldwell, Directress ..... | | 1 | 24 | White and black lace from Tinnevelly, showing considerable aptitude for this class of manufacture, and that with perseverance great progress would likely be made |
| Trichinopoly.. | Rosalingum, Assary .. | 1 | | 33 | For gold chain and bracelet and general merit of workmanship |
| | Government of Madras . | 1 | | 4c | For a collection of vegetable products |
| | Do. | 1 | | 24 | For a beautiful embroidered dress, with gold and beetle wings; black muslin dress embroidered with ditto; also a white muslin dress, with other articles of the same class, showing excellence of design and workmanship ‖ |
| Mysore ... | Astagram Sugar Company | 1 | | 3b | For excellence of manufacture of their sugar |
| | Do. | 1 | | 3c | For general excellence of rum |
| | Onslow, Col. W. Campbell | 1 | | 3b | For excellent quality of coffee |
| | Puckle, Capt. J. .... | 1 | | 4c | For a series of 26 specimens of wood of Mysore, showing the strength of each |
| | Government of Mysore . | 1 | | 4c | For sandalwood carvings, and for cleaned cotton from Mysore, valued at 12¼d. per lb. |
| | Do. | | 1 | 4c | For cotton from *Gossypium herbaceum*, value 8¾d. per lb. |
| | Do. | | 1 | 3c | For leaf tobacco |
| | Do. | | 1 | 9 | For models of agricultural implements from Bangalore |
| NATIVE DIGNITARY | | | | | |
| Vizianagram .. | H. H. the Rajah ... | 1 | | 1 | For the interest attaching to his graphite, found in a new locality |
| | Do. | | 1 | 27d | For native work in native shoes and sandals |
| | Do. | 1 | 1 | 30 | For inlaid work, being of good design and workmanship |

* Awarded to 'Collector of Chingleput,' but exhibited by Dr. Shortt.
† Awarded, under head of 'Madras Manufacturers,' to the Ellore carpet weavers.
‡ This cotton, forwarded by the Collectors of Salem, is from Messrs. Fisher & Co., whose specimens of Bourbon cotton are also referred to in the Jury's award of a medal for indigo.
§ The silk rug here referred to was forwarded for exhibition by Saccharam Sahib, son-in-law of the late Rajah of Tanjore.
‖ These embroideries were executed to order by Hoossain Khan.

| Presidency or Locality | To whom awarded | Medal | Hon. Mention (Certificate) | Class | Objects awarded and reasons for the award |
|---|---|---|---|---|---|
| BURMAH AND STRAITS SETTLEMENTS Rangoon . . . | Phayre, Colonel . . . . | 1 | | 3c | For sound, well-grown, and fragrant leaf tobacco |
| | Local Committee . . . | 1 | | 4c | For a collection of the useful vegetable products of Burmah |
| | Brandis, Dr. . . . . . | 1 | | 4c | For an excellent scientifically named collection of 120 woods of Burmah |
| Malacca . . . | De Wind, A. A. . . . . | 1 | | 4c | For Gutta terble |
| | Baumgarten, J. . . . . | | 1 | 4c | For specimens of Malacca cotton, value 12½d. per lb. |
| | H. H. Inche Wan Aboo Bakar | | 1 | 4c | For good specimens of Malaccan woods |
| | Neubronner, T. . . . . | 1 | | 4c | For several varieties of gutta percha from Malacca |
| Penang and Province Wellesley | Glugor Estate (the Proprietors) | 1 | | 4c | For Sea Island cotton, at 24d. per lb., and for Pernambuco, at 13d. per lb. |
| | Horsman, Hon. E., M.P. . | 1 | | 3B | For excellent quality of sugars |
| | Hutchinson, A. . . . . | 1 | | 4c | For Washington cotton from Province Wellesley |
| | Man, Hon. Major . . . | | 1 | 4c | For 25 specimens of Penang woods |
| | Nairn, Lawrence . . . | | 1 | 3B | For good tapioca, and novelty of preparation |
| | Scott, G. . . . . . . | 1 | | 4D | For excellence of quality of fragrant oils |
| Singapore . . | Local Committee . . . | | 1 | 4A | For good collection of oils |
| | Do. | 1 | | 3B | For excellent quality of spices, tapioca, and sago |
| | Do. | 1 | | 4c | For series of mats and mat-making materials, from Penang and Province Wellesley |
| | Angus, G. . . . . . . | 1 | | 4c | For a valuable collection of resins, Dammar gums, and tanning materials |
| | Brown, G. H. . . . . | 1 | | 4c | For Sea Island cotton, valued at 14d. per lb. |
| | Cavanagh, Hon. Col. . . | | 1 | 4c | For cleaned cotton from Singapore, Pernambuco seed, 12½d. |
| | Do. | 1 | | 4c | For cotton grown at Government House Garden |
| | Do. | | 1 | 11c | For his interesting collection of arms |
| | D'Almeida, José. . . . | 1 | | 3B | For excellency of peppers |
| | Do. | 1 | | 3c | For general excellence of rum |
| | Thompson, D. C.* . . . | 1 | | 3B | For excellent quality of tapioca, and novelty of preparation |
| | Tan Kim Sing . . . . | 1 | | 3c | For general excellence of arrack |
| | Do. | 1 | | 27D | For native boots, shoes, clogs, &c., well made |
| Tringanu . . . | H. H. the Rajah . . . | | 1 | 11c | For his interesting collection of arms |
| | Do. | 1 | | | For turnery in wood, supposed to be tamarind |
| ENGLAND | Montgomery Martin . . | 1 | | 1 | For his illustration of the hydrographical basins of India |
| | Do. | 1 | | 29 | For the merit of a topographical map of India |
| | Cautley, Col. Sir P., K.C.B. | | 1 | 13 | For a copy of Troughton's level; considered as a good specimen of native workmanship |
| | Reid, Colonel, C.B. . . . | 1 | | 25A | For showing two large and fine specimens of Bengal tiger furs |
| | Do. | | 1 | 29 | For two stuffed tigers |
| | Halliday, Fox, & Co. . . | 1 | | 4c | For cleaned cotton from Arracan, valued at 13d. per lb. |
| | Do. | 1 | | 3B | For excellent quality of spices from Rangoon |
| | Do. | | 1 | 3A | For paddy and rice from Rangoon, comparatively good |
| | Do. | 1 | | 4c | For an extensive collection of the dyeing and tanning materials of Rangoon |

* Manager on the estate of R. Wilson, Esq.

| Presidency or Locality | To whom awarded | Medal | Hon. Mention (Certificate) | Class | Objects awarded and reasons for the award |
|---|---|---|---|---|---|
| ENGLAND— *cont.* | Watson, Dr. J. Forbes . | 1 | | 1 | For an extensive collection of soils, and the scientific labour bestowed on the analyses* |
| | Do. | 1 | | 4c | For his arrangements of the fibres of India in an admirably practical manner, showing the relative capabilities of the various materials |
| | India Museum . . . . | 1 | | 3A | For excellence of Indian grains and pulses, with analyses by Dr. J. Forbes Watson |
| | Moore, F. . . . . . . | 1 | | 4B | For a very complete collection of the various species of silkworms |
| | Do. | 1 | | 29 | For his collection of Asiatic silk-producing moths |

\* The soils here referred to were selected for examination from the collection, now in the India Museum, made by Messrs. de Schlagintweit. The analyses were executed for my department by Dr. Albert Bernays, Professor of Chemistry, St. Thomas's Hospital; by Mr. W. Valentine, Senior Assistant, Royal College of Chemistry, and by Mr. F. A. Manning, of Leadenhall Street.

---

In consequence of an accidental oversight the excellent specimens of canvas and bagging exhibited by the 'Borneo Company' did not come under the notice of the Jury.

J. F. W.

# INDEX.

### ADA

**PAGE**

ADAPTATION of metals to special purposes . . 10
Agricultural implements (4229–4262) . . . . 193
Agricultural produce . . 45
Alkalies, earths, and their compounds . . . . 27
Animal and vegetable substances used in manufactures 85
Animal food products . . 76
Animal oils (1859–73) . . 85
Animal substances used in manufacture . . . 97
Apparel, miscellaneous . . 244
Arms, small (4326–36) . . 196
Arrowroot (1633–43) . . 71

BATTLE-AXES, spears, arrows (4387–4413) . . 198
Beadwork . . . 232
Bidri ware (6815–7) . . 264
Bituminous bodies and native naphtha . . . 14
Blanketings, etc. (5000–17) . 218
Bone, horn, etc. . . . 103
Books, etc. (6107–11) . . 248
Boots and shoes (6018–54) . 245
Building stones (326–336) . 16

CALIGRAPHY, &c. (6112–6131) . . . . 248
Canvas (4682–95) . . 208
Caoutchouc (2431–33) . . 112
Carpets, rugs, and mats . . 222
Carved furniture (6226–60) . 252
Carvings in sandalwood, etc. (3843–3939) . . . 182
Cashmere shawls . . 219
Cellular substances (3147–8) 152
Cements and artificial stones (371–4) . . . . 17
Cereals (1023–1335) . . 46
Charcoal (3745–50) . . 180
Chemical substances used in manufacture . . . 27
Chemical substances and products . . . . 27
Chemical substances used in medicine and pharmacy . 30
Chickun work (5304–67) . 228
Chogas, etc. (5825–33) . 240
Cholees, etc. materials for . 244

### COT

**PAGE**

Churrus, or hemp resin (1803–1815). . . . . 81
Civil engineering and building 194
Clays (392–425) . . 18
Clothing, articles of . . 238
Coals, series of Indian, (224–254) . . . . . 11
Coffee, table of exports . . 69
Coins (6803–10) . . . 264
Coir, and coir rope, table of exports . . . 150
Committees, Indian . v.–xvi.
Copper ores (125-7, 131-3) . 8
Cordage of all kinds (4808–4818). . . . . 212
Cornelian, agate, etc. manufactures . . . . 268
Cotton, table of exports . . 123
Cotton, from
  Arracan, Pegu, Tenasserim and Straits' Provinces (2908–25) . . . 132
  Assam (2766–70) . . 127
  Bancoorah (2726–31) . 125
  Behar (2745–6) . . . 126
  Bogra (2735) . . . 126
  Burdwan (2712–25) . . 125
  Cachar (2772–5) . . 127
  Chittagong (2707–11) . 125
  Chota Nagpore (2781–90) . 128
  Cuttack and Sumbulpore (2693–2706) . . . 122
  Darjeeling (2780) . . 128
  Garrow Hills (3778–9) . 128
  Gwalior and Ulwar (2810–2822) . . . . 130
  Madras (2885–99) . . 131
  Midnapore (2732–4) . . 126
  Monghyr (2737–44) . . 126
  Mysore (2900–7) . . 132
  North-Western Provinces—
    Bundelkund . . . 129
  Patna (2751–2). . . 127
  Pubna (2736) . . . 126
  Punjaub and Sinde (2823–2841) . . . 130
  Sarun (2747–50) . . 126
  Shahabad (2753–62) . . 127
  Silhet (2776–7) . . 128
  Tirhoot (2763–4) . . 127
  Western India and Berar (2842–84) . . . 131
Cotton goods, table of exports 205

### FUR

**PAGE**

Cottons, silk . . . . 133
Cotton yarn and thread (4556–4568) . . . . 204
Counterpanes (5258–60) . 227
Cutch, or Catechu (Acacia Catechu) (2682–92) . . 122
Cutlery (6516–27) . . . 257

DATE-SUGAR, manufacture of . . . . . 75
Dhotees (5834–63) . . 241
Drawings for teaching . . 249
Dressing-cases, etc. . . 271
Dressing-cases and fittings (7083–7113) . . . 272
Drugs, Indian (613–952) . 30
Dye stuffs and tanning materials . . . . . 112
Dyes, table of exports . . 112
Dyeing, samples of . . 226

EARTHY and semi-crystalline minerals . . . 22
Edible algæ, etc. (1673–81) . 73
Educational works and appliances . . . . 248
Elastic and plastic gums (2431–2449) . . . . 112
Embroidery, miscellaneous . 232
Enamelling and Damascene work . . . . . 264
Ezarbunds and Cummerbunds 242

FEATHERS, table of exports . . . . 234
Feathers and feather work (5625–67) . . . 235
Female attire . . . 243
Fibres (2943–3148) . . 139
Fibrous substances . . 122
Figures, plastic, etc. (7233–7344) . . . . . 275
Filigree work . . . 261
Fine Arts Division . . 273
Fish maws and shark fins (1726–40) . . . . 76
Fish maws, table of exports . 77
Fossil Cephalopoda, etc. (6150–6225) . . . . . 249
Fruits, dried, (1336–53) . 61
Fuel . . . . . 11
Furniture, upholstery, etc. . 252

## GAM

| | PAGE |
|---|---|
| GAMES and Toys (6134–6149) | 249 |
| Geological maps, plans and sections | 1 |
| Geological survey of India, maps of | 2 |
| Glass inlaid with gold | 265 |
| Gold and silver, articles in | 258 |
| Gold embroidery | 229 |
| Grains, Indian, in the ear, (962–1022) | 45 |
| Grocery, or preparations of food | 61 |
| Gums (2256–2341) | 104 |
| Gums, table of exports | 105 |
| Gunny bags, table of exports | 143 |
| Gutta percha (2438–44) | 112 |
| HAIR, table of exports | 234 |
| Half-stuff (6055–60) | 246 |
| Handkerchiefs, silk (4950–2) | 217 |
| Hardware, metal work, &c. | 255 |
| Hats, caps, etc. | 239 |
| Head coverings | 238 |
| Hemp, and hemp rope, table of exports | 146 |
| Hides and skins, table of exports | 236 |
| Hookahs (1854–8) | 84 |
| Horns, table of exports | 103 |
| Horological instruments (4538) | 203 |
| INDIA cotton, tabular synopsis of | 134 |
| India rubber | 112 |
| Indigo (2450–2503) | 113 |
| Indigo, table of exports | 114 |
| Inlaid work (5306–6422) | 253 |
| Iron and general hardware | 255 |
| Iron ores (26–38, 83–112) | 356 |
| Ivory carvings (7190–7232) | 275 |
| Ivory, paintings on (7114–59) | 274 |
| Ivory, table of exports | 250 |
| JADE and rock crystal, manufactures | 265 |
| Japanned work (6293–6304) | 253 |
| Jewellery | 258 |
| Jewellery, table of exports | 258 |
| Jury awards | 279 |
| Jute, and jute rope, table of exports | 142 |
| KASHMERE shawls (5018–5073) | 220 |
| Kashmere shawls, table of exports | 219 |
| Kerchiefs for the head (5948–5957) | 243 |
| Kincobs (5261–80) | 227 |
| Koftgari work (6822–33) | 265 |
| LACE (5281–5303) | 278 |
| Lace, gold | 230 |
| Lacs, crude, stick lac (2349–2372) | 108 |

## MUS

| | PAGE |
|---|---|
| Lac, table of exports | 109 |
| Lead ores (137–141) | 8 |
| Leather, with saddlery and harness | 235 |
| Lichens (2657–9) | 121 |
| Lignite and peat (255–261) | 13 |
| Limestones (299–325) | 15 |
| Linseed, table of exports | 91 |
| Lithographic stones (543) | 23 |
| Loongees, or scarfs (5768–5824) | 239 |
| MADDER, table of exports | 116 |
| Manufactures in stone, alabaster, etc. (4264–4325) | 194 |
| Manufacturing machines and tools (4190–4225) | 190 |
| Manufactures in flax, hemp, and other fibres (478–96) | 211 |
| Manufactures from straw, grass and other similar materials (3949–4017) | 184 |
| Marbles (337–342) | 16 |
| Materials for baskets, brooms, etc. (3099–3113) | 151 |
| Mat-making materials (3091–8) | 15 |
| Mats (5144–5207) | 225 |
| Mats, table of exports | 225 |
| Matting of hemp, etc. | 224 |
| Medals, jury awards, etc. | 279 |
| Medicinal gum resins (2330–2348) | 107 |
| Medicinal substances from the vegetable and animal kingdom (613–952) | 30 |
| Metallic preparations employed in medicine (602–612) | 30 |
| Mica (366–9) | 17 |
| Mica, paintings on, etc. (7160–7185) | 274 |
| Military engineering | 196 |
| Millets | 48 |
| Mineral waters (497–520) | 22 |
| Minerals used in the manufacture of pottery and glass | 17 |
| Minerals used in manufacture (550–6) | 23 |
| Minerals used for ornament (433–464) | 19 |
| Mining and quarrying operations | 1 |
| Miscellaneous drugs (953–961) | 44 |
| Mixed fabrics, etc. (5080–5092) | 221 |
| Models, topographical, of India | 1 |
| Models carved in wood (3940–3948) | 183 |
| Molasses, table of exports | 75 |
| Munjeet or madder root | 115 |
| Musical instruments (4539–4555) | 204 |
| Musk, table of exports | 186 |
| Muslins (4569–98) | 206 |

## RUG

| | PAGE |
|---|---|
| NAPKINS and doyleys (4637–55) | 207 |
| Native metals | 9 |
| Natural history, illustrations | 249 |
| Naval architecture and ships' tackle (4414–24) | 198 |
| Non-metallic mineral products | 11 |
| OIL PAINTINGS (7186–7189) | 274 |
| Oils, table of exports | 86 |
| Oils and oil seeds, table of exports, Bombay | 92 |
| Oleo-resins (2413–30) | 111 |
| Opium (1763–1802) | 78 |
| Opium, table of exports | 79 |
| Ores and metallurgical operations | 3 |
| PAPER, printing, etc. | 246 |
| Papier maché, etc. (6265) | 253 |
| Perfumery | 185 |
| Perfumery, table of exports | 185 |
| Perfumes of animal origin (4018–9) | 185 |
| Perfumes derived from plants (4020–4184) | 186 |
| Personal ornaments in tinsel, ivory, horn, hair, etc. | 269 |
| Petroleum | 14 |
| Philosophical instruments (4425–6) | 200 |
| Photographic pictures and likenesses (4427–4537) | 200 |
| Physical education appliances | 249 |
| Pickles, preserves, dried fruits, etc. (1354–99) | 61 |
| Pigments, dyes, and various other chemical manufactures (595–600) | 29 |
| Pigments and dyes (2251–2555) | 104 |
| Pine apple (Ananassa sativa) (3027–32) | 147 |
| Plastic models and figures (7233–7344) | 275 |
| Portraits, ivory (7152–9) | 274 |
| Potstone (358–365) | 17 |
| Pottery (7000–7065) | 170 |
| Precious metals, working in | 258 |
| Precious stones | 19 |
| Precious stones, table of exports | 19 |
| Prepared woods (3734–6) | 180 |
| Printing | 248 |
| Printing and dyeing, specimens of (5208–57) | 226 |
| Pulses | 56 |
| RAILWAY plant (4185–7) | 190 |
| Rattans and canes (3146) | 152 |
| Rattans and canes, table of exports | 152 |
| Raw silk | 99 |
| Resins (2373–2430) | 109 |
| Rice | 52 |
| Rugs (5106–24) | 223 |

## SAD

| | PAGE |
|---|---|
| SADDLERY, harness, etc. (5675–84) | 235 |
| Sago (1658–65) | 72 |
| Saleps (1666–72) | 72 |
| Saltpetre (563–582) | 27 |
| Saltpetre, table of exports | 28 |
| Salt | 20 |
| Sandstones (273–291) | 14 |
| Sarees (5876–5925) | 242 |
| Scarfs for skirts (5926–5947) | 243 |
| Sculpture, models, etc. | 275 |
| Seeds employed for bracelets, and other ornamental purposes (3741–4) | 180 |
| Shark fins, table of exports | 77 |
| Shawls, Kashmere (5018–73) | 220 |
| Shawls, silk (5077–9) | 221 |
| Sheetings (4677–81) | 208 |
| Shellac | 108 |
| Silk cottons (2926–37) | 133 |
| Silk, various specimens of | 99 |
| Silk, table of exports | 100 |
| Silk embroidery | 230 |
| Silk goods | 214 |
| Silk goods, table of exports | 215 |
| Silks from Bombay and Southern India | 216 |
| Silks, plain and fancy (4885–4990) | 214 |
| Silk and velvet | 214 |
| Silk yarns (4879–84) | 214 |
| Silver, articles in | 261 |
| Silver embroidery | 230 |
| Skins and furs (5621–4) | 235 |
| Skins, fur, feathers, and hair | 234 |
| Skirts, etc. (5958–69) | 243 |
| Slate (292–8) | 14 |
| Soap and candles (1874–6) | 85 |
| Soils, Indian | 23 |
| Analysis of a series of | 24 |
| Soosees, etc. (5080–92) | 221 |
| Spices, etc. (1400–65) | 62 |
| Spirits, intoxicating or stimulating drugs (1741–62) | 78 |
| Starches, etc. (1633–65) | 71 |
| Statuettes, etc., in ivory (7190–7232) | 275 |
| Steel, cutlery, and edge tools | 257 |
| Substances used as food | 45 |
| Sugars (1682–1725) | 73 |
| Sugar, table of exports | 74 |
| Suttringees (5125–43) | 223 |
| Swords, daggers (4337–86) | 196 |
| TABLE CLOTHS (4616–4636) | 207 |
| Table showing export of Coffee | 69 |

## TEA

| | PAGE |
|---|---|
| Table showing export of | |
| Coir and coir-rope | 150 |
| Cotton | 123 |
| Cotton goods | 205 |
| Dyes | 112 |
| Feathers | 234 |
| Fish maws | 77 |
| Gums | 105 |
| Gunny bags and gunnies | 143 |
| Hair | 234 |
| Hemp and hemp rope | 146 |
| Hides and skins | 236 |
| Horns | 103 |
| Indigo | 114 |
| Ivory | 250 |
| Jewellery | 258 |
| Jute and jute rope | 142 |
| Kashmere shawls | 219 |
| Lac | 109 |
| Linseed | 91 |
| Madder (munjeet) | 116 |
| Mats and matting | 225 |
| Molasses | 75 |
| Musk | 186 |
| Oils | 86 |
| Oils and oil seeds, Bombay | 92 |
| Opium | 79 |
| Perfumery | 185 |
| Precious stones | 19 |
| Rattans and canes | 152 |
| Rice | 51 |
| Saltpetre | 28 |
| Shark fins | 77 |
| Silk goods | 215 |
| Silk | 100 |
| Sugar | 74 |
| Tea and coffee plantations, Darjeeling | 66 |
| Tobacco | 83 |
| Wax and wax candles | 87 |
| Wool | 98 |
| Tables, showing results of experiments on the transverse strength of woods from Mysore (3398–3427) | 162 |
| Tanning substances | 121 |
| Tapestry, lace, and embroidery | 227 |
| Tapestry, ornamental | 227 |
| Tapioca (1649–56) | 72 |
| Tea, collection of, from | |
| Assam (1466–1505) | 64 |
| Burmah (1616) | 68 |
| Cachar (1506–25) | 65 |
| Darjeeling (1526–61) | 66 |
| Deyrah Dhoon (1562–69) | 67 |
| Kangra, Punjaub (1600–15) | 68 |
| Kumaon and Gurwhal (1570–99) | 67 |

## WRI

| | PAGE |
|---|---|
| Tents (4682) | 208 |
| Timber and fancy woods used for construction and ornament, and prepared for dyeing (3149–3736) | 153 |
| Tincal | 22 |
| Tinctorial gums (2310–29) | 106 |
| Tobacco, table of exports | 83 |
| Topees (5731–6) | 239 |
| Towellings (4656–74) | 208 |
| Toys and games (6134–49) | 249 |
| Turban pieces (5685–5730) | 238 |
| Turnery, plain and lacquered (3753–3842) | 181, 182 |
| UMBRELLAS, walking-sticks, etc. (6001–17) | 245 |
| VEGETABLE oil series, (1877–2107) | 85 |
| Vegetable substances used in manufacture | 104 |
| Vertu, articles of | 265 |
| WAISTBANDS, etc. (5864–5875) | 242 |
| Walking-sticks, umbrellas, etc. (6001–17) | 245 |
| Wax and wax candles, table of exports | 87 |
| Wood, general manufactures from, (3753–4017) | 181 |
| Woodcut blocks (6102–6) | 248 |
| Woods from— | |
| Assam (3202–18) | 156 |
| Burmah, (3428–3540) | 167 |
| Chittagong (3277–3296) | 157 |
| Chota Nagpore (3229–39) | 156 |
| Cuttack (3149–62) | 153 |
| Darjeeling (3219–28) | 156 |
| Jubbulpore (3163–3201) | 154 |
| Madras (gun carriage manufactory) (3297–3311) | 158 |
| Malacca (3632–72) | 177 |
| Malayan Peninsula (3673–3699) | 179 |
| Moulmein (3541–3631) | 175 |
| Oude (3252–76) | 156 |
| Pinang Hill, Forest of (3700–33) | 179 |
| Wood used for making charcoal | 3 |
| Wool (2108–64) | 97 |
| Wool, table of exports | 98 |
| Woollen and worsted fabrics | 218 |
| Writing cases and fittings (7066–82) | 271 |
| Writing, native (6112–31) | 248 |

PRINTED FOR HER MAJESTY'S COMMISSIONERS

BY

SPOTTISWOODE AND CO., NEW-STREET SQUARE, LONDON.

*THE*

ILLUSTRATED CATALOGUE

OF THE

INTERNATIONAL EXHIBITION.

𝔉𝔬𝔯𝔢𝔦𝔤𝔫 𝔇𝔦𝔳𝔦𝔰𝔦𝔬𝔫.

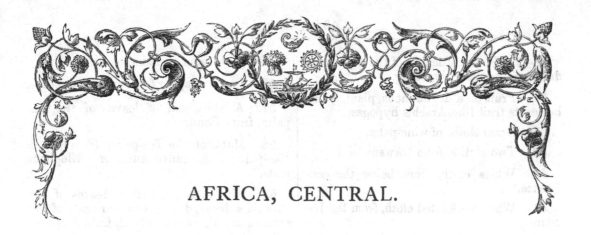

# AFRICA, CENTRAL.

NORTH-EAST COURT, UNDER STAIRCASE, NEAR CENTRAL ENTRANCE TO

HORTICULTURAL GARDENS.

BAIKIE, DR. W. BALFOUR, R.N.—

1-2. Striped cloth for men, from Hausa.

3-4. Cloth made of fibres of the wine-palm and cotton, from the right bank of Kwarra.

5. A tobe, poorest quality, made in Nupe.

6. A tobe of finer quality.

7. A white tobe, with plaits, from Nupe.

8. Striped trowsers, Nupe or Hausa make.

9-10. Common cloth, for women, from Bonu.

11. A woman's wrapper, made in Nupe.

12. A woman's wrapper, from Nupe.

13. A woman's wrapper, not made up, called "Locust's tooth."

14. A wrapper containing red silk, called Maizha'n baki, or "red mouth."

15. An inferior wrapper, from Nupe.

16. Blue and white cloth, from Nupe.

17-18. Cloth made in Yoruba.

19-20. Cloths from Nupe.

21-25. Cloths from Yoruba.

26. Small cloth for girls, from Nupe.

27. Bag from Onitsha.

28. Mat, from right bank of Kwarra.

29. Tozoli (sulphuret of lead), applied to the eyelids.

31. Man's wrapper, from Ki, in Bonu.

32. Woman's head-tie, or alfuta, from Nupe.

33. Bags for gunpowder, from Onitsha.

34-35. A calabash and ladle.

37. Red silk, or "Al harini," of Hausa.

38. Sword hangings, or "Amila," made at Kano, in Hausa.

39. Siliya, or red silk cord, from Kano.

40. Rope, from Onitsha.

41-42. Bags.

43. White cloth, or fari, made in Nupe and Hausa.

44. White cloth, from below the confluence.

45. A white tobe, from Nupe.

46. Four calabashes, for pepper, &c.

47. A small calabash and lid, for food.

48-49. Pinnæ of leaves of the wine-palm, dried and used for thatching.

50. Fruit of a leguminous plant, which buries its fruit like Arachis hypogæa.

51. Grass cloth, of wine-palm.

52. Two cloths, from Okwani.

53. White cloth, from below the confluence.

54. White perforated cloth, from the Ibo country.

55. Mats from Onitsha.

56. Large wrapper for a man, from Nupe.

57. A white mat of leaves of the fan-palm, from Bonu.

58. Mats of the fan-palm, from Bonu. Fan-palm mats, called guva, or "Elephant mats."

59. Fine mats and hats of leaves of the Phœnix spinosa, dyed. Circular mats of the same material, used by chiefs, from Nupe.

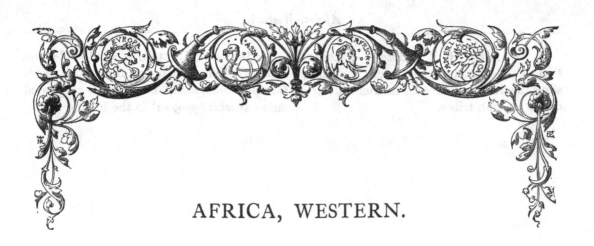

# AFRICA, WESTERN.

COMMERCIAL ASSOCIATION OF ABEOKUTA. —1. Oils: Of beni seed, obtained by fermentation and boiling. 2. Of Egusi, from wild melon seed. 3. Of palm, for home consumption; 4. For exportation, obtained by beating, pressing, and boiling the fruit. 5, 7. Of palm-nut, for home consumption; 6. For exportation. 11. Shea butter. 10. Egusi, or wild melon, fruit. 8. Beni seed. 9. Fruit of the Shea butter tree.

1. White cotton thread; 2. Dyed; 3. Blue. 4. Fine spun cotton. 5. Coarse strong spun cotton, called "Akase." 6. Akase cotton, cleaned and bowed; 7. In seed. 8. Seed itself of Akase cotton. 9, 10. Ordinary native cotton. 11, 12, 13. Ordinary green, black, and brown seeded cottons. 14. Silk cottons. 16. Country rope of bark. 17. Palm fibre. 18. Red dyed native silk, from Illorin. 20. Fibre used for native sponge. 23, 24, 25. Native silk, from a hairy silk-worm at Abeokuta. 26. Leaves of the cotton tree. 27. Pine-apple fibre. 29. Bow-string fibre. 30. Jute.

15. Long black pepper. 22. Senna. 21. A sample of native antimony, from Illorin.

## Sundry native manufactures.

N.B.—Cotton is obtainable in any quantity, and is now grown extensively throughout the Yoruba country, especially to the east and north. Great quantities of cotton cloths, of a strong texture, are annually made, finding their way to the Brazils, and into the far interior. To obtain a largely increased supply of cotton, it is only necessary to open roads, and bring money to the market. Upwards of 2,000 bales have been exported this year, and the quantity would have been doubled or trebled if the country had been at peace. The present price is 4½d. per lb. The other fibres are not at present made for exportation, though doubtless, some of them—jute, for instance—would be, if in demand. Of the native manufactures, the grass cloths, made from palm fibre, and the cotton cloths, are most prominent. Very nice leather work is done. The art of dyeing Morocco leather different colours has been introduced from the interior. Indigo is almost the only dye which can be obtained in considerable quantities. The natives manufacture all their own iron implements, and the quality of the metal is considered good.

2. MCWILLIAM, THE LATE DR. C. B.— 1. Cloth, from the confluence of the Niger and Tchadda. 2. Raw silk from Egga. 3. Cotton from the confluence. 4. Fishing spear, used by the natives of Kakunda. 5. Spoons, from Gori market. 6. A curved horn for holding galena, used to paint the eyelids. 7. Cloths, from towns on the Gambia. 8. Grass mat, from Angola. 9. Grass mat, from Binguela.

3. WALKER, R. B. *Gaboon.*—A collection of mats, fibres, commercial products, skins, native arms, musical instruments, &c., of the Ba Fan tribes.

4. BARNARD, JOHN A. L. 8, *Alfred-villas, Dalston.*—Tallicoonah or Kundah oil, from *Carapa Tallicoonah*, and a bundle of ground nuts (*Arachis hypogæa*) in the haulm.

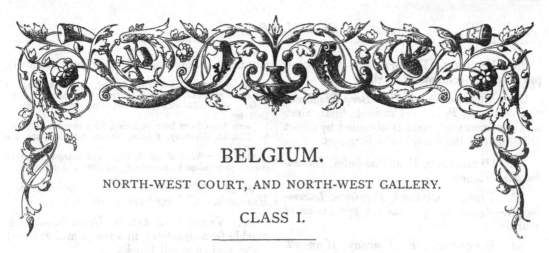

# BELGIUM.

## NORTH-WEST COURT, AND NORTH-WEST GALLERY.

## CLASS I.

1. THE MINISTRY OF PUBLIC WORKS, *Brussels.*—Constituent rocks, and mineral products of Belgium, collected by M. Jules Van Scherpenzeel-Thim.

2. AMAND, E. *Mettet, Namur.*—Hydrated iron ores; charcoal iron, and castings.

3. BRINCOURT-ANDRÉ, L. *Herbeumont, Luxemburgh.*—Various kinds of slate.

4. COUPERY DE SAINT-GEORGES, E. *Dinant, Namur.*—Black marble in polished slabs and blocks.

5. DE JAER & CO. *Antwerp.*—Alluvial pyrites.

6. DEJAIFFE-DEVROYE, T. *Saint-Martin-Bálatre, Namur.*—Black marble from Golzinnes.

7. DE MERCX DE CORBAIS, MRS.—Lead ores and potters' clay.

8. DASSONVILLE DE SAINT HUBERT, L. *Namur.*—Belgian millstones (siliceous).

9. DESCAMPS, J. & CO. *Saint-Josse-ten-Noode, near Brussels.*—Grit-stone pavement.

10. DESMANET DE BIESME, VISCOUNT, *Biesme, Namur.*—Black marble from Golzinnes.

11. DE THIER, A. *Theux.*—Black marble from Theux.

12. DEVILLIERS & CO. *Brussels.*—Sainte-Anne marbles, polished.

13. DE WYNDT, J. & CO. *Antwerp.*—Refined sulphur, in rolls and flowers.

14. DUPIERRY, *Viel-Salm, Luxemburgh.*—Whet and grinding-stones.

EXHIBITORS' COMMITTEE FOR THE DISTRICT OF VERVIERS :—

16. SOCIÉTÉ ANONYME DE CORPHALE, *Antheit, near Huy.*—Ores of zinc, lead, pyrites; refined lead; crude zinc.

17. SOCIÉTÉ ANONYME DE LA NOUVELLE-MONTAGNE, *Verviers.*—Ores of lead, zinc, iron; pyrites; metals; sulphur.

18. SOCIÉTÉ ANONYME DE ROCHEUX ET D'ONEUX, *Theux.*—Pyrites; ores of zinc, lead, iron.

19. SOCIÉTÉ ANONYME DES HAUTS-FOURNEAUX ET LAMINOIRS DE MONTIGNY-SUR-SAMBRE, *Montigny-sur-Sambre, Hainault.*—Iron ores, coke-castings, puddled steel, &c.

20. SOCIÉTÉ ANONYME DES HAUTS-FOURNEAUX, USINES, ET CHARBONNAGES DE CHÂTELINEAU, *Châtelineau, Hainault.*—Coke castings, &c.

21. SOCIÉTÉ ANONYME DES HAUTS-FOURNEAUX, USINES, ET CHARBONNAGES DE MARCINELLE ET COUILLET, *Couillet, Hainault.*—Pit-coal, iron ores, coke castings, puddled steel, &c.

22. SOCIÉTÉ ANONYME DES HAUTS-FOURNEAUX ET CHARBONNAGES DE SCLESSIN, *Sclessin, Liége.*—Iron ores, coke castings, pyrites.

23. SOCIÉTÉ DES MINIÈRES DE HONTHEM, *Dolhain.*—Ores of lead; pyrites.

24. SOCIÉTÉ ANONYME DE VEZIN-AULNOYE, *Huy.*—Iron ores, coke castings.

25. SOCIÉTÉ ANONYME DU BLEYBERG, *Bleyberg ès-Montzen.*—Zinc and lead ores, lead-pigs, zinc ingots, regulus of silver, glazed pottery, crystals.

26. LA PLUME-ROUXHE, J. N. *Salm-Château, Luxemburgh.*—Whet and grinding-stones.

27. LEBENS-SCHUL, E. *Brussels.*—Quartzite paving-stones from Hal.

28. MARCHAL, D. *Brussels.*—Grit-stone pavement, specimens of marble.

29. MULLER, A. & Co. *Berg-Gladbach, near Cologne, Prusse.*—Crude Belgian zinc, and specimens of products obtained by direct treatment of the ores, in the furnaces.

30. OFFERGELD, P. J. *Viel-Salm, Luxemburgh.*—Hones.

32. PIERLOT-QUARRÉ, *Forrières, Luxemburgh.*—Quartzite pavement; specimens of marble.

33. SACQUELEU, F. *Tournay, Hainault.*—Specimens of marble; flags, slabs; mangers, manger fronts, &c. Black marbles, for the pavement of porches, churches, and public monuments; obtained at Basècles (Hainault).

Besides black marble in blocks and sawn slabs, flags of the same material, of all sizes, from 0·10 metres to 0·60 metres, and 0·70 metres square, may be obtained at Basècles.

These flags have been exported, for a number of years, to Holland, Germany, France, Spain, America, and the Levant.

Also, every kind of lintel, step, and manger, whether smooth or polished by machinery, or merely chiselled.

34. TACQUENIER, A. C. & Bros.—*Lessines, Hainault.*—Chlorophyre pavement.

35. VERBIST-LAMAL, R. *Brussels.*—Black marble from Basècles, in a rough and finished state, and in small blocks.

36. WATRISSE, L. *Dinant, Namur.*—Polished black marble slabs.

## CLASS II.

37. BARBANSON, P. *Brussels.*—Animal-black and bone-dust.

38. BORTIER, P. *Ghistelles, West Flanders.*—A mixture of manures, that produces in the soil a nitrification of calcareous matters, which renders the effect more energetic and lasting.

39. BRASSEUR, E. *Ghent.*—White lead, and ultramarine.

40. BRUNEEL, J. J. & Co. *Ghent.*—Chemical products extracted from wood, and adapted for domestic purposes and the arts: including acids, acetates, vinegar, oil, and alcohol.

41. CAPPELLEMANS, J. B. SEN. DEBY, A. & Co. *Brussels.*—Chemical products.

42. COOSEMANS & Co. *Berchem, Antwerp.*—Naphtha, photogene; lubricating, paraffine, and other oils, obtained by the distillation of bituminous schists, or Scotch Boghead coal.

43. DE CARTIER, A. *Auderghem, near Brussels.* — "*Minium de fer d'Auderghem,*" a preservative paint for iron and wood, which answers as a substitute for minium of lead, &c.

His "minium de fer" obtained the first bronze medal at the Exhibition of Dort in 1861.

The "minium de fer d'Auderghem," a preservative paint, has superseded minium, lead, and other colours, on account of its solidity and most valuable property of completely preserving iron from oxidations, its hardening wood, and, above all, its great economy.

Twelve years of extraordinary and deserved success have caused its adoption and use by all great industrial establishments, sugar refineries, railway and steamboat companies of Europe, as is proved by the collection of certificates which may be had at the manufactory. Prospectuses and specimens can be procured at the following dèpots :—

*Rotterdam*—WANDEN, HELM, and SON; L. H. GARCIA.
*Paris*—J. DROUVIER, DARCHE, and PUISSANT.
*Lille*—FONTAINE and GRANDEL.
*Havre*—H. SONDHEIM.
*Nantes*—A. DUREL and Co.
*Bordeaux*—L. BAY and DUPRAT.
*Marseilles*—JULES CAMAN and Co.
*Lyons*—LORNOT and LESSIEUX.
*Metz*—A. ROUSSEAUX.
*Strasbourg*—L. WILHELM.
*Algiers*—JOLY Bros.
*London*—L. FLOERSHEIM.
*Hull*—JOHN FOSTER and Co.
*Newcastle-on-Tyne*—BENJ. PLUMMER.
*Leith*—MITCHELL, SOMERVILLE, and Co.
*Glasgow*—A. G. KIDSTON and Co.
*Liverpool*—F. E. B. SCOTT
*Manchester*—FR. BUTTERFIELD.
*Birmingham*—R. and F. J. ADAMS.
*Bristol*—ROBERT BRUCE.
*Dublin*—CHARLES PALGRAVE.
*Cork*—W. and M. GOULDING.
*Belfast*—WALKINGTON and SON.
*Cologne*—J. W. WEIBER.
*Manheim*—ED. EISENHARDT.
*Stutgard*—AUG. HEVINGER.
*Heilbron*—FR. ED. MAYER.
*Dresden*—GEHE and Co.
*Berlin*—AHREND and VEIT.
*Stettin*—J. G. WEISS.
*Bremen*—ROHLIG and Co.
*Hambourg*—VAN STRAALEN URLINGS.
*Vienna*—BRUDNER Bros.
*Trieste*—LUIGI HESS.
*Winterthur*—H. JAEGGLI.
*St. Petersburg*—H. PEPINSTER.

*Warsaw*—JULIAN SIMLEN.
*Odessa*—AD. WENDELSTEIN.
*Copenhagen*—J. L. MADSEN.
*Stockholm*—AUT. BENDIX.
*Valence*—MALTS. VELTERS.
*Madrid*—EMILIO LESTGARENS.
*Lisbon*—AUG. SCHMITZ.
*Turin*—FRED. SCHMID.
*Naples*—ACHARD and Co.

44. DELMOTTE-HOOREMAN, C. *Maria-kerke, near Ghent.*—White lead.

45. DELTENRE-WALKER, *Brussels.*—Fine varnishes for various purposes : collodion.

46. DE MOOR, A. *Brussels.* — Elastic copal, and other varnishes.

47. DE SAEGHER, H. *Brussels.*—A chemical product which removes incrustations from steam-boilers.

48. GENNOTTE, L. *Brussels.*—Vegetable powder for the destruction of insects, troublesome to man or domestic animals, or mischievous to woven fabrics, furs, &c.

49. MATHYS, M. *Brussels.*—Thirty fine specimens of varnishes, for external and internal painting, ornamental wood work, metals, paper, sculpture, &c.

50. MERTENS, B. & Co. *Lessines, Hainault.*—Preservative blacking, lucifer-matches.

51. MERTENS, G. *Overboelaere, East Flanders.*—Inodorous lucifer-matches, preservative blacking.

52. KAYSER, A. & POPELEMON, J. *Brussels.*—"Cellulose," a powder made from linen cloths by chemical disintegration of the fibres, supplying the place of lycopodium, and answering instead of gun-cotton in the making of collodion.

53. RAVE & Co. *Court-Saint-Etienne, Brabant.*—Alkal-oxide, and an oxygenated compound: both of them substitutes for powder in mining operations, and inexplosive in ordinary circumstances.

54. SEGHERS, B. *Ghent.*—Bone black and ivory black.

55. VANDER ELST, P. D. *Brussels.*—Sulphuric and nitric acids, sulphate of soda, copperas, bleaching-powder.

56. VANSETTER, CONINCKX & Co. *Neder-Overheembeek, near Brussels.*—Turpentine, animal black.

57. VERSTRAETEN, E. *Ghent.* — Animal black.

# CLASS III.

AGRICULTURAL ASSOCIATION OF THE ARRONDISSEMENT OF YPRES :—

58. AGRICULTURAL ASSOCIATION.—Hops, wheat, rye, Indian wheat, pease, colza; *œillette*, a variety of poppy ; leaf tobacco, &c.

59. COEVOET, L. F. *Poperinghe.*—Hops grown in 1861.

60. DE GRYSE, W. *Poperinghe.*—Hops.

61. DELBAERE, P. *Poperinghe.* — Wheat and pease.

62. DEMOOR, B. *Passchendaele.*—Kidney-beans.

63. GOMBERT & CAMERLYNCK, *Reninghelst.*—Hops.

64. LEBBE-BEERNAERT, B.—*Poperinghe.*—Hops, wheat, oats.

65. LESAFFRE, A. *Gheluwe.*—Tobacco.

66. MALOU, J. B. *Dickebusch.* — Hops grown in 1861.

67. PATTYN, C.—Rye.

68. PEENE BROS. *Elverdinghe.*—Hops.

69. QUAGHEBEUR - VERDONCK, P. *Poperinghe.*—Hops.

70. RICQUIER, L. *Warneton.*—Œillette, a variety of poppy ; colza.

71. ROMMENS, F. *Poperinghe.* — Hops grown in 1861.

72. VANDERGHOTE, E. *Elverdinghe.* — Hops grown in 1861.

73. VANDERMEERSCH, J. B. *Bas-Warneton.*—Tobacco.

74. VANDROMME, P. *Westoutre.* — Hops grown in 1861.

AGRICULTURAL SOCIETY OF EAST FLANDERS :—

75. DE BERLAERE, KN. *Vinderhaute.*—Hops.

76. DE CROESER, Baron, ED. *Mooreghem.*—Leaf tobacco ; wheat, oats, kidney beans, pease.

77. DEMEULDER, J. F. *Poesele.* — Rye, Australian white wheat.

78. DERORE, J. *Mooreghem.*—Wheat and oats.

79. GHENT COMMITTEE.—Cereals, hops.

80. GUEQUIER, J. *Wachtebeke.* — Buck-wheat, sorgho.

81. LATEUR, L. *Mooreghem.* — Wheat, Polish oats.

82. VAN BUTSELE, G. *Nuckerke.*—Australian wheat in ears.

83. VAN PELT, J. F. *Tamise.*—Rye and barley.

83A. MONTON & AUTHOMPEN, *Herstel, Liége.*—Starch, &c.

84. BALCAEN, P. *Peteghem.*—Chicory.

85. BEERNAERT, L. *Thourout, West Flanders.* — Wheat, rye, oats, buck-wheat, kidney-beans, colza ; tobacco.

86. BELPAIRE & OOMEN, *Antwerp.* — Cigars.

87. BENOIT, A. *Chermont, sous St. Hubert, Luxemburgh.* — Summer wheat, winter rye, summer barley and black oats.

88. BLAESS, C. B. *Borgerhout, near Antwerp.*—Vinegars, from grain.

These products, obtained from vegetable substances of the very best kind, are remarkable for their great strength, and are adapted to general use, on account of their perfect freedom from any kind of impurity.

89. BORGHS, *Turnhout.* — Wheat and oats.

90. BORTIER, P. *Ghistelles.*—Giant wheat, in ear.

92. CAPOUILLET, P. *Brussels.* — Sugar-loaves, raw beetroot sugar.

93. DE BISEAU D'HAUTEVILLE, *Entre-Monts, sous Buvrinnes, Hainault.*—Wheat.

94. DELANNOY, N. *Tournay, Hainault.*—Chocolate, cocoa, racahout, and fancy articles.

95. D'ELPIER, C. *Castle of Mielen, Saint-Trond.*—Giant wheat, potatoes.

96. DE MARNEFFE-VAN PETEGHEM, *Alost, East Flanders.*—Alost hops, grown in 1861.

97. DENOTER, R., *Laerne, East Flanders.*—Leaf tobacco.

98. DEWYNDT-AERTS, J. & Co. *Antwerp.*—Samples of sugar-candy.

99. DEYMANN, J. H. *Charleroi.* — Dey-mann-bitter, a liqueur.

100. DIERT DE KERKWERVE, BARON, *Castle of Hemixem, Antwerp.*—Wheat, rye, and oats, in sheaves, and grain.

101. DUBUS & DESCAMPS, *Brussels.* — Extract of tobacco, for the manufacture of cigars.

102. HANSSENS, B. & SON (TRITHART, Director), *Vilvorde, Brabant.*—Fecula, starch, gommeline, dextrine, leiogomme, gum arabic, various pastes.

103. HEIDT-CUITIS, J. *Chokier, Liége.*—Starch, and finishing materials.

104. JORISSEN, L. *Liége.*—Alcohol rectified, concentrated, and chemically pure.

105. JOVENEAUX, A. *Tournay, Hainault.*—Chocolate for ordinary and hygienic purposes.

106. LE HON, F. SEN. *Brussels.*—Curaçao, and other liqueurs.

107. MERKEL, G. *Tournay, Hainault.*—Vinegar made of grain alcohol, and not strengthened by pyroacetic acid.

Manufacturer of vinegar, containing 12 per cent. acetic acid, made from grain alcohol, no pyroacetic acid or wood vinegar being used to strengthen it, although this is employed everywhere, even at Orleans, for the purpose.

Price, for quantities weighing not less than 500 kilogrammes, 38 francs the 100 litres, delivered at the Tournay Station, barrels not included.

Common vinegar contains only 3 per cent., and the best Orleans vinegar only 7 per cent. acetic acid. All kinds of vinegar warranted pure may therefore be obtained from it by the mere addition of water.

108. MIRLAND & Co. *Pecy, near Tournay.*—Apple paste in various states.

109. NORTHERN AGRICULTURAL SOCIETY, *Antwerp.*—Wheat, rye, oats, hops, &c.

110. PAILLET-JONEAU, A. *Ville-en-Hesbaye, Liége.*—Syrups, prepared from fruit, and beetroot.

111. PATRON-JOLY, *Huy.* — Sparkling wine, made of indigenous wine.

112. PENITENTIARY OF SAINT HUBERT (MARINUS, Director), *Luxemburgh.*—Wheat, rye, barley and black oats ; the products of the Audennes ; grown nearly 1400 feet above the level of the sea.

114. REMY, E. & Co. *Wygmael, Louvain, Brabant.*—Rice starch.

115. SCHALTIN-DUPLAIS & Co. *Spa, Liége.*—Liqueurs, Elixir of Spa.

116. SCHOOFFS, J. B. *Brussels.*—Extracts used in the manufacture of liqueurs.

117. SERRÉ, L. *Hal, near Brussels.*—Beer.

119. SOCIÉTÉ ANONYME DES MOULINS A VAPEUR DE BRUXELLES, *Molenbeek, near Brussels.* — Starch, the product of wheat; for purposes requiring very great purity.

120. STEENS, H. *Schooten.*—Wheat, rye, oats, buck-wheat, colza, pease.

121. STEIN, A. & Co. *Antwerp.*—Cigars made with Havannah and other tobaccos.

122. TINCHANT, L. *Antwerp.*—One thousand different kinds of cigars, made of genuine and imitation Havannah, &c.

123. ULLENS, C. F. *Schooten.*—Wheat, barley, rye, and oats.

125. VAN BERCHEM & Co. *Brussels.*—Cigars.

126. VANDEN WYNGAERT, *Wilmarsdonck.*—Barley.

127. VANDEVELDE, N. *Ghent.*—Liqueurs, champagne beer, rectified gin, &c.

128. VAN GEETERUYEN - EVERAERT, *Hamme, East Flanders.* — Eight specimens of starch produced from damaged wheat.

129. VAN PUT, *Antwerp.*—Hay, gathered in irrigated meadows.

130. VANSTRAELEN, H. *Hasselt, Limburgh.*—Gin from grain.

131. VAN VOLSEM, P. *Hal, Brabant.*—Rye and oats grown on heaths.

132. VERGOUTS, F. *Lillo, Antwerp.* — Australian white wheat, Polders oats.

133. VERHEYDEN, DILBECK, *Brabant.*—Wheat, rye, oats, hops.

134. VERHEYEN, P. J. *Turnhout, Antwerp.*—Hops grown on the heaths of the Antwerp Campine.

135. VERTONGEN, BROS. *Raegels.*—Wheat, rye, oats, trefoil.

136. WINCQ, J. B. *Ochamps, Luxemburgh.*—Oats, gathered on heaths.

# CLASS IV.

AGRICULTURAL ASSOCIATION OF THE ARRONDISSEMENT OF YPRES :—

137. AGRICULTURAL ASSOCIATION. — Steeped and unsteeped flax, madder.

138. BEERNAERT, L. *Thourout.* — Raw and unsteeped flax, peeled flax.

139. HERMAN, J. *Becelaere.* — Steeped flax, peeled wax.

140. VAN LEENE, D. *Dickebusch.* — Steeped flax.

141. VAN WALLEGHEM. C. *Zannebeke.*—Unsteeped flax.

142. VERMEULIN, A. *Becelaere.* — Unsteeped flax.

___

AGRICULTURAL SOCIETY OF EAST FLANDERS :—

143. REYNIERS, J. A. *Seveneeke.*—Flax and hemp in different stages of preparation.

144. VAN PELT, J. F. *Tamise.*—Hemp, raw, steeped, in filaments, carded.

145. VAN RANTERGHEM, L. *Drongen.*—Flax.

___

146. BERTOU, BROS. *Liége.*—Waterproof grease for shoes and boots.

147. BIHET, H. *Huy.*—Glue.

148. BISSÉ, E. & Co. *Cureghem, near Brussels.*—Lubricating oils for manufacturing purposes and locomotives, for lamps, and for dyeing Turkey red, &c.

149. BRUGELMAN & HALSTEAD, *Cureghem, near Brussels.*—Artificial wools.

150. CLAUDE, L. *Brussels.*—Pure colza oil.

151. DAVID, C. *Antwerp.*—Flemish flax, peeled by hand.

152. DE BEHAULT, *Buggenhout, East Flanders.*—-Raw flax.

153. DE BRUYN, J. *Thermonde, East Flanders.*—Peeled flax, raw flax and hemp.

154. DE CATERS, BARON, *Antwerp.* — Flax.

155. DE COCK, BROS. *Brussels.*—Watch and clock makers' oil.

156. DE CONINCK, BROS. *Brussels.* — Household soap, purified colza oil for carcel lamps.

157. DE CURTE, V. *Ghentbrugge, near Ghent.*—Distilled stearine and candles.

158. DE MOOR, E. & Co. *Antwerp.*— Stearine candles, and oleine, produced from the waters in which wool had been washed.

159. DENAEYER, P. *Lebbeke, East Flanders.*—Artificial wools (mungo-shoddy) of all colours and degrees of fineness.

160. DENS, *Putte.*—Flax.

161. DE ROUBAIX-JENAR & Co. *Brussels.* —Stearine, candles, raw materials, oleic acid, products of distillation.

162. DE ROUBAIX-OEDENKOVEN & Co. *Borgerhout, near Antwerp.*—Stearine candles, fat-acids.

163. DE SAINT-HUBERT BROS. *Warnant-Moulins, Namur.*—Flax steeped by a manufacturing process, and peeled by machine.

164. DES CRESSONNIÈRES, WIDOW & SON, *Brussels.*—Soaps for Turkey red dyeing, household and toilet soaps, oil for turning.

166. DUBOIS-CREPY, *Mons.* — Perfumed and household soaps.

167. EECKELAERS, L. *Saint - Josse - ten - Noode.*—Toilet and household soaps.

168. FELHOEN, BROS. (FELHOEN-PECQUE-RIAU), *Courtrai.*—Courtrai flax, hemp, and jute, peeled by machine.

169. GAUCHEZ, L. *Brussels.* — Oil for carding-engines and for felting threads.

170. GENNOTTE, T. *Brussels.* — Night lights, porcelain floats, waxed wicks.

171. HANSOTTE-DELLOYE, V. G. *Huy, Liége.*—Glue.

173. LEFEBURE, J. *Brussels.* — Preparation of flax and hemp by machinery.

174. MARTIN, *Saint - Josse - ten - Noode.*— Watchmakers' oil prepared without acid.

175. MECHANT, H. *Hamme-Saint-Anne, East Flanders.*—Specimens of flax in all stages of preparation.

176. MULLENDORFF, *Ixelles, near Brussels.*—Vegetable oil, purified, and completely freed from acid, by a new process; for the lubrication of machinery.

177. NORTHERN AGRICULTURAL SOCIETY, *Antwerp.*—Specimens of flax.

178. PEERS, BARON E. *Oostcamp, West Flanders.* — Raw flax grown in a heathy soil.

179. QUANONNE, C. & MIDDAGH, P. *Molenbeek-St.-Jean, near Brussels.*—Distilled stearic acid and other candles.

180. REYNAERT, CH. *Reninghe, West Flanders.*—Raw flax cultivated by a new method.

181. ROMBOUTS-VREVEN, *Hasselt, Limburgh.*—Bleached wax, tapers.

182. STEENS, H. *Schooten.* — Flax and hemp.

183. TAULEZ-BOTTELIER, C. *Bruges.* — Samples of peeled flax.

184. VAN DEN PUT, V. *Brussels.*—Soaps and perfumery.

185. VANDERPLASSE, BROS. *Brussels.* — Oil for horological purposes.

186. VANDERSCHRIECK, BROS. *Antwerp (succur-saal at Saint-Denis, near Paris).*— Woollen rags, suited for unravelling.

187. VAN ROYE, G. & H. BROS. *Brussels.* —Purified colza oil.

188. VAN-SETTER-CONINCKX & Co. *Brussels.*—Neat's-foot and other oils for lubricating machinery.

189. VERBESSEM, C. *Ghent.*—Glue, size, and gelatine.

190. VERCRUYSSE - BRACQ, F. *Deerlyk, near Courtrai.*—Flax, raw and peeled.

191. VERPOORTEN, *Bruges.*—Colza, and other seed; Colza, and other oils, common and refined; oil cakes.

This manufactory, worked by water and steam power, has been established for the treatment of oleaginous seeds :

Indigenous Colza seed, refined Colza oil for carcel lamps, Colza oil-cakes; indigenous linseed, linseed oil, linseed oil-cakes—an excellent food for cattle; indigenous cameline seed, refined cameline oil, cameline oil-cakes; indigenous poppy seed, superfine poppy seed oil, poppy seed oil-cake.

As the value of these articles is subject to great variation, the exhibitor cannot mention any prices.

Those desirous of it will obtain the fullest information at the above address.

192. VERSCHEURE, J. *Oyghem, West Flanders.*—Raw, steeped and peeled flax.

193. VERTONGEN BROS. *Raevels.*—Flax.

194. WINNEN BROS. & SISTERS, *Brussels.* —Shoddy; artificial whalebone.

195. ZOUDE, L. *Val de Poix, Luxemburgh.* —Beech-wood gun-stocks, and fellies.

## CLASS V.

196. ARNOULD, G. *Mons.*—Forged iron railway chairs; new kind of fish-plates.

197. BLONDIAUX & Co. *Thy-le-Château, Namur.*— Rails, splints, and breakings of various kinds of rail.

198. CEURVORST, S. P. *Antwerp.*—Patent railway break, J. Briere's principle.

199. COMPAGNIE GÉNÉRALE DE MATÉRIELS DE CHEMINS DE FER, *Brussels.*—Railway carriage; trophy of wheels and iron fittings for railway waggons, &c.

200. GOFFIN, C. & J. *Brussels.*—Axles for locomotive engines.

201. HEINDRYCKX, *Ixelles, near Brussels.*—Wrought iron railway chairs; model of crossing, for preventing danger of every kind.

202. SOCIÉTÉ DES HAUTS-FOURNEAUX, USINES, ET CHARBONNAGES DE CHATELINEAU, *Châtelineau, Hainault.*—Rails.

203. SOCIÉTÉ ANONYME DES HAUTS-FOURNEAUX, USINES, ET CHARBONNAGES DE MARCINELLE ET COUILLET, *Couillet, Hainault.*—A six-wheeled locomotive; rails.

204. SOCIÉTÉ ANONYME DE LA FABRIQUE DE FER D'OUGRÉE, *Seraing, Liége.*—Unwelded wheel tires for waggons and steam engines, axles and wheels for waggons.

204A. SOCIÉTÉ ANONYME DES HAUTS-FOURNEAUX ET LAMINOIRS DE MONTIGNY SUR SAMBRE, *Hainault.*—Rails of puddled steel, &c.

205. SOCIÉTÈ DES FORGES ET LAMINOIRS DE L'HEURE, *Marchienne-au-Pont, Hainault.*—Axles.

206. SOCIÉTÉ ANONYME DES FORGES DE LA PROVIDENCE, *Marchienne-au-Pont, Hainault.*—Waggon wheels forged in one piece, by a new process.

207. THIRION, *Aische-en-Refail, Namur.*—Model of a new kind of waggon; the load is sustained on moveable spheres, that do not touch the axle.

208. SOCIÉTÉ ANONYME DE L'USINE VANDENBRANDE, *Schaerbeek, near Brussels.*—Patent double-acting excentric; railway crossings, as used in Belgium.

209. VAN DER ELST, L. & Co. *Braine-le-Comte, Hainault.* — Weigh-bridge for railways, having an isolating apparatus; intended to bear a weight of 30,000 kilogrammes.

## CLASS VI.

210. DE RUYTTER, J. *Bruges.*—A clarence.

211. JONES BROS. *Brussels.* — Various carriages.

Distinctions obtained at various Exhibitions :—
  1825. *Harlem*—First bronze medal.
  1840. *Antwerp*—Silver medal.
  1841. *Brussels*—Silver-gilt medal.
  1847. *Brussels*—Gold medal.
  1851. *London*—Prize medal.
  1855. *Paris*—Medal of the first-class.

And decorations of the Order of Leopold, by a royal decree, dated 16th December, 1855.

Carriages exhibited in London in 1862 :—
  A coupé d'Orsay, the pole of iron, fitted with a steel plate. An extremely light carriage.—Price 4,800 francs.
  A phaeton cabriolet, mounted on elliptic springs.—Price 2,750 francs.
  A calash for ordinary use, mounted on elliptic springs (the kind made for exportation).—Price 2,750 francs.

212. VAN AKEN BROS. *Antwerp.* — A calash with double suspension.

213. VAN AKEN, C. B.—Calash, double suspension.

# CLASS VII.

214. CAIL, J. F. HALOT, A. & Co. *Brussels.*—Various apparatus for the manufacturers and refiners of beet-root and other sugar.

215. COMPAGNIE GÉNÉRALE DE MATÉRIELS DE CHEMINS DE FER, *Brussels.*—A mortising machine for wood.

216. DAUTREBANDE, H. *Huy.*—Machine for manufacturing endless paper.

217. DE BRUYNE, E. *Hamme, East Flanders.*—Hair-cloth bags for oil presses.

218. DE BRUYNE & SON, *Waesmunster, East Flanders.*—Hair-cloth bags for the extraction of oil from seeds; also, one to be used with an hydraulic press.

219. DE GROOTE, C. *Brussels.*—Bottle corker with glass tube, with or without needle.

220. DEHAYNIN, F. *Gosselies et Marcinelle, Hainault.*—Two drawings of a machine for agglomerating coal; and coal-bricks made by it.

221. DEKEYSER-DUMORTIER, S. *Eecloo, East Flanders,*—Hair-cloth for pressing oleaginous seeds.

224. DUMONT, E. *Liége.*—Endless mechanical sieve for preparing ores, coal, &c.

EXHIBITORS' COMMITTEE FOR THE DISTRICT OF VERVIERS —Machinery for the manufacture of wool:—

225. BOVY, J. D. & VANDERMAESEN, L. C. *Verviers*—Machine for making velvet.

226. HÔUGET, J. D. & TESTON, C. *Verviers.*—Various machines used in the manufacture of woollen cloth.

229. MARTIN, C. *Pepinster.*—Machine for oil-pressing wool; carding machine; articulated pads.

230. MARTIN, T. *Verviers.*—Backs and ribbons for cards; cards for wool in leather and felt, in artificial leather and felt, and in leather.

231. NEUBARTH & LONGTAIN, *Verviers.*—Longitudinal shearing implements, on a new principle.

232. TROUPIN, J. P. *Verviers.*—Blades, tables and rulers in cast steel for shearing cloths, shawls, and stuffs in wool, silk, and cotton.

Manufacturer of blades, tables, and guides for shearing cloth, stuffs, and shawls, of wool, silk, and cotton.

He obtained the following distinctions :—
In 1847, National Exhibition of Belgium—1st Prize.
1851, Universal Exhibition of London—6th „
1855, Paris Exhibition—2nd Class.
1861, Exhibition of the Arrondissement of Verviers—The Decoration of his Majesty Leopold I.

233. WANKENNE ET DEBIAL, *Verviers.*—Shearing or smoothing implements with right and left-handed blades.

235. FETU, A. & DELIÉGE, *Liége.*—Specimens of cards for spinning wool and cotton.

236. GENNOTTE, L. *Brussels.*—Apparatus for making gaseous beverages instantaneously.

237. GÉRARD, D. *Charleroi.*—Steam-engine for mines, without beam, &c., requiring only one cylinder, and a single rope instead of two and their accessories; a mode of raising waggons from mines with one rope, and without the use of two cages, the waggons attaching and liberating themselves; a parachute always acting instantly but gradually, at the pleasure of the miners who are being conveyed, or when the rope breaks, &c.

239. LAROCHE & Co. *Brussels.*—Machine for the manufacture of paper.

239A LLONG, C. & BISCOP, J. B. *Wiers, Hainault.*—Waggon for coal mine.

240. LEROY, A. *Brussels.*—Six sewing-machines, on different principles.

241. LIBOTTE, N. *Gilly, Hainault.*—Two miners' cages with parachutes and waggons.

242. MEERENS, *Brussels.*—Flower bleaching apparatus, preserving the lace makers from the effects of the white lead; the latter being placed in an hermetically sealed box.

243. MERTENS, *Gheel, Antwerp.* — Flax and hemp scutching and peeling machines.

245. NYST, F. *Liége.*—Friction parachute for miners' cages, &c.

246. PERRIN, N. *Brussels.*—Drill stock, on a new principle, with an Archimedean screw, worked with one hand only.

247. PREVOT, C. *Haine St. Pierre, Hainault.*—Machine for making pegs; applicable to boat-building.

248. RYCX, A. & SON, *Ghent.*—Patent cards for cotton.

249. SACRÉ, A. *Brussels.*—Flax-drawing apparatus, with double spiral system; intended to facilitate the manufacture of the

thread, and to render it more uniform and better in quality.

250. VALLÉE, F. *Molenbeek-St.-Jean, near Brussels.*—Small working model for spinning flax, wool, and cotton (the invention of the exhibitor).

251. VANDER ELST, L. & Co. *Braine-le-Comte, Hainault.* — Drawing for a paper-manufacturing machine.

252. VANGINDERTAELEN & Co. *Brussels.* —Distilling apparatus, constructed on scientific principles, for the production and rectification of spirits; and consisting of a still, an analyser which may be cleaned instantaneously, and a refrigerator. Refrigerators for various uses. Pumps and taps for ordinary and hygienic purposes.

253. VAN GOETHEM, C. & Co. *Brussels.* —Centrifugal machines for purifying sugar.

254. VERMEULEN, C. *Roulers.*—Shuttles with rollers for weaving various fabrics.

255. VINCENT, J. *Alost.* — A Jacquard machine with 700 hooks.

256. WERGIFOSSE, *Brussels.*—Liege mangle, washing and calendering machines.

257. WINNEN, *Brussels.*—Mill for unravelling rags.

258. WISSAERT, J. *Brussels.*—Embossing and gilding plates for bookbinding purposes.

259. WYNANTS & MACKINTOSH, *Brussels.* —Frames for locking up printing formes, without wedges or feather edges, or the use of hammers.

## CLASS VIII.

260. ARNOULD, G. *Mons.*—A water-level; free-air manometer; miner's safety-lamp.

261. BERTIEAUX, H. *Antwerp.*—Steam-engine.

262. CAIL, J. F. HALOT, A. & Co. *Brussels.*—Giffard-injectors; tubular steam-boiler, &c.

COMMITTEE OF THE EXHIBITORS OF VERVIERS:—
263. HOUGET, J. D. & TESTON, C. *Verviers.*—Portable steam-engine.

264. CUNGNE, U. *Langhemarcq, West Flanders.*—Weighing scales, with pans above.

265. DE LANDTSHEER, *Brussels.*—Horizontal steam-engine, on Woolf's principle.

268. FONDU, J. B. *Lodelinsart, Hainault.* —Economic fire-bars perforated horizontally, forming tubes under the ignited fuel.

270. GOUTEAUX, P. J. *Gilly, Hainault.*— Check chains and safety apparatus, applicable to mining engines.

271. LIBOTTE, N. *Gilly, Hainault.* — A fire-grate and accessories.

272. OBACH, N. *Brussels.* — Weighing-machine with double mechanism and square platform.

273. PERARD, L. *Liége.* — Horizontal blowing machine with two cylinders (Fossey's principle), 200-horse power.

274. PETIT, H. J. & Co. *Brussels.* — Level, with one air bulb and check screw.

275. PIROTTE, L. & SISTERS, *Brussels.*— Balance, on Roberval's principle. Bronze and marble stand.

276. REGUILÉ, JUN. & BEDRIVE, *Liége.*—
Exhibitors obtained the following medals:—
*F. Reguilé,* the Silver Prize Medal at the Brussels Exhibition of 1847.
*J. Bedrive,* a Second-class Medal at the Paris Exhibition of 1855, and the Gold Medal of the National Academy of Paris in 1857.
Articles exhibited by them in London in 1862—
A watering cart.—Price 3,000 francs.
A double action (suction and forcing) fire-engine.—Price 3,200 francs.
They manufacture all kinds of fire-engines, draining-pumps for mines, hydraulic pumps, &c.; riveted leather tubing, hose, and buckets.
Prospectuses sent on application.

277. SACRÉ, C. *Brussels.* — Hydrometer for alcohol, giving the quantity and strength of the liquor manufactured.

278. SCRIBE, G. *Ghent.*—Patent horizontal engine (Woolf's principle), with connected cylinders, 30 horse power.

279. THIRION, A. L. *Aische-en-Refail, Namur.*—Model of a windmill, in which a helix transmits the motion from the sails to the stones, without shock, and without a necessity for lubrication: the sails regulate themselves, and close up in a storm.

280. VANDERHECHT, E. *Brussels.*—Model of an apparatus for preventing shocks from sudden communication of motion; as when an engine is first started, or a horse first begins to draw; applicable to mines and traction.

281. WINAND, F. *Goffontaine, Liége.*— Patent safety screw-jack.

## CLASS IX.

283. BORTIER, P. *Ghistelles, West Flanders.*—Plan, in relief, of Britannia farm, at Ghistelles.

284. DAMS, *Tilleur, Liége.*—Unalterable enamelled labels for Botanic and Zoological gardens, inscriptions, sun-dials, &c.

285. D'AUXY, MARQUIS G. *Frasnes, near Leuze, Hainault.*—A granary.

286. DE GREEF, E. *Hal, Brabant.* —Agricultural implements.

287. DELSTANCHE, *Marbais, Brabant.*—Improved plough.

288. DE SOER, O. *Ben-Ahin, Liége.*—A skeleton roller.

291. LECOMTE, *Pont-à-Celles, Hainault.*—Iron plough with double mould-board.

292. MARIE, L. J. *Marchienne-au-Pont, Hainault.*—Apparatus for cleaning grain.

293. ODEURS, J. M. *Marlinne, Limburgh.*—Common plough. A plough, with sub-soil apparatus and balance.

294. PAS, P. A. *Londerzeel, Brabant.*—Churn, on a new principle.

295. PEERS, BARON E. *Oostcamp, West Flanders.*—A plan for a farm.

297. ROMEDENNE, A. J. *Erpent, Namur.*—Agricultural implements.

298. TIXHON, J. *Fléron, Liége.*—Agricultural implements.

300. VAN MAELE, E. *Thielt.*—Ploughs, straw chopper, bread cutter, sowing machine.

## CLASS X.

301. ADEN, L. *Brussels.*—Patent door, opening and closing on four sides.

302. BEERNAERT, A. *Brussels.*—Marble chimney-pieces, &c.

303. BOCH BROS. *La Louvière, Hainault.*—Mosaic slabs for pavements.

304. BOUCHER, T. *Saint-Ghislain, Hainault.* — Refractory substances, bricks, crucible, retorts, stone for spreading melted glass.

He obtained Medals at the Brussels and Mons Exhibitions of 1847 and 1851, and at the universal Exhibitions of London and Paris in 1851 and 1855 ; and a First-class Honorary Medal from the Paris Academy of Agriculture, Manufactures, and Commerce.

His establishment is of the very highest order, and is the oldest on the Continent. It is singularly well-situated, being in the centre of his own works, for obtaining primary substances and fuel, and in immediate communication with railways and steam vessels.

His processes are of a peculiar kind; they have been brought to great perfection, and are patented. And his attention is specially devoted to products that do not shrink, and are of the best quality, equal, indeed, to what are commonly termed extra, and are used for puddling furnaces and linings.

His business is on a very large scale; and he manufactures articles of great size. His customers are of the most respectable class, and he exports to all parts of the world.

Trade mark

Names of the objects exhibited, and their prices, when delivered at a railway, or placed on board a steam-vessel :—

1st Series.—Bricks, both rectangular and of other forms ; all, as nearly as possible, of the same length and size.—From 20 to 23 francs the 1,000 kilogrammes.

2nd Series.—Rectangular bricks, and articles of a shape not included among those to be described.—From 30 to 35 francs the 1,000 kilogrammes, models included.

3rd Series.—Large articles for the construction of all kinds of furnaces conformable to the plans which may be furnished.—From 40 to 50 francs the 1,000 kilogrammes, models included.

4th Series.—Crucibles and fittings for smelting furnaces. From 55 to 65 francs the 1,000 kilogrammes.

Articles not comprised among the above—

Pulverized cement, packing not included.—18 francs.

Retorts of all shapes and sizes.—From 50 to 70 francs, according to the dimensions.

Slabs or plates for spreading glass.—The price in proportion to the size.

He exhibits part of the crucible of a smelting furnace, weighing 700 kilogrammes ; and a slab for spreading glass, 1·60 metres long and 1·10 metres wide.

305. BOUCNEAU, L. *Brussels.*—A marble chimney-piece, Renaissance style.

306. BOUWENS, *Mechlin, Antwerp.*—Music-desk, door-lock.

307. CHAUDRON, J. *Brussels.*—Model of cast-iron lining, for the formation of shafts in humid soils, &c.

308. DEFUISSEAUX, MRS. *Bandour, near Mons.*—Articles in fire-clay.

310. DELPERDANGE, V. *Brussels.*—New method of joining water-pipes, gas-pipes, &c.

311. DEWYNDT, J. & Co. *Antwerp.*—Cedar wood veneer.

312. GODEFROY, J. *Brussels.*—A room door of rich woods, in carefully-selected shades; a room door of oak, with carvings.

313. GUIBAL, T. *Mons.*—Ventilator for mines, capable of displacing more than 100 cubical yards of air per second.

314. JACOBS, *Mechlin.*—A chimney-piece in portor marble, with interior and flooring.

315. JOSSON, N. & DELANGLE, *Antwerp.*—Hydraulic cements, mastic, terra cotta, bricks, tiles, and flags.

316. KELLER, A. *Ghent.*—Gas-retorts in refractory clay.

317. LAMBRETTE, J. *Brussels.* — Zinc roofing on a new principle.

318. LECLERQ, A. J. *Brussels.*—Chimney-pieces of statuary marble, &c. in the Flemish style, suited to English fire-places.

319. SIEGLITZ, J. *Brussels.*—Chimney-piece, with statuary work.

320. VANDER ELST-BOURGOIS, *Brussels.* Black chimney-pieces.

321. VAN NEUSS, M. *Brussels.*—Inodorous water-closet.

322. WYNEN, G. *Schaerbeek, near Brussels.*—A specimen of flooring.

# CLASS XI.

323. BAYET BROS. *Liége.*—Ornamented fire-arms; Lefaucheux guns, and guns with ramrods; Swiss carbine and revolver.

324. BERNIMOLIN BROS. *Liége.*—Lefaucheux-Bernimolin guns, pistols, carbines on Flobert's principle.

325. COOPPAL & Co. (Director: C. VAN CROMPHAUT), *Wetteren, East Flanders.*—Gunpowder of various kinds, refined saltpetre.

326. DANDOY, C. *Liége.*—Fire-arms of all kinds.

327. DE LEZAACK, A. *Liége.*—Fowling-pieces, &c.

328. DITS, A. J. *Saint-Gilles, near Brussels.*—Cartridges for Lefaucheux guns and revolvers: patent balls.

329. DUHENT, L. *Brussels.* — Wheelbarrow convertible into a camp-bed, ambulance, tent, boat, or bridge.

330. DUMOULIN-LAMBINON, G. *Liége.*—Guns, revolvers, pistols, carbines, &c

331. FAFCHAMPS, *Brussels.*—New kinds of fire-arms; new system of defence, &c.

332. FUSNOT, C. & Co. *Brussels.*—Cartridges for Lefaucheux guns and revolvers: ball cartridges of gun-cotton and fulminating powder.

This establishment exhibits bushes for Lefaucheux fowling-pieces, copper bushes for six-barrelled revolver pistols, and ball charges for needle guns.

The revolution which is taking place in the construction of breech-loading fire-arms has created a branch of trade that, from the perfection at which it has arrived, has contributed to the spread of new principles of construction. Sometimes metal only, sometimes a combination of metal and paper, are used in the formation of a case which is capable of resisting a powerful charge.

This case is intended to prevent the fire from issuing behind when the charge is ignited and goes off. As a consequence of this arrangement, the method of charging is simplified to an extraordinary degree, and the discharge is effected with great rapidity.

The more perfect the bushing of the cartridges the more fully these results are attained. The exhibitors show their productions as possessing a superiority altogether exceptional, which they attribute to the care with which they are executed. If the various details are examined and tested they will be found arranged in such a manner as to produce a combination that leaves nothing to be desired.

They offer their copper bushes for revolvers as articles which have never been surpassed. Their bushes for fowling-pieces possess a flexibility, and at the same time a strength which are inimitable.

Notwithstanding those excellent qualities they have succeeded in producing them at a very moderate price, of which any one may satisfy himself by application to them.

Mr. Charles Fusnot obtained a medal at the Belgium Exhibition of 1847 for the new contrivances which he invented. Among them was a ball-charge which he devised. This invention has been confirmed to him by judicial decisions, condemning those who pirated it.

333. HERMAN, J. *Liége.*—Designs for the manufacture of fancy fire-arms.

334. HUBAR, *Herstal, Liége.* — Fancy guns, and the various pieces used in making them.

335. JANSEN, A. *Brussels.* — Double-barrel guns, &c.

336. JONGEN BROS. *Liége.*—Fire-arms, for military purposes, sporting, &c.

337. LADRY, F. *Brussels.*—A rest, for taking correct aim with portable fire-arms; instrument for measuring the distance of the bullet-marks from the centre of a target.

338. LARDINOIS, N. C. *Liége.*—Breech-loading carbine.

339. LEMAIRE, J. B. *Liége.*—Fowling-pieces, revolver pistol, &c.

340. LEXIN, C. *Ghent.*—Cuirasses for infantry, cavalry, artillery, &c., in hammered steel, ball proof at twenty-five yards distance.

341. MALHERBE, P. J. & Co. *Liége.*—Guns, musketoons, pistols, gun-barrels.

342. MASU BROS. *Liége.*—Breech-loading fowling-pieces, each made in a different way.

343. SIMONIS, N. & Co. *Val-Benoît, near Liége.*—Gun-barrels.

344. TINLOT, J. M. *Herstal, Liége.*—A carbine on Flobert's principle.

## CLASS XII.

345. VAN BELLINGEN, A. J. *Antwerp.* — Proved chain cables, and rigging chains of Belgian iron.

## CLASS XIII.

346. BULTINCK, E. *Ostend.* — Portable electro-galvanic apparatus with inodorous acid, giving any required current.

347. DUSAUCHOIT, E. *Ghent.* — Signal speaking-trumpets and whistles.

348. GÉRARD, A. *Liége.*—Electric clock, electric battery, and electro-magnet; plans of instruments and machinery.

349. GLOESENER, M. *Liége.*—Electric chronoscopes, registering multipliers, electric clock, and electric telegraph apparatus, &c.

350. JASPAR, *Liége.* — Chronoscope on Major Navez's principle; Doctor Stacquez's "electro-medical;" a regulator of electrical light.

351. LIPPENS, P. *Brussels.*—Telegraph apparatus, &c.

352. SACRÉ, E. M. *Brussels.*—Philosophical balance, eclemeter-compass, circle-level.

353. VANDEVELDE, N. *Ghent.*—A saccharometer.

## CLASS XIV.

355. DAVELUY, *Bruges.* — Photographic views of Bruges.

356. DUPONT, *Antwerp.*—Photographs : portraits selected from the collection named "The Antwerp School."

357. FIERLANTS, ED. *Brussels.*—Photographs, representing the master-pieces and monuments of Belgium; executed by order of the Government.

358. GHÉMAR BROS. *Brussels.*—Photographs, natural size, and others ; visiting cards.

359. MASCRÉ, J. *Brussels.*—Photographs from pictures, plaster casts, &c.

360. MICHIELS, J. J. *Brussels.*—Photographs : copies of pictures.

361. NEYT, A. L. *Ghent.* — Photographic micrography (obtained through the agency partly of solar and partly of electric light).

362. NEYT, CH. *Brussels.*—Photographs : portraits, and copies after Vander Hecht.

# CLASS XV.

363. GÉRARD, A. J. *Liége.*—Clocks and watches.

# CLASS XVI.

364. AERTS, F. G. *Antwerp.*—Oblique-trichord seven-octave pianos.

365. ALBERT, E. *Brussels.* — Clarinets, flutes, hautboys, bassoons.

266. BERDEN, F. & Co. *Brussels.*—Upright pianos, with oblique and vertical strings.

368. DARCHE, C. F. *Brussels.*—Tenor-violins, violins, violoncello, &c.

371. JASTRZEBSKI, F. *Brussels.*—Grand pianoforte ; upright transpositional pianoforte.

372. MAHILLON, C. *Brussels.*—A complete collection of musical instruments.

373. STERNBERG, L. & Co. *Brussels.*—Four pianos of different kinds.

374. VUILLAUME, N. F. *Brussels.*—Violins, violoncello, counter-bass.

# CLASS XVII.

375. GLITSCHKA, H. *Ghent.* — Surgical instruments ; artificial limb.

376. KAYSER, *Brussels.*—Case of instruments for the royal railway trains.

377. ODEURS, J. M. *Marlinne, Limburgh.*—Speculum uteri ; a mouth-opener.

378. WAERSEGERS, J. *Antwerp.*—Herniary trusses ; ventral and hypogastric belts ; orthopœdic apparatus ; artificial limbs.

# CLASS XVIII.

BELGIAN GOVERNMENT. — Cotton goods produced in the Flemish Apprentice Schools :—

## WEST FLANDERS.

379. APPRENTICE SCHOOL OF BECELAERE.—Cotton goods for summer.

380. —— OF MOORSEELE. — Cotton checks.

381. APPRENTICE SCHOOL OF MOORSLEDE.—Summer goods : stuffs for furniture.

382. —— OF POPERINGHE.—Cotton checks.

383. —— OF ROULERS. — Cotton checks.

384. —— OF RUDDERVOORDE.—Cotton woven fabrics.

385. APPRENTICE SCHOOL OF YPRES.—Cotton checks.

386. LATE APPRENTICE SCHOOL OF BRUGES (owners: MM. DE RANTERE & CO. *Bruges*).—Dimity, and frame-embroidered muslins.

387. —— OF COURTRAI (owner: M. SISENLUST, *Courtrai*).—Cotton velvet.

### EAST FLANDERS.

388. APPRENTICE SCHOOL OF CALCKEN. — Cotton stuffs for dresses and window blinds.

389. —— OF OLSENE.—Dimity; cotton-satin.

390. —— OF OORDEGHEM.—Cotton velvet, stuffs for window blinds.

391. —— OF SINAY (owner: M. VERELLEN-RODRIGO, *Saint Nicolas*).—Cotton stuffs for dresses; cravats.

392. LATE APPRENTICE SCHOOL OF LEDE (owner: M. V. DERCHE, *Brussels*).—Dimity and frame embroidered muslins; muslins; "royaumont" dimity.

393. —— OF NAZARETH (owner: M. VANDEN BOSSCHE-VERVIER, *Nazareth*). — "Leather-dimity."

394. —— OF NEDERBRAKEL (owner: M. DE PROOST, *Opbrakel*).—Beverteens and cotton-satins.

395. —— OF SLEYDINGE (owners: MM. CEUTERICK & DE COCK, *Ghent*).—White Jacquard cotton stuffs.

396. —— OF WAESMUNSTER (owner: M. VAN HOOF, *Lokeren*).—Cravats; fine cotton checks.

———

397. DE BACKER, L. & N. *Braine-le-Château, Brabant.*—Short staple Georgia cotton yarn.

398. DE BAST, C. *Ghent.*—Woven goods from raw cotton.

399. DE BLOCK-DELSAUX, *Termonde, East Flanders.*—Cotton bed-covers.

401. DE SMET BROS. *Ghent.*—Cotton warp, dyed and dressed; plain and printed fabrics.

402. DE SMET, E. & CO. *Ghent.*—Dressed and dyed raw cotton warp for mixed fabrics.

403. DIERMAN-SETH, F. *Ghent.*—Woven fabrics of Surat cotton.

404. DUCHAMPS, G. *Brussels.*—Cotton stuffs for trousers and other garments.

405. DUJARDIN, J. E. & L. *Bruges.*—Raw cotton spun, warp and weft.

406. DUPREZ & CO. *Dottignies, near Courtrai.*—Cotton stuffs for trousers and other garments.

407. HOOREMAN-CAMBIER & SON, *Ghent.*—Cotton fabrics for trousers and other garments.

409. LEMAIRE-DUPRET & SON, *Tournay.*—Cotton stuffs for trousers.

410. MOUSCRON, CITY OF, DISTRICT OF COURTRAI, WEST FLANDERS, COMMITTEE:—

| | |
|---|---|
| Desprets Bros. | Labis-Delecoeillerie. |
| Dujardin, L. | |

—Cotton stuffs for trousers.

411. PHILIPS-GLAZER, J. *Termonde.*—Cotton bed-covers, calicoes, pilous, Belgian leather-cloth, dimity half linen.

412. PIRON, J. *Tournay.*—Stuffs for trousers, all cotton.

413. ROELANDTS, F. *Courtrai.*—Cotton stuffs for trousers and other garments.

414. ROOS & VAN BELLE, *Termonde.*—Cotton bed-covers.

415. RYCX, A. & VERSPEYEN, *Ghent.*—Cotton spools; cotton fabrics.

316. SAEYS BROS. *Termonde.*—Cotton bed-covers.

416A. SCHMIDT & CO. *Courtrai.*—Cotton stuffs for trousers.

417. STAELENS, P. & CO. *Ghent.*—Surat cotton yarns.

418. VAN HEE BROS. *Mouscron, West Flanders.*—Cotton stuffs for trousers, &c.

419. VAN HEUVERSWYN, F. & CO. *Ghent.*—Counterpanes, petticoats, dimity, damasks (white and coloured), calicoes.

420. VANNESTE, P. & CO. *Rolleghem, near Courtrai.*—Cotton stuffs for trousers, dresses, and waistcoats.

# CLASS XIX.

BELGIAN GOVERNMENT. — Linen goods produced in the Flemish Apprentice Schools:—

## WEST FLANDERS.

421. APPRENTICE SCHOOL OF AERSEELE. —Half-bleached linens.

422. —— OF AERTRYCKE.—Linen.

423. —— OF ANSEGHEM.—Linen.

424. —— OF ARDOYE.—Linens.

425. —— OF AVELGHEM.—Linens.

426. —— OF BECELAERE. — Linens and handkerchiefs.

427. —— OF CLERCKEN.—Linens.

428. —— OF CORTEMARCQ.—Linens.

429. —— OF COURTRAI.—Linens and damasks.

430. —— OF DEERLYK.—Linens.

431. —— OF DENTERGHEM.—Linens.

432. —— OF DESSELGHEM. — Linens and cambric handkerchiefs.

433. —— OF GHISTELLES.—Linens.

434. —— OF HEULE.—Linens.

435. —— OF HOOGHLEDE.—Linens.

436. —— OF HULSTE. Linens.

437. —— OF INGOYGHEM.—Linens.

438. —— OF LANGHEMARCQ.—Linens, bleached and unbleached; linens for mattresses.

439. —— OF LENDELEDE.—Linens.

440. —— OF LICHTERVELDE. — Linens.

441. —— OF MENIN. — Linens and handkerchiefs.

442. —— OF MEULEBEKE.—Linen.

443. —— OF MOORSEELE.—Linens.

444. —— OF MOORSLEDE.—Linens.

445. —— OF OOSTNIEUWKERKE.—Linens.

446. APPRENTICE SCHOOL OF OOST-ROOSEBEKE.—Linens.

447. —— OF OUCKENE.—Linen; cambric handkerchiefs.

448. —— OF OYGHEM. — Woven fabrics, in flax and hemp.

449. —— OF PASSCHENDAELE. — Linens.

450. —— OF PITTHEM.—Plain linen; linens for napkins and towels.

451. —— OF POPERINGHE.—Linen for mattresses, diaper, handkerchiefs.

452. —— OF ROULERS.—Linens and damasks.

453. —— OF RUDDERVOORDE.—Linen fabrics.

454. —— OF RUYSSELEDE—Linen.

455. —— OF STADEN.—Linens.

456. —— OF SWEVEGHEM.—Linens.

457. —— OF SWEVEZEELE.—Linens.

458. —— OF THIELT.—Linens.

459. —— OF THOUROUT.—Ticks and diapers.

460. —— OF WAEREGHEM.— Linens made of raw, bleached, and half-bleached yarn, linen for towels.

461. —— OF WESTROOSEBEKE. — Bleached linens.

462. —— OF YPRES.—Linens, plain and damasked, for mattresses.

463. LATE APPRENTICE SCHOOL OF BLANKENBERGHE (owner: M. L. DE LESCLUZE, *Bruges*).—Ticks, of flax only; and of flax and cotton, English mode of manufacture.

464. —— OF BRUGES (owner: M. MARLIER, *Bruges*).—Blue linens; diaper linens, blue and white.

465. —— OF BRUGES (owner: M. C. POPP, *Bruges*).—Linens and cambrics.

466. —— OF BRUGES (owner: M. ARDRIGHETTI, *Bruges*).—Fore-parts of shirts, with moveable breasts made of flax-yarn.

467. LATE APPRENTICE SCHOOL OF ISE-GHEM (owner: M. MAES-VAN-CAMPENHANDT). —Linens, ticks, and handkerchiefs.

## EAST FLANDERS.

468. APPRENTICE SCHOOL OF BAELEGEM (M. ROBYNS, *Baelegem*).—Linens, handkerchiefs, and linen ticks.

469. ——— OF CALCKEN.—Flax yarn fabrics:

470. ——— OF EYNE (MM. L. & A. VAN DE PUTTE, *Ghent*).—Linens and handkerchiefs made of flax yarn, linens for mattresses.

471. ——— OF NEDERBRAKEL (M. DE PROOST, *Opbrakel*).—Napkins.

472. ——— OF OLSENE.—Linens.

473. ——— OF OORDEGEM.—Damasked linens for mattresses, &c.

474. ——— OF SYNGEM.—Unbleached linens.

475. ——— OF URSEL.—Table linens, &c.

476. LATE APPRENTICE SCHOOL OF ALOST (owners: MM. J. & P. NOËL, BROS.) —Damask and diaper table-linen, linen for mattresses.

477. ——— OF BEILEM (owner: M. MOERMAN-VAL-LAERE, *Gand*).—Sail cloth, &c.

478. ——— OF SLEYDINGE (owner: M. DOBBELAERE-HULIN, *Ghent*.)—Sail cloth, plain linens, diapers for mattresses.

———

480. CAESENS, V. & SON, *Zele, East Flanders.*—Bolting cloth, cloth for stopping up casks, and other purposes.

482. DE BRANDT, J. *Alost, East Flanders.* —Damask and diaper table-linen, &c.

483. DE BROUCKERE BROS. *Roulers.*—Tow-yarn.

483A DEVOS, F. & Co. *Courtrai.*—Plain unbleached hand-spun flax-yarn.

484. FRANCHOMME, L. *Brussels.*—Various sorts of ticks.

485. JELIE, J. B. *Alost.*—Flax sewing thread, hand and machine spun.

487. LEFEBVRE, F. F. *Alost.*—Plain linens, &c.

[*The Exhibitor received eleven First-class Medals, at the Exhibitions of 1859, and 1861, in France and Belgium.*]

Flax, hemp, &c., prepared by machinery.

Class A, from 1·50f. to 1·75f the kilogramme.
„ B, „ 2·50f. „ 3·00f the do.

The price varies with the cost of the raw material in the place where it grows.

Nature of the process, its cost, and relation to health—

1st. The products are obtainable immediately after the gathering of the textile matters.

2nd. The work, which is easy and regular, may be done at all seasons without any chance of loss, or any offensive or disagreeable operation.

3rd. The flax and hemp gathered off many hectares may be manipulated in one day, the operation being more regular in proportion to the greatness of the scale according to which it is carried on.

4th. The refuse, which constitutes 70 per cent. of the raw flax, is used for fuel; and its ashes afford 20 per cent. potash.

5th. All the textile materials of the flax and hemp are obtained.

6th. The fibres are separated completely, regularly, and with certainty, without injury to their strength.

7th. The silvery or slightly golden shade natural to flax is preserved.

8th. A different tint may be given to the raw flax.

9th. A greater value is imparted to every kind of flax.

10th. All kinds of flax and hemp may be spun in any way with cold water.

11th. The thread is strong, regular, and clean, exhibiting the shades which are natural to it.

12th. As the fibres of the flax contain no resinous matter, it is not necessary to prepare the thread for creaming.

13th. Bleaching is effected with ease and rapidity.

14th. The thread and stuff are dyed directly with water, the most delicate tints being given to the raw flax.

15th. The process is carried on with a great economy of labour, without any chance of loss. The operations are merely routine. The material is very simple when pure, and is not subject to deterioration. The machinery, which is inexpensive, has been improved by an experiment continued during three years of practical working.

A manufactory is in operation at Brussels.

488. MAES-VAN-CAMPENHOUDT, *Iseghem, West Flanders.*—Linens, tickings, cambric handkerchiefs.

489. SAINT BERNARD HOUSE OF CORRECTION, *Hemixem, Antwerp.*—Linens of various kinds.

491. SIREJACOB, E. & COUCKE, C. *Brussels.*—Diaper and damasked napkins, with crests, towels, &c.

492. SOCIÉTÉ LINIÈRE DE BRUXELLES, *St. Gilles, near Brussels.*—Machine-spun flax and tow yarns; flax and hemp fabrics, woven by hand and power looms.

493. SOCIÉTÉ LINIÈRE GANTOISE, *Ghent.* —Bleached and unbleached flax and tow yarns.

494. SOCIÉTÉ LINIÈRE DE SAINT LÉONARD, *Liége.*—Flax and tow yarns.

495. TANT-VERLINDE, *Roulers.* — Flax, flax-yarns and linens, unbleached, &c.

496. THIENPONT, L. & SUNAERT, A, *Ghent.*—Damask and diaper table-linen, linen cloth for mattresses, towels.

498. VAN ACKERE, J. C. *Wevelghem and Courtrai, West Flanders.*—Unbleached and bleached linens; linen and cambric hand-kerchiefs.

499. VAN DAMME BROS. *Roulers.*—Un-bleached linen.

500. VAN DE WYNCKELE BROS. & ALS-BERGE, J. *Ghent.*—Flax-yarns, in every stage of bleaching.

501. VAN MELDERT, *Haeltert, near Alost.*—Unbleached linen, table-linen, &c.

502. VAN OOST, P. *Hooghlede, West Flanders.*—Linens made of machine and hand-spun yarn.

503. VAN ROBAYS, A. J. *Waereghem, West Flanders.*—Sail cloths, russias, sack-ings made of jute.

504. VAN TIEGHEM & Co. *Courtrai.*—Linen made of machine and hand spun yarn.

505. VERRIEST, P. *Courtrai.*—Diaper and damask table-linen and cloths for mat-tresses.

506. VERTONGEN-GOENS, C. S. *Termonde.*—A piece of manille-hemp flat cable with eight strands.

## CLASS XX.

BELGIAN GOVERNMENT.—Silk and velvet goods produced in Flemish Apprentice Schools:—

### WEST FLANDERS.

507. LATE APPRENTICE SCHOOL OF BRUGES (M. AVANZO), *Brussels.*—Ribbons for hats and caps, cravats, &c.

### EAST FLANDERS.

508. LATE APPRENTICE SCHOOL OF

ALOST (owner: M. LEVIONNOIS-DEKENS, *Alost.*)—Articles in plain black silk.

509. LATE APPRENTICE SCHOOL OF DEYNZE (owners: MM. LAGRANGE BROS. *Deynze.*)—Various articles in silk.

510. THYS, C. *Brussels.*—Thrown, un-bleached, and dyed silks, for mercers' and lace-makers' goods.

## CLASS XXI.

BELGIAN GOVERNMENT.—Woollen and mixed fabrics produced in the Apprentice Schools of Flanders:—

### WEST FLANDERS.

511. APPRENTICE SCHOOL OF BECELA-ERE.—Articles of wool and cotton.

512. —— OF BRUGES (MM. KAUWERZ & Co. *Brussels.*)—Bournous, half-wool, tar-tans, galaplaids, and goats' hair cloth.

513. —— OF COURTRAI.—Stuffs for trousers.

514. —— OF DEERLYK.—Roubaix cloths, for trousers; materials for dresses; fancy stuffs of silk mixed with wool and cotton.

515. APPRENTICE SCHOOL OF HULSTE.—Woollen fabrics.

516. —— OF LANGHEMARCQ.—Black paramatta.

517. —— OF MENIN.—Stuff for trou-sers.

518. —— OF MOORSLEDE.—Siamese, an article of Roubaix.

519. —— OF MOUSCRON.—Woven fa-brics, wool and cotton.

520. —— OF POPERINGHE.—Siamese, plain and twilled.

521. —— OF ROULERS.—Orléans.

522. APPRENTICE SCHOOL OF THIELT (MM. SCHEPPERS, *Loth, near Brussels*).—Thibets, lastings, serges, &c.

523. ——— OF YPRES.—Molletons.

### EAST FLANDERS.

524. APPRENTICE SCHOOL OF CALCKEN.—Stuffs for dresses.

525. ——— OF OLSENE.—Orléans, paramattas.

526. ——— OF RUYEN.—Mixed fabrics for dresses and trousers; fabrics manufactured on Jacquard's principle.

527. ——— OF SINAY (M. VERELLEN-RODRIGO, *St. Nicolas*).—Materials for dresses, &c., in wool, cotton, and silk.

528. ——— OF URSEL.—Stuffs for trousers in wool and cotton.

529. ——— OF WICHELEN (M. F. VAN BRABANDER, *Wichelen*). — Stuffs for mattresses.

530. LATE APPRENTICE SCHOOL OF NAZARETH (owner: M. VAN DEN BOSSCHE-VERVIER, *Nazareth*).—Tweed, corded stuff, satin, satin-reps (wool and cotton).

531. ——— OF WAESMUNSTER (owner: M. VAN HOOFF, *Lokeren*).—Woven goods, in wool, and in wool and cotton.

532. ANDRIES & WAUTERS, *Mechlin, Antwerp.*—Woollen blankets.

533. BEGASSE, CH. *Liége.* — Woollen blankets; felts for paper factories; woollen stuffs.

COMMITTEE FOR THE EXHIBITION OF THE DISTRICT OF VERVIERS.—Woollen yarns and fabrics:—

535. BARAS-NAVAUX, *Hodimont, near Verviers.*—Light woollen stuffs for suits, and caps.

536. BERCK, CH. *Aerve.*—Spun goods for borders.

537. BIOLLEY, F. & SON. *Verviers.*—Cloths, satins, cashmeres, fancy woollen cloths.

538. BRULS-RIGAUX, *Goffontaine, Cornesse.*—Mixed wool and cotton thread.

539. CHANDELLE-HANNOTTE, *Dison.*—Beavers, and knitted articles.

540. CHAUDOIR & HOUSSAT, *Hodimont.*—Fancy stuffs for winter and summer.

541. COMMISSION VERVIÉTOISE.—Corded stuffs, billiard cloths, tweeds, satins, &c.

542. DEBEFVE-BLAISE, *Dison.* — Hangings, fancy cloths, military cloths.

543. DEHESELLE, *Thimister, near Verviers.*—Flannels, domets, gauzes, and swanskins.

543A. DELEVAL & SON, *Dison, Verviers.*—Fancy stuffs, &c.

544. DEL MARMOL, F. *Francomont, near Verviers.*—Domets, and flannels.

545. DORET, V. (LÉONARD DORET), *Verviers.*—Woollen cloth, dyed and undyed.

546. DUBOIS, GÉRARD, & CO. *Verviers.*—Stuffs of wool, and of silk and wool; wool-satin, and velvet cloths.

547. FLAGONTIER, J. J. *Verviers.*—Stuffs of wool, and of wool and silk for trousers, &c.

548. GAROT, J. *Hodimont, near Verviers.*—Stuffs of wool, and of wool and silk; fancy cloths for trousers, great-coats, cloaks, &c.

549. GRANDJEAN, H. J. *Verviers.*—Stuffs of wool, and of wool and silk.

550. GRÉGOIRE & PELTZER, *Dison.* — Woollen stuffs for great-coats and trousers.

551. HAUZEUR, P. & VIGAND, BROS. *Ensival, near Verviers.*—Stuffs of wool, and of wool and silk.

552. HAZEUR, GÉR. & SON, *Verviers.*—Thread of carded wool, for weaving.

552A. HENROTTY, MARÈCHAL, *Ensival, Verviers.*—Woollen stuffs.

553. HENRION, J. J. *Hodimont, Verviers.*—Cloths, and woollen stuffs for trousers, &c.

554. LAHAYE, M. & CO. *Verviers.*—Cloth and woollen stuffs for suits.

555. LAOUREUX, G. J. *Verviers,* Cloth plain and twilled; woollen stuffs.

556. LECLERCQ, N. *Dison.* — Beavers, duffels, satins, moscows, &c.

557. LEJEUNE-VINCENT, H. S. *Dison.*—Fancy stuffs, ladies' cloaks, &c.

558. LEJEUNE-VINCENT, J. C. *Dison.*—Woollen stuffs, moscows, wool-satins.

559. Lieutenant & Peltzer, *Verviers.* — Thread, cashmeres, beavers, wool-satins, stuffs of fancy wool, and of wool and silk.

560. Lincé, Widow H. & Son, *Dison.* — Moscows, fancy stuffs, stuffs of wool and silk.

561. Marbaise & Son, *Hodimont.* — Military and other cloth, woollen stuffs.

562. Masson, L. *Verviers.* — Hangings, manufactured stuffs, black and coloured.

562A. Mathieu, J. F. *Dison, Verviers.* — Woollen stuffs.

563. Modion, A. & Bertrand M. *Verviers.*—Moscows, stuffs of wool and silk.

564. Mullendorff & Co. *Verviers.* — Thread made of carded wool for fancy cloths, stuffs, shawls, &c.

565. Navaux, R. & Son, *Hodimont.* — Reps and summer goods.

566. Olivier, J. J. & Son, *Verviers.* — Drapery and woollen stuffs.

567. Pirenne & Duesberg, *Verviers.* — Wool-satins, fancy and other woollen stuffs, military cloth.

568. Piron-Thimister, *Francomont.* — Stuffs of wool and silk, double-milled cloth used in garments for the Belgian army.

569. Rahlenbek & Co. *Verviers.* — Cloths, fancy stuffs, fabrics for gloves, cloth gloves.

570. Sagehomme-Lutaster, S. *Dison.* —Moscows, cotelines, corded stuffs, &c.

571. Sauvage, A. J. *Francomont, near Verviers.*—Woollen stuffs.

572. Seret & Pirard, *Verviers.*—Thread made of white and other wool, washed and unwashed.

572A. Simar, Dréze, *Dison, Verviers.*— Billiard cloths.

573. Simon, J. & Diet, *Chaineux, near Verviers.*—Thread made of carded wool, unbleached, and mixed in different shades.

574. Simonis, I. *Verviers.* — Hangings, stuffs of wool, and of wool and silk.

575. Sirtaine, F. *Verviers.*—Cloths and woollen stuffs, fancy goods.

577. Snoek, E. *Charneux, near Verviers.* —Cloths, zephyrs, cashmeres, wool-satins, corded stuffs, moscows, and other woollen goods.

578. Van der Maesen, L. C. *Verviers.* —Fancy stuffs for great-coats, ladies' cloaks, &c.; stuffs in wool and silk.

578A. Suhs, J. A.—Wool-satin.

579. Vervier & Gregoire, *Verviers.*— Fancy stuffs, velvets, stuffs in wool and silk.

580. Voos, J. J. *Verviers.*—Fancy stuffs, cloths, hangings, double-milled cloth, &c.

580A. Winandy-Veuster, *Dison.*—Coteline.

581. Xhibitte, *Charneux, near Verviers.* —Carded wool for fancy cloths, stuffs, &c.

581A. Xhoffray, C. & Bruls, C. *Doltrain.*—Carded wool.

---

582. Duchamps, G. *Brussels.*—Stuffs of cotton and wool mixed, for trousers, &c.

583. Duprez & Co. *Dottignies, near Courtrai.*—Stuffs of wool and cotton for trousers, &c.

584. Gauchez, L. *Brussels.*—Blankets, felted threads, fancy and mixed fabrics woven from felted threads.

585. Kauwerz, P. & Co. *Brussels.*— Tartan shawls.

They exhibit these shawls, chiefly with the view of showing the superiority of their establishment, taken in its entirety; the finishing process, which is the most difficult portion of the manufacture, being executed by their own workmen, and not by finishers who devote themselves to nothing else.

They are engaged also in the production of small shawls, stuffs for dresses and other purposes, comforters, chatelaines, &c. They offer for sale goods which have been made with great care, and whose merits have already been several times officially acknowledged by medals, which have been obtained at various Exhibitions, and are represented on their cards.

587. Lemaire-Dupret & Son, *Tournay.* —Stuffs for trousers, of silk and cotton.

588. Mouscron (City of), *District of Courtrai.*—Stuffs in cotton and wool, for trousers, &c.

589. Piron, J. *Tournay.* — Stuffs for trousers in wool and cotton; ticks in thread and cotton.

590. Roelandts, F. *Courtrai.*—Stuffs in wool and cotton, for trousers, &c.

591. Rolin, H. Son & Co. *Saint-Nicolas, East Flanders.*—Tartan shawls, fabrics all wool, or wool cotton and silk.

592. Schmidt & Co. *Courtrai.* Fabrics for trousers, in wool and cotton, and cotton and thread.

593. VAN HEE BROS. *Mouscron, West Flanders.*—Stuffs, in wool and cotton, for trousers, &c.

594. VAN NESTE & VANDER MERSCH, *Rolleghem, West Flanders.*—Stuffs, in wool and cotton, for trousers, dresses, &c.

595. WAUTERS, A. & A. *Tamise, East Flanders.*—Wool-poplin, silk poplin, shawls.

# CLASS XXII.

596. BRAQUENIÉ BROS. & CO. *Ingelmunster, West Flanders.*—Carpeting, Flanders tapestry for furniture and hangings.

597. MOYERSOEN-CAMMAERTS, R. *Brussels.*—Pilous tapestry, carpetings, rugs, &c.

598. SCHEPENS, L. *Ghent.* — Pattern, drawn for a high-warp carpeting,

599. SOCIÉTÉ DE LA MANUFACTURE ROYALE DE TAPIS DE TOURNAY, *Brussels.*—Carpets.

600. TIMMERMANS, MISS M. *Ixelles, near Brussels.*—Tapestry done with the needle, on canvas, in silk and wool; two drawings on canvas for the same purpose.

# CLASS XXIII.

601. DEWOLF & DE MEY, *Rouge-Cloter-under-Auderghem, near Brussels.*—Turkey red, and other cotton yarns in fast colours.

603. IDIERS, E. *Auderghem, near Brussels.*

—Cotton yarns in Turkey red, and other fast colours.

604. RAVE, N. SEN. *Curreghem. near Brussels.*—Dyed goods; wool, silk, cotton spun and raw.

# CLASS XXIV.

BELGIAN GOVERNMENT.—Embroidered articles, manufactured in the Apprentice Schools of Flanders :—

## WEST FLANDERS.

605. APPRENTICE SCHOOL OF SWEVEGHEM (GIRLS) —Embroidered articles, style of St. Gall.

606. LATE APPRENTICE SCHOOL OF BRUGES (owner: M. AVANZO, *Brussels*) — Laces; galloons.

## EAST FLANDERS.

607. APPRENTICE SCHOOL OF CALLOO. —Embroidery on lace.

609. BOETEMAN, A. J. *Bruges.*—Handkerchief and collars in Valenciennes lace.

610. BONNOD, P. *Brussels.*—Designs for all kinds of lace.

611. BRUYNEEL, SEN. *Grammont, East Flanders.*—Black silk lace.

612. BUCHHOLTZ & CO. *Brussels and Valenciennes.*—Point de Venise, and other laces, &c.—

Application, gauze, Valenciennes, and Chantilly lace; embroideries, cambrics, and lawns.

613. CHRISTIAENSEN, G. H. J. *Antwerp.* Embroidered lace.

614. CUSTODI-BESME, J. *Brussels.*—Handkerchief in gauze point, Chantilly veil.

[*His laces obtained the Medal at the Florence Exhibition.*]

Established for the manufacture of point and appliqué laces.

Trousseaus, complete, from 50*l.* upwards. Specimens forwarded on application.

615. DAIMERIES-PETITJEAN, *Brussels.*— White and black lace, antique style, &c.

616. DE CLIPPÈLE, Mrs. C. *Brussels.*— Point de Venise, and other laces.

617. DELAPORTE, Mrs. *Brussels.* — Galloons for carriages and livery lace.

618. DE RANTERE & Co. *Bruges.*—Embroidered articles, in the style termed " Plumetis," &c.

619. EVERAERT, J. & SISTERS, *Brussels.* Black and white lace.

620. GEFFRIER-DELISLE BROS. & Co.— Brussels lace, gauze point, &c.

621. GHYSELS, V. & Co. *Brussels.* — Brussels application lace, gauze point, guipure.

622. GREGOIR-GELOEN, N. J. *Brussels.*— Brussels and Valenciennes lace.

623. HANSSENS-HAP, B. *Vilvorde, near Brussels.*—Lace; galloons for carriages.

624. HOORICKX, E. J. *Brussels.*—Articles in lace.

625. HOUTMANS, A. J. *Brussels.*—Designs for lace, &c.

626. HOUTMANS, C. C. *Brussels.*—Designs for lace, &c.

627. HUTELLIER, *Brussels.*—Application lace, in point, cushion work, and gauze.

Lace articles exhibited—
Volants, handkerchiefs, sets comprising collars and sleeves, in Brussels appliqué, point gauze, and point laces, à l'aiguille and plat gauze.

628. KEYMEULEN, H. *Brussels.*—Flemish black lace, and lace articles.

629. LEPAGE-KINA, J. G. *Grammont.*— Ladies' apparel in black lace, and other lace gold articles.

630. MELOTTE, E. *Brussels.*—Banner of the Brussels Tennis Club, embroidered in gold on velvet.

631. MINNE-DANSERT, C. *Brussels.* — Various descriptions of lace, and lace articles.

632. MULLIE-TRUYFFAUT, P. *Courtrai.*— Valenciennes lace, and articles made of it.

633. NAETEN, J. *Brussels.*—Designs for various articles in lace.

634. PHILIPPE, L. *Brussels.* — An embroidery in gold and silk, representing the royal arms of Belgium.

635. REINHEIMER, C. (MAISON SOPHIE DEFRENNE), *Brussels.* — Brussels lace, in Brussels point, and point and plat.

Patent Brussels point and plat lace volants and handkerchiefs.
Articles in lace, both at a low price and of the richest and finest qualities.
C. Reinheimer obtained First-class Medals for his goods at the following Exhibitions :—

| | |
|---|---|
| Brussels, 1847. | London, 1851. |
| New York, 1853. | Paris, 1855. |

636. ROOSEN, H. (MAISON SECLET-VANCUTSEM), *Brussels.*—Articles in lace.

637. SALIGO-VANDENBERGHE, *Grammont.*—Articles enriched with lace.

638. SASSE, MRS. P. F. *Brussels & London.*—Lace articles.

639. SCHUERMANS & THRO, *Brussels.*— Brussels application lace, gauze point, English point, embroidery, imitations.

640. STOCQUART BROS. *Grammont.* — Black lace, and lace articles.

641. STREHLER, J. *Brussels.*—Lace, applied to gauze; Valenciennes lace, and embroidery.

642. VAN CAULAERT-STIÉNON, E. *Brussels.*—Articles in Brussels lace; head-dress in black lace.

643. VAN DER DUSSEN, B. J. *Brussels.* —Designs for various articles in lace.

644. VANDERHAEGEN & Co. *Brussels.*— Brussels application lace.

645. VAN DER PLANCKE, SISTERS, *Courtrai.*—Valenciennes lace.

646. VANDER SMISSEN-VANDEN BOSSCHE, *Alost.*—Specimens of Brussels and Valenciennes application lace, &c.

647. VANDER SMISSEN, V. *Brussels.*— Brussels application lace, and embroidery on net; various articles.

648. VAN ROSSUM, J. B. *Hal, Brabant.* —Gauze point lace, handkerchiefs, collars, sleeves, and lappets.

649. WASHER, V. *Brussels.* — Imitation lace, and articles in lace.

650. WITTOCKX, H. *Saint-Josse-ten-Noode, near Brussels.* — Black silk lace, tunic, flounces, pelerine, &c.

## CLASS XXV.

651. BERTOU BROS. J. J. & A. P. *Liége.* —Various tanned skins.

652. BULTER, CH. *Brussels.*—A collection of articles in furs of various kinds.

653. DELMOTTE, H. *Ghent.* — Belgian hog's bristles.

654. DEVACHT, G. A. *Brussels.*—Articles in hair.

655. HANSSENS-HAP, B. *Vilvorde, near Brussels.*—Hair cloth; hog's bristles; painter's brushes.

656. HESNAULT, A. & BROTHER, *Ghent.* —Rabbit and cat skins finished; hare and rabbit fur.

657. JONNIAUX, E. & Co. *Brussels.* — Tawed skins.

658. LONCKE-HAESE, *Roulers.*—Brushes, hog's bristles.

659. MOTTIE, *Brussels.*—Wigs on a new principle.

660. SCHMITZ, F. A. *Brussels.*—Morocco dressed sheep-skins; bands of cut leather for hat-making.

661. SOMZÉ, H. JUN. *Liége.*—Brushes, hog's bristles.

662. SOMZÉ-MAHY, H. *Liége.*—Brushes, hog's bristles.

663. VERRYCK-FLEETWOOD, *Brussels.*— Perukes, and hair fronts.

## CLASS XXVI.

664. ARRETZ-WUYTS, G. *Aerschot, Brabant.*—Leather, vamps, &c.

665. BOONE, A. J. *Alost.*—Tanned, curried, and japanned skins.

666. BOONE, J. & Co. *Cureghem, near Brussels.*—Curried calf-skins.

667. BOUVY, A. *Liége.*—Calf-skins, and leather for various purposes.

668. COLLET, L. J. *Brussels.*—Leather and hides, japanned and plain, for saddlery, &c.

669. D'ANCRÉ, P. *Louvain, Brabant.*— Buenos-Ayres hides tanned but not beaten.

670. DAVID, P. *Stavelot, Liége.*—Strong sole leather.

671. DECLERQ-VANHAVERBEKE, L. *Iseghem, West Flanders.*—Tanned and curried skins, calf-skins, vamps, leather for soles.

672. DE CLIPPÈLE, CH. & Co. *Brussels.* —Engine-straps, joined on a new principle.

Patent leather straps, with permanent joints and with lace-holes.

The straps made by their method may be considered as free from all risk of lengthening—

1st. Because the leather employed receives only the quantity of oil or grease which is absolutely required to give it the necessary pliancy; and therefore it is not spongy, like that which is ordinarily used for the purpose.

2nd. Because they are made only from the centre of the hide, the flanks, neck, and other parts which are inferior in strength being carefully excluded.

3rd. And especially because they are severely tested before being handed over to the finishers.

4th. And lastly, the method of junction used with the different stripes of leather of which these straps consist causes them to be of a perfectly uniform thickness throughout their whole length, which prevents any jolt on the pullies.

673. EVERAERTS, C. *Wavre, Brabant.*— Leather: calf-skin boot-legs and fronts.

675. FETU, J. G. J. & Co. *Brussels.*— Straps for machinery, leathern hose.

676. FONTEYNE, J. *Bruges.*—Foreign and native leather for soles, calf-skin, curried horse-hide, &c.

677. HEGH, F. & DUGNIOLLE, A. *Mechlin.*—Curried goods, morocco-leather, varnished leather and articles for hat-making.

678. HOUDIN & LAMBERT, *Brussels.*— Sole-leather, calf-skins; French and Belgian military accoutrements.

679. JOREZ, L. & SON, *Brussels.*—Oilcloths, American linen cloths, gummed taffetas, varnished leathers, American cloth panels.

680. LUYTEN, C. F. & J. *Cureghem, near Brussels.*—Leather, imitation leather, varnished linens and cottons.

681. Maréchal, V. J. *Brussels.*—Harness.

682. Massange, A. *Stavelot, Liége.* — American leather for soles and engine-straps; polished native cow-hides.

683. Mouthuy, A. *Brussels.*—Engine-straps.

684. Perleau-Taziaux, Mrs. *Saint-Hubert, Luxemburgh.*—Brazil tanned hides for soles.

685. Piret-Pauchet, E. *Namur.*—Sole-leather, &c. tanned with oak bark.

687. Roussel, E. *Tournay.*—Strong leather for cylinder-packings, sole-leather.

689. Van Molle, L. L. *Lennick Saint Quentin, Brabant.*—Harness for a draught horse.

690. Van Schoonen, E. *Ghent.*—Straps for machinery on an improved principle.

## CLASS XXVII.

691. Canisius, G. *Huy, Liége.* — Silk hats, caps, &c.

692. César, A. & Co. *Brussels.*—Shoes and boots, ordinary and with wooden soles.

693. Colin Renson, H. *Brussels.*—Kid and leather gloves.

694. Deblock, Mrs. *Antwerp.*—Elastic hygienic corsets.

695. De Coster, H. *Brussels.*—Shoes and boots for various purposes.

697. Fagel-Vallaeys, B. *Ypres, West Flanders.*—Silk hats, invented by the exhibitor.

698. Frenay Bros. *Roclenge, Limburgh.* —Straw-plats, straw bonnets, and hats.

699. Hansen, F. G. *Liége.*—Boots of morocco and varnished leather.

700. Jonniaux, Ed. & Co. *Brussels.*— Kid gloves.

701. Lainglet, J. *Brussels.*—Silk corsets.

702. Leclercq, N. *Bruges.* — Boots, shoes, half-boots, fishing-boots, &c.

703. Liévain, L. *Mechlin.*—Silk and felt hats.

704. Masson-Fouquet, Mrs. A. *Brussels.*—Horse-hair corsets.

705. Soitoux, Et. *Saint Gilles, near Brussels.*—Galoches with wooden soles.

706. Somzé-Mahy, H. *Liége.*—Shoes and boots.

707. Troostenberghe, D. *Bruges.* — Leather half-boots, waterproof shoes without seams, leather gaiters in a single piece.

708. Valentyns & Vander Plaetsen, *Saint-Josse-ten-Noode, near Brussels.* — Kid gloves, " gants duchesse " gloves.

709. Vanden Bos-Poelman, *Ghent.* — Waterproof sporting boots; other boots, fancy and plain.

710. Vanderoost, M. *Brussels.*—Boot-trees and lasts; half-boots, &c.

711. Vimenet & Son, *Brussels.*—Hats of various kinds, in felt.

712. Watrigant, late Allard, *Brussels.*—Boots and shoes of all sorts, but particularly the fancy kinds.

## CLASS XXVIII.

713. Asselberghs-Lequime, *Brussels* —Letter-paper.

714. Barbier-Hanssens, L. E. *Brussels.* —Packing and wrapping-paper.

715. Brepols, Dierckx, & Son, *Turn-*

hout, *Antwerp.*—Playing-cards, fancy-paper; bound books.

716. Briard, J. H. *Brussels.* — Specimens of Bibliography.

717. Bruck, P. A. *Arlon, Luxemburgh.*

—Scientific works on the manufacture of paper.

719. CALLEWAERT BROS. *Brussels.* — Stationery.

720. DAVELUY, *Bruges.*—Playing-cards, chromo-lithographs.

The exhibitor obtained the Bronze Medal at the Brussels Exhibition of 1847; the Silver Medal at the Flanders Exhibition of 1849; and Honourable Mention at the Paris Universal Exhibition.
Patent-playing-cards of all kinds, both plain and ornamented, for exportation.
Thirty-two different patterns and qualities.—Prices from 15 to 200 francs.

721. DEMAEGT, J. *Saint-Josse-ten-Noode, near Brussels.*—Paper and pulp, made without rags or straw.

722. DESSAIN, H. *Mechlin.*—Liturgical, theological, and devotional works.

723. GLÉNISSON & SON, *Turnhout.* — Fancy paper, playing cards.

724. GOUWELOOS, A. *Brussels.*—Samples of account books, railway tickets, lithography, &c.

725. GREUSE, C. J. A. *Schaerbeek, near Brussels.*—Folio and quarto illustrated works, &c.

726. HAYEZ, M. J. F. *Brussels.*—Books.

727. HENRY, P. *Dinant, Namur.*—Pressing boards, paste-board.

728. JERVIS, G. *Brussels.*—Diagram to illustrate a new method of printing chromo-lithographically with four impressions.

729. LELONG, C. *Brussels.*—Typographical specimens.

730. MORREN, ED. *Liége.*—An horticultural and botanical review, with chromo-lithographed illustrations of flowers.

731. MUQUARDT, C. *Brussels.* — Illustrated works.

732. OLIN & DEMEURS, G. *Brussels.*—Printing and packing paper.

733. PARENT, W. & SONS, *Brussels.*—Illustrated, and other works.

734. POISSONNIEZ, J. B. *Brussels.* — Pasteboard and cards.

735. SCHAVYE, J. C. E. *Brussels.*—Ancient and modern bookbindings, designs, &c.

736. SEVEREYNS, G. M. C. *Saint-Josse-ten-Noode, near Brussels.* — Scientific and chromo-lithographic drawings, &c.

737. SOCIETY OF BELGIAN PAPER MANUFACTURES, *Basse-Wavre, Brabant.* — Writing and printing paper, paste-board.

739. TARDIF BROS. *Brussels & Paris.*—Tracing and photographic paper.

740. TIRCHER, J. B. *Brussels.*—"History of Glass Staining."

741. VAN CAMPENHOUT, *Brussels.*—Specimens of account books.

742. VAN DOOSSELAERE, J. S. *Ghent.*—Typography; wood-cuts printed on vellum, silk, and enamelled paper.

743. VAN GENECHTEN, A. *Turnhout.*—Playing-cards, fancy papers, enamelled pasteboard; typography, lithography, registers.

744. VAN VELSEN, E. F. *Mechlin.*—Illustrated, and other books.

746. WEISSENBRUCH, MISS, *Brussels.*—Books.

## CLASS XXIX.

748. BELGIAN GOVERNMENT. — Collection of educational objects, formed under the superintendence of Professor Braun.

749. BRAUN, CH. *Rivelles, Brabant.* — Pedagogical and classical works.

750. CALLEWAERT BROS. *Brussels.* —

Method of writing adopted in the Belgian schools, &c.; atlases.

751. CAMPION, J. J. *Brussels.*—"Journal of Popular Education."

752. GÉRARD, JOSEPH, *Brussels.*—Tablets, for teaching history, &c.

## CLASS XXX.

753. DAEMS-SCHOY, J. B. *Brussels.*—Furniture in sculptured wood, framings, fancy articles.

754. DEBASIN-SCHMIDT, CH. *Namur.*—Painted imitations of woods and marbles.

755. DE GOBART, EM. *Ghent.*—Dining-room furniture.

756. DEKEYN BROS. *Saint-Josse-ten-Noode, near Brussels.*—Inlaid flooring, in wood of different colours.

757. DELEÉUW-DEMARÉE, *Brussels.*—A gilt frame.

758. DERENNE, L. J. *Evelette, Namur.*—A sofa, table, chairs, &c.

759. DERUDDER, SON, & CO. *Brussels.*—Frames and console, gilt, &c.

760. GODEFROY, J. 14, *Rue Haute, Brussels,* Joiners to His Majesty the King of the Belgians; and, by appointment, maker of inlaid floors to His Majesty the Emperor of the French.—Parquetry, joinery, and carpenters' work.

Besides the establishment founded in 1820 by his father, Mr. Godefroy has, since 1853, owned and carried on the old-established manufactory of inlaid floors of Messrs. Couvert and Lucas, which received the Prize Medal at the Universal Exhibition of London in 1851.

He has many and important customers, not only in Belgium but in France, Holland, and Portugal.

His productions have merited for him in Belgium the Silver Medal of the National Industrial Exhibition of 1835, and a First-class Medal of the Universal Exhibition of Paris in 1855; and, in bestowing the latter, the international jury gave the following as the reasons of its decision:—

"Messrs. Godefroy Bros. have attained at once the very highest position, and their manufactures have met with universal approbation. The inlaid floors of Messrs. Godefroy are executed by a process which is as simple as it is ingenious, and secures a perfect solidity, while it allows the application of the richest and most varied designs. They have, besides, the advantage of being made at the factory in detached portions, which can be fixed in their places in a very short space of time, and without any difficulty. Messrs. Godefroy have also exhibited a species of door, in which all the ordinary iron work is dispensed with, being replaced merely by a swing movement, which is invisible externally. This principle possesses many advantages, and is a real improvement, since it gives to doors a more pleasing appearance, and greatly simplifies their fittings. Moreover it can be applied to wood of any thickness.

"The uncommon elegance of design and tasteful richness of decoration which distinguish the works of Messrs. Godefroy Bros. have been appreciated by the international jury, which, in awarding to them a Medal of the first-class, bestows upon them the recompense justly due to the efforts they have made, and the extraordinary results which they have attained. —Report of the Mixed International Jury (Paris, 1856, page 1131.

761. GOYERS BROS. J. & H. *Louvain.*—A pulpit in the Gothic style (14th century.)

762. HODY, J. J. *Aubel, Liége.*—Piece of furniture serving four purposes: wash-hand stand, dressing-table, praying-desk, and writing-table.

763. LEARCH, A. *Brussels.*—Panels, in imitation of ancient leather.

765. LUPPENS, H. *Brussels.* — Clocks, vases, bronze model of a monument.

767. MARLIER, *Brussels.*—Buffet or cupboard.

768. OLIN & DEMEURS, *Brussels.*—Paper-hangings.

769. PEETERS-VIERING, J. *Mechlin.*—Dining-room furniture in antique style.

770. POHLMANN, G. & DALK, A. *Brussels.*—Mouldings for panel and frame works, specimens of frames.

771. REISSE, CH. *Brussels.* — Gothic chimney-piece and clock, in carved oak.

772. RYCKERS, E. & SON, *Brussels.*—Buffet in the style of Louis XIII.

773. VAN DEN BRANDE BROS. *Mechlin.*—Carved and inlaid drawing-room furniture.

774. VAN DEN BROECK, D. *Brussels.*—A gilt wood toilet-console, in the style of Louis XV.

775. VAN HOOL, J. F. *Antwerp.*—A sculptured altar, and a crucifix.

776. WARIN, J. *Brussels.*—A book-case in imitation ebony.

777. WATRISSE, L. *Dinant.* — Round claw-tables of Belgian marble.

778. VAN DE LAER, P. *Brussels.*—Panels, specimens of stained and gilt papers.

779. WAHLEN-FIERLANTS, MRS. *Brussels.*—Paper-hangings.

———

863. GERMON-DIDIET, A. *Brussels.*—Artificial flowers in paper and muslin, wax fruits.

## CLASS XXXI.

781. BAYARD, M. *Herstal, Liége.*—Bolts, screw-wrenches, compasses, squares, iron fittings for carriages, &c.

782. BECQUET BROS. *Brussels.*—Samples of forged nails used in different countries.

783. BOGAERTS, ALP. *Antwerp.*—Works of art in bronze.

784. BROERMANN, F. G. SEN. *Brussels.* —Iron bedsteads, an aviary flower-stand, and garden chairs.

785. CANIVEZ, J. B. *Ath, Hainault.*— Zinc letters in relief, gilt, &c.

786. CARLIER, F. *Chênée, Liége.* — Wrought-iron anvil.

787. CHAUDOIR, CH. & H. *Liége.*—Unsoldered copper tubes for locomotive engine boilers, steam-boats, &c.

788. COMPANY FOR THE MANUFACTURE OF BRONZE AND ZINC, *Brussels.*—Statues, works of art, &c.

789. DARDENNE, T. & SON, *Chimay, Hainault.*—Screw-iron, for saddle and harness horses.

790. DAWANS, A. & ORBAN, H. *Liége.*— Nails of various kinds.

791. DE BAVAY, P. & Co. *Brussels.*— Common iron, iron for wire, nails, &c.

793. DELLOYE-MASSON, E. & Co.—*Laeken, near Brussels.*—Forged iron tinned and galvanized, enamelled cast-iron.

794. DELLOYE-MATHIEU, C. *Huy, Liége.* —Sheet-iron, polished and unpolished.

795. FABRIQUE DE FER D'OUGRÉE, *Seraing, Liége.*—Sheet-iron, specimens of iron.

796. - FAUCONIER-DELIRE, WIDOW, *Châtelet, Hainault.*—Hand-wrought iron nails.

797. FRAIGNEUX BROS. *Liége.* — Fire and thief proof safes.

798. GAILLIARD, L. C. C. *Brussels.*— Models in chiseled and chased metal.

799. GÉRARD, H. & DIDIER, *Bouillon, Luxemburgh.*—Hooks, hinges; iron-work, for buildings and furniture.

800. GOFFIN, C. & J. *Brussels.*—Cast-iron tubes.

801. GROTHAUS BROS. *Gosselies, Hainault.*—Wrought-iron nails.

802. HOORICKX, G. *Brussels.*—Iron safes.

803. LALMAND-LEFORT, F. J. *Bothey, Namur.*—Iron safe, with invisible key-hole.

804. LAMAL, P. & Co. *Brussels.*—Lead and tin pipes.

805. LAMBERT, W. G. J. *Charleroi.*— Rivets of all kinds.

807. LECHERF, IS. DE, *Brussels.*—Bronze articles.

808. LESAGE, V. *Saint-Josse-ten-Noode, near Brussels.*—Nails, rivets, springs, telegraph-wires.

809. MATHYS-DECLERCK, *Brussels.*—A lock, having 629 fixed and 414 moveable pieces, and a key with 84 different divisions.

812. NICAISE, P. & N. *Marcinelle, near Charleroi.*—Bolts, nuts, &c.

814. RAIKEM-VERDBOIS, H. J. *Liége.*— Sheet-iron, polished.

815. REMACLE, J. & PERARD, *Liége.*— Sheet-iron.

816. SIÉRON, L. *Brussels.*—Iron, copper, and zinc nails.

817. SOCIÉTÉ ANONYME DE CORPHALIE, *Antheit, near Huy.*—Sheet zinc.

818. SOCIÉTÉ ANONYME DES FORGES DE LA PROVIDENCE, *Marchienne au-Pont, Hainault.*—A collection of iron, manufactured specially for building purposes.

819. SOCIÉTIÉ ANONYME DES HAUTS-FOURNEAUX ET LAMINOIRS DE MONTIGNY-SUR-SAMBRE, *Montigny, near Charleroi.*— Round bar-iron, for various purposes; rolled iron for doors, windows, &c.

820. SOCIÉTÉ ANONYME DES HAUTS-FOURNEAUX, USINES, ET CHARBONNAGES DE CHÂTELINEAU, *Châtelineau, near Charleroi.* —Rolled iron.

821. SOCIÉTÉ ANONYME DES HAUTS-FOURNEAUX, USINES, ET CHARBONNAGES DE MARCINELLE ET COUILLET, *Couillet, near Charleroi.*—Bar and sheet iron.

822. SOCIÉTIÉ ANONYME DES HAUTS-FOURNEAUX, USINES, ET CHARBONNAGES DE

SCLESSIN, *Tilleur, Liége.* — Rolled and wrought-iron.

823. SOCIÉTIÉ DES FORGES ET LAMINOIRS DE L'HEURE, *Marchienne-au-Pont, Hainault.*—Rolled iron, sheet-iron.

824. SOCIÉTIÉ DES LAMINOIRS DE HAUTPRÉ, *Ougrée, Liége.*—Sheet-iron.

825. SOCIETY FOR MANUFACTURING OF

NAILS BY MACHINE, *Fontaine-l'Evêque, Hainault.*—Machine-made nails and tacks.

826. TREMOROUX BROS. & DE BURLET, *Saint Gilles, near Brussels.*—Forged iron, household articles tinned and glazed.

827. VANDERMILEN, CH. *Brussels.*—Iron safes.

828. VAN NEUSS, M. *Brussels.*—A grilling oven.

# CLASS XXXII.

829. BOMBOIR, G. *Houffalize, Luxemburgh.*—Sickles, scythes, axes, &c.

831. MONNOYER, P. J. *Namur.*—Knives, razors, scissors, &c.

832. NOTTE, F. *Gembloux, Namur.*—Cutlery of all kinds.

833. OLIVIER, A. *Enghien, Hainault.*

—Cast-steel hammers for dressing millstones.

834. ROBERT, J. & DE LAMBERT, *Liége.* —Files for watchmakers, jewellers, armourers, &c.; gravers.

835. SOCIÉTIÉ ANONYME DES HAUTS-FOURNEAUX ET LAMINOIRS DE MONTIGNY-SUR-SAMBRE, *near Charleroi.* — Knife of puddled steel, for paper-makers.

# CLASS XXXIII.

836. DEHIN, J. J. *Liége.*—A silver monstrance.

837. DUFOUR, J. & BROTHER, *Brussels.* —Various articles of jewelry; tea-service, style of Louis XVI., &c.

838. GOUVERNEUR, C. & SON, *Brussels.*

—Gold and silver wire; lace for Belgian military purposes.

839. HOKA, A. *Liége.*—Engraved bracelets, pins, brooches, &c.

840. PETERS, L. *Tongres, Limburgh.*— Silver-gilt pyxes and communion-cups, in the Gothic style; fire-gilt articles, &c.

# CLASS XXXIV.

841. ANDRIS-LAMBERT & Co. *Marchienne-au-Pont, Hainault.*—Window-glass of various kinds.

842. BENNERT & BIVORT, *Jumet, Hainault.*—Bottles; window-glass.

844. BOURDON, J. & Co. *Chénée and Liége.*—Wine and liqueur bottles, demijohns; patent wickered glass pots for butter, syrup, honey, &c.; patent glass milk-pans.

845. CAPPELLEMANS, J. B. SEN., BEBY, A. & Co. *Brussels and Saint-Vaast, Hainault.*—Bottles; window-glass, and glass tiles.

846. CAPRONNIER, *Brussels.* — Stained glass window, for Howden Church, Yorkshire.

847. DAUBRESSE BROS. *La Louvière, Hainault.*—Window-glass.

848. De Dorlodot de Moriamé, L. & Son, *Lodelinsart, Hainault.*—Window-glass.

849. Floreffe Co. *Floreffe, near Namur.* —Silvered and unsilvered plate-glass; window-glass and flint-glass.

851. Jambers, J. *Liége.* — Engravings on glass.

852. Jonet, D. & Co. *Charleroi.*—Window-glass, coloured, polished, engraved, &c.

853. Ledoux, J. B. & C. *Jumet, Hainault.*—Window-glass.

854. Mondron, J. *Lodelinsart, Hainault.* —Window-glass of various thicknesses, &c.

855. Société Anonyme d'Herbatte, *Herbatte, near Namur.*—One complete cut crystal service; various articles in crystal, demi-crystal, and gobeletterie, plain, cut, and coloured.

John Berry and Co., 1 New Broad Street, City, London E.C., are the sole agents of the above company, and will give any information required. Illustrated catalogues of the society's manufactures may be obtained at their office.

856. Société des Manufactures de Glaces, Verres a Vitres, Cristaux et Gobeleteries, *Brussels.*—Plate and window glass, bottles, drinking-glasses, silvered glass; a sheet of glass sixteen and a half feet high, and nearly ten feet wide, &c.

857. Vanderpoorten, J. L. *Molenbeek-Saint-Jean, near Brussels.*—Painted church windows.

# CLASS XXXV.

858. Barth, D. *Andenne, Namur.*—Clay smoking-pipes.

859. Boch Bros. *Kéramis, Hainault.*— Coarse and fine crockery-ware, plain and ornamented.

860. Cappellemans, J. B. Sen. *Hal.*— Table, coffee and tea services, and other articles.

861. De Fuisseaux, Mrs. *Baudour, near Mons.*—Articles of crockery and porcelain.

862. Demol, *Brussels.*—Painted crockery-ware and porcelain.

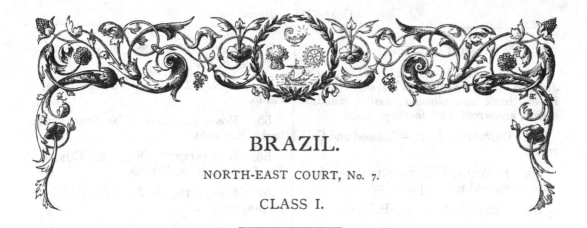

# BRAZIL.

## CLASS I.

1. BURLAMAQUE, DR. F. L. C.—A collection of gold, diamonds, emeralds, topazes, and various Brazilian minerals.

2. TASSARA, A.—Slates from Minas Geraes.

3. MOULEVADE.—Asbestos and kaolin, from Minas Geraes.

4. COPPET.—Limestone, &c., from Rio Janeiro.

5. BARBACENA, VISCOUNT — Coal from Laguna, province of Santa Catharina.

6. LEÃO, J. A. F.—Malachite, iron ores, coal, and various other minerals.

7. LUZ, M. M. DA.—Diamond, in gravel, from Minas Geraes.

8. SOUZA, M. S. DA.—Diamond, in gravel, from Diamantina, Minas Geraes.

9. TEXIER, J. I. JUN.—Sandstone from the banks of the Cahy, S. Pedro.

10. ROHAN, H. DE B.—Amethysts from S. Paulo.

11. BELLO, O.—Quartz crystals from San Pedro.

12. BOULIECH, G.—Jasper and kaolin from S. Pedro; minerals from the Jaquaro Mines, S. Pedro.

13. M. J. P.—Lignite from Ouro Preto, Minas Geraes.

14. CARVALHO. J. P. D. DE.—Gold, in quartz, from Ouro Preto, Minas Geraes.

15. ANCHIETA, J. DE.—Iron ore from Cocaes, Minas Geraes.

16. REIS, J. M. DOS. — Quartz from Goyas.

## CLASS II.

17. PECKOLT, T. *Cantagallo, Rio Janeiro.* —Vegetable acids, and essential extracts from native plants.

18. SANTOS, M. E. C. DOS, & SON, *Rio Janeiro.*—Organic and inorganic chemicals.

19. MAGALHÃES, M. DA C. — Ipecacuanha.

20. GARY, M. M. ALEIXO, & Co. *Rio Janeiro.*—A collection of organic and inorganic chemicals.

21. CASTRO, M. M. & Mendes, *Nitherohy, Rio Janeiro.*—Chemical products.

22. BLANC, J. F. A. *Rio Janeiro.* — Chemical products.

## CLASS III.

23. PECKOLT, T. *Cantagallo, Rio Janeiro.* --Starch from various plants.

24. PIRAQUARA, BARON DE, *Rio Janeiro.* —Sugar-cane rum ; coffee.

25. HUET, D. D. H. *Rio Janeiro.*— Rum.

26. COATS, R. *Rio Janeiro.* — Rum, Hollands, and loaf sugar.

27. WENTEN, J. R.—Rum.

28. FARO, J. P. D. & J. D. DE, *Rio Janeiro.*—Orange rum, and loaf sugar; coffee in the husk and cleaned; maize, mandioc starch, arrowroot, and Jacatupè flour.

29. CALDERON, L. B.—Aniseed and Cajá rum.

30. RABELLO, J. H. DA SILVA.—Cashew wine, rum, and pine-apple syrup.

31. PEREIRA, A. J. G.—Hollands.

32. RIBEIRO, M. R. J.—Rum.

33. HEWLER, S. *Campo, Rio Janeiro.*—Anhydrous alcohol, and sugar-cane vinegar.

34. MARCHADO & REDONDO, *Rio Janeiro.*—Alcohol; white and brown vinegar.

35. COUTINHO, J. DE A.—Alçohol.

36. COSTA, F. G. DA, & SONS.—Paddy, rice, coffee, and tapioca.

37. GOMES, A. & CUNHA, A. DA.—Rice.

38. NITHEROHY SUGAR-REFINERY & DISTILLERY Co.—Refined sugar.

39. DOUS DE JULHO FACTORY, *Bahia.*—Crystallized sugar.

40. SOUZA, S. DE, & SILVA, *Pernambuco.*—Refined and loaf sugar.

41. MONTEIRO MANUFACTORY, *Pernambuco.*—White sugar.

42. GEREMOABO, T. P. *Bahia.*—White sugar.

43. LOURENÇO, BARON S. *Bahia.*—Refined sugar.

44. CARVALHO, J. P. D. DE, *Gavia, Rio de Janeiro.*—Specimens of coffee.

45. DIAS, H. J. *Rio Janeiro.*—Coffee.

46. CRUZ, J. B. DA, *Cantagallo, Rio Janeiro.*—Coffee.

47. FARO, A. P. DE, *Rio Janeiro.*—Coffee.

48. TAVARES, J. P. *Itaguahy, Rio Janeiro.*—Coffee.

49. ANDRADE, F. DE P.—Coffee.

50. ALMEIDE, DR. C. M. DE.—Coffee.

51. MUNIZ, H. F.—White and yellow carimam, prepared from the mandioc; mandioc starch.

52. TREASURER'S ESTATE, *Minas Geraes.*—Varieties of tea.

53. BOTANIC GARDENS, *Minas Geraes.*—Varieties of tea.

54. SILVA, C. I. DA.—Tea from Itú, S. Paulo.

55. ROSA, J. C. DA.—Tea from Constitucâo, S. Paulo.

56. BITTANCOURT, M. J. DA CUNHA.—Tea from Coritiba, Paraná.

57. FROUGETH, Dr. J. F. *Rio Janeiro.*—Paquequer tea.

58. SILVA, J. J. DA, *S. Roque, S. Paulo.*—Green tea.

59. AMARAL, J. V. DE ARRUDA, *S. Paulo.*—Scented green tea.

60. BLANC, J. F. A. *Rio Janeiro.*—Chocolate.

61. BERRINI, G. *Rio Janeiro.*—Chocolate.

62. SRA. V. CASTAGNIER, *Rio Janeiro.*—Preserves.

63. DEROCHE & Co. *Rio Janeiro.*—Pine-apple preserve.

64. VASCONCELLOS, F. P. DE, *Bahia.*—Preserved vegetables.

65. FREITAS, J. DA COSTA, *Rio Janeiro.*—Mandioc starch.

66. SOUZA, A. C. DE.—Tapioca.

67. FURTADO, J. C.—Tapioca.

68. OLIVEIRA, V. J. DE.—Wheaten flour.

69. AZEVEDO, J. F. DE.—Thirty varieties of Theresopolis beans, from the province of Rio Janeiro.

70. LAGOS, M. F.—Eleven varieties of Theresopolis beans, from the province of Ceará.

71. NATIONAL FACTORY, *Gamboa, Rio Janeiro.*—Various liqueurs.

72. FERREIRA, A. J. BRAGA, & ISMÃO, *Rio Janeiro,*—Liqueurs.

73. GOMEZ, A. J. *Rio Janeiro.*—Liqueurs, and barley wine.

74. BASTOS, A. J. G. P. *Rio Janeiro.*—Liqueurs and syrups.

75. TAVEIRA, A. M.—Specimens of Mandioc starch.

76. LEÃO, J. C. DE M. JUN.—Mandioc starch.

77. GAMBOA FACTORY, *Rio Janeiro.*—Rose vinegar.

78. BRASIL, P. A. *Rio Janeiro.*—Coloured vinegar.

79. LOBO, J. F.—Brazilian wines.

80. RENDON, J. A. DE T.—Sweet grape wine.

81. AGUIAR, A. P. DE.—Orange wine.

82. BITTENCOURT, J. DE.—Orange wine.

83. PINHEIRO, J. H.—Mogy das Cruzes wine.

84. MARSE, C. *S. Leopoldo, S. Pedro.*—Grape wine.

85. BAUN & CASTANÊRA, *Rio Janeiro.*—Cigars.

86. PALOS, D. *Rio Janeiro.*—Cigars.

87. MONTES, J. J. & Co.—*Rio Janeiro.*—Cigars.

88. SOUZA FLORES, J. J. DE, *Rio Janeiro.*—Cigars.

89. PALOS, P. *Rio Janeiro.*—Cigars.

90. MACHADO, F. A.—Cigarettes.

91. GONÇALVES, J.—Cigarettes.

92. PERES, S.—Cigarettes.

93. SILVEIRA, P. *Rio Janeiro.*—Snuff.

94. CORDEIRO, J. P. *Rio Janeiro.*—Snuff.

95. JAGUARARY, BARON DE.—Leaf tobacco.

## CLASS IV.

96. PECKOLT, S. *Cantagallo.* — Fruits, seeds, roots, barks, vegetable fibres, gums, resins and dye stuffs, vegetable oils, &c.

97. PIMPARDE, H. *Rio Janeiro.*—Aloe-water and oil of aloes.

98. PINTO, J. DE A.—Indigo from Pernambuco.

99. M. C. O. *Rio Janeiro.*—Tallow oil.

100. STRAUSS, H. A. — Manufactured Indian rubber.

101. HERBST & ROSSITER, *Rio Janeiro.*—Mexican variety of vanilla.

102. CASANOVA, —. —Charcoal and potash, from the coffee husk.

103. CARNEIRO, J. M. DOS S.—Wax.

104. SIQUEIROS, M. J. P. DE.—Wax.

105. LAGE, M. P. F.—Wax.

106. RAMOS, A. DA SILVA.—Wax.

107. ALBUQUERQUE, G. A. G. DE.—Black wax.

108. LAGOS, M. F.—Barks, and leaf tobacco: a collection of bees with their wax and honey.

109. GONÇALVES, J. A. *Rio Janeiro.* — Extract of Brazil-wood.

110. LEÃO, J. A. F.—Leaf tobacco.

111. SOARES, J. J.—Leaf tobacco.

112. STEARINE CANDLE COMPANY, *Rio Janeiro.*—Glycerine, soap, and candles.

113. BRELAZ, L. *Pará.*—Vegetable oils.

114. ARAUJO, J. A. DE, *Rio Janeiro.*—Oils.

115. MARIA, S., *S. Pedro.*—Castor oil.

116. ARAUJO, J. M. DE, & Co. *Penedo, Alagoas.*—Vegetable oils.

117. MAUÁ, BARON DE, CRUZ, M. D. DA, CRUZ, J. B. & OTHERS.—Woods of Brazil, comprising 410 varieties.

118. BRUSQUE, F. C. DE A.—Mosaic of the woods of the province of Pará.

119. MUNICIPAL CHAMBER OF DESTERRO.—Mosaic of the woods of the province of Santa Catharina.

120. CARVALHO, A. L. P. *Rio Janeiro.*—Soap.

121. MONTEIRO, J. F. C. *Aracaty, Ceará.*—Soap.

122. ARÊDE, J. B. DE, & Co. *Pará.*—Soap.

123. REGO, DR. P. DA S. *Bahia.*—Soap.

124. MARTELET, R. & Co. *Rio Janeiro.*—Soap.

125. BARCELLOS, A. P. S. *Pernambuco.* Carnauba palm-oil candles.

126. ARAUJO & IRMÃO, *Rio Janeiro.*—Tallow candles.

## CLASS VII.

127. SILVA, J. Y. DA.—A blacksmith's bellows.

## CLASS IX.

128. SANTOS, M. C. DOS, *Rio Janeiro.*—Agricultural implements.

## CLASS XI.

129. REAL, C. & PINTO, *Rio Janeiro.*—Embroidered scarf, and gold epaulettes.

130. MILITARY ARSENAL, *Rio Janeiro.*—Carbine, pistol, and Minié rifle.

131. MILITARY ARSENAL, *Pernambuco.*—A pistol.

## CLASS XII.

132. NAVAL ARSENAL, *Rio Janeiro.*—Models of ships, &c.

133. MIERS IRMÃO, & MAYLOR, *Rio Janeiro.*—Models of ships, &c.

134. PONTA DA ARÉA COMPANY.—Models of ships, &c.

135. NAVAL ARSENAL, *Pernambuco.*—Model of a ship.

## CLASS XIII.

136. MASCARENHAS, A. M. DE, *Rio Janeiro.*—A ship's compass.

137. REIS, J. M. DOS, *Rio Janeiro.*—Spectacles, reading glasses, &c.

## CLASS XIV.

138. PACHECO, J. I. *Rio Janeiro.*—Photographic portraits of the Imperial Family of Brazil, &c.

139. DAER, —. *Rio Janeiro.*—Photographic views of the Botanic Gardens, Rio Janeiro.

## CLASS XV.

140. GONDOLO & Co. *Rio Janeiro.*—A gold watch.

## CLASS XVII.

141. BLANCHARD, —. *Rio Janeiro.*—A set of surgical instruments.

## CLASS XVIII.

142. PEREIRA, M. N. B.— White and yellow cotton.

143. RODRIGUES, C. J. A.—Raw cotton from Rio Janeiro.

144. MELLO, L. C. DE.—Raw cotton from Pernambuco.

145. REZENDE, L. R. DE S.—Cotton in the pod from Alagôas.

146. MASCARENHAS, D. L. DE A.—Cotton counterpanes.

147. ALBUQUERQUE, A. P. DE.—Cotton piece goods from Todos os Santos, Bahia.

148. USMAR, J. C. M. DE.—Cotton piece goods from Andarahy, Rio Janeiro.

149. FILGUEIRAS, J. A. DE A. & Co.—Cotton piece goods from Magé, Rio Janeiro.

150. ANDRADE, J. DAS C.—Cotton piece goods from Passa Tempo, Minas Geraes.

151. COSTA, M. DE A. — Cotton piece goods from Campo Grande, Rio Janeiro.

152. JUMBEBA, F. R. DA C.—Cotton piece goods from Brumado de Suassuhy, Rio Janeiro.

153. PADUA, F. N. N. DE.—Cotton piece goods from Queluz, Minas Geraes.

154. LAGOS, M. F.—Cotton piece goods from Crato, Ceará.

## CLASS XIX.

155. BARBACENA, VISCOUNT DE.—Guaxima from Pilar, Rio Janeiro.

156. BURLAMAQUE, GENERAL F. L. C.—Aloe fibre cloth.

157. MOTTA, F. L. DA, *Rio Janeiro.*—Aloe fibre cloth, embroidered with gold.

## CLASS XX.

158. UBATUBA, DR. M. P. DA S., ARAUJO, D. C. R. DE, and CAPBDEBILA, V. F.— Cocoons; raw and manufactured silks from the province I. Pedro.

## CLASS XXI.

159. SILVA, S. V. DA.—Woollen counterpanes from Minas Geraes.

160. PADUA, F. N. N. DE. — Woollen counterpanes from the same province.

## CLASS XXV.

161. PINGARILHO, J. M. DA S.—Skin of the red socuryú snake, tanned.

162. THE IMPERIAL MORDOMIA.—Feather flowers.

## CLASS XXVI.

163. GUIMARAES, L. & SOUZA, *Rio Janeiro.*—Coloured morocco and other leathers.

164. ROMANN, BRET, & KILIAN, *Rio Janeiro.* — Coloured morocco and other leathers.

165. GUIMARAES, C. J. DE A. *Rio Janeiro.*—A saddle.

166. GUIMARAES, A. DE A. *Rio Janeiro.*—A saddle.

167. JANSEN, G. *Rio Janeiro.*—A saddle.

168. PEIXE, G. DE S. *Pernambuco.*—A saddle.

169. SILVA, J. M. DA, & Co. *Rio Janeiro.*—A saddle.

170. DIAS, J. R. *Pernambuco.*—A saddle.

171. GUIMARAES, T. T. DE A. *Rio Janeiro.*—A saddle.

## CLASS XXVII.

172. MURIAMÉ, A. M. *Rio Janeiro.* — Boots.

173. CARREIRO, J. C. *Rio Janeiro.* — Boots.

174. PINGARILHO, J. M. DA S. *Pará.*—Boots.

175. QUEIRÓS, J. M. DE, *Rio Janeiro.*—Boots.

176. CAMPAS, J. & SON, *Rio Janeiro.*—Boots.

177. GUILHERME, P. A. & SON. *Rio Janeiro.*—Boots.

178. THER, P. *Porto Alegre, S. Pedro.* Boots.

179. PINHEIRO, J. DE L. *Rio Janeiro.*—Felt hats.

180. COSTA, F. A. DA, & Co. *Rio Janeiro*—Felt hats.

181. GOMES, V. J. & Co. *Rio Janeiro.*—Felt hats.

182. BARCELLOS & VIANNA, *Rio Janeiro.*—Felt hats, made from hares' fur.

183. ALMEIDA, R. A. DE, *Rio Janeiro.*—Felt hats.

184. CHASTEL & Co. *Rio Janeiro.*—Silk hats.

185. MELLO & ALMEIDA, *Rio Janeiro.*—Silk hats.

186. CASTRO, P. DE, & Co. *Rio Janeiro.*—Silk hats.

## CLASS XXVIII.

187. RENSBURG, E. *Rio Janeiro.* — An Atlas and Report on the S. Francisco river.

188. LEUZINGER, G. *Rio Janeiro.*—Merchants' ledgers, &c.

189. LAEMMERT, E. & H. *Rio Janeiro.*—Merchants' office books.

190. LOMBAERTS, —. —Merchants' office books.

191. OLIVEIRA, M. J. DE, JUN. *Rio Janeiro.*—Vegetable writing ink.

192. AZEVEDO, J. V. R. DE, *Rio Janeiro.* Writing ink.

# CLASS XXX.

193. JOHN, A. *Santa· Isabel, Espirito Santo.*—Workbox, of various Brazilian woods.

194. ZANCHI, —, *Rio Janeiro.*—Workbox of various Brazilian woods.

195. CAPÓTE, J. A. *Rio Janeiro.*—An inlaid workbox.

196. NASCENTES DE AZAMBUJA, B. A.—An inlaid work-table.

197. VALLIM, M. DE A.—A work-table of Candêa wood.

198. QUINTANILHA, B. *Rio Janeiro.*—Flowers made of insects' wings.

199. GARCIA, C. A. G. *Rio Janeiro.*—Paper-hangings.

200. PEREIRA, G. G. *Rio Janeiro.*—Printed paper-hangings.

201. HORN, F. DE.—Mosaic of Brazilian woods.

202. G. B. S.—A mosaic of flooring woods.

203. LANDOT, J. B. S. *Rio Janeiro.*—A mosaic of Vinhatico wood.

204. LEITE, J. A. JUN.—A vase of shell-flowers.

205. FERRAZ, A. M. DA S.—A vase of fish scale and shell flowers.

206. SILVA, E. F. DA. — A vase with artificial rose tree.

# CLASS XXXI.

207. ANDRADE, A. R. DE.—A bar of wrought iron from Minas Geraes.

208. BARROS, L. A. M. DE.—A bar of wrought iron from Congonhas do Campo, Minas Geraes.

209. COTTA, M. P.—A bar of wrought iron from Antonio Pereira, Minas Geraes.

210. ANDRADE, J. C. DA C.—A bar of wrought iron from Itabira de Mato Dentro, Minas Geraes.

211. MOULEVADE, J. A. DE.—A bar of iron, from the same locality.

212. SANTOS, M. C. DOS. *Rio Janeiro.*—A lock.

213. FERREIRA, J. V. *Rio Janeiro.*—A secret door-lock.

214. URBACH, A. *Rio Janeiro.* — Cast-iron medallion.

215. HARGREAVES, —. *Rio Janeiro.* — Cast-iron medallion.

216. SANTOS, M. C. DOS, *Rio Janeiro.*—Cast-iron ornaments and panel.

217. BEUCHON, —. *Rio Janeiro.*—Nails, screws, &c.

# CLASS XXXII.

218. PRADINES, J. *Pernambuco.*—Knives, and other cutlery.

219. BLANCHARD.—Specimens of razors.

# CLASS XXXIII.

220. DOMINGOS, FARINI, & IRMÃO, *Rio Janeiro.*—The Imperial arms of Brazil; a silver medallion.

221. LOPES, A. J.—Gold lace.

222. REIS, J. M. DOS.—Eye-glass, the property of His Imperial Majesty.

## CLASS XXXIV.

223.   FONSECA, S. DA E, SÁ, *Rio Janeiro.* Engraved flint-glass.

224.   CASTRO, PAES, & Co. *Praia, Formosa, Rio Janeiro.*—Ornamental glass.

225.   LOMBOS, M. & ROQUE, S. *Rio Janeiro.*—Ornamental glass.

## CLASS XXXV.

226.   ESBERARD, F. *Rio Janeiro.* — Earthenware.

227.   SARVILLO, P. A. & Co. *Nitherohy, Rio Janeiro.*—Tiles, bricks, and pipes.

228.   LAGE, M. P. F.—Tiles from the União e Industria Co.

229.   FEREIRA, J. S. *Bahia.*—Tiles.

## CLASS XXXVI.

230.   FORESTE, A. *Rio Janeiro.*—Jewelry-cases.

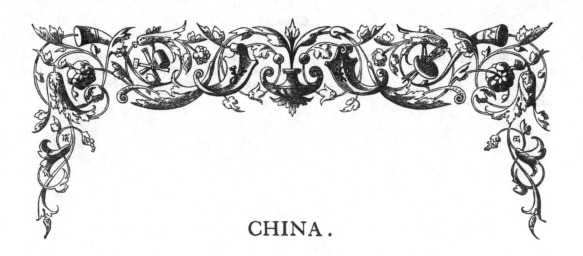

# CHINA.

NAVE, NORTH SIDE, NEAR HORTICULTURAL
SOCIETY'S ENTRANCE.

1. MICHEL, GEN. SIR JOHN.—A carved screen, from behind the Emperor's throne in the Summer Palace ; jars.

2. COPLAND, C.—Backgammon board.

3. DUNCANSON, E. J.—Chinese paper, and manufactured goods.

4. SWINHOE, R. *H.M. Vice-Consul Tai-wan-foo, Formosa.* — Various articles from Formosa.

5. FORREST, R. J. *H.M. Acting Vice-Consul at Kiukiang.*—Autograph of first rebel chief ; coins made by the rebel authorities at Nanking.

6. LEGGE, REV. J., D.D.—Specimens of Chinese types.

7. MERCER, W. T.—Two screens.

8. KANE, DR. — Porcelain vases and stands.

9. ROWLAND, J. C. — A table, vases, bronzes

10. RENNIE, W. H.—Bath tub in porphyry.

11. JACOB, CAPT. 99*th Regt.* — Sundry articles.

12. WALKINSHAW, W.—A pagoda-stand, model of scaffolding, and carved ivory ball.

13. MURRAY, DR. *Chairman of Hong Kong Committee.*—Silver vase, and ivory articles.

14. MALCOLM, CAPT. C. D., R.E. — Carved ivory chessmen.

15. HOACHING, MR. — Carved ivory casket.

16. FLETCHER, ANG.—Jade ornaments and Chinese medicines.

17. MONTEIRO, MR.—Jewelled cups.

18. AN ARTILLERY OFFICER.—Ancient bronze incense burner and two candlesticks.

19. TAIT, CAPT. W.—A human skull richly set in gold ; reported to be the skull of Confucius.

20. HEWITT, W. & CO. 18, *Fenchurch Street, E.C.*—Mandarin jars ; tea, &c. services, enamels, &c.

21. HALL, CAPT. R.N.—Chinese pictures.

22. CAMPBELL, P.—A jade-stone sceptre, bowl, vase, &c.

23. OLDING, J. A.—The Emperor's jade-stone seal, used to stamp documents certifying literary proficiency.

24. ROSARIO, R. A. — Pharmaceutical articles.

25. CAREY, H. W.—Chinese drugs, and miscellaneous articles.

26. RÉMI, SCHMIDT, & Co.—Raw silks, vases, bronzes, lacquer ware, cups of jade and agate, carpets from the Summer Palace, &c.

# COSTA RICA.

## NORTH-EAST COURT, No. 1.

1.   EXHIBITORS, THE GOVERNMENT OF COSTA RICA :—

## CLASS I.

2.   Ores of gold, silver, copper, and lead, from various mines; gold, after separation of the mercury, and after having been melted; volcanic sulphur.

## CLASS II.

3.   Sarsaparilla; balsams; medicinal and other roots; gums; medical and chemical substances.

## CLASS III.

4.   Fruits, beans, rice, coffee, sugar, tobacco, cacao, rum, &c.

## CLASS IV.

5.   Twine, &c. made of the Agave leaf, and other vegetable fibres; dye-stuffs, and matters used for tanning; nuts for the production of oils, &c.; numerous specimens of indigenous woods; tortoise and mother-of-pearl shells; cotton, caoutchouc, ocre, &c.

## CLASS XXII.

6.   A tule mat.

## CLASS XXV.

7.   Bird skins.

## CLASS XXVI.

8.   Tanned tapir skins; otter and jaguar skins; whips of deer skin and tapir skin.

## CLASS XXVII.

9.   Articles manufactured with English yarn; palm-leaf hats.

## CLASS XXXIII.

10.   Gold, silver, and filigree work.

## CLASS XXXVI.

11.   Cigar-cases and purses, made of pita and tule; halters of Cabuga; calabashes; cocoa-nut goblet.

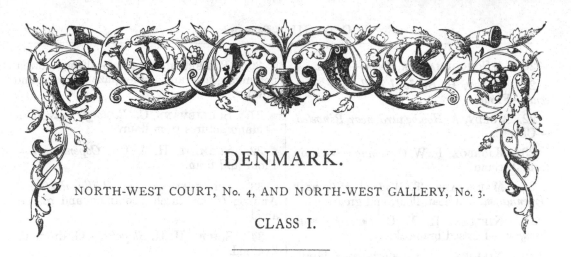

# DENMARK.

NORTH-WEST COURT, No. 4, AND NORTH-WEST GALLERY, No. 3.

## CLASS I.

1. FORCHAMMER, G. *Copenhagen.* — Minerals from Denmark and her colonies.

2. SOUTH GREENLAND MINING CO.—

Tin, copper, lead, cryolite, and other minerals.

3. WEBER, TH. & CO. *Copenhagen.* — Cryolite and its products.

## CLASS II.

### SUB-CLASS A.

4. BENZON, A. *Copenhagen.*—Chemical, photographic, technical, and economical preparations and articles.

5. FREUDENREICH, A. G. *Flensborg.*— Chrome-colours.

6. HEYMANN & RÓNNING, *Sophiehaab, near Copenhagen.* — Chemical preparations and colours.

7. KEDENBERG & BLECKER, *Uetersen.*— Manures, superphosphate, and crushed bones.

8. MEIER, F. C. S. *Copenhagen.*—Linseed oil, varnishes, and drying extracts.

9. MÖLLER, H. C. *Kiel.*—Crushed bones, chemically clean and pulverized.

13. NISSEN & VOLKENS, *Heide.* — Bituminous sand (raw produce), asphalte tar (half manufactured), solar oil, asphalte oil, and mineral asphalte.

10. WEIL, M. & CO. *Copenhagen.*—Phosphate of lime.

### SUB-CLASS B.

13. BENZON, A. *Copenhagen.*—Pharmaceutical preparations.

14. ERIKSEN, J. *Copenhagen.*—Artificial mineral-waters.

15. RIISE, A. H. *St. Thomas.*—Oil of lemon-grass (of *Andropogon Citratum,* D.C.).

16. ROSENBORG MANUFACTORY OF MINERAL WATERS, *Copenhagen.*—Artificial mineral waters.

17. STOLTZENBERG & UFFHAUSEN, *Altona.* — Hydro-chloride of ammonia, camphine, glycerine, citric chromate of potass, cariophyl oil, and nitric ether.

## CLASS III.

### SUB-CLASS A.

18. ROYAL AGRICULTURAL HIGH SCHOOL, *Copenhagen.*—Grain and sheaves.

19. AGRICULTURAL SOCIETY, *Kiel.*—Holstein agricultural produce.

20. A SCHOOLMASTER, *Fyen.*—Hops.

21. HOLST, H. *Bredvad Mill, near Horsens.*—Flour.

22. JORDY, A. *Hómbgaard, near Ringsted.*—Cheese.

23. KJÓRBOE, F. W. *Copenhagen.*—Grain from Jutland.

24. MARSTRAND, T. *Wodroff Mill, near Copenhagen.*—Wheat, flour, and groats.

25. NIELSEN, H. M. C. *Lystofte, near Lyngby.*—Pressed bran-cakes.

26. NIELSEN, C. G. *Flensborg.*—Grain from Slesvig.

27. PASCHKOWSKY, G. *Flensborg.*—Starch (amidon) and potato-flour.

28. PUGGAARD & HAGE, *Nakskov.*—Grain and seeds.

29. PUGGAARD, H. & Co. *Copenhagen.*—Chevalier barley.

30. RADBRUCH, H. *Kiel.*—Flour starch, common-glaze and blue-glaze starch and powder.

31. SCHLIEMANN, C. *Rastorff, near Kiel.*—Manufactures from flour.

32. SCHMIDT, H. & Co. *Copenhagen.*—Flour and bran.

33. SCHOENFELDT, A. *Heiligenhafen.*—Amidon (flour starch), common and glazed amidon.

34. SCHOU, H. H. *Slagelse.*—Grain from Zealand.

35. TESDORFF, E. *Ourupgaard, Falster.*—Grain from Falster.

36. VEIS, A. *Aarhuus-mill.*—Flour and bran.

37. WINNING & Co. STEAM-MILL, *Horsens.*—Flour and groats.

## SUB-CLASS B.

38. BEAUVAIS, J. B. D. *Copenhagen.*—Hermetically sealed boxes, containing meat and fish.

39. HANSEN, A. N. & Co. *Copenhagen.*—India pork and mess pork.

40. HANSEN, J. J. & Co. *Copenhagen.*—Danish West India sugar.

41. HILL, MISS R. *St. John's.*—Arrowroot.

42. JÜRGENSEN, D. *Flensborg.* — India pork, mess pork, lard in bladders, sausages, and hams.

43. MEYER, J. C. F. & SON, *Altona.*—Cocoa paste, cocoa, and preparations of cocoa, vanille, and powder chocolate.

44. NEWTON, F. R. *St. Croix.*—Muscovado sugar.

45. PARTSCH, J. W. F. *Flensborg.*—India pork, mess pork, India beef, and lard in bladders.

46. PLASKETT, W. *St. Croix.* — Muscovado sugar.

47. RESTORFF, M. C. & Co. *Thorshavn, Faro Isles.*—Dried fish.

48. ROTHE, L. *St. Croix.* — Muscovado sugar.

49. ROTHE, MRS. C. *St. Croix.*—Guava jelly, limes preserved in pickled vinegar.

50. STEVENS, J. Y. *St. Croix.*—Muscovado sugar.

51. WENDT, MISS C. DE, *St. John's.*—Pickles.

## SUB-CLASS C.

53. HERRING, P. F. *Copenhagen.* — Cherry cordial.

54. HILL, MISS R. *St. John's.*—Guaverberry rum, old rum, and shrub.

55. PETERSEN, L. E. *Kolding.*—Danish corn spirits.

56. RIISE, A. H. *St. Thomas.* — Bay spirits.

57. ROTHE, MRS. C. *St. Croix.*—Old rum.

58. WILMS, H. B. *Flensborg.*—Vinegar.

300. MAACK, F. *Flensborg.*—Vinegar.

# CLASS IV.

## SUB-CLASS A.

59. ASMUS, G. E. A. *Kiel.* — Raw and refined rape oil.

60. BENZON, A. *Copenhagen.*—Stearine candles.

61. CLAUSEN, H. A. C. *Copenhagen.*— Fish oil.

62. DRIESHAUS, *Altona.*—Wax and composite candles.

63. HOLM, J. & SONS, *Copenhagen.*— Oils, oil cake, and composite candles.

64. HOLMBLAD, L. F. *Copenhagen.* — Stearine candles and oil cakes.

65. KRACKE, C. W. *Flensborg*—Linseed cakes and linseed oils.

66. NIELSEN, J. *Frederiksborg.*—Dzierson's beehives, improved.

67. RESTORFF, M. C. & Co. *Thorshavn, Faro Isles.*—Whale oil, and cod-liver oil.

68. UFFHAUSEN, J. F. *Mölln.*—Brilliant blacking, deep black varnish for preserving leather.

## SUB-CLASS B.

69. THE GREENLAND TRADING Co. *Copenhagen.*—Raw products, skins, &c., from Greenland.

70. CLAUSEN, H. A. C. *Copenhagen.*— Iceland produce, wool, and eider-down.

72. A FARMER, *Iceland.*—Spoons of horn.

74. HOSKIER, F. *Copenhagen.* — Wools, and specimens of Greenland industry.

76. LUND, J. *Iceland.*—Spoons of horn.

77. MAGNUSSON, G. *Stokkholt, Iceland.*— Travelling knife and fork, the handles of whale tooth, mounted in brass.

79. THORSTEINSSON, J. *Vindás, Iceland.* —Travelling bottle of horn, mounted with brass.

## SUB-CLASS C.

80. ANDERSEN, ORLOW, *Frederiksborg.*— Flax in different stages of dressing.

81. BEVENSEE, T. *Seegeberg.*—Turning in grey alabaster.

82. BORNHÓFT, T. *Seegeberg.*—Turning in grey alabaster.

83. CHRISTENSEN & KJELDSEN, *Copenhagen.*—Works in cork and cork-shavings.

84. HILL, A. C. *St. John's.*—Baskets.

85. JEBSENS, WIDOW P. H. *Seegeberg.*— Turning in grey alabaster.

86. LANGMAACK, E. *Plöen.* — Turning and carving in meerschaum.

87. LUND, J. *Iceland.*—Snuff-box (baukr) of mahogany, mounted with brass.

88. PETERS, J. F. C. *Windloch, near Flensborg.*—Improved trough.

89. ROSENÓRN LEHN, BARON, *Guldborgland.*—Samples of wood.

90. ROTHE, MRS. C. *St. Croix.*—Box containing wild cotton, in buds and blown, miniature fish-pots, the one in a conic form is for catching eels.

91. SCHWARTZ & Son, J. G. *Copenhagen.* —Specimens of turning, manufactures of whalebone, umbrellas, and combs.

92. SKIFBÓGGER, JÓRGEN, *Elstrup, near Sönderborg.*—Wood articles for ship, domestic, and dairy purposes.

93. TAYLOR, J. W. *Greenland.*—Baskets of Greenland grass.

94. VESSUP, *St. John's.*—Fish-pot.

95. WEBER, TH. & Co. *Copenhagen.*— Paper pulp made from wood.

# CLASS VI.

96. FIFE, HENRY, *Copenhagen.*—A carriage.

97. SCHRÓDER, H. A. *Flensborg.*—A phaeton.

# CLASS VII.

## SUB-CLASS A.

98. DITTMANN & BRIX, *Flensborg.*—Turned and polished case-hardened cast-iron roller, and piece of a smaller one.

99. MARSTRAND, TH. *Copenhagen.*—Weaving appliances, shuttles, &c.

100. NÓRHOLM, NIELS, *Copenhagen.*—An apparatus for measuring and cutting out clothes.

## SUB-CLASS B.

101. DALHOFF, J. B. *Copenhagen.*—Machine for making files.

102. HAMMER & SÓRENSEN, V. *Copenhagen.*—Lasts and boot-trees.

103. RÜINNING & KROLL, *Preetz.*—Specimens of cooperage.

# CLASS VIII.

104. GAMEL & WINSTRUP, *Copenhagen.*—Fire-engines, constructed for large farms and small towns.

# CLASS IX.

105. ALLERUP, M. P. *Odense.*—Agricultural implements.

106. MARSTRAND, TH. *Copenhagen.*—Agricultural, domestic, and gardening implements.

# CLASS X.

110. DITHMER, H. H. *Renneberg.*—Tile-work, polished flower-vase of burnt clay.

111. NIELSEN, P. E. (SCHELLERS, Suc.) *Copenhagen.*—Sepulchral monuments.

301. MULZENBECHER, T. H. *Rensing.*—Moulded and pressed bricks.

302. VIDAL, C. *Fernsicht.*—Tiles, &c.

# CLASS XI.

## Sub-Class B.

112.   COHEN, I.   *Copenhagen.*—Camp-kettle and appurtenance.

## Sub-Class C.

113.   KRONBORG MANUFACTORY OF ARMS, *Hellebek, near Elsinore.*—Rifles.

# CLASS XII.

## Sub-Class A.

114.   THE NAVY-YARD, *Copenhagen.*—Models of ships.

115.   WILDE, CAPT. A., R.N. *Copenhagen.*—Drawing and model of a line-of-battle ship.

## Sub-Class B.

116.   THE HOME DEPARTMENT, *Copenhagen.*—Model of a life-boat.

117.   MÜLLER, H. C. *Thorshavn, Faro Isles.*—Whale and fishing boat, with weapons.

118.   SOUTH GREENLAND MINING CO.—A cajak for seal-hunting, with weapons.

## Sub-Class C.

119.   BRÜTZ, *Rendsburg.*—Ropes, &c.

120.   HOLM, JACOB, & SONS, *Copenhagen.*—Ropes and sail-cloth.

121.   THÖL, W. *Rendsburg.*—Ornamental work for ship's stern; wheel, and specimen of clock turning.

122.   WINGE, P. W. *Randers.*—Ropes.

# CLASS XIII.

123.   DANISH STATE TELEGRAPH, *Copenhagen.*—Isolators and galvanic battery.

124.   FAXÓ, *Stubbekjöbing.* — Machine worked by heated air.

125.   KHYL, C. C. *Copenhagen.*—Relays and translators for telegraphic purposes.

126.   ORNSTRUP, L. *Copenhagen.*— Gas boiling apparatus for chemical purposes.

## CLASS XIV.

127. HANSEN, G. E. *Copenhagen.*—Photographs.

129. KIRCHHOFF, A. W. *Copenhagen.*—Photographs.

130. KRIEGSMANN, M. *Flensborg.*—Photographs.

131. LANGE, E. *Copenhagen.*—Photographs.

132. MOST, P. H. C. *Copenhagen.*—Photographs.

133. STRIEGLER, R. *Copenhagen.*—Photographs.

## CLASS XV.

134. FUNCH, A. *Copenhagen.* — Tower clock (improved construction) and case chronometer.

135. JÚRGENSENS, URBAN, & SONS, *Copenhagen.* — Sea and portable chronometers.

136. KRILLE (KESSEL'S successors), *Altona.*—Pendulum clock with quicksilver compensation; a chronograph (galvanic registering apparatus), and galvanic interrupter.

137. RANCH, CARL, *Copenhagen.*—Chronometers.

———

303. TENSEN, T. C. *Bornholm.*—Clocks.

## CLASS XVI.

139. ALPERS, O. F. *Copenhagen.*—Pianoforte.

140. CARLSEN, D. & Co. *Uetersen.*—Grand piano, brass tuning instrument, and iron sounding-board.

141. HANSEN, O. *Flensborg.*—Upright piano.

142. HORNUNG & MÓLLER, *Copenhagen.*—Grand and upright pianos.

143. JACOBSEN, J. *Haderslev.* — Organ æolodicon; æolodicon with one stop; upright pianoforte.

144. KNUDSEN, CHR. *Copenhagen.*—Pianoforte.

145. LARSEN, CHR. *Odense.*—Stringed instruments.

146. MARSCHALLS, A. & SON, *Copenhagen.*—Demi-oblique upright piano.

147. PETERSEN, P. & SUNDAHL, *Copenhagen.*—Pianoforte.

148. SCHMIDT, P. E. *Copenhagen.*—Brass instruments.

149. SÓRENSEN, J. P. *Copenhagen.*—Pianoforte.

150. WULFF, L. *Copenhagen.*— Pianoforte.

## CLASS XVII.

151. NYROP, PROF. CAMILLUS, *Copenhagen.*—Surgical instruments, bandages, orthopœdical machines, and apparatus.

152. RASMUSSEN, A. *Copenhagen.*—Bandages, surgical, orthopœdical, and electro-galvanic apparatus.

153. WULFF, CARL, *Copenhagen.*—Artificial leeches.

## CLASS XVIII.

154. BIERFREUND, LOR. *Odense.*—Specimens of cotton manufactures.

## CLASS XIX.

156. OLSEN, O. F. *Wintersbölle, near Vordingborg.*—Damask and drill, all linen.

## CLASS XXI.

157. USSERÖD FACTORY, *near Hilleröd.*—Military cloth, blankets, and horse-cloth for the army.

158. BECH, MARCUS, *Aarhuus.*—Shoddy, &c.

159. CHRISTIANSEN, H. *Thorshavn, Faro Isles.*—Woollen and worsted goods.

160. CLAUSEN, H. A. C. *Copenhagen.*—Woollen goods.

161. DAVIDSEN, I. & Co. *Thorshavn, Faro Isles.*—Woollen and worsted goods.

162. EHLEN, MARIE, *Lutterbeck.*—Hand-spun wool.

163. FISCHER, C. *Vestbirk, near Horsens.*—Woollen fabrics and worsted.

164. MODEVEG, J. C. & SON, *Brede, near Lyngby.*—Broad cloths.

165. RESTORFF, M. C. & Co. *Thorshavn, Faro Isles.*—Woollen and worsted goods.

166. SCHLIEMANN, CHR. *Rastorff, near Kiel.*—Shoddy of various kinds.

————

304. ALBECH, C. E. & SON, *Copenhagen.*—Shoddy.

## CLASS XXII.

167. GROTH & SONS, *Flensborg.*—Wax cloth table-cover, oil-cloth, and lacquered calf-skins.

168. IÓNSDÓTTIR, T. *Iceland.*—Sewed carpet, old fashioned.

169. MAGNÚSDÓTTIR, MISS H. *Reykjavik, Iceland.*—Sewed carpet, old fashioned.

170. MEYER, J. E. *Copenhagen.*—Oil-cloth, and lacquered goods.

171. STEPHENSEN, MRS. *Videy, Iceland.*—Woven carpet, old fashioned.

## CLASS XXIII.

172. SCHRIEVER, *Rendsburg.*—Dyed yarns.

## CLASS XXIV.

173. BRIX, MISS, *Industrial Depôt, Copenhagen.*—Specimens of sewing by the country-women, Hedeboerne.

174. HANSEN, DETLEV, *Mógeltónder, near Tónder.*—Specimens of lace, trimmings, and collars.

175. KRAGH, MISS EMILIE, *Frederiksborg.*—Specimens of sewing by the country-women, Hedeboerne.

176. LEVISOHN, J. C. *Copenhagen.*—Embroidery in wool.

177. LOHSE, MISS HENRIETTE, *Copenhagen.*—Ladies' sets in point lace.

178. RICHTER, MRS. S. *Holstebro.*—White embroidery.

179. TOPP, MISS MATHILDE, *Copenhagen.*—White embroidery.

———

305. BOIESEN, MRS. M. *Copenhagen.*—Embroidery.

## CLASS XXV.

### SUB-CLASS A.

180. BANG, J. C. *Copenhagen.*—Carpet and fur manufactures.

181. BRINCKMANN, FR. *Copenhagen.*—Fur manufactures.

182. SCHMID, *Kiel.*—Fox and cat furs.

183. TAYLOR, J. W. *Greenland.*—Female Greenlanders' costumes; Esquimaux hunting dress, with sundry Esquimaux articles; seal and dog skin mat; footstool, and seal-skin gloves; white haired skin of seal fœtus; prepared bird-skins for articles of dress; hand-spun yarn, from hair of the white hare; sinews, from which thread is made.

184. TROLLE, C. A. *Copenhagen.*—Fur coat for travelling; carpets; trimming and lining for ladies' dress.

### SUB-CLASS C.

185. LANGE, MISS HENRIETTE, *Altona.*—Articles worked in human hair.

## CLASS XXVI.

### SUB-CLASS A.

186. BORCH, BROS. *Copenhagen.*—Skins and leather.

188. ERIKSEN, S. *Horsens.* — Tanned lambskins.

191. MESSERSCHMIDT, E. *Copenhagen.*—Tanned hides and skins.

193. WIENGREEN & FIRJAHN, *Slesvig.*—Lacquered leather, skins, &c.

## SUB-CLASS B.

194. BARTH, MAJOR S. C. cavalry, *Copenhagen.*—Riding equipage for cavalry.

195. DAHLMANN, F. & L. *Copenhagen.*—Set of double harness, saddles, &c.

196. HINTZ, C. O. *Kiel.*—Set. of double harness.

197. SÖRENSEN, C. P. *Copenhagen.*—A stuffed horse.

## SUB-CLASS C.

198. JENSEN, H. C. *Flensborg.*—Copper-clinched fire-engine hose.

# CLASS XXVII.

## SUB-CLASS A.

199. BODECKER, A. F. *Copenhagen.*—Silk and felt hats.

200. BRET, A. H. *Copenhagen.*—Hats.

## SUB-CLASS C.

201. CHRISTENSEN, PETER, *Copenhagen.*—Frock coats, vest, and trousers.

202. COHEN, H. *Copenhagen.*—Gentlemen's linen, underclothing, and neckties.

203. LANDER, P. JUN. *Copenhagen.*—Gloves.

204. LARSEN, H. C. *Copenhagen.*—Gloves.

205. LORENTZEN, P. J. *Flensborg.*—Gentlemen's linen.

207. RASMUSSEN, HANS, *Copenhagen.*—A uniform.

208. RUBEN, M. M. *Copenhagen.*—Gentlemen's linen.

209. SCHOTTLÆNDER & GOLDSCHMIDT, *Copenhagen.*—Shirts, shirt-fronts, surtout, and vest.

## SUB-CLASS D.

210. BENGAHL, J. V. *Copenhagen.*—Spring shoes, with pasteboard bottoms, for flat-footed persons.

211. CORDWAINERS' GUILD, *Preetz.*—Boots and shoes.

212. DÜRING, N. P. *Copenhagen.*—Boots and shoes.

213. HJORTH, M. H. *Copenhagen.*—Boots and goloshes.

214. RUSCHE, *Altona.*—Boots.

215. SCHWARZ, *Altona.*—Specimens of boot and shoe making.

216. VOGES, J. C. *Altona.*—Specimens of bootmaking.

# CLASS XXVIII.
## Sub-Class A.

217. Drewsen & Sons, *Silkeborg.*—Colombier, chart, and writing, printing, and cartoon paper.

219. Holmblad, L. F. *Copenhagen.*—Playing-cards.

220. Rosenberg, Caroline, *Hoffmansgave, Fyen.*—Writing-paper decorated with moss and fern.

## Sub-Class B.

222. Göttsch, Wilhelmine, *Kiel.*—Specimens of cutting in leather and paper.

## Sub-Class C.

223. Bærentzen, Em. & Co. *Copenhagen.*—Chromo-lithograph, mezzotints, and lithographs.

224. Henneberg & Rosenstand, *Copenhagen.*—Frames containing woodcuts.

225. Klein, Louis, *Copenhagen.*—A book set by Sórensen's compositor.

226. Luno, Bianco, *Copenhagen.*—Printed books.

## Sub-Class D.

228. Clement, D. L. *Copenhagen.*—Bound books, typography, xylography, and copper and steel engravings.

229. Junge, Chr. *Copenhagen.*—Picture books.

# CLASS XXIX.
## Sub-Class A.

230. Royal Ordnance Survey, *Copenhagen.*—Maps.

231. Direction of Public Schools, *Copenhagen.*—Maps.

232. Director of Public Schools, *Slesvig.*—Maps.

233. Nissen, J. V. *Ramten, near Grenaa.*—A Bible historical map.

234. Steen, Chr. & Son, *Copenhagen.*—Maps.

## Sub-Class B.

235. Director of Public Schools, *Slesvig.*—Books and apparatus for instruction from the Duchy of Slesvig.

236. Direction of Public Schools, *Copenhagen.*—Books and apparatus for instruction.

237. Conradsen, Rudolf, *Copenhagen.*— Apparatus for educational purposes; stuffed animals.

238. Hestermann, *Altona.*—Model for educational purposes, specially adapted for natural philosophy in elementary schools.

239. Schiött, *Copenhagen.*—New writing apparatus for the blind.

240. Steen, Chr. & Son, *Copenhagen.*—Globe.

241. Thornam, J. C. *Copenhagen.*—Zoological drawings for educational purposes.

306. Galberg, *Copenhagen.* — Writing apparatus for the blind.

# CLASS XXX.

## SUB-CLASS A.

242. ART AND INDUSTRIAL UNION, *Copenhagen.*—Furniture and domestic utensils.

243. DAHL, EMANUEL, *Haderslev.* — Carved table inlaid with German silver.

244. FREESE, H. C. *Kiel.*—Furniture of wicker-work; chairs easily taken to pieces.

246. GRIMM, *Neustadt.* — Furniture in rosewood.

247. HANSEN, F. DUMONT, *Copenhagen.* —A commode.

248. HEINSEN, N. H. *Altona.*—Furniture.

249. HELLMANN, S. D. *Altona.*—Wicker-work furniture.

251. LARSEN, L. *Copenhagen.* — Couch, armchair, and chair.

252. LUND, J. G. *Copenhagen.* — China and plate cupboard, and chairs.

253. NIELSEN, O. *Odense.* — Sideboard, and model of a secretaire.

254. RAMCKE, H. H. *Altona.*—Furniture of rosewood and mahogany.

255. SCHIRMER, F. *Kiel.*—Veneering.

## SUB-CLASS B.

257. CLAUDIUS, S. *Kiel.*—Wall-painting (new invention), Pompeian style.

258. DAHL, A. *Copenhagen.* — Printed blinds, Venetian blinds, and Persiennes.

259. FJELDSKOV, W. *Copenhagen.* — A figure (Christian IV.) tankard, with stand.

260. FREESE, F. *Kiel.*—Tapestry from wood-shavings.

261. FRÓLICH, L.—Decorative paintings illustrative of Northern mythology; allegory of " Morning."

262. HARBOE, J. O. *Copenhagen.*—Blinds, Venetian blinds, Persiennes, and floor paper.

263. HENRICHSEN, J. *Copenhagen.* — Frames.

264. HULBE, C. *Kiel.*—Beading.

265. MASSMANN, F. *Kiel.*—Rough and polished beading.

266. NIELSEN, O. *Odense.*—Mirror frames.

268. WARNHOLZ, H. D. *Neumünster.*—Paper-hangings, designs, roller and hand stencilling.

# CLASS XXXI.

## SUB-CLASS A.

269. BUHLMANN, C. & Co. *Heide.*—Gas-meters and water-cistern which do not require constant pressure.

270. HOLLER & Co. *Iron Foundry, Carlshütte, near Rensburg.*—Enamelled milkpans, with appendages.

271. MARTIN, L. A. *Copenhagen.*—Pattern card of buttons.

272. RAMES, C. A. *Copenhagen.*—Nails.

273. UNION IRON WORKS, *Pinneberg.*—Tinned cooking-apparatus, currycombs, lacquered iron sugar-loaf moulds.

———

307. SCHWEFFEL & KOWALD, *Kiel.* — Iron milk-dishes, for large dairies.

## SUB-CLASS B.

274. CRUSAA COPPER WORKS, *Flensborg.*—Yellow metal, copper in plates, and brass pans.

275. HALLVARD, *Thvera, Iceland.* —

Padlock of brass, with two appertaining keys.

276. UNION IRON WORKS, *Pinneberg.*—Brass goods.

## SUB-CLASS C.

277. HÓY, Hans, *Copenhagen.*—Pewter utensils.

278. JRGENS, C. & SON, *Copenhagen.*—Tin plate goods.

279. MEYER, F. *Copenhagen.*—Brass, tin, and japanned goods.

280. RASMUSSEN, L. *Copenhagen.*—Figures cast in zinc.

# CLASS XXXIII.

281. ART & INDUSTRIAL UNION, *Copenhagen.*—Silver plate.

282. CHRISTESEN, V. *Copenhagen.*—Silver plate.

283. CLAUSEN, N. C. *Odense.*—Spoons and forks.

284. DAHL, E. F. *Copenhagen.*—Works in gold and silver.

285. DAHLHOFF, J. B. *Copenhagen.*—Chased bust.

286. DIDRICHSEN, JUL. *Copenhagen.*—Chased figures and animals in gold and silver.

287. DRAGSTED, *Copenhagen.*—Drinking horns in the northern antique; silver plate.

288. DREWSEN, H. O. *Copenhagen.*—Electro plate.

289. FERSLEV, O. & Co. *Copenhagen.*—Seals and arms.

290. HERTZ, P. *Copenhagen.*—Epergne with figures.

291. MAYENTZHUSEN, H. C. V. *Copenhagen.*—Articles in gold and silver.

292. MÓLLER, CASPAR, *Copenhagen.*—Galvano plastic works, plated and bronzed.

293. THORNING, J. C. *Copenhagen.*—Jewelry.

294. VIGFUSSON, S. *Reykjavik, Iceland.*—Works in silver.

_____

308. HOLM, C. *Copenhagen.*—Chased bronze.

# CLASS XXXIV.

## SUB-CLASS B.

295. JENSEN, H. *Flensborg.*—Articles in glass.

# CLASS XXXV.

296. ROYAL PORCELAIN MANUFACTORY, *Copenhagen.*—Porcelain, table services, bisquit figures, table ornaments, vases, &c.

297. BING & GRÓNDAHL'S PORCELAIN WORKS, *Copenhagen.*—Bisquit figures, bas-reliefs, domestic and apothecary utensils, and telegraph insulators.

298. MATZENBECHER, J. H. *Rensing.*—Modelled and pressed bricks.

299. VIDAL, C. *Fernsicht.*—Stoves, architectural ornaments, tiles, figures, and vases of baked clay.

_____

309. SONNE, F. M. *Rónne, Bornholm.*—Stone ware.

# ECUADOR.

## NORTH-WEST GALLERY.

1. Gold dust from the mines of Cachabi.

2. Gold ornaments found in different parts of the country.

3. Silver ore; ditto roasted and crushed; ditto crushed.

4. Copper ore mixed with emeralds.

5. Set of emeralds mounted in gold by native workmen; exhibited by Mrs. Prichard.

6. Cotton.

7. Cacao.

8. Leaf tobacco and cigars.

9. Coffee.

10. Orchella weed.

11. Ivory nuts.

12. Cinchona bark, flat and round.

13. Caoutchouc.

14. Silk produced by Senor Chiriboia.

15. A collection of woods exhibited by the Ecuador Land Company.

16. Pita, or the fibre of aloe.

17. Panama, or palm-leaf hats; Panama straw.

18. Embroidery work by Indians, exhibited by his Excellency Senor Flores.

19. Antiquities of pottery, found six feet below the level of the sea, at the Pailon.

20. Paintings from churches at Quito, by native Indians, exhibited by Señor Sanquirico y Ajesa.

21. Paintings representing views of Ecuador, by native Indians, exhibited by Mr. Mocatta.

22. Head of an Inca, reduced to tenth part of its natural size by an unknown process; an idol, from the Temple of Jivaros.

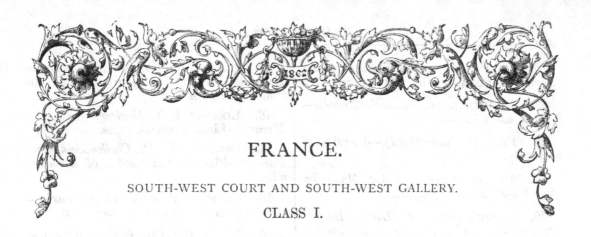

# FRANCE.

SOUTH-WEST COURT AND SOUTH-WEST GALLERY.

## CLASS I.

1. DUPONT & DREYFUS, *Ars-sur-Moselle* (*Moselle*).—Iron in bars, and for special purposes; angle-iron; &c.

2 SCHOOL OF THE MASTER-MINERS OF ALAIS (*Gard*).—Mineralogical collection of the Department of the Gard.

3. COLLECTIVE EXHIBITION OF THE DEPARTMENT OF CORSICA.—Ores and marbles (4 Exhibitors).

4. JAMES JACKSON, SON & CO. *St. Seurin-sur-l'Isle* (*Gironde*).— Steel, by Bessemer's process; bars, springs, &c.

5. DE DIÉTRICH & CO. *Niederbronn* (*Bas-Rhin*).—Ornamental and other castings; enamelled cast-iron articles; charcoal iron; puddled steel, in bars and sheets; waggon and locomotive wheels, forgings for steam engines.

6. BONNOR, DEGROND, & CO. *Eurville* (*Haute-Marne*).—Wood castings, rolled iron, wire, and chains.

7. MARTIN, E. O. *Sireuil* (*Charente*).— Manganiferous iron ores of Perigord and La Charente; wood, and anthracite castings; refined and puddled iron, rolled cast steel, puddled and cast steel rails and tyres, bent axle.

8. LALOUËL DE SOURDEVAL & MARGUERITTE, *Paris*.—Converted and cast steel tools.

9. BARON DE ROSTAING & BAUDOUIN BROS. *Paris*.—Metals granulated by centrifugal force: with steel, oxides, and salts, their products.

10. BAUDRY, A. & COTTREAU, *Athis-Mons* (*Seine-and-Oise*).—Iron, and cast steel.

11. JOINT-STOCK MINING AND RAILWAY CO. OF CARMAUX, *Avalats* (*Tarn*).—Iron and puddled steel; axles.

12. DURAND, JUN. & GUYONNET, P. *Perigeux* (*Dordogne*). — Cast-iron, refined and welded iron, iron-wire, &c.

13. JACQUINOT, F. & CO. *Solenzara* (*Corsica*).—Iron ores, wood-charcoal, cast iron, &c.

14. CHENOT, A. & E. BROS. *Clichy-la-Garenne* (*Seine*).— Plans and models of furnaces; sponge iron, steel.

15. GUILLEM & CO. *Marseilles* (*Bouches-du-Rhône*).— Lead pigs, and pipes; silver plate, copper nails.

16. OESCHGER, MESDACH, & CO. *Biache-St. Vaast* (*Pas-de-Calais*).—Ores; lead, copper, zinc; unsoldered copper tubes, &c.

17. COMMITTEE OF COAL PROPRIETORS OF THE DEPARTMENT OF THE LOIRE, *St. Etienne* (*Loire*).—Pit-coal; coke, agglomerates, and other derivatives of coal.

18. JOINT-STOCK ARGENTIFEROUS LEAD MINE AND FOUNDRY CO. OF PONTGIBAUD (*Puy-de-Dôme*).—Lead-ores, lead-pigs.

19. CHAMBER OF COMMERCE OF CHAMBERY (*Savoy*).—Marbles, ores, cements.

20. DELMAS, E. *St. Capraix* (*Dordogne*).— Alluvial iron ore.

21. TAMISIER & CO. *St. Gervais-les-Bains* (*Haute-Savoie*).—Red jasper.

22. FOMMARTY & CO. *Perigueux* (*Dordogne*).—Hydraulic lime.

23. MAGNEUR, R. *Hautefort* (*Dordogne*). —Manganesiferous alluvial iron ore.

24. GUÉRIN, DR. J. *Paris*.—Marls, lime, hydraulic cements; plan of a new kind of limekiln.

25. BICKFORD, DAVEY, CHANU, & CO. *Rouen* (*Seine-Inf.*).—Safety fusees.

26. DE PAGÈZE DE LAVERNÈDE, *Salles* (*Gard*).—Pit-coal, regulus of antimony.

27. DUBRULLE, A. N. *Lille* (*Nord*).— Miners' safety-lamps.

28. BEAU D. *Alais* (*Gard*).—Regulus of antimony.

29. MICHEL, ARMAND, & Co. *Marseilles* (*Bouches-du-Rhône*).—Lignite.

30. LEBRUN-VIRLOY, A. *Lanty* (*Haute-Marne*).—Plan of a portable apparatus for drying and carbonizing wood and peat, and collecting the volatile products.

31. SPIERS, *Paris.*—Coal agglomerated with, and without, bituminous matter.

32. MATHIEU BROS. *Anzin* (*Nord*). — Air and water counter-pressure apparatus for mines, &c.

33. MARQUIS DE CHAMBRUN, *Marvéjols* (*Lozère*).—Argentiferous lead ore.

34. CROS, J. *Albi* (*Tarn*).— Manganese ore.

35. NICOLI, J. B. *Ajacccio* (*Corsica*). — Rocks and ores.

36. COUVRAT - DESVERGUES - GEOFFROY, *Excideuil* (*Dordogne*).—Manganesiferous alluvial iron ore.

37. MINING Co. OF BÉDOUÈS AND CO-CURÈS, *Meyrueis* (*Lozère*).—Ores of argentiferous lead, and of copper.

38. MINING Co. OF PALLIÈRES, *Alais* (*Gard*).—Sulphates of lead.

39. MINING Co. OF RICHALDON, *Collet-de-Dèze* (*Lozère*).— Argentiferous lead ore, and products obtained from it by mechanical means.

40. MULOT, SON, & DRU, *Paris.*—Sounding and boring tools.

41. JACQUET, N. J. SEN. *Arras* (*Pas-de-Calais*).—Parachute for coal mines.

42. DEPLAYE, JULLIEN, & Co. *Paris.*— French lithographic · stone six feet seven inches long, and three feet four inches wide.

43. COLETTES KAOLIN Co. *Château de Veauce* (*Allier*).— Natural china clay : vases and other articles made of it.

44. MINING Co. OF MARSAC, *Coussac-Bonneval* (*Haute-Vienne*)—Kaolin : sulphate of alumina.

45. PLANTIÉ & SON, *Bayonne* (*Basses-Pyrénées*) Kaolin, crude, ground, and elutriated ; felspath.

46. LIÉNART, L. T. *Mortcerf* (*Seine-and-Marne*).—Limo, cements, pipes.

47. PARQUIN, L. P. *Chelles Seine-and-Marne*).—Plaster, and model of a plaster mill.

48. CHAPUIS, P. & A. BROS. *Paris.*— Ores of platina, and articles made of it.

49. MORIN, P. & Co. *Nanterre* (*Seine*).— Aluminium, and aluminium bronze, and a variety of articles made of them.

50. LÉTRANGE, L. & Co. *Paris.*—Rolled and hammered copper, lead, and zinc.

51. HERNIO, E. *Clohars* (*Finistèrre*). — Kaolin.

52. MAIRE, E. *Plessis-en-Coësmes* (*Ille-and-Vilaine*).—Slates.

53. MAIRE, X. (*Moisdon Loire - Inf.*). Roofing-slate.

54. DEGOUSÉE & LAURIENT, C. *Paris.*— Model of sounding apparatus ; sounding tubes and tools.

55. JOINT-STOCK MINING Co. OF LA GRAND-COMBE (*Gard*). — Coal, coke, and agglomerates.

56. GAILLARD, SEN., PETIT & HALBOU, A. *La Ferté-sous-Jouarre* (*Seine-and-Marne*).— Millstone.

57. DANGREVILLE-CHERRON & VALLOND, J. *La Ferté-sous-Jouarre* (*Seine-and-Marne*).—Millstone.

58. CHASSIANG-PEYROT & Co. *Domme* (*Dordogne*).—Millstones, slabs.

59. ALLORD, SON, & Co. *Sarlat* (*Dordogne*).—Millstones, nut oil.

60. BAILLY & Co. *La Ferté-sous-Jouarre* (*Seine-and-Marne*).—Millstones.

60A. THIERRION, *Épernay* (*Marne*).—Circular guide for dressing millstones.

61. GILQUIN, P. S. *La-Ferté-sous-Jouarre* (*Seine-and-Marne*).—Millstones, slabs, and parts of millstones.

62. DESMOUTIS, CHAPUIS, & QUENNESSEN, *Paris.*—Platina, and apparatus made of it ; metals obtained from platina ores.

63. DUPETY, THEUREY-GUEUVIN, BOUCHON & Co. *La-Ferté-sous-Jouarre* (*Seine-*

*and-Marne*).—Millstones, slabs: and parts of millstones.

64. GAILLARD, J. F. *La-Ferté-sous-Jouarre (Seine-and-Marne)*.—Millstones.

65. LEVEAU-BAUDRY, *Villaine-la-Gonais (Sarthe)*.—Millstones.

66. CHAUVEAU, *Villaine - la - Gonais (Sarthe)*—Millstones.

67. DELÉPINE, C. & T., A. *La-Ferté-sous-Jouarre (Seine-and-Marne)*.—Millstones.

68. ROGER, SON, & Co. *La-Ferté-sous-Jouarre (Seine-and-Marne)*.—Millstones, and parts of millstones.

69. BARDEAU, E. *Fleury (Yonne)*.—Millstones.

70. TIGER & JONQUET, *Cloyes (Eure-and-Loire)*.—Millstones.

71. BESNARD, *Epernon (Eure-and-Loire)*.—Millstones.

72. MATHER & SON, *Toulouse (Haute-Garonne)*.—Ingot, sheet, and cupola copper.

73. VISCOUNT A. N. DESSERES, *Caylus (Tarn-and-Garonne)*.—Lithographic stones.

74. MALBEC, A. A. *Paris.*—Artificial stones, for mills and other purposes.

75. MESNET, T. A. *Cinq-Mars-la-Pile (Indre-and-Loire)*.—Millstones.

76. DESPAQUIS, P. A. *Harol (Vosges)*.—Lithographic stones.

77. DELESSE, A. *Paris.*—Hydrological map of the Department of the Seine.

78. DORMOY, E. *Valenciennes (Nord)*.—Subterranean map of the coal basin of Valenciennes, and of the coal field at Mons: intended to guide miners to the best points for sinking shafts.

79. SENS, E. *Arras (Pas-de-Calais)*.—Typographical map of the coal basin of the Department of the Pas-de-Calais.

80. POUGNET, M. & Co. *Landroff (Moselle)*.—Model of shaft-lining, for coal mines.

81. CABANY, A. *Valenciennes (Nord)*.—Waggon, and plan of a mine.

82. LECOQ, H. *Clermont-Ferrand (Puy-de-Dôme)*.—Geological map of the Department of the Puy-de-Dôme.

83. DEHAYIN, F. *Paris.*— Agglomerates, and plan of the machine used for agglomerating small coal.

84. CHALAIN, E. *Riaden (Ille-and-Vilaine)*.—Slates for flooring, roofing, and billiard tables; a slate thirteen feet long, and three feet four inches wide.

85. BORDE, RAYMOND, PALAZZI, *Corte (Corsica)*.—Streaked copper ore.

86. BARON O. DE BARDIES, *Oust (Ariége)*.—Argentiferous galena ores.

87. CHALLETON, J. F. F. *Montauger (Seine-and-Oise)*.—Peat, purified, condensed, and carbonized; essential oils, ammoniacal compounds, manures.

88. BRIÈRE, A. *Brassac-les-Mines (Puy-de-Dôme)*.—Ore, and regulus of antimony; arsenious acid (Schweinfurth green).

89. CHAPERON, PERRIGAULT, & Co. *Libourne (Gironde)*.—Millstones.

90. LAVALLÉE, E. *Fontenay (Seine)*.—Moulding sand, for founders.

————

90A. TERQUEM, *Metz (Moselle)*.—Geological section of Mount St. Quentin, near Metz.

90B. BONHOMÉ, *Paris.*— Designs having reference to mines and metallurgical establishments.

90C. CHUART, M. *Paris.*—Safety-lamp, gazoscope for preventing explosions in mines.

90D. POUYAT BROTHERS. *Limoges (Haute-Vienne)*.—Kaolin, felspath.

90E. ROSIER WIDOW, & BAROCHE, *Tain (Drôme)*,—Kaolin.

90F. VERDIÉ, F. F. & Co. *Firminy (Loire)*.—Iron, puddled and cast steel.

90G. VIEILLARD, J. & Co. *Bordeaux (Gironde)*.—Kaolin, and felspath.

# CLASS II.

91. CALLOU, A. & VALLÉE, *Paris.*—Salts, &c. obtained from the waters of Vichy.

92. CHERBOUQUET-BADOIT & CHAMPAGNON, *St. Galmier (Loire)*.—Mineral waters, from the springs of Badoit, and André de Saint-Galmier.

93. BOULOUMIE, L. *Vittel (Vosges)*.—Mineral waters and ferruginous products of the springs of Vittel; corks for preventing the decomposition of mineral waters.

94. THE CITY OF BAGNÈRES-DE-LUCHON (*Haute-Garonne*).—Plans of a Thermal establishment.

95. FRANÇOIS, J. *Paris.*—Collection of rocks accompanying the mineral waters of Bagnères-de-Luchon, Cauterets, and Lamalou-l'Ancien.

96. THE PROPRIETORS OF THE MINERAL WATERS OF FRANCE.—Mineral waters, from forty-six localities.

97. CROC, L. *Aubusson (Creuse)*.—Telegraphic and writing ink.

The exhibitor has received the Honorary Medal for his indestructible ink, which has been patented in France, England, Belgium, Austria, Prussia, Holland, &c.

This ink has the fluidity, tone, and brilliancy of the best inks known.

It is indestructible—an important quality which the most numerous and unceasing researches have never before been able to attain.

It possesses those properties which have been so long sought after, for the purpose of placing public and private documents out of all danger of being tampered with by the cleverest forger, or injured by the action of time, which destroys every other kind of ink.

The experiments made in the chemical laboratory of the Imperial Polytechnic School of Paris, that have demonstrated its resistance to the action of chemical re-agents, have been confirmed by the international jury, which has awarded an Honorary Medal to this inestimable discovery.

98. CHARVIN, F. *Lyon (Rhône)*.—Green dye-stuff.

100. GERTOUX, J. *Bagnères-de-Bigorre (Hautes Pyrénées)*.—Labassère water, containing sulphuret of sodium.

101. BURGADE & SISTERS, *Garost (Hautes-Pyrénées)*.—Iodurated sulphurous water of Garost.

102. MANINAT, Jun. *Ossan (Hautes-Pyrénées)*.—Sulphurous mineral water from the springs of Nabias, La Rallière, and César.

103. BRUN, M. *Puteaux (Seine)*.—Mordants for dyeing.

104. ROSELEUR, A. *Paris.*—Chemica products.

105. ARRAULT, H. *Paris.*—Chemical products, and medicine chests for military purposes, &c.

106. LALOUËL DE SOURDEVAL & MARGUERITTE, *Paris.*—Alkaline cyanides obtained by means either of atmospheric nitrogen, or ammonia; ammoniacal salts for agricultural purposes.

107. PENNÈS, J. A. *Paris.*—Mineral salts for baths.

108. LEFRANC & Co. *Paris.*—Colours, varnishes, and typographical ink.

109. LE PERDRIEL & MARINIER, *Paris.*—Pharmaceutical products.

110. FUMOUZE-ALBESPEYRES, *Paris.*—Dressings for blisters, &c.

111. DUROZIEZ, M. E. A. *Paris.*—Artists' materials, photographic chemicals.

412. GARZEND, A. *Paris.*—Prepared woods for dyeing.

113. USÈBE, C. J. *St. Ouen (Seine)*.—Carmine of saffron, as a liquid, paste, &c.

114. ROCQUES & BOURGEOIS, *Ivry (Seine)*.—Chemical products derived from the carbonization of wood.

115. POIRRIER & CHAPPAT, JUN. *Paris.*—Chemical products for dyeing.

116. PETERSEN, F. & SICHLER, *Villeneuve-la-Garenne, near St. Denis (Seine)*.—Chemical products, dye-stuffs, and colours.

117. MALLET, A. A. P. *Paris.*—Caustic ammonia, crude, and refined; sulphate and hydrochlorate of ammonia; pure sulphate of iron.

118. PARISIAN GAS-LIGHTING AND HEATING Co. *Paris.*—Chemical products obtained during the destructive distillation of coal.

119. CAMUS, C. & Co. *Paris.*—Chemical products.

120. PIVER & RONDEAU, A. *Paris.*—Colours, and varnishes.

121. DEROCHE, C. *Paris.*—Chemical products.

122. KUHLMANN & Co. *Lille (Nord).*—Chemical products, and specimens of new modes of applying them to industrial purposes.

123. DRION-QUÉRITÉ PATOUX, & DRION, A. *Aniche (Nord).*—Chemical products.

124. DESESPRINGALLE, A. *Lille (Nord).*—Chemical products obtained from alcohol, and tar; salts of cadmium.

125. PÉRUS, J. & Co. *Lille (Nord).*—White lead.

126. RICHTER, B. & F. *Lille (Nord).*—Ultramarine blue.

127. DORNEMANN, G. W. *Lille (Nord).*—Ultramarine blue and green.

128. CHAPUS, A. *Lille (Nord).*—Ultramarine blue.

129. SERRET, HAMOIR, DUQUESNE, & Co. *Valenciennes (Nord).*—Chemical products: alcohol, sugar.

130. DEHAYNIN, M. G. *Valenciennes (Nord).*—Chemical products and dye-stuffs derived from tar.

131. SERBAT, L. *St. Saulve (Nord).*—Mastic, for steam joints; oils and fats for industrial purposes.

132. GAUTIER-BOUCHARD, L. J. *Paris.*—Chemical products, colours and varnish.

133. BONZEL BROS. *Haübourdin (Nord).*—White-lead, chicory.

134. JOINT-STOCK MINING Co. OF SAMBRE-AND-MEUSE, *Hautmont (Nord).*—Chemical products.

135. MINING Co. OF BOUXWILLER (*Bas-Rhin*).—Chemical products; alum; prussiates of potash; sulphate of copper; gelatine.

136. KESTNER, C. *Thann (Haut-Rhin).*—Chemical products and dye-stuffs.

137. SCHAAFF & LAUTH, *Strasbourg (Bas-Rhin).*—Madder, its extracts and lakes.

138. RIESS, M. *Dieuze (Meurthe).*—Gelatine, and phosphate of lime.

139. MERLE, H. & Co. *Alais (Gard).*—Various salts derived from sea-water; chemicals, and dye-stuffs.

140. PLANCHON, S. *St. Hippolyte (Gard).*—Glue and gelatine.

141. CHIRAUX, L. *Cambrai (Nord).*—Blacking.

142. CAZALIS, H. and Co. *Montpellier Hérault.*—Sulphuric and other acids; salts.

143. LE BEUF, F. *Bayonne (Basses-Pyrénées).*—Pharmaceutical products, &c.

144. TACHON, SON, & Co. *Roanne (Loire).*—White lead, obtained directly with carbonic acid evolved from the mineral springs of Saint-Alban.

145. BERJOT, F. *Caen, (Calvados).*—Pharmaceutical extracts prepared and dried in vacuo; flowers and medicinal plants preserved so as to retain their form and appearance. Hermetically sealed flasks. Apparatus for determining the amount of oil obtained from oleaginous seeds. Models of apparatus for producing gaseous waters.

146. PARQUIN, LEGUEUX ZAGOROWSKI, & SONNET, *Auxerre (Yonne).*—Ochres, raw and manufactured.

147. CAROF, A. & Co. *Portsal-Ploudalmézeau (Finistèrre).*—Chemical products, obtained from sea-wrack.

148. COURNERIE, SON, & Co. *Cherbourg (Manche).*—Chemical products, obtained from sea-wrack.

149. MAUMENÉ & ROGELET, *Reims (Marne).*—Potassa, and its salts, obtained from sheep-grease.

150. HUILLARD & GRISON, *Deville-lez-Rouen (Seine-Inf.).*—Products for dyeing.

151. DELACRETAZ & CLOUET, *Hâvre.*—Chrome oxide; chromate and bichromate of potash.

152A. DELACRETAZ, *Paris.*—Chemical products; sulphuric acid, made with sulphur, and with pyrites; stearic and oleic acid; &c.

152. MULLER, P. *Rouen (Seine-Inf.).*—Gelâtine.

153. LAURENTZ, P. C. *Rouen (Seine-Inf.).*—Chlorides.

154. TISSIER & SON, *Conquet (Finistèrre).*—Chemical products, obtained from sea-wrack.

155. PICARD & Co. *Granville (Manche).*—Chemical products, obtained from sea-wrack.

156. JOINT-STOCK GLASS AND CHEMICAL Co. OF S. COBAIN, CHAUNY, & CIREY, *Paris.*—Chemical products.

158 CHEVENEMENT, L. *Bordeaux (Girónde).*—Blacking, black and coloured inks.

159. FOURNIER-LAIGNY & Co. *Courville (Eure-and-Loire).*—Products obtained from pyroligneous acid.

160. DESCHAMPS, BROS. *Vieuz-Jean-d'heures (Meuse).*—Ultramarine blue and green.

161. BARTHE, DURRSCHMIDT, PORLIER, & Co. *Pont-St. Ours (Nièvre).*—Acetic acid, alum, and sulphate of alumina.

162. BAZET, HAPPEY, & Co. *Paris.*—Gazogène apparatus.

163. LUTTON, A. LOLLIOT, & Co. *Neuvy-sur-Loire (Nièvre).*—Pyroligneous acid, and its products.

164. BRUZON, J. & Co. *Portillon (Indre-and-Loire).*—Chemical products, &c.

165. BERTRAND & Co. (*Dijon Côte-d'or).*—Ultramarine blue.

166. DANIEL, H. *Paris.*—Blacking and inks.

157. GUINON, MARNAS, & BONNET, *Lyon (Rhône).*—Chemical products, obtained during the destructive distillation of pit-coal.

168. ALESMONIÈRES, A. *Lyon (Rhône).*—Chemical products and dye-stuffs obtained from pit-coal.

169. GILLET & PIERRON, *Lyon (Rhône).*—Crystals of soda; palm oil, and olive oil, soap, &c.

170. MONNET & DURY, *Lyon (Rhône).*—Products for dyeing; aniline and its derivatives.

171. FAYOLLE & Co. *Lyon (Rhône).*—Chemical products for dyeing and printing; aniline colours.

172. PLATEL L. J. & BONNARD, J. *Lyon (Rhone).*—Products for dyeing; tannic acid, obtained from chestnut wood.

173. BRUNIER, JUN. & Co. *Lyon (Rhône).*—Prussiates of potash.

175. RENARD, BROS. & FRANC, *Lyon (Rhône).*—Chemical products and dye-stuffs.

176. GUIMET, J. B. *Lyon (Rhône).*—Ultramarine blue.

177. BLUM-GAY & Co. *Lyon (Rhône).*—Chemical products, &c.

178. COIGNET, SON & Co. AND COIGNET, BROS. & Co. *Lyon (Rhône).*—Chemical products obtained from bone; amorphous phosphorus matches.

179. MESSIER, *Paris.*—Lakes for paper staining.

180. LANGE-DESMOULIN, J. B. C. *Paris.*—Colours.

181. BOYER & Co. *Paris.*—Chemical products, &c.

182. CHEVÉ, L. J. JUN. *Paris.*—Chemical products.

183. FOURCADE, A. & Co. *Paris.*—Chemical products.

184. JAVAL, J. *Paris.*—Products for dyeing; aniline, and colours derived from it.

185. MATHIEU-PLESSY, E. *Paris.*—Chemical products, for photography; colours for printing, &c.

186. JACQUES-SAUCE, *Paris.*—Cochineal carmine, &c.

187. ROQUES E. & Co. *Paris.*—Chemical products.

188. POMMIER & Co. *Paris.*—Chemical products for dyeing, and the manufacture of paper-hangings.

189. DALEMAGNE, L. *Paris.*—Silicate of potash, for preserving calcareous stone.

190. DURET, SEN. & BOURGEOIS, *Paris.*—Non-poisonous colours.

191. PERRA, B. *Petit-Vanvres (Seine).*—Pharmaceutical and dyeing products; phenic acid, picric acid, and picrates.

192. BOBŒUF, P. A. F. *Paris.*—Chemical products; picric and phenic acids; picrates, and alkaline phenates.

193. DEISS, E. *Paris.*—Fatty substances from refuse matter.

194. ADVIELLE, L. B. *Paris.*—Liquid for silvering.

195. ACCAULT, C. *Paris.*—Calcined magnesia.

196. ARMET DE LISLE, J. *Nogent-sur-Marne (Seine).*—Ultramarine blue, salts of quinine.

197. LAROCQUE, A. *Paris.*—Benzine, nitro-benzine; products obtained from apple peel.

198 GÉLIS, A. *Paris.*—Lactate of iron; salts of gold; pyrodextrine; process for the preparation of prussiate of potash by means of sulphuret of carbon.

199. BLANCARD, H. *Paris.*—Iodide of iron pills.

200. BURDEL & Co. *Paris.*—Liquids for cleansing and reviving cloth.

201 DEFAY, J. B. & Co. *Paris.*—Dried blood, for manure, &c; albumen, from blood, for printing textile fabrics.

202. JORET, E. M. F. & HOMOLLE, E. *Paris.*—Apiol, the active principle in parsley seed.

202A. HOMOLLE & DEBREIUL, *Paris.*—Digitaline.

203. COLLAS, C. & Co. *Paris.*—Benzine, &c.

204. MENIER, E. J. *Paris.*—Pharmaceutical and chemical products. Organic substances, obtained artificially by chemical synthesis.

205. COËZ E. & Co. *St. Denis (Seine).*—Extracts from dyewoods, and lakes obtained from colouring matters.

206. LAURENT, F. & CASTHÉLAZ, *Paris.*—Chemical and pharmaceutical products, &c.

207. POULENC-WITTMANN, E. J. *Paris.*—Chemical and pharmaceutical products, &c.

208. DUBOSC, F. & Co. *Paris.*—Chemical products, for pharmacy and photography; quinine, and its salts.

209. FREZON, J. B. *Neuilly (Seine).*—Mordants.

210. LAURENT, C. & LABÉLONYE, C. *Paris.*—Pharmaceutical extracts, prepared in vacuo.

211. SCHOEN & REUTER, *Paris.*—Colouring substances, derived from archil, &c.

212. BEZANÇON BROS. *Paris.*—White lead.

213. FERRAND, M. *Paris.*—Artist's colours.

214. STRAUSS-JAVAL & Co. *Paris.*—Dry extracts from dye-woods.

215. HARDY-MILORI, G. *Montreuil-sous-Bois (Seine).*—Colours in the dry and pasty state.

216. LATRY, A. & Co. *Paris.*—Zinc-white.

217. GELLÉ, J. B. A. SEN. & Co. *Paris.*—Perfumes, and toilet soap.

218. MOLLARD, A. A. *Paris.*—Toilet soap.

220. SARDOU, L. *Cannes (Alpes-Maritimes).*—Perfumes, pomades.

221. HUGUES, SEN. *Grasse (Alpes-Maritimes).*—Perfumes; alcoholic extracts.

222. MÉRO, J. D. *Grasse (Alpes-Maritimes).*—Essences, perfumed oils, &c.

223. ISNARD-MAUBERT, *Grasse (Alpes-Maritimes).*—Essences, orange-flower water.

224. RANCÉ, F. & LAUTIER, JUN. *Grasse (Alpes-Maritimes).*—Essences, orange-flower water, &c.

225. ARDISSON & VARALDI, *Cannes (Alpes-Maritimes).*—Essences, orange-flower water; perfumed oils, and pomades.

226. COUDRAY, P. E. *Paris.*—Perfumes pomades, and toilet soap.

227. CLAYE, V. L. *Paris.*—Perfumes, and toilet soap.

228. PINAUD, E. & MEYER, E. *Paris.*—Perfumes, pomades, and toilet soap.

229. SICHEL, J. *Paris.*—Perfumes and toilet soap.

230. BLEUZE-HADANCOURT, *Paris.*—Perfumes, pomades, and toilet soap.

231. DELABRIERE - VINCENT, *Paris.*—Perfumes, and toilet soap.

232. MOUILLERON, A. *Paris.*—Vinegar, toilet soap, &c.

233. PORTE, F. X. L. *Paris.*—Eau de Cologne, tooth-powder, &c.

234. LANDON-LEMERCIER, *Paris.*—Vinaigre de Bully.

236. GUERLAIN. P. F. P. *Paris.*—Essences, cosmetics, &c.

237. BOUTRON - FAGUER, *Paris.* — Essences, perfumes, &c.

238. MAILLY, F. *Paris.*—Perfumes and toilet soap.

239. PIVER, A. *Paris.*—Perfumes and toilet soap.

240. TITARD, J. L. *Paris.*—Vegetable rouge and blanc, for the toilet, &c.

241. JASSAU-RAIMOND, *Paris.*—Rouge for the toilet.

242. GIRAUD BROS. *Paris.*—Perfumed oils, extracts, &c.

# CLASS III.

## SUB-CLASS A.

251. DELAFONTAINE & DETTWILLER, *Paris.*—Cocoa, chocolate, &c.

252. MÉNIER, E. J. *Paris.*—Chocolate.

253. DEVINCK, F. J. *Paris.*—Chocolate.

254. GUÉKIN-BOUTRON, M. L. A. *Paris.* —Cocoa and chocolate.

255. CHOQUART, C. F. *Paris.*—Chocolate.

256. IBLED BROS. & Co. *Paris.*—Chocolate.

257. LEGUERRIER, C. L. M. *Paris.*—Chocolate and roasted coffee.

258. HERMANN, G. *Paris.*—Chocolate.

259. ALLAIS, E. *Paris.*—Ordinary and ferruginous chocolate.

260. LABRIC, P. E. *Paris.*—Cocoa and chocolate.

261. PELLETIER, E. & Co. *Paris.*—Butter of cacao and chocolate.

262. POTIN, L. E. *Paris.*—Chocolate.

263. TREBUCIEN, BROS. *Paris.*—Coffee, chocolate, tapioca.

264. FAGALDE, P. *Bayonne (Basses-Pyrénées).*—Chocolate.

265. PÉNIN, C. & Co. *Bayonne (Basses Pyrénées).*—Chocolate.

266. RUBIÑO, A. *Nice (Alpes-Maritimes).* —Cocoa and chocolate.

267. LOUIT BROS. & Co. *Bordeaux (Gironde).*—Chocolate, alimentary pastes, &c.

268. ASSOCIATION OF THE PASTE MANUFACTURERS, &C. OF AUVERGNE, *Clermont-Ferrand (Puy-de-Dôme).*—Alimentary pastes (13 Exhibitors).

269. BOUDIER, F. *Paris.*—Alimentary pastes.

270. GROULT, JUN. *Paris.*—Pastes, flour, fecula; food for infants and invalids; &c.

271. FRELUT & Co. *Clermont-Ferrand (Puy-de-Dôme).*—Preserved fruits.

272. NOEL-MARTIN & Co. *Paris.*—Alimentary pastes; gluten gluten flour for diabetic patients; starch.

273. COLLECTIVE EXHIBITION OF THE CITY OF EPINAL *(Vosges).*— Potato-fecula, and alimentary pastes (8 Exhibitors).

274. COMBIER-DESTRE, *Saumur (Maine-and-Loire).*—Elixir of "Raspail."

275. LERVILLES, J. *Lille (Nord).* — Roasted chicory.

276. BÉRIOT, C. *Lille (Nord).*—Chicory, and varnishes.

277. BOYER, A. *Paris.*—Carmelite water of Melissa.

278. GIRAUD, BROS. *Paris.*—Olive and perfumed oils; extracts, essences, and distilled waters.

279. ROUSSIN, ELIAS, *Rennes (Ille-and-Vilaine).*—Groats and pearled barley.

280. ARNAUD, SEN. & Co. *Voiron (Isère).* —Liqueurs.

281. OLIBET, *Bordeaux (Gironde).*—Sea-biscuits.

282. CAUSSEROUGE BROS. *Paris.*—Liqueurs, syrups; fruits preserved in brandy, &c.

283. DAVID, J. & Co. *Orléans (Loiret).* "French bitter."

284. GOURRY & Co. *Cognac (Charente).*— Liqueurs.

285. ROBIN, L. P. JUN. *L'Ille d'Espagnac (Charente).*—Coffee and chicory.

286. ROCHER BROS. *Côte St. André (Isère).*—Liqueurs.

287. LEGIGAN & LEFÉVRE, *Paris.*—Liqueurs.

288. SAINTOIN BROS.—*Orléans (Loiret).* Curaçoa, chocolate, &c.

289. MARIE BRIZARD & ROGER, *Bordeaux (Gironde).*—Liqueurs.

290. JOURDAN-BRIVE, G. SEN. *Marseilles (Bouches-du-Rhône).*—Wines, liqueurs, &c.

291. COLLECTIVE EXHIBITION OF THE DEPARTMENT OF THE CÔTE D'OR (5 Exhibitors).—Liqueurs.

292. HOFFMANN-FORTY, F. *Phlasbourg* (*Meurthe*).—Liqueurs.

293. PAULIN-FORT, DESPAX, & BACOT, *Toulouse* (*Haute-Garonne*).—Wines, liqueurs, and syrups.

294. ROUSSEAU & LAURENS, *Paris.*—Liqueurs, preserved fruits, &c.

295. TESSON, A. *Pantin-lez-Paris* (*Seine*). —Wines and liqueurs; preserved fruits.

296. COLLECTIVE EXHIBITION OF THE CITY OF DUNKERQUE (*Nord*).—Juniper-berry liqueurs.

297. GALLIFET & Co. *Grenoble* (*Isère*).— Liqueurs.

298. FAIVRE, DOCTOR C. *Paris.*—" Mont Carmel" liqueur.

299. LASSIMONNE, C. *Paris.*—Liqueurs, syrups, preserved fruits, &c.

300. MAGNÉ, A. *Rouen* (*Seine-Inf.*).— Jellies, and apple-preserve.

301. POURCHIER, J. B. *Avignon* (*Vaucluse*).—Chocolate, cocoa nuggets.

302. NÉGRE, J. *Grasse* (*Alpes-Maritimes*). —Preserved fruits; orange-flower water.

303. BAUDOT-MABILLE, *Verdun* (*Meuse*). —Sugar-plums and liquorice.

304. CAIZERGUES, A. *Montpellier* (*Hérault*).—Preserved fruits, confectionery.

305. MALSALLEZ, C. *Paris.*—Liqueurs, syrups, chocolate, sweetmeats.

306. AUVRAY, JUN. *Orléans* (*Loiret*).— Bon-bons.

307. BRUNET, L. *Paris.*—Concentrated extracts, for the manufacture of liqueurs by mere mixture.

308. GELLER, G. *Marseilles* (*Bouches-du-Rhône*).—Fruits preserved in sugar, and in brandy; sweetmeats.

309. MUSSO, *Nice* (*Alpes-Maritimes*).— Fruits, liqueurs, and syrups.

310. BONFILS BROS. & Co. *Carpentras* (*Vancluse*).—Conserve of truffles.

311. JACQUIN, WIDOW, & SON, *Paris.*— Sugar-plums and crisped almonds.

312. CHOLLET & Co. *Paris.*—Vegetables preserved by drying and compression; chocolate, alimentary pastes.

313. REY, F. A. *Paris.*—Preserved fruits.

314. DEMEURAT, DOCTOR L. *Tournan* (*Seine-and-Marne*).—Meats and vegetables preserved by drying and compression; biscuits.

315. CORMIER, E. *Neuilly* (*Seine*).— Preserved eggs and vegetables; sardines in oil.

316. CARNET & SAUSSIER, *Paris.*—Preserved alimentary substances.

317. GALOPIN, P. JUN. *Paris.*—Preserved truffles.

318. CHEVET, C. J. *Paris.*—Preserved alimentary substances.

319. HENRY, L. *Strasbourg* (*Bas-Rhin*)· —Goose-liver pasty.

320. GUILLOUT, E. *Paris.*—Gingerbread, biscuits, dry confectionery.

321. SIGAUT, J. J. *Paris.*—Gingerbread biscuits, dried confectionery.

322. DRIOTON, *St. Seine-l Abbaye* (*Côte d'Or*).—Barberry preserve.

323. BLANC, *Perigueux* (*Dordogne*).— Preserved alimentary substances, containing truffles.

325. SAUSOT, J. *Bordeaux* (*Gironde*).— Preserved alimentary substances containing truffles; preserved vegetables, fish, &c.

327. REBOURS-GUIZELIN, DIONE, & Co. *Paris.*—Preserved alimentary substances, &c.

328. QUILLET, A. & SON, *Paris.*—Mustard, vinegar, pickles.

329. BORDIN-TASSART, A. *Paris.*—Mustard, vinegar, and pickles.

330. MAILLE ET SEGOND, *Paris.*—Mustard, vinegar, and pickles.

331. DUBOSC, *Paris.*—Mustard.

332. AMAND-GUENIER, *Auxerre* (*Yonne*). —Mustard.

333. COLLECTIVE EXHIBITION OF THE DEPARTMENT OF THE CÔTE D'OR.—Mustard. (7 Exhibitors).

334. DIETRICH BROS. *Strasbourg* (*Bas-Rhin*).—Mustard.

335. ROUZÉ, H. *Paris.*—Preserved fruits.

336. PERRIER, J. P. F. *Crest (Drôme).*—Preserved truffles.

337. BATTENDIER, A. F. JUN. *Paris.*—Preserved truffles.

338. GALLOIS, H. *Paris.*—Ground pepper.

339. JOURDAIN, E. *Paris.*—Preserved fruits and sweetmeats.

340. PHILIPPE, C. & CANAUD WIDOW, *Nantes (Loire-Inf.).*—Preserved alimentary substances.

341. PELLIER BROS. *Mans (Sarthe).*—Sardines in oil.

342. RODEL & SONS, *Bordeaux (Gironde).*—Preserved alimentary substances.

343. SAUCEROTTE & PARMENTIER, *Lunéville (Meurthe).*—Preserved fruits and vegetables.

344. HÉRON, *Paris.*—Fish, preserved, without the bones; essence of coffee.

345. BALESTRIÉ, R. *Cancarneau (Finisterre).*—Sardines in oil.

347. LIREUX, S. *Le Havre (Seine-Inf.).*—Caramel.

348. CONNIÉ & MARTIN, *La Rochelle (Charente-Inf.).*—Sardines.

349. VOISIN, A. *Paris.*—Chestnut conserve.

350. SALLES, A. & SON, *Paris.*—Preserved alimentary substances.

350A. ALLARD, SON, & CO. *Sarlat (Dordogne).*—Nut oil.

## SUB-CLASS B.

[REGION I.—Producing Wheat, but neither Wine for Exportation, nor Silk.]

(*Nord*).
351. AGRICULTURAL ASSOCIATION OF LILLE.—Collection of cereals, forage, oils, alcohols, &c. (19 Exhibitors).

352. AGRICULTURAL SOCIETY OF BOURBOURG.—Collection of cereals, forage, colza, flax, &c. (21 Exhibitors).

353. AGRICULTURAL SOCIETY OF HAZEBROUCK.—Cereals, forage, flax, hops, tobacco, &c. (19 Exhibitors).

354. COMMUNE OF REXPOÈDE.—Wheat, oats, flax, beans, (11 Exhibitors).

355. VANDERCOLME, A. *Rexpoède.*—Wheat, oats in the sheaf, forage, &c.

356. HAMOIR, G. *Saultain.*—Soils, cereals, forage, sugar.

357. GOUVION-DEROY, *Denain.*—Cereals, sugar, alcohol: potash, obtained by calcination of the residue left after distilling beet-root juice.

358. FIEVET, *Masny.*—Wheat, oats in the sheaf, flax, sugar, &c.

359. CHEVAL, B.—Agricultural products.

360. VARDAELE, F. *Warhem.*—Corn, beet-root, flax, oil, oil-cake.

361. PORQUET-DOURIN, *Bourbourg.*—Wheat, oats, flax, pease.

362. RYCKELYNCK. *Beaudignies.*—Agricultural products.

363. SPIERS, J. A. *Valenciennes.*—Natural and artificial guano, made of the refuse of fish.

364. MESSERSCHMIDT, *St. Amand-les-Eaux.*—Strong vinegar, obtained by a new process.

(*Pas de Calais.*)
365. DECROMBECQUE, G. *Lens.*—Cereals, loaf-sugar, alcohol.

366. DELABY, A. & CO. *Courcelles-lez-Lens.*—Wheat, flax, beet-root sugar.

367. DELAUNE, A. *Courrières-lez-Lens.*—Sugar from beet-root molasses; alcohol from cane-sugar molasses; barytic products belonging to the manufacture.

368. MARQUIS D'HAVRINCOURT.—*Corbehem.*—Fleece; plan of a manure pit.

369. DE PLANCQUE, *St. André-lez-Gouy.*—Colza seed.

370. PROYART, *Hendecourt-lez-Gagnicourt.*—Cereals in sheaves, forage, flax.

(*Aisne.*)
371. COLLECTIVE EXHIBITION BY THE DEPARTMENT OF THE AISNE.—Agricultural products; indigenous raw opium; fleeces,

honey, flax; cocoons, wine, sugar, alcohol; chemical and pharmaceutical products used in agriculture; bricks, and refractory clays; millstones, &c. (52 Exhibitors).

### (Oise.)

373. NORMAL AGRICULTURAL INSTITUTE OF BEAUVAIS. — Agricultural products including 276 varieties of wheat, 5 of rye, 16 of barley, 23 of oats, 30 of maize, 160 of potatoes, 12 of carrots, 22 of turnips, 38 of pease, 120 of kidney beans, 16 of onions, 20 of beet-roots, &c. Synoptical tables of the produce obtained from the chief varieties exhibited of wheat, carrots, &c.

374. FAULTE DE PUYPARLIER, A.—Compressed bread, a substitute for military and naval biscuit.

375. FLAMAND-SEZILLE, *Noyon.*—Shelled and husked peas.

376. COLLECTIVE EXHIBITION OF THE WHEAT AND WINE DISTRICTS.—Flour (14 Exhibitors).

### (Somme.)

377. COLLECTIVE EXHIBITION OF THE DEPARTMENTS OF SOMME.—Cereals, forage, oils, sugars, wood, leather, wool, ligneous fibres, clays, bricks, &c. (52 Exhibitors).

### (Seine-and-Marne.)

378. COLLECTIVE EXHIBITION OF THE DEPARTMENT OF SEINE-AND-MARNE.—Cereals, forage, alcohols, vegetables, honey, wax, cheese, &c. (55 Exhibitors).

### (Seine-and-Oise.)

379. COLLECTIVE EXHIBITION OF THE AGRICULTURAL ASSOCIATION OF THE DEPARTMENT OF SEINE-AND-OISE.—Fecula, oils, alcohol, cereals, &c. (23 Exhibitors).

### (Seine.)

380. CHODZKO, *Neuilly, near Paris.*—Model of a drying apparatus for fecal matters.

382. GRIVEL, CHATEAU, & BAYLE, *Paris.*—Manure from sewerage, &c.

383. KRAFFT, L. *Paris.*—Manure made from offal of an abattoir.

384. ROHART & SON, *Paris.* — Animal matters for manures.

385. ROUILLIER, E. *Paris.*—Beer, of various kinds.

386. VOLLIER, J. B. A. *Paris.*—Malt, hops, beer.

387. LABADY, *Paris.*—Beer.

388. BOUCHEROT, *Puteaux.*—Table and Bavarian beer.

389. L'HOMME-LEFORT, *Paris.*—Mastic for grafting and for curing unhealthy trees and shrubs.

390. DESCROIX, *Paris.*—Wine-vinegar.

391. VOIRIN, *Paris.*—Liqueurs.

392. FENAILLE & CHATILLON, *Paris.*—Fatty matters, and resinous oils.

394. THOURET, E. *Paris.*—Model of preservative granary.

395. VICAT, *Paris.*—Insect-killing powder, and apparatus used with it.

396. BEAUSSIER, *Paris.*—Indigenous tea.

397. BIGNON, *Paris.*—Products obtained by improved cultivation, and notices of the method pursued.

### (Seine-Inf.)

398. AGRICULTURAL SOCIETY OF THE ARRONDISSEMENT OF HÀVRE (7 Exhibitors). —Cereals, flax, fleeces, cyder, &c.

399. DESMAREST, *Bully.*—Cereals.

400. MOISSON, *Luzy.*—Oats in the sheaf.

401. MULLOT, *St. Aubin-Cilloville.*—Oats in the sheaf.

402. Rasset, *Minterollier.*—Wheat, oats, barley.

403. SÉMICHON. JUN. *Vieux-Rouen.*—Wheat in the sheaf and in grain.

404. MAMBOUR DELAGRAVE, *Foucarmont.*—Hops.

405. DUVIVIER, *St. Martin.* — Cyder, cheese.

406. JOLY, *La Mobraye.*—Cyder.

407. LESUEUR, *Forges-les-Eaux.*—Refractory clays.

### (Manche.)

408. SOCIETY OF THE POLDERS OF THE WEST.—Specimens of the soil, and products of the Polders.

409. MOSSELMANN & Co. *La Rocque-Genest.*—Limestone, lime; ridge tiles, hollow bricks, draining pipes.

410. LAJOYE, *St. Lô.*—Animal manure.

411. LEMOIGNE - DULONGPRÉ. — Cyder, kaolin.

(*Calvados.*)

412. DELAUNEY, A. *St.-Désir.*—Liquid resin.

413. DE VILADE, L. C. *Surire.*—Cyder-brandy.

(*Allier.*)

414. AGRICULTURAL ASSOCIATION OF MONTLUÇON.— Rye in the sheaf; casket made of different kinds of wood, &c.

415. AGRICULTURAL ASSOCIATION OF EBREUIL.—Wines of the country.

416. BARON DE VEAUCE, *Château de Veauce.*—Wines.

416. DE FINANCE, *Trévelles.* — Wheat, oats, hemp, and wool.

(*Corrèze.*)

418. AGRICULTURAL ASSOCIATION OF THE CANTON OF MEYSSAC.—Wines (8 Exhibitors).

419. COUNT J. DE COSNAC, *Château du Pui.*—Cereals, nuts, hemp, cyder, wine, &c. Samples of the soil.

420. MAVIDAL, *Bronceilles.*—Wine.

(*Puy de Dôme.*)

422. MARQUIS DE LA SALLE, *St. Germain-Lembrou.*—Red wines.

423. AUBERGIER, *Clermont.*—Indigenous opium, &c.

424. CHESNEAU, *Clermont.*—Vinegar.

425. DELMAS, *Besse-en-Chandèse.*—Liqueurs.

426. DUMAS-GIRAUD, *Courpiere.*—Artificial guano of Dumas.

(*Seine-and-Oise.*)

427. IMPERIAL AGRICULTURAL SCHOOL AND AGRONOMIC SOCIETY OF GRIGNON.—Collection of cereals, honey; specimens of soils and manures, &c.

(*Sarthe.*)

428. SOCIETY OF AGRICULTURE, SCIENCES, AND ARTS, OF THE DEPARTMENT OF SARTHE.—Cereals, and other agricultural products.

(*Ille-and-Vilaine.*)

429. DEPARTMENTAL AGRICULTURAL SOCIETY AND GENERAL COMMITTEE OF THE ASSOCIATIONS OF THE DEPARTMENT OF ILLE-AND-VILAINE.—Flax, hemp, Linen.

430. RITTER, *Fougères.*—Kirschwasser.

(*Côtes-du-Nord.*)

431. COLLECTIVE EXHIBITION OF THE DEPARTMENT OF THE COTES-DU-NORD.—Wheat, oats, flax, hemp, &c. (10 Exhibitors).

432. LECOQ, *Dinan.*—Flax seed.

(*Finisterre.*)

433. COMMUNE OF ROSNOEN.—Cereals, roots, forage, types of animals.

434. BRIOT DE LA MALLERIE, *Kerlogotu.*—Corn, buck-wheat, wines.

435. COMTE DU COUEDIC, DIRECTOR OF THE IRRIGATION SCHOOL OF LÉZARDEAU.—Plans in relief, &c.

436. HERTEL, *Kerbourg.*—Agricultural products.

437. COLLECTIVE EXHIBITION OF THE THREE REGIONS OF FRANCE.—Sugars (8 Exhibitors).

438 COLLECTIVE EXHIBITION OF THE WHEAT REGION.—Alcohol (6 Exhibitors).

439. ———Fecula, and starch (6 Exhibitors).

440. CHIRADE, P. P. *Paris.* — Eggs, butter.

441. BOURDOIS & SON, *Paris.* — Cattaert's mode of preserving cheese.

442. COLLECTIVE EXHIBITION OF THE WHEAT DISTRICT.—Cheese (7 Exhibitors).

443. COLLECTIVE EXHIBITION. — Oils, and oil-cake, from linseed, colza, &c. (4 Exhibitors).

444. IMPERIAL AND CENTRAL HORTICULTURAL SOCIETY OF PARIS.—Fruits grown in the wheat region and round Paris—done from nature by M. Buchetet.

(*Aisne.*)

445. VICOMTE DE COURVAL, *Pinon.*—Specimens of wood cut by the new and old methods.

446. ROBERT, DR. E. *Bellevue (Seine-and-Oise)*.—Treatment and cure of diseased elms illustrated, &c.

447. COLLECTIVE EXHIBITION OF THE WHEAT DISTRICT.—Wools (3 Exhibitors.)

448. ———Flax (2 Exhibitors).

449. BOURSIER-DELAPLACE, *Chevrières (Oise)*.—Hemp.

450. LÉONI & COBLENZ, *Vaugenlieu (Oise)*.—Hemp, mechanically prepared by a new process.

451. DEMOLON & COCHERY, *Paris.*— Phosphate of lime.

452. DURIVAU, *St. Jean-de-la-Motte (Sarthe)*.—White and red Brouassin wine.

453. PERS, A. *Paris.*—Artificial manures.

454. CAILLEAUX, *Melun (Seine-and-Oise)*.—Plans for drainage.

455. RICHARD DE JUVENCE, *Versailles (Seine-and-Oise)*.—Plans, agricultural statistics, &c.

456. HEUZE, G. *Grignon (Seine-and-Oise)*.—Plain and coloured engravings of cereals, and of the plants required for manufacturing processes, and for forage—after the drawings of M. Rouyer; agricultural maps of France.

457. ABOILARD, C. *Paris.* — Proposed methods of draining.

458. DUVILLERS, *Paris.*—Plans of parks and gardens.

459. BRUNIER, *Rouen (Seine-Inf.).*— Plan of a distillery.

———

[REGION II.—Producing Wheat and Wine for Exportation, but no Silk.]

(*Ardennes.*)
481. CHANAL, *Mézières.*—Hops.

482. GOSSIN, C. *Latour-Audry.*—Osiers.

(*Meuse.*)
483. JURY OF THE MEUSE.—Wines.

484. MAUPAS & SCHLAÏSSE, *Bar-le-Buc.* —Fossil phosphate of lime.

485. MÉRION, *Bar-le-Buc.* — Sparkling wines.

(*Moselle.*)
486. MANGUIN, C. E. *Metz.*—Fecula.

487. ST.-JACQUES, *Metz.*—Starch.

488. CHAMPIGNEUL, *Metz.*—Soft corn.

489. HIRT, J. *Sarreguemines.*—Wines.

490. MACHETAY, JUN. *Metz.*—Wines.

(*Meurthe.*)
491. BLOCH & SON, *Tomblaine.*—Tapioca, sago, colourless glucose, fecula and starch. A table of the chief substances derived from the potato.

492. DERMIER, *Nancy.*—Plants prepared for exportation.

493. VOIRIN, JUN. *Nancy.*—Liqueurs.

(*Vosges.*)
494. COLLECTIVE EXHIBITION BY THE AGRICULTURAL ASSOCIATION OF THE ARRONDISSEMENT OF EPINAL.—Cereals, oils, honey, tiles (10 Exhibitors).

495. AGRICULTURAL ASSOCIATION OF RAMBERVILLERS.—Hops, farina, groats, &c. (20 Exhibitors).

496. COLLECTIVE EXHIBITION BY THE ARRONDISSEMENT OF REMIREMONT.— Kirschwasser and gentian brandy.

497. COLLECTIVE EXHIBITION BY THE DEPARTMENT OF VOSGES.—Fecula (15 Exhibitors).

498. CUNY, GERARD, *St.-Dié.*—Corn.

499. FLEUZOT & THIERRY, *Val d'Ajol.*— Kirschwasser.

500. LEMASSON, *Val d'Ajol.*—Kirschwasser.

501. PARIS, *Remiremont.*—Kirschwasser.

(*Bas-Rhin.*)
502. AGRICULTURAL COLONY OF OSTWALD.—Agricultural products; tobacco.

503. SCHATTENMANN, *Rouxvillers.*—Geological specimens of the Eastern region; cereals, leguminous, tuberous and oleaginous plants; forage; textile matters; tobacco; wines; types of animals, &c.

504. VOCLKER, *Strasbourg.*—Grain, &c.

505. ANDÉOUD, *Avolsheim.*—Wines.

506. DANTHIN, *Oltrott.* Red wine.

507. PASQUAY BROS., *Wasselonne.*— Wine.

508. PROST, *Strasbourg.*—Wine.

509. REISSER, *Oltrott.*—Kirschwasser.

510. REYSZ, *Traenheim.*—Wine.

511. SPIELMANN, *Werthoffen.*—Wine.

512. STOLZ, SEN. *Andlau.*—Wines and Kirschwasser.

513. ZEYSSOLFF, *Strasbourg.*—Wines.

514. ZIMMER, *Wangen.*—Wines.

515. COLLECTIVE EXHIBITION BY THE DEPARTMENT OF THE HAUT-RHIN, INCLUDING THE ARRONDISSEMENT OF COLMAR.—Wines, brandy, and liqueurs (18 Exhibitors).

(*Marne.*)

516. AGRICULTURAL SOCIETY, AND CENTRAL ASSOCIATION OF THE DEPARTMENT OF MARNE.—Agricultural products.

517. ROQUEPLAN, N. *Reims.*—Sparkling wine of Champagne.

518. RICBOUR MEUNIER, *Avenay.*—Semoule, farina of groats.

519. CHEMERY, *Noirmont.*—Agricultural products.

(*Haute-Marne.*)

521. PASSY, A. & Co. *Arc-en-Barrois.*—Wrought indigenous wood.

522. DELETTRE-COURTOIS, *Arc-en-Barrois.*—Preserved truffles and other eatables.

(*Haute-Saône.*)

523. LOCAL COMMITTEE OF GREY.—Wines.

524. JURY OF LURE.—Wines.

525. MARQUIS D'ANDELARRE, *Lure.*—Plans in relief, of a stable, with a granary, for forage.

526. LUZET, *Luxeuil.*—Wines.

(*Jura.*)

527. SOCIETY OF AGRICULTURE, SCIENCE, AND ARTS, OF POLIGNY.—Claret; brandy made from the husks of the grape, gentian brandy (6 Exhibitors).

528. BURY, *Lons-le-Saunier.*—Wines.

529. GAUDARD, *Courbouzon.*—Wines.

530. GENOT BROS. *Lons-le-Saunier.*—Wines.

531 MANGIN & GIROD, *Lons-le-Saunier.* Wines.

532. MONARD, *Lons-le-Saunier.*—Wines.

533. MOREAU, *Quintigny.*—Wines.

534. RENAUD, *Lons-le-Saunier.*—Wines.

(*Aube.*)

535. BEAU, SEN. *Riceys.*—Red and pink wine.

536. GRATTEPAIN, *Loches-sur-Ource.*—Wines.

(*Cote-d'Or.*)

538. STRONG WINES CO. OF BURGUNDY.—Wines of the Romanée-conti, the Clos de Vogeot, and Chambertin.

539. BOUTON, E. *Montigny-sur-Aube.*—Preserved truffles, wines.

540. COUQUAUX-JOLY & Co. *Dijon.*—Liqueurs.

541. DEVILLEBICHOT WIDOW J. *Dijon.*—Liqueurs.

542. MARQUIS DE LAGARDE.—Wines of the Romanée-conti, the Clos de Vogeot, and Chambertin.

543. HUAN & FONTAGNY, *Dijon.*—Vinegar.

544. SAGLIER, *Dijon.*—Truffles.

545. VIEILHOMME, H. *Paris.*—Wines of Musigny, and Petits-Vougeots.

546. CHOLET-LHUILLIER, *Fixin.*—Wines of Chambertin, Corton and Volnay.

547. CRÉTIN-CHOLET, *Fixin.*—Wine of the Clos Napoleon.

548. GRAY, M. *Dijon.*—Mustard.

(*Yonne.*)

549. AGRICULTURAL SOCIETY OF JOIGNY.—Corn and various agricultural products.

550. BARDEAU, E. *Fleury.*—Wheat, and oats in the sheaf.

551. ROY, *Tonnerre.*—Alcohol.

552. BONNEVILLE, A. A. *Villeneuve-sur-Yonne.*—Confection of grapes; wines of Chablis, Côte, Moutonne, Saint-Julien, and Chaumont.

553. LE PÈRE, C. *Auxerre.*—Wines.

(*Saône-and-Loire.*)

554. COLLECTIVE EXHIBITION OF THE MÂCONNAIS.—Wines (44 Exhibitors)..

555. THE COMMUNE OF ROMANÉCHE.— Wines of Thorins, Romanéche, and Moulin-a-vent.

556. DESMARQUEST & Co. *Macon.*— Wines of Moulin-a-vent, Moriers, and Fuissé-Pouilly.

557. ANDELLE, G. *Epinac.*—Wines.

558. COMTE DE BÉTHUNE, *Macon.*— Wines of Sommeré.

559. COLLECTIVE EXHIBITION OF MA-CON.—Wines (8 Exhibitors.)

560. DE MURARD, *Macon.*—Wines of Juilliénas.

561. RUFFARD, *Macon.*—Vinegar.

562. BEAUPÉRE & Co. *Chalons-sur-Saône.* —Beet-root sugar.

(*Rhône.*)

563. ASSOCIATION OF BEAUJEN.—Wines (107 Exhibitors.)

564. BLAIN, *Lyon.*—Wine.
565. TREVOUX, E. *Lyon.* — Artificial guano.

(*Loire.*)

566. AGRICULTURAL SOCIETY OF PER-REUS.—Wines of the Roannais.

567. MARQUIS DE VOUGY, *Roanne.*— Wines.

(*Haute-Loire.*)

568. COLLECTIVE EXHIBITION OF THE, SOCIETY OF AGRICULTURE, SCIENCE, ART, AND COMMERCE OF THE PUY.—Cereals, leguminous plants, forage, draining tiles (3 Exhibitors).

(*Ain.*)

569. GL. BAR. GIROD DE L'AIN, *Gex.*— Merino fleeces.

570. JACQUIN, *Seyssel.*—White wine of Seyssel.

(*Savoie.*)

571. ROUX-VOLLON, *St.-Jean-de-Belle-ville.* Gruyère cheese.

572. TATOUT, J. *St. Bon.* — Gruyère cheese.

573. CHRISTIN, *St.-Pierre-de-Belleville.*— Wine.

(*Eure-and-Loir.*)

574. AGRICULTURAL AND HORTICUL-TURAL SOCIETY OF THE DEPARTMENT OF EURE-AND-LOIR.—Results of felling, and forest culture by a new method; wool.

575. RICOUR, *Chartres.*—Cereals in the sheaf and in the ear.

(*Loiret.*)

576. COLLECTIVE EXHIBITION BY THE DEPARTMENT OF LOIRET.—Wines (17 Exhibitors.)

577. ——Vinegar (6 Exhibitors).

578. ——Honey and Wax (5 Exhibitors).

579. ——Saffron (6 Exhibitors).

580. DE BÉHAGUE, *Dampierre.*—Fecula.

581. ANSELMIER, DIRECTOR OF THE FARM SCHOOL OF MAUBERNEAUME.—Cereals and roots.

582. DAVID, *Orléans.*—Bitters.

583. HOARAU, *Orléans.*—Prunes.

(*Loire-Inf.*)

584. IMPERIAL AGRICULTURAL SCHOOL OF GRAND-JOUAN. — Cereals, plants for forage, angelica; liqueurs.

585. LIAZARD, A. *Tréguel.*—Collection of cereals, oleaginous plants, forage; wine, cyder, brandy, vinegar; oak bark, woods; fleeces; wax, honey, eggs; &c.

586. JOUBERT, *St. Herblon.*—Wines.

587. LEROUX & Co. *Nantes.*—Manures.

588. DERRIEN, E. *Chantenay-Nantes.*— Manure.

(*Maine-and-Loire.*)

589. INDUSTRIAL AND AGRICULTURAL SOCIETY OF ANGERS.—Wines.

590. HENNEQUIN, D. *Angers.*—Grains for soups and forage.

591. COMBIER-DESTRE, *Saumur.* — Brandy.

592. BOURDON & JAGOT, C., *Saumur.*— Wines.

593. BOLOGNÈSI, *Saumur.* — Elixir "Raspail."

(*Loir-and-Cher.*)

594. EXHIBITION BY THE DEPARTMENT OF LOIR-AND-CHER.—Wines, vinegars and alcohols (11 Exhibitors).

595. SOYER, *Nouan*.—Wooden poles.

596. DESVAUX-SAVOURÉ, *Beauchéne*.—Cyder.

597. BRETHEAU-AUBRY, *Meusnes*.—Flints.

(*Cher*.)

598. AGRICULTURAL SOCIETY OF THE DEPARTMENT OF THE CHER.—Wines of Sancerre, Ricardes, La Pincette, Côteau, &c. (7 Exhibitors).

599. AGRICULTURAL ASSOCIATION OF AUBIGNY.—Twenty-four kinds of grain; prunes.

600. LALOUEL DE SOURDEVAL, *Laverdines*.—Soils, cereals, sugar, alcohol.

(*Indre*.)

601. COUSIN-MONOURY, *Issoudun*.—Vinegar.

602. GODEFROY, M. *Reuilly*.—Red wine.

603. AGRICULTURAL SOCIETY OF CHÂTEAUROUX.—Specimens of soils and their products; wines of Lamoustière, Châteauroux, Lagnys, Argenton, Veuil-la-Tourdubreuil; beet-root alcohol; artificial fruits (13 Exhibitors).

(*Indre-and-Loire*.)

604. AGRICULTURAL ASSOCIATION OF THE ARRONDISSEMENT OF CHINON.—Agricultural products, wines, liqueurs, fruits, &c. (17 Exhibitors).

605. ASSOCIATION OF THE PROPRIETORS OF VOUVRAY.—Wines.

606. HÉBERT, A. *Athée, near Tours*.—Starch, fecula, flour.

607. DELABROUSSE, *Civray-sur-Cher*.—Wine from the slopes of the Cher.

608. HARDY, *Joué-lès-Tours*.—Wine.

609. PETIT DE VAUZELLES WIDOW.—Wines.

610. ROUILLÉ-COURBÉ, *Tours*.—Red and white wine.

611. VAUGONDY, *Rochecorbon*.—Wine.

612. DESBORDES & VOISIN, *Chinon*.—Vinegar.

(*Deux-Sèvres*.)

613. HORTICULTURAL SOCIETY OF NIORT. — Plants of fruit and forest trees (9 Exhibitors).

614. APERCÉ, *Giffont*.—Wheat, maize, barley, nuts, trefoil, colza seed, œilette; a fleece.

615. DE MESCHINET.—Specimens of soils, wheat, oats, barley.

616. MICHCAUD, *La Charrière*.—Trefoil seed.

617. PINARD, *St. Étienne*.—Wines and brandy.

618. DAVID, *Niort*.—Brandy

619. FONTAINE, *Greffier*.—Brandy.

620. DESCOLLARD, *Epannes*. — Ray-grass.

621. PRIEUR, *Epannes*.—Hemp.

622. GRIFFIER-VERRASSON, *Niort*. — Osiers.

(*Vienne*.)

623. AGRICULTURAL ASSOCIATION OF THE ARRONDISSEMENT OF CHATELLERAULT.—Wheat in the sheaf and in the grain; leguminous plants and forage; textile and oleaginous plants; feathers, down; honey, wax; wines of Vaux, and Saint-Romain, vinegar, oil, mustard; truffles (26 Exhibitors).

624. ASSOCIATION OF CIVRAY.—Collection of wheat in the grain and in sheaves, &c.

625. DE LARCLAUSE, DIRECTOR OF THE FARM SCHOOL OF MONTS.—Agricultural products.

(*Haute Vienne*.)

626. BRUCHARD, DIRECTOR OF THE FARM SCHOOL OF CAVAIGNAC.—Geological specimens.

(*Dordogne*.)

627. DE LENTILLAC, DIRECTOR OF THE FARM SCHOOL OF LAVALLADE.—Collection cereals, leguminous plants, tobacco, silkworms, eggs, silk cocoons.

628. LASALVÉTAT, H. *Périgueux*.—Alimentary preserves.

629. HOARAU, DE LA SOURCE, *Chateau de Ponthet*.—Prunes.

630. GOURSALLE, *Périgueux*. — Yellow wax.

631. Bourson, E. *Farcies.* — Leaf tobacco, prunes, and red wine.

632. Blanc, *Périgueux.* — Alimentary preserves.

633. Allard, Son, & Co.—Nut oil, nut-bread.

634. Collective Exhibition of the Department of the Dordogne.—Wines and liqueurs (34 Exhibitors).

(*Gironde.*)

635. Agricultural Society of the Gironde.—Cones, seeds, &c. of the maritime pine; resin, tar, oils, hops, &c. (10 Exhibitors).

636. Clamargeran, *La Lambertie, near St. Foy.*—Agricultural products.

637. Constantin, *Bordeaux.* — Rich wines of Bordeaux.

638. Rousse, J. *Bordeaux.*— Alchohol, &c., obtained by a new method of distillation.

(*Lozère.*)

639. Society of Agriculture, Industry, Science, and Arts of the Department of the Lozère.—Agricultural products.

(*Vendée*).

640. Jury of the Vendée.— Agricultural products.

(*Charente-Inf.*)

641. Chamber of Commerce of Rochefort.—Wood, wheat, and other agricultural products; building and moulding sand, sulphuret of iron, refractory clay, &c.

642. Bouscasse, Director of the Farm School of Puilboreau.—Beet-root seed, brandy, red wine.

643. Guillon-Desamis, *La Côte, near Nieul-sur-Mer.*—Oysters.

644. Dr. Kemmerès, *La Côte de Rivadoux, Ile de Ré.*—Oysters, &c.

645. Lem, Widow, *St. Martin, Ile de Ré.* —Honey.

646. Collective Exhibition by the Department of the Charente Inf.— Wines and brandy (10 Exhibitors).

647. Dr. A. Menudier, *Pleaud - Chermignac.*—Wine.

648. Conte & Co. *St. Pierre d'Oleron.*— White wine vinegar.

649. Olivier, *La Flotte, Ile de Ré.*— Vinegar.

650. Robineau, P. & Co. *La Tremblade.* —Strong and clarified vinegar.

(*Charente.*)

651. Collective Exhibition by the Department of the Charente.—Brandy, and alcohol (9 Exhibitors).

652. E. Thiac, *Puyréaux.*— Specimens of the soil, potatoes, beet-root, plan of the farm, wine, &c.

653. Brumauld des Allées, *St.Cloud.*— Cement and hydraulic lime.

654. Galland, *Ruffec.*—Corn for poultry.

660. Collective Exhibition.— Agricultural implements and produce (37 Exhibitors).

661. Collective Exhibition. — The wines of Champagne (19 Exhibitors).

662. Collective Exhibition. — The wools of the wine region (12 Exhibitors).

663. Collective Exhibition. — The wines of Burgundy (248 Exhibitors).

664. General Administration of French Tobaccos. — Indigenous tobacco, &c.

665. Collective Exhibitors.—Indigenous tobacco (2 Exhibitors.)

666. Voeleker, *Benfield (Bas-Rhin).*— Products obtained from chicory root.

667. Sengenwald, *Strasbourg (Bas-Rhin).*—Madder and its products.

668. Collective Exhibition.—Hops of the wine region (4 Exhibitors).

669. Collective Exhibition. — The wines of Bordeaux (289 Exhibitors.)

670. Normal School of the Department of the Haut-Rhin.—Cereals and farinaceous grain (5 Exhibitors.)

671. Toilard. P. *Paris.*—Collection of grains and forage.

672. Tamiset, C. *Plombières-lez-Dijon (Côte d'Or).*—Wheat and bean flour.

673. PERTHUY - MARTINEAU, *Nantes* (*Loire-Inf.*).—Wines.

674. ROUCHIER, SEN. *Ruffec* (*Charente*). —Preserves, liqueurs, and biscuits.

———

[REGION III.—Producing Wheat, Wine for Exportation, and Silk.]

(*Lot.*)

690. BOUTAREL - MEMBRY, *Luzech.*— Wine.

691. CAPMAS, *Prayssac.*—Wine.

692. IZARN, C. *Cahors.*—Wine.

693. LABICHE, C. *Cahors.*—Wines.

694. VIEULS, JUN. *Gaillac.*—Wines.

695. CABANÈS & MALGOUISARD, *Gourdon.* —Liqueurs, and nut oil.

(*Lot-and-Garonne.*)

696. DÉFFEZ, C. G. A. *Nerac.*—Wheat, maize, red wine, brandy.

697. DUCOS-BERNARD, *Beauziac.*— Ears of corn.

698. NADAU, *St. Livrade.*—Prunes.

699. CUZOL, SON, & CO. *Castelmoron-sur-Lot.*—Prunes.

700. TRUANT, E. *Domaine de Pader.*— Red and white wine.

701. MARGUES & DUVIGNEAU, *Nérac.*— Armagnac brandy.

702. DÉHOC, LAROZE, & CO. *Mézin.*— Brandy.

703. SIGAUD, A. *Nérac.*—Liqueurs and fruits.

(*Tarn-and-Garonne.*)

704. HORTICULTURAL AND ACCLIMATIZATION SOCIETY OF MONTAUBAN.—Cocoons and silk.

705. AGRICULTURAL ASSOCIATION OF MONTAUBAN.—Wheat, millet, maize, giant rye (7 Exhibitors).

706. SOCIETY OF SCIENCE AND AGRICULTURE OF MONTAUBAN.—Cocoons, and silk.

707. COUDERC, & SOUCARET, JUN. *Montauban.*—Raw silk.

708. GASCOU, NEPH. & ALBRESPY, *Montauban.*—Cocoons, and silk.

709. AGRICULTURAL ASSOCIATION OF NÈGREPELISSE.—Montricoux marble.

710. AGRICULTURAL ASSOCIATION OF MONCLAR.—Wheat, maize, large chestnuts, raw hemp.

711. SOL, *Verdun.*—Ears of corn, and maize.

712. COLLECTIVE EXHIBITION OF MONTAUBAN.—Wine and brandy (13 Exhibitors).

(*Tarn.*)

713. ASSOCIATION OF THE WINE GROWERS OF GAILLAC (5 Exhibitors).— Wine.

714. THE MAYOR OF GAILLAC.—Wine.

715. THE MAYOR OF GRAULHET.—Trefoil seed.

716. RAYNAL & SON, *Gaillac.*— Trefoil seed, aniseed, prunes.

717. MARAVAL & CO. *Lavaur.*— Raw silk.

(*Landes.*)

718. COLLECTIVE EXHIBITION OF ST. SEVER.—Red and white wines of Chalosse.

719. COLLECTIVE EXHIBITION OF PARLEBOSCQ, AND THE SURROUNDING COMMUNES. — Low Armagnac brandy (16 Exhibitors).

720. DUPRAT, *Hontaux.* — Wine and brandy.

721. LABADIE, P. *Arthez*—Brandy.

722. DUPUY, *Mont-de-Marsan.*—Oil.

723. DIVES, H. *Mont-de-Marsan.*—Resinous products.

724. COLLECTIVE EXHIBITION OF ARMAGNAC.—Brandies (22 Exhibitors).

725. DARQUIER, *Lectoure.*—Red wine.

726. LAFFITTE, J. *Castres.*—White vinegar.

(*Haute-Garonne.*)

727. FORT DESPAX & BACOT, *Toulouse.*— Wine and liqueurs.

728. DELORME & CO. *Toulouse.*—Vegetable horse-hair, the produce of the dwarf palm.

(*Basses-Pyrénées.*)

729. PÉCAUT, *Salies.* — Wine, refined salt.

(*Hautes-Pyrénées.*)

733. FONTAN, *Bernadets-Debat.*—Wine.

734. NABONNE, *Madiran.*—Wine.

735. MANINAT, JUN. *Ossun.* — Mineral wafers.

(*Pyrénées-Orientales.*)

736. BONET-DESMARES, *St. Laurent-de-la-Salanque.*—Wines.

737. SALLENS, P.—Liqueur wine. Manure.

(*Ardèche.*)

738. MALLET-FAURE & SON, *St. Péray.*—Wines of St Péray, and Châteaubourg.

739. ROY, *Privas.*—Model of a farm waggon.

740. RICHARD, H. *Tournon-sur-Rhône.*—Wine.

741. PRADIER, J. *Annonay.*—Agricultural products, raw and prepared silk.

742. CHANGEA, *Lamastre.* — Raw and prepared silk.

743. NICOD & SON, *Annonay.*—Silkworms' eggs, and cocoons.

744. BUISSON, C. *La Tronche.*—Raw and prepared silk.

(*Isère.*)

745. AGRICULTURAL AND HORTICULTURAL SOCIETY OF THE ARRONDISSEMENT OF GRENOBLE.—Wheat, beet-root, nuts, hemp, oil, brandy, kirschwasser cocoons, honey, wax, hides, resinous products, manures, marbles, anthracite, charcoal (12 Exhibitors).

746. ARNAND, SEN. & Co. *Voiron.*—Liqueurs.

747. HEURARD D'ARMIEU, *Armieu St. Gervais.*—Nuts, and nut-oil.

(*Drôme.*)

749. COMBRIER, BROS. *Livron.* — Raw silk.

750. GAUTHIER A. *Chabeuil.*—Raw and prepared silk.

751. HELME, A. *Loriol.*—Raw and prepared silk.

752. LACROIX, P.— Raw and prepared silk.

753. LASCOUR, *Crest.*—Raw and prepared silk.

754. LEYDIER BROS. *Buis-lès-Barronies.*—Raw and prepared silk.

755. NOYER BROS. *Dieulefit.*—Raw and prepared silk.

756. SAUVAGEON, *Valence.*—Cocoons, obtained under the influence of electricity.

757. COLLECTIVE EXHIBITION (26 Exhibitors.—Hermitage wines.

758. COLLECTIVE EXHIBITION BY THE DEPARTMENT OF THE DRÔME.—Wines (12 Exhibitors.)

759. CHARRAS & SON, *Nyons.*—Liqueurs.

760. CHEVALLIER - ROBERT, & CUILLERIER, *Romans.*—Cherry liqueurs and ratafias.

761. MARKERT, G. *Tain.*—Wine, crême de l'Hermitage.

762. BLANC-MONTBRUN, *Chateau de la Rolière.*—White wine, raw silk.

THE CURE OF CHARVAT, PRESIDENT OF THE ASSOCIATION OF RÉAUVILLE.—Cereals, almonds, madder.

764. AGRICULTURAL ASSOCIATION OF RÉAUVILLE.—Cereals, wines.

765. BRUN, JUN. *Réauville.*—Maize, oats, French beans, madder.

766. BOUTAREL - MAUBRY, *Valence.* — Nuts, prunes, wines.

767. GIRARD.—Yellow wax.

768. GUERBY, V.—Agricultural products. Crozes wine.

769. DELHOMME, *Larnage.*—Wine, kaolin.

770. MARRON-STOUPANI, *Montelimart.*—Nugget, &c. of Provence.

771. PREMIER, & SON, *Romans.*— Preserved fruits, liqueurs.

772. ROBEUX, *Valence.*—Liqueurs.

773. GALOPIN.—Alimentary preserves.

774. PERRIER, J. *Crest.*—Alimentary preserves.

775. CHARBONNET & SON, *Montelimart.*—Preserved black truffles.

(*Gard.*)

776. Collective Exhibition by the Department of the Gard.—Wines and liqueurs (15 Exhibitors).

777. Lacombe, I. *Alais.*—Raw silk, &c.

778. Vernet Bros. *Beaucaire.*—Raw and prepared silk.

779. De Fournès, *Remoulins.* — Long-stapled upland cotton, grown in the domain of the Exhibitors, from Algerian seed.

780. Charenon, Bonifas, & Co. *Moussac.*—Liquorice juice.

781. David, P.—Liquorice wood and juice.

(*Herault.*)

783. Collective Exhibition by the Department of Herault — Wines and brandy (11 Exhibitors).

784. Jury of the Arrondissement of Montpellier.—Wine, oil, wool.

785. Nourrgiat, *Lunel.*—Raw silk, vegetable and animal substances; silkworms fed on the leaves of a sulphured mulberry tree.

786. Boyer & Heil, *Gignac.*—Preserved truffles, olives, aromatic essences.

(*Pyrénées-Orient.*)

787. Jury of Prades.—Honey.

(*Aude.*)

788. Collective Exhibition by the Department of Aude.—Wines (5 Exhibitors).

789. Delcasse, G. *Limoux.*—Fleeces, woollen-yarn, and wines.

790. Débosque, *Espéraza.*—Ferruginous water.

791. Denille, Director of the Farm School of Besplas.—Teasels, wheat, maize, forage, &c.

792. De Martin, J. *Narbonne.*—Sea salt, wine.

(*Vaucluse.*)

794. Chabaud, A. *Avignon.*—Raw and prepared silk.

795. Berton Bros. *Avignon.*—Wines of different growths.

796. Comte de Maleyssie, *Chateauneuf.*—Wine of Lanerthe.

797. Sautet, A. *Sorgues.*—Alcohol and sulphuric ether.

798. Faure, P. *Avignon.*—Madder, and wine.

799. Julian, Jun. & Hoquer, *Sorgues.*—Alizarine, madder, and its derivatives.

800. Leplay, H. & Co. *Avignon.*—Alcohol, derived from sorgo, beet-root, various kinds of grain, fruits of different kinds, indigenous and foreign molasses, madder, and husks of the grape.

801. Reynaud, *Pertuis.*—Alimentary preserves.

802. Bonnet, *Aps.*—Bark of the green oak.

(*Boúches du Rhône.*)

803. Agricultural Society of the Bouches-du-Rhône.—Wheat, forage, madder; oils, nuts, teasels; wine, wool (12 Exhibitors).

804. Agricultural Association of the Arrondissement of Aix.—Wheat, beans, resinous products, tobacco, teasels, madder, &c. (3 Exhibitors).

805. Agricultural Association of Aubagne.—Corn, maize, farina, oils.

806. Brunet, *Marseilles.*—Wheat, flour, Semoule.

807. Be Bec, P. Director of the Farm-School of Montauronne—Collection of almonds.

808. Aubert, F. *Aix.*—Wheat, teasels, farina.

809. Pomirol, *Marseilles.*—Wines.

810. Monier, *Aubagne.*—Wines.

811. Reinaud, Chappaz, & Co. *Marseilles.*—Liqueurs.

812. Jourdan, G. & Brive, Sen.—Wines, liqueurs, preserved fruits, &c.

813. Olive, Nephew, & Michel, *Marseilles.*—Liquorice juice.

(*Var.*)

814. Agricultural and Commercial Society of the Department of the Var.—Cereals, woods, cork, tobacco, wine.

815. Agricultural Association of the Arrondissement of Toulon.—Wines.

816. CORNEILLE & FABRE, *Trans.*—Cocoòns, raw silk, &c.

(*Corsica.*)

817. JURY OF AJACCIO.—Specimens of rocks and minerals; animal and vegetable products (14 Exhibitors).

818. JURY OF CALVI.—Wines, oils, tobacco.

819. COLLECTIVE EXHIBITION OF CORSICA.—Corsican wine (10 Exhibitors).

820. BATTIONI, *Bastia.*—Myrtle liqueurs, mulberry alcohol.

821. GASPARINI, I. *L'ille-Rousse.*—Alimentary pastes.

822. LINGÉNIEUR.—Cedrates, preserves.

823. GARINI & MARIOTTI, *Campele.*—Dried fruits, chestnuts, mulberry alcohol.

824. CAFFARELLI, J. *Bastia.*—Italian pastes.

825. BREGANTI, J.—Cigars.

826. LICCIA, *Monticello.*—Leaf tobacco.

827. JURY OF BASTIA.—Marbles and ores.

(*Hautes-Alpes.*)

828. COLLECTIVE EXHIBITION BY THE DEPARTMENT OF THE UPPER ALPS.—Madder, teasels, honey, wines, wax, cocoons. (10 Exhibitors).

(*Basses-Alpes.*)

829. RAYBAUD-L-ANGE DIRECTOR OF THE FARM SCHOOL OF PAILLEROLS.—Collection of grains specially cultivated in the silk region; flour, teasels, olives, honey, &c.

# EXHIBITION OF SILK, &c., FROM VARIOUS LOCALITIES.

830. DE BAILLET, *St. Germain-et-Mons* (*Dordogne*).—Silk.

831. BÉRARD & BRUNET, *Lyon* (*Rhône*).—Silks, raw and prepared.

832. CHABOD, JUN. *Lyon* (*Rhône*).—Cocoons.

833. COUNTESS C. DE CORNEILLAN, *Paris.*—Cocoons, raw silk, &c.

834. DUSEIGNEUR, P. *Lyon* (*Rhône*).—A collection of cocoons.

835. FARA, JUN. *Bourg-Argental* (*Loire*).—Silks, of Bourg-Argental, for Caen lace.

836. FRIGARD, *Bourg-Argental* (*Loire*).—Raw silk, white and yellow.

837. GUERIN-MENEVILLE, DOCTOR F. *Paris.*—Silkworms.

838. COLLECTIVE EXHIBITION.—Products of the olive tree (11 Exhibitors).

839. BLANCHON, L. *St. Julien-en-St. Alban* (*Ardèche*).—Yellow cocoons, raw and prepared silk.

840. BARRÈS BROS. *St. Julien-en-St. Alban* (*Ardèche*).—Raw and prepared silk.

841. SÉRUSCLAT, L. *Etoile* (*Drôme*).—Prepared silk.

842. MONESTIER, SEN. *Avignon* (*Vaucluse*).—Raw and prepared silk.

843. BLANCHON, SEN. *Flaviac* (*Ardèche*).—Raw and prepared silk.

844. PALLUAT & Co. *Lyon* (*Rhône*).—Prepared silks.

845. BOISRAMEY, JUN. *Caen* (*Calvados*).—Raw and prepared silk, for laces, &c.

847. TEISSIER DU CROS, *Valleraugue* (*Gard*).—Cocoons; raw and prepared silk, white and yellow.

848. CHAMBON WIDOW, *St. Paul-Lacoste* (*Gard*).—Raw silk, and organzine.

849. BONNET & BOUNIOLS, *Vigan* (*Gard*).—Cocoons, raw silk.

850. BROUILHET & BAUMIER, *Vigan* (*Gard*).—Cocoons, white and yellow; raw and prepared silk.

851. MARTIN, L. & Co. *Lasalle* (*Gard*).—Raw silk.

852. BOUDET, F. *Uzès* (*Gard*).—Raw silk.

853. CHAMPANHET-SARGEAS BROS. *Vals* (*Ardèche*).—Cocoons, raw and prepared silk.

854. FOUGEIROL, A. *Ollières* (*Ardèche*).—Raw silk, organzine, cocoons.

855. REGARD BRO. *Privas* (*Ardèche*).—Raw and prepared silk.

856. BISCARRAT, P. *Bouchet* (*Drôme*).—Raw and prepared silk.

857. CHARTRON & SON, *St. Vallier* (*Drôme*).—Silk.

858. FRANQUEBALME & SON, *Avignon* (*Vaucluse*).—Wrought Chinese and Japanese silk.

859. BANNETON, *St. Vallier* (*Drôme*).—Raw silk, organzine.

862. MAHISTRE, A. JUN. *Vigan* (*Gard*).—Raw silk.

863. COLLECTIVE EXHIBITION.—Rough and manufactured cork (8 Exhibitors).

## SPECIAL EXHIBITIONS.

880. IMPERIAL SOCIETY OF ACCLIMATIZATION.—Results of its labours, with reference to six species of mammifers, twenty-five of birds, eleven of silkworms, and ten of vegetables, acclimatized, or in process of being so; and to the industrial products derived from them.

881. ROUYER, L. *Paris.*—Paintings, engravings, and lithographs, of animals, and agricultural products.

882. VILMORIN-ANDRIEUX & Co. *Paris.*—Agricultural products, and industrial and economic products derived from them.

883. CHAMBRELENT, *Bordeaux* (*Gironde*).—Forest products of the Landes or the Gironde.

884. JAVAL, *Arès* (*Gironde*).—Forest products of the Landes of Gascony, at Arès.

885. FLORENT-PREVOST, F. *Paris.*—Part of a collection of preparations, to determine the food of French birds: they consist of stomachs, with analyses of their contents at the time they were examined, and the dates of such examinations.

886. MUSEUM OF NATURAL HISTORY, *Paris.*—The principal types of mammifers and birds of the three agricultural regions of France, both useful and mischievous.

887. A COLLECTION of the principal kinds of game of the three regions of France; made with the concurrence of the Museum of Natural History of Paris.

888. ELOFFE & Co. *Paris.*—Geological and Botanical collection: soils and subsoils of the three agricultural regions of France.

## CLASS IV.

941. CUSINBERCHE, JUN. *Paris.*—Stearic and oleic acids, wax candles, soap.

942. LEROY, C. & DURAND.—*Gentilly* (*Seine*).—Stearic and oleic acids, wax and other candles, soap.

943. DE MILLY, L. A. *Paris.*—Stearic and oleic acids, glycerine, wax candles, soda-soap.

944. PETIT BROS. & Co. *Paris.*—Oleic acid, glycerine, and wax candles.

945. TREMEAU & MALEVAL, *Vienne* (*Isère*).—Stearic and oleic acids, wax candles, and oleine soap.

946. GAILLARD BROS. *Paris.*—Stearic and oleic acids, wax candles, wax, and tapers.

947. BUREAU, C. *Bordeaux* (*Gironde*).—Wax; wax, and other candles.

948. AUTRAN, L. *Paris.*—Candles, some of them made of the tallow recovered after the manufacture of stearic acid.

949. AMENE, L. *Clermont-Ferrand* (*Puy-de-Dôme*).—Animal oil; olive oil, and saponine for lubrication.

950. BLANCHARD, G. *Lyon* (*Rhône*).—Purified oils and soap.

951. FAULQUIER-CADET & Co. *Montpellier* (*Hérault*).—Stearic and oleic acids, tapers, wax and tallow candles.

952. LE TAROUILLY, A. & Co. *Rennes* (*Ille-and-Vilaine*).—Bleached wax, tapers.

953. GOHIN, SEN. *Vire* (*Calvados*).—Teazles.

954. ROBERT GALLAND & Co. *Paris.*—Bituminous schist and its products.

955. D'AMBLY, C. & Co. *Paris.*—Buffalo-horn, imitating whalebone.

956. D'ENFERT BROS. *Paris.*—Gelatine and glue.

957. COGNIET, C. MARÉCHAL, & Co. *Paris.*—Spermaceti, paraffine, and lubricating oils.

958. ROUSSEAU DE LAFARGE, L. & Co. *Persan-Beaumont* (*Seine-and-Oise*).—Vulcanised india-rubber.

959. AUBERT, A. & GÉRARD, *Paris.*—Articles in caoutchouc, both hard and flexible.

960. ARNAVON, H. *Marseilles* (*Bouches-du-Rhône*).—White and marbled soaps.

961. DELATTRE & Co. *Dieppe* (*Seine-Inf.*).—Oil from the liver of the Squalus: manure from the refuse of fish.

962. MONTALAND, C. & Co. *Lyon* (*Rhône*).—Stearic and wax candles.

963. FOURNIER, F. *Marseilles* (*Bouches-du-Rhône*).—Stearic candles and soap.

964. CAUSSEMILLE, JUN. & Co. *Marseilles* (*Bouches-du-Rhône*).—Lucifer matches of wood and wax.

965. MILLIAU, JUN. *Marseilles* (*Bouches-du-Rhône*).—White soap.

966. ROUX. C. JUN. *Marseilles* (*Bouches-du-Rhône*).—Marbled soap.

967. ROCCA BROS. & NEPHEWS, *Marseilles* (*Bouches-du-Rhône*).—Marbled soap, oil from seeds, oil-cake.

968. GRESLAND, C. *Paris.* — Candle wicks.

969. ROULET, C. H. & CHAPONNIÈRE, *Marseilles* (*Bouches-du-Rhône*). — Marbled soap, oil from seeds, oil-cake.

970. GOUNELLE, C. *Marseilles* (*Bouches-du-Rhône*).—Marbled soap, and seed-oils.

971. SEMICHON, J. JUN. *Paris.*—Lamp-black.

972. JACQUEMART & Co. *Paris.*—Varnishes for carriages, metals, lithography, photography, and pictures.

973. NOIROT & Co. *Paris.*—India-rubber tubing without joint, and apparatus for making it.

974. VANSTEENKISTE WIDOW (DORUS), *Valenciennes* (*Nord*).—Starch.

975. SOCHNÉE BROS. *Paris.* — Spirit varnish for metals, oil paintings, and photography.

976. MICHAUD, C. H. *Paris.*—Seed-oils, and purified animal oils.

977. MORISOT, C. T. *Vincennes* (*Seine*).—Lubricating oils.

. 978. SAUVAGE & Co. *Paris.*—Illuminating and lubricating oils, tar, paraffine, and lamp-black.

979 CAHOUET & MORANE, *Paris.* — Candle moulds and candles.

980. TESTON, J. *Nyons,* (*Drôme*).—Oil for clocks and telegraph apparatus.

981. RIOT, L. M. T. *Paris.*—Soap made without heat.

982. GONTARD, A. & Co. *Paris.*—Marbled soap.

983. STEINBACH, J. J. *Petit-Quevilly, near Rouen* (*Seine Inf.*).—Starch and gums, for printing and finishing woven fabrics.

984. PLICHART & CUVELIER, *Valenciennes* (*Nord*).—Bone, and animal blacks.

985. LARTIGUE, J. *Bayonne* (*Basses-Pyrénées*).—Animal black and manure.

986. BRIEZ, F. JUN. *Arras* (*Pas-de-Calais*).—Fabrics in horse-hair, for the manufacture of oils, and of stearic acid, &c.

987. FERMIER DE LA PROVOTAIS & GAUMONT, *Paris.*—Pulp, fibrous substances, and paper, from the common broom.

988. CABANIS, F. *Paris.*— Sprigs and bark of the mulberry, for making paper-pulp, and for the manufacture of thread or woven fabrics.

989. DETHAN, A. *Paris.*—Lubricating oils.

990. FENAILLE & CHATILLON, *Paris.*—Fats, with resinous base for carriages; resin-oils.

991. PASQUIER, DE RIBAUCOURT, & Co. *Paris.*—Oils, and grease for lubrication, &c.

992. BARBIER & DAUBRÉE, *Clermont-Ferrand (Puy-de-Dôme).*—Products resulting from the manufacture of caoutchouc.

993. DELACRETAZ, *Paris.*—Stearic and oleic acids, soap, and candles.

994. MULER, P. *Rouen (Seine Inf.).*—Gelatine.

995. PLANCHON, S. *St. Hippolite (Gard).*—Glue, gelatine.

996. SERBAT, L. *St. Saubie (Nord).*—Putty for making the joints of steam-engines; oils, and grease for manufacturing purposes.

# CLASS V.

1011. CASTOR, A. *Paris.*—Collection of steam apparatus, used by exhibitor in the formation of canals and railways, and the construction of the bridges of Kehl and Argenteuil.

1012. ORLEANS RAILWAY Co. *Paris.*—Smoke-consuming locomotive; tender, and first-class carriage, &c.

1013. SAGNIER, L. & Co. *Paris.*—Sextuple weigh-bridge for locomotives, and other apparatus for weighing.

1014. GENERAL RAILWAY PLANT Co. *Paris.*—Carriage, switches, &c.

1015. BARANOWSKI, J. J. *Paris.*—Automatic signals to prevent the collision of trains.

1016. LYONS RAILWAY Co. *La Croix Rousse, Paris.*—Self-acting break, for stopping a waggon on an incline, steeper than about one in eight.

1017. JOINT-STOCK IRON-MASTERS' Co. OF MAUBEUGE (*Nord*).—Turn-table, cast-steel rails, hydraulic crane, &c.

1018. MAZILIER, *Paris.*—Designs and models of an iron road, without cast-iron or wood.

1019. EVRARD, *Douai (Nord).*—Axle; specimen of a new method of lubrication.

1020. MATHIEU, *Anzin (Nord).*—Miner's waggon.

1021. CABANY, A. *Anzin (Nord).*—Miner's waggon of galvanized iron, with patent axle.

1022. CAIL, J. F. & Co. *Paris.*—Locomotive.

1023. ACHARD, A. *Paris.*—Application of electrical apparatus to securing safety on railways, and a uniform water-level in steam-boilers.

1024. DE JOANNES, E. *Valenciennes (Nord).*—Design for an American sleeping carriage.

1025. MEYER & SON, J. J. & A. *Vienna (Austria).*—Designs for an articulated locomotive, cut off apparatus, &c.

1026. DELANNOY, A. F. *Paris.*—Grease box; model of locomotive.

1027. COQUATRIX, J. B. *Paris.*—Self-acting lubricating apparatus.

1028. RASTOUIN, A. *Château-Renault (Indre-and-Loire).*—Grease box, for waggon axles; model of lever for facilitating operations at railway stations; articulated waggon axle, for turning sharp curves.

1029. FONTENAY, T. *Grenoble (Isère).*—Plan of a smoke-consuming furnace for a locomotive.

1030. GUERIN, E. *Paris.*—Self-acting break for a railway waggon.

1031. GARGAN & Co. *Paris.*—Cistern-waggon for transport of liquids and manure; lever for working vehicles on railways, by hand; feed apparatus for steam boilers; designs for forgings, and parts of boilers, &c.

1032. NORTHERN RAILWAY Co. *Paris.*—Locomotive for steep inclines, &c.

1033. CATENOT-BERANGER & Co. *Lyon (Rhône).*—Decuple weigh-bridge for locomotives; and other apparatus for weighing.

1034. ARBEL, L., DEFLASSIEUX BROS. and PEILLON, *Rive-de-Gier (Loire)*.—Wheels for locomotives.

1035. VERDIÉ, J. F. & Co. *Firminy (Loire)*.— Wheels, mounted for waggons; tires of combined iron and cast-steel; springs, for railway and other carriages; iron, puddled and cast-steel.

1036. VIGNIER, *Paris*.— Designs for a safety apparatus for railways.

1037. ALEXANDRE - LESEIGNEUR & Co. *Paris*.—Tires and switches.

1038. DEYEUX, N. T. *Liancourt (Oise)*.— Cast-steel tires.

1039. DÉZELU & GUILLOT, *Paris*.—Railway carriage lamp.

1040. POMME DE MIRIMONDE, L. *Paris*. —Railway grease boxes.

———

1040A. POLONCEAU WIDOW, *Paris*.— Model of locomotive with eight coupled wheels, for curves of small radius.

1040B. DIDIER, *Paris*.—Model of a break.

# CLASS VI.

1041. DESOUCHES - TOUCHARD & SON, *Paris*.—Chariot, coupé d'Orsay, model of a handle for a coach door.

1042. PARIS GENERAL OMNIBUS COMPANY.—An omnibus.

1043. BELVALLETTE BROS. *Paris*.— A landau; designs for carriages.

1044. PERRET, C. *Paris*.—A Victoria vis-à-vis, designs for carriages.

1046. BECQUET, J. F. *Paris*.—A calash.

1047.' BENOIST BROS. *Nogent-sur-Seine (Aube)*.— Axle-tree ends, which may be greased without taking off the wheels.

1049. POITRASSON, P. *Paris*.—A chariot.

1050. MOINGEARD BROS. *Paris*.—Eight-springed Berline.

1051. MOUSSARD & Co. *Paris*.—Coupé d'Orsay.

1052. MUHLBACHER BROS. *Paris*.— Four-wheel carriage.

1053. COLAS,. DELONGUEIL, & COMMUNAY, *Courbevoie (Seine)*.—Carriage wheels.

1054. DUFOUR BROS. *Perigueux (Dordogne)*.—Carriage.

1055. CLIQUENNOIS BROS. *Lille (Nord)*. —Calash.

———

1056. ALEXANDRE - LESEIGNEUR & Co. *Paris*.—Wheel tires of cast steel.

1057. DEYEUX, N. T. *Liancourt (Oise)*.— Wheel tires of cast steel.

1058. VERDIÉ, J. F. & Co. *Firminy (Loire)*.—Tires formed of a combination of iron and cast steel; carriage-springs.

# CLASS VII.

1061. ONFROY & Co. *Paris.*—Mechanism for printing woven fabrics.

1062. BRISSET WIDOW, *Paris.*—Lithographic press.

1064. CALVET-ROGNIAT & Co. *Louviers* (*Eure*).—Ribbons and backs for cards.

1065. DE CELLES, *Paris.*—Sewing machines, which fasten each stitch by a weaver's knot.

1066. CALLEBAUT, C. *Paris.*— Sewing machines.

1067. JOURNAUX-LEBLOND, J. F. *Paris.* —Sewing machines.

1068. TAILBOUIS, E. *St. Just-en-Chaussée* (*Oise*).—Machine for making netting.

1069. BRUNEAUX, L. JUN. *Rethel* (*Ardennes*).—Power-loom for wool.

1070. BACOT, P. *Sedan* (*Ardennes*).— Power-loom for wool.

1071. TRIQUET, JUN. *Lyon* (*Rhône*).— Glazing machine.

1072. DELCAMBRE, *Lille* (*Nord*).—Composing machine.

1073. LANEUVILLE, J. B. V. *Paris.*— Machine for making watch-guards and purses.

1074. DESHAYS, A. *Paris.*—Machine for making various kinds of cords, &c.

1075. LEMAIRE, E. F. *Paris.*—Automatic machine for making silk lace.

1076. SCRIVE, H. *Lille* (*Nord*).—Requisites for carding cotton, wool, &c.

1077. HARDING-COCKER, *Lille* (*Nord*).— Combs for spinning mills, and for the combing of textile matters.

1078. RONZE, R. *Lyon* (*Rhône*).—Economical Jacquard-loom.

1079. GILLET, F. *Troyes* (*Aube*).—Circular looms for making nets.

1080. BETHELOT, N. *Troyes* (*Aube*). — Circular and rectilinear looms for hosiery.

1081. PESIER, E. *Valenciennes* (*Nord*).— Model of apparatus for refining beet-root juice, by the application of alcohol.

1082. HERMANN-LACHAPELLE&GLOVER, *Paris.* — Apparatus for making gaseous drinks. Portable steam-engine, with furnace inside the boiler.

1084. FRANÇOIS, E. S. *Paris.*—Apparatus for making seltzer water.

1085. BARIL BROS. *Amiens* (*Somme*).— Machine for cutting velvet.

1086. ANDRÉ & GUILLOT, *Paris.*—Apparatus for making seltzer water.

1087. MONDOLLOT BROS. *Paris.*—Apparatus for making seltzer water.

1088. FÈVRE, G. D. *Paris.*—Apparatus for making seltzer water.

1089. CAIL, J. F. & Co. *Paris.*—Apparatus for distilling in vacuo, at a low temperature; sugar-cane mill and steam-engine; animal charcoal filters; centrifugal apparatus, for refining sugar, &c.

1090. CHOUILLOU & JAEGER, *Paris.*— Currying machine, acting by means of helical blades.

1091. DE COSTER WIDOW, *Paris.* — Punching machine; machine for making sugar by centrifugal force; portable expansive engine.

1092. BERNIER, SEN. & ARBEY, *Paris.*— Machines for working wood.

1093. DE DIETRICH & Co. *Reichshoffen* (*Bas-Rhin*).—Machine for bending metals.

1094. PERIN, J. L. *Paris.*—Endless saw, and fittings.

1095. TOUGARD, E. F. JUN. *Bapaume-lez-Rouen* (*Seine Inf.*).—Machine for preparing printing plates.

1096. SCHMERBER BROS. *Mulhouse* (*Haut-Rhin*).—Mechanical pestle.

1097. FREY, P. A. & SON, *Paris.*—Portable saw, for forest work.

1098. CORSEL, *Paris.*—Machine for reeling cocoons.

1099. MARESCHAL, J. *Paris.*—Machine for shaping wood, by means of helical blades.

1100. MERCIER, A. *Louviers (Eure).*—Spinning and weaving apparatus for wool; machine for opening the wool before it is carded; cards of various kinds; spinning machine with 210 spindles; continuous machine with 210 spindles, Vimaut's principle; winding machine with 20 bobbins; machine with 20 bobbins, for making woof; loom for glazed fabrics; machine for making bobbins; circular carding engine; felting machine, &c.

1101. FRESNE, C. *Louviers (Eure).* — Plates and ribbons for cards.

1102. BAUDOUIN BROS. & JOUANIN, *Paris.*—Machine for making fishing nets.

1104. DURAND, F. & PRADEL, *Paris.*—Jacquard-loom, in which paper is substituted for pasteboard.

1105. BERNARD, F. *Bourges (Cher).*—A jack.

1106. DAVID BROS. & CO. *St. Quentin (Aisne).*—Dynamometer, for measuring the tension of threads for warp.

1107. VILLAIN, E. P. *Montmartre, Paris.*—Machine for making twisted fringes.

1108. CRESSIER, E. *Gras (Doubs).*—Horological tools.

1109. THOUROT BROS. *Vandoncourt (Doubs).*—Horological tools.

1110. CAMBRAY, *Valenciennes (Nord).*—Apparatus for the grouping of frames of different sizes.

1110A. ALAUZET, P. *Paris.*—Mechanical press, with variable movement.

1110B. BAZET, HAPPEY, & CO. *Paris.*—Apparatus for making gaseous waters.

1110C. BERJOT, F. *Caen (Calvados).* — Model of an apparatus for gaseous waters.

1110D. BOURGEOIS BOTZ, *Reims (Marne).*—Plates and ribbons for cards.

1110E. DUTARTRE, P. *Paris.*—Typographic press for two colours.

1110F. LEGAL, F. *Nantes (Loire-Inf.).*—Model of an apparatus for refining sugar in vacuo.

1110G. MATHIEU BROS. *Anzin (Nord).*—Sugar apparatus.

1110H. SILBERMANN, J. J. *Paris.*—Universal presses.

## CLASS VIII.

1111. CHÉRET, M. J. *Paris.*—Model of a movement for a beam-engine.

1112. BOUILLON, MÜLLER & CO. & MÜLLER, E. *Paris.*—Washing and bath apparatus.

1113. EGROT, E. A. *Paris.*—Continuous distilling apparatus.

1114. MOUQUET, H. *Lille (Nord).*—Model of an apparatus for concentrating syrups in vacuo.

1115. DROUOT, E. *Paris.*—Steam bread-making machine, moved by the waste heat of the oven.

1116. BOLAND, O. J. JUN. *Paris.*—Steam kneading machine, with helicoidal blades, without internal arbour, and capable of being reversed.

1117. LESOBRE, C. *Paris.*—Rolland's revolving hot-air oven, and mechanical kneading machine.

1118. MALBEC, A. A. *Paris.*—Apparatus for sharpening saws; grinding-stones, &c.

1119. BEZIAT, J. C. M. *Paris.*—A jack for racking wines.

1120. GLEUZER, J. L. *Paris.*—Cork-cutting machine.

1121. LEMERCIER, E. *Paris.*—Machine for screwing shoes, &c.

1122. RATEL, *Saulieu (Cote d' Or).*—Machine for beating out scythes.

1123. TUSSAUD, F. *Paris.*—Machine for hashing meat; machine for making bricks, and draining pipes; new mode of mounting a screw-propeller, permitting the vessel to be tacked rapidly in a small space.

1124. VANÇON, J. A. *La Bresse (Vosges).*—Apparatus for the carriage of fish alive.

1125. VANDEVILLE BROS. *Ferin (Nord).*—Hammers for dressing millstones.

1126. GODIN-CHAMBRIAND, *Guise (Aisne).*—Hammers for dressing millstones.

1127. DUFOURNET & Co. *Clichy-la-Garenne (Seine).*—Pasteboard sugar moulds.

1128. LECOQ, E. F. *Paris.*—Machine for making and printing railway tickets.

1129. GERVAIS, A. *Paris.*—Apparatus for warming green-houses.

1130. FAUCONNIER, F. L. *Paris.*—Mill for pounding and sifting dry substances.

1131. HÉDIARD, A. *Paris.*—Inexplosive instantaneous steam-boiler.

1132. NILLUS, E. *Havre (Seine-Inf.).*—Screw propeller apparatus.

1133. FAIVRE BROS. *Nantes (Loire-Inf.).*—Removable screw propeller, with variable speed.

1134. SUCKFULL, L. *Déville-lez-Rouen (Seine-Inf.).*—Pistons.

1135. LEMIELLE, T. *Valenciennes (Nord).*—Ventilators for mines.

1136. TOUAILLON, *Paris.*—Model of a mill; stove for drying flour; model of a machine for cleaning corn, and glazing rice; machine for dressing mill-stones; model of a machine for amalgamating auriferous matters, &c.

1137. ZAMBEAUX, *St. Denis (Seine).*—Portable vertical steam-boiler.

1138. PARENT, SCHAKEN, CAILLET, & Co. *Oullins (Rhône).*—Crane.

1139. DESBORDES, L. & RONDAULT, E. A. *Paris.*—Metallic dial barometers, metallic manometers and safety valves; metallic vacuum-indicators for low-pressure engines.

1140. LETHUILLIER-PINEL, *Rouen (Seine-Inf.).*—Magnetic indicators of the level of water in boilers, and automatic feeders.

1141. RENAUD, P. *Nantes (Loire-Inf.).*—Whistle-float for steam-boiler.

1142. PERREAUX, L. G. *Paris.*—Caoutchouc valves for pumps.

1143. POUGAULT, A. *Decize (Nièvre.)*—Steam cleansing machine.

1144. CAIL, J. F. & Co. *Paris.*—Steam apparatus for washing pit-coal; crane; tubular boiler, with fire-place of steel, and tubes of drawn iron with copper ends; horizontal condensing engine, with variable expansion, &c.

1145. HERMANN, G. *Paris.*—Chocolate grinding machine, &c.

1146. BERLIOZ, A. & Co. *Paris.*—Magneto-electric machine, for the production of light.

1147. BAUCHET-VERLINDE, & Co. *Lille (Nord).*—Ruling machines for account-books.

1148. THIERS, A. *Paris.*—Electric lamp, and light regulator.

1149. DEVINCK, F. J. *Paris.*—Machines for weighing, grinding, and wrapping chocolate.

1150. VORUZ, SEN. *Nantes (Loire-Inf.).*—Machines for moulding bricks and projectiles.

1151. LAURENS & THOMAS, *Paris.*—Tubular boiler with moveable furnace; demi-fixed steam engine; plans, and models.

1152. FARCOT & SONS, *Port St. Ouen (Seine).*—Horizontal condensing steam-engine, exhausting pump, pestle, &c.

1153. BRÉVAL, *Paris.*—Portable steam-engines.

1154. SAGEBIEN, A. *Amiens (Somme).*—Drawing of a water-wheel.

1155. CHENAILLER, P. C. *Paris.*—Evaporating apparatus.

1156. BOURDON, E. *Paris.*—Steam-engine, &c.

1157. LÉGAL, F. *Nantes (Loire-Inf.).*—Model of an apparatus for refining sugar in vacuo.

1158. CLERC, E. *Lyon (Rhône).*—Metallic manometers.

1159. DEDIEU, C. *Lyon (Rhône).*—Metallic manometers.

1160. DESBORDES, L. J. F. *Paris.*—Manometers, safety valves.

1161. FAUCONNIER, C. *Paris.*—Cranes: hydraulic crane for feeding locomotives.

1162. SILBERMANN, J. J. *Paris.*—Universal presses acting by water pressure, and capable of printing any kind of surface, &c.

1163. DURENNE, J. F. *Courbevoie (Seine).*—"Hydratmopurificateur," for purifying the water of steam boilers.

1164. FLAUD, H. *Paris.*—Fire-engine, Giffard's injectors.

1165. BOLLÉE & SON, *Mans (Sarthe).*—Hydraulic ram.

1166. LECOINTE, J. *St. Quentin (Aisne).* Apparatus for working a certain number of hydraulic presses under a continuous pressure.

1167. LETESTU, M. A. *Paris.*—Various pumps.

1168. DESPLAS, H. *Elbeuf (Seine Inf.).*—Fulling machine.

1169. BARRÉ, ROUGNON, & Co. *Paris.*—Machine for washing minerals.

1170. GAUTRON, B. J. *Paris.*—Hydro-extractor, for manufacturing fecula and starch.

1171. MOISON, F. T. *Mouy (Oise).*—Regulator with differential movement for steam engines; regulator for hydraulic engines; dynamometer for testing motive machines, when power is let out to hire; apparatus for removing the grease from wool.

1172. BARRÉ & BESNARD, *Paris.*—Tubular boiler, on a new principle.

1173. FONTAINE & BRAULT, *Chartres (Eure-and-Loir).*—Turbines, helical ventilator.

1174. GIGNOUX, G. G. *Lige, near Andenge (Gironde).*—Drawing of a machine for the transport of spawn.

1175. LECOUTEUX, H. *Paris.*—Double cylinder beam-engine.

1176. MIGNOT, H. *Paris.*—Metallic manometers.

1177. ALAUZET, P. *Paris.*—Mechanical press, with variable movement.

1178. DUTARTRE, A. B. *Paris.*—Typographic press for two colours.

1179. NORMAND, F. *Paris.*—Photograph of a press; model of universal joint, &c.

1180. WORMS DE ROMILLY, M. *Paris.*—Lifting machine.

1181. ARMENGAND, SEN. *Paris.*—Drawings of steam-engines, &c.

1182. VERNAY, L. *Paris.*—Machine for lifting and weighing heavy articles.

1183. POIRIER DE ST. CHARLES, *Gentilly (Seine).*—Type-founding machine.

1184. JUTTEAU, A. *Orléans (Loiret).*—Specimens in relief and drawings of a system of stone veneering, for the restoration of buildings.

1185. DE CHODSKO, N. F. B. *Paris.*—Smoke-consuming apparatus, for steam-boilers.

1186. PALAZOT, *Bordeaux (Gironde).*—Smoke-consuming apparatus for steam-boilers.

1187. DUMERY, C. J. *Paris.*—Plan of a smoke-consuming apparatus, and of an apparatus for preventing the incrustation of steam boilers.

1188. LENOIR & Co. *Paris.*—Expanded air-engine, acting by means of coal-gas inflamed by electricity.

1189. FORTIN-HERMANN BROS. *Paris.*—Apparatus for the distribution of water for public use, &c.

1190. DARDONVILLE, V. *Paris.*—Apparatus for the purification of water; charcoal filters, &c.

1191. COOLING APPARATUS CO. CARRÉ & Co. 149 *Rue de Menilmontant (Paris).*—Ma-chine for making ice by successive evapora-tions and liquefactions.

Fig. 1.

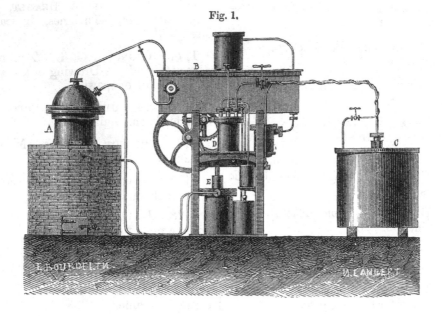

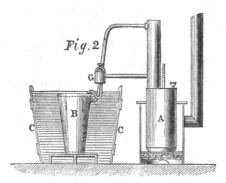

Fig. 2

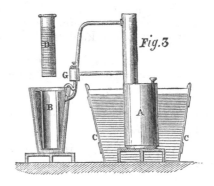

Fig. 3

M. Carré's apparatus is of two kinds—

One, intermittent for household use, for the production of artificial ice and sherbets, and icing champagne, and other drinks. It requires no special preparation, it being sufficient to warm it for a short time, and then leave it to cool to produce the desired effect. The sole expense, therefore, is the fuel; and one pound of coal can produce three or four pounds of ice.

The other, continuous, for the wholesale production of ice, and combined so as to insure the greatest economy of fuel. It is principally used for icing sherbets, decanters, but especially for freezing in general on a large scale. It may be used also—in breweries, for the precipitation of sulphate of soda from the waste water of salt reservoirs; for distilling sea-water by congelation; for concentrating dilute solutions; in candle manufactories, for solidifying fatty or glutinous substances, &c. It will be found invaluable in all trades generally which are obliged to stop working during summer, or which require a stock of ice, necessitating a considerable outlay of capital and ware-house room. One pound of coal can produce with this apparatus from ten to fifteen pounds of ice, according to size.

MR. FLOREAU, *Agent,*
9, *Fulham Road, Brompton.* ]

1192. PROFESSIONAL SCHOOL OF DOUAI (*Nord*).—Small horizontal steam-engine.

1193. VARRAL, EHVELL, & POULOT, *Paris.*—Steam-engine, saws, &c.

1194. QUILLACQ, L. A. *Anzin (Nord).*—Horizontal double cylinder steam-engine for working coal mines.

1195. SOCIÉTÉ NOUVELLE DES FORGES ET CHANTIERS DE LA MEDITERRANÉE, *Paris.*—400-horse-power marine engines for screw propeller.

1196. PRINCE DE POLIGNAC, *Paris.*—Curved-cylinder steam-engine.

1197. MAUZAIZE, J. N. *Chartres (Eure-and-Loir).*—Apparatus for flour mills.

1198. HOËL-RENIER, F. *Lille (Nord).*— Manometer.

1199. MARQUIS DE MONTAGU, *Paris.*— Apparatus for preventing chimneys from smoking.

1200. ANTONY MASSON, *Paris.*—Smoke-box.

———

1200A. HUBERT, H. *Paris.*—Feed-pump for the fountains placed in the gardens of the Horticultural Society.

1200B. FELDTRAPPE BROS. *Paris.*—Design for a cylinder for printing stuffs and paper.

1200C. COMTE DE EPRÉMESNIL, *Bernay (Eure).*—Design for a system of transmission to great distances.

1200D. DE COSTER WIDOW A. *Paris.*— Expansive portable engine.

1200E. DEHAYNIN, F. *Paris.*—Design for a machine to agglomerate small pit-coal.

1200F. HERMANN - LACHAPELLE & GLOVER, *Paris.* — Portable engine, with furnace within the boiler.

# CLASS IX.

1201. MAZIER, DR. *L'Aigle (Orne).*— Two-horse reaping and mowing machine.

1203. BELLA, F. *Grignon (Seine-and-Oise).* —Agricultural instruments.

1204. AGRICULTURAL ASSOCIATION OF SEINE-AND-OISE, *Versailles.*—Agricultural implements (10 Exhibitors).

1205. BARBIER & DAUBRÉE, *Clermont-ferrand (Puy-de-Dôme).*—Portable four-horse engine, reaping machine, &c.

1206. CUMMING J. *Orleans (Loiret).*— Portable engine, and thrashing machine.

1207. ALBARET & Co. *Liancourt (Oise).* —Portable steam-engine, thrashing machine, &c.

1208. GANNERON, E. *Paris.*—Agricultural machines.

1209. PINET, J. JUN. *Abilly (Indre-and-Loire).*—Horse-gin, thrashing machine, winnowing machines.

1210. JACQUET-ROBILLARD, *Arras (Pas-de-Calais).*—Sowing machines.

1211. AGRICULTURAL COLONY OF METTRAY *(Indre-and-Loire).*—Ploughs.

1212. PERRIGAULT, J. *Rennes (Ille-and-Vilaine).*—Apparatus for the graduated aëration of millstones while grinding, and the preservation of the mill dust.

1213. LANET & SON, *Cette (Hérault).*— Round and oval oak tuns.

1214. DOYÈRE & Co. *Paris.*—Model of subterranean cellars.

1215. PAVY, E. *Chemillé-sur-Dême (Indre-and-Loire).*—Model and drawings of a preservative granary.

1216. ROBIN, H. *Nantes (Loire-Inf.).*— Steam-reaping machine.

1217. RADIDIER & SIMONEL, JUN. *Paris.* —Root-cutters, &c.

1218. GRANDVOINET, J. A. *Paris.*— Plough.

1219. LECACHEUX, A. *Pieux (Manche).* —Plough, with instantaneous regulator.

1220. PARQUIN, L. V. *Villeparisis (Seine-and-Marne).*—Ploughs.

1221. CARDEILHAC & SON, *Toulouse (Haute-Garonne).*—Agricultural instruments.

1222. HOOK, J. M. *Paris.* Apparatus for the manufacture of potato-fecula.

1223. DENET, E. *Paris.*—Moulds, &c. for alimentary pastes.

1224. CREUZÉ DES ROCHES, R. *Chateau-de-Grandmaison (Indre).*—Model of a horse-gin, with a vertical movement.

1225. CLAUZEL & Co. *Sauves (Gard).*—Pitchforks of wood, shovel and scythe handles, &c., obtained naturally by giving the lote-tree and the green oak a particular form.

1226. DORLÉANS, E. *Paris.*—Machine for making mats for vines.

1227. REDIER, A. & Co. *Paris.*—Machines for sowing seed and spreading manure at the same time.

1229. HERBEAUMONT, P. F. *Paris.*—Designs for iron greenhouses.

1230. VANDERCOLME, A. *Rexpoëde (Nord).*—Plans of the commune of Rexpoëde.

1231. COMTE DU COUËDIC, *Quimperlé (Finistèrre).*—Plan of the domain of Lèzardeau, and the mode of irrigating it.

1232. BOURGERIE, E. *Remilly, Ardennes.*—Plough mounted on a fore carriage.

1233. DUMONT C. *St. Ouen Railway-station (Seine).*—Machine for decorticating colza.

1234. GUIGUET, *Paris.*—Drawings of machines and agricultural operations.

1235. BARBIER, C. *Paris.*—Plans of experiments made in the application of liquid manures; drainage pipes made to overcome the obstructions offered by roots.

1236. LAVOISY, *Paris.*—Mechanical churn.

1237. O'REILLY & DORMOIS, *Paris.*—Plans of iron greenhouses.

1239. HUART, H. *Cambrai (Nord).*—Plan of a preservative granary.

1240. LEBŒUF, *Paris.*—Venetian blinds for greenhouses.

1241. HAMOIR, G. *Saultain (Nord).*—Horse rake; metallic nave for the wheels of chariots and carriages.

1242. DEVAUX, A. C. L. *London.*—Model of a ventilating and preservative granary.

1243. BUISSON, *Tullins (Isère).*—Flour-mill.

1244. CAIL, J. F. & Co. *Paris.*—Straw and root cutters.

———

1245. ROY, *Privas (Ardèche).*—Model of an agricultural wheel-barrow.

1246. THOURET, E. *Paris.*—Model of a preservative granary.

## CLASS X.

1251. THE MINISTER OF AGRICULTURE, COMMERCE, AND PUBLIC WORKS.—Collection of models and drawings relating to the public works of the French empire :—

PASCAL, ANDRÉ, & DUSSAUD.—Model of a portion of the pier of the Napoleon Basin, Port of Marseilles; a plan of that port, with drawings of the means used in constructing it (1).

BOUNICEAU, LEMAITRE, ESCARRAGUEL, J., MONET, ESCARRAGUEL, A., ROULET, DUFFIEU, & PERRIN.—Sluice-gates of the citadel of Havre (2).

BOUNICEAU, BELLOT, COUCHE, E., ESCARRAGUEL, A., ROULET, DUFFIEU, & BATTAILLE.—Model of a graving-dock of the basin of the Eure, at Havre (3).

WATIER, CHATONEY.—Sluice-gate, &c. of St. Nazaire (4).

REYNAUD, L., FORESTIER, & MARIN.—Light-house of Barges, built on a rock, four miles from the the Port of Sables d'Olonne (5).

LEPAUTE, H., REYNAUD, L., & ALLARD.—Lenticular light-house apparatus; white light varied alternately by red and green flashes, without obscuration (6).

SAUTTER, REYNAUD, & ALLARD.—Lenticular light-house apparatus; red light with obscurations (7).

CADIAT, OUDRY, MATHIEU, MAITROT DE VARENNES, AUMAITRE, & ROSSEAU —A turning-bridge of sheet-iron, constructed at Brest (8).

PERRIER, GENDARME, DE BAVOTTE, & CONTE.—Collection of drawings relating to the irrigation Canal at Carpentras. (Vaucluse) (9).

CHANOINE & DE LAGRENÉ.—Model of a moveable dam, constructed on the Upper Seine (10).

POIRÉE, CAMBUZAT & MARINI.—Model of a moveable dam, constructed on the Yonne, near Auxerre (11).

LOUICHE - DEFONTAINES, CARRO, AND HOLLEAUX.—Model of a moveable dam constructed on the Marne (12).

ROZAT DE MANDRES.—Model of a portion of the dam of the reservoir of Settons, for supplying the navigation of the Yonne (13).

VERON-DUVERGER & SCIAMA.—Model of a part of the dam of the reservoir of Montaubry, for feeding the Canal du Centre (14).

MARX & BRUNIQUEL.—Photographs of the Napoleon Bridge, at St. Saveur, on the road from Paris to Spain (15).

MATHIEU, JOLY, A., & VIGOUROUX.—Model of the Bridge of St. Just, on the Ardèche (16).

TARBE RUINET, REGNIER, & DOURDET.—Chart of the Pools of the Dombes (Ain), in two colours (17).

BOURDALOUE.—Map of the levels taken throughout France (18).

FRANCOIS, J., & DUBRIEU.—The Baths of Ussat: subterranean works, searches for, and discovery of the sources of the mineral waters of that place (19).

FRANCOIS, J., & CHAMBERT.—Subterranean works, searches for, and discovery of the sources of the thermal waters of Bagnères-de-Luchon (20).

DAUBRÉE & JUTIER.—Plans and drawings of the search for, and conduction of, the waters of Plombières (21).

GRUNER.—Geological and mineralogical map of the department of the Loire (22).

ELIE DE BEAUMONT & DE CHANCOURTOIS.—Geological map of the Department of the Haute Marne (23).

LEVALLOIS, M.—Geological map of the Department of the Meurthe (24).

DORMOY, M.—Map of the coal-basins of Valenciennes and Mons (25).

JAQUOT, M.—Agricultural map of the Arrondissement of Toul (Meurthe) (26).

REIBELE, G., VIRLA, MAHYER, & BONNIN.—Cherbourg breakwater (27).

NOEL & CALMAN.—Models of the docks of Castigneau, Nos. 1, 2, and 3 (28).

CHATONEY, LAROCHE, LE BOUËDEC, & KIEZELL.—Dam, with vertical shafts, used in laying the foundations of the new basins of the Port of Lorient (29).

DEHARNE & VERRIER, M.—Process of gradual removal of rock La Rose, at Brest (30).

GARNIER, A., ANGIBOUST, & CHARVIN.—New dry-dock at the military port of Rochfort (31).

REYNAUD, L., & ALLARD, E.—Metallic lighthouse for New Caledonia (32).

VUIGNIER, FLEUR-ST.-DENIS, & SAPPEL.—Lattice railway bridge over the Rhine at Kehl; caisson used in laying the foundation, fragment of a pile in process of driving, and of a pile in its place (33).

ZEILLER & DECOMBLE. — Viaduct of Chaumont, on the Paris and Mulhouse Railway (34).

VUIGNER, COLLET-MEYGRET, & PLUYETTE.—Viaduct of Nogent on the Paris and Mulhouse Railway (35).

BOMMART, A., SURELE, DE LA ROCHE-TOLAY, REGNAULT, PAUWELLS, NEPVEU, & EIFFEL.—Sheet-iron railway bridge, over the Garonne, at Bordeaux (36).

MAGUES & SIMONNEAU.—The Canal du Midi, at the level of the torrent of Librons (37).

MICHAL, BELGRAND, DELAPERCHE, ROUSSELLE, & VALLÉE.—Map of the sewers of Paris, section of the sewer of Sebastopol, with its adjuncts; section of the great receptacle of the waters of Paris, with its adjuncts. Statistics of the water supply and sewerage of Paris, and of the works of the canal of St. Martin (38).

MICHAL, BELGRAND, & ROZAT DE MANDRES.—Map of the distribution of waters in the City of Paris; section of the reservoir of Passy, constructed for the supply of a part of Paris (39).

HERDEVIN, GUINIER, & FORTEIR HER-MANN.—Apparatus used in connection with the Parisian fountains (40).

MICHAL, ALPHAND, DARCEL, & KIND.—The artesian wells of Passy, near Paris; geological section of the strata crossing them; boring tools and apparatus used with them; plan of the workshop (41).

LAUDET.—Machine for making the paving-stones used in Paris (42).

MICHAL, ALPHAND, DARCEL, GRÉGOIRE, & FOULARD.—A collection of drawings of the works executed by the City of Paris in the Bois de Boulogne, the wood of Vin-cennes, and various squares of the capital (43).

LEFEBURE DE FOURGY.—Maps of the subterranean quarries of the city of Paris (44).

DELESSE.—Subterranean geological map of the City of Paris (45).

LEFEBURE DE FOURGY. — Hydrological map of the City of Paris (46).

MATHIEU, JACQMIN, & DURBACH.—Sheet-iron railway bridge constructed at Fribourg (47).

CHAUBART.—Self-regulating sluice-gate (48).

BEAUDEMOULIN & BOUZIAT.—Improved apparatus for striking the centres of bridges, by means of sand, on the principle of Baude-moulin, used for the Bridge of Austerlitz (49).

1252 GARNAUD, E. F. JUN. *Paris.*—Objects in terra cotta.

1253. MIGNOT, L. *Paris.*—Objects in vegeto-mineral mastic.

1254. FACONNET, CHEVALLIER, & Co. *Paris.*—Tiles.

1255. GUICESTRE & Co. *Paris.*—Non-bitumenized pasteboard for roofs; coating and colours which resist moisture.

1256. FERARY, C. A. *Grenoble (Isère).* —Cement, and articles made of it.

1257. LIPMANN, SCHNECKENBURGER, & Co. *Paris.*—Artistic articles, in factitious marble and stone.

1258. COIGNET, F. BROS. & Co. *Paris.*—Plain and decorated artificial stone, called agglomerated beton.

1259. MONDUIT, N. & BÉCHET, *Paris.*—Ornaments in lead and copper for building and decorative purposes.

1260. CRAPOIX, J. *Paris.*—Decorative objects in stucco.

1261. GÉRUZET, L. *Bagnères-de-Bigorre* (*Hautes Pyrénées*). — Works in sculptured marble, for ornamentation and furniture.

1262. DE TINSEAU, P. *St. Ylie (Jura).* —Articles in Jura marble.

1263. SOYER, A. *Mareuil-lez-Meaux* (*Seine-and-Marne*).—Objects in factitious stone.

1264. ARNAUD, J. VENDRE, & CARRIERE. *La Porte de France (Isère).*—Cements, and fountain-pipes made of it.

1265. LINGÉE & Co. *Paris.*—Cement.

1266. GODIN-LEMAIRE, J. B. A. *Guise* (*Aisne*).—Cast-iron heating apparatus, &c.

1267. GASTELLIER, C. A. *Montauglaust* (*Seine - and - Marne*).—Compressed bricks, tiles, drainage pipes, hollow bricks, and mosaic squares.

1268. ZELLER & Co. *Ollwiller* (*Haute-Rhin*).—Glazed earthenware pipes for con-veying water and gas.

1269. DAVID, L. *Uzés (Gard).*—Refrac-tory clay and bricks.

1270. ALGOUD BROS. DEPUY DE BORDES, & Co. *Grenoble (Isère).*—Cement, and cement tubes for fountains.

1271. VICAT, J. *Grenoble (Isère).*—Arti-ficial cements.

1272. COUISSINIER, A. *Saint - Henry* (*Bouches-du-Rhône*).—Tiles for flooring.

1273. TROUILLIET, P. *Sens (Yonne).*—Refractory bricks.

1274. DUPONT, P. H. *Cherbourg* (*Manche*).—Metallic varnish for preserving wood and metals.

1275. LÉDIER, A. *Auffay (Seine-Inf.).*—Bricks.

1276. AGOMBART, A. *St. Quentin* (*Aisne*).—Hydraulic lime, &c.

1277. BIGOT-DUVAL, WIDOW, & Co. *Manceliére* (*Eure-and-Loir*).—Natural hydraulic lime.

1278. DESFEUX, P. A. *Paris.*—Waterproof pasteboard for roofing.

1279. DUMONT, E. H. *Roanne* (*Loire*).—Tiles of various kinds.

1280. MACHABÉE, L. *Paris.*—Hydraulic mastic.

1281. BAUDON, F. & SON, *Paris.*—Economic cooking apparatus.

1282. BREBAR, C. R. *Lille* (*Nord*).—Siliceous paintings.

1283. GRADOS, L. *Paris.* — Repoussé ornaments in zinc and lead, for exterior decoration.

1284. MICHELET, H. F. *Paris.*—Stamped articles in zinc and lead, for exterior decoration.

1285. PLANIER, E. & Co. *Paris.*—Elastic pads for stopping chinks in apartments.

1286. CANDELOT, L. F. SEN. *Paris.*—Various kinds of cement.

1287. VIEILLARD, G. (LACROIX), *Compeigne* (*Oise*).—Waterproof coating for wood, plaster, and stone.

1288. DERENUSSON, C. *Paris.*—Safety apparatus for scaffolding.

1289. LE BLANC, C. *Rennes* (*Ille-and-Vilaine*).—Drawings of the Viaduct of Corbinières, on the Rennes and Redon railway.

1290. MOULIN, A. *Bonsecours-lez-Rouen* (*Seine - Inf.*). — Imitation of Florentine mosaics.

1291. CAIL, J. F. & Co. *Paris.*—Model of a sheet-iron bridge, &c.

1292. BARTHÉLEMY, H. *Paris.*—Designs for theatres.

1293. FONTENELLE, C. C. *Paris.* — Cement for flagging.

1294. BOULANGÉ, *Auneuil* (*Oise*). — Enamelled mosaic tiles.

1295. JALOUREAU, A. *Paris.*—Pipes of bitumenized paper, for water or gas.

———

1296. PARISIAN GAS - LIGHTING AND HEATING Co. *Paris.*—Refractory bricks.

1297. DUPRAT, M. C. V. *Canéjan* (*Gironde*).—Refractory bricks.

1298. JUTTEAU, A. *Orléans* (*Loiret*).—Specimen and drawings of veneering in stone for the restoration of buildings.

1299. ROSIER, WIDOW, & BAROCHE, *Tain* (*Drôme*).—Refractory bricks.

1300. VIEILIARD, J. & Co. *Bordeaux* (*Gironde*).—Refractory bricks.

## CLASS XI.

1301. BOCHE-TORDEUX, A. *Paris.*—Requisites for sporting.

1302. CARON, A. *Paris.*—Breech-loading cannon, carbines; apparatus for cleaning arms.

1303. BRUN, A. *Paris.*—Breech-loading and other guns, pistols, &c.

1304. LESPIAUT, A. *Paris.*—Requisites for sporting.

1305. THOMAS, J. M. *Paris.*—Guns, pistols, and carbines.

1306. FLOBERT, L. N. A. *Paris.*—Sporting guns, revolver pistols, cartridges, &c.

1307. GÉVELOT, *Paris.*—Percussion caps, and cartridges.

1308. JAVELLE-MAGAUD & SON, *St. Etienne* (*Loire*).—Damasked double barrels, for fowling-pieces.

1309. RONCHARD-SIAUVE, *St. Etienne* (*Loire*).—Damasked double barrels, for fowling-pieces; cast steel barrels for guns and carbines.

1310. DIDIER - DREVET, *St. Etienne* (*Loire*).—Double barrels for fowling-pieces.

1311. BLACHON, J. *St. Etienne* (*Loire*).—Double barrels: cast steel barrels.

1312. JAVELLE - MICHEL, *St. Etienne* (*Loire*).—Revolvers, cartridges.

1313. MURGUES, *St. Etienne* (*Loire*).—Fowling-pieces, pistols, &c.

1314. GABION - FOURNEL, *St. Etienne* (*Loire*).—Double-barrel fowling-pieces.

1315. BERGER, F. *St. Etienne* (*Loire*).—Fowling-pieces.

1316. VERNEY - CARRON, *St. Etienne* (*Loire*).—Fowling-pieces.

1317. ESCOFFIER, F. *St. Etienne* (*Loire*).—Military fire-arms.

1318. PONDEVAUX & JUSSY, *St. Etienne* (*Loire*).—Double-barrel fowling-pieces.

1319. AURY, L. *St. Etienne* (*Loire*).—Double-barrel fowling-pieces.

1320. GEREST, A. *St. Etienne* (*Loire*).—Fancy guns.

1321. BOURGAUD & Co. *St. Etienne* (*Loire*).—Fowling-pieces, &c.

1322. GIRARD, L. *Chatellerault* (*Vienne*).—Seamless leather sword-sheaths.

1323. BLANC - MARTY & Co. *Paris.*—Military accoutrements.

1324. BERNARD, A. *Paris.*—Gun-barrels.

1325. GÉERINCKX, F. E. *Paris.*—Hunting guns and knives, &c.

1326. MARQUIS, F. *Paris.*—Weapons for the chase.

1327. PERRIN, L. *Paris.* — Fowling-pieces, guns, carabines, pistols, and revolvers.

1328. DELVIGNE, H. G. *Paris.*—Duck-guns; rifled barrels; pistols on a new principle.

1329. TARDY & BLANCHET, *Paris.*—Percussion caps.

1330 GAUPILLAT, SON, & ILLIG, *Paris.*—Percussion caps and metallic cartridges for revolvers, &c.

1331. SUDRE, F. *Paris.* — Telephonic system, for the transmission of sound to a distance, with the aid of the clarion, the drum, and the cannon: a system of telegraphic correspondence between two armies, by means of three signs.

1332. GRANGER, E. *Paris.*—Copies of ancient arms and weapons.

1333. DELACOUR, L. F. *Paris.* — Burnished weapons.

1334. HOULLIER - BLANCHARD, C. H. *Paris.*—Sporting guns: pistols.

1335. GASTINNE-RENETTE, L. J. *Paris.*—Guns, carbines, and pistols, &c.

1336. CLAUDIN, F. *Paris.* — Fowling-pieces and pistols.

1337. BERNARD, L. *Paris.*—Barrels of fire-arms.

1338. DEVISME, L. F. *Paris.* — Gun-barrels, fancy guns; weapons for whaling and the chase.

1339. LEPAGE-MOUTIER, M. L. *Paris.*—Sporting-guns, carbines, revolvers, &c.

1340. CORDIER & Co. *Paris.*—Revolvers and carbines.

1341. CHAUDUN, A. SON, & DERIVIÈRE, N. *Paris.*—Cartridges.

———

1342. GEIGER, Z. *Paris.*—Requisites for sporting.

1343. WALCKER, W. *Paris.*—Requisites for sporting.

# CLASS XII.

1361. Léfevre Bros. *Sotteville - lez - Rouen (Seine-Inf.).*—Rigging.

1362. Lafaye, G. & Co. *Bordeaux (Gironde).*—Hemp and wire rigging.

1363. Delvigne, H. G. *Paris.*—Howitzers for carrying a line to a shipwrecked vessel.

1364. Salette, J. *Marseilles (Bouches-du-Rhône).*—Model of apparatus for weighing anchors.

1365. Pécoul, A. *Marseilles (Bouches-du-Rhône).*—Sounding log, for measuring the speed of vessels, and sounding without stopping, and for giving warning of approach to land or to reefs.

1366. Cavayé, F. *Montpellier (Hérault).*—Safety and swimming belt.

1367. Labat, T. *Bordeaux (Gironde).*—Apparatus for hauling vessels on land.

1368. Tisserant, J. G. *Orléans (Loiret).*—Apparatus for saving from shipwreck.

1369. Cabirol, J. M. *Paris.* — Cork jacket; submarine lamp.

1370. Gallois-Foucault, *St. Martin (Charante-Inf.).*—Floating fog-bells, &c.

1371. Bouquié, F. *Paris.*—Model of a system of towing by steam; chains made without joining.

1373. Broquant, Hochard, & Co. *Dunkerque (Nord).*—Machine-made fishing nets.

1374. Delage-Montignac, *Paris.*—Implements for fishing.

1375. David & Co. *Hâvre (Seine Inf.).*—Chain-cables, anchors, &c.

1376. Ouarnier - Mathieu, *Compiêgne (Oise).*—Chain and hemp cables, for marine, or mining purposes.

1377. Besnard, F. Richou, & Genest, *Angers (Maine-and-Loire).*—Hemp and wire-rope, for marine or mining purposes; fishing requisites.

# CLASS XIII.

1391. Naudet & Co. *Paris.*—Metallic barometer.

1392. Fastré, J. T. Sen. *Paris.*—Philosophical instruments; barometers; thermometers; hygrometers.

1393. Hardy, E. *Paris.*—Philosophical instruments; chronographs, and inductive chronoscope, for measuring the speed of projectiles; polytrope, for examining electric currents; apparatus for measuring the secular deviations of the needle.

1394. Kœnig, R. *Paris.*—Acoustic instruments.

1395. Colombi, C. Jun. *Paris.*—Surveying, levelling, and geodesic instruments.

1396. Molteni, J. *Paris.*—Mathematical and other instruments.

1397. Silbermann, Jun. *Paris.*—Concave celestial and terrestrial hemispheres; chromatic chronometer for astronomical instruments; diagrams.

1398. Gavard, A. *Paris.*—Pantographs producing direct, or inverted copies, compass for tracing ovals.

1399. Burdon, J. A. *Paris.*—Ruler for calculation.

1400. Bonis, P. F. *Paris.* — Insulated wire for electric and telegraphic apparatus.

1401. Biloret, A. *Paris.* — Insulated wire for electric apparatus, &c.

1402. Prud'homme, P. D. *Paris.* — Electric bells and signals, for private houses, for naval and railway purposes.

1403. Poullot, V. N. *Paris.* — Eye-glasses, spectacles, &c.

1404. Lemaire, *Paris.*—Opera-glasses, &c.

1405. Breton Bros. *Paris.* — Philosophical instruments.

1406. Mabru, *Paris.* — Apparatus for maintaining a vacuum.

1407. Thomas, C. X. *Paris.* — Calculating machines.

1408. Bardou, P. G. *Paris.* — Opera-glasses, and glasses of various kinds.

1409. Lebrun, A. *Paris.* — Optical glasses, opera-glasses, microscopes, telescopes, &c.

1410. Dutrou, E. P. *Paris.*—Philosophical instruments; levels, areometers; barometers, thermometers, and hygrometers; metallic pyrometer.

1411. Santi, A. *Marseille (Bouches-du-Rhône).*—Mathematical and other instruments.

1412. Balbreck, M. *Paris.*—Geodesic and other instruments.

1413. Breguet, L. C. F. *Paris.*—Philosophical and horological instruments; electric telegraph apparatus.

1414. Digney Bros. & Co. *Paris.*— Telegraphic apparatus.

1415.. Brunner & Son, *Paris.*—Optical instruments, &c.

1416. Nachet & Son, *Paris,*—Microscopes.

1417. Hartnack, E. F. *Paris.*—Microscopes.

1418. Mirand, A. Sen. *Paris.*—Microscopes.

1419. Mouchet, A. *Rochefort.*—Apparatus for instantly finding again, in any preparation, a microscopic object already observed.

1420. Duboscq, L. J. *Paris.*—Optical apparatus.

1421. Deleuil, J. A. *Paris.*—Delicate instruments for weighing, &c.

1422. Bourette, E. H. *Paris.*—Thermometers.

1423. Digeon, R. H. *Paris.*—Scientific diagrams.

1424. Tiffereau, T. *Paris.*—Nautical hour-glasses; apparatus for receiving, measuring, and transferring gas.

1425. Sagey, *Boulogne-sur-Mer (Pas-de-Calais).*—Gas-meter.

1426. Ricourt, C. *La Machine (Nièvre).* —Calculating cylinder.

1427. Collot, M. & A. Bros. *Paris.*— Balances of aluminium, &c.

1428. Rattier & Co. *Paris.*—Submarine and subterranean telegraph cables.

1429. Perreaux, L. G. *Paris.*—Delicate instruments for measurements of various kinds.

1430. Taurines, J. M. H. A. *Paris.*— Balances; dynamometers.

1431. Prudent, L. *Paris.* — Opera-glasses.

1432. Cam, J. C. *Paris.*—Opera-glasses, spectacles, &c.; block of rock crystal.

1433. Coiffier, A. *Paris.* — Opera-glasses, &c.

1434. Lafleur, A. *Paris.* — Opera-glasses.

1435. Chuard, M. *Paris.*—Safety-lamp; gazoscope, to give warning of the danger of explosion in mines; apparatus to show the motion of the earth, &c.

1436. Warren Thomson, *Paris.*—Telegraph printing instruments.

1437. Serrin, V. *Paris.* — Self-acting regulator for electric light.

1438. Mouilleron & Vinay, *Paris.*— Electric telegraph apparatus.

1439. Dujardin, P. A. J. *Lille (Nord).* —Printing telegraph instrument.

1440. Hoffmann, J. G. *Paris.*—Optical instruments and articles.

1441. Lamotte-Lafleur, C. G. *Paris.* —Boxes of mathematical instruments.

1442. BERTAUD, A. S. *Paris.*—Photographic object lenses, astronomical and nautical glasses, &c.

1443. MOREAU, A. *Paris.* — Electro-medical apparatus.

1444. GUYOT, D'ARLINCOURT, L. C. *Paris.*—Alphabetical printing telegraph instrument, by Breguet.

1445. LARGEFEUILLE, *Châlons-sur-Saône (Saône-and-Loire).*—Wire covered with an insulating enamel for submarine and subterranean telegraphy; elastic submarine cables without joint.

1446. ROBERT, H. *Paris.*—Astronomical clock, &c.

1447. SORTAIS, T. *Lisieux (Calvados).*—Morse's telegraph apparatus.

1448. DU PUY DE PODIO, *Courbevoie (Seine).*—Stadiometer.

———

1449. BERLIOZ, A. & Co. *Paris.*—Magneto-electric machine, for producing light.

1450. COLLECTIVE EXHIBITION OF THE CITY OF CLUSES (*Haute-Savoié*).—Metallic thermometers.

1450A. DESHAYS, A. *Paris.* — Delicate compass.

1450B. PATRY, E. L. *Paris.*—Spectacles, eye-glasses, &c., of silver, aluminium, &c.

# CLASS XIV.

1451. TITUS-ALBITÈS, *Paris.* — Photographic apparatus and photographs.

1452. SILBERMANN, J. JUN. *Paris.* — Table of the photogenic effects of the principal colours, on different substances, &c.

1453. ROBIN, A. *Paris.*—Photographic copies of plans, &c.

1454. DU MONT, H. *Paris,*—Photographic apparatus, representing the different phases of motion.

1455. ANTHONI, G. *Paris.* — Portable photographic laboratory.

1456. VILLETTE, E. *Paris.*—Large photographs on collodion, obtained by amplification, electric light, Duboscg's apparatus being used.

1457. DUBOSCG, L. J. *Paris.*—Photographic apparatus.

1458. BERTAUD, *Paris.* — Automatic camera and portable laboratory for obtaining small photographs which are intended to be enlarged; heliographic megascope, for the production of enlarged photographs; autographic apparatus for changing the two

negatives of the automatic camera into a single positive on glass, with moist collodion; automatic microscopic camera; large photographs obtained by means of exhibitor's apparatus.

1459. BERTSCH, A. *Paris.*—Automatic camera, and portable laboratory, for obtaining photographs which are to be enlarged; heliographic megascope for enlarging photographs; automatic apparatus for changing the two negatives of the automatic camera into a single positive on glass with moist collodion; automatic microscopic camera; enlarged photographs, obtained by exhibitor's apparatus.

1460. BILORDEAUX, A. *Paris.*—Photographs, specimens of prepared paper, &c.

1461. DISDÉRI, *Paris.* — Photographs, enlarged photographs of the natural size, &c.

1462. ALOPHE, M. *Paris.*—Photographs of the natural size, and obtained directly, &c.

1463. BISSON BROS. *Paris.* — Photographs from nature, &c.

1464. DELBARRE, P. J. & LELARGE, A.

*Paris.*—Portraits of the natural size, obtained directly, &c.

1465. Cammas, *Paris.* — Photographs: views in Egypt.

1466. Yvon, *Paris.*—Specimens of photography obtained rapidly.

1467. Delessert, E. *Paris.* — Large photographs obtained directly, &c.

1468. Delton, *Paris.* — Photographs taken quickly.

1469. Baldus, E. *Paris.*—Large photographs obtained directly.

1470. Potteau, *Paris.*—Application of photography to the sciences; types of the different races of men; anatomical, and natural history objects.

1471. Tournachon, A. Jun. *Paris.*—Photographs obtained rapidly.

1472. Jamin, *Paris.*—Photographs.

1473. Rolloy, Jun. *Paris.* — Photographic chemicals, paper, and apparatus.

1474. Marrion, *Paris.*—Photographic paper, &c.

1475. Puech, L. *Paris.*—Photographic chemicals and apparatus.

1476. Briois, C. A. *Paris.*—Photographic chemicals and apparatus.

1477. Richardin, J. B. *Paris.*—Machine for polishing daguerreotype plates, and cleaning glass plates.

1478. Poirier, *Paris.*—Press for glazing photographs.

1479. Lecu, F. N. *Paris.*—Photographic requisites.

1480. De Poilly, Sen. *Boulogne (Pas-de-Calais).*—Photographic apparatus for the country.

1481. Dumonteil, *Paris.*—Photographic apparatus, &c.

1482. Koch, *Paris.*—Large photographic apparatus.

1483. Relandin, *Paris.*—Photographic apparatus.

1484. Lefêvre, *Paris.*—Polishing requisites, and colours for photographers.

1485. Deriveau, *Paris.*—Photographic alembic for travelling.

1486. Garin & Co. *Paris.*—Photographic chemicals and paper.

1487. Quinet, A. M. *Paris.* — Photographic apparatus; sensitive glass plates and paper, which may be kept without any special precautions.

1488. Derogy, *Paris.*—Photographic apparatus, &c.

1489. Millet, A. *Paris.*—Photographic apparatus.

1490. Hermagis, *Paris.*—Photographic apparatus, solar microscope, &c.

1491. Darlot, *Paris.*—Photographic apparatus, &c.

1492. Laverdet, *Paris.* — Collodion photographs; photographs coloured by exhibitor's process.

1493. Mathieu-Plessy, E. *Paris.*—Photographic chemicals and paper.

1494. Numa-Blanc & Co. *Paris.*—Photographs, plain and coloured.

1495. D'Orzagh, *Paris.* — Transferred collodion and retouched photographs.

1496. Plaisant, *Paris.* — Oil-coloured photographic portraits.

1497. Mayer & Pierson, *Paris.*—Photographs, some of them retouched or painted.

1498. Nadar, *Paris.* — Photographs, many of them taken by electric light.

1499. Pesme, *Paris.*—Photographs, plain and painted.

1500. Ken, A. *Paris.* — Photographs, some of large size, obtained directly.

1501. Lemercier, *Paris.*—Litho-photographs, &c.

1502. Couvez, H. & Colombat, *Paris.*—Photographs on wood; heliographic metallic plates.

1503. Niepce, de St. Victor, *Paris.*—Photographs on glass and heliographic engravings on steel; heliochromic photographs, fixed instantaneously by means of chloride of lead, &c.

1504. Nêgre, C. *Paris.* — Heliographs obtained on steel, &c.

1505. DUFRESNE, *Paris.*—Photographic engraving and damascening on steel.

1506. LAFOND DE CAMARSAC, *Paris.*—Unchangeable photographs on enamels and porcelain, &c.

1507. JOLY-GRANGEDOR, *Paris.*—Artistic photographs, &c.

1508. POITEVIN, A. *Paris.* — Carbon photographs and photo-lithographs.

1509. VIDAL, L. *Marseilles (Boûches-du-Rhône).*—Carbon photographs, &c.

1510. PETIT, P. *Paris.*—Photographs.

1511. CORBIN, H. *Paris.*—Photographs, &c.

1512. TAUPENOT (dec.).—Photographs.

1513. GAUMÉ, *Mans (Sarthe).*—Photographs, &c.

1514. FARGIER, *Lyon (Rhône).*—Photographs.

1515. CHARAVET, *Paris.*—Carbon photographs.

1516. GARNIER & SALMON, *Paris.*—Carbon photographs, heliographic engraving, &c.

1517. ROBERT, *Sèvres (Seine).*—Photographs, taken from Sèvres articles, &c.

1518. DAVANNE, A. *Paris.*—Photographs.

1519. GIRARD, A. *Paris.*—Photography of the eclipse of 18th July, 1860, observed at Batna, Province of Constantine, by a Commission of the Polytechnic School.

1520. DAVANNE & GIRARD, *Paris*—Specimens of photography.

1521. MAGNY, *Paris.*—Photographs.

1522. BRETON, MADAME, *Rouen (Seine-Inf.).*—Photographs.

1523. MARVILLE, *Paris.*—Archæological and other photographs.

1524. BAYARD & BERTALL, *Paris.*—Photographic portraits and copies.

1525 RENARD — *Bourbonne-les-Bains (Haute-Marne).*—Photographs from nature, &c.

1526. JEANRENAUD, *Paris.*—Views obtained with dry collodion.

1527. BRAUNN, A. *Dornach (Haut-Rhin).*—Photographs.

1528. JOUET, E. *Paris.*—Photographs.

1529. MAILAND, E. *Paris.*—Waxed paper photographs.

1530. DE BRÉBISSON, *Falaise (Calvados).*—Photographs; specimens of different photographic processes.

1531. ADAM SALOMON, *Paris.* — Photographs from nature.

1532. GAILLARD, P. *Paris.*—Moist collodion photographs.

1533. BINGHAM, R. *Paris.*—Moist collodion photographs, &c.

1534. MICHELEZ, C. *Paris.*—Moist collodion photographs.

1535. CARJAT & Co. *Paris.*—Moist collodion photographs.

1536. LAFFON, J. C. *Paris.* — Photographs; photographs on silk and glass.

1537. MAXWEL-LYTE, *Bagniéres-de-Bigorre (Hautes Pyrénées).*—Views in the Pyrenees.

1538. ALEO, *Nice (Alpes-Maritimes).* — Photographs by various processes.

1539. MUZET, *Grenoble (Isère).* — Views of the Isère and Savoy,

1540. BERTHIER, P. *Paris.* — Photographs.

1541. COMTE O. AGUADO, *Paris.*— Enlarged photographs obtained by solar and by electric light.

1542. VICOMTE O. AGUADO, *Paris.*—Enlarged photographs obtained by solar and by electric light.

1543. SILVY, *Paris.*—Views in Algeria, &c.

1544. BRAQUEHAIS, *Paris.*—Stereoscopic photographs, coloured, with specimens of the colours used.

1545. DAGRON, E. *Paris.*—Microscopic photographs mounted on jewels.

1546. FERRIER & SON, *Paris.*—Large photographs on glass; views taken instantaneously in Paris while passing through the streets, the carriages and passengers represented being in motion.

1547. MARLÉ, C. A. *Paris.*—Moist collodion photographs.

1548. WARNOD, *Hâvre (Seine Inf.)*.—Photographs.

1549. MASSON, *Seville (Spain) and Paris.*—Photograph views in Spain; &c.

1550. BOUSSETON & APPERT, *Paris.*—Portraits.

1551. CREMIERE. *Paris.*—Instantaneous portraits, &c.

1552. MARQUIS DE BÉRENGER, *Paris.*—Photograph views of the Department of the Isère

1553. DELONDRE, P. *Paris.* — Photographs obtained by means of dry waxed paper.

1554. DE CLERCQ, L. *Paris.* — Photographs obtained on dry iodized waxed paper.

1555. DE VILLECHOLLE, F. *Paris.*—Photographs.

1556. ANTHONY-THOURET, JUN. *Paris.*—Photographs from nature, and copies.

1557. CHARNAY, D. *Mâcon (Saône-and-Loire)*.—Photographs.

1558. LACKERBAUER, *Paris.* — Natural history and anatomical photographs, of the natural size, &c.

1558. ROMAN, D. *Arles (Bouches-du-Rhône)*.—Photographs.

1569. TILLARD, F. *Bayeux (Calvados)*.—Photographs.

1561. COLLARD, *Paris.* — Photographic views, &c.

1562. RICHEBOURG, *Paris.* — Photographs.

1563. DE LUCY, L. G. *Paris.*— Portraits and groups.

1564. DE CHAMPLOUIS, *Paris.*—Views in Syria.

1565. DUVETTE & ROMANET, *Amiens (Somme)*.—Photographs of the cathedral of Amiens.

1566. MOULIN, F. *Paris.*—Photographs.

1567. AUTIN, *Caen (Calvados)*.—Photographs.

1568. GUESNÉ, *Paris.*—Photographs.

1569. BACOT, *Caen (Calvados).* — Portraits and studies from nature, with collodion.

# CLASS XV.

1581. DESFONTAINE, LEROY, & SON, *Paris.*—Watches, chronometers; time-pieces, one of them marking the hours, minutes, and seconds, the day of the week and month, the month, the phases of the moon, and sun's place.

1582. ROBERT, H. *Paris.*—Time-pieces, chronometers, gravers' tools, oils, &c.

1583. COLIN, A. *Paris.*—Clocks, regulators, travellers' time-pieces; bronzes; tools.

1584. ANQUETIN, M. *Paris.*—Watches; universal time-pieces, giving the hour at the chief cities of the world.

1585. DETOUCHE, C. L. *Paris.*—Time-pieces, watches, regulators; electric clocks; astronomical regulators, marking seconds, and indicating the true time, the day of the week and month, the month and year, the phases and age of the moon, the sun's rising, setting, and altitude, the signs of the zodiac, and the hour at different parts of the earth.

1586. MONTANDON BROTHERS, *Paris.* — Springs for clocks, watches, &c.

1587. FARCOT, H. A. E. *Paris.*—Time-pieces, alarms.

1588. BROCOT, L. A. *Paris.*—Regulators, half-second time-pieces, and perpetual calendars.

1589. REDIER, A. *Paris.*—Time-pieces, alarms; applications of the conical pendulum, &c.

1590. CHARPENTIER, P. A. *Paris.* — Watches, time-pieces, chronometers, regulators, &c.

1591. GUYARD, F. V. *Dieppe (Siene-Inf.).* —Electric clocks.

1592. GONTARD, *Besançon (Doubs).* — Watches, &c.

1593. FERNIER, N. *Besançon (Doubs).*— Watches, &c.

1594. BERTHELOT, *Besançon (Doubs).*— Watches, &c.

1595. MONTANDON, *Besançon (Doubs).*— Watches, &c.

1596. JEANNOT-DROZ, *Besançon (Doubs).* —Watches, &c.

1597. RICHARDEY, *Besançon (Doubs).*— Watches, &c.

1599. SAVOYE BROS. *Besançon (Doubs).* —Watches, &c.

1600. CRESSIER, E. *Besançon (Doubs).*— Watches, &c.

1601. GILET, E. *Besançon (Doubs).*— Watches, &c.

1602. BOSSY, *Besançon (Doubs).* — Watches, &c.

1603. BOUTEY & SON, *Besançon (Doubs).* —Watches, &c.

1604. ADLER, N. *Besançon (Doubs).*— Watches, &c.

1605. FAVRE - HEINRICK, *Besançon (Doubs).*—Watches, &c.

1606. PERROT, E. *Besançon (Doubs).*— Watches, &c.

1607. GROZ BROS. *Besançon (Doubs).*— Watches, &c.

1608. HUMBERT, A. *Besançon (Doubs).* —Watches, &c.

1609. CHALONS, V. *Besançon (Doubs).*— Watches, &c.

1610. LAMBERT, H. *Besançon (Doubs).*— Watches, &c.

1611. STERKY, A. *Besançon (Doubs).*— Watches, &c.

1612. BERTHET, *Besançon (Doubs).*— Watches, &c.

1613. PIGUET, *Besançon, (Doubs).*— Watches, &c.

1614. FUMEY, *Besançon (Doubs).*— Watches, &c.

1615. GONDELFINGER & BICHET, *Besançon (Doubs).*—Watches, &c.

1616. AMET, F. *Besançon (Doubs).* — Watches, &c.

1617. PONSOT & HÉRIZÉ, *Besançon (Doubs).* Watches, &c.

1618. BRÉDIN, *Besançon (Doubs).* — Watches, &c.

1619. CROUTTE, A. & Co. *Rouen (Seine-Inf.).*—Time-pieces; alarms.

1620. LÉGER, P. J. *Paris.*—Watches.

1621. GINDRAUX, A. & SONS, *Paris.*— Rubies and diamonds for wire drawers, and diamond points for engravers on steel; files.

1622. VISSIÈRE, S. *Hâvre (Seine-Inf.).*— Chronometers, &c.

1623. LEROY, T. *Paris.*—Chronometers, &c.

1624. JACOT, H. L. *Paris.*—Traveller's time-pieces.

1625. FLÉCHET, P. *Paris.*—Sun-dials.

1626. COUËT, L. C. *Paris.*—Watches, travellers' time-pieces, regulators.

1627. PATAY, P. *Paris.* — Traveller's time-pieces.

1628. DROCOURT, P. *Paris.*—Traveller's time-piece

1629. PIERRET, V. *Paris.* — Watches, portable time-pieces, &c.

1630. COLLECTIVE EXHIBITION OF THE CITY OF CLUSES (*Haute-Savoie*).—Parts of watches and clocks, tools, &c.

1631. SCHARF, B. *St. Nicolas d'Aliermont (Seine-Inf.).* — Astronomical clock, chronometers, &c.

1632. STRÉBET, L. *Paris.* — Steeple-clock.

1633. PERREAUX, L. G. *Paris.*—Public clocks.

1634. SCHIRRMANN, E. *Paris.* — Time-pieces, &c. in carved wood.

1635. BREQUET, L. C. F. *Paris.*—Time-pieces, regulators, chronometers, repeaters, &c.

## CLASS XVI.

1641. MUSTEL, C. V. *Paris.* — Organ-harmonium.

1642. LABBAYE, J. C. *Paris.* — Brass instruments.

1643. RODOLPHE, A. *Paris.* — Organ-harmonium.

1644. HENRY E. & MARTIN, J. *Paris.*—Military musical instruments, in brass, &c.

1645. BEAUCOURT, H. C. *Lyon (Rhône).* Organ-harmonium.

1646 DAVID, L. *Paris.*—Brass military instruments.

1647. MAYER-MARIX, *Paris.*—Harmoni-flutes.

1648. LECOMTE, A. & Co. *Paris.*—Wind instruments, of brass and wood.

1649. KASRIEL, L. M. *Paris.* — Flute-harmonium, harmoni-flutes, &c.

1650. ALEXANDRE & SON, *Paris.* — Organs.

1651. REMY & GROBERT, *Mirecourt (Vosges).*—Non-metallic wind instruments; an organ.

1652. BUSSON, C. *Paris.*—Accordions; harmoni-flutes.

1653. DERAZEY, J. J. H. *Mirecourt (Vosges).*—Stringed instruments, &c.

1654. MANGEOT BROS. & Co. *Nancy (Meurthe).*—Upright piano, with oblique strings.

1655. À CAUDÈRES, J. J. *Bordeaux (Gironde).*—Upright piano, with a triple clavier.

1656. POIROT, D. *Mirecourt (Vosges).*—Organ, and stringed instruments.

1657. CAVAILLÉ-COLL, A. *Paris.*—Plan of the great organ of the church of St. Sulpice; chamber organ; models of pneumatic movements with simple and double actions; regulators of air and gas pressure; bellows for different pressures, &c.

1658. KELSEN, E. *Paris.*—Self-acting organ, moved by a spring or weight.

1659. FARRENC, J. H. A. *Paris.*—Piano music, comprising choice pieces of the masters of all countries and periods from the sixteenth to the middle of the nineteenth century.

1660. GRANDJON, J. *Mirecourt (Vosges).*—Stringed instruments.

1661. GÉRARD, E. & Co. *Paris.*—Select music.

1662. VINCENT, *Paris.*—Organ, having a double clavier with quarter tones.

1663. DEBAIN, A. F. *Paris.*—Mechanical piano, harmonichords, &c.

1664. MARTIN, P. & SON, *Toulouse (Haute-Garonne).*—Upright piano, with demi-oblique strings.

1665. KRIEGELSTEIN, J. G. *Paris*—Upright piano with oblique strings, and grand piano with repeating movement.

1666. HUSSON-BUTHOD, & THIBOUVILLE, *Paris.*—Wind and stringed instruments.

1667. COURTOIS, ANTONY, *Paris.*—Brass instruments.

1668. ELCKÉ, F. *Paris.*—Piano.

1669. LEMOINE, H. A. P. *Paris.* — Music.

1670. VUILLAUME, J. B. *Paris.*—Violins and bows.

1671. SAVARESSE, H. *Paris.*—Strings for instruments.

1672. KLEINJASPER, *Paris.* —Upright piano.

1673. BOISSELOT & SON, *Marseille (Bouches-du-Rhône).*—Pianos.

1674. WIART, F. S. *Châteauroux (Indre).*—Piano mechanism.

1675. GEHRLING, C. *Paris.*—Mechanism for pianos.

1676. GAUDONNET, P. *Paris.*—Upright piano.

1677. MIRMONT, C. A. *Paris.*—Stringed instruments.

1678. MONTAL, C. *Paris.*—Upright and grand pianos.

1679. DE ROHDEN, F. *Paris.*—Mechanism for pianos.

1680. BONNET, C. *Marseilles (Boûches-du-Rhône).*—Stringed instruments.

1681. FAVRE, J. *Lyon (Rhône).*—Harmonichords.

1682. JAULIN, L. J. *Paris.*—Harmonichord.

1683. BARBIER, V. *Paris.*—Requisites for the manufacture of pianos.

1684. BORD, A. *Paris.*—Pianos.

1685. AUCHER, L. & J. BROS. *Paris.*—Pianos.

1686. PLEYEL, WOOLF, & Co., pianoforte makers, 22 *Rue Rochechouart, Paris.* Warerooms, &c., 95 *Rue Richelieu;* Manufactory, *Rue des Récollets;* Timber Yard and Saw Works, *Rue des Portes Blanches (Montmartre), à Paris.*

Sole agents for the United Kingdom—METZLER & Co., 35, 37, 38, and 16, *Great Marlborough Street, London.*

[PRIZE MEDAL AWARDED 1862.]

"For excellence in every kind of piano, power and equality of tone, precision of mechanism, and solidity." Price lists will be forwarded on application to Metzler and Co., Great Marlborough Street, London.

1687. JACQUOT, C. *Nancy (Meurthe).*—Violins, violoncellos, &c.

1688. PAPE, J. H. *Paris.*—Pianos.

1689. HERZ, H. *Paris.*—Pianos.

1690. BLANCHET, P. A. C. *Paris.*—Pianos.

1691. WÖLFEL, F. *Paris.*—Pianos.

1692. COTTIAU, P. F. J. *Paris.*—Reeds, for organ-harmoniums and accordions.

1693. LOT, L. *Paris.* — Flutes with cylindric bore.

1694. GODEFROY, C. SEN. *Paris.* — Flutes with cylindric and conical bore, &c.

1695. BUFFET, L. A. JUN. *Paris.*—Clarinets, hautboys, &c.

1696. TRIÉBERT, F. *Paris.*—Wind instruments; mouth-pieces for clarinets; reeds and tools for making them.

1697. BUFFET-CRAMPON & Co. *Paris.*—Wind instruments.

1698. BRETON, J. D. *Paris.*—Wind instruments; clarinets and flutes of wood, silver, aluminium, and crystal; crystal mouthpieces, &c., for clarinets.

1699. THIBOUVILLE, SEN. *Paris.*—Wind instruments.

1700. BARBU, J. P. *Paris.*—Reeds.

1701. SAX, A. J. *Paris.*—Wind instruments, &c.

1702. GAUTROT, P. L. SEN. *Paris.*—

Wind and percussion instruments, of brass and wood.

1703. DE TILLANCOURT, E. *Paris.*—Silk acribelles-strings.

1704. FRELON, L. F. A. *Paris.*—Apparatus for teaching music.

1705. BAUDASSÉ-CAZOTTES, *Montpellier* (*Hérault*).—Harmonic strings.

1706. GICHENÉ, L'ABBÉ, *St. Medard* (*Landes*).—Symphonista.

1707. SAX, A. JUN. *Paris.*—Wind instruments.

## CLASS XVII.

1711. MERICANT, E. *Paris.* — Instruments for veterinary surgeons.

1712. CHARRIÈRE, J. J. *Paris.*—Surgical instruments.

1713. LUËR, G. G. A. *Paris.*—Surgical instruments.

1714. MATHIEU, L. J. *Paris* —Surgical apparatus, artificial limbs, &c.

1715. SALES-GIRONS, DR. *Paris.*—Apparatus for medicating mineral waters.

1716. CHARLES, G. *Paris.*—Baths, with heating apparatus, &c.

1717. FOUQUET, A. *Paris.*—Bathing apparatus.

1718. MARTRÈS, DR. A. *St. Cyr* (*Seine-and-Oise*).—Tent answering for a litter.

1719. LÉCUYER, F. J. *Paris.*—Bath, with interior warming apparatus.

1720. PAQUET, DR. F. *Roubaix* (*Nord*).—Ferruginated gutta percha, for surgical uses.

1721. GALANTE H. & CO. *Paris.*—Vulcanized india-rubber surgical apparatus.

1722. REBOLD, E. *Paris.* — Electro-medical apparatus.

1723. JUNOD, V. T. *Paris.*—Cupping-glasses.

1724. GRANCOLLOT, L. P. *Paris.*—Orthopœdic and herniary apparatus.

1725. TOLLAY, MARTIN, & LEBLANC, *Paris.*—Shower-baths.

1726. PICHOT, J. A. & MALAPERT, *Poitiers* (*Vienne*).—Carboniferous disinfecting paper, for dressing wounds.

1727. THIERS, L. P. T. *Paris.*—Hygienic, &c., apparatus.

1728. BURG, *Paris.*—Metallo-therapeutic apparatus for the treatment of diseases of the nerves.

1729. MORIN, F. J. *Paris.*—Electro-medical apparatus ; electric apparatus for the ignition of gunpowder.

1730. NACHET & SON, *Paris.*—Optical apparatus for anatomy.

1731. LEPLANQUAIS, P. F. *Paris.* — Trusses, &c.

1732. LE PERDIEL, C. & MARINIER, J. *Paris.*—Trusses, &c.

1733. WICKHAM BROS. *Paris.*—Trusses.

1734. LORIEL, H. F. *Paris.*—Trusses.

1735. LE BELLEGUIC, P. J. *Paris.*—Orthopœdic apparatus, trusses, &c.

1736. BÉCHARD, R. L. *Paris.*—Trusses, artificial limbs, orthopœdic apparatus.

1737. VASSEUR, P. N. *Paris.*—Osteological, and comparative anatomical preparations.

1738 BAILLIÈRE, J. B. & SON, *Paris.*—Works on natural history, anatomy, &c.

1739. DUCHENNE, DR. G. *Paris.*—Photographs from nature, representing the different expressions of the countenance under the action of electricity.

1740. LAKERBANER, P. *Paris.*—Drawings, lithographs, engravings, and photographs, for the natural and medical sciences.

1741. GION, D. J. *Paris.* — Artificial teeth; articles for the repair of injuries to the mouth.

1742. LÉGER, DR. E. V. *Paris.*—Anatomical and pathological models in paper.

1743. SIMON, P. *Paris.*—Dental instruments.

1744. LAVEZZARI, E. *Montreuil-sur-Mer (Pas-de-Calais).*—Plans of an hospital, and baths for warm sea-water in the winter.

1745. TALRICH, J. V. J. *Paris.*—Anatomical wax models, &c.

1746. GUÉRIN, J. J. B. *Paris.*—Comparative osteological preparations.

1747 LEFEVRE, A. A. *Paris.*—Taxidermic preparations of animals' heads.

1748. AUZOUX, DR. L. *Paris.*—Elastic anatomical preparations.

1749. BOISSONNEAU, A. SEN. *Paris.*—Moveable artificial eyes.

1750. COULOMB, J. *Paris.* — Moveable artificial eyes.

1751. LUCAS, L. P. *Paris.*—Preserved plants.

1752. JULIENNE, MADAME M. J. E. *Paris.*—Bathing-belt for infants.

1753. DARBO, F. *Paris.*—Hygienic and other instruments.

1754. PARZUDAKI, E. *Paris.*—Taxidermic preparations.

1755. BOURGOGNE, J. SEN. *Paris.*—Microscropic, anatomical, &c. objects.

1756. PRÉTERRE, P. A. *Paris.*—Articles for the repair of injuries to the mouth.

1757. BOURGOGNE BROS. *Paris.* — Microscopic preparations of animal and vegetable anatomy.

1758. LAMI, A. *Paris.* — Anatomical model, representing the human muscular system.

1759. LÉTHO, F. H. *Paris.*—Artificial eyes for taxidermic preparations.

1760. DESJARDINS DE MORAINVILLE, DR. J. B. L. *Paris.*—Artificial eyes; specimens of diseases of the eye.

1761. BOISSONNEAU, A. P. JUN. *Paris.*—Moveable artificial eyes; specimens of diseases of the eye.

1762. MAREY, DR. J. *Paris.*—Sphygmographic apparatus, for recording the beatings of the pulse; the hemo-manometer, an instrument for measuring the pressure of the blood in the arteries; apparatus representing the physical phenomena of the circulation of the blood.

1763. DAMOISEAU, A. *Alençon (Orne).*—The zérabdella, an instrument intended as a substitute for leeches.

1764. FRANÇOIS, J. CONTE-GRANDCHAMP, & DESBUISSONS, *Paris.*—Plans of a proposed thermal establishment at Amélie-les-Bains.

1765. FRANÇOIS, J. & DURRIEU.—Plans of an establishment at the baths of Ussat.

1766. BILLET, FRANÇOIS, J., & PELLEGRINI, B.—Plans and views of the thermal establishment of Marlioz.

1767. FRANÇOIS, J., & NORMAND.—Plans of the baths of Baréges.

1768. CARON, A. *Paris.*—Medical spoon, for administering disagreeable medicine.

1769. MOUILLARD, P. F. V. *Paris.*—Process for purifying tobacco smoke, or combining it with matters intended to be inhaled.

# CLASS XVIII.

1771. MALLET BROS. *Lille (Nord).*—Cotton thread, plain, twisted, &c.

1772. LOYER, H. *Lille (Nord).*—Twisted thread of American and Algerian cotton.

1773. COLLECTIVE EXHIBITION OF THE CITY OF ROANNE (*Loire*).—Grain-dyed cotton fabrics (13 Exhibitors).

1774. COLLIN & CO. *Bar le Duc (Meuse).*—Cotton, and wool and cotton fabrics.

1775. HUMBERT & CO. *Gamaches (Somme).*—Cotton thread, plain and twisted.

1776. LEJEUNE, L. & VOITRIN, *Paris.*—Wadding, white and coloured.

1777. DUPONT-POULET, *Troyes (Aube).*—Spun and dyed cotton.

1778. DURET, *Brionne (Eure).* — Spun and dyed American and Algerian cotton.

1779. RITAUD, C. FLEURY, V. & CO. *Paris.*—Thread made of cotton dyed before being spun.

1780. CARTIER-BRESSON, *Paris.* — Twisted cotton thread.

1782. FAUQUET-LEMAITRE & PREVOST, *Pont-Audemer (Eure).*—Cotton thread.

1783. TABOUEL & LÉMERY, *Darnetal (Seine-Inf.).*—Dressed lustrings and fabrics.

1784. PIMONT, P. *Rouen (Seine-Inf.).*—Fabrics made of bleached and prepared cotton.

1785. FANQUET & LHEUREUX, *Rouen (Seine-Inf.).*—Cotton handkerchiefs.

1786. FANQUET, O. & CO. *Oissel-sur-Seine (Seine-Inf.).*—Spun cotton.

1788. DERLY, A. & CHABOY, *Bellemcombre (Seine-Inf.).*—Spun cotton.

1789. PEYNAUD & CO. *Remilly-sur-Andelle (Seine-Inf.).*—Spun cotton and calico.

1792. ROUSSELIN, S. *Darnetal (Seine Inf.).*—Spun cotton.

1793. VAUSSARD, *Nôtre-Dame-de-Boudeville (Seine-Inf.).*—Spun cotton, and cotton fabrics.

1794. BOUCLY-MARCHAND, *St. Quentin (Aisne).*—Plaited shirt-fronts, embroidered, &c., by machine.

1795. HUGUES-CAUNIN, *St. Quentin (Aisne).*—Figured, bleached, and dyed muslins for furniture.

1796. LEDOUX-BEDU, & CO. *St. Quentin (Aisne).*—Cotton fabrics, handkerchiefs of silk and cotton.

1797. TROCMÉ, P. L. *Hervilly (Somme).*—White and figured quilting.

1798. LEROY-DAUPHIN, *St. Quentin (Aisne).*—Plain cotton fabrics, &c.

1799. HUET-JACQUEMIN, A. *St. Quentin (Aisne).*—Worked muslins, &c.

1800. DERCHE-GIRARDE, *St. Quentin (Aisne).*—White worked muslins, &c.

1801. ROLAND, C. *St. Quentin (Aisne).*—White and coloured quilting for petticoats.

1802. DELACOURT, C. *Epehy (Somme).*—Quilting.

1803. COLOMBIER BROTHERS, *St. Quentin (Aisne).*—Fine quilting, for various purposes.

1804. CARCENAC and ROY, *Paris.*—Plain and other cotton fabrics, white and coloured.

1805. ODERIEU, C. and CHARDON, L. *Rouen (Seine Inf.).*—White and figured quilting.

1807. MOTTE-BOSSUT, & CO. *Roubaix (Nord).*—Various articles in cotton.

1808. BOIGEOL-JAPY, *Giromagny (Haut-Rhin).*—Spun and woven articles in Indian and American cotton.

1809. GROS, ODIER, ROMAN, & CO. *Wesserling (Haut-Rhin).*—White cotton fabrics.

1810. DOLFUS, MIEG, & CO. *Mulhouse (Haut-Rhin).*—Plain and twisted cotton-thread; raw and bleached cotton fabrics.

1811. FÉROUELLE & ROLLAND, *St. Quen-*

*tin (Aisne)*.—Figured window blinds and curtains.

1812. MENNET-POSSOZ, DAVID, & TROULLIER, *Paris*.—Worked muslins and gauzes, for curtains.

1813. DUBOIS, V. *Paris*.—Window curtains and blinds, &c.

1814. CHATELUS - DUBOST, *Tarare (Rhône)*.—Light cotton fabrics, tarlatans, &c.

1815. THIVEL-MICHON, *Tarare (Rhône)*. —Tarlatans.

1816. COLLECTIVE EXHIBITION OF THE CITY OF TARARE (*Rhône*)—Figured muslins, &c., made with a Jacquard machine (8 Exhibitors.)

1817. RUFFIER - LEUTNER, *Tarare (Rhône)*.—Muslins, plain and figured.

1818. MAC-CULLOCH BROTHERS, *Tarare (Rhône)*—Dyed cotton fabrics.

1819. PATUREAU, L. *Paris*.—Cases for sewing cotton.

## CLASS XIX.

1831. CASSE, J. & SON, *Lille (Nord)*.— Damasked table-linen, mixed flax and cotton fabrics for toilet covers.

1832. LAUWICK BROTHERS, *Commines (Nord)*.—Flax and cotton tapes.

1833. CORNILLEAU, L. SEN. & CO. *Mans (Sarthe)*.—Machine-made hempen cloth.

1834. BARY, JUN. & CO. *Mans (Sarthe)*. —Hemp-yarn, unbleached and bleached cloth.

1835. RÉVEILLÈRE, *Mans (Sarthe)*.— Raw and manufactured hemp; hemp-yarn for naval purposes; packing cloths.

1836. JOURNÉ, P. *Paris*.—Ticking of flax, of flax and cotton, &c., for trousers.

1837. DICKSON & CO. *Dunkerque (Nord)*. —Duck.

1838. FROMAGE, L. *Darnetal (Seine Inf.)*. —Ribbed sail-cloths.

1839. LEVEAU, A. *Evreux (Eure)*.— Ticking of cotton, and of flax and cotton.

1840. DEQUOY, J. *Lille (Nord)*.—Flax-yarn, tow-yarn, and cloth.

1841. DRUMMOND, BARTES, & CO. *Moulins-Lille (Nord)*.—Jute-yarn.

1842. VRAU, P. *Lille (Nord)*.—Sewing thread.

1843. MONCHAIN, Z. *Lille (Nord)*.— Flax-yarn.

1844. POUCHAIN, V. *Armentières (Nord)*. —Raw, bleached and dyed flax fabrics; wrought and damasked table linen.

1845. GRASSOT & CO. *Lyon (Rhône)*.— Damask table linen; fabrics of wool and cotton, for furniture.

1847. DENEUX BROTHERS, *Hallencourt (Somme)*.—Table linen.

1848. BUCHHOLTZ & CO. *Valenciennes (Nord)*.—Cambrics and lawns.

1849. DELAME-LELIÈVRE & SON, *Valenciennes (Nord)*.—Cambrics, and lawns; handkerchiefs.

1850. GUYNET, L. H. *Paris & Cambrai (Nord)*.—Linens of Cambrai, &c.

1851. LUSSIGNY BROTHERS, *Paris & Valenciennes (Nord)*.—Cambrics, of hand and machine-spun yarn, raw, bleached and printed; white handkerchiefs.

1852. GODARD, A. & BONTEMPS, *Valenciennes (Nord)*.—Cambrics, of hand and machine-spun yarn, &c.

1853. THOURY BROTHERS, & FLAIR, *Mans (Sarthe)*.—Carded hemp, thread, and packthread.

1854. VERSTRACTE & CO. *Lille (Nord).* —Flax-yarn, and undressed tow-yarn; twisted sewing-thread.

1855. HEUZE, RADIGUET, HOMON, GOURY, & LEROUX, *Landerneau (Finistèrre).* —Flax-yarn and tow-yarn; duck, &c.

1856. DEFREY, HOUSSIER, & LEPRÊTRE, *Alençon (Orne).*—Waterproof fabrics, and cordage.

1857. LEONI & COBLENZ, *Vaugenlieu (Oise).*—Hemp, peeled mechanically without steeping, by a new process.

# CLASS XX.

1861. MATHEVON & BOUVARD, *Lyon (Rhône).*—Silk fabrics, for furniture.

1862. CARQUILLAT, M. M. *Lyon (Rhône).* —Silk fabrics.

1863. HECKEL, SEN., BROSSET, *Lyon (Rhône).*—Silk fabrics.

1864. BLACHE & CO. *Lyon (Rhône).*— Silk velvet.

1865. BRUNET-LECOMTE, DEVILLAINE, & Co. *Lyon (Rhône).*—Silk fabrics, plain or printed.

1866. PONCET, L., LENOIR, V. & Co. *Lyon (Rhône).*—Silk fabrics.

1867. TEILLARD, C. M. *Lyon (Rhône).*— Plain silk fabrics.

1868. BROSSET, SEN. & BOISSIEU, *Lyon (Rhône).*—Plain and figured silks.

1869. GIRARD, NEPHEW, QUINZON & Co. *Lyon (Rhône).*—Plain velvet.

1870. BONNET, C. J. & Co. *Lyon (Rhône).* —Taffetas and satins.

1871. SCHULZ BROTHERS & BERAUD, *Lyon (Rhône).*—Silk fabrics for dresses; silk shawls.

1872. MARTIN, J. B. & P. *Tarare (Rhône).* —Dyed and milled silks; plush for hats; velvet.

1873. GANTILLON, D. *Lyon (Rhône).*— Dressed silk handkerchiefs.

1874. VILLARD & JACKSON, *Lyon (Rhône).*—Black and coloured velvet.

1875. ROUGIER & CO. *Lyon (Rhône).*— Plain and worked silk fabrics.

1876. RIBOUD, J., PRAVAZ, H. & Co. *Lyon (Rhône).*—Silk crapes and stuffs.

1877. GAUTHIER, J. & Co. *Lyon (Rhône).* —Plain velvet.

1878. SÉVÈNE, BARRAL, & Co. *Lyon (Rhône).*—Silk fabrics, plain and figured.

1879. DONAT, A. & Co. *Lyon (Rhône).* —Silk, velvet, and stuffs for waistcoats; grenadines for dresses and shawls.

1880. FONT, CHAMBEYRON, & BENOIT, *Lyon (Rhône).*—Plain velvet.

1881. CAQUET-VAUZELLE, & COTE, *Lyon (Rhône).*—Plain and figured silks for dresses.

1882. BOCOUP, VILLARD, & CO. *Lyon (Rhône).*—Silk fabrics, plain and figured; Chambery gauze.

1883. BARBEQUOT, CHENAUD, & Co. *Lyon (Rhône).*—Silk shawls.

1884. BOYRIVEN BROS. *Lyon (Rhône).*— Carriage lace.

1885. ALGOUD BROS. *Lyon (Rhône).*— Black taffetas.

1886. ARAUD BROS. *Lyon (Rhône).*— Silk fabrics for umbrellas.

1887. YÉMÉNIZ, *Lyon (Rhône).*—Silk fabrics for furniture.

1888. TAPISSIER, JUN. & HUTET, P. *Lyon (Rhône).*—Plain and figured silk fabrics; velvet.

1889. SÈVE & CO. *Lyon (Rhône).*—Plain, black, and coloured velvet.

1890. VANEL, L. & Co. *Lyon (Rhône).* —Worked silk fabrics.

1891. MILLION, J. P. & Co. *Lyon (Rhône).* —Plain silk fabrics; velvet.

1892. TAPISSIER, JUN. & DEBRY, *Lyon* (*Rhône*).—Black taffetas and stuffs of silk.

1893. SILO COUSINS, *Lyon* (*Rhône*).—Silk fabrics for fancy goods.

1894. RIGOD, *Lyon* (*Rhône*).—Velvet.

1895. ROSSET, RENDU, & Co. *Lyon* (*Rhône*).—Silk fabrics, shawls.

1896. BRISSON BROS. *Lyon* (*Rhône*).—Plain velvets; plushes; galoons.

1897. BAYARD BROS. *Lyon* (*Rhône*).—Plain and figured silk fabrics; galoons for hats.

1898. BELLON BROS. & CONTY, *Lyon* (*Rhône*).—Plain silk fabrics.

1899. COCHAUD, ADAM, & Co. *Lyon* (*Rhône*).—Plain and figured silks.

1900. CHARBIN & TROUBAT, *Lyon* (*Rhône*).—Plain velvet.

1901. HAMELIN, A. *Paris.*—Twisted silk for sewing and embroidery, plain and dyed.

1902. FABRE, C. *Paris.*—Twisted silk, for sewing and embroidery, plain and dyed.

1903. CHILLIAT, E. *Paris.*—Silks, milled, undyed, and dyed.

1904. PLAILLY, N. P. *Paris.*—Sewing silk for glove making.

1905. HOCK, A. *Strasbourg* (*Bas-Rhin*).—Cottons and silks, yellow, and in imitation of straw and wood.

1906. BALLY, J. *Paris.* — Ribbons for orders of knighthood, French and foreign.

1907. DE BARY-MERIAN, *Guebwiller* (*Haut-Rhin*).—Plain and figured silk ribbons.

1908. WILLIAM, S. *Soultz* (*Haut-Rhin*).—Black silk ribbons, satined and figured taffetas.

1910. DENIS, A. *St. Etienne* (*Loire*).—Velvet, and articles in lace.

1911. GIRINON, JUN. *St. Etienne* (*Loire*).—Articles in lace.

1912. FAURE, E. *St. Etienne* (*Loire*).—Articles in lace.

1913. LARCHER, FAURE, & Co. *St. Etienne* (*Loire*).—Plain and figured silk ribbons.

1914. REBOURG, C. *St. Etienne* (*Loire*).—Figured silk ribbons.

1915. CALEMARD, J. *St. Etienne* (*Loire*).—Plain and figured silk ribbons.

1916. EPITALON BROTHERS, *St. Etienne* (*Loire*).—Plain satin ribbons.

1917. JOUCERAND, C. *St Etienne* (*Loire*).—Plain and figured silk ribbons.

1918. COLLECTIVE EXHIBITION OF THE CITY OF ST. CHAMOND (*Loire*).—Silk laces and braids (4 Exhibitors).

1919. RENARD BROS. *Sarreguemines* (*Moselle*).—Silk plushes for hats.

1920. MASSING BROS. & Co. *Puttelange* (*Moselle*).—Silk plushes for hats.

1921. LACOUR & WALTER, *Sarreguemines* (*Moselle*).—Silk and cotton plushes for hats.

1922 MASSING, P. & Co. *Sarreguemines* (*Moselle*).—Silk plushes for hats.

1923. BARALLON & BROSSARD, J. *St. Etienne* (*Loire*).—Ribbons made of raw silk.

1924. BALAY, J. & Co. *St. Etienne* (*Loire*).—Ribbons made of raw silk, and dyed in the piece.

1925. NEYRET, J. *St. Etienne* (*Loire*).—Ribbons for belts and decorations.

1926. DUGNAT-GAUTHIER, *St. Etienne* (*Loire*).—Silk and velvet ribbons.

1927. GÉRENTET & COIGNET, *St. Etienne* (*Loire*).—Plain and figured silk ribbons.

1928. DONZEL, L. *St. Etienne* (*Loire*).—Silk and velvet ribbons.

1929. DESCOURS, A. *St. Etienne* (*Loire*).—Silk and velvet ribbons.

1930. AVRIL, A. & Co. *St. Etienne* (*Loire*).—Plain and figured velvet ribbon.

1931. GIRON BROS. *St. Etienne* (*Loire*).—Velvet fabrics and ribbons.

1932. DAVID, J. B. *St. Etienne* (*Loire*).—Ribbons of silk and velvet.

1933. PILLET-MEAUZÉ & SON, *Tours* (*Indre-and-Loire.*)—Silk fabrics for furniture and carriages.

1934. FEY, MARTIN, EUDE & VIEUGUÉ, *Tours* (*Indre-and-Loire*).—Silk fabrics for furniture and carriages.

1935. LACHARD & BESSON, *Lyon (Rhône).* —Silk for dresses; shawls.

1936. BERNARD-JOLY & CHAPPET, *Lyon (Rhône).*—Velvet.

1937. GALLAND, F. *Lyon (Rhône).*— Figured silks.

1938. BADOIL, G. and Co. *Lyon (Rhône).* —Plain and figured silks; grenadine shawls.

1939. BERLIE, A. SON, & Co. *Lyon (Rhône).*—Plain and figured silks.

1940. BARDON & RITTON, *Lyon (Rhône).* —Plain and figured silks.

1941. SERVANT DEVIENNE & Co. *Lyon. (Rhône).*—Plain and figured silks for waist-coats and cravats.

1942. PONCET, JUN. & Co. *Lyon (Rhône).* —Silk fabrics for umbrellas.

1943. NICOLAS, F. & Co. *Lyon (Rhône).* —Fabrics of silk, of silk and wool, of wool and cotton, and of silk and cotton.

1944. GOURD, CROIZAT, SON, & DUBOST, *Lyon (Rhône).*—Plain and figured silks.

1945. MAUVERNAY & DUBOST, *Lyon (Rhône).*—Black and coloured silks.

1946. GUISE & Co. *Lyon (Rhône).*—Plain silk and velvet fabrics.

1947. MERCIER, VUILLEMOT, & NEYRET, *Lyon (Rhône).*—Silk velvets and fabrics, for waistcoats.

1948. CHARBONNET & VILLATTE, *Lyon (Rhône).*—Silk fabrics.

1949. VERPILLAT, J. *Lyon (Rhône).*— Plain silk fabrics.

1950. BRÉBANT, SALOMON, & Co. *Lyon (Rhône).*—Silk fabrics.

1951. BELMONT-TERRET, & Co. *Lyon (Rhône).*—Plain silk fabrics.

1952. BÉRARD E. PONCET, & Co. *Lyon (Rhône).*—Silks for dresses.

1953. FAVROT BROTHERS, *Lyon (Rhône).* —Plain, printed, and figured silk handker-chiefs.

1954. LYON, A. & Co. *Lyon (Rhône).*— Black silk fabrics; silk handkerchiefs; velvet.

1955. KUPPENHEIM, *Lyon (Rhône).* — Printed handkerchiefs.

1956. EMERY, L. *Lyon (Rhône).*—Plain and figured silks.

1957. TRAPADOUX, A. & Co. *Lyon (Rhône).*—Silks for handkerchiefs.

1958. RÉROLLE, G. & Co. *Lyon (Rhône).* —Printed handkerchiefs.

1959. RONZE & VACHON, *Lyon (Rhône).* —Silk fabrics.

1960. MENET, J. H. & DURINGE, S. *Lyon (Rhône).*—Silk fabrics.

1961. LABORÉ, RODIER, & Co. *Lyon (Rhône).*—Figured silks.

1962. VALANSOT, M. *Lyon (Rhône).*— Plain silk fabrics.

1963. BOUVARD & SON, *Lyon (Rhône).*— Silk fabrics for furniture, and church orna-ments.

1964. TABART, G. F. & Co. *Lyon (Rhône).* —Plain, black, and coloured velvet.

1965. SAVOYE, RAVIER & CHANU, *Lyon (Rhône).*—Plain and figured silks.

1966. ROCHE & DIME, *Lyon (Rhône).*— Shawls; figured silks.

1967. PONSON, C. *Lyon (Rhône).*—Silk fabrics; velvets.

1968. MEYNIER, P. *Lyon (Rhône).*—Fi-gured silks.

1969. JANDIN, C. & DUVAL, A. *Lyon (Rhône).*—Printed silk handkerchiefs.

1970. GINDRE & Co. *Lyon (Rhône).*— Plain silk fabrics.

1971. MONTESSUY, A. & CHOMER, A. *Lyon (Rhône).*—Silk fabrics; crape; mus-lin, &c.

1972. LAMY, A. *Lyon (Rhône).*—Figured silk for dresses.

1973. GONDRE & Co. *Lyon (Rhône).*— Plain silk fabrics; velvet.

1974. FRANC, SON, & MARTELIN, *Lyon (Rhône).*—Fancy silk thread.

1975. COLLET - LEFRANCQ, *Amiens (Somme).*— Raw silk waste, carded, and spun.

1976. RÉVIL, C. & Co. *Amilly (Loiret).* —Spun floss silk.

1977. BINDSCHEDLER, LEGRAND, & FAL-

LOT, *Thann* (*Haut - Rhin*). — Spun floss silk.

1978. BLONDEAU-BILLIET, *Lille* (*Nord*). Spun floss silk.

1979. HENNECART, J. F. *Paris.*—Silk fabrics for bolting flour, and fecula, gunpowder, pharmaceutical powders, and porcelain earths.

1980. GASCOU, NEPHEW, & ALBRESPY, A. *Montauban* (*Tarn-and-Garonne*).— Silk fabrics for bolting flour; raw silk.

1981. COUDERC, A. & SOUCARET, JUN. *Montauban* (*Tarn-and-Garonne*).—Silk fabrics for bolting flour; raw silk.

1982. REYBAUD, J. *Lyon* (*Rhône*).—Portraits on silk, executed by a Jacquard loom.

---

The following appear also in Class III., as contributors of raw materials:—

830. DE BAILLET, *St. Germains-es-Mons* (*Dordogne*).—Silk.

859. BANNETON, *St. Vallier* (*Drôme*).— Raw silk, organzine.

840. BARRÉS BROS. *St.-Julien-en-St.-Alban* (*Ardèche*).—Raw and prepared silk.

832. BERARD & BRUNET (*Rhône*).—Raw and prepared silk.

856. BISCARRAT, P. *Bouchet* (*Drôme*).— Raw and prepared silk.

839. BLANCHON, L. *St.-Julien-en-St.-Alban.*—Raw and prepared silk.

843. BLANCHON, JUN. *Flaviac* (*Ardèche*). —Raw and prepared silk.

762. BLANC - MONTBRUN, *La Rolière* (*Drôme*).—Raw and prepared silk.

845. BOISRAMEY, JUN. *Caen* (*Calvados*). —Raw and prepared silk, for blondes and net. Grenadines for lace.

849. BONNET & BOUNIOLS, *Vigan* (*Gard*). —Raw silk.

852. BONDET, F. *Uzès* (*Gard*).—Raw silk.

850. BROUILHET & BAUMIER, *Vigan* (*Gard*).—Raw and prepared silk.

744. BUISSOU, C. *Tronche* (*Isère*).—Raw and prepared silk.

794. CHABAUD, A. *Avignon* (*Vaucluse*).— Raw and prepared silk.

848. CHAMBON, WIDOW, *St.-Paul-Lacoste* (*Gard*).—Raw silk, and organzine.

853. CHAMPANOHET - SARGEAS BROS. *Vals.* (*Ardèche*).—Raw and prepared silk.

742. CHANGEA, *Lamastre* (*Ardèche*).— Raw and prepared silk.

857. CHARTRON & SON, *St. Vallier* (*Drôme*) —Silk.

749. COMBIER BROS. *Livron* (*Drôme*).— Raw silk.

833. COMTESSE C. DE CORNEILLAN, *Paris.*—Raw silk.

816. CORNEILLE & FABRE, *Trans* (*Var*). —Raw silk, &c.

835. FARA, JUN. *Bourg-Argental* (*Loire*). —Silk for laces.

854. FOUGEIROL, A. *Ollières* (*Ardèche*). —Raw silk, organzine.

858. FRANQUEBALME & SON, *Avignon* (*Vaucluse*).—Worked Chinese and Japanese silk, for woof and nap.

836. FRIGARD, *Bourg-Argental* (*Loire*). —White and yellow raw silk.

750. GAUTIER, F. *Chabeuil* (*Drôme*).— Raw and prepared silk.

751. HELME, A. *Loriol* (*Drôme*).—Raw and prepared silk.

777. LACOMBE, J. *Alais* (*Gard*).—Raw silk.

752. LACROIX, P. (*Drôme*).—Raw and prepared silk.

753. LASCOUR, *Crest* (*Drôme*).—Raw and prepared silk.

754. LEYDIER BROTHERS, *Buis-lès-Baronnies* (*Drôme*).—Raw and prepared silk.

862. MAHISTRE, A. JUN. *Vigan* (*Gard*). —Raw silk.

717. MARAVAL & CO. *Lavaur* (*Tarn*).— Raw silk.

851. MARTIN, L. & CO. *Lasalle* (*Gard*). —Raw silk.

842. MONESTIER, SEN. *Avignon* (*Vaucluse*).—Raw and prepared silk.

704. AGRICULTURAL ASSOCIATION OF MONTAUBAN (*Tarn-and-Garonne*).—Silk.

706. SOCIETY OF SCIENCE AND AGRICULTURE OF MONTAUBAN (*Tarn-and-Garonne*). —Silk.

785. NOURRIGAT, *Lunel (Herault)*.—Raw silk.

755. NOYER BROS. *Dieulefit (Drôme).*— Raw and prepared silk.

844. PALLUAT & Co. *Lyon (Rhône.*— Prepared silk.

2045. PASQUAY BROTHERS, & Co. *Wasselonne (Bas-Rhin)*.—Spun floss silk.

741. PRADIER, J. *Annonay (Ardèche)*.— Raw and prepared silk.

829. RAIBAUD-L'ANGE, DIRECTOR OF THE FARM SCHOOL OF PAILLEROLS.—

855. REGARD BROTHERS, *Privas (Ardèche)*.—Raw and prepared silk.

841. SÉRUSLAT, L. *Etoile (Drôme).*— Prepared silk.

847. TESSIER DU CROS, *Valleraugue (Gard)*.—Raw and prepared silk.

778. VERNET BROS. *Beaucaire (Gard)*. —Raw and prepared silk.

# CLASS XXI.

1991. LELARGE, F., & AUGER, A. *Reims (Marne)*.—Plain and twilled flannels for hygienic purposes.

1992. BOUFFARD, FERRIER, & Co. *Paris and Reims.*—Plain and twilled flannels; various kinds of cloth.

1994. PRADINE & Co. *Reims (Marne).*— Merinos wove by power-loom.

1995. LUCAS BROTHERS, *Reims (Marne).* —Yarn from carded wool; merinos.

1996. PHILIPPOT, J. M. *Reims (Marne).* —Merinos, figured fabrics.

1998. VILLEMINOT, *Reims (Marne).*— Yarn of combed wool; merinos made by power-loom.

1999. ROGELET, C., GAND BROS., GRAND-JEAN, IBRY, & Co. *Reims (Marne).*—Yarn of combed wool, merinos.

2000. GILBERT & Co. *Reims (Marne).*— Yarns of combed wool; merinos made by power-loom.

2001. CROUTELLE, ROGELET, GAND, & GRANDJEAN, *Reims (Marne).* — Yarns of carded wool; plain and twilled flannels.

2002. SAUTRET, A. T. *Bétheniville (Marne).*—Merinos made by power-loom.

2003. ROBERT-GALLAND, *Pontfaverger (Marne).*—Merinos.

2004. OUDIN BROS. *Reims (Marne).*— Merinos made by power-loom.

2005. MAUGRAS, H.—Mérinos made by power-loom.

2006. CHATELAIN-FÉRON, *Reims (Marne).* —Cloth and flannel, plain and mixed.

2007. DESTEUQUE, BOUCHEZ, & QUE-NOBLE BROS. *Reims (Marne).*—Cloth, flannel, figured tissues.

2008. JOLTROIS, C. *Reims (Marne).*— Cloth, flannel.

2009. BENOIST - MALOT, & WALBAUM; *Reims (Marne).*—Cloth for fancy articles; merinos made by power-loom.

2011. BENOIST & GREVIN, *Reims (Marne).* —Flannel, fabrics of cotton, and of wool, figured.

2012. APPERT-TARTAT, *Reims (Marne).* —Cloth.

2014. VIÉVILLE & Co. *Reims (Marne).*— Cloth.

2017. HARMEL BROS. *Warmeriville (Marne).*—Yarn of combed, carded, and both combed and carded wool.

2018. PIERRARD - PARPAITE, *Reims (Marne).*—Combed wool.

2019. HOLDEN, *Reims (Marne)*.—Combed wool.

2020. LEGRAND, T. & SONS, *Fourmies (Nord)*.—Combed wool with woollen yarn, and woollen fabrics made of it.

2021. COLLECTIVE EXHIBITION BY THE DEPARTMENT OF THE SOMME. — Cotton, linen, and woollen yarn; Utrecht velvet; various fabrics (20 Exhibitors).

2022. COLLET-DUBOIS & CO. *Amiens (Somme)*.—Fabrics of wool, of wool and cotton, and of wool and silk.

2023. CRIGNON, SON, & HUE, *Amiens (Somme)*.—Combed wool, yarn for hosiery, embroidery, and lace.

2024. GAUTHIER, E. *Amiens (Somme)*.—Shawls of wool, of wool and cotton, and of wool and silk; fabrics of wool, and of wool and silk, for dresses; stuffs for waistcoating.

2025. VULFRAN-MOLLET, *Amiens (Somme)*, *and Paris*.—Plain and figured fabrics of wool and silk.

2026. PAYEN & CO. *Amiens (Somme)*.—Utrecht velvet.

2027. HORDÉ, E. & CO. *Amiens (Somme)*.—Plain and worked fabrics, for dresses and shawls; satins, for boots and shoes; fabrics of wool and silk, termed *barrepours*.

2028. BARIL, SON, & CO. *Amiens (Somme)*.—Utrecht velvet made with the power-loom, cotton velvet.

2029. GAMONNET-DEHOLLANDE, *Amiens (Somme)*.—Satins for shoes and boots.

2030. BOQUET, J. & CO. *Amiens (Somme)*.—Utrecht velvet, satins, figured fabrics.

2031. FÉRY, J. JUN. *Metz (Moselle)*.—Circular felts, for paper-making.

2032. CRÉTIEN, DEBOUCHAUD, MATTARD, VERIT, & CO. *Nersac (Charente)*.—Circular felts, for paper-making.

2033. VOUILLON, F. *Louviers (Eure)*.—Felted thread, and cloth made of it.

2038. DAVID-LABBEZ & CO. *Saint-Richaumont (Aisne)*.—Twisted yarn of wool and cotton; merinos made by power-loom.

2039. AUDRESSET, SON, & MENUET, *Paris*.—Cachemire wool and down, combed, spun, and woven.

2040. NOIRET, CHOPPIN, & CO. *Paris*.—Combed wool; plain and twisted yarn; double thread of wool and silk, for Cashmeres.

2041. BELLOT, C. & COLLIÈRE, O. *Angecourt (Ardennes)*.—Yarn of carded wool.

2043. COLBECK & GREVEN, *Douai (Nord)*.—Unravelled wool.

2044. VERNET BROS. *Nimes (Gard)*.—Carded wool.

2045. PASQUAY BROS. & CO. *Wasselonne (Bas-Rhin)*.—Spun wool, and floss silk; plaits of waste silk for covering steam-pipes.

2046. PALLIER, P. *Nimes (Gard)*.—Laces and cords of cotton, silk and caoutchouc; plaits of silk, alpaca, &c.

2047. DE FOURMENT & SON, *Cercamp-lez-Frévent (Pas-de-Calais)*.—Yarn of combed wool.

2048. ARRECKX-COLLETTE, WIDOW, *Tourcoing (Nord)*.—Woollen yarn for making poplin and laces.

2050. BUIRETTE-THIAFFAIT, & FARAGUET, *Dijon (Côte-d'Or)*.—Carded and spun wool.

2051. ROBIN, A. *Dijon (Côte d'Or)*.—Woollen yarns.

2052. KOECHLIN-DOLLFUS, *Mulhouse (Haut-Rhin)*.—Unbleached wool.

2053. MASSE, H., & CRESSIN, JUN. *Corbie (Somme)*.—Twisted yarns.

2054. CRESSIN, C., DEWAILLY, E. & CO. *Fouilloy (Somme)*.—Twisted yarn of cotton, of wool, of silk, &c.

2055. OUIN, A. & CO. *Paris*.—Dyed unravelled wool.

2056. HARTMANN, SCHMALZER, & CO. *Malmerspach (Haut-Rhin)*.—Woollen yarn, unbleached, plain, twisted, and milled.

2057. CHEGUILLAUME, P. & CO. *Cugand (Vendée)*.—Cloth, druggets, and fustians.

2058. HONNORAT, E. JUN. *St. André (Basses-Alpes)*. — Common cloth; fleeces, washed and unwashed.

2059. SIGNORET, P. *Vienne (Isère)*.—Drapery, and fancy articles for trousers.

2061. POIX-COSTE, WIDOW, BARBARIN, & CO. *Vienne (Isère)*.—Drapery.

2062. GOUET, NEPHEW, & SON, *Vienne (Isère)*.—Drapery.

2063. LAFONT & GAY, *Vienne (Isère)*.—Drapery, and fancy articles for trousers.

2065. PERTUS & JULLIEN, A. *Vienne (Isère).*—Cloth and fancy articles for trousers.

2066. BONON, & ALEX. BROS. *Vienne (Isère).*—Drapery.

2067. REYMOND & BAYARD-BARON, *Vienne (Isère).*—Drapery.

2068. SAVOYE, BLANC, & Co. *Vienne (Isère).* — Double-milled cloth ; cloth and fancy articles for trousers.

2069. PONCHON, SEN. *Vienne (Isère).*—Double-milled cloth ; castor.

2070. GAUDCHAUX-PICARD SONS, *Nancy (Meurthe).*—Plain and figured cloth for men's clothes.

2071. BLIN, SON, & BLOC, *Bischwiller (Bas-Rhin).*—Black and coloured cloth.

2072. ROUSTIC, WIDOW A. *Carcasonne (Aude).*—Plain and figured cloth.

2073. COLLECTIVE EXHIBITION OF THE CITY OF CASTRES (*Tarn*).—Fabrics of wool, of wool and cotton, of wool and silk ; fancy drapery.

2074. COLLECTIVE EXHIBITION OF THE CITY OF MAZAMET (*Tarn*).—Flannel, figured woollen fabrics, &c.

2075. VERNAZOBRES BROS. *Bédarieux (Hérault).*—Plain and figured cloth.

2077. DAUTRESME, D. JUN. *Rouen (Seine-Inf.).*—Figured cloth.

2078. MAUZIÈRES. SEN. & Co. *Castres (Tarn).*—Flannel ; figured cloth.

2079. BATUT-LAVAL, P. & Co. *Castres (Tarn).*—Drapery.

2080. HOULÈS, SON, & CORMOULS, *Mazamet (Tarn).*—Velvet, figured cloth for trousers and paletots.

2081. OLOMBEL BROS. *Mazamet (Tarn).*—Woollen fabrics, wool velvets ; cloths and fancy articles for trousers.

2082. JUHEL-DESMARES, J. *Vire (Calvados).*—Double-milled cloths ; satins ; cloth for uniforms.

2083. LENORMAND, A. *Vire (Calvados).*—Cloth, and wool velvet.

2084. DEMAR, L. & Co. *Elbeuf (Seine-Inf.).*—Fancy cloth for various purposes.

2085. THILLARD, J. & Co. *Elbeuf (Seine-Inf.).*—Fancy articles for trousers.

2086. LEGRIX & MAUREL, *Elbeuf (Seine-Inf.).*—Fancy stuffs for garments.

2087. GERIN-ROSE, H. *Elbeuf (Seine-Inf.).*—Figured fabrics.

2088. CHARY, F. & LAFENDEL, J. *Elbeuf (Seine Inf.).*—Black and coloured cloth.

2089. VAUQUELIN, F. *Elbeuf (Seine-Inf.).*—Fancy stuffs for trousers.

2090. COSSE, L. *Elbeuf (Seine-Inf.).*—Stuffs for paletots ; drapery.

2091. BELLEST, E. BENOIST, & Co. *Elbeuf (Seine-Inf.).*—Cloth, satins.

2092. MIGNART & Co. *Elbeuf (Seine-Inf.).*—Stuffs for garments.

2093. DUCLOS, L. & NEPHEW, *Elbeuf (Seine-Inf.).*—Cloth for uniforms, &c.

2094. LEVIEUX, F. A. *Elbeuf (Seine-Inf.).*—Fancy goods.

2095. OSMONT, A. & LHERMUZEAX, *Elbeuf, (Seine-Inf.).*—Fancy goods.

2096. POUSSIN, A. & SON, *Elbeuf (Seine-Inf.).*—Cloth, black satins.

2097. CHENNEVIÈRE, WIDOW, & SON, *Elbeuf (Seine-Inf.).*—Fancy goods.

2098. DESBOIS, P. & Co. *Elbeuf (Seine-Inf.).*—Black and coloured cloth ; stuffs for paletots.

2099. BORDEREL, JUN. *Sédan (Ardennes).*—Black cloth, wool velvet.

2100. DE MONTAGNAC, E. & SON, *Sédan (Ardennes).* — Fabrics for garments, wool velvet.

2101. LECOMTE BROS. *Sédan (Ardennes).*—Cloth, satins, velvet.

2102. LABROSSE BROS. *Sédan (Ardennes).*—Black and coloured cloths ; plain and figured velvets.

2103. CUNIN-GRIDAINE, SON, & Co. *Sédan (Ardennes).*—Cloth, satin, velvet, fancy drapery.

2104. BERTÉCHE, BOUDOUX-CHESNON & Co. *Sédan (Ardennes).*—Cloth, satins, and cashmeres, black, and in colours ; plain and figured stuffs for paletots ; wool velvet.

2105. RENARD, A. *Sédan (Ardennes).*—Cloth, satins, &c.

2106. HULIN, E. G. & H. PAQUIN, JUN. *Sédan (Ardennes).*—Cloth, satin, &c.

2107. BACOT, P. & SON, *Sédan (Ardennes).*—Cloths, fancy goods, &c.

2108. BACOT, F. & SON, *Sédan (Ardennes).*—Cloth, satins, &c.

2109. POITEVIN & SON, *Louviers (Eure).* —Cloth and fancy goods.

2110. NOUFFLARD, C. & Co. *Louviers (Eure).*—Cloth, and fancy articles for paletots.

2111. BRUGUIÈRE, H. & Co. *Louviers (Eure).*—Cloth, stuffs, &c.

2112. JEUFFRAIN & SON, *Louviers (Eure).* —Black cloths and satins; blue cloths and satins for military uniforms; stuffs for men's use.

2113. PELLIER & TRUBERT, *Louviers (Eure).*—Fancy goods.

2114. REMY, E. & PICARD, P. *Louviers (Eure).*—Stuffs; fancy goods.

2115. GASTINNE, H. *Louviers (Eure).*— Fancy goods.

2116. POITEVIN, C. *Louviers (Eure).*— Cloth, and stuffs, for paletots.

2117. DANNET & Co. *Louviers (Eure).*— Fancy goods, and fabrics of carded wool.

2118. RENAULT, R. & Co. *Louviers (Eure).*—Black cloth and satins; blue cloth for uniforms.

2119. BRETON, L. & BARBE, L. *Louviers (Eure).*—Cloth, and fancy goods.

2120. BERTIN, J. & PENNELLE, J. *Louviers (Eure).*—Fancy goods.

2121. HURSTEL BROS. *Ribemont (Aisne).* —Combed and spun wool, white, and dyed.

2122. LEFEBVRE-DUCATTEAU BROS. *Roubaix (Nord).*—Combed, carded, and spun wool; stuffs for waistcoats and shoes and boots; plain and figured merinos.

2123. CATTEAU, P. *Roubaix (Nord).*— Stuffs for dresses.

2124. DELATTRE, H. & SON. *Roubaix (Nord).*—Combed and spun wool; plain and mixed fabrics:

2125. CORDONNIER, L. *Roubaix (Nord).* —Mixed fabric for dresses, made by power-loom.

2126. HARINKOUCK & CUVILLIER, *Roubaix (Nord).*—Fabrics for furniture.

2127. LAGACHE, J. *Roubaix (Nord).*— Fabrics for waistcoats.

2128. BULTEAU BROS. *Roubaix (Nord).*— Fabrics for dresses, and shawls.

2129. SCRÉPEL-LEFEBVRE, *Roubaix (Nord).*—Fabrics of wool, of wool and cotton, of wool and silk.

2130 SCREPEL, L. & SON, *Roubaix (Nord).* —Fabrics of wool, and of wool and silk.

2131. SCREPEL-ROUSSEL, *Roubaix (Nord).* —Fabrics of wool, of wool and cotton, of wool and silk.

2132. DILLIES BROS. *Roubaix (Nord).* —Plain and figured fabrics mixed with cotton and silk.

2133. SADON & Co. *Roubaix (Nord).*— Figured stuffs of wool and silk; shawls, railway wrappers.

2134. CAZIER, E. *Roubaix (Nord).* — Woollen yarn, for poplins and shawls.

2135. PROUVOST A. & Co. *Roubaix (Nord).*—Combed wool.

2136. ERNOULT-BAYART & SON, *Roubaix (Nord).*—Yarns of carded wool.

2137. CATTEAU-LEPLAT, *Roubaix (Nord).* —Woollen stuffs for dresses; satins for boots and shoes.

2138. SCRÉPEL, C. *Roubaix (Nord).*— Woollen fabrics for various purposes:

2139. PIN-BAYART, *Roubaix (Nord).*— Fabrics of wool, of wool and cotton, and of wool and silk, for dresses.

2140. VAN-DONGHEN, E. & V. *Roubaix (Nord).*—Fancy goods for dresses.

2141. POULLIER-DELERUE, *Roubaix (Nord).*—Woollen fabrics, fancy goods.

2142. HEYNDRICKX-DORMEUIL, WIDOW, *Roubaix (Nord).*—Fabrics for waistcoats.

2143.—MOURCEAU, C. H. *Paris.*—Tapestries and reps for furniture.

2144. PIN, A. & Co. *Lyon (Rhône).*— Worked shawls.

2145. RIVOIRON, PERRAUD, & GUIGNARD, *Lyon (Rhône).*—Worked shawls of wool and cashmere.

2146. GELLIN, C. & Co. *Lyon (Rhône.)* —Worked shawls of wool and cashmere.

2147. DAVIN, F. *Paris.*—Combed and spun wool; and products obtained with the soft wool termed indigenous cashmere of Graux de Mauchamps; products obtained with the hair of the Angora goat acclimatized in France, and with the hair of the Asiatic and African camel.

2148. BOUTEILLE, COUSINS, *Paris.* — Worked long, and square, woollen shawls.

2149. CHAPUSOT, PREVOST, & BOING, *Paris.*—Worked woollen shawls.

2150. RIBES, ROUX, & DURAND, *Nîmes (Gard.)*—Worked woollen shawls.

2151. BIÉTRY, L. *Paris.* — Cashmere fabrics for dresses; and plain shawls; French cashmere shawls.

2152. SEYDOUX, A. SIEBER, & Co. *Le Cateau (Nord).*—Merino wool combed and spun, pure and mixed fabrics.

2153. LAIR, H. & LAIR, H. *Paris*— Worked woollen shawls.

2154. HESS, G. *Paris.*—Fabrics of goosedown; fancy goods.

2155. DUCHÉ, A., DUCHÉ, JUN., BRIÈRE, & Co. *Paris.*—Worked woollen and cashmere shawls.

2156. FABART & Co. *Paris.* — Shawls, imitations of Indian cashmere.

2157. CHAMPION, L. *Paris.* — Worked shawls of cashmere and wool.

2158. AUBÉ, NOURTIER, & Co. *Paris.*— Shawls of cashmere and wool; shawls embroidered with gold and silver.

2159. BOUTARD & LASSALLE, *Paris.*— Figured shawls of wool and cashmere.

2160. BOAS BROS. & Co. *Paris.*—Figured shawls of wool and cashmere.

2161. HÉBERT, E. F. & SON, *Paris.*— Shawls of wool and cashmere, &c.

2162. GÉRARD, C. & CANTIGNY, *Paris.* —Shawls of wool and cashmere.

2163. CATHERINE, V. & Co. *Paris.*— Shawls of wool and cashmere.

2164. DACHÈS & DUVERGER, *Paris.*— Worked shawls of wool and cashmere.

2165. BOURGEOIS BROS. *Paris.*—Worked shawls of wool and cashmere.

2166. ROBERT & GOSSELIN, JUN, *Paris.* —Worked shawls of wool and cashmere.

2167. HOOPER, G., CARROZ, & TABOURIER, *Paris.*—Pure and mixed woollen fabrics; silk fabrics; net, laces.

2168. LEGRAND, A. *Paris.* — Worked shawls.

2169. DREYFOUS, F. *Paris.*—Mixed fabrics.

2170. VATIN, F. JUN. & Co. *Paris.*— Fabrics of silk, and of wool and silk; gauze, and bareges for dresses and shawls.

2171. ZADIG, J. B. *Paris,*—Gauze, and bareges for dresses, shawls, &c.

2173. PLANCHE, L. & Co. *Paris.*—Fabrics of combed wool, muslins, &c.; plain, figured, and printed shawls; shawls and tissues of gauze and barege.

2174. SABRAN, V. & JESSÉ, G. *Paris.*— Woollen and mixed fabrics.

2175. DUNCAN & CHARPENTIER, *Paris.* —Plain and figured fabrics for shawls and dresses.

2176. WEISGERBER BROS. & KIENER, *Paris.*—Mixed fabrics.

2177. COCHETEUX, SON, & Co. *Templeuve- (Nord).*—Fabrics of wool and silk for furniture.

2178. JOURDAIN-HERBERT, JUN. *Amiens (Somme).*—Goats' hair velvet for furniture.

2179. MAZURE-MAZURE, *Roubaix (Nord).* —Fabrics for dresses and furniture; reps, and damask.

2180. BERCHOUD, L. & GUERREAU, *Paris.* —Stuffs for furniture; Belleville tapestry.

2181. DAGER, E., MENAGER, H., & WALMEZ, H. *Paris.*—Reps for furniture; imitation Aubusson tapestry.

2182. ARNAUD-GAIDAN, J. & Co. *Nîmes (Gard).*—Stuffs for furniture.

2183. DAMIRON & Co. *Lyon (Rhône).*— Shawls of wool, and cashmere.

2184. CHANEL, J. *Lyon (Rhône)*.—Worked woollen shawls.

2185. MANTELIER, P. & Co. *Lyon (Rhône)*. —Worked woollen shawls.

2186. DESPRÉAUX, A. *Versailles (Seine-and-Oise)*.—Stuffs for furniture.

2187. GRASSOT & Co. *Lyon (Rhône)*.— Fabrics of wool and cotton for furniture.

# CLASS XXII.

2191. IMPERIAL MANUFACTORIES OF THE GOBELINS, THE SAVONNERIE, & BEUAVAIS. —Gobelins tapestry; the Assumption of the Virgin after Titian; portrait of Louis XIV., after Rigault. La Savonnerie tapestry: specimens in the style of the Rennaissance, and of the time of Louis XVI. Beauvais tapestry: characteristics of the chase; still life after Desportes; furniture in the style of Louis XV. and XVI., the composition of H. Chabal-Dessurgey, and intended for imperial palaces.

2192. BRAQUENIÉ BROS. *Aubusson (Creuse)*.—Aubusson tapestry; tapestry for hangings, &c.

2193. REQUILLART, ROUSSEL, & CHOCQUEEL, *Aubusson (Creuse)*.—Aubusson and Moquette tapestry, tapestry panels, and tapestry for furniture.

2195. ARNAUD-GAIDAN, J. & Co. *Nimes (Gard)*.—High-napped tufted carpet.

2196. GRAVIER, C. & Co. *Nimes (Gard)*. Tufted carpet.

2197. PLANCHON, F. & Co. *Neuilly (Seine)*. —Neuilly tapestry.

2198. IMBS, *Brumath (Bas-Rhin)*.—Carpets.

2199. GADRAT, P. *Meaux (Seine-and-Oise)*.—High-napped carpets; hearth rugs.

2200. SALLANDROUZE, J. J. & SON, *Aubusson (Creuse)*.—Tufted carpets, moquettes.

2201. POLONCEAU, *Paris.* — Moquettes, with coating to make the nap adhere to the warp.

# CLASS XXIII.

2211. GROS, ODIER, ROMAN, & Co. *Wesserling (Haut-Rhin).* — Cotton yarn; fabrics of white and printed cotton; fabrics of wool and silk; mixed printed fabrics.

2212. PARAF-JAVAL BROS. & Co. *Thann (Haut-Rhin)*.—Dyed and printed fabrics.

2213. ECK, D. *Cernay (Haut-Rhin)*.— Printed cotton and mixed fabrics.

2214. STEINBACH, KOECHLIN, & Co. *Mulhouse (Haut-Rhin)*.—Printed fabrics.

2215. THIERRY, MIEG, & Co. *Mulhouse (Haut-Rhin)*.—Printed fabrics of cotton and of wool, for furniture; printed woollen shawl.

2216. HUGUENIN-SCHWARTZ, & CONILLEAU, *Mulhouse (Haut-Rhin)*.—Printed cotton fabrics; printed fabrics of cotton, and of wool for furniture.

2217. KOECHLIN BROS. *Mulhouse (Haut-Rhin)*.—Printed fabrics.

2218. DOLLFUS, MIEG, & Co. *Mulhouse (Haut-Rhin)*.—Printed cotton fabrics.

2219. LEFEBVRE, C. *Paris.* — Printed cotton fabrics.

2220. CANTEL & Co. *Rouen (Seine-Inf.)*. — Leather-cloths, plain, printed, and repoussé.

2221. LAMY-GODARD BROS. *Darnetal* (*Seine-Inf.*). — Printed cotton fabrics, for cravats and furniture.

2222. GRUEL, *Rouen* (*Seine-Inf.*). — Printed cotton fabrics.

2223. PETEL, *Darnetal* (*Seine-Inf.*). — Dyed cotton and chintzes.

2224. BAYLE - TABOURET, *Darnetal* (*Seine-Inf.*).—Dyed cottons.

2225. LEGRAS, F. *Rouen* (*Seine-Inf.*).— Dyed cottons.

2226. GUEROULT, N. *Rouen* (*Seine-Inf.*). —Dyed yarns of wool, silk, and cotton.

2227. DREVON, E. SEN. *Lyon* (*Rhône*). —Black silk.

2228. FREPPEL, F. A. *Saint - Mandé* (*Seine*).—Dressing materials, dressed fabrics.

2229. LACUFFER, SEN. *Annecy* (*Haut-Savoie*).—Cotton-yarn, fabrics, and prints.

2230. HENRY & SON, *Savonnières-devant-Bar* (*Meuse*).—Dyed cotton yarn.

2231. MESSIER, A. *Paris.*—Dyed wools in powder, for the manufacture of velvet paper. Dyed wool and cotton in powder, for printing on stuffs.

2232. JACQUES-SAUCE, *Paris.* — Dyed wools and cottons in powder, for making velvet paper, and printing on stuffs.

2233. DESSAINT & DALIPHAR, *Radepont* (*Eure*).—Printed cotton fabrics.

2234. PARENT, *Saint - André - lez - Lille* (*Nord*).—Dyed flax and cotton yarns.

2235. FEAU-BÉCHARD, A. *Paris.*—Dyed yarns of wool and cashmere, for dresses and shawls.

2236. ROUQUÈS, A. *Clichy-la-Garenne* (*Seine*).—Fabrics of wool and cashmere, dyed and dressed.

2237. POURCHELLE, F. *Amiens* (*Somme*). —Dyed cotton velvets.

2238. GUILIAUMET, A. *Puteaux* (*Seine*). Dyed and dressed fabrics of wool, and of wool and silk.

2239. BOULOGNE & HOUPIN, *Reims* (*Marne*).—Dyed fabrics.

2240. VEISSIÈRE, *Puteaux* (*Seine*). — Dyed woollen fabrics.

2241. SEIB, HOFFMAN, & CO. *Strasbourg* (*Bas-Rhin*).—Plain and printed oil cloths.

2242. BAUDOUIN BROS. *Paris.* — Plain and printed oil-cloths.

2243. LE CROSNIER, M. L. *Paris.* — Fabrics of cotton, of flax, and of silk gummed or waxed, plain and printed.

2244. DESCAT BROS. *Flers* (*Nord*). — Dyed, printed, and dressed fabrics.

2245. MOTTE, A. & CO. *Roubaix* (*Nord*). —Dyed and dressed fabrics.

2246. DUPARET, MADAME A. *Paris.*— Printed fabrics.

2247. PETITDIDIER, *Paris.* — Dyed fabrics of wool, of cashmere, and silk.

2248. GUINON, MARNAS, & BONNET, *Lyon* (*Rhône*).—Dyed silks and cottons.

2249. STEINER, C. *Ribeauvillé* (*Haut-Rhin*). — Turkey-red dyed cotton fabrics, plain and printed, for furniture.

2250. KLOTZ, M. *Paris.*—Light fabrics for ball-dresses, ornamented by the application of cut and printed designs.

2251. WERNER & MICHNIEWICZ, *Paris.* —Laces, printed fabrics imitating lace and fur.

2252. HOFER-GROSJEAN, E. *Mulhouse* (*Haut-Rhin*).—Printed fabrics for dresses.

2253. ONFROY & CO. *Paris.*—Fabrics of cotton, wool, and silk, printed by hand, by plates and rollers.

2254. TROESTER, B. & CO. *Bourgoin* (*Isère*).—Printed silk fabrics.

2255. RENARD BROS. *Lyon* (*Rhône*). — Silks and dyed silken fabrics.

2256. BRUNET-LECOMTE, H. *Bourgoin* (*Isère*).—Printed silks.

2257. CHOCQUEEL, L. *Puteaux* (*Seine*). —Shawls, dresses, and printed fabrics for furniture.

2258. GUILLAUME & SON, *Saint Denis* (*Seine*).—Printed fabrics.

2259. DELAMOTTE & FAILLE, *Reims* (*Marne*).—Fabrics of washed merino-wool, dyed and dressed.

2260. FRANCILLON & CO. *Puteaux* (*Seine*).—Fabrics of wool, of wool and cotton, and of wool and silk, dyed and dressed.

2261. PALLOTEAU-GUYOTIN & MARQUANT, F. *Reims (Marne).*—Dressing material, used by the manufacturers of Rheims in preparing the articles exhibited.

2262. BERNOVILLE BROS. LARSONNIER BROS. & CHENEST, *Paris.* — Combed and spun wool, woollen fabrics raw and dyed, woollen and mixed printed fabrics.

2263. KŒCHLIN-DOLPHUS & Co. *Mul-house (Haut-Rhin).*—Wools, raw and dyed, for embroidery, hosiery, and fancy purposes.

2264. JAPUIS, KASTNER, & CARTERON, *Claye (Seine-and-Marne).* — Printed fabrics for furniture.

2265. DOPTER, A. *Paris.*—Stuffs printed by lithography.

2266. WEISGERBER BROS. & KIENER, *Paris.*—Dyed wool, cotton, and silk yarn.

# CLASS XXIV.

2271. CHAMPAILLER, A. *Lyon (Rhône).* —Machine-made black lace.

2272. BURNIER, A. *Lyon (Rhône).* — Fancy net.

2273. BOURRY BROS. *Paris.* — Muslins, cambrics, worked velvet, and taffetas; worked fabrics.

2274. SAJOU, J. J. *Paris.*—Designs for tapestry and embroidery; articles in tapestry, crochet, and net.

2275. DUBUS, T. *Paris.* — Embroidered church ornaments.

2276. FERGUSON & SON, *Paris.* — Machine-made lace; black Cambrai lace; lama lace.

2277. KEENAN BROS. *Paris.*—Machine-made lace and net, for furniture.

2278. MONARD, F. T. *Paris.*—Machine-made imitation Chantilly lace.

2279. LEFORT, WIDOW L. P.—*Grand-Couronne (Seine-Inf.).*—Imitations of silk lace.

2280. FOULQUIÉ, MISS A. & Co.—*Paris.* —Articles in net.

2281. LECOMTE, C. & Co. *Paris.*—Woven net, of silk and cotton.

2282. GALOPPE, H. & Co. *Paris.*—Silk net, embroidered by hand, an imitation of Chantilly lace.

2283. RAYMOND, H. *Paris.* — Machine-made silk net and lace.

2284. DOGNIN & Co. *Paris.*—Black silk imitation Chantilly lace; damasked Lyons lace; lama, and pusher lace.

2285. PANNIER, A. RAMIBERT, & Co. *Paris.*—Articles in lace.

2286. GIRARD-THIBAULT, *Paris.*—Plain and worked articles in net, &c.

2287. HELBRONNER, MADAME S. *Paris.* —Tapestry worked with the needle, for furniture.

2288. GARAPON, L. *Paris.*—Aluminium thread; laces, &c. made of it.

2289. TRUCHY & VANGEOIS, *Paris.* — Embroideries and laces in gold and silver, for fancy articles, furniture, and military accoutrements.

2290. BROUILLET, A. *Paris.*—Laces, for fancy articles, and furniture.

2291. CHEVALIER, L. *Paris.*—Laces, for fancy articles, and furniture.

2292. GLÉNARD, E. & Co. *Paris.* — Worked articles in net.

2293. PARIOT-LAURENT, *Paris.* — Silk buttons and lace articles.

2294. COLLECTIVE EXHIBITION OF THE CITY OF ST. PIERRE-LÈZ-CALAIS (*Pas-de-Calais*).—Machine-made lace and net (10 Exhibitors).

2295. BRUNOT & LEFÈVRE, *Calais* (*Pas-de-Calais*).—Machine-made net.

2296. L'HEUREUX BROS. *St.-Pierre-lez-Calais.*—Imitation Chantilly lace, silk net.

2297. HOUETTE, L. *St.-Pierre-lez-Calais.* —Imitation of lace and guipure, in silk, thread, and cotton.

2298. HERBELOT, JUN. & GENET-DUFAY, *Calais* (*Pas-de-Calais*).—Silk net in imitation of Chantilly lace, and silk blondes imitating those of Caen.

2299. DUBOUT & SONS, *Calais* (*Pas-de-Calais*). — Black silk imitation Chantilly lace.

2300. MALLET BROS. *Calais* (*Pas-de-Calais*).—Machine-made silk blondes.

2301. REBIER, L. & VALOIS, F. *St.-Pierre-lez-Calais.* — Valenciennes net, and machine-made silk net.

2302. TROPHAM BROS. *St.-Pierre-lez-Calais* (*Pas-de-Calais*).—Machine-made lace and blondes.

2303. GEFFRIER, DELISLE BROS. & Co. *Paris.*—Lace made with spindles, and with the needle.

2304. LECORNU, *Caen* (*Calvados*).—Spindle-made lace.

2305. LACHEZ-BLEUZE, *Paris.* — White embroidery on muslin and cambric.

2306. LALLEMANT, *Paris.*— White embroideries.

2307. DRIOU, MOREST, & Co. *Paris.*— White embroideries.

2308. MOREAU BROS. *Paris.* — Worked shirt-fronts.

2309. CHAPRON, L. *Paris.* — Worked handkerchiefs.

2310. HUSSON - HEMMERLÉ, *Paris.* — Worked muslins and cambrics.

2311. PLANCHE, L., LAFON, G., & SIVAL, D. *Paris.*—Machine-made spindle lace.

2312. BACOUEL - TROUSSEL, A. *Arras* (*Pas-de-Calais*).—Spindle-made Arras lace.

2313. BERR, D. & SON, *Nancy* (*Meurthe*). —White embroidery; plain and worked shirt fronts.

2314. AUBRY-FEBVREL, *Mirecourt* (*Vosges*).—Spindle-made white and black guipures.

2315. LOISEAU, J. B. *Paris.*— Spindle-made lace.

2316. SEGUIN, J. *Paris.*—Spindle-made lace.

2317. ROBERT-FAURE, *Paris.*—Spindle-made lace.

2318. POIRET BROS. & NEPHEW, *Paris.* — Needle-made tapestry; dyed wool and canvas for tapestry; cottons for embroidery.

2319. ROGUIER, C. *Paris.*—Needle-made tapestry; dyed wools for tapestry. The *quadrilateur*, an instrument for copying designs on canvas.

2320. BLAZY BROS. *Paris.*— Wools for tapestry and net work; cotton canvas; needle-made tapestry.

2322. RAFFARD, E. & GONNARD, *Lyon* (*Rhône*).—Figured net.

2323. IDRIL, *Lyon* (*Rhône*). — Figured net, and machine-made lace.

2324. DOLLFUS-MOUSSY, *Lyon* (*Rhône*). —Figured net, and machine-made lace.

2325. TARPIN & SON, *Lyon* (*Rhône*).— Embroidery in gold.

2326. CLIFF BROS. *St. Quentin* (*Aisne*). —Machine-made blondes.

2327. HINAUT-COL. & Co. *St. Quentin* (*Aisne*).—White embroidery.

2328. CARPENTIER, A. *Caudry* (*Nord*). —Plain and fancy net.

2329. MAISON DE ST. MARIE, *Cherbourg* (*Manche*).—Black spindle-made lace.

2330. COSTALLAT-LAFORQUE, *Bagneres-de-Bigorre* (*Hautes-Pyrénées*). — Articles in woollen net.

2331. POIGNAT, G. *Paris.*—Borders for hats.

———

2332. HOOPER, G., CARREZ, & TABOURIER, *Paris.*—Laces and nets.

2333. PALLIER, P. *Nimes* (*Gard*).—Articles in lace.

2334. WERNER & MICHNIEWIER, *Paris.* —Articles in lace.

## CLASS XXV.

2351. BENEDICK, *Paris.*—Buffalo-horn brushes.

2352. DUPONT & DESCHAMPS, *Beauvais (Oise).*—Brushes, buttons, bristles.

2353. PITET, SEN. & LIDY, *Paris.*—Hair-pencils and brushes.

2354. HÉNOC, *Paris.*—Dusters.

2355. RENNES, A. J. M. *Paris.*—Brushes for various purposes.

2256. ROMANCEY & RICARD, *Paris.*—Shaving and hair brushes.

2357. LAURENÇOT, J. E. *Paris.*—Toilet brushes.

2358. HASSE, F. *Lyon (Rhône).*—Skins and furs.

2359. BEAUFOUR-LEMONNIER, *Paris.*—Jewelry, designs, and fancy articles in hair; application of photography to works in hair.

2360. LHUILLIER, *Paris.* — Skins and manufactured furs.

2361. PASQUIER, V. *Poitiers (Vienne).*—Dressed goose-skins.

2362. CAVY, A. *Nevers (Nièvre).*—Garments made with indigenous furs.

## CLASS XXVI.

2371. CHANEY, J. A. *Nantes (Loire-Inf.).*—Black and white calf-skin, for boots and shoes.

2372. COURVOISIER P. *Paris.*—Kid-skin, and kid-skin gloves.

2373. BUNEL BROS. *Pont-Andemer (Eure).*—Leather.

2374. COLLECTIVE EXHIBITION OF THE CITY OF ANNONAY (*Ardèche*).—Tawed skins (6 Exhibitors).

2375. ROQUES, *Montpellier (Herault).*—Tanned sheep-skins.

2376. AMIC, L. & Co. *Toulon (Var).*—Tanned and polished leather for soles; morocco-dressed goat-skin; bark of the green oak, and indigenous sumac of the Department of the Var.

2377. SCELLOS, E. *Paris.*—Tanned and curried leather: straps for machinery.

2378. FORTIER-BEAULIEU, JUN. *Roanne (Loire).*—Legs of military boots.

2379. SUSER, H. *Nantes (Loire-Inf.).*—Leather of various kinds, for boots, shoes, &c.

2380. CAMUSAT-GUYON, *Auxerre (Yonne).*—Bridles.

2381. PINAULT-BRISOU, E. *Rennes (Ille-and-Vilaine).*—Tanned leather.

2382. ROBAUT, A. *Valenciennes (Nord).*—Leather for cards, driving belts, and saddlery.

2383. LEGAL, *Chateaubriant (Loire-Inf.).*—Tanned and tawed leather.

2384. PRIN, A. *Nantes (Loire-Inf.)*—Black calf-skin.

2385. LES FILS COPPIN-LEJUNE, *Douai (Nord).*—Leather for cards, and spinning mills.

2387. BOUCHON, V. *Niort (Deux-Sevres).*—Cow, horse, and calf-skins.

2388. HOCHEDÉ, *Montdidier (Somme).*—Strong leathers, for military shoes and boots.

2389. SUEUR G. & STOFFT, *Amiens (Somme).*—Tanned and curried leather.

2390. LES FILS D'HERRENSCHMIDT, G. F. *Strasbourg (Bas-Rhin).*—Strong leathers, for various purposes.

2391. POULLAIN-BEURRIER, P. I. E. *Paris.* —Leather for spinning mills, cards, &c.

2392. CHEVILLOT BROS. *Paris.* — Cow-skin, for uppers.

2393. MASSERANT & ROULLEAU, *Paris.* —Harness traces.

2394. JODOT, J. *Paris.*—Calf-skin, and boot legs.

2385. VINCENT, J. *Nantes* (*Loire-Inf.*).— Boot legs.

2396. SUEUR, T. & LUTZ, *Paris.*—Varnished leather.

2397. OGEREAU BROS. *Paris.*—Leather for boots and shoes: chamois dressed sheep-skins; leather for saddlery.

2398. MARTEAU, A. & Co. *Paris.*—Black and white calf-skin; boot-legs.

2399. WALTZ, B. BROS. *Paris.*—Saddlery, military boots and shoes, driving belts.

2400. GLATARD, L. *Roanne* (*Loire*).— Harness for unyoking instantaneously.

2401. LAMBIN & LEFEVRE, *Paris.*—Saddles, bridles, and harness.

2402. PATUREL & BOYER, *Paris.*—Whips, and sticks.

2403. LEROUX, F. *Paris.*—Saddlery.

2404. DURAND, A. & L. BROS. *Paris.*— Tanned, curried, and polished leather for various purposes.

2405. PELTEREAU, A. *Château-Renault* (*Indre-and-Loire*). — Polished leather for soles.

2406. PELTEREAU, P. *Château - Renault* (*Indre-and-Loire*). — Polished leather for soles; driving belts; metal tubes with leather joints, for conveying water.

2407. PAILLART, G. *Paris.*—Leather for spinning mills, cards, &c.

2408. LEROUX, E. & LE BASTARD, A. *Rennes* (*Ille-and-Vilaine*).—Strong, and polished leather.

2409. SENDRET, R. & SON, *Metz* (*Moselle*). —Leather for soles and uppers, for military accoutrements, &c.

2410. GALLIEN & Co. *Longjumeau* (*Seine-and-Oise*).—Strong leather, &c.

2411. FORTIER-BEAULIEU, C. A. *Paris.*— Tanned and curried leather for saddles and harness; pig skin for saddles.

2412. STERLINGUE, *Bourges* (*Cher*).— Strong leather.

2414. TREMPÉ & SON, *Paris.*—Leather for shoes and boots.

2415. LOIGNON & CASSE, *Amiens* (*Somme*). —Muzzle for horses.

2416. COURTOIS, E. C. *Paris.*—Varnished calf-skin for boots and shoes.

2417. LEVEN, M. & SON, *Paris.*—Curried, varnished, and black calf-skin.

2418. ALDEBERT, L. *Milhau* (*Aveyron*). Black and white calf-skin.

2419. CARRIÈRE, P. BROS. DUPONT, *Milhau* (*Aveyron*).—Black and white calf-skin.

2420. CORNEILLAN BROS. *Milhau* (*Aveyron*) Tanned calf-skin.

2421. ALDEBERT, A. SON & BROS. *Milhau* (*Aveyron*).—Black calf-skin; chamois dressed sheep-skin; tanned lamb-skin.

2422. HOUETTE, A. & Co. *Paris.*—Varnished calf-skin, for boots and shoes.

2423. JUMELLE, F. C. *Paris.*—Coloured and varnished leathers, for boots and shoes; calf and buffalo skins for military accoutrements; grained calf, goat, and sheep-skins for boots and shoes, carriages and travelling requisites.

2424. NYS & Co.—*Paris.*—Varnished leather, for boots and shoes.

2425. LANDRON, JUN. *Orleans* (*Loiret*). —Strong leather.

2426. LATOUCHE - ROGER, *Avranches* (*Manche*).—Skins and leather.

2427. BAYVET BROS. *Paris.*—Morocco-dressed and coloured sheep-skin, and calf-skin for binding, &c.

2428. EHMANN, HERING, & GOERGER, *Strasbourg* (*Bas-Rhin*). — Morocco-dressed and coloured sheep-skin, tawed kid-skin.

2429. CERF-LANGENBERG, *Strasbourg* (*Bas-Rhin*). — Morocco-dressed sheep-skin, &c.

2430. VERGER, G. & C. BROS. *Pont-Audemer* (*Eure*).—Strong leather.

2430A. ROULLIER, *Paris.*—Driving belts.

2430B. DORÉ & Co.—*Bordeaux* (*Gironde*). —Black calf-skin, boot-legs.

2430C. FANIEN & SONS, *Liliers* (*Pas-de-Calais*).—Tanned leather and curried skins.

# CLASS XXVII.

2431. BEZON, J. *Lyon* (*Rhône*).—Gloves made by power-loom; net termed China lace.

2432. DUVELLEROY, P. *Paris.*—Fans and hand-screens; a collection of Chinese fans and screens, with their prices in China.

2433. VANDEVOORDE, A. *Paris.*—Carved fan-mountings.

2434. ALEXANDRE, P. F. V. *Paris.*— Fans.

2435. MOLÉON, L. A. *Paris.*— Ready-made garments for men.

2436. MALLET BROS. *Lille* (*Nord*).— Ready-made garments for men.

2437. LELEUX, A. *Paris.*—Ready-made garments for men.

2438. GIBORY, *Paris.*—Court dresses, &c.

2439. WUY, A. *Paris.* — Ready-made garments for men.

2440. LEMANN, & Co. *Paris.*—Ready-made garments for men, &c.

2441. MASSEZ, M. *Paris.* — Boots and shoes.

2442. BROE, *Paris.* — Boots and half-boots, &c.

2443. ROND, T. *Paris.*—Bottines and shoes.

2444. MÉLIÈS, L. S. *Paris.*—Boots and shoes.

2445. BESSARD, E. J. B. *Paris.*—Bottines for ladies and children.

2446. PRADEL-HUET, SEN. *Paris.*— Boots and shoes, for city wear, hunting, &c.

2447. PROUST, H. G. *Paris.*—Boots, half-boots, shoes, and slippers.

2448. DUPUIS, S. & Co. *Paris.*—Screwed boots and shoes.

2449. PELLETIER, L. *Paris.*—Boots and shoes.

2450. DELAIL, G. *Paris.* — Hunting boots.

2451. PICARD BROS. *Paris.*—Children's boots and shoes.

2452. PINET, F. J. L. *Paris.*—Boots and shoes for ladies.

2453. CABOURG, T. *Paris.* — Screwed-boots and shoes.

2454. TOUZET, J. C. *Paris.*—Boots and shoes, for men and women.

2455. LATOUR, P. *Paris.*—Riveted-boots and shoes.

2456. WALTER, F. *Paris.* — Leather breeches, gloves, &c.

2457. GEIGER, Z. *Paris.*—Articles for the chase, and travelling.

2458. NOËL, E. *Paris.*—Requisites for boots and shoes.

2459. CHAZELLE, 'E. *Tours* (*Indre-and-Loire*).—Boots and shoes, screwed and clouted by machine.

2460. MONTEUX & GILLY, *Paris.*—Boots and shoes, for men and women.

2461. ROUSSET & Co. *Blois (Loir-and-Cher).*—Boots and shoes, for men and women.

2462. POIRIER, P. *Châteaubriant (Loire-Inf.).*—Boots, half-boots, &c.

2463. HAULON, S. JUN. *Bayonne (Basses Pyrénées).*—Boots and shoes.

2464. PETIT, J. A. *Paris.*—Boots and shoes, for various purposes.

2465. ETANCHAUD, F. *Paris.* — Fancy boots and shoes, for ladies.

2466. DORÉ & Co. *Bordeaux (Gironde).*—Riveted boots and shoes for men, boot-legs, &c.

2467.—FANIEN & SONS, *Lillers (Pas-de-Calais).*—Sewed, riveted, and screwed boots and shoes, for men, &c.

2468. SUSER, H. *Nantes (Loire-Inf.).*—Boots and shoes, &c.

2469. CLEROX, A. M. *Paris.* — Fancy boots and shoes, &c.

2470. MAYER, *Paris.*—Fancy boots and shoes.

2471. BARRÉ, WIDOW, *Paris.* — Fancy boots and shoes, for ladies.

2472. BERNARD, A. L. *Paris.*—Flowers for head-dresses.

2473. MEUREIN, A. *Paris.* — Artificial foliage and fruit.

2474. PERROT, PETIT, & Co. *Paris.*—Artificial flowers, and feathers.

2475. BARDIN, J. L. F. *Paris.*—Flowers and fancy articles, made of materials obtained from quills.

2476. CHAGOT, D. A. *Paris.*—Artificial flowers, and feathers.

2477. HERPIN-LEROY, *Paris.*—Artificial flowers, and ornaments.

2478. KRAFFT, MADAME E. *Paris.*—Artificial flowers.

2479. TOURNIER, C. *Paris.*—Pistils, for artificial flowers.

2480. DEVRIÈS & Co. *Paris.*—Feathers.

2481. MARIENVAL-FLAMET, L. *Paris.*—Artificial flowers, and feathers.

2482. JAVEY & Co. *Paris.*—Stuffs, and coloured papers for artificial flowers, foliage, and fruit.

2483. GÉROLD, L. A. *Paris.*—Artificial flowers.

2484. CHAGOT, A. *Paris.*—Feathers and artificial flowers.

2485. VANDEREYKEN, H. *Paris.*—Feathers.

2486. TRAVERSIER, A. *Paris.*—Bonnets and head-dresses, for ladies ; mechanical hat, fitting into a box about three inches high.

2487. LANGEVIN-COULON, *Paris.*—Bonnets, caps, &c.

2488. DUFOUR, F. *Paris.* — Bonnets, head-dresses, &c.

2489. BRUN, MADAME M. *Paris.*—Head-dresses.

2490. BOUILLET, J. B. *Paris.*—Mantles, mantillas, scarfs, shawls.

2491. GOSSEIN-JODON, MADAME, *Paris.*—Linen drapery, children's dresses.

2492. BAPTEROSSES, *Briare (Loiret).*—Enamelled buttons, &c.

2493. LAVILLE, PETIT & CRESPIN, *Paris.*—Silk and felt hats.

2494. COUPIN, J. *Aix (Boûches-du-Rhône).*—Felt hats.

2495. BESSON BROS. *Bordeaux (Gironde).*—Silk and felt hats.

2497. MOCH, F. *Paris.*—Head-dresses, for children.

2500. VIEL, WIDOW, & VALLAGNOSE, J. *Marseilles (Boûches-du-Rhône).*—Pliable machine-made hats.

2501. HAAS, Y. *Paris.*—Hats, caps.

2502. MONROY, A. *Paris.*—Elastic hats, umbrella hats.

2503. HISPA & BOUQUET, *Toulouse (Haute-Garonne).*—Felt and woollen hats.

2504. VINCENDON, E. JUN. *Bordeaux (Gironde).*—Beaver, silk, and woollen hats ; fancy hats.

2505. MOSSANT, C. & SON, *Bourg-du-Péage (Drôme).*—Hats, mantles, &c., of felt.

2506. QUENOT & LEBARGY, *Paris.*—Hats of felt, beaver, and silk.

2507. DURST-WILD & Co. *Paris.*—Straw hats.

2508. BAMMES & Co. *Paris.*—Straw hats.

2509. LES FRERES AGNELLET, *Paris.*—Requisites for dresses, &c.

2510. GALLOT, A. SEN. *Paris.*—Straw and horse-hair hats.

2511. CHAUMONT & Co. *Paris.*—Straw hats.

2512. WILD, J. U. *Nancy (Meurthe).*—Straw hats.

2513. OPPENHEIM, WEILL, & DAVID, *Paris.* — Collars, shirt fronts, plain and worked.

2514. WERLY, R. *Bar-le-Duc (Meuse).*—Seamless corsets.

2515. CUNY, MADAME, S. J. *Paris.*—Hygienic corsets, &c.

2516. GRINGOIRE, MADAME V. *Paris.*—Elastic corsets, &c.

2517. BOCQUET, J. T. *Paris.*—Corsets and belts.

2518. JOSSELIN, J. J. & Co. *Paris.*—Corsets.

2519. COSTALLAT-BOUCHARD, *Paris.*—Articles in crochet-work.

2520. HEMARDINGUER, *Paris.*—Shirts.

2521. DAUGARD, A. *Paris.*—Moveable busts for hairdressers and milliners.

2522. HAYEM, S. SEN. *Paris.*—Collars, cravats, shirts, &c.

2523. ALLAIN-MOULARD, L. A. F. *Paris.*—Articles in crochet-work, &c.

2524. RAVAUT, BOCKAIRY BROS. & Co. *Paris.*—Ready-made garments; laces, &c.

2525. OPIGÈZ-GAGELIN & Co. *Paris.*—Court-dresses, &c.

2526. MATHIEU, F. & GARNOT, S. *Paris.* Garments for ladies; worked shawls.

2527. CHARAGEAT, G. E. *Paris.*—Walking-sticks, umbrellas.

2528. CAZAL, R. M. *Paris.*—Walking-sticks, whips, umbrellas.

2529. ALLAMAGNY, P. *Paris.* — Brass tubing for umbrellas.

2530. ELLUIN, F. *Paris.* — Walking-sticks, whips, &c.

2531. LIPS, C. *Paris.*—Carved handles, for umbrellas.

2532. SALLES, JUN. & Co. *Paris.* — Gloves of woollen webs, cotton, and silk.

2533. TRÉFOUSSE, J. L. & D. *Chaumont (Haute-Marne).*—Kid-skin gloves.

2534. CALVAT, F. *Grenoble (Isère).* — Kid and lamb-skin gloves.

2535. VIAL, M. *Grenoble (Isère).*—Kid-skin gloves.

2536. CALVAT, E. & NAVIZET, H. *Grenoble (Isère).*—Kid-skin gloves.

2537. BAYOUD, A. *Grenoble (Isere).*—Dyed skins for gloves.

2538. GUIGNIÉ, *Grenoble (Isère).*—Kid-skin gloves.

2539. CURTIS, L. B. & Co. *Grenoble (Isère).*—Kid-skin gloves.

2540. BERTHOIN, JUN. *Grenoble (Isère).*—Kid-skin gloves.

2541. PLAUD BROS. & Co. *Grenoble (Isère).*—Kid-skin gloves.

2542. MORIQUAND, SEN. *Grenoble (Isère).*—Kid-skin gloves.

2543. FRANCOZ, *Grenoble (Isère).*—Kid-skin gloves.

2544. ROUILLON, F. *Grenoble (Isère).*—Kid-skin gloves.

2545. CHEILLEY, JUN. & Co. *Paris.*—Kid-skin gloves.

2546. FONTAINE, A. & L. *Paris.*—Leather gloves.

2547. DESCHAMPS, P. J. T. *Paris.*—Kid-skin gloves.

2548. MARIOTTE, P. V. *Paris.*—Gloves, of tawed skins.

2549. GUÉRIMEAU-AUBRY, *Paris.* — Leather gloves

2550. BONNEVEY, B. *Paris.*—Kid-skin gloves.

2551. BILLION, A. *Paris.* — Leather gloves.

2552. JOUVIN, WIDOW, X. & Co. *Paris.*—Leather gloves.

2553. ALEXANDRE, *Paris.*—Tawed and dyed skins; gloves.

2554. COURVOISIER P. *Paris.*—Kid-skin gloves, &c.

2555. COMPÈRE, E. & DUFORT, *Paris.*—Kid-skin, and gloves.

2556. ROUQUETTE, MEUNIER, & Co. *Paris.*—Kid, beaver, and lamb-skin gloves.

2557. PRÉVILLE, A. L. *Paris.*—Kid-skin gloves.

2558. BOUDIER, MADAME, M. A. *Paris.*—Long Victoria gloves.

2559. JOUVIN & Co. *Paris.*—Tawed skins; gloves.

2560. PARENT, A., & HAMET, T., & Co. *Paris.*—Silk and metal buttons.

2561. BAGRIOT, F. A. *Paris.*—Metal buttons.

2562. MARIE, E., & DUMONT, A. *Paris.*—Mother-of-pearl buttons.

2563. GRELLOU, H. *Paris.*—Silk buttons, &c.

2564. LEMESLE, *Paris.*—Buttons.

2565 ROBINEAU, JUN. *Paris.*—Chaplets, medals.

2566. GOURDIN & Co. *Paris.*—Fancy buttons.

2567. RENÉE-BERNIER, *La Neuville-lèz-Corbie (Somme).*—Hosiery.

2568. BOULY-LEPAGE, *Moreuil (Somme).*—Cotton hosiery.

2569. BAIL BROS. *Villers-Bretonneux (Somme).*—Stockings.

2570. DORÉ & Co. *Troyes (Aube).*—Stockings and under-stockings.

2571. VERJEOT-GOMIER, *Châtres (Aube).*—Cotton hosiery.

2572. MEURICE, A. *Arcis-sur-Aube (Aube).*—Hosiery.

2573. LAFAIST, V. *Aix-en-Othe (Aube).*—Wool and cotton hosiery.

2574. QUINQUARLET, A. *Aix-en-Othe (Aube).*—Stockings, with sides respectively of wool and cotton.

2575. QUINQUARLET, H. *Aix-en-Othe (Aube).*—Stockings, with sides respectively of wool and cotton.

2576. MAROT, SEN. & JOLLY, *Troyes, (Aube).*—Cotton hosiery.

2577. BAZIN ARGENTIN, *Troyes (Aube).*—Cotton hosiery.

2578. FRÉROT, A. *Troyes (Aube).*—Cotton and wool under-stockings.

2579. DOUÉ & ROSENBURGER, *Troyes (Aube).*—Wool and cotton hosiery.

2580. EVRARD, E. *Troyes (Aube).*—Cotton hosiery.

2581. PORON BROS. *Troyes (Aube).*—Cotton stockings, and under-stockings.

2582. BARADUC, J. *Paris.*—Steel buttons and buckles; polished steel articles for furniture.

2583. PHILIPPON-DEGOIS, *Romilly-sur-Seine (Aube).*—Stockings and under-stockings.

2584. GORNET & BAZIN, *Romilly-sur-Seine (Aube).*—Cotton hosiery.

2585. LACOUR, *Romilly-sur-Seine (Aube).*—Cotton hosiery.

2586. COGNON, *Romilly-sur-Seine (Aube).*—Children's stockings, and under-stockings of cotton.

2587. BELLEMÈRE-VOISEMBERT, *Romilly-sur-Seine (Aube).*—Cotton stockings, and under-stockings.

2588. TAILBOUIS, E. & Co. *Paris.*—Hosiery of various materials.

2589. MEYRUEIS BROS. *Paris.*—Hosiery gloves.

2590. CONTOUR, A. F. *Paris.*—Hosiery, &c.

2591. MILON, SEN. *Paris.*—Hosiery, &c.

2592. BLANCHET, J. B. *Paris.*—Hosiery of various materials.

2593. PASQUAY, L. *Wasselonne (Bas-Rhin).*—Woollen socks and nets.

2596. DUCOURTIOUX, *Paris.* — Elastic stockings for varicose legs.

2597. LEDUC & CHARMENTIER, *Nantes (Loire Inf.).*—Cotton and wool hosiery.

2598. COANET, *Nancy (Meurthe).*—Leather gloves.

2599   BRUNET, MADAME L. & Co. *Paris.* Artificial flowers.

2600.   GRUYER, *Paris.*—Umbrellas.

2601.   GUYOT, C. V. *Paris.*—Braces and garters.

2602.   DUPILLE, H. *Paris.*—Corsets.

2603.   HUET & Co. *Rouen (Seine Inf.).*—Elastic caoutchouc fabrics for braces, belts, garters, &c.

2604.   CHANDELIER, *Méru (Oise).*—Buttons.

2605.   COURVOISIER, P. *Paris.*—Gloves and kid skins.

# CLASS XXVIII.

2631.   IMPERIAL PRINTING OFFICE OF FRANCE, *Paris.*—Printed works of various kinds, maps, electrotype matrices, &c.

2632.   GRUEL - ENGELMANN, *Paris.* — Fancy and artistic bindings.

2633.   LORTIC, P. M. *Paris.*—Artistic bindings.

2634.   BLANCHET BROTHERS, & KLÉBER, *Paris.*—Various kinds of paper.

2635.   ENGELMANN & GRAF, *Paris.*—Chromo - lithographs ; chromo - lithographic imitations of stained-glass windows.

2636.   LEGRAND, *Paris.*—Letter-paper envelopes.

2637.   OUTHENIN-CHALANDRE, *Paris.*—Various kinds of paper.

2638.   BOULARD, G. *Chamary (Nièvre).*—Paper for copper-plate engraving and lithography.

2639.   APPEL, F. A. *Paris.*—Fancy tickets, &c.

2640.   OLIER, J. P. *Paris.*—Paper for bank notes and playing-cards.

2641.   BOULARD, G. *Corvol-l' Orgueilleux (Nièvre).* Paper for copper-plate engraving, and lithography,

2642.   BRETON BROTHERS. & Co. *Pont-de-Claix (Isère).*—Various kinds of paper.

2643.   NISSOU, G. *Paris.*—Fancy cards ; chromo-lithographs.

2644.   JOINT-STOCK PAPER MILLS Co. OF THE SOUCHE, *near St. Dié (Vosges).*—Various kinds of paper.

2645.   ROMAIN & PALYART, *Paris.*—Fancy cards.

2646.   BADOUREAU, M. P. L. *Paris.*—Lithographs, chromo-lithographs, &c.

2647.   DUPUY, T. *Paris.*—Lithographs.

2648.   LATUNC & Co. *Blacons (Drôme).*—Various kinds of paper.

2649.   VORSTER, *Montfourat, near Angoulême.*—Various kinds of paper.

2650.   DAMBRICOURT BROTHERS, *S. Omer (Pas-de-Calais).*—Various kinds of paper.

2651.   OLLION, E. A. *Paris.* — Tracing paper.

2652.   TALLE & Co. *Paris.*—Blacks, ink, varnishes, colours, paper.

2653.   JOINT-STOCK PAPER MILL Co. OF ESSONNE *(Seine-and-Oise).*—Various kinds of paper.

2654.   BATAILLE, H. *Paris.*—Fancy paper for boxes, and fancy cards.

2655.   BÉCOULET, C. & Co. *Angoulême (Charente).*—Paper for writing, lithography, and photography.

2656.   PEULVEY, A. *Paris.* — Ink, carmine, &c.

2657.   HÉRARD. L. *Paris.*—Copper-plate engravings, done by machine.

2658.   LANDA, *Chalons (Saone-and-Loire).*

—Diagrams and cards for the Arts and Sciences; tradesmen's cards.

2659. NEWELL, T. F. *Paris.*—Letter-paper, envelopes, copying-paper.

2660. DESROZIER, *Moulins (Allier).*—Printed books.

2661. BARBAT, *Chalons (Marne)*—Lithographs and chromo-lithographs.

2662. BERNARD, J. & Co. *Prouzel(Somme).*—Various kinds of paper.

2663. JOINT-STOCK PAPER MILL CO. OF THE MARAIS AND ST. MARIE, *Paris.*—Various kinds of paper.

2664. DOPTER, J. A. V. *Paris.*—Lithographic impressions on stuff, &c.

2665. COLLECTIVE EXHIBITION OF THE CITY OF ANNONAY (*Ardéche*).—Writing-paper (2 Exhibitors).

2666. COLLECTIVE EXHIBITION OF THE CITY OF ANGOULÊME, (*Charente*).—Writing-paper (3 Exhibitors).

2667. LEBLOND, J. D. *Paris.*—Mannequin photographs.

2668. COSQCIN, J. *Paris.*—Topographic cards.

2669. TARDIE, M. J. A. *Paris.*—Stuffs and paper, crayons, and stamps for chalk drawing.

2670. AVRIL BROTHERS, *Paris.*—Plan of Paris; cards and plans.

2671. WIESENER, P. F. *Paris.*—Specimens of typography, bank notes, &c.

2672. JACOMME, C. & LELOUP, *Paris.*—Artistic lithographs, black and coloured.

2673. DE MOURGUES, C. BROTHERS, *Paris.*—Albums and printed books.

2674. TAMBON, A. *Paris.*—Engraved copper-plates.

2675. COBLENCE, S. V. *Paris.*—Electrotypes in copper for the production of engraved plates; illustrated works.

2676. BOUDIN, L. A. E. *Paris.*—Black and coloured inks; sealing-wax.

2677. SALLE, E. *Levallois (Seine).*—Engravings in relief, on copper.

2678. DESJARDINS, J. L. *Paris.*—Imitations of oil and water-colour paintings, by copper-plate engraving.

2679. CARLES, *Paris.*—Lithographs.

2680. PETITFRÈRE, T. P. *Paris.*—Composing sticks and moveable letters for gilders and bookbinders.

2681. PORTHAUX, *Paris.*—Engravings and vignettes, &c.

2682. BERTAUTS, J. V. *Paris.*—Lithographs of paintings.

2683. SCHMAUTZ, C. *Paris.*—Rollers for lithographers and engravers.

2684. KNAPP, *Strasbourg (Bas-Rhin).*—Bronze powder for paper.

2685. MATHIEU, C. *Paris.*—Chromo-lithographs.

2686. POUSSIN, *Paris.*—Cards.

2687. HARO, E. F. *Paris.*—Restored pictures; the paint having been separated from the wood or cloth. Colours, varnish, &c.

2688. VINCENT & FOREST, *Paris.*—Sealing-wax and wafers; office ink.

2689. BAULANT, SEN. *Paris.*—Fancy papers.

2690. FORESTIÉ, T. JUN. *Montauban (Tarn-and-Garonne).*—Printed books.

2691. BEAU, E. *Paris.*—Lithographs and chromo-lithographs.

2692. LEFRANC & CO. *Paris.*—Black and coloured inks.

2693. PÉQUÉGNOT, A. *Paris.*—Etchings.

2694. CREMNITZ, M. *Paris.*—Fancy cards, &c.

2695. DUNAND-NARAT, P. B. *Paris.*—Paintings on cloth; copies of water-colour drawings and engravings, by topography.

2696. TROUILLET, A. *Paris.*—Mechanical counters.

2697. RIESTER, M. *Paris.*—Steel engravings, for ornamentation.

2698. POUPART-DAVYL. *Paris.*—Books and engravings.

2699. LORILLEUX, C. & SON, *Paris.*—Black and coloured inks.

2700. LANÉE, *Paris.*—Maps.

2701. SAGANSAN, L. *Paris.*—Maps.

2702. SILBERMANN, G. *Strasbourg* (*Bas-Rhin.*)—Specimens of typography.

2703. CHATELAIN, A. *Paris.*—Cards of the modes of communication, by steam and telegraph, throughout the world.

2704. DUNOD, P. C. *Paris.*—Architectural and other plates, &c.

2705. ARMENGAUD, J. E. SEN. *Paris.*—Printed works and diagrams.

2706. TURGIS, L. JUN. *Paris.*—Prints: sketches with lace borders.

2707. TANTENSTEIN, J. C. *Paris.*—Printed music books and sheets.

2708. CRETÉ, L. & SON, *Corbeil* (*Seine-and-Oise*).—Printed books.

2709. MOREL, A. & Co. *Paris.*—Treatises on the arts.

2710. BEST, J. & Co. *Paris.*—Illustrated works, &c.

2711. RENOUARD, WIDOW, J. *Paris.*—Books and engravings.

2712. BERGER-LEVRAULT, WIDOW, *Strasbourg* (*Bas-Rhin*).—Printed books.

2713. BARRE, A. *Paris.*—Coins, medals, &c.

2714. DULOS, C. *Paris.*—Engravings in intaglio and relief.

2715. DALY, C. *Paris.* — Architectural engravings.

2716. ARNOLD, WIDOW, & SON, *Lille* (*Nord*).—Bookbinding.

2717. BASSET, J. A. *Paris.*—Bordered prints, &c.

2718. BOUCHER-MOREAU, *Anzin* (*Nord*).—Account books, printed and ruled.

2719. COURT & Co.—*Renage* (*Isère*).—Writing and drawing paper.

2720. GÉRAULT, H. *Paris.* — Account books.

2121. GONTHIER-DREYFUS, *Paris.*—Account books.

2722. CURMER, H. L. *Paris.*—Illustrated works.

2723. DUCROQUET, V. *Paris.* — Account books, &c.

2724. DARRAS, A. D. *Paris.*—Account books.

2725. COUTTENIER - PRINGUET, *Lille* (*Nord*).—Specimens of bookbinding.

2726. BRISSE, L. *Paris.*—Album of the French Exhibition of 1855.

2727. BRY, M. E. A. *Paris.*—Lithographs, black and coloured.

2728. FOSSEY, J. *Paris.*—Pasteboards.

2729. LATRY, A. *Paris.*—Cards and enamelled paper.

2730. CHARPENTIER, H. D. *Nantes* (*Loire-Inf.*).—Illustrated works.

2731. HULOT, A. A. *Paris.*—Specimens of postage stamps, &c.

2732. DIDRON, V. *Paris.*—Archæological and other works.

2733. DUPONT, P. & Co. *Paris.*—Specimens of fancy printing, &c.

2734. BANCE, B. *Paris.*—Treatises on Architecture, and the Arts.

2735. CLAYE, J. *Paris.*—Classical works, gift-books, &c.

2736. LENÈGRE, *Paris.*—Albums for drawings and photographs, &c.

2737. BARON I. S. J. TAYLOR, *Paris.*—Books of travels.

2738. MALLET - BACHELIER, A. L. J. *Paris.*—Scientific works.

2739. DÉRRIEY, J. C. *Paris.*—Specimens of type, &c.

2740. MAME, A. & Co. *Tours* (*Indre-and-Loire*).—Printed books; engravings.

2741. ANDRIVEAU-GOUGON, E. *Paris.*—Maps.

2742. ALMIN, A. S. *Paris.*—Pasteboard boxes.

2743. PAGNERRE, C. A. *Paris.*—Literary and popular publications, &c.

2744. CHARPENTIER, G. H. *Paris.*—Literary works.

2745. LEMERCIER, R. J. *Paris.*—Lithographs.

2746. GUILLEMOT, C. A. *Paris.*—Stamps.

2747. REPOS, E. *Paris.* — Liturgical works.

2748. PLON, H. *Paris.*—Matrixes, electrotype plates, &c.

2749. DANEL, L. *Lille (Nord).*—Specimens of typography.

2750. PERRIN, L. *Lyon (Rhône).*—Printed books.

2751. MARION, A. & Co. *Paris.*—Letter-paper, &c.

2752. MONCOURT, *Amiens (Somme).*—Lithographs.

2753. CABASSON, *Paris.*—Letter-paper with envelopes attached.

2754. REGNIER & DOURDET, *Paris.*—Maps on stone.

2755. HUMBLOT, CONTE & Co. *Paris.*—Crayons.

2756. BOUTON, N. V. *Paris.*—Imitation of old manuscripts, &c.

2757. LEFEVRE, T. *Paris.*—Educational works.

2758. MARICOT & VAQUEREL, *Paris.*—Coloured and fancy paper; pasteboard.

2759. ERHARD SCHIEBLE, *Paris.*—Maps on stone.

2760. FROMONT, *Paris.*—Wafers.

2761. CHARDON, F. C. SEN. *Paris.*—Copper-plate engravings.

2762. RAPINE, *Paris.*—Copper-plate relievo engravings.

2763. CORNILLAC, E. & Co. *Chatillon-sur-Seine (Côte-d'Or).*—Specimens of fancy binding.

2764. JUNDT, J. & SON, *Strasbourg (Bas-Rhin).*—Pasteboard and enamelled paper.

2765. LEMAIRE-DAIMÉ, *Andrecy (Seine-and-Oise).*—Cigarette paper.

2766. GARNIER, H. & SALMON, A. *Paris.*—Steeled copper-plates; plates engraved by autographic and photographic processes.

2767. DE CHOISEUL-BEAUPRÉ, *Château de Dury (Somme).*—Frame with specimens of paper cutting.

2768. WILLOUGHBY, J. *Paris.*—Pencil case with perpetual almanack.

2769. LEPRINCE, T. *Paris.*—Paper for the copying-press.

2770. GODCHAUX, A. *Paris.*—Specimens of writing done in copper-plate.

# CLASS XXIX.

PUBLICATIONS EXHIBITED BY THE IMPERIAL COMMISSIONER :—

2771. Works on the art of teaching.

2772. Journals of education.

2773. Works on law, rules, and discipline.

2774. Works on statistics, and reports.

2775. Works on school buildings.

2776. Information regarding infant schools and nurseries.

2777. Works on Roman Catholic religious instruction.

2778. Works on Protestant religious instruction.

2779. Works on Jewish religious instruction.

2780. Works on the art of reading.

2781. Reading books.

2782. Works on writing.

2783. Works on the French language.

2784. Works on arithmetic.

2785. Elementary works on geometry and algebra.

2786. Historical works.

2787. Geographical works.

2788. Works on the physical and natural sciences.

2789. Works on agriculture.

2790. Industrial treatises.

2791. Treatises on commerce, and the keeping of accounts.

2792. Works on living languages.

2793. Works on drawing.

2794. Works on vocal and instrumental music.

2795. Works on amusements and rewards.

2796. Works on school and communal libraries.

2797. Works on the special instruction of the blind.

2798. Works on the special instruction of deaf-mutes and idiots.

2799. Plans, &c. of school buildings and furniture.

2800. Apparatus for teaching reading and writing.

2801. Apparatus for teaching arithmetic and geometry.

2802. Apparatus for teaching geography and cosmography.

2803. Apparatus for teaching the physical and natural sciences.

2804. Apparatus for teaching drawing.

2805. Apparatus for teaching vocal and instrumental music.

2806. Gymnastic apparatus.

2807. Apparatus for the special instruction of the blind, and of deaf-mutes.

2808—60. Specimens of what has been done by the pupils of the primary schools, in various places.

## CLASS XXX.

2861. AHRENS, H. *Paris.*—Specimens of marquetry: table tops, panels.

2862. GOUAULT, A. F. *Paris.*—Marble chimney-piece, style of Louis XIII.

2863. DELAPIERRE, P. & Co. *Paris.*—Frame with plate-glass, gilt and ornamented with carton-pierre, article in carved oak.

2864. SAUVREZY, A. H. *Paris.*—Carved furniture in various styles.

2865. FOURNIER, A. M. E. *Paris.*—Sculptured and gilt furniture in various styles; hangings in the style of Louis XIV.; Venetian glass.

2866. CHARMOIS, C. *Paris.*—Ebony furniture, ornamented with bronze and mosaics.

2867. KNECHT, E. *Paris.*—Furniture in carved wood, &c.

2868. CHAIX, P. A. *Paris.*—Ebony bookcase.

2869. HUBER BROS. *Paris.* — Cartonpierre decorations.

2870. LAURENT, E. & Co. *Paris.*—Inlaid floors and mouldings, &c.

2871. PALLU A. & Co. *Paris.*—Article in Algerian onyx.

2872. LAROQUE, P. *Paris.*—Billiard table and requisites.

2873. FOURDINOIS, H. *Paris.*—Furniture of carved ebony, decorated with marquetry, gilt, &c.

2874. GROHÉ, G. *Paris.*—Furniture of carved mahogany, ebony, &c., decorated with mosaics and gilt bronze.

2875. DUVAL BROS. *Paris.*—Transformable couch; fancy seats.

2876. GUÉRET BROS. *Paris.* — Carved furniture, &c.

2877. MAZAROZ RIBAILLIER, P. *Paris.*—Furniture of carved walnut and oak, &c.

2878. GROS, J. L. B. *Paris.*—Ebony furniture, inlaid, decorated with mosaics, gilt bronze, &c.

2879. RIVART, J. N. *Paris.* — Artistic furniture, decorated with marquetry and painted porcelains.

2880. ROUX, F. *Paris.*—Inlaid furniture, decorated with gilt bronze.

2881. CRAMER BROS. *Paris.*—Bedsteads of wood, conveniently resolvable into many pieces.

2882. DEXHEIMER, P. *Paris.*—Buhl furniture, decorated with bronzes and Florentine mosaics.

2883. RAMONDENE, D. *Paris.*—Tables, capable of extension.

2884. BADÍN, MADAME R. *Paris.* — Articles in basket-work, &c.

2885. MUTET, E. *Paris.*—Rush and cane furniture for gardens.

2886. JEANSELME, SON, & GODIN, *Paris.*—Cabinet furniture.

2887. MELLI, J. L. *Paris.*—Marbles and bronzes.

2888. MEYER, L. F. A. *Paris.*—Lacquered furniture.

2889. WIRTH BROS. *Paris.* — Fancy carved furniture.

2890. QUIGNON, N. J. A. *Paris.*—Chairs, arm-chairs, book-case.

2891. DESPRÉAUX, A. *Versailles (Seine-and-Oise).*—Stuffs for furniture.

2892. ARMÉ MÉDARD, *Lyon (Rhône.)*—Carved wood-articles.

2893. ZIMMERMANN, P. *Paris.*—Gildings on wood, &c.

2894. FONTAINE, H. *Valenciennes (Nord).*—Curved oak furniture.

2895. GANSER, L. G. *Paris.*—Imitation bamboo chairs, &c.

2896. ZUBER, J. & Co. *Rixheim (Haut-Rhin).*—Paper hangings.

2897. LEGLAS-MAURICE, F. *Nantes (Loire-Inf.).*—Common and fancy furniture.

2898. DULUD, J. M. *Paris.*—Leather for hangings, &c.

2899. GALLAIS, C. A. *Paris.*—Panels; lacquered furniture.

2900. BEAUFILS, *Bordeaux (Gironde).*—Fancy furniture.

2901. PECQUEREAU & SON, *Paris.*—Carved furniture.

2902. BALNY, J. P. *Paris.*—Side-boards, gun cases, &c.

2903. LÉROUX, C. H. F. *Paris.*—Sofa bed.

2904. LÉONARD, C. *Paris.*—Iron bedsteads, &c.

2905. DE LATERRIÈRE, J. & Co. *Paris.*—Iron bed, &c.

2906. RIBAL, F. *Paris.*—Dining tables, capable of extension.

2907. VANLOO, P. *Paris.*—Bed, chest of drawers, work-table, &c.

2908. EVETTE, H. & C. L. YON, *Paris.*—Elastic hair quilts, mattresses, and pillows.

2909. DUBREUIL, SEN. *Paris.*—Paper-hangings.

2910. REBEYROTTE, F. *Paris.*—Chairs, and armchairs, of lacquered wood, gilt.

2911. FOSSEY, JUN. *Paris.*—Carved and decorated furniture.

2912. LEMOINE, H, *Paris.*—Carved and decorated rose-wood furniture.

2913. CREMER, J. *Paris.*—Furniture, enriched with marquetry, mosaics, &c.

2914. LOREMY & GRISEY, *Paris.*—Gilt frames and cabinets.

2915. JOSSE, C. L. *Paris.*—Paper-hangings, gilt to imitate embroidery.

2916. GRANTIL, JUN. & DIDION, C. *Montigny-lez-Metz (Moselle).*—Paper-hangings.

2917. POLGE, C. & BEZAULT, *Paris.*—Paper-hangings.

2918. TURQUETIL & MALZARD, *Paris.*—Paper-hangings in panels.

2919. DESFOSSÉ, J. *Paris.*—Paper-hangings.

2920. BARTHELEMY & DUBREUIL, *Paris.*—Paper-hangings.

2921. SEEGERS, A. *Paris.*—Paper-hangings.

2922. GENOUX & Co. *Paris.*—Paper-hangings.

2923. BINANT, L. A. *Paris.*—Stuffs for the decoration of ceilings, &c.

2924. LEROY, J. *Paris.*—Paper-hangings made by machine.

2925. RIOTTOT, J., J. CHARDON & PACON, *Paris.*—Paper-hangings.

2926. LEROY, C. V. *Paris.*—Painted window blinds.

2927. BACH-PÈRÈS, A. *Paris.*—Painted window blinds.

2928. GERFAUX, J. JUN. *Paris.*—Picture frames.

2929. THIAULT & CORNET, *Paris.*—Paper imitating oak.

2930. WAASER, C. A. *Paris.*—Kiosque, and specimens of trellis-work decoration.

2930A. CORNU, T. JUN. *Paris.*—Furniture of ebony, enriched with bronze.

2930B. PHILOUÉ, A. *Paris.*—Chairs and arm-chairs.

2930C. MURE, C. *Paris.*—Steel nails and ornaments for furniture, &c.

2930D. BARBEDIENNE, F. *Paris.*—Artistic articles in wood, and sculptured marble.

2930E. BLUMER, C. *Strasbourg (Bas-Rhin).*—Inlaid floors, artistic furniture.

2930F. DELACOUR, L. F. *Paris.*—Screen-curtains.

2930G. PLANIER, E. & Co. *Paris.*—Elastic pads for keeping apartments warm.

2930H. TRONCHON. N. J. *Paris.*—Iron seats and furniture.

# CLASS XXXI.

2931. BRICARD & GAUTHIER, *Paris.*—Locks and hardware.

2932. HERDEVIN, J. M. *Paris.*—Cocks, and steam apparatus, &c.

2933. DOUCHAIN, C. *Paris.*—Apparatus for the distribution of water, &c.

2934. LEBOUC, J. L. J. *Paris.*—Ornamental, and other apparatus connected with artificial light.

2935. CRIBIER, E. JUN. *Paris.*—Steel springs for petticoats.

2936. DUCROS, L. J. *Paris.*—Wrought-iron balustrade, &c.

2937. ROUY, P. A. *Paris.*—Fecula for the moulding of metals.

2938. FRAISSINET, E. & Co. *Alais (Gard).*—Metallic articles and fabrics.

2939. BARBOU & SON, *Paris.*—Iron bottle-cases.

2940. LEPAN, T. *Lille (Nord).*—Foil and tubes of lead and tin.

2941. PASSAGER, J. D. D. *Paris.*—Lamp for congealed fats.

2942. AROUX, *Melun (Seine-and-Marne).*—Hardware.

2943. LETURC, A. M. *Paris.*—Venetian blinds, of moveable sheet-iron plates, &c.

2944. BAUDRIT, A. *Paris.*—Iron work.

2945. CHAUVEL, E. J. B. *Evreux (Eure).*—Buckles, thimbles, &c.

2946. JUBERT BROS. *Charleville (Ardennes).*—Iron fittings for carriages.

2947. Miette Bros. *Braux prés Charleville* (*Ardennes*).—Ironwork for.waggons, &c.

2948. Demans & Co. *Chambon-Feugerolles* (*Loire*).—Bolts and files.

2949. Demandre, C. *Aillevillers* (*Haute-Saône*).—Iron wire for cards and combs.

2950. De Poilly, Jun. *Boulogne-sur-Mer* (*Pas-de-Calais*).—Locks.

2951. De la Haye de Barbezières, S. N. *Paris.*—Iron sashes.

2952. Valéry & Son, & Seroux, *Beauchamps* (*Somme*).—Keys, having the numbers, &c., of the apartments to which they belong on the rings.

2953. Goldenberg, G. *Zornhoff* (*Bas-Rhin*).—Iron-work, cutting instruments, &c.

2954. Bourgeois-Botz, *Reims* (*Marne*). —Plates and ribbons for cards.

2955. Truche, J. M. J. *Lyon* (*Rhone*). —Mirrors for catching larks.

2956. Gaillard, C. C. *Paris.*—Metallic cloths.

2957. Weber, *Nancy* (*Meurthe*).—Metallic cloths.

2958. Massiere, E. F. P. *Paris.*—Tin foil; metallic paper for damp walls, &c.

2959. Cribier, *Viroflay* (*Seine-and-Oise*). —Hair pins.

2960. Laperche, A. Jun. *Paris.*—Military accoutrements.

2961. Bohin, B. F. *L'Aigle* (*Orne*).— Hardware, &c.

2962. Delage, J. P. *Paris.*—Articles in malleable cast-iron; requisites for perambulators and copying-presses.

2963. André J. L. *Paris.* — Lever-buckles for driving belts, &c.

2964. Tronchon, N. J. *Paris.*—Garden locks, iron furniture, &c.

2965. Fichet, A. *Paris.*—Iron safes.

2966. Sauve, L. & Magaud, *Marseilles* (*Boûches-du-Rhône*).—Iron safes, and locks for iron safes.

2967. Delacour, L. F. *Paris.*—Screens, &c.

2968. Lhermitte, B. *Paris.*—Polished steel safe, &c.

2969. Haffner Bros. *Paris.* — Iron safes.

2970. Lang, L. *Schelestadt* (*Bas-Rhin*). Metallic cloths.

2971. Franck & Co. *Schelestadt* (*Bas-Rhin*).—Metallic cloths.

2972. Tronchon, A. *Paris.*—Models of portable iron buildings.

2973. Gandillot, *Paris.*—Iron tubing.

2974. Estivant Bros. *Givet* (*Ardennes*). —Unsoldered copper tubes.

2975. Bouttevillain, L. F. *Paris.*— Iron tubes, for boilers, &c.

2976. Vicaire, L. F. *Paris.*—Copper tubing.

2977. Maquemneheu, E. & Imbert, *Escarbotin* (*Somme*). — Common and fancy locksmiths' work.

2978. Mage, Sen. *Lyon* (*Rhône*).—Metallic cloths.

2979. Roswag, V. *Schelestadt* (*Bas-Rhin*). —Metallic thread and cloth.

2980. Tailfer, A. & Co. *L'Aigle* (*Orne*). —Galvanized iron pins, &c.

2981. Sirot-Wagret, *Trith-St.-Leger* (*Nord*).—Pegs for boots and shoes.

2982. Cantagrel, F. and Co. *Paris.*— Instrument for finding leaks in gas apparatus, &c.

2983. Fournier, C. A. *Paris.*—Safety apparatus for finding leaks in gas pipes, &c.

2984. Rebour, C. J. N. *Paris.*—Locks.

2985. Brodin, C. J. *Soissons* (*Aisne*).— Apparatus for cleaning lamps.

2986. Lambert, S. *Paris.*—Tin foil, &c.

2987. Hubert-Regnault & Co. *Charleville* (*Ardennes*).—Machine-made nails.

2988. Gailly, Sen. *Charleville* (*Ardennes*).—Machine-made nails for boots and shoes.

2989. Fergusson, E. *Paris.*—Regulators of gas pressure.

2990. Boubilla, J. R. *Paris.*—Apparatus for securing bags for despatches.

2991. GAUSSERAN & Co. *Paris.*—Smoothing irons.

2992. MONIER, H. & Co. *Paris.*—Crystal gas jets.

2993. PRADEL, P. *Paris.* — Mode of fastening without a key.

2994. NOS-D'ARGENCE, *Rouen Seine-Inf.).*—Metallic teazles, and brush.

2995. DUPUCH, G. *Paris.*—Copper and bronze articles.

2996. BOUCHER, E. *Fumay (Ardennes).*—Kitchen articles, copper-coated iron wire, &c.

2997. PAJOT & Co. *Randonnay(Orne).*—Cast-iron household requisites.

2998. SICARD, J. A. *Paris.*—Copper kitchen requisites.

2999. FONDET, J. B. *Paris.*—Presmatic tubing for heating apparatus.

3000. POTTECHER, B. *Bussang (Vosges).*—Curry-combs, &c.

3001. DALIFOL & DALIFOL, A. *Paris.*—Articles in malleable cast-iron.

3002. BRUN, P. *Lyon (Rhône).*—Portable forge.

3003. MASSON, J. B. F. *Paris.*—Tin foil.

3004. MENAND, F. R. *Paris.*—Lanterns of paper and stuff, for illuminations.

3005. CHÊNE, J. & SON, *Paris.*—Heating apparatus.

3006. PAROD, E. *Paris.*—Cooking utensils.

3007. AULON, C. L. E. *Darney (Vosges).*—Sheet-iron covers.

3008. PARET, N. E. A. *Paris.*—Portable kitchen furnaces.

3009. PIAT, J. J. D. *Paris.*—Mechanism for numbering cotton threads.

3010. JAVAL, E. *Paris.*—Hardware.

3011. COLETTE, *Paris.*—Hardware.

3012. PÉNANT, *Paris.*—Coffee pots.

3013. MORISOT, N. J. *Paris.*—Bronze fittings for grates.

3014. GRANDRY-GRANDRY & SON, *Nouzon (Ardennes).*—Fittings for grates.

3015. BROQUIN & LAINÉ, *Paris.*—Ornamental and other bronzes.

3016. BION, V. *Paris.*—Bronze fittings for grates.

3017. CLAVIER, E. *Paris.*—Bronze and iron andirons.

3018. BESNARD, H. *Paris.*—Ornamental articles in zinc and bronze.

3019. BLOT & DROUARD, *Paris.*—Ornamental articles in zinc, imitating bronze.

3020. ROBIN BROS. *Paris.*—Candelabras; lustres and gas fittings.

3021. LEFEVRE, E. J. *Paris.*—Time-pieces, &c., of bronzed or gilt zinc.

3022. GAUTROT & LATOUR, *Paris.*—Ornamental zinc articles.

3023. BRUNEL, L. *Paris.*—Ornamental zinc articles.

3024. LAMBIN, SAGUET, & FOUCHET, *Paris.*—Ornamental zinc articles.

3025. RIGOLET, R. M. *Paris.*—Imitation bronzes.

3026. GRILLOT, C. *Paris.*—Imitation-bronze groups, statuettes, &c.

3027. FOUBERT, *Paris.* — Time-pieces, groups in bronzed or gilt zinc.

3028. HOTTOT, L. *Paris.*—Time-pieces, &c., in zinc.

3029. LECLERCQ, F. P. *Paris.*—Time-pieces, and ornamental articles in zinc, bronzed or gilt.

3030. BOY, J. *Paris.* — Ornamental articles in zinc and bronze.

3031. MOURCY, P. *Paris.*—Articles in gilt aluminium, and in gilt zinc.

3032. MIROY FRÈRES, *Paris.*—Articles in zinc and bronze.

3033. MARCHAND, L. *Paris.*—Bronzes.

3034. DUCEL, J. J. *Paris.*—Ornamental articles in cast-iron.

### 3035. DURENNE, A., Iron-master, *Sommevoire (Haute-Marne)*.    *Paris: 30 Rue de la Verrerie.*

The Exhibitor's works at Sommervoire are specially devoted to ornamental castings for constructive purposes and for the decoration of churches, squares, and gardens. The collection he shows at the London Exhibition is both very complete and very diversified. It consists of six parts :—

1st. The great monumental fountain in the Gardens of the Horticultural Society, forty-five feet high, and fifty-two feet in diameter at the base of the basin.

It comprises in the lower portion, four sea-horses ridden by Amazons, each group thirteen feet long and nine feet high.

In the next portion, eight Syrens pouring water, in the form of cascades, from shells and basins, and supporting a large vase measuring twenty-three feet in diameter.

In the third portion, four females in a sitting posture, representing Painting, Poetry, Science, and Industry.

In the upper portion, Abundance crowning the whole.

This fountain, which at present throws out two hundred cubic metres of water every hour, is capable of producing a much greater effect.

2nd. A great gate, intended for a square or a park, and made in imitation of hammered iron. Each compartment is a single casting, four feet wide and thirteen and a half feet high. The two pilasters, by a happy introduction of two panels placed in juxtaposition, and of the thickness of each pilaster, are rendered light, and the entire is finished above by a decoration eleven feet wide and five feet high.

3rd. A statue representing an undecided victory, bespoke by his Imperial Highness Prince Napoleon, and cast in a single jet, the arms not taken into account: it is seven feet and a half high, and as it came from the mould, without any mending.

4th. A collection of statues, vases, candelabra, busts, bas-reliefs, and groups, to be seen in the French Court.

All these castings are as they came from the moulds, and without any dressing, to show the care expended in the moulding and execution. A few are bronzed, to afford a means of comparing articles in the rough state, and those which are finished, that it may be seen how little was required to render them complete.

The ornaments for religious purposes are—

A Christ, in one piece, six feet high, and weighing five hundred kilogrammes; rough from the mould.

The Virgin, six feet high, bronzed.

Do.        „        „

A medallion of the Virgin, two feet and a half, rough from the mould.

Christ in the Sepulchre, six feet long and three feet wide, also rough from the mould.

The same, three feet long and two feet wide.

The ornaments for a garden porch are—

Statues for lights, with cornucopias, in the Florentine style, four feet high; rough from the mould.

Statues for lights—Cornelia and Antiphona—Greek style, three feet high, rough from the mould.

The Four Seasons, Renaissance style, four feet high, rough from the mould.

Two groups—Amazons on horseback, one of them as it was cast, the other bronzed—each four feet and a half long and three feet high.

Three groups of animals—

A Dog defending its Young, bronzed : six feet long, and three feet wide.

A Wolf ready to devour a Dog, as cast; six feet long and three feet wide.—These latter groups form a pair.

A Boar Hunt, rough from the mould, six feet long and three feet wide; which, with a Wolf Hunt, also rough from the mould, forms a pair.

The following, which, likewise, are as they were cast, have not been exhibited, for want of space :—

A Light-bearer, consisting of a child supporting girandoles.

Statuettes, comprising—Paris; a Female bathing; Jupiter and Leda.

An Aurora vase, four feet and a half high.

Busts of Madame Du Barry; Mademoiselle De Ligny, an artiste of Louis XV.'s time; Diana De Gabie; Ceres; Germain Pilons; Three Graces.

And a plateau, consisting of a cup of Benvenuto Cellini.

5th. A series of castings for constructive purposes, rough from the moulds, such as window balconies, balustrades for large balconies, requisites for Venetian sunshades, friezes, flower-stands, crosses, requisites for gates, grates, &c.

6th. A collection of articles, consisting of vases, tazzas, statues, a fountain, and groups, placed in the Gardens of the Horticultural Society.

DURENNE'S FOUNTAIN.

3035.—DURENNE, A.—*continued.*

A BOAR HUNT.

3035.—DURENNE, A.—*continued.*

STATUETTES—THE SEASONS—AND GARDEN VASE IN CAST IRON.

3036. GRAUX-MARLEY, L. J. B. *Paris.*—Bronzes.

3037. BARBEZAT & Co. *Paris.*—Ornamental articles in cast-iron.

3038. THIÉBAUT, V. *Paris.* — Ornamental articles in bronze; monumental fountain in the Horticultural Society's Garden.

3039. DELAFONTAINE, A. M. *Paris.*—Bronzes.

3040. CAIN, A. *Paris.*—Bronzes.

3041. DUPLAN & SALLES, *Paris.* — Bronzes.

3042. MOIGNIEZ, J. JUN., *Paris.* — Bronzes.

3043. PEYROL, H. *Paris.*—Bronzes.

3044. MÈNE, P. J. *Paris.*—Bronzes.

3045. GALOPIN, A. *Paris.*—Bronze and porcelain lamps.

3046. GONON, E. *Nantes (Loire-Inf.).*—Bronzes.

3047. POILLEUX, J. B. *Paris.*—Bronzes.

3048. ROLLIN, F. G. *Paris.* — Bronze time-pieces, &c.

3049. MAGE, F. *Paris.*—Plaster groups for models of bronzes.

3050. FÉTU, J. L. *Paris.*—Bronzes.

3051. CHAUMONT & LANGEREAU, *Paris.*—Bronzes.

3052. POPON, N. *Paris.*—Bronzes.

3053. MARERCHAL & BERNAND, *Paris.*—Porcelain lamps mounted in bronze.

3054. DAUBRÉE, A. *Nancy (Meurthe).*—Bronzes.

3055. BARBEDIENNE, F. *Paris.*—Bronzes.

3056. GAGNEAU, E. *Paris.*—Bronzes.

3057. SCHLOSSMACHER, J. & Co. *Paris*—Bronze and other lamps.

3058. HADROT, L. JUN. BONNET, C. & BORDIER, *Paris.*—Bronze lamps.

3059. LACARRIÈRE, A. SON, & Co. *Paris.*—Gas-fittings, in cast-iron, zinc, and bronze.

3060. CARLHIAN & CORBIÈRE, *Paris.*—Lamps.

3061. HADROT, L. N. *Paris.* — Bronze lamps, &c.

3062. LEROLLE, L. *Paris.*—Articles in bronze.

3063. LEVY, J. F. *Paris.* — Artistic bronzes.

3064. CHARPENTIER & Co. *Paris.* — Groups, statues, &c. in bronze.

3065. DENIÈRE, G. JUN. *Paris.*—Bronzes.

3066. GAUTIER, F. *Paris.*—Bronzes.

3067. RAINGO BROS. *Paris.*—Bronzes.

3068. PAILLARD. A. V. *Paris.*—Bronzes.

3069. HOUDEBINE, C. H. A. *Paris.*—Bronzes.

3070. BOYER, V. & SON, *Paris.*—Bronzes.

3071. LIONNET BROS. *Paris.*—Objects of art, &c., obtained by the electrotype process.

3072. VITTOZ, E. JUN. *Paris.*—Bronzes.

3073. MERCIER, D. *Paris.*—Bronzes.

3074. PERROT, H. *Paris.*—Fancy bronzes.

3075. SUSSE BROS. *Paris.*—Bronzes.

3076. SERVANT, JUN. & DEVAY, *Paris.*—Bronzes.

3077. ZIER, A. J. *Paris.*—Bronzes.

3078. LEMAIRE, A. *Paris.*—Bronzes.

3079. PICKARD, C. & PUNANT, A. *Paris.*—Bronze chimney ornaments, &c.

3080. VANVRAY BROS. *Paris.*—Bronzes.

3081. HERMANN, G. *Paris.*—Artistic objects mounted in bronze.

3082. FENQUIÈRES, J. *Paris.*—Bronzes, silver plate, gold jewelry, enamels.

3083. LEVY BROS. *Paris.*—Bronze chimney ornaments, flower stands, and vases mounted in gilt bronze.

3084. OUDRY, L. *Paris.*—Objects in cast-iron, coated with copper, &c. by the electrotype.

3085. MATIFAT, C. S. *Paris.*—Bronzes.

3086. NADAULT DE BUFFON, *Paris.*—Tubular filters.

3087. BURG, DR. V. *Paris.*—Apparatus for purifying and cooling water for drinking.

3088. DESCOLE, P. *Paris.*—Apparatus for giving light.

3089. SERVAU, J. J. *Paris.*—Articles in chiselled and repouséd copper.

3090. GONSSE, C. L. *Paris.*—Fancy articles in bronze, crystal, and porcelain; baskets for flowers, fruits, &c.

3091. LARGAUT, WIDOW, *Paris.*—Polishing powder.

3092. FARJAT, *Rouen* (*Sen. Inf.*).—Scrapers of iron, and leather.

3093. CUBAIN, R. & Co. *Verneuil* (*Eure*).—Laminated copper, &c.

3094. BRUNT & Co. *Paris.*—Gas-meters.

# CLASS XXXII.

3131. VITRY BROTHERS, *Nogent* (*Haute-Marne*).—Cutlery of all kinds; surgical instruments.

3132. THUILLIÉR - LEFRANC, *Nogent* (*Haute-Marne*). — Tailors' shears, pruning knives, &c.

3133. PERRAY, F. C. *Nogent* (*Haute-Marne*).—Cutlery.

3134. GIRARD, C. *Nogent* (*Haute-Marne*).—Kitchen knives, choppers, garden instruments.

3135. MARESCHAL - GIRARD, *Nogent* (*Haute-Marne*).—Cutlery of all kinds.

3137. PLUMEREL-DAGUIN, *Nogent* (*Haute-Marne*).—Common cutlery.

3138. GUILLEMIN-RENAUT & Co. *Nogent* (*Haute-Marne*).—Cutlery.

3140. LÉCOLLIER, V. *Nogent* (*Haute-Marne*).—Razors.

3141. GIRARD, M. *Nogent* (*Haute-Marne*).—Knife-boxes, razor strops, emery.

3142. ROZE, C. *St. Dizier* (*Haute-Marne*).—Cutlery.

3143. COLLECTIVE EXHIBITION BY THE CITY OF THIERS (*Puy-de-Dôme*).—Cutlery (9 Exhibitors).

3144. ROMMETIN, C. M. *Paris.*—Dentists' instruments.

3145. GUERRE, C. JUN. *Langres* (*Haute-Marne*).—Common and fancy cutlery.

3146. PIAULT, J. *Nogent* (*Haute-Marne*).—Cutlery.

3147. TOURON-PARISOT, E. *Paris.* —Fancy cutlery.

3148. PAGÉ BROS. *Châtellerault* (*Vienne*).—Table and kitchen cutlery, razors.

3149. MERMILLIOD BROTHERS, *Prieuré* (*Vienna*).—Cutlery.

3150. CARDEILHAC, A. E. *Paris.*—Fancy cutlery; silver articles for the table.

3151. PICAULT, G. F. *Paris.*—Common and fancy cutlery; razors made without heat, &c.

3152. CHARRIERE, J. J. *Paris.*—Common and fancy cutlery, garden instruments, toilet requisites.

3153. CREMIÈRE, *Portillon* (*Indre-et-Loire*).—Cast steel files.

3154. PROUTAT, MICHOT, & THOMERET, *Arnay-le-Duc* (*Cote-d'Or*).—Steel files and tools.

3158. CLICQUOT, R. M. *Courbevoie* (*Seine*). Gravers' tools, &c.

3159. PETREMENT, F. *Paris.*—Gauges of cast steel, for measuring sheet iron and wire, the hundredth of a millimetre (0·003937 in.) in thickness.

3160. MAILLET, C. P. A. *Compiègne* (*Oise*).—Shoemakers' tools.

3161. DELAHAYE, E. *Paris.* — Screws, broaches, &c.

3162. PERRIN, A. *Nogent (Haute-Marne).* —Cutlery.

3163. LEPAGE, J. H. *Paris.*—Cast steel files.

3164. MANGIN, SEN. & Co. *Paris.*—Files, rasps, and gravers' tools.

3165. LANGUEDOCQ, C. F. *Paris.*—Cutlery.

3166. BLANZY & Co. *Boulogne (Pas-de-Calais).*—Steel pens and holders, &c.

3167. MALLAT, J. B. *Paris.*—Steel, pens, gold and platina pens, with iridium and ruby points.

3168. LIBERT & Co. *Boulogne (Pas-de-Calais).*—Steel pens; steel laminated without heat; steel springs.

3169. MONCHICOURT, V. *Paris.*—Steel pens and holders.

3170. PARYS, A. *Paris.*—Plates for wire-drawers.

3171. RENARD, J. A. *Paris.*—Steel tools for artistic and industrial engraving; engraving machine.

3172. LIMET, LAPAREILLÉ, & Co. *Paris.* —Steel files and cutlery; magnets; flatting machines of tempered cast steel; azotized, cyanuretted, and carburetted steel.

3173. HUËT, J. *Paris.*—Steel beads; clasps and ornaments for purses; buckles, combs, and jewelry of steel.

3174. ESSIQUE, L. *Paris.*—Steel thimbles, and steel jewelry.

3175. JACQUEMIN & SORDOILLET, *Paris.* —Jewelry and ornaments of steel.

3176. BOURGAIN, J. B. *Paris.*—Caskets, jewels, and ornaments of steel.

3177. GOLDENBERG, G. & Co. *Lornhoff (Bas-Rhin).*—Steel tools, spring steel.

3178. CRIBIER, E. JUN. *Paris.*—Steel springs for petticoats.

# CLASS XXXIII.

3201. ODIOT, *Paris.*—Goldsmith's work; toilet requisites, table services, &c.

3202. DEBAIN, A. *Paris.*—Silver candelabras, tea services, &c.

3203. CHERTIER, A. *Paris.* —Church plate.

3204. CHRISTOFLE, C. & Co. 56 *Rue de Bondy, Paris,* and *Carlsrhue, Grand Duchy of Baden.*

Their establishment comprises—

1st. The manufacture of table ornaments in maillechort and brass.
2nd. Of articles in silver.
3rd. Of articles in electrotype, both hollow and massive.
4th. The gilding and plating of goods, whether made by themselves or sent to them for the purpose.
They employ 1,200 workpeople.

The following are specimens of those different branches :—

### PLATED ARTICLES.

Two CUPS, emblematic of the North and West of France. They form a portion of the large set of table ornaments which were made for his Majesty the Emperor, and were exhibited at Paris in 1855.
The sculpture is by Messrs. Gilbert, Diebott, Daumas, Caudron, Briant Boros, Montaguy, Rouillard, and Dernay.
The CENTRE-PIECE of a set of table ornaments, made for fètes at the Hotel de Ville of Paris, after a design by Baron Haussman, senator, and Prefect of the Seine.
The sculpture of this fine composition has been executed

under the direction of Mr. V. Ballard, architect, director of the public works at Paris, and Inspector of Fine Arts, by the sculptors, Diebott, Maillet, Thomas, Germery, and Mathurin-Moreau, who obtained the first prizes at Rome; Rouillard, Capy; and Auguste Madroux, ornamental modeller.
The Centre-piece consists of a large plateau of plate-glass, the framework of which is relieved by a rich frieze moulding, gilt in different shades: four great candelabra, worked into this moulding, form a connection between its principal portions.
The centre is occupied by a vessel symbolical of the arms of the city of Paris. On its deck the statue of the city is raised upon a shield, supported by four Caryatides, representing the Sciences, the Arts, Industry, and Commerce—emblems of her glory and power.
At the bow is an eagle drawing the vessel onward to her future destiny; the genius of Progress lights the way; Prudence is at the stern holding the helm.
Round the vessel are groups of Tritons and Dolphins, sporting in the water.
The two extremities of the composition are occupied by

3204. CHRISTOFLE, C., & Co.—*continued.*

groups of sea-horses, whom Genii and Tritons are endeavouring to tame.

A drawing, placed along with the portion exhibited, shows the entire of the details : only the centre-piece could be finished in time for the Exhibition.

PLATED AND GILT TABLE ORNAMENTS, after the antique, from designs by Rossigneux, architect, and decorator for the Exhibition.

A DESSERT SERVICE, executed after drawings by Mr. Dieterle, formerly head of the art department at the manufactory of Sèvres.

A PAIR OF CANDELABRA, consisting of female Caryatides, in the style of the silver service of his Majesty the Emperor (to be noticed below). The sculpture by Capy.

A PLATED CENTRE-PIECE, used by the Duke of Magenta during the fêtes given at the coronation of the King of Prussia at Konigsberg.

Two CANDELABRA, formed by the figures of children, and other portions of a dinner service, in the style of Louis XIV. are exhibited as specimens of the entire service.

A LARGE SET OF GILT TABLE ORNAMENTS, in the style of Louis XIV.

Two CANDELABRA, each supported on three Caryatides, and other articles belonging to a desert service, in the style of Louis XVI., are exhibited as specimens of the entire service.

A SMALL SET OF TABLE ORNAMENTS, and a GILT DESSERT SERVICE, decorated with children, from Nature.

A SMALL SET OF TABLE ORNAMENTS and a DESSERT SERVICE, silvered, decorated with fruits from Nature, inlaid.

A SMALL SET OF TABLE ORNAMENTS and a DESSERT SERVICE, guilloched and silvered, in the Greek style.

FOUR SERVICES in plate—consisting of warmers, dishes, stew-pans, covers, salt-cellars, &c., in four different styles :—

The first, in that of Louis XVI., is guilloched and chased.
The second, in the same, is chased.
The third, in the style of Louis XIV., is engraved.
And the fourth is in the style of Louis XV.

A GREAT NUMBER OF ARTICLES IN PLATE, &c., such as ice-pails, bread-baskets, soup tureens, oil cruets, menagères, salad-bowls, salt-cellars, mustard-pots, egg-cups, liqueur-flasks, tea-trays, branch candlesticks—articles for the Levant. Among these we direct attention to the following :—

No. 1. A plate-warmer and cover, Greek style.
No. 2. A plate-warmer and cover, decorated with ram's-heads, style of Louis XVI.
No. 3. A bread-basket, style of Louis XIV.
No. 4. A bread-basket, decorated with ears of corn.
No. 5. A salt-cellar, Greek style.
No. 6. A toothpick-holder—decoration, a bird.
No. 7. A toothpick-holder, in the form of a vase, decorated with swallows.
No. 8. A Chinese sugar-bowl.
No. 9. A sugar-bowl, Arab style.
No. 10. A candlestick with two branches, style of Louis XIII.
No. 11. Another, in the Greek style.
No. 12. Another, in the Pompeian style.
No. 13. An ice pail, style of Louis XIV.
No. 14. Another, in the style of Louis XVI., and enriched with medallions.
No. 15. A cake-plate, style of Louis XVI.

### SILVER PLATE.

A SILVER-GILT SERVICE, executed for his Majesty the Emperor, and consisting of warmers, dishes, stew-pans, covers, salt-cellars, fruit-dishes, étagères, compotiers, &c.

A SPECIMEN of the twelve silver cups offered as prizes by the Minister of Agriculture, Commerce, and Public Works in the district competitions throughout France.

AN ALLEGORICAL PIECE, offered by the City of Carlsrhue to the Grand Duchess Mary of Baden, on the occasion of her marriage with the Russian Grand Duke Michael Nicolajevitch.

A COFFEE SERVICE, Arab style, consisting of a large enamelled plateau, a coffee-pot, and ten enamelled cans.

A TEA SERVICE, style of Louis XV., with a plateau, having outline decorations.

A TEA SERVICE, enriched with bas-reliefs of birds, in repoussée chasing.

A SILVER-GILT AND ENAMELLED TEA SERVICE.

No. 16 is an engraved and guilloched tea service.

3204. Christofle, C., & Co.—*continued*.

3204.   CHRISTOFLE, C., & Co.—*continued.*

3204.  CHRISTOFLE, C., & Co.—*continued.*

No. 17.  A ewer and basin, Arab style.
No. 18.  An enamelled cup, with cover.
No. 19.  A silver perfume burner, enamelled.

### ALUMINIUM.

VARIOUS GROUPS.
A CENTRE-PIECE.
BASKETS IN ALUMINIUM, both solid and with a bronze foundation, gilt and silvered.
A HELMET in aluminium bronze.

### ARTICLES IN ELECTROTYPE.

#### HOLLOW WORK.

BUSTS OF THE EMPEROR AND EMPRESS, from models by De Meenwerkerke : the plaster casts are those of Mr. Susse.
THE SPRING-TIME OF LIFE, an electrotype silver-gilt copy of a statue by Maillet, the sculptor who obtained the first prize at Rome.
THE FAUNUS AND KID, a group by Fesquet, sculptor.
TWO STATUETTES of gladiators, copies from studies by Gerôme the painter.
THE VENUS OF ALLEGRAIN, the copy of a model in bronze belonging to Mr. Susse.
CINCINNATUS, a copy on plaster, from a reduced copy, belonging to M. Sauvage.

### ELECTROTYPE MASSIVE WORK.

TWO CASES, in rosewood, enriched with electrotype ornaments.
A SPECIMEN OF LOCKSMITHS' WORK, belonging to the apartments of her Majesty the Empress, at the Tuilleries, after models by Dolessany.
BATH COCKS, for the Tuilleries, after models by M. Lepretre.
Seven pictures, illustrating the different models which they have had made for cabinetmakers and other artisans.

### WORKS IN ELECTROTYPE NOT EXHIBITED.

THE POPE'S RAILWAY CARRIAGE.
It has been made under the direction of Emile Trelat, architect, and Professor to the Conservatoire des Arts et Metiers.
The exterior covering, as well of the main portions, as of the decorative, and the figures, are in electrotype.
The photograph which is exhibited is an exact representation of the work.
They have applied the electrotype process to MONU-MENTAL DECORATIONS, under the direction of Lefuel, architect to the Louvre.  The following are examples of this—
The chimneypieces of the Tuilleries.
The ramp of the staircase at the residence of the Minister of State.
The chandelier.
The gates of the riding-school of the Louvre.
The imperial stables.
Il Penseroso, the copy of a statue by Michael Angelo.

17

---

3205.  MARRET & BEAUGRAND, *Paris.*— Jewelry in gold and silver.

3206.  PETITEAU, E. *Paris.*—Jewelry.

3207.  BOUVENAT, L. *Paris.*—Jewelry.

3208.  LHOMME, L. *Paris.*—Bracelets.

3209.  BELLEAU, F. A. *Paris.*—Work-requisites.

3210.  PHILIPPI, A. *Paris.*—Jewelry.

18

19

3211. PAYEN & SON, *Paris.*—Jewelry enriched with unpolished filagree work; specimens of the filagree work of all countries.

3212. LEMOINE, V. *Paris.*—Gold chains and bracelets.

3213. OGEZ, WIDOW, & CADET-PICARD, *Paris.*—Gold jewelry; enamelled fancy articles.

3214. JARRY, L. G. SEN. *Paris.*—Agate cup, mounted in silver; gold jewelry; fancy articles in gold and silver.

3215. CONSTANT-VALÈS, & CO. *Paris.*—Imitation pearls; necklaces and bracelets.

3216. JARDIN - BLANCOUD, *Paris.*—Metals, inlaid stones, &c.

3217. PURPER, L. *Paris.*—Onyx cameos.

3218. LANGEVIN, H. *Paris.*—Copper and silver jewelry; imitations of niello jewelry.

3219. TOPART BROS. *Paris.*—Imitation pearls and coral.

3220. GUEYTON, A. *Paris.*—Gilt jewelry.

3221. BON, L. A. *Paris.*—Imitation lapis-lazuli, &c.

3222. CHEVALIER, A. E. *Paris.*—Seals, coins, &c.

3223. STUBLER, A. *Paris.*—Diamonds, for graving tools, wire-drawing, and jewelry.

3224. VILLEMONT, C. H. *Paris.*—Gilt jewelry.

3225. FAASSE, *Paris*—Imitation jewelry.

3226. BENDER, L. A. E. *Paris.*—Imitation jewelry.

3227. TRUCHY, C. E. *Paris.*—Imitation pearls.

3228. GRANGER, E. *Paris.*— Historic jewelry, for the theatre; imitation jewelry.

3229. BACHELET, L. C. *Paris.*—Church plate and bronzes.

3230. CHANUEL, WIDOW, *Paris.*—Ornamental repoussé articles in silver.

3231. HUBERT, L. J. L. *Paris.*—Repoussé and enamelled silver articles.

3232. RUDOLPHI, F. J. *Paris.*—Ornamental plate.

3233. BALAINE, C. *Paris.* — Plated articles.

3234. GOMBAULT A. & CO. *Paris.*—Maillechort table service.

3235. POUSSIELGUE-RUSAND, *Paris.* — Church plate and bronzes.

3236. DOTIN, A. C. *Paris.*—Enamelled articles, imitations of Limoge and Byzantine enamels.

3237. FLORANGE, E. H. *Paris.*—Plated articles.

3238. ARMAND-CALLIAT, *Lyon (Rhône).*—Devotional articles in gold and silver.

3239. TRIOULLIER, E. C. *Paris.*—Church plate and bronzes.

3240. COMTE, H. C. C. DE RUOLZ & FONTENAY, *Paris.*—Various articles made of a new alloy.

3241. SAVARD, A. F. *Paris.*—Double-gold jewelry.

3242. DOBBÉ, V. & HÉMON, *Paris.*—Double-gold jewelry.

3243. MURAT, C. *Paris.*—Double-gold jewelry.

3244. CORBEELS, S. *Paris.*—Inlaid tortoiseshell jewelry.

3245. SAVARY, A. *Paris.* — Imitation jewelry, trinkets mounted on gold and on silver, gems rough and cut.

3246. CASALTA, L. & ISLER, L. *Paris.*—Carved corals, cameos.

3247. DECAUX, E. *Paris.*—Jewelry.

3248. BONNET-GROLLIER, *Bourg (Ain).*—Enamelled silver jewelry.

3249. CELLIER, L. *Marseilles (Bouches du-Rhône).*—Gold and silver jewelry, &c.

3250. DE LAURENCEL, *Paris.* — Auriferous agate of San Francisco.

2351. JOUANIN, C. V. *Paris.*—Cameos cut on stones and shells.

3252. BOURET & FERRÉ, *Paris.*—Jewelry obtained by stamping.

3253. BARBARY, E. *Paris.*—Ivory workboxes, &c., furnished in gold or silver.

3254. THÉNARD, F. *Paris.*—Plate, jewelry.

2255. FRIBOURG, G. A. *Paris.*— Gold jewelry.

3256. LOBJOIS, A. *Paris.*—Jewelry.

3257. BRUNEAU & Co. *Paris.*—Jewelry, plate.

3258. MAGNIADAS, MADAME, SEN. *Paris.* —Fancy jewelry.

3259. COFFIGNON BROS. *Paris.*—Silver and plated jewelry.

3260. GENTILHOMME, J. H. *Paris.* — Gold and silver jewelry.

3261. CAILLOT, JUN., JECK & Co. *Paris.* —Jewelry; ornaments enriched with diamonds and precious stones; imitations of antique jewelry.

3262. JACTA, E. & Co. *Paris.* — Jewelry.

3263. MELLERIO (MELLER), BROTHERS, *Paris.*—Jewelry, &c.

3264. WIESE, J. *Paris.*—Jeweller's and artistic goldsmith's work.

3265. DURON, C. *Paris.*—Artistic jewelry; agate cups, mounted in gold for enamel.

3266. DESURY, A. *St. Briene* (*Côtes-du-Nord*).—Plate; silver plateau.

3267. CARMANT, T. A. NORMAND, P. *Paris.*—Gilt jewelry.

3268. BRISSON, T. A. *Paris.*—Designs for bronze and goldsmith's work.

3269. FANNIERE, *Paris.*—Silver plate.

3270. GUEYTON, A. *Paris.* — Artistic goldsmith's work.

3271. DUPONCHEL, H. *Paris.* — Candelabra, plate.

3272. ROUCOU, *Paris.*—Articles damascened and inlaid with gold and silver.

3273. AUCOE, L. SEN. *Paris.*—Table ornaments.

3274. FEUQUIÈRES, J. *Paris.* — Gold, silver, and enamelled jewellery.

## CLASS XXXIV.

3281. JOINT-STOCK CO. OF THE MANUFACTURES OF GLASS AND CHEMICAL PRODUCTS OF SAINT-GOBAIN, CHAUNY, AND CIREY, *Paris.*—Flint glass.

3282. ROUGET DE LISLE, T. A. *Paris.*— Bottles, and alimentary cases of glass.

3283. WALTER, BERGER, & Co. *Paris.*— Watch glasses, &c.

3284. DE POILLY, DE FITZJAMES, & LABARBE, *Follembray* (*Aisne*).—Bottles, glasses, decanters, garden bell-glasses.

3285. PASQUES, A. & Co. *Blanzy* (*Saône and Loire*).—Bottles, garden bell-glasses, &c.

3286. VAN LEEMPOEL, *Quiquengronne* (*Aisne*).—Champagne bottles.

3287. GOBBE, O., FOGT, A. & Co. *Aniche* (*Nord*).—Window glass.

3288. DELHAY, H. *Aniche* (*Nord*).— Glass for windows and photography.

3289. RAABE, C. & Co. *Rive-de-Gier* (*Loire*).—Bottles, window glass, &c.

3290. CHAPPUY, L. *Frais-Marais-lez-Douai* (*Nord*).—Bottles, of large and small sizes.

3291. CHARTIER, P. *Douai* (*Nord*).— Large bottles.

3292. HOUTARD, F. & Co. *Lourches* (*Nord*).—Bottles.

3293. RENARD & SON, *Fresnes* (*Nord*).— Window glass, and glass for optical purposes.

3994. DEVIOLANE BROS. *Vauxrot* (*Aisne*).—Champagne bottles.

3295. JOINT-STOCK CO. OF THE GLASS WORKS OF EPINÀC (*Saône and Loire*).— Bottles.

3296. DEROCHE, C. *Paris.*—Glass, porcelain and chemical ware.

3297. MAËS, L. J. *Clichy-la-Garenne* (*Seine*).—Flint glass, white and coloured, cut, engraved, and enriched.

3298. OUDINOT, E. *Paris.*—Glass windows, styles of the 12th and 14th centuries; window glass.

3299. LAFAYE, P. *Paris.*—Glass windows.

3300. NICOD, *Paris.* — Glass cases, church windows.

3301. HONER, *Nancy* (*Meurthe*).— Stained glass, styles of the 13th and 15th centuries; a window.

3302. BOURGEOIS, G. *Reims* (*Marne*).— Glass cases, style of the 13th century: church windows.

3303. DRION-QUÉRITÉ, PATOUX, & DRION, A. *Aniche* (*Nord*).—Glass windows, glass for roofs, &c.

3304. MARÉCHAL, C. L. *Metz* (*Moselle*). —Windows.

3305. LUSSON, A. *Paris.*—Painted windows, in the styles of the 13th and 16th centuries.

3306. DIDRON, A. N. SEN. *Paris.*— Painted windows, styles of the 13th, 15th, and 16th centuries.

3307. LAURENT & GSELL, *Paris.*— Church window.

3308. MONOT, E. S. *Pantin* (*Seine*).— Flint-glass, moulded, cut, and engraved, mounted on gilt bronze, &c.

3309. THOMAS, KUHLIGER BROS. & CO. *Paris.*—Plated glass.

3310. ALEXANDRE, JUN. *Paris.*—Flint-glass.

3311. SIMEON & SON, *Paris.*—Glass ornamented by transference of a painting on cloth.

3312. COFFETIER, N. *Paris.* — Stained glass, in the styles of the 12th and 13th centuries.

3313. MARÉCHAL, C. R. JUN. *Metz* (*Moselle*).—Windows.

3314. VEILLARD. J. & CO. *Bordeaux* (*Gironde*).—Bottles.

3315. PICARD BROTHERS, *Sarrebourg* (*Meurthe*).—Watch glasses.

3316. CATTAERT, *Paris.* — Decanters closing hermetically.

3317. DORSY, A. *Paris.* — Flasks and cases.

3318. LAHOCHE & PANNIER, *Paris.*— Cut flint glass for the table, &c.

## CLASS XXXV.

3321. IMPERIAL MANUFACTORY OF SÈVRES.—Articles in porcelain, crockery, and terra cotta.

3322. ROZIER, WIDOW, & BAROCHE, *Tain* (*Drôme.*)—Kaolin, refractory bricks.

3323. DUMERIL, C. & LEURS, H. *St. Omer* (*Pas de Calais*).—Earthenware pipes.

3324. LAUDET (BEAUFAY), SOUCHART, & CO. *Paris.*—Laboratory articles in refractory clay.

3325. DEYEUX, N. T. *Liancourt* (*Oise*). —Refractory crucibles.

3326. DUPRAT, M. C. V. *Canéjan* (*Gironde*).—Refractory crucibles and bricks.

3327. FIOLET, L. *St. Omer* (*Pas de Calais*).—Clay pipes.

3328. GISCLON, *Lille* (*Nord*). — Clay pipes.

3329. GOSSE, F. A. *Bayeaux* (*Calvados*). —Hard porcelain articles, made to stand the fire, &c.

3330. HAVILAND, C. F. & CO. *Limoges* (*Haute-Vienne*).—Decorated porcelain.

3331. POUYAT BROS. *Limoges* (*Haute-Vienne*).—Articles in hard porcelain.

**3332.** ARDANT, H. & Co. *Limoges (Haute-Vienne).*—Fancy porcelain articles.

**3333.** JULLIEN, *Saint-Leonard (Haute-Vienne).*—Porcelain vases and table services.

**3334.** BATIER, H. *Limoges (Haute-Vienne).*—Ornamented porcelain.

**3335.** VIEILLARD, J. & Co. *Bordeaux (Gironde).*—Crockery ware and porcelain, &c.

**3336.** MÉNARD, C. *Paris.*—Artistic porcelain articles, &c.

**3337.** ROUSSEAU, E. *Paris.*—Porcelain crockery ware, and enamels.

**3338.** LAHOCHE & PANNIER, Maison de l'Escalier de Cristal, 162, 163, and 164, *Galerie de Valois (Côte de la Banque), Palais Royal;* and 13 *Rue de Valois* (for carriages), *Paris.*

[OBTAINED SIX MEDALS AT LONDON, NEW YORK, AND PARIS.]

Established for the manufacture of porcelain and cut-glass table services, gilt and plated bronze *surtouts*, time-pieces, lustres, candelabra, artistic and fancy articles.

From the French Official Catalogue :—

"The *Escalier de Cristal* comes forward with an exhibition which is of an unusual kind, and confers the greatest honour on French ceramic art. M. Lahoche has aimed at richness, elegance, and cheapness, and has reached them all.

"Let the millionaire, let the crowned-head pause before that masterly glass case : each of them will find abundant opportunities for selecting the most magnificent articles contributed by French industry. Let the wealthy member of the aristocracy, the visitor nice and hard to be pleased, one in search of productions classically conceived and skilfully contrived, they will discover exquisite gems, and works altogether above the ordinary standard. Or let him whose nobler aspirations dream of the beautiful—condemned though he be to the practice of economy—long for one of those favourable purchases which are followed by no regrets, and which a moderate price shall place within his reach : he too may stop before that glass case, which is accessible to all.

"It is for table services that the *Escalier de Cristal* is particularly distinguished—complete services, that include *surtouts*, which vary through every degree of economy or splendour. In our opinion, the great triumph of the exhibition is the magnificent surtout, the bronzes of which are themselves little chefs-d'œuvres of sculpture. From the lofty figures in the midst of it, constituting a charming group, which supports a basket of flowers, to the graceful reclining figures on which rest the compotiers and the cups, each detail differs from all the rest : there is no mere reproduction of the same idea, enlarged or diminished, according to the circumstances; but a purpose exactly suited to the form, the magnitude, and the importance of each particular object.

"Vases of great value, time-pieces of new patterns and arrangements—the former deserving a place in the palaces of Emperors, the latter fit to attract admiration on the chimney-piece of a Parisian boudoir. The vase, the flower-stand, the tazza, the bouquet-holder, all add to the decorations of this industrial trophy. In leaving this glass case it is impossible to determine which claims our admiration most—the immense articles of cut glass—the porcelain, decorated by the most celebrated pencils—or the mountings in bronze, enriched with fine gold or oxidised silver — classic and artistic forms, which M. Lahoche knows how to use so profusely for the general purposes of his manufacture, or to impart to it a perfection peculiar to itself.

"The Industry, at the head of which stands the *Escalier de Cristal*, is an art, and one that depends on taste, and perhaps also, in some degree, on fashion. Every grade, every class possessing a capacity to appreciate, comes to Paris, from every quarter of the globe, to seek the gratification of its ruling propensity. At the gathering of the human race, at the meeting place of the inhabitants of every country, M. Lahoche holds aloft, and firmly, the French standard of ceramic art."

3339. LAURIN, F. *Bourg-le-Reine (Seine).* —Crockery ware.

3340. DECK, T. *Paris.*—Artistic crockery ware, panels, medallions, &c.

3341. PINART, H. A. *Paris.* — Ornamented plates and dishes.

3342. AVISSEAU, E. *Tours (Indre-and-Loire).*—Artistic articles in pottery.

3344. CHABLIN, N. L. *Paris.*—Porcelain, flint-glass, enamels encrusted with engraved gold and silver.

3346. LYONS, G. *Nevers (Nièvre).*—Common and artistic crockery, &c.

3347. DANIEL, S. *Paris.*—Porcelain imitations of precious stones.

3348. DE ST. ALBIN, MADAME, C. H. *Paris.*—Fruits and flowers painted on porcelain.

3349. BARBIZET, V. *Paris.*—Glazed pottery, style of Palissy.

3350. MACÉ, L. A. C. *Paris* — Ornamented porcelain.

3351. LAVALLE, J. *Premières (Cote d' Or).*

—Artistic pottery painted on coarse enamel, imitation of Etruscan.

3352. DEVERS, *Paris.*—Enamelled earthenware.

3353. JEAN, A. *Paris.*—Artistic decorations for apartments in porcelain..

3354. GILLET & BRIANCHON, *Paris.*—Enamelled ceramic pastes, imitating mother-of-pearl, ivory, and emerald.

3355. GILLE, JUN. *Paris.*—White and ornamented porcelain statues, groups, &c.

3356. LÉTU & MAUGER, *Isle Adam (Seine-and-Ouse).*—Statue of the Virgin in biscuit.

3357. HACHE, A. & PEPIN-LEHALLEUR, *Paris.* — White and ornamented porcelains.

3359. PILLIVUYT, C. & Co. *Paris*— White and ornamented porcelain.

3359. PREVOST, J. R. *Paris.* — Ornamented porcelain.

3360. DE BETTIGNIES, M. *St. Amand-les-Eaux (Nord).*—Soft porcelain, imitating old Sèvres.

# CLASS XXXVI.

3371. POISSON, P. L. M. *Paris.*—Fancy articles in ivory, &c.

3372. HEMERY, J. J. L. *Paris.*—Porcelain, &c. mounted in bronze ; malachite jewelry, &c.

3374. GIRAUDON, S. A. *Paris.*—Cases, &c.

3375. COLLECTIVE EXHIBITION OF THE CITY OF DIEPPE *(Seine-Inf.).*—Articles in ivory (11 Exhibitors).

3376. JEANTET-DAVID, *St. Claude (Jura).* —Snuff-boxes, pipes, combs, &c.

3377. SCHOTTLANDER, H. *Paris.*—Photograph albums.

3378. LEFORT, V. M. *Paris.*—Articles of carved ivory, &c.

3379. DROUARD BROS. *Paris.*— Photograph albums.

3380. MARX, W. *Paris.*—Albums, portfolios, &c.

3381. LATRY, SEN. & SON, *Paris.* — Fancy articles made of the sawdust of exotic wood, hardened and compressed.

3382. GERSON & WEBER, *Paris.*—Requisites and fancy articles in carved wood.

3383. LERUTH, F. L. *Paris.*—Small cabinet-work articles in morocco, &c.

3384. MERCIER, C. V. *Paris.* — Snuff-boxes in exotic wood, &c.

3385. TRIEFFUS & ETTLINGER, *Paris.*— Morocco articles, tortoise-shell and ivory jewelry.

3386. HORCHOLLE, A. *Paris.*—Articles in engraved ivory.

3387. HOCHAPFEL BROS. *Strasbourg (Bas-Rhin).*—Pipes, enriched with meerschaum, ivory, and amber.

3388. NŒTINGER, C. *Mutzeg (Bas-Rhin).* —Stone balls.

3389. CHANDELIER, *Méru (Oise).*—Buttons.

3390. BEUGNOT, C. H. *Paris.*—Requisites in plain, carved, and encrusted ivory; fancy articles.

3391. BONHOMÉ, *Paris.*—Designs for metallurgical establishments.

3392. BONDIER-DONNINGER, & ULBRICH, *Paris.*—Pipes, cigar-cases.

3393. PATRY, E. L. *Paris.*—Spectacles, &c. of aluminium, tortoise-shell, gold, &c.

3394. BEAUDOIRE-LEROUX, & Co. *Paris.* —Albums, note books.

3395. GRUMEL, F. R. *Paris.*—Photograph-albums and frames.

3396. BLUMER, C. *Strasbourg (Bas-Rhin).*—Inlaid floors; artistic furniture.

3397. WALCKER, *Paris.* — Travellers', sportsmen's, and military requisites.

3398. MOREAU, J. L. *Paris.*—Carved ivory casket; small Gothic monument in ivory, ornamented with statues.

3399. BONTEMS, B. *Paris.*—Mechanical toys and automata.

3400. DUCLOS & RUBALTO, *Paris.*—A fume-cigarre.

3401. BÉREUX, MISS J. *Paris.*—Dolls, and dresses for them.

3402. LEMAIRE-DAIMÉ, *Andrèsy (Seine-and-Oise).*—Atmospheric playthings.

3903. THÉROUDE, A. N. *Paris.*—Automatic toys; self-acting organ.

3404. PEZET, C. *Paris.*—Articles in morocco, and toilet requisites.

3405. MIDOCQ, N. E. & GAILLARD, E. A. *Paris.*—Morocco articles; requisites for travelling, &c.

3406. SORMANI, P. *Paris.*—Travelling requisites; small and fancy cabinet work.

3407. GELLÉE BROS. *Paris.*—Sheaths, caskets, &c.

3408. SCHLOSS, S. & NEPHEW, *Paris.*— Morocco articles, cigar-cases, &c.

3409. AUCOC, L. SEN. *Paris.*—Morocco articles requisite for the toilet; tea and coffee services, &c.

3410. ULMANN, P. *Paris.*—Designs for fabrics, shawls, &c.

3411. DELAYE, P. V. *Paris.*—Designs for shawls, &c.

3412. NAZE, SON, & Co. *Paris.*—Designs for shawls, dresses, &c.

3413. BERRUS, A. BROS. *Paris.*—Designs for long shawls.

3414. AUBRY, C. H. *Paris.*—Designs for lace, &c.

3415. GONELLE BROS. *Paris.*—Designs for long shawls.

3416. HENRY, H. F. *Paris.*—Designs for paper-hangings, and stuffs for furniture.

3417. CAPTIER, E. V. *Fontainebleau (Seine-and-Marne).*—Designs for silks, &c.

3418. MATHIEU, E. *Paris.*—Designs for shawls.

3419. VAILLANT BROTHERS, *Paris.*— Designs for shawls, &c.

3420. FAURE, E. *Paris.* — Designs for lace.

4521. MADELEINE BROTHERS, *Paris.*— Designs for lace.

3422. GATTIKER, G. *Paris.* — Designs for dresses.

3423. GOURDET & ADAN, *Paris.*—Designs for carpets, &c.

3424. GUICHARD, E. A. D. *Paris.*— Industrial designs.

3425. BOISSIER, *Paris.* — Designs for worked shawls.

3426. FAUVELLE-DELEBARRE, *Paris.*— Combs.

3427. DESRUELLES, A. *Paris.*—Combs.

3428. JEUNIAUX, D. *Paris.*—Combs.

3429. MASSUE, E. *Paris.*—Combs.

3430. LEGAVRE, G. B. *Paris.*—Combs.

3431. PICARD, F. A. *Paris.*—Combs.

3432. CASELLA, E. *Paris.*—Combs.

3433. HEUDE, *Paris.*—Combs.

3434. PINSON BROTHERS, *Paris.*—Panel imitating tortoise-shell, mother-of-pearl, and ivory.

3435. RAMBERT, C. D. *Paris.*—Designs for the goldsmith and cabinet-maker.

3436. LIENARD, *Paris.*—Industrial designs.

3437. ALLAIN-MOULARD, L. A. F. *Paris.* —Morocco articles.

3438. DUPONT & DESCHAMPS, *Beauvais* (*Oise*).—Fancy articles.

# FRENCH COLONIES.

SOUTH-WEST COURT AND SOUTH-WEST GALLERY.

---

## GUIANA.

### CLASS II.

3451. HERARD.—Citrate of lime.

### CLASS III.

3452. Flour and fecula (2 Exhibitors).

3453. Sugar (3 Exhibitors).

3454. Alcohol, liqueurs, syrups, and preserves (5 Exhibitors).

3455. Cocoa (1 Exhibitor).

3456. Spices (2 Exhibitors).

3457. Tobacco (2 Exhibitors).

3458. Wood (3 Exhibitors).

3459. Organic industrial products — cocoons, cotton, dye-stuffs (5 Exhibitors).

3460. Animal products—salt fish, fish-glue, tortoise-shell (3 Exhibitors).

3461. Collection of colonial products (1 Exhibitor).

### CLASS XVIII.

3462. Thread and woven fabrics made of Guiana cotton (5 Exhibitors).

### CLASS XXV.

3463. Articles made of feathers (2 Exhibitors).

### CLASS XXVII.

3464. Costumes, arms, and garments (1 Exhibitor).

## ST. PIERRE AND MIQUELON.

### CLASS III.

3465. Products of the Fisheries—salt cod, salt herrings, cod-liver oil (7 Exhibitors).

3466. Cereals, and medicinal plants (1 Exhibitor).

## COCHIN-CHINA.

3467.   Collection of colonial products (1 Exhibitor).

## NEW CALEDONIA.

### CLASS I.

3468.   Pit-coal (1 Exhibitor).

### CLASS II.

3469.   Medicinal substances (2 Exhibitors).

### CLASS III.

3470.   Tuberous roots and alimentary grain (3 Exhibitors).

3471.   Sugar and coffee (2 Exhibitors).

3472.   Wool (2 Exhibitors).

3473.   Products of the fisheries—tortoiseshell, tripang, pearl-oysters, &c. (4 Exhibitors).

3474.   Specimens of wood (5 Exhibitors).

3475.   Industrial organic products—cotton and textile fibres, gums, resins, oleaginous plants and oils, dye-stuffs (9 Exhibitors).

### CLASS XXVII.

3476.   Arms and garments (2 Exhibitors).

## TAHITI AND ITS DEPENDENCIES.

### CLASS III.

3477.   Alimentary substances — fecula, sugar, alcohol, coffee, vanilla, &c. (7 Exhibitors).

3478.   Organic industrial products—wood, cotton, textile fibres, oleaginous substances, tobacco, pearl-oysters, and pearls (7 Exhibitors).

## MAYOTTE AND NOSSI-BE.

### CLASS III.

3479.   Alimentary substances — rice, sugar, coffee, rum (4 Exhibitors).

3480.   Organic industrial products—wax, tortoise-shell (1 Exhibitor).

# ST. MARY OF MADAGASCAR.

## CLASS III.

3481.   Collection of colonial products (1 Exhibitor).

# EAST INDIES.

## CLASS II.

3482.   Medicinal substances (3 Exhibitors).

## CLASS III.

3483.   Alimentary substances — cereals, fecula, sugar, spices, models of fruit from nature, preserved fish (4 Exhibitors).

3484.   Organic industrial products—silk, cotton, wood, oleaginous substances, indigo and other dye-stuff, tanning matters (9 Exhibitors).

## CLASS IV.

3485.   Articles in straw and cane (3 Exhibitors).

## CLASS XX.

3486.   Mixed cotton and silk fabrics (2 Exhibitors).

## CLASS XXVII.

3487.   Boots, shoes, &c. (1 Exhibitor).

# WEST COAST OF AFRICA.

## CLASS III.

3488.   Alimentary substances — Rio Nuñez coffee (1 Exhibitor).

3489.   Organic industrial products—cotton and textile fibres, medicinal plants, oleaginous substances, colouring matters, peltry (4 Exhibitors).

3490.   Collection of the products of the country (1 Exhibitor).

## CLASS XXVII.

3491.   Arms and costumes (1 Exhibitor).

# GUADALOUPE.

### CLASS I.

3492.  Mineral products (1 Exhibitor).

### CLASS II.

3493.  Medicinal substances (1 Exhibitor).

### CLASS III.

3494.  Alimentary substances—flour and fecula, sugar, coffee, alcohol, preserves and liqueurs, spices, sea salt (10 Exhibitors).

3495.  Organic industrial products—silk, cotton, oleaginous substances, dye-stuffs, tobacco (19 Exhibitors).

3496.  Collection of colonial products (1 Exhibitor).

3497.  Woven fabrics made with colonial cotton (4 Exhibitors).

# ISLE OF REUNION.

### CLASS I.

3498.  Mineral products (4 Exhibitors).

### CLASS II.

3499.  Medicinal substances (3 Exhibitors).

### CLASS III.

3500.  Alimentary substances — alimentary grain, flour, and fecula, sugar, alcohol, preserves, coffee, cocoa, spices, wax, honey, &c. (59 Exhibitors).

3501.  Organic industrial products — wood, cotton and textile fibres; gums, balsams and resins; oleaginous substances, dyestuffs, and tanning matters; tobacco (25 Exhibitors).

### CLASS XXX.

3502.  Furniture and ornamental articles (13 Exhibitors).

# MARTINIQUE.

### CLASS I.

3503.  Mineral products (3 Exhibitors).

### CLASS III.

3504.  Alimentary substances—flour and fecula, sugar, alcohol, liqueurs, preserves, coffee, cocoa, salt meats (33 Exhibitors).

3505.  Organic industrial products — wood, cotton, gums, oleaginous substances, manures (8 Exhibitors).

3506.  Collection of colonial products (1 Exhibitor).

3507.  Birds—group of **Martinique** birds (1 Exhibitor).

# ALGERIA.

**3601. COLLECTIVE EXHIBITION. — Metalliferous minerals.**

Province of Algiers: ores of iron, copper, and lead (3 Exhibitors).

—— of Oran: ores of iron, copper, lead, zinc, and nickel (5 Exhibitors).

—— of Constantine: ores of iron, copper, lead, zinc, antimony, manganese, and mercury (13 Exhibitors).

**3602. COLLECTIVE EXHIBITION. — Mineral non-metalliferous products.**

Province of Algiers: marbles, calcareous substances, onyx marbles, gypsum, phosphyroidal diorite (7 Exhibitors).

—— of Oran: marbles, calcareous substances, lime and mortar, alabaster, plaster, saltpetre, sea salt (9 Exhibitors).

—— of Constantine: marbles, calcareous substances, lime, plaster, alabaster, porphyry, granite, grit stone, sal gemma (11 Exhibitors).

**3603. Metallurgical products—**
Province of Constantine: cast iron from the ores of Allelick (1 Exhibitor).

## CLASS II.

**3604. COLLECTIVE EXHIBITION. — Mineral and thermal waters—**
Province of Constantine (2 Exhibitors).

**3605. COLLECTIVE EXHIBITION. — Medicinal substances.**

Province of Algiers: lichens, poppies, umbelliferous aromatic plants, mallows, borage, liquorice, mint, indigenous pepper and senna, oak-galls, opium (6 Exhibitors).

—— of Oran: absynthe, hops, hachish, pimenta, liquorice, cumin, anise, pastes, syrups (10 Exhibitors).

—— of Constantine: kef, a kind of hemp used instead of tobacco; indigenous tea, fernugreek, pimenta, liquorice, coriander seed, poppies, opium, &c. (12 Exhibitors).

**3606. COLLECTIVE EXHIBITION. — Essences and essential oils for perfumery.**

Province of Algiers: essential oil of geranium, peppermint water (2 Exhibitors).

—— of Constantine: orange-flower water; the leaves and flowers of odoriferous plants (2 Exhibitors).

## CLASS III.

**3607. COLLECTIVE EXHIBITION. — Cereals.**

Province of Algiers: wheat, barley, rice, maize, oats, &c. (9 Exhibitors).

—— of Oran: wheat, barley, oats, maize, sergho (23 Exhibitors).

—— of Constantine: wheat, barley, rice, maize, oats, millet, sergho (65 Exhibitors).

**3608. COLLECTIVE EXHIBITION.—Farinaceous vegetables, and plants for forage.**

Province of Algiers: French beans, peas, lentils, beans, &c. (2 Exhibitors).

—— of Oran: French beans, peas, chick-peas, lentils, beans (9 Exhibitors).

—— of Constantine: French beans, peas, chick-peas, lentils, beans (12 Exhibitors).

**3609. COLLECTIVE EXHIBITION.—Tuberous roots.**

Province of Algiers: Tubers, colocases, &c. (1 Exhibitor).

—— of Oran: Tubers, beet-root, potatoes, &c. (5 Exhibitors).

—— of Constantine: Bulbs of the taronda, potatoes, &c. (4 Exhibitors).

**3610. COLLECTIVE EXHIBITION.—Fruits.**
Province of Algiers: almonds (1 Exhibitor).

—— of Oran: various fruits, raisins, capers, &c. (10 Exhibitors).

—— of Constantine: various fruits, jujubes, raisins, &c. (6 Exhibitors).

**3611. COLLECTIVE EXHIBITION.—Flour and alimentary pastes.**

Province of Algiers: wheat-flour, semoule, macaroni, small pastes, biscuit for the army (4 Exhibitors).

—— of Oran : wheat and maize flour, semoule (3 Exhibitors).

—— of Constantine : wheat-flours, fecula of the various tubers, semoules, and semouleka, pastes (15 Exhibitors).

Metropolitan France : flour, alimentary pastes, and other articles, prepared with the wheat and maize of Algeria (3 Exhibitors).

3612. COLLECTIVE EXHIBITION.—Honey and wax.

Province of Algiers : honey, wax, bee-hive of a new construction (1 Exhibitor).

—— of Oran : honey, candles made of indigenous wax (4 Exhibitors).

—— of Constantine : honey, wax, candles made of indigenous wax (7 Exhibitors).

3613. COLLECTIVE EXHIBITION.—White and red wine.

Province of Algiers (24 Exhibitors).

—— of Oran (37 Exhibitors).

—— of Constantine (17 Exhibitors).

3614. COLLECTIVE EXHIBITION.— Alcohol, and alcoholic drinks, liqueurs, preserves, and confectionery.

Province of Algiers (3 Exhibitors).

—— of Oran (9 Exhibitors).

—— of Constantine (2 Exhibitors).

3615. COLLECTIVE EXHIBITION. — Leaf and manufactured tobacco.

Province of Algiers (5 Exhibitors).

—— of Oran (10 Exhibitors).

—— of Constantine (21 Exhibitors).

3616. COLLECTIVE EXHIBITION.—Plants and oleaginous substances.

Province of Algiers : flax, &c. ; olives, oil of olives, &c. (7 Exhibitors).

—— of Oran : olives, and oil of olives, flax, colza, &c. (21 Exhibitors).

—— of Constantine : olives, and oil of olives, flax, white mustard, oil of sweet almonds, &c. (15 Exhibitors).

3617. COLLECTIVE EXHIBITION.—Wools.

Province of Algiers : Sheep's wool and camel's hair (7 Exhibitors).

Province of Oran : sheep's wool (13 Exhibitors).

—— of Constantine : sheep's wool, camel's hair (20 Exhibitors).

3618. COLLECTIVE EXHIBITION. — Raw silk and cocoons.

Province of Algiers (10 Exhibitors).

—— of Oran (12 Exhibitors).

—— of Constantine (17 Exhibitors).

3619. COLLECTIVE EXHIBITION.—Woods and forest products.

Province of Algiers : 47 specimens of woods, &c. (4 Exhibitors).

—— of Oran : 18 specimens of woods (6 Exhibitors).

—— of Constantine : 104 specimens of woods (3 Exhibitors).

3620. COLLECTIVE EXHIBITION.— Raw and prepared corks.

Province of Algiers (1 Exhibitor).

—— of Oran (2 Exhibitors).

—— of Constantine (5 Exhibitors).

3621. COLLECTIVE EXHIBITION.—Vegetable textile fibres.

Province of Algiers : fibres of the Nain palm, termed vegetable horse-hair, hemp, flax, agave, &c. (12 Exhibitors).

—— of Oran : fibres of the Nain palm, agave, flax, &c. (18 Exhibitors).

—— of Constantine : flax, hemp, and various other textile fibres (8 Exhibitors).

3622. COLLECTIVE EXHIBITION. — Cotton.

Province of Algiers (24 Exhibitors).

—— of Oran (19 Exhibitors).

—— of Constantine (13 Exhibitors).

3623. COLLECTIVE EXHIBITION. — Dye stuffs, and other industrial organic substances.

Province of Algiers : cochineal, indigo, kermes, madder, carthamum, glunes of sorgho, and teasels (5 Exhibitors).

—— of Oran : hemp, carthamum, kermes, sumach, &c. (15 Exhibitors).

—— of Constantine : madder, indigo, nutgalls, tanning bark, pine-resin, sugar (7 Exhibitors).

## CLASS IX.

3624. Agricultural instruments (3 Exhibitors).

## CLASS XII.

3625. Instrument ascertaining the rate at sea (1 Exhibitor).

## CLASS XVIII.

3626. Yarns, and woven fabrics made with Algerian cotton (10 Exhibitors).

## CLASS XIX.

3627. Thread, and woven fabrics made of Algerian flax (1 Exhibitor).

## CLASS XX.

3628. Woven fabrics, made of Algerian silk (1 Exhibitor).

## CLASS XXI.

3629. COLLECTIVE EXHIBITION.—Woven fabrics, made of Algerian wool (3 Exhibitors).

## CLASS XXII.

3630. COLLECTIVE EXHIBITION.—Indigenous carpets.
Province of Algeria (3 Exhibitors).
—— of Oran (4 Exhibitors).
—— of Constantine (2 Exhibitors).

## CLASS XXVI.

3631. COLLECTIVE EXHIBITION. — Leather, peltry, and other matters derived from animals.
Province of Oran: goose down, coral (3 Exhibitors).
—— of Constantine: calf, sheep, and goat-skins; skins of the lion, jackall, hyæna; feathers and down of birds; coral (3 Exhibitors).

3632. COLLECTIVE EXHIBITION. — Saddlery and harness.
Province of Constantine (5 Exhibitors).

## CLASS XXVII.

3633. COLLECTIVE EXHIBITION. — Costumes, arms, garments.
Province of Algiers (8 Exhibitors).
—— of Oran (15 Exhibitors).
—— of Constantine (21 Exhibitors).

3634. Flowers and head-dresses made of Algerian products (1 Exhibitor).

## CLASS XXVIII.

3635. COLLECTIVE EXHIBITION.—Books and manuscripts.
Province of Algiers (3 Exhibitors).
—— of Oran (3 Exhibitors).
—— of Constantine (1 Exhibitor).

## CLASS XXX.

3636. COLLECTIVE EXHIBITION. — Articles of furniture and decoration.
Province of Algiers (3 Exhibitors).
—— of Oran (3 Exhibitors).
—— of Constantine (4 Exhibitors).

3636. COLLECTIVE EXHIBITION. — Furniture and articles made of indigenous wood (8 Exhibitors).

Printed in the United States
By Bookmasters